Applied Regression Analysis: A Research Tool

THE WADSWORTH & BROOKS/COLE STATISTICS/PROBABILITY SERIES

Series Editors
O. E. Barndorff-Nielsen, Aarhus University
Peter J. Bickel, University of California, Berkeley
William S. Cleveland, AT&T Bell Laboratories
Richard M. Dudley, Massachusetts Institute of Technology

R. Becker, J. Chambers, A. Wilks, *The New S Language: A Programming Environment for Data Analysis and Graphics*
P. Bickel, K. Doksum, J. Hodges, Jr., *A Festschrift for Erich L. Lehmann*
G. Box, *The Collected Works of George E. P. Box, Volumes I and II*, G. Tiao, editor-in-chief
L. Breiman, J. Friedman, R. Olshen, C. Stone, *Classification and Regression Trees*
J. Chambers, W. S. Cleveland, B. Kleiner, P. Tukey, *Graphical Methods for Data Analysis*
R. Durrett, *Lecture Notes on Particle Systems and Percolation*
F. Graybill, *Matrices with Applications in Statistics, Second Edition*
L. Le Cam, R. Olshen, *Proceedings of the Berkeley Conference in Honor of Jerzy Neyman and Jack Kiefer, Volumes I and II*
P. Lewis, E. Orav, *Simulation Methodology for Statisticians, Operations Analysts, and Engineers*
H. J. Newton, *TIMESLAB*
J. Rawlings, *Applied Regression Analysis: A Research Tool*
J. Rice, *Mathematical Statistics and Data Analysis*
J. Romano, A. Siegel, *Counterexamples in Probability and Statistics*
J. Tanur, F. Mosteller, W. Kruskal, E. Lehmann, R. Link, R. Pieters, G. Rising, *Statistics: A Guide to the Unknown, Third Edition*
J. Tukey, *The Collected Works of J. W. Tukey*, W. S. Cleveland, editor-in-chief
Volume I: Time Series: 1949–1964, edited by D. Brillinger
Volume II: Time Series: 1965–1984, edited by D. Brillinger
Volume III: Philosophy and Principles of Data Analysis: 1949–1964, edited by L. Jones
Volume IV: Philosophy and Principles of Data Analysis: 1965–1986, edited by L. Jones
Volume V: Graphics: 1965–1985, edited by W. S. Cleveland

Applied Regression Analysis: A Research Tool

John O. Rawlings

North Carolina State University

Wadsworth & Brooks/Cole Advanced Books & Software
Pacific Grove, California

Wadsworth & Brooks/Cole Advanced Books & Software
A Division of Wadsworth, Inc.

Printed in the United States of America

10 9 8 7 6 5 4 3 2 1

Library of Congress Cataloging-in-Publication Data
Rawlings, John O., [date]
Applied regression analysis.
(The Wadsworth & Brooks/Cole statistics/probability series)
Bibliography: p.
Includes index.
1. Regression analysis. I. Title. II. Series.
QA278.2.R38 1988 519.5′36 88-20638
ISBN 0-534-09246-2

Sponsoring Editor: *John Kimmel*
Editorial Assistant: *Maria Tarantino*
Production Editor: *Joan Marsh*
Manuscript Editor: *Susan Reiland*
Permissions Editor: *Carline Haga*
Interior and Cover Design: *Flora Pomeroy*
Art Coordinator: *Sue C. Howard*
Interior Illustration: *John Foster*
Typesetting: *Asco Trade Typesetting Ltd; Hong Kong*
Printing and Binding: *Arcata Graphics/Fairfield*

To my wife, Janie, and our loving family.

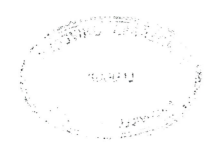

Preface

This text is the outgrowth of several years of teaching an applied regression course to graduate students in the sciences. Most of the students in these classes had taken a two-semester introduction to statistical methods that included experimental design and multiple regression at the level provided in texts such as Steel and Torrie (1980) or Snedecor and Cochran (1980). For most, the multiple regression had been presented in matrix notation.

The basic purpose of the course and this text is to develop an understanding of least squares and related statistical methods without becoming excessively mathematical. The emphasis is on regression concepts, rather than on mathematical proofs. Proofs are given only to develop facility with matrix algebra and comprehension of mathematical relationships. Good students, even though they may not have strong mathematical backgrounds, quickly grasp the essential concepts and appreciate the enhanced understanding. The learning process is reinforced with continuous use of numerical examples throughout the text and with several case studies.

The first four chapters of the book provide a review of simple regression in algebraic notation (Chapter 1), an introduction to key matrix operations and the geometry of vectors (Chapter 2), and a review of ordinary least squares in matrix notation (Chapters 3 and 4). Chapter 5 is a case study giving a complete multiple regression analysis based on the methods reviewed in the first four chapters. Chapter 6 then gives a brief geometric interpretation of least squares, illustrating the relationships among the data vectors, the link between the analysis of variance and the lengths of the vectors, and the role of degrees of freedom. Chapter 7 discusses the methods and criteria for determining which independent variables should be included in the models. Class variables and the analysis of variance of designed experiments (models of less than full rank) are introduced in Chapter 8.

Chapters 9 through 13 address some of the problems that might be encountered in regression analysis. A general introduction to the various kinds of problems is given in Chapter 9. This is followed by discussions of regression diagnostic techniques (Chapter 10), and scaling or transforming variables to rectify some of the problems (Chapter 11). Analysis of the correlational structure of the data and biased regression are discussed as techniques for dealing with the collinearity problem common in observational data (Chapter 12). Chapter 13 is a case study illustrating the analysis of data in the presence of collinearity.

Response curve modeling, including models that are nonlinear in the parameters, is presented in Chapter 14. Chapter 15 is another case study using polynomial response models, nonlinear modeling, transformations to linearize, and analysis of residuals. Chapter 16 addresses the analysis of unbalanced data. The final section of Chapter 16 introduces linear models that have more than one random effect. The ordinary least squares approach to such models is given. This is followed by the definition of the variance–covariance matrix for such models and a brief introduction to the use of iterative maximum likelihood estimation of both the variance components and the fixed effects. The final chapter, Chapter 17, is a case study of the analysis of unbalanced data.

I am grateful for the assistance of many in the development of this book. Of particular importance have been the dedicated editing by my daughter Gwen Briggs and her many suggestions for improvement. It is uncertain when the book would have been finished without her support. A special thanks goes to my former student Virginia Lesser for her many contributions in reading parts of the manuscript, in data analysis, and in the enlistment of many data sets from her graduate student friends in the biological sciences. I am indebted to my friends, both faculty and students, at North Carolina State University for bringing me many interesting consulting problems over the years that have stimulated the teaching of this material. I am particularly indebted to those (acknowledged in the text) who have generously allowed the use of their data. In this regard, Rick Linthurst warrants special mention for his stimulating discussions as well as the use of his data. I acknowledge the encouragement and valuable discussions of colleagues in the Department of Statistics at NCSU, and I thank Matthew Sommerville for checking answers to the exercises. Finally, I want to express appreciation for the critical reviews and many suggestions provided by the Wadsworth Brooks/Cole reviewers: Mark Conaway, University of Iowa; Franklin Graybill, Colorado State University; Jason Hsu, Ohio State University; Kenneth Koehler, Iowa State University; B. Lindsay, The Pennsylvania State University; M. B. Rajarshi, University of Poona (India); Muni Stribastava, University of Toronto; and Patricia Wahl, University of Washington.

Acknowledgment is given for the use of the material in the appendix tables. I am grateful to the Literary Executor of the late Sir Ronald A. Fisher, F.R.S., to Dr. Frank Yates, F.R.S., and the Longman Group Ltd, London for permission to reprint Table III from their book *Statistical Tables for Biological, Agricultural and Medical Research* (6th edition, 1974). This material is reproduced in Appendix Table A. Appendix Table B has been reproduced in part with permission of the Biometrika Trustees from *Biometrika Tables for Statisticians*, Volume II, Table 5, edited by E. S. Pearson and H. O. Hartley, published for Biometrika Trustees, Cambridge University Press, Cambridge, England, 1972. Appendix Table C, is reproduced in part from Tables 4, 5, and 6 of Durbin and Watson (1951) with permission of the Biometrika Trustees. Appendix Table D is reproduced with permission from Shapiro and Francia (1972), *Journal of the American Statistical Association.* I gratefully acknowledge permission of other authors and publishers for use of material from their publications as noted in the text.

Note to the Reader

Most research is aimed at quantifying relationships among variables that either measure the end result of some process or are likely to affect the process. The process in question may be any biological, chemical, or physical process of interest to the scientist. The quantification of the process may be as simple as determining the degree of association between two variables or as complicated as estimating the many parameters of a very detailed nonlinear mathematical model of the system.

Regardless of the degree of sophistication of the model, the most commonly used statistical method for estimating the parameters of interest is the method of **least squares**. The criterion applied in least squares estimation is simple and has great intuitive appeal. The researcher chooses the model that is believed to be most appropriate for the project at hand. The parameters for the model are then estimated such that the predictions from the model and the observed data are in as good agreement as possible as measured by the **least squares criterion**, minimization of the sum of squared differences between the predicted and the observed points.

Least squares estimation is a powerful research tool. Few assumptions are required and the estimators obtained have several desirable properties. Inference from research data to the true behavior of a process, however, can be a difficult and dangerous step due to unrecognized inadequacies in the data, misspecification of the model, or inappropriate inferences of causality. As with any research tool, it is important that the least squares method be thoroughly understood in order to eliminate as much misuse or misinterpretation of the results as possible. There is a distinct difference between understanding and pure memorization. Memorization can make a good technician, but it takes understanding to produce a master. A discussion of the **geometric interpretation** of least squares is given to enhance your understanding. You may find your first exposure to the geometry of least squares somewhat traumatic but the visual perception of least squares is worth the effort. I encourage you to tackle the topic in the spirit in which it is included.

The general topic of least squares has been broadened to include statistical techniques associated with **model development and testing**. The backbone of least squares is the classical multiple regression analysis using the linear model to relate several independent variables to a response or dependent variable. Initially, this classical model is assumed to be appropriate. Then methods for detecting inadequacies in this model and possible remedies are discussed.

The connection between the analysis of variance for designed experiments and multiple regression is developed to build the foundation for the analysis of **unbalanced data**. (This also emphasizes the generality of the least squares method.) Interpretation of unbalanced data is difficult. It is important that the application of least squares to the analysis of such data be understood if the results from computer programs designed for the analysis of unbalanced data are to be used correctly.

The objective of a research project determines the amount of effort to be devoted to the development of realistic models. If the intent is one of prediction only, the degree to which the model might be considered realistic is immaterial. The only requirement is that the predictions be adequately precise in the region of interest. On the other hand, realism is of primary importance if the goal is a thorough understanding of the system. The simple linear additive model can seldom be regarded as a realistic model. It is at best an approximation of the true model. Almost without exception, models developed from the basic principles of a process will be nonlinear in the parameters. The least squares estimation principle is still applicable but the mathematical methods become much more difficult. You will be introduced to **nonlinear least squares regression methods** and some of the more common nonlinear models.

Least squares estimation is controlled by the correlational structure observed among the independent and dependent variables in the data set. Observational data, data collected by observing the state of nature according to some sampling plan, will frequently cause special problems for least squares estimation because of strong correlations or, more generally, near-linear dependencies among the independent variables. The seriousness of the problems will depend on the use to be made of the analyses. Understanding the correlational structure of the data is most helpful in interpreting regression results and deciding what inferences might be made. Principal component analysis is introduced as an aid in characterizing the correlational structure of the data. A graphical procedure, Gabriel's biplot, is introduced to help visualize the correlational structure. Principal component analysis also serves as an introduction to **biased regression methods**. Biased regression methods are designed to alleviate the deleterious effects of near-linear dependencies (among the independent variables) on ordinary least squares estimation.

Least squares estimation is a powerful research tool and, with modern low-cost computers, is readily available. This ease of access, however, also facilitates misuse. Proper use of least squares requires an understanding of the basic method and assumptions on which it is built, and an awareness of the possible problems and their remedies. In some cases, alternative methods to least squares estimation might be more appropriate. It is the intent of this text to convey the basic understanding that will allow you to use least squares as an effective research tool.

Contents

Applied Regression Analysis: A Research Tool

CHAPTER ONE

Review of Simple Regression

This chapter reviews the elementary regression results for a linear model in one variable. The primary purpose is to establish a common notation and to point out the need for matrix notation. A light reading should suffice for most students.

Modeling refers to the development of mathematical expressions that describe in some sense the behavior of a random variable of interest. This variable may be the price of wheat in the world market, the number of deaths from lung cancer, the rate of growth of a particular type of tumor, or tensile strength of metal wire. In all cases, this variable will be called the **dependent variable** and denoted with Y. A subscript on Y will identify the particular unit from which the observation was taken, the time at which the price was recorded, the county in which the deaths were recorded, the experimental unit on which the tumor growth was recorded, and so forth. Most commonly the modeling is aimed at describing how the **mean** of the dependent variable, $E(Y)$, changes with changing conditions; the variance of the dependent variable is assumed to be unaffected by the changing conditions.

Other variables which are thought to provide information on the behavior of the dependent variable are incorporated into the model as predictor or explanatory variables. These variables will be called the **independent variables** and will be denoted by X with subscripts as needed to identify different independent variables. Additional subscripts will denote the observational unit from which the data were taken. The X's are assumed to be known constants. In addition to the X's, all models will involve unknown constants called **parameters**, which control the behavior of the model. These parameters will be denoted by Greek letters and are to be estimated from the data.

The mathematical complexity of the model and the degree to which it is a realistic model will depend on how much is known about the process being studied and on the purpose of the modeling exercise. In preliminary studies of a process or in cases where prediction is the primary objective, the models will almost always fall in the class of models that are **linear in the parameters**. That is, the parameters enter the model as simple coefficients on the independent variables or functions of the independent variables. Such models will be referred to loosely as **linear models**. The more realistic models, on the other hand, are often **nonlinear in the parameters**. Most growth models, for example, are nonlinear models. Nonlinear models fall into two categories: **intrinsically linear models**, which can be linearized by an appropriate transformation on the dependent variable, and those which cannot be so transformed. Most of the discussion will be devoted to the linear class of models and to those nonlinear models that are intrinsically linear. Nonlinear models are discussed in Section 11.2 and Chapter 14.

1.1 The Linear Model and Assumptions

Model

The simplest linear model involves only one independent variable and states that the true mean of the dependent variable changes at a constant rate as the value of the independent variable increases or decreases. Thus, the functional relationship between the true mean of Y_i, $\mathcal{E}(Y_i)$, and X_i is the equation of a straight line:

$$\mathcal{E}(Y_i) = \beta_0 + \beta_1(X_i) \qquad [1.1]$$

β_0 is the intercept, the value of $\mathcal{E}(Y_i)$ when $X = 0$, and β_1 is the slope of the line, the rate of change in $\mathcal{E}(Y_i)$ per unit change in X.

Assumptions

The observations on the dependent variable, Y_i, are assumed to be random observations from populations of random variables with the mean of each population given by $\mathcal{E}(Y_i)$. The deviation of an observation Y_i from its population mean $\mathcal{E}(Y_i)$ is taken into account by adding a random error ε_i to give the statistical model

$$Y_i = \beta_0 + \beta_1 X_i + \varepsilon_i \qquad [1.2]$$

The subscript i indicates the particular observational unit, $i = 1, 2, \ldots, n$. The X_i are the n observations on the independent variable and are assumed to be measured without error. That is, the observed values of X are assumed to be a set of known constants. The Y_i and X_i are paired observations; both are measured on every observational unit.

The random errors, ε_i, have zero mean and are assumed to have common variance σ^2 and to be pairwise independent. Since the only random element in the model is ε_i, these assumptions imply that the Y_i also have common variance σ^2 and are pairwise independent. For purposes of making tests of significance, the random errors are assumed to be normally distributed, which implies that the Y_i are also normally distributed. The assumptions about the random error are frequently stated as

$$\varepsilon_i \sim \text{NID}(0, \sigma^2) \qquad [1.3]$$

where "NID" stands for "normally and independently distributed." The quantities in parentheses denote the mean and the variance, respectively, of the normal distribution.

1.2 Least Squares Estimation

The simple linear model described in Section 1.1 has two parameters, β_0 and β_1, which are to be estimated from the data. If there were no random error in Y_i, any two data points could be used to solve explicitly for the values of the parameters. The random variation in Y, however, causes each pair of observed data points to give different results. (All estimates would be identical only if the observed data fell exactly on the straight line.) A method is needed that will combine all the information to give one solution which is "best" by some criterion.

The Least Squares Criterion

The **least squares estimation procedure** uses the criterion that the solution must give the smallest possible sum of squared deviations of the observed Y_i from the estimates of their true means provided by the solution. Let $\hat{\beta}_0$ and $\hat{\beta}_1$ be numerical estimates of the parameters β_0 and β_1, respectively, and let

$$\hat{Y}_i = \hat{\beta}_0 + \hat{\beta}_1 X_i \qquad [1.4]$$

be the estimated mean of Y for each X_i, $i = 1, \ldots, n$. Note that $\hat{Y}_i$ is obtained by substituting the estimates for the parameters in the functional form of the model relating $\mathscr{E}(Y_i)$ to X_i, given by equation 1.1. The least squares principle chooses $\hat{\beta}_0$ and $\hat{\beta}_1$ that minimize the sum of squares of the residuals, SS(Res):

$$\begin{aligned} \text{SS(Res)} &= \sum_{i=1}^{n} (Y_i - \hat{Y}_i)^2 \\ &= \sum e_i^2 \end{aligned} \qquad [1.5]$$

where $e_i = (Y_i - \hat{Y}_i)$ is the observed residual for the ith observation. The summation indicated by $\sum$ is over all observations in the data set as indicated by the index of summation, $i = 1$ to n. (The index of summation will be omitted when the limits of summation are clear from the context.)

The Normal Equations

The estimators for β_0 and β_1 are obtained by using calculus to find the values that minimize SS(Res). The derivatives of SS(Res) with respect to $\hat{\beta}_0$ and $\hat{\beta}_1$ in turn are set equal to zero. This gives two equations in two unknowns called the **normal equations**:

$$\begin{aligned} n(\hat{\beta}_0) + \left(\sum X_i\right)\hat{\beta}_1 &= \sum Y_i \\ \left(\sum X_i\right)\hat{\beta}_0 + \left(\sum X_i^2\right)\hat{\beta}_1 &= \sum X_i Y_i \end{aligned} \qquad [1.6]$$

Solving the normal equations simultaneously for $\hat{\beta}_0$ and $\hat{\beta}_1$ gives the estimates of β_1 and β_0 as

$$\hat{\beta}_1 = \frac{\sum(X_i - \bar{X})(Y_i - \bar{Y})}{\sum(X_i - \bar{X})^2} = \frac{\sum x_i y_i}{\sum x_i^2}$$
$$\hat{\beta}_0 = \bar{Y} - \hat{\beta}_1 \bar{X} \qquad [1.7]$$

Note that $x_i = (X_i - \bar{X})$ and $y_i = (Y_i - \bar{Y})$ denote observations expressed as deviations from their sample means, $\bar{X}$ and $\bar{Y}$, respectively. The more convenient forms for hand computation of sums of squares and sums of products are

$$\sum x_i^2 = \sum X_i^2 - \frac{(\sum X_i)^2}{n}$$
$$\sum x_i y_i = \sum X_i Y_i - \frac{(\sum X_i)(\sum Y_i)}{n} \qquad [1.8]$$

Thus, the computational formula for the slope is

$$\hat{\beta}_1 = \frac{\sum X_i Y_i - \frac{(\sum X_i)(\sum Y_i)}{n}}{\sum X_i^2 - \frac{(\sum X_i)^2}{n}} \qquad [1.9]$$

These estimates of the parameters give the regression equation

$$\hat{Y}_i = \hat{\beta}_0 + \hat{\beta}_1 X_i \qquad [1.10]$$

Example 1.1 The computations for the linear regression analysis are illustrated using treatment mean data from a study conducted by Dr. A. S. Heagle at North Carolina State University on effects of ozone pollution on soybean yield (Table 1.1). Four dose levels of ozone and the resulting mean seed yield of soybeans are given. The dose

Table 1.1 Mean yields of soybean plants (gm per plant) obtained in response to the indicated levels of ozone exposure over the growing season. (Data courtesy of Dr. A. S. Heagle, USDA and North Carolina State University.)

X *Ozone (ppm)*		Y *Yield (gm/plt)*
.02		242
.07		237
.11		231
.15		201
$\sum X_i = .35$		$\sum Y_i = 911$
$\bar{X} = .0875$		$\bar{Y} = 227.75$
$\sum X_i^2 = .0399$		$\sum Y_i^2 = 208{,}495$
	$\sum X_i Y_i = 76.99$	

of ozone is the average concentration (parts per million, ppm) during the growing season; yield is reported in grams per plant.

Assuming a linear relationship between yield and ozone dose, the simple linear model described by equation 1.2 is appropriate. The estimates of β_0 and β_1 obtained from equations 1.7 and 1.9 are

$$\hat{\beta}_1 = \frac{76.99 - \frac{(.35)(911)}{4}}{.0399 - \frac{(.35)^2}{4}} = -293.531 \qquad [1.11]$$

$$\hat{\beta}_0 = 227.75 - (-293.531)(.0875) = 253.434$$

The least squares regression equation characterizing the effects of ozone on the mean yield of soybeans in this study, assuming the linear model is correct, is

$$\hat{Y}_i = 253.434 - 293.531X_i \qquad [1.12]$$

The interpretation of $\hat{\beta}_1 = -294$ is that the mean yield is expected to decrease, since the slope is negative, by 294 grams per plant with each 1 ppm increase in ozone, or 2.94 grams with each .01 ppm increase in ozone. The observed range of ozone levels in the experiment was .02 ppm to .15 ppm. Therefore, it would be an unreasonable extrapolation to expect this rate of decrease in yield to continue if ozone levels were to increase, for example, to as much as 1 ppm. It is safe to use the results of regression only within the range of values of the independent variable. The intercept, $\hat{\beta}_0 = 253$ grams, is the value of Y where the regression line crosses the Y-axis. In this case, since the lowest dose is .02 ppm, it would be an extrapolation to interpret $\hat{\beta}_0$ as the estimate of the mean yield when there is no ozone. ■

1.3 Predicted Values and Residuals

The regression equation from Example 1.1 can be evaluated to obtain estimates of the mean of the dependent variable Y at chosen levels of the independent variable. Of course, the validity of such estimates is dependent on the assumed model being correct, or at least a good approximation to the correct model within the limits of the pollution doses observed in the study.

Estimates and Predictions

Each quantity computed from the fitted regression line, $\hat{Y}_i$, will be used as both (1) the **estimate** of the population mean of Y for that particular value of X and (2) the **prediction** of the value of Y one might obtain on some future observation at that level of X. Hence, the $\hat{Y}_i$ will be referred to both as **estimates** and as **predicted values**. Different degrees of confidence and different variances are associated with the two uses of $\hat{Y}_i$.

Residuals

If the observed values Y_i in the data set are compared with their corresponding values of $\hat{Y}_i$ computed from the regression equation, a measure of the degree of agreement between the model and the data is obtained. Remember that the least squares principle makes this agreement as "good as possible" in the least squares sense. The residuals

$$e_i = Y_i - \hat{Y}_i \qquad [1.13]$$

measure the discrepancy between the data and the fitted model. The results for Example 1.1 are shown in Table 1.2. Notice that the residuals sum to zero, as they always will when the model includes the constant term β_0. The least squares estimation procedure has minimized the sum of squares of the e_i. That is, there is no other choice of values for the two parameters β_0 and β_1 that will provide a smaller $\sum e_i^2$.

Example 1.2 A plot of the regression equation and the data from Example 1.1 (Figure 1.1) provides a visual check on the arithmetic and the adequacy with which the equation characterizes the data. The regression line crosses the Y-axis at the value of

Table 1.2 Observed values, estimated values, and residuals for the linear regression of soybean yield on ozone dosage.

Y_i	$\hat{Y}_i$	e_i	e_i^2
242	247.563	−5.563	30.947
237	232.887	4.113	16.917
231	221.146	9.854	97.101
201	209.404	−8.404	70.627
		$\sum e_i = 0.0$	$\sum e_i^2 = 215.592$

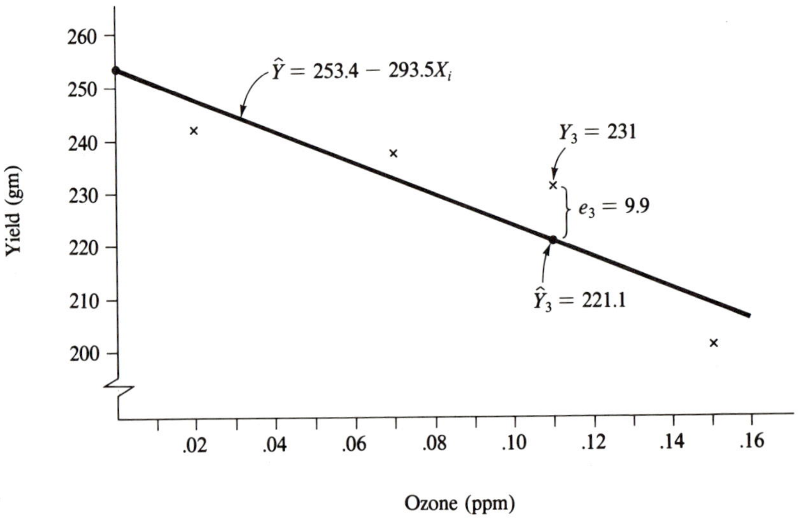

Figure 1.1 Regression of soybean yield on ozone level.

$\hat{\beta}_0 = 253.4$. The negative sign on $\hat{\beta}_1$ is reflected in the negative slope. Inspection of the plot shows that the regression line decreases to approximately $Y = 223$ when $X = .1$. This is a decrease of 30 grams of yield over a .1 ppm increase in ozone, or a **rate** of change of -300 grams in Y for each unit increase in X. This is reasonably close to the computed value of -293.5 grams per ppm. Figure 1.1 shows that the regression line "passes through" the data as well as could be expected from a straight-line relationship. The pattern of the deviations from the regression line, however, suggests that the linear model may not adequately represent the relationship. ■

1.4 Analysis of Variation in the Dependent Variable

The residuals are defined in equation 1.13 as the deviations of the observed values from the estimated values provided by the regression equation. Alternatively, each observed value of the dependent variable Y_i can be written as the sum of the estimated population mean of Y for the given value of X and the corresponding residual:

$$Y_i = \hat{Y}_i + e_i \qquad [1.14]$$

$\hat{Y}$ is the part of the observation Y_i "accounted for" by the model, whereas e_i reflects the "unaccounted for" part.

SS(Model) ***and*** **SS(Res)**

The **total uncorrected sum of squares** of Y_i, SS(Total$_{\text{uncorr}}$) $= \sum Y_i^2$, can be similarly partitioned. Substitute $\hat{Y}_i + e_i$ for each Y_i and expand the square. Thus,

$$\begin{aligned}\sum Y_i^2 &= \sum(\hat{Y}_i + e_i)^2 \\ &= \sum \hat{Y}_i^2 + \sum e_i^2 \\ &= \text{SS(Model)} + \text{SS(Res)}\end{aligned} \qquad [1.15]$$

(The cross-product term, $\sum \hat{Y}_i e_i$, is zero, as can be readily shown with the matrix notation of Chapter 3.) The term SS(Model) is the sum of squares "accounted for" by the model; SS(Res) is the "unaccounted for" part of the sum of squares. The forms SS(Model) $= \sum \hat{Y}_i^2$ and SS(Res) $= \sum e_i^2$ show the origins of these sums of squares. The more convenient computational forms are

$$\begin{aligned}\text{SS(Model)} &= n\bar{Y}^2 + \hat{\beta}_1^2 \sum(X_i - \bar{X})^2 \\ \text{SS(Res)} &= \text{SS(Total}_{\text{uncorr}}) - \text{SS(Model)}\end{aligned} \qquad [1.16]$$

The partitioning of the total uncorrected sum of squares can be reexpressed in terms of the **corrected sum of squares** by subtracting the sum of squares due to correction for the mean, the correction factor, $n\bar{Y}^2$, from each side of equation 1.15:

SS(Regr)

$$\text{SS}(\text{Total}_{\text{uncorr}}) - n\bar{Y}^2 = [\text{SS}(\text{Model}) - n\bar{Y}^2] + \text{SS}(\text{Res})$$

or, using equation 1.16,

$$\begin{aligned}\sum y_i^2 &= \hat{\beta}_1^2 \sum (X_i - \bar{X})^2 + \sum e_i^2 \\ &= \text{SS(Regr)} + \text{SS(Res)} \qquad [1.17]\end{aligned}$$

Notice that lowercase y is the deviation of Y from $\bar{Y}$ so that $\sum y_i^2$ is the **corrected total sum of squares**. Henceforth, SS(Total) will be used to denote the corrected sum of squares of the dependent variable. Also notice that SS(Model) denotes the sum of squares attributable to the entire model, whereas SS(Regr) denotes only that part of SS(Model) that exceeds the correction factor. The correction factor is the sum of squares for a model that contains *only* the constant term β_0. Such a model postulates that the mean of Y is a constant, or is unaffected by changes in X. Thus, SS(Regr) measures the *additional* information provided by the independent variable.

Degrees of Freedom

The **degrees of freedom** associated with each sum of squares is determined by the sample size n and the number of parameters p' in the model. [We will use p' to denote the number of parameters in the model and p (without the prime) to denote the number of independent variables; $p' = p + 1$ when the model includes an intercept as in equation 1.2.] The degrees of freedom associated with SS(Model) is $p' = 2$; the degrees of freedom associated with SS(Regr) is always 1 less to account for subtraction of the correction factor, which has 1 degree of freedom. SS(Res) will contain the $(n - p')$ degrees of freedom not accounted for by SS(Model). The mean squares are found by dividing each sum of squares by its degrees of freedom.

Example 1.3

The partitions of the degrees of freedom and sums of squares for the ozone data from Example 1.1 are given in Table 1.3. The definitional formulas for the sums

Table 1.3 Partitions of the degrees of freedom and sums of squares for yield of soybeans exposed to ozone.

Source of Variation	*Degrees of Freedom*	*Sum of Squares*	*Mean Square*
$\text{Total}_{\text{uncorr}}$	$n = 4$	$\sum Y_i^2 = 208{,}495$	
Corr. factor	1	$n\bar{Y}^2 = 207{,}480.25$	
$\text{Total}_{\text{corr}}$	$n - 1 = 3$	$\sum y_i^2 = 1{,}014.75$	
Due to model	$p' = 2$	$\sum \hat{Y}_i^2 = 208{,}279.39$	
Corr. factor	1	207,480.25	
Due to regr.	$p' - 1 = 1$	$\sum \hat{Y}_i^2 - n\bar{Y}^2 = 799.14$	799.14
Residual	$n - p' = 2$	$\sum e_i^2 = 215.61$	107.81

Table 1.4 Analysis of variance of yield of soybeans exposed to ozone pollution.

Source	d.f.	SS	MS
Total	3	1,014.75	
Due to regr.	1	799.14	799.14
Residual	2	215.61	107.81

of squares are included. An abbreviated form of Table 1.3, omitting the total uncorrected sum of squares, the correction factor, and SS(Model), is usually presented as the analysis of variance table (Table 1.4). ■

Coefficient of Determination

One measure of the contribution of the independent variable(s) in the model is the **coefficient of determination**, denoted by R^2:

$$R^2 = \frac{\text{SS(Regr)}}{\sum y_i^2} \tag{1.18}$$

This is the proportion of the (corrected) sum of squares of Y attributable to the information obtained from the independent variable(s). The coefficient of determination ranges from zero to one and is the square of the product moment correlation between Y_i and $\hat{Y}_i$. If there is only one independent variable, it is also the square of the correlation coefficient between Y_i and X_i.

Example 1.4

The coefficient of determination for the ozone data from Example 1.1 is

$$R^2 = \frac{799.14}{1{,}014.75} = .7875$$

The interpretation of R^2 is that 79% of the variation in the dependent variable, yield of soybeans, is "explained" by its linear relationship with the independent variable, ozone level. Caution must be exercised in the interpretation given to the phrase "explained by X." In this example, the data are from a controlled experiment where the level of ozone was being controlled in a properly replicated and randomized experiment. It is therefore reasonable to infer that any significant association of the variation in yield with variation in the level of ozone reflects a causal effect of the pollutant. If the data had been observational data, random observations on nature as it existed at some point in time and space, there would have been no basis for inferring causality. Model-fitting exercises can only reflect associations in the data. With observational data there are many reasons for associations among variables, only one of which is causality. ■

Expected Mean Squares

If the model is correct, the residual mean square is an unbiased estimate of σ^2, the variance among the random errors. The regression mean square is an unbiased estimate of $\sigma^2 + \beta_1^2(\sum x_i^2)$, where $\sum x_i^2 = \sum(X_i - \bar{X})^2$. These are referred to as the **mean square expectations** and are denoted by $\mathscr{E}$[MS(Res)] and $\mathscr{E}$[MS(Regr)]. Notice that MS(Regr) is estimating the same quantity as MS(Res) *plus* a positive quantity that depends on the magnitude of β_1 and $\sum x_i^2$. Thus, any linear relationship between Y and X, where $\beta_1 \neq 0$, will on the average make MS(Regr) larger than MS(Res). Comparison of MS(Regr) to MS(Res) provides the basis for judging the importance of the relationship.

Example 1.5 The estimate of σ^2 will be denoted by s^2. For the data of Example 1.1, MS(Res) $= s^2 = 107.81$ (Table 1.4). MS(Regr) $= 799.14$ is much larger than s^2, which suggests that β_1 is not zero. Testing of the null hypothesis that $\beta_1 = 0$ will be discussed in Section 1.6. ■

1.5 Precision of Estimates

Any quantity computed from random variables is itself a random variable. Thus, $\bar{Y}$, $\hat{Y}$, e, $\hat{\beta}_0$, and $\hat{\beta}_1$ are random variables computed from the Y_i. Measures of precision, variances or standard errors of the estimates, provide a basis for judging the reliability of the estimates.

Variance of a Linear Function

The computed regression coefficients, the $\hat{Y}_i$, and the residuals are all **linear functions** of the Y_i. Their variances can be determined using the basic definition of the variance of a linear function. Let $U = \sum a_i Y_i$ be an arbitrary linear function of the random variables Y_i, where the a_i are constants. The general formula for the variance of U is

$$\text{Var}(U) = \sum a_i^2 \,\text{Var}(Y_i) + \sum_{i \neq j}\sum a_i a_j \,\text{Cov}(Y_i, Y_j) \qquad [1.19]$$

where the double summation is over all possible $n(n-1)$ pairs of terms where i and j are not equal. Cov$(\cdot, \cdot)$ denotes the covariance between the two variables indicated in the parentheses. (Covariance measures the tendency of two variables to increase or decrease together.) When the random variables are independent, as is assumed in the usual regression model, all of the covariances are zero and the double summation term disappears. If, in addition, the variances of the random variables are equal, again as in the usual regression model where $\text{Var}(Y_i) = \sigma^2$ for all i, the variance of the linear function reduces to

$$\text{Var}(U) = \left(\sum a_i^2\right)\sigma^2 \qquad [1.20]$$

Variances of linear functions play an extremely important role in every aspect of statistics. Understanding the derivation of variances of linear functions will prove valuable; for this reason, we now give several examples.

Example 1.6 The variance of the sample mean of n observations will be derived. The coefficient a_i on each Y_i in the sample mean is $1/n$. If the Y_i have common variance σ^2 and zero covariances (that is, they are independent), equation 1.20 applies. The sum of squares of the coefficients is

$$\sum a_i^2 = n\left(\frac{1}{n}\right)^2 = \frac{1}{n}$$

and the variance of the mean becomes

$$\text{Var}(\bar{Y}) = \frac{\sigma^2}{n} \qquad [1.21]$$

which is the well-known result for the variance of the sample mean. ■

Example 1.7 In this example, the variance is derived for a linear contrast of three treatment means,

$$C = \bar{Y}_1 + \bar{Y}_2 - 2\bar{Y}_3 \qquad [1.22]$$

If each mean is the average of n independent observations from the same population, the variance of each sample mean is equal to $\text{Var}(\bar{Y}_i) = \sigma^2/n$ and all covariances are zero. The coefficients on the $\bar{Y}_i$ are 1, 1, and -2. Thus,

$$\begin{aligned}\text{Var}(C) &= (1)^2\text{Var}(\bar{Y}_1) + (1)^2\text{Var}(\bar{Y}_2) + (-2)^2\text{Var}(\bar{Y}_3)\\ &= (1 + 1 + 4)\left(\frac{\sigma^2}{n}\right) = 6\left(\frac{\sigma^2}{n}\right) \qquad [1.23]\end{aligned}$$

■

Variance of $\hat{\beta}_1$

We now turn to deriving the variances of $\hat{\beta}_1$, $\hat{\beta}_0$, and $\hat{Y}_i$. The variance of $\hat{Y}_i$ will be derived for the two situations where $\hat{Y}_i$ is used as an estimate of the mean and where it is used as a prediction. When $\hat{Y}_i$ is used as a prediction, its variance will be labeled $\text{Var}(\hat{Y}_{\text{pred}_i})$. To determine the variance of $\hat{\beta}_1$, express

$$\hat{\beta}_1 = \frac{\sum x_i y_i}{\sum x_i^2} \qquad [1.24]$$

as

$$\hat{\beta}_1 = \left(\frac{x_1}{\sum x_i^2}\right)Y_1 + \left(\frac{x_2}{\sum x_i^2}\right)Y_2 + \cdots + \left(\frac{x_n}{\sum x_i^2}\right)Y_n \qquad [1.25]$$

The coefficient on each Y_i is $x_i/\sum x_i^2$, which is a constant in the regression model. The Y_i are assumed to be independent and to have common variance σ^2. Thus, the variance of $\hat{\beta}_1$ is

$$\text{Var}(\hat{\beta}_1) = \left(\frac{x_1}{\sum x_i^2}\right)^2 \sigma^2 + \left(\frac{x_2}{\sum x_i^2}\right)^2 \sigma^2 + \cdots + \left(\frac{x_n}{\sum x_i^2}\right)^2 \sigma^2$$

$$= \frac{\sum x_i^2}{(\sum x_i^2)^2}\sigma^2 = \frac{\sigma^2}{\sum x_i^2} \qquad [1.26]$$

Variance of $\hat{\beta}_0$

Determining the variance of the intercept

$$\hat{\beta}_0 = \bar{Y} - \hat{\beta}_1 \bar{X} \qquad [1.27]$$

is a little more involved. The random variables in this linear function are $\bar{Y}$ and $\hat{\beta}_1$; the coefficients are 1 and $(-\bar{X})$. Equation 1.19 can be used to obtain the variance of $\hat{\beta}_0$:

$$\text{Var}(\hat{\beta}_0) = \text{Var}(\bar{Y}) + (-\bar{X})^2 \text{Var}(\hat{\beta}_1) + 2(-\bar{X})\text{Cov}(\bar{Y}, \hat{\beta}_1) \qquad [1.28]$$

It has been shown that $\text{Var}(\bar{Y}) = \sigma^2/n$ and $\text{Var}(\hat{\beta}_1) = \sigma^2/\sum x_i^2$, but $\text{Cov}(\bar{Y}, \hat{\beta}_1)$ remains to be determined.

Covariances of Linear Functions

The covariance between two linear functions is only slightly more complicated than the variance of a single linear function. Let U be the linear function defined earlier with a_i as coefficients and let W be a second linear function of the same random variables using d_i as coefficients:

$$U = \sum a_i Y_i \qquad \text{and} \qquad W = \sum d_i Y_i$$

The covariance of U and W is given by

$$\text{Cov}(U, W) = \sum a_i d_i \text{Var}(Y_i) + \sum\sum a_i d_j \text{Cov}(Y_i, Y_j) \qquad [1.29]$$

where the double summation is again over all possible $n(n-1)$ combinations of different values of the subscripts. If the Y_i are independent, the covariances are zero and equation 1.29 reduces to

$$\text{Cov}(U, W) = \sum a_i d_i \text{Var}(Y_i) \qquad [1.30]$$

Note that products of the corresponding coefficients are being used, whereas the squares of the coefficients were used in obtaining the variance of a linear function.

Variance of $\hat{\beta}_0$ (cont.)

Returning to the derivation of $\text{Var}(\hat{\beta}_0)$, where U and W are $\bar{Y}$ and $\hat{\beta}_1$, we note that the corresponding coefficients for each Y_i are $1/n$ and $x_i/\sum x_i^2$, respectively. Thus, the covariance between $\bar{Y}$ and $\hat{\beta}_1$ is

$$\text{Cov}(\bar{Y}, \hat{\beta}_1) = \sum\left(\frac{1}{n}\right)\left(\frac{x_i}{\sum x_i^2}\right)\text{Var}(Y_i)$$

$$= \left(\frac{1}{n}\right)\left(\frac{\sum x_i}{\sum x_i^2}\right)\sigma^2 = 0 \qquad [1.31]$$

since $\sum x_i = 0$. Thus, the variance of $\hat{\beta}_0$ reduces to

$$\begin{aligned} \text{Var}(\hat{\beta}_0) &= \text{Var}(\bar{Y}) + (\bar{X})^2\text{Var}(\hat{\beta}_1) \\ &= \frac{\sigma^2}{n} + \bar{X}^2\frac{\sigma^2}{\sum x_i^2} \\ &= \left(\frac{1}{n} + \frac{\bar{X}^2}{\sum x_i^2}\right)\sigma^2 \end{aligned} \qquad [1.32]$$

Variance of $\hat{Y}_i$

Recall that $\hat{\beta}_0$ is the estimated mean of Y when $X = 0$, and thus $\text{Var}(\hat{\beta}_0)$ can be thought of as the $\text{Var}(\hat{Y})$ for $X = 0$. The formula for $\text{Var}(\hat{\beta}_0)$ can be used to obtain the variance of any $\hat{Y}_i$ for any given value of X_i by replacing $\bar{X}$ with $(X_i - \bar{X})$. Thus,

$$\text{Var}(\hat{Y}_i) = \left[\frac{1}{n} + \frac{(X_i - \bar{X})^2}{\sum x_i^2}\right]\sigma^2 \qquad [1.33]$$

The variance of the fitted value attains its minimum of σ^2/n when the regression equation is being evaluated at $X_i = \bar{X}$, and increases as the value of X at which the equation is being evaluated moves away from $\bar{X}$.

Variance of Predictions

Equation 1.33 gives the appropriate variance when $\hat{Y}_i$ is being used as the **estimate** of the true mean of Y at the specific value of X. Recall that $\hat{Y}_i$ is also used as a **predictor** of some future observation. The variance for prediction must take into account the fact that the quantity being predicted is itself a random variable. The success of the prediction will depend on how small the difference is between $\hat{Y}_i$ and the future observation, say Y_0. The average squared difference between $\hat{Y}_i$ and Y_0, $\mathscr{E}(\hat{Y}_i - Y_0)^2$, is called the **mean squared error of prediction.** If the model is correct and prediction is for an individual in the same population from which the data were obtained, so that $\mathscr{E}(\hat{Y}_i - Y_0) = 0$, the mean squared error is also the variance of prediction. Assuming this to be the case, the **variance for prediction**, $\text{Var}(\hat{Y}_{\text{pred}_i})$, is the variance of the difference between $\hat{Y}_i$ and the future observation, Y_0:

$$\begin{aligned} \text{Var}(\hat{Y}_{\text{pred}_i}) &= \text{Var}(\hat{Y}_i - Y_0) \\ &= \text{Var}(\hat{Y}_i) + \sigma^2 \\ &= \left[1 + \frac{1}{n} + \frac{(X_i - \bar{X})^2}{\sum x_i^2}\right]\sigma^2 \end{aligned} \qquad [1.34]$$

Notice that the variance for prediction is the variance for estimation *plus* the variance of the quantity being predicted.

The derived variances are the *true* variances; they depend on knowledge of σ^2. $\text{Var}(\cdot)$ and σ^2 are used to designate true variances. *Estimated* variances are obtained by replacing σ^2 in the variance equations with an estimate of σ^2. The residual mean square from the analysis provides an estimate of σ^2 if the

Table 1.5 Summary of important formulas in simple regression.

Formula	*Estimate of*
$\hat{\beta}_1 = \sum x_i y_i / \sum x_i^2$	β_1
$\hat{\beta}_0 = \bar{Y} - \hat{\beta}_1 \bar{X}$	β_0
$\hat{Y}_i = \hat{\beta}_0 + \hat{\beta}_1 X_i$	$\mathcal{E}(Y_i)$
$e_i = Y_i - \hat{Y}_i$	ε_i
$\text{SS(Total}_{\text{uncorr}}) = \sum Y_i^2$	Total uncorrected sum of squares
$\text{SS(Total)} = \sum Y_i^2 - (\sum Y_i)^2/n$	Total corrected sum of squares
$\text{SS(Model)} = n\bar{Y}^2 + \hat{\beta}_1^2(\sum x_i^2)$	Sum of squares due to model
$\text{SS(Regr)} = \hat{\beta}_1^2(\sum x_i^2)$	Sum of squares due to X
$\text{SS(Res)} = \text{SS(Total)} - \text{SS(Regr)}$	Residual sum of squares
$R^2 = \text{SS(Regr)}/\text{SS(Total)}$	Coefficient of determination
$s^2(\hat{\beta}_1) = s^2/\sum x_i^2$	Variance of $\hat{\beta}_1$
$s^2(\hat{\beta}_0) = [1/n + \bar{X}^2/\sum x_i^2]s^2$	Variance of $\hat{\beta}_0$
$s^2(\hat{Y}_i) = [1/n + (X_i - \bar{X})^2/\sum x_i^2]s^2$	Variance of estimated mean
$s^2(\hat{Y}_{\text{pred}_i}) = [1 + 1/n + (X_i - \bar{X})^2/\sum x_i^2]s^2$	Variance of prediction

correct model has been fitted. As will be seen later, estimates of σ^2 that are not dependent on the correct regression model being used are available in some cases. The estimated variances obtained by substituting s^2 for σ^2 will be denoted by $s^2(\cdot)$, with the quantity in parentheses designating the random variable to which the variance applies.

Table 1.5 provides a summary to this point of the important formulas in linear regression with one independent variable.

Example 1.8 For the ozone data from Example 1.1, $s^2 = 107.81, n = 4$, and $\sum x_i^2 = [.0399 - (.35)^2/4] = .009275$. Thus, the estimated variances for the linear functions are as follows:

$$s^2(\hat{\beta}_1) = \frac{s^2}{\sum x_i^2} = \frac{107.81}{.009275} = 11{,}623.281$$

$$s^2(\hat{\beta}_0) = \left(\frac{1}{n} + \frac{\bar{X}^2}{\sum x_i^2}\right)s^2$$

$$= \left[\frac{1}{4} + \frac{(.0875)^2}{.009275}\right](107.81) = 115.942$$

$$s^2(\hat{Y}_1) = \left(\frac{1}{n} + \frac{(X_1 - \bar{X})^2}{\sum x_i^2}\right)s^2$$

$$= \left[\frac{1}{4} + \frac{(.02 - .0875)^2}{.009275}\right](107.81) = 79.91$$

Making appropriate changes in the values of X_i gives the variances of the remaining $\hat{Y}_i$:

$$s^2(\hat{Y}_2) = 30.51$$

$$s^2(\hat{Y}_3) = 32.84$$

$$s^2(\hat{Y}_4) = 72.35$$

If the $\hat{Y}_i$ are to be used as predictions of future random observations at the corresponding levels of ozone, each of the above variances would be increased by the amount $s^2 = 107.81$. Thus, the variances of the four *predictions* would be 187.72, 138.32, 140.65, and 180.16, respectively. ■

1.6 Tests of Significance and Confidence Intervals

Tests of Significance

The most common hypothesis of interest in simple linear regression is the hypothesis that the true value of the linear regression coefficient, the slope, is zero. This says that the dependent variable Y shows neither a linear increase nor decrease as the independent variable changes. In some cases, the nature of the problem will suggest other values for the null hypothesis. The computed regression coefficients, being random variables, will never exactly equal the hypothesized value even when the hypothesis is true. The role of the test of significance is to protect against being misled by the random variation in the estimates. Is the difference between the observed value of the parameter, $\hat{\beta}_1$, and the hypothesized value of the parameter greater than can be reasonably attributed to random variation? If so, the null hypothesis is rejected.

To accommodate the more general case, the null hypothesis is written as H_0: $\beta_1 = m$, where m is any constant of interest and of course can be equal to zero. The alternative hypothesis is H_a: $\beta_1 \neq m$, H_a: $\beta_1 > m$, or H_a: $\beta_1 < m$, depending on the expected behavior of β_1 if the null hypothesis is *not* true. In the first case, H_a: $\beta_1 \neq m$ is referred to as the two-tailed alternative hypothesis (interest is in detecting departures of β_1 from m in either direction) and leads to a two-tailed test of significance. The latter two alternative hypotheses, H_a: $\beta_1 > m$ and H_a: $\beta_1 < m$, are one-tailed alternatives and lead to one-tailed tests of significance.

If the random errors in the model, the ε_i, are normally distributed, the Y_i and any linear function of the Y_i will be normally distributed (see Searle, 1971). Thus, $\hat{\beta}_1$ is normally distributed with mean β_1 ($\hat{\beta}_1$ is shown to be unbiased in Chapter 3), and variance $\text{Var}(\hat{\beta}_1)$. If the null hypothesis that $\beta_1 = m$ is true, then $\hat{\beta}_1 - m$ is normally distributed with mean zero. Thus,

$$t = \frac{\hat{\beta}_1 - m}{s(\hat{\beta}_1)} \qquad [1.35]$$

is distributed as Student's t with degrees of freedom determined by the degrees of freedom in the estimate of σ^2 in the denominator. The computed t-value is compared to the appropriate critical value of Student's t (Appendix Table A), determined by the Type I error rate α and whether the alternative hypothesis is one-tailed or two-tailed. The critical value of Student's t for the two-tailed alternative hypothesis places probability $\alpha/2$ in each tail of the distribution. The critical values for the one-tailed alternative hypotheses place probability α in only the upper or lower tail of the distribution, depending on whether the alternative is $\beta_1 > m$ or $\beta_1 < m$, respectively.

Example 1.9 The estimate of β_1 for Heagle's ozone data from Example 1.1 was $\hat{\beta}_1 = -293.53$ with a standard error of $s(\hat{\beta}_1) = \sqrt{11{,}623.281} = 107.81$. Thus, the computed t-value for the test of H_0: $\beta_1 = 0$ is

$$t = \frac{-293.53}{107.81} = -2.72$$

The estimate of σ^2 in this example has only 2 degrees of freedom. Using the two-tailed alternative hypothesis and $\alpha = .05$ gives a critical t-value of $t_{(.05,2)} = 4.303$. Since $|t| < 4.303$, the conclusion is that the data do not provide convincing evidence that β_1 is different from zero.

In this example one might expect the increasing levels of ozone to *depress* the yield of soybeans; that is, the slope would be negative if not zero. The appropriate one-tailed alternative hypothesis would be H_a: $\beta_1 < 0$. For this one-tailed test, the critical value of t for $\alpha = .05$ is $t_{(.10,2)} = 2.920$. Although the magnitude of the computed t is close to this critical value, strict adherence to the $\alpha = .05$ size of test leads to the conclusion that there is insufficient evidence in these data to infer a real (linear) effect of ozone on soybean yield. (From a practical point of view, one would begin to suspect a real effect of ozone and seek more conclusive data.) ■

In a similar manner, t-tests of hypotheses about β_0 and any of the $\hat{Y}_i$ can be constructed. In each case, the numerator of the t-statistic is the difference between the estimated value of the parameter and the hypothesized value, and the denominator is the standard deviation (or standard error) of the estimate. The degrees of freedom for Student's t is always the degrees of freedom associated with the estimate of σ^2.

The F-statistic can be used as an alternative to Student's t for two-tailed hypotheses about the regression coefficients. It was indicated earlier that MS(Regr) is an estimate of $\sigma^2 + \beta_1^2 \sum x_i^2$ and that MS(Res) is an estimate of σ^2. If the null hypothesis that $\beta_1 = 0$ is true, both MS(Regr) and MS(Res) are estimating σ^2. As β_1 deviates from zero, MS(Regr) will become increasingly larger (on the average) than MS(Res). Therefore, a ratio of MS(Regr) to MS(Res) appreciably larger than unity would suggest that β_1 is not zero. This ratio of MS(Regr) to MS(Res) follows the F distribution when the assumption that the residuals are normally distributed is valid and the null hypothesis is true.

Example 1.10 For the ozone data of Example 1.1, the ratio of variances is

$$F = \frac{\text{MS(Regr)}}{\text{MS(Res)}} = \frac{799.14}{107.81} = 7.41$$

which can be compared to the critical value, $\alpha = .05$, of the F distribution with 1 degree of freedom in the numerator and 2 degrees of freedom in the denominator, $F_{(.05,1,2)} = 18.51$ (Appendix Table B), to determine whether MS(Regr) is sufficiently larger than MS(Res) to rule out chance as the explanation. Since $F = 7.41 < 18.51$, the conclusion is that the data do not provide conclusive evidence of a linear effect of ozone. The F-ratio with 1 degree of freedom in the numerator is the square of the corresponding t-statistic. Therefore, the F and the t are equivalent tests for this two-tailed alternative hypothesis. ■

Confidence Intervals

Confidence interval estimates of parameters are more informative than point estimates because they reflect the precision of the estimates. The 95% confidence interval estimates of β_1 and β_0 are, respectively,

$$\hat{\beta}_1 \pm t_{(.05,\nu)} s(\hat{\beta}_1) \tag{1.36}$$

and

$$\hat{\beta}_0 \pm t_{(.05,\nu)} s(\hat{\beta}_0) \tag{1.37}$$

where ν is the degrees of freedom associated with s^2.

Example 1.11 The 95% confidence interval estimate of β_1 for Example 1.1 is

$$-293.53 \pm (4.303)(107.81)$$

or $(-757, 170)$.

The confidence interval estimate indicates that the true value may fall anywhere between -757 and 170. This very wide range conveys a high degree of uncertainty (lack of confidence) in the point estimate $\hat{\beta}_1 = -293.53$. Notice that the interval includes zero. This is consistent with the conclusions from the t-test and the F-test that $H_0\colon \beta_1 = 0$ cannot be rejected.

The 95% confidence interval estimate of β_0 is

$$253.43 \pm (4.303)(10.77)$$

or (207.1, 299.8).

The value of β_0 might reasonably be expected to fall anywhere between 207 and 300 based on the information provided by this study. ■

In a similar manner, interval estimates of the true mean of Y for various values of X are computed using $\hat{Y}_i$ and their standard errors. Frequently, these

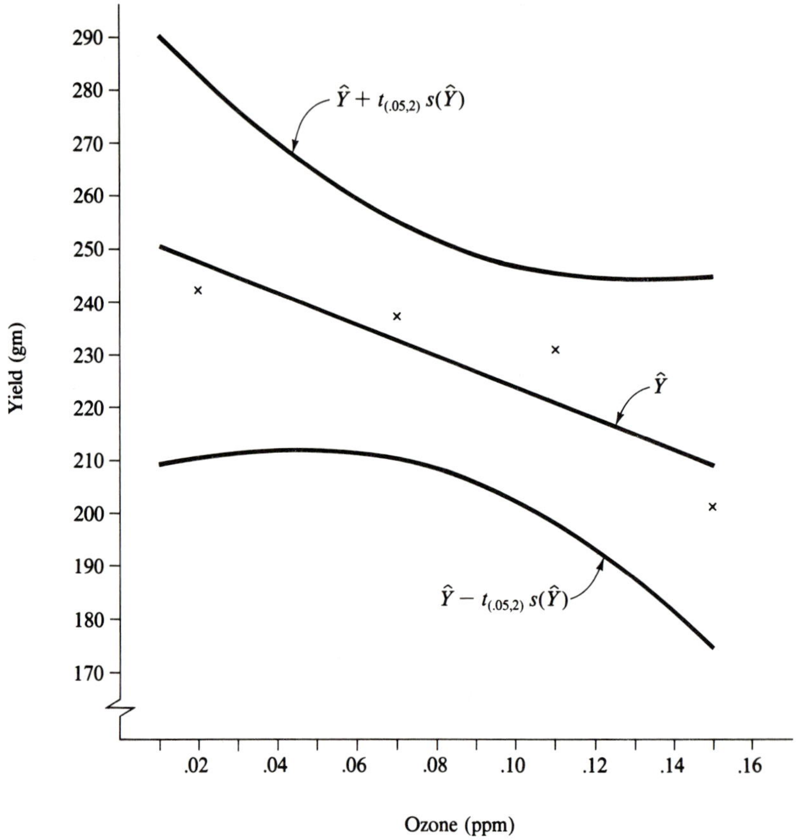

Figure 1.2 The regression of soybean mean yield (grams per plant) on ozone (ppm) showing the confidence interval estimates of the mean response.

confidence interval estimates of $\mathscr{E}(Y_i)$ are plotted with the regression line and the observed data. Such graphs convey an overall picture of how well the regression represents the data and the degree of confidence one might place in the results. Figure 1.2 shows the results for the ozone example. The confidence coefficient of .95 applies individually to the confidence intervals on each estimated mean. Simultaneous confidence intervals will be discussed in Section 4.6.

The failure of the tests of significance to detect an effect of ozone on the yield of soybeans is, in this case, a reflection of the lack of power in this small data set. This lack of power is due primarily to the limited degrees of freedom available for estimating σ^2. In defense of the research project from which these data were borrowed, we must point out that only a portion of the data (the set of treatment means) is being used for this illustration. The complete data set from this experiment provides for an adequate estimate of error and the effects of ozone are highly significant. The complete data set will be used at a later time.

1.7 Regression Through the Origin

In some situations the regression line is expected to pass through the origin. That is, the true mean of the dependent variable is expected to be zero when the value of the independent variable is zero. Many growth models, for example, would show zero growth at time zero. Or the amount of chemical produced in a system requiring a catalyst would be zero when there is no catalyst present. The linear regression model is forced to pass through the origin by setting β_0 equal to zero. The linear model then becomes

$$Y_i = \beta_1 X_i + \varepsilon_i \qquad [1.38]$$

There is now only one parameter to be estimated and application of the least squares principle gives

$$\beta_1\left(\sum X_i^2\right) = \sum X_i Y_i \qquad [1.39]$$

as the only normal equation to be solved. The solution is

$$\hat{\beta}_1 = \frac{\sum X_i Y_i}{\sum X_i^2} \qquad [1.40]$$

Both the numerator and denominator are now uncorrected sums of products and squares.

The regression equation becomes

$$\hat{Y}_i = \hat{\beta}_1 X_i \qquad [1.41]$$

and the residuals are defined as before,

$$e_i = Y_i - \hat{Y}_i \qquad [1.42]$$

Unlike the model with an intercept, in the no-intercept model the sum of the residuals is not zero.

The uncorrected sum of squares of Y can still be partitioned into the two parts

$$\text{SS(Model)} = \sum \hat{Y}_i^2 \qquad [1.43]$$

and

$$\text{SS(Res)} = \sum (Y_i - \hat{Y}_i)^2 = \sum e_i^2 \qquad [1.44]$$

Since only one parameter is involved in determining $\hat{Y}_i$, SS(Model) has only 1 degree of freedom and cannot be further partitioned into the correction for the mean and SS(Regr). For the same reason, the residual sum of squares has $(n - 1)$ degrees of freedom. The residual mean square is an estimate of σ^2 if the model is correct. The expectation of MS(Regr) is $\mathscr{E}[\text{MS(Regr)}] = \sigma^2 + \beta_1^2(\sum X_i^2)$. This is the same form as for a model with an intercept except the sum of squares for X is the uncorrected sum of squares.

The variance of $\hat{\beta}_1$ is determined using the rules for the variance of a linear function (see equations 1.25 and 1.26). The coefficients on the Y_i for the no-intercept model are $X_i/\sum X_i^2$. With the same assumptions of independence of the Y_i and common variance σ^2, the variance of $\hat{\beta}_1$ is

$$\begin{aligned}\text{Var}(\hat{\beta}_1) &= \left[\left(\frac{X_1}{\sum X_i^2}\right)^2 + \left(\frac{X_2}{\sum X_i^2}\right)^2 + \cdots + \left(\frac{X_n}{\sum X_i^2}\right)^2\right]\sigma^2 \\ &= \frac{\sigma^2}{\sum X_i^2} \qquad [1.45]\end{aligned}$$

The divisor on σ^2, the *uncorrected* sum of squares for the independent variable, will always be larger (usually much larger) than the corrected sum of squares. Therefore, the estimate of $\hat{\beta}_1$ will be much more precise when a no-intercept model is appropriate.

The variance of $\hat{Y}_i$ is most easily obtained by viewing it as a linear function of $\hat{\beta}_1$:

$$\hat{Y}_i = X_i\hat{\beta}_1 \qquad [1.46]$$

Thus, the variance is

$$\begin{aligned}\text{Var}(\hat{Y}_i) &= X_i^2\text{Var}(\hat{\beta}_1) \\ &= \left(\frac{X_i^2}{\sum X_i^2}\right)\sigma^2 \qquad [1.47]\end{aligned}$$

Estimates of the variances are obtained by substitution of s^2 for σ^2.

Example 1.12

Regression through the origin will be illustrated using data on increased risk incurred by individuals exposed to a toxic agent. Such health risks are often expressed as relative risk, the ratio of the rate of incidence of the health problem for those exposed to the rate of incidence for those not exposed to the toxic agent. A relative risk of 1.0 implies no increased risk of the disease from exposure to the agent. Table 1.6 gives the relative risk to individuals exposed to differing levels of dust in their work environments. Dust exposure is measured as the average number of particles/ft^3/year scaled by dividing by 10^6. By definition, the expected relative risk is 1.0 when exposure is zero. Thus, the regression line relating relative risk to exposure should have an intercept of 1.0 or, equivalently, the regression line relating $Y =$ (relative risk $-$ 1) to exposure should pass through the origin. The variable Y and key summary statistics on X and Y are included in Table 1.6.

Table 1.6 Relative risk of exposure to dust for nine groups of individuals. Dust exposure is reported in particles/ft^3/year and scaled by dividing by 10^6.

X = *Dust Exposure*	*Relative Risk*	Y = *Relative Risk* − 1
75	1.10	.10
100	1.05	.05
150	.97	−.03
350	1.90	.90
600	1.83	.83
900	2.45	1.45
1,300	3.70	2.70
1,650	3.52	2.52
2,250	4.16	3.16
$\sum X_i = 7{,}375$		$\sum Y_i = 11.68$
$\sum X_i^2 = 10{,}805{,}625$		$\sum Y_i^2 = 27.2408$
	$\sum X_i Y_i = 16{,}904$	

Table 1.7 Y_i, $\hat{Y}_i$, and e_i from linear regression through the origin of increase in relative risk (Y = relative risk − 1) on exposure level.

Y_i	$\hat{Y}_i$	e_i
.10	.1173	−.0173
.05	.1564	−.1064
−.03	.2347	−.2647
.90	.5475	.3525
.83	.9386	−.1086
1.45	1.4079	.0421
2.70	2.0337	.6663
2.52	2.5812	−.0612
3.16	3.5198	−.3598
$\sum Y_i^2 = 27.2408$	$\sum \hat{Y}_i^2 = 26.4441$	$\sum e_i^2 = .7967$

Assuming a linear relationship and zero intercept, the point estimate of the slope β_1 of the regression line is

$$\hat{\beta}_1 = \frac{\sum X_i Y_i}{\sum X_i^2} = \frac{16{,}904}{10{,}805{,}625} = .00156$$

The estimated increase in relative risk is .00156 for each increase in dust exposure of 1 million particles per cubic foot per year. The regression equation is

$$\hat{Y}_i = .00156 X_i$$

When $X_i = 0$, the value of $\hat{Y}_i$ is zero and the regression equation has been forced to pass through the origin.

The regression partitions each observation Y_i into two parts, that accounted for by the regression through the origin, $\hat{Y}_i$, and the residual or deviation from the regression line, e_i (Table 1.7). The sum of squares attributable to the model,

Table 1.8 Summary analysis of variance for regression through the origin of increase in relative risk on level of exposure to dust particles.

Source	d.f.	SS	MS	$\mathscr{E}$(MS)
$\text{Total}_{\text{uncorr}}$	$n = 9$	27.2408		
Due to model	$p = 1$	26.4441	26.4441	$\sigma^2 + \beta_1^2(\sum X_i^2)$
Residual	$n - p = 8$	.7967	.0996	σ^2

$$\text{SS(Model)} = \sum \hat{Y}_i^2 = 26.4441$$

and the sum of squares of the residuals,

$$\text{SS(Res)} = \sum e_i^2 = .7967$$

partition the total uncorrected sum of squares,

$$\sum Y_i^2 = 27.2408$$

In practice, the sum of squares due to the model is more easily computed as

$$\begin{aligned}\text{SS(Model)} &= \hat{\beta}_1^2(\sum X_i^2)\\ &= (.00156437)^2(10{,}805{,}625) = 26.4441\end{aligned}$$

The residual sum of squares is computed by difference. The summary analysis of variance, including the mean square expectations, is given in Table 1.8.

If the no-intercept model is correct, MS(Res) is an estimate of σ^2. MS(Model) is an estimate of σ^2 plus a quantity that is positive if β_1 is not zero. The ratio of the two mean squares provides a test of significance for H_0: $\beta_1 = 0$. This is an F-test with 1 and 8 degrees of freedom, if the assumption of normality is valid, and is significant beyond $\alpha = .001$. There is clear evidence that the linear regression relating increased risk to dust exposure is not zero.

The estimated variance of $\hat{\beta}_1$ is

$$s^2(\hat{\beta}_1) = \frac{s^2}{\sum X_i^2} = \frac{.09958533}{10{,}805{,}625} = 92.161 \times 10^{-10}$$

or

$$s(\hat{\beta}_1) = 9.6 \times 10^{-5} = .000096$$

Since each $\hat{Y}_i$ is obtained by multiplying $\hat{\beta}_1$ by the appropriate X_i, the estimated variance of a $\hat{Y}_i$ is

$$\begin{aligned}s^2(\hat{Y}_i) &= X_i^2[s^2(\hat{\beta}_1)]\\ &= (92.161 \times 10^{-10})X_i^2\end{aligned}$$

if $\hat{Y}_i$ is being used as an estimate of the true mean of Y for that value of X. If $\hat{Y}_i$ is to be used for prediction, the variance is

$$s^2(\hat{Y}_{\text{pred}_i}) = s^2 + s^2(\hat{Y}_i)$$
$$= .09958 + (92.161 \times 10^{-10})X_i^2$$

The variances and the standard errors provide measures of precision of the estimate and are used to construct tests of hypotheses and confidence interval estimates.

The data and a plot of the fitted regression line are shown in Figure 1.3. The 95% confidence interval estimates of the mean response $\hat{Y}_i$ are shown as bands on the regression line in the figure. Notice that with regression through the origin the confidence bands go to zero as the origin is approached. This is consistent with the model assumption that the mean of Y is *known* to be zero when $X = 0$. While the fit appears to be reasonable, there are suggestions that the model might be improved. The three lowest exposures fall below the regression line and very near zero; these levels of exposure may not be having as much impact as linear regression through the origin would predict. In addition, the largest residual, $e_7 = .6663$, is particularly noticeable. It is nearly twice as large

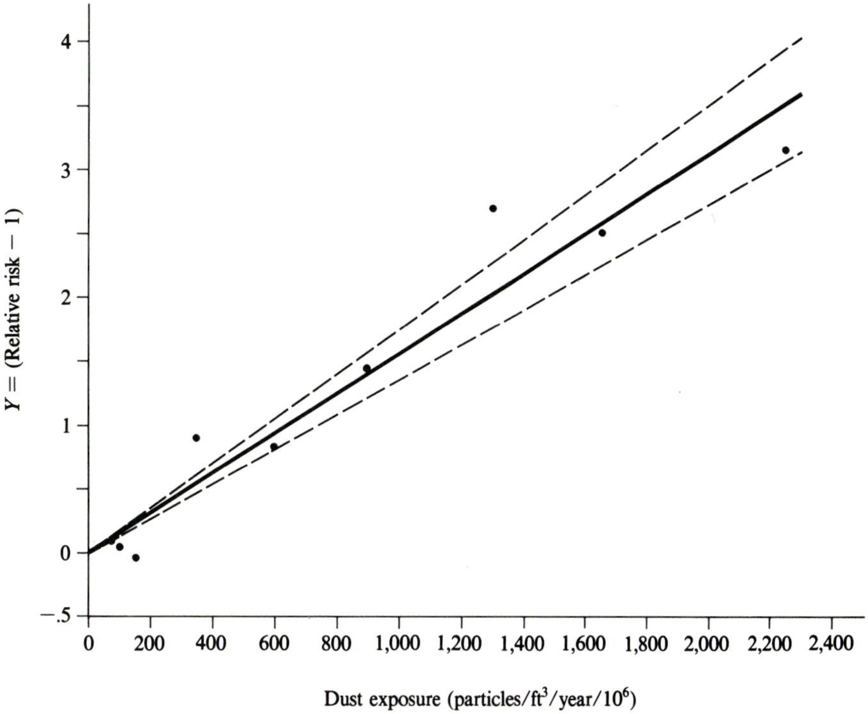

Figure 1.3 Regression of increase in relative risk on exposure to dust particles with the regression forced through the origin. The bands on the regression line connect the limits of the 95% confidence interval estimates of the means.

as the next largest residual and is the source of over half of the residual sum of squares (see Table 1.7). This large positive residual and the overall pattern of residuals suggest that a curvilinear relationship without the origin being forced to be zero would provide a better fit to the data. In practice, such alternative models would be tested before this linear no-intercept model would be adopted. We will forgo testing the need for a curvilinear relationship at this time (fitting curvilinear models will be discussed in Chapter 14) and continue with this example to illustrate testing the appropriateness of the no-intercept model assuming the linear relationship is appropriate.

The test of the assumption that β_0 is zero is made by temporarily adopting a model that allows a nonzero intercept. The estimate obtained of the intercept is then used to test the null hypothesis that β_0 is zero. Including an intercept in this example gives $\hat{\beta}_0 = .0360$ with $s(\hat{\beta}_0) = .1688$. (The residual mean square from the intercept model is $s^2 = .1131$ with 7 degrees of freedom.) The t-test for the null hypothesis that $\hat{\beta}_0$ is zero is

$$t = \frac{.0360}{.1688} = .213$$

and is not significant; $t_{(.05,7)} = 2.365$. There is no indication in these data that the no-intercept model is inappropriate. (Recall that this test has been made assuming the linear relationship is appropriate. If the model were expanded to allow a curvilinear response, the test of the null hypothesis that $\beta_0 = 0$ might become significant.) An equivalent test of the null hypothesis that $\beta_0 = 0$ can be made using the difference between the residual sums of squares from the intercept and no-intercept models. This test will be discussed in Chapter 4. ■

1.8 The Linear Model with Several Independent Variables

Most models will use more than one independent variable to explain the behavior of the dependent variable. The linear additive model can be extended to include any number of independent variables:

$$Y_i = \beta_0 + \beta_1 X_{i1} + \beta_2 X_{i2} + \beta_3 X_{i3} + \cdots + \beta_p X_{ip} + \varepsilon_i \qquad [1.48]$$

The subscript notation has been extended to include a number on each X and β to identify each independent variable and its regression coefficient. There are p independent variables and, including β_0, $p' = p + 1$ parameters to be estimated.

The usual least squares assumptions apply. The ε_i are assumed to be independent and to have common variance σ^2. For constructing tests of significance or confidence interval statements, the random errors are also assumed to be normally distributed. The independent variables are assumed to be measured without error.

The least squares method of estimation applied to this model requires that estimates of the $p + 1$ parameters be found such that

$$SS(Res) = \sum(Y_i - \hat{Y}_i)^2$$
$$= \sum(Y_i - \hat{\beta}_0 - \hat{\beta}_1 X_{i1} - \hat{\beta}_2 X_{i2} - \cdots - \hat{\beta}_p X_{ip})^2 \qquad [1.49]$$

is minimized. The $\hat{\beta}_j$, $j = 0, 1, \ldots, p$, are the estimates of the parameters. The values of $\hat{\beta}_j$ that minimize SS(Res) are obtained by setting the derivative of SS(Res) with respect to each $\hat{\beta}_j$ in turn equal to zero. This gives $(p + 1)$ normal equations which must be solved simultaneously to obtain the least squares estimates of the $(p + 1)$ parameters.

It is apparent that the problem is becoming increasingly difficult as the number of independent variables increases. The algebraic notation becomes particularly cumbersome. For these reasons, matrix notation and matrix algebra will be used to develop the regression results for the more complicated models. The next chapter is devoted to a brief review of the key matrix operations needed for the remainder of the text.

1.9 Summary

This chapter has reviewed the basic elements of least squares estimation for the simple linear model containing one independent variable. The more complicated linear model with several independent variables was introduced and will be pursued using matrix notation in subsequent chapters. The student should understand the following concepts:

- The form and basic assumptions of the linear model
- The least squares criterion, the estimators of the parameters obtained using this criterion, and measures of precision of the estimates
- The use of the regression equation to obtain estimates of mean values and predictions, and appropriate measures of precision for each
- The partitioning of the total variability of the response variable into that explained by the regression equation and the residual or unexplained part

EXERCISES

[*Note*: Answers to exercises preceded by an asterisk are given in the back of the book.]

1.1. Use the least squares criterion to derive the normal equations, equation 1.6, for the simple linear model of equation 1.2.

1.2. Solve the normal equations, equation 1.6, to obtain the estimates of β_0 and β_1 given in equation 1.7.

1.3. Use the statistical model

$$Y_i = \beta_0 + \beta_1 X_i + \varepsilon_i$$

to show that $\varepsilon_i \sim \text{NID}(0, \sigma^2)$ implies each of the following:

a. $\mathscr{E}(Y_i) = \beta_0 + \beta_1 X_i$

b. $\sigma^2(Y_i) = \sigma^2$

c. $\text{Cov}(Y_i, Y_{i'}) = 0$

For parts (b) and (c), use the following definitions of variance and covariance:

$$\sigma^2(Y_i) = \mathscr{E}\{[Y_i - \mathscr{E}(Y_i)]^2\}$$

$$\text{Cov}(Y_i, Y_{i'}) = \mathscr{E}\{[Y_i - \mathscr{E}(Y_i)][Y_{i'} - \mathscr{E}(Y_{i'})]\}$$

* 1.4. The data in the accompanying table relate heart rate at rest, Y, to kilograms body weight, X.

X		Y
90		62
86		45
67		40
89		55
81		64
75		53
$\sum X_i = 488$		$\sum Y_i = 319$
$\sum X_i^2 = 40{,}092$		$\sum Y_i^2 = 17{,}399$
	$\sum X_i Y_i = 26{,}184$	

a. Graph these data. Does it appear that there is a linear relationship between body weight and heart rate at rest?
b. Compute $\hat{\beta}_0$ and $\hat{\beta}_1$ and write the regression equation for these data. Plot the regression line on the graph from part (a). Interpret the estimated regression coefficients.
c. Now examine the data point (67, 40). If this data point were removed from the data set, what changes might occur in the estimates of β_0 and β_1?
d. Obtain the point estimate of the mean of Y when $X = 88$. Obtain a 95% confidence interval estimate of the mean of Y when $X = 88$. Interpret this interval statement.
e. Predict the heart rate for a particular subject weighing 88 kg using both a point prediction and a 95% confidence interval. Compare these predictions to the estimates computed in part (d).
f. Without doing the computations, for which measured X would the corresponding $\hat{Y}$ have the smallest variance? Why?

* 1.5. Use the data and regression equation from exercise 1.4 and compute $\hat{Y}_i$ for each value of X. Compute the product moment correlations between
a. X_i and Y_i
b. Y_i and $\hat{Y}_i$
c. X_i and $\hat{Y}_i$

Compare these correlations to each other and to the coefficient of determination, R^2. Can you prove algebraically the relationships you detect?

1.6. Show that

$$\text{SS(Model)} = n\bar{Y}^2 + \hat{\beta}_1^2 \sum (X_i - \bar{X})^2 \quad \text{(equation 1.16)}$$

1.7. Show that

$$\sum (Y_i - \hat{Y}_i)^2 = \sum y_i^2 - \hat{\beta}_1^2 \sum (X_i - \bar{X})^2$$

Note that $\sum y_i^2$ is being used to denote the *corrected* sum of squares.

1.8. Show algebraically that $\sum e_i = 0$ when the regression equation includes the constant term β_0. Show algebraically that this is not true when the regression does not include the intercept.

1.9. The accompanying data relate biomass production of soybeans to cumulative intercepted solar radiation over an 8-week period following emergence. Biomass production is the mean dry weight in grams of independent samples of four plants. (Data courtesy of Virginia Lesser and Dr. Mike Unsworth, North Carolina State University.)

X Solar Radiation	*Y* Plant Biomass
29.7	16.6
68.4	49.1
120.7	121.7
217.2	219.6
313.5	375.5
419.1	570.8
535.9	648.2
641.5	755.6

a. Compute $\hat{\beta}_0$ and $\hat{\beta}_1$ for the linear regression of plant biomass on intercepted solar radiation. Write the regression equation.
b. Place 95% confidence intervals on β_1 and β_0. Interpret the intervals.
c. Test H_0: $\beta_1 = 1.0$ versus H_a: $\beta_1 \neq 1.0$ using a t-test with $\alpha = .1$. Is your result for the t-test consistent with the confidence interval from part (b)? Explain.
d. Use a t-test to test H_0: $\beta_0 = 0$ against H_a: $\beta_0 \neq 0$. Interpret the results. Now fit a regression with $\beta_0 = 0$. Give the analysis of variance for the regression through the origin and use an F-test to test H_0: $\beta_0 = 0$. Compare the results of the t-test and the F-test. Do you adopt the model with or without the intercept?
e. Compute $s^2(\hat{\beta}_1)$ for the regression equation without an intercept. Compare the variances of the estimates of the slopes $\hat{\beta}_1$ for the two models. Which model provides the greater precision for the estimate of the slope?
f. Compute the 95% confidence interval estimates of the mean biomass production for $X = 30$ and $X = 600$ for both the intercept and the no-intercept models. Explain the differences in the intervals obtained for the two models.

*1.10. A linear regression was run on a set of data using an intercept and one independent variable. You are given only the following information:

1. $\hat{Y}_i = 11.5 - 1.5X_i$
2. The t-test for H_0: $\beta_1 = 0$ was nonsignificant at the $\alpha = .05$ level. A computed t of -4.087 was compared to $t_{(.05,2)}$ from Appendix Table A.
3. The estimate of σ^2 was $s^2 = 1.75$.

a. Complete the analysis of variance table using the given results.
b. Compute and interpret the coefficient of determination R^2.

*1.11. An experiment has yielded sample means for four treatment regimes, $\overline{Y}_1$, $\overline{Y}_2$, $\overline{Y}_3$, and $\overline{Y}_4$. The numbers of observations in the four means are $n_1 = 4$, $n_2 = 6$, $n_3 = 3$, and $n_4 = 9$. The pooled estimate of σ^2 is $s^2 = 23.5$.
a. Compute the variance of each treatment mean.
b. Compute the variance of the mean contrast $C = \overline{Y}_3 + \overline{Y}_4 - 2\overline{Y}_1$.

c. Compute the variance of $(\overline{Y}_1 + \overline{Y}_2 + \overline{Y}_3)/3$.
d. Compute the variance of $(4\overline{Y}_1 + 6\overline{Y}_2 + 3\overline{Y}_3)/13$.

*1.12. Obtain the normal equations and the least squares estimates for the model

$$Y_i = \mu + \beta_1 x_i + \varepsilon_i$$

where $x_i = (X_i - \overline{X})$. Compare the results to equation 1.6. (The model expressed in this form is referred to as the "centered" model; the independent variable has been shifted to have mean zero.)

1.13. Recompute the regression equation and analysis of variance for the Heagle ozone data (Table 1.1) using the centered model,

$$Y_i = \mu + \beta_1 x_i + \varepsilon_i$$

where $x_i = (X_i - \overline{X})$. Compare the results with those in Tables 1.2–1.4.

1.14. Derive the normal equation for the no-intercept model, equation 1.39, and the least squares estimate of the slope, equation 1.40.

1.15. Derive the variance of $\hat{\beta}_1$ and $\hat{Y}_i$ for the no-intercept model.

1.16. Show that

$$\sum(X_i - \overline{X})(Y_i - \overline{Y}) = \sum(X_i - \overline{X})Y_i = \sum X_i(Y_i - \overline{Y})$$

*1.17. The variance of $\hat{Y}_{\text{pred}_i}$ as given by equation 1.34 is for the prediction of a single future observation. Derive the variance of a prediction of the *mean* of q future observations all having the same value of X.

1.18. An experimenter wants to design an experiment for estimating the rate of change in a dependent variable Y as an independent variable X is changed. He is convinced from previous experience that the relationship is linear in the region of interest, between $X = 0$ and $X = 11$. He has enough resources to obtain 12 observations. Use $\sigma^2(\hat{\beta}_1)$, equation 1.26, to show the researcher the best allocation of the design points (choices of X-values). Compare $\sigma^2(\hat{\beta}_1)$ for this optimum allocation with an allocation of one observation at each integer value of X from $X = 0$ to $X = 11$.

*1.19. The data in the table relate seed weight of soybeans, collected for six successive weeks following the start of the reproductive stage, to cumulative seasonal solar radiation for two levels of chronic ozone exposure. Seed weight is mean seed weight (grams per plant) from independent samples of four plants. (Data courtesy of Virginia Lesser and Dr. Mike Unsworth.)

Low Ozone		*High Ozone*	
Radiation	*Seed Weight*	*Radiation*	*Seed Weight*
118.4	.7	109.1	1.3
215.2	2.9	199.6	4.8
283.9	5.6	264.2	6.5
387.9	8.7	358.2	9.4
451.5	12.4	413.2	12.9
515.6	17.4	452.5	12.3

a. Determine the linear regression of seed weight on radiation separately for each level of ozone. Determine the similarity of the two regressions by comparing the confidence interval estimates of the two intercepts and the two slopes and by visual inspection of plots of the data and the regressions.

b. Regardless of your conclusion in part (a), assume that the two regressions are the same and estimate the common regression equation.

1.20. A hotel experienced an outbreak of Pseudomona dermatis among its guests. Physicians suspected the source of infection to be the hotel whirlpool-spa. The data in the table give the number of female guests and the number infected by categories of time (minutes) spent in the whirlpool.

Time (Minutes)	*Number of Guests*	*Number Infected*
0–10	8	1
11–20	12	3
21–30	9	3
31–40	14	7
41–50	7	4
51–60	4	3
61–70	2	2

a. Can the incidence of infection (number infected/number exposed) be characterized by a linear regression on time spent in the whirlpool? Use the midpoint of the time interval as the independent variable. Estimate the intercept and the slope, and plot the regression line and the data.

b. Review each of the basic assumptions of least squares regression and comment on whether each is satisfied by these data.

*1.21. Hospital records were examined to assess the link between smoking and duration of illness. The data reported in the table are the number of hospital days (per 1,000 person-years) for several classes of individuals, the average number of cigarettes smoked per day, and the number of hospital days for control groups of nonsmokers for each class. (The control groups consist of individuals matched as nearly as possible to the smokers for several primary health factors other than smoking.)

# Hospital Days (Smokers)	*# Cigarettes Smoked/Day*	*# Hospital Days (Nonsmokers)*
215	10	201
185	5	180
334	15	297
761	45	235
684	25	520
368	30	210
1,275	50	195
3,190	45	835
3,520	60	435
428	20	312
575	5	590
2,280	45	1,131
2,795	60	225

a. Plot the logarithm of number of hospital days (for the smokers) against number of cigarettes. Do you think a linear regression will adequately represent the relationship?

b. Plot the logarithm of number of hospital days for smokers minus the logarithm of number of hospital days for the control group against number of cigarettes. Do you think a linear regression will adequately represent the relationship? Has subtraction of the control group means reduced the dispersion?

c. Define $Y = \ln(\#\text{ days for smokers}) - \ln(\#\text{ days for nonsmokers})$ and $X = (\#\text{ cigarettes})^2$. Fit the linear regression of Y on X. Make a test of significance to determine if the intercept can be set to zero. Depending on your results, give the regression equation, the standard errors of the estimates, and the summary analysis of variance.

CHAPTER
TWO

Introduction to Matrices

Chapter 1 reviewed simple linear regression in algebraic notation and showed that the notation for models involving several variables is very cumbersome.

This chapter introduces matrix notation and all matrix operations that will be used in this text. Matrix algebra greatly simplifies the presentation of regression and will be used throughout the text. Sections 2.7 and 2.8 will not be used until later in the text and can be omitted for now.

Matrix algebra is extremely helpful in multiple regression in simplifying notation and algebraic manipulations. You must be familiar with the basic operations of matrices in order to understand the regression results to be presented. A brief introduction to the key matrix operations is given in this chapter. You are referred to matrix algebra texts [for example, Searle (1982), Searle and Hausman (1970), or Stewart (1973)] for more complete presentations of matrix algebra.

2.1 Basic Definitions

Matrix

A **matrix** is a rectangular array of numbers arranged in orderly rows and columns. Matrices will be denoted with boldface capital letters. The following are examples:

$$\boldsymbol{Z} = \begin{bmatrix} 1 & 2 \\ 6 & 4 \\ 5 & 7 \end{bmatrix} \qquad \boldsymbol{X} = \begin{bmatrix} 1 & 5 \\ 1 & 6 \\ 1 & 4 \\ 1 & 9 \\ 1 & 2 \\ 1 & 6 \end{bmatrix}$$

$$\boldsymbol{B} = \begin{bmatrix} 15 & 7 & -1 & 0 \\ 15 & 5 & -2 & 10 \end{bmatrix}$$

Elements

The numbers that form a matrix are called the **elements** of the matrix. A general matrix could be denoted as

$$A = \begin{bmatrix} a_{11} & a_{12} & \cdots & a_{1n} \\ a_{21} & a_{22} & \cdots & a_{2n} \\ \vdots & \vdots & & \vdots \\ a_{m1} & a_{m2} & \cdots & a_{mn} \end{bmatrix}$$

The subscripts on the elements denote the row and column, respectively, in which the element appears. For example, a_{23} is the element found in the second row and third column. The row number is always given first.

Order

The **order** of a matrix is its size given by the number of rows and columns. The first matrix given, $\boldsymbol{Z}$, is of order (3, 2) or $\boldsymbol{Z}$ is a 3×2 matrix, since it has three rows and two columns. Matrix $\boldsymbol{A}$ is an $m \times n$ matrix.

Rank

The **rank** of a matrix is defined as the number of linearly independent columns (or rows) in the matrix. Any subset of columns of a matrix are **linearly independent** if no column can be expressed as a linear combination of the others. The matrix

$$A = \begin{bmatrix} 1 & 2 & 4 \\ 3 & 0 & 6 \\ 5 & 3 & 13 \end{bmatrix}$$

contains a linear dependency among its columns. The first column multiplied by 2 and added to the second column produces the third column. In fact, any one of the three columns of $\boldsymbol{A}$ can be written as a linear combination of the other two columns. On the other hand, any *two* columns of $\boldsymbol{A}$ are linearly independent since one cannot be produced as a multiple of the other. Thus, the rank of the matrix $\boldsymbol{A}$, denoted by $r(\boldsymbol{A})$, is 2.

Full-Rank Matrices

If there are no linear dependencies among the columns of a matrix, the matrix is said to be of **full rank**, or **nonsingular**. If a matrix is not of full rank it is said to be **singular**. The number of linearly independent rows of a matrix will always equal the number of linearly independent columns. The linear dependency among the rows of $\boldsymbol{A}$ is shown by 9(row 1) + 7(row 2) = 6(row 3). The critical matrices in regression will almost always have fewer columns than rows and, therefore, rank is more easily visualized by inspection of the columns.

2.2 Special Types of Matrices

Vector

A **vector** is a matrix having only one row or one column, and is called a row or column vector, respectively. Although vectors are often designated with boldface lowercase letters, this convention will not be followed rigorously in this text. A boldface capital letter will be used to designate a data vector and a boldface Greek letter will be used for vectors of parameters. Thus, for example,

$$v = \begin{pmatrix} 3 \\ 8 \\ 2 \\ 1 \end{pmatrix} \quad \text{is a } 4 \times 1 \text{ column vector.}$$

$$\boldsymbol{\mu} = (\mu_1 \ \mu_2 \ \mu_3) \quad \text{is a } 1 \times 3 \text{ row vector.}$$

(Vectors will usually be defined as column vectors but they need not be.) A single number such as 4, −2.1, or 0 is called a **scalar**.

Square Matrix

A **square matrix** has an equal number of rows and columns.

$$\boldsymbol{D} = \begin{bmatrix} 2 & 4 \\ 6 & 7 \end{bmatrix} \quad \text{is a } 2 \times 2 \text{ square matrix.}$$

Diagonal Matrix

A **diagonal matrix** is a square matrix in which all elements are zero except the elements on the main diagonal, the diagonal of elements, $a_{11}, a_{22}, \ldots, a_{nn}$, running from the upper-left cell to the lower-right cell.

$$\boldsymbol{A} = \begin{bmatrix} 5 & 0 & 0 \\ 0 & 4 & 0 \\ 0 & 0 & 8 \end{bmatrix} \quad \text{is a } 3 \times 3 \text{ diagonal matrix.}$$

Identity Matrix

An **identity matrix** is a diagonal matrix having all the diagonal elements equal to 1; such a matrix is denoted by $\boldsymbol{I}_n$. The subscript identifies the order of the matrix and will be omitted when the order is clear from the context.

$$\boldsymbol{I}_3 = \begin{bmatrix} 1 & 0 & 0 \\ 0 & 1 & 0 \\ 0 & 0 & 1 \end{bmatrix} \quad \text{is the } 3 \times 3 \text{ identity matrix.}$$

After matrix multiplication is discussed, it can be verified that multiplying any matrix by the identity matrix will *not* change the original matrix.

Symmetric Matrix

A **symmetric matrix** is a square matrix in which element a_{ij} equals element a_{ji} for all i and j. The elements form a symmetric pattern around the diagonal of the matrix.

$$\boldsymbol{A} = \begin{bmatrix} 5 & -2 & 3 \\ -2 & 4 & -1 \\ 3 & -1 & 8 \end{bmatrix} \quad \text{is a } 3 \times 3 \text{ symmetric matrix.}$$

Note that the first row is identical to the first column, the second row is identical to the second column, and so on.

2.3 Matrix Operations

Transpose

The **transpose** of a matrix $\boldsymbol{A}$, designated $\boldsymbol{A}'$, is the matrix obtained by using the rows of $\boldsymbol{A}$ as the columns of $\boldsymbol{A}'$. If

$$A = \begin{bmatrix} 1 & 2 \\ 3 & 8 \\ 4 & 1 \\ 5 & 9 \end{bmatrix}$$

the transpose of $\boldsymbol{A}$ is

$$A' = \begin{bmatrix} 1 & 3 & 4 & 5 \\ 2 & 8 & 1 & 9 \end{bmatrix}$$

If a matrix $\boldsymbol{A}$ has order $m \times n$, its transpose $\boldsymbol{A}'$ has order $n \times m$. A symmetric matrix is equal to its transpose: $\boldsymbol{A}' = \boldsymbol{A}$.

Addition

Addition of two matrices is defined if and only if the matrices are of the same order. Then, addition (or subtraction) consists of adding (or subtracting) the corresponding elements of the two matrices. For example,

$$\begin{bmatrix} 1 & 2 \\ 3 & 8 \end{bmatrix} + \begin{bmatrix} 7 & -6 \\ 8 & 2 \end{bmatrix} = \begin{bmatrix} 8 & -4 \\ 11 & 10 \end{bmatrix}$$

Addition is commutative: $\boldsymbol{A} + \boldsymbol{B} = \boldsymbol{B} + \boldsymbol{A}$.

Multiplication

Multiplication of two matrices is defined if and only if the number of *columns* in the first matrix equals the number of *rows* in the second matrix. If $\boldsymbol{A}$ is of order $r \times s$ and $\boldsymbol{B}$ is of order $m \times n$, the matrix product $\boldsymbol{AB}$ exists only if $s = m$. The matrix product $\boldsymbol{BA}$ exists only if $r = n$. Multiplication is most easily defined by first considering the multiplication of a row vector by a column vector. Let $\boldsymbol{a}' = (a_1 \ a_2 \ a_3)$ and $\boldsymbol{b}' = (b_1 \ b_2 \ b_3)$. (Notice that both $\boldsymbol{a}$ and $\boldsymbol{b}$ are defined as column vectors.) Then, the product of $\boldsymbol{a}'$ and $\boldsymbol{b}$ is

$$\begin{aligned} \boldsymbol{a}'\boldsymbol{b} &= (a_1 \ a_2 \ a_3)\begin{pmatrix} b_1 \\ b_2 \\ b_3 \end{pmatrix} \\ &= a_1b_1 + a_2b_2 + a_3b_3 \end{aligned} \qquad [2.1]$$

The result is a scalar equal to the sum of products of the corresponding elements. Let

$$\boldsymbol{a}' = (3 \ 6 \ 1) \qquad \text{and} \qquad \boldsymbol{b}' = (2 \ 4 \ 8)$$

The matrix product is

$$\boldsymbol{a}'\boldsymbol{b} = (3 \ 6 \ 1)\begin{pmatrix} 2 \\ 4 \\ 8 \end{pmatrix} = 6 + 24 + 8 = 38$$

Matrix multiplication is defined as a sequence of vector multiplications. Write

$$A = \begin{bmatrix} a_{11} & a_{12} & a_{13} \\ a_{21} & a_{22} & a_{23} \end{bmatrix} \quad \text{as} \quad A = \begin{pmatrix} \boldsymbol{a}'_1 \\ \boldsymbol{a}'_2 \end{pmatrix}$$

where $\boldsymbol{a}'_1 = (a_{11}\ a_{12}\ a_{13})$ and $\boldsymbol{a}'_2 = (a_{21}\ a_{22}\ a_{23})$ are the 1×3 row vectors in $\boldsymbol{A}$. Similarly, write

$$B = \begin{bmatrix} b_{11} & b_{12} \\ b_{21} & b_{22} \\ b_{31} & b_{32} \end{bmatrix} \quad \text{as} \quad B = (\boldsymbol{b}_1\ \boldsymbol{b}_2)$$

where $\boldsymbol{b}_1$ and $\boldsymbol{b}_2$ are the 3×1 column vectors in $\boldsymbol{B}$. Then the product of $\boldsymbol{A}$ and $\boldsymbol{B}$ is the 2×2 matrix

$$AB = C = \begin{bmatrix} \boldsymbol{a}'_1\boldsymbol{b}_1 & \boldsymbol{a}'_1\boldsymbol{b}_2 \\ \boldsymbol{a}'_2\boldsymbol{b}_1 & \boldsymbol{a}'_2\boldsymbol{b}_2 \end{bmatrix} = \begin{bmatrix} c_{11} & c_{12} \\ c_{21} & c_{22} \end{bmatrix} \qquad [2.2]$$

where

$$c_{11} = \boldsymbol{a}'_1\boldsymbol{b}_1 = \sum_{j=1}^{3} a_{1j}b_{j1} = a_{11}b_{11} + a_{12}b_{21} + a_{13}b_{31}$$

$$c_{12} = \boldsymbol{a}'_1\boldsymbol{b}_2 = \sum_{j=1}^{3} a_{1j}b_{j2} = a_{11}b_{12} + a_{12}b_{22} + a_{13}b_{32}$$

$$c_{21} = \boldsymbol{a}'_2\boldsymbol{b}_1 = \sum_{j=1}^{3} a_{2j}b_{j1} = a_{21}b_{11} + a_{22}b_{21} + a_{23}b_{31}$$

$$c_{22} = \boldsymbol{a}'_2\boldsymbol{b}_2 = \sum_{j=1}^{3} a_{2j}b_{j2} = a_{21}b_{12} + a_{22}b_{22} + a_{23}b_{32}$$

In general, element c_{ij} is obtained from the vector multiplication of the ith *row* vector from the first matrix and the jth *column* vector from the second matrix. The resulting matrix $\boldsymbol{C}$ has the number of rows equal to the number of rows in $\boldsymbol{A}$ and number of columns equal to the number of columns in $\boldsymbol{B}$.

Example 2.1 Let

$$T = \begin{bmatrix} 1 & 2 \\ 4 & 5 \\ 3 & 0 \end{bmatrix} \quad \text{and} \quad W = \begin{pmatrix} -1 \\ 3 \end{pmatrix}$$

The product $\boldsymbol{WT}$ is not defined because the number of columns in $\boldsymbol{W}$ is not equal to the number of rows in $\boldsymbol{T}$. The product $\boldsymbol{TW}$, however, is defined:

$$TW = \begin{bmatrix} 1 & 2 \\ 4 & 5 \\ 3 & 0 \end{bmatrix} \begin{pmatrix} -1 \\ 3 \end{pmatrix}$$

$$= \begin{pmatrix} (1)(-1) + (2)(3) \\ (4)(-1) + (5)(3) \\ (3)(-1) + (0)(3) \end{pmatrix} = \begin{pmatrix} 5 \\ 11 \\ -3 \end{pmatrix}$$

The resulting matrix is of order 3×1 with the elements determined by multiplication of the corresponding row vector from $\boldsymbol{T}$ by the column vector in $\boldsymbol{W}$. ■

Matrix multiplication is not commutative; $\boldsymbol{AB}$ does not necessarily equal $\boldsymbol{BA}$. As for the matrices $\boldsymbol{W}$ and $\boldsymbol{T}$ in Example 2.1, the matrices will often not be of the proper order for multiplication to be defined in both ways. The first step in matrix multiplication is to verify that the matrices do conform (have the proper order) for multiplication.

The transpose of a product is equal to the product *in reverse order* of the transposes of the two matrices. That is,

$$(\boldsymbol{AB})' = \boldsymbol{B}'\boldsymbol{A}' \qquad [2.3]$$

The transpose of the product of $\boldsymbol{T}$ and $\boldsymbol{W}$ from Example 2.1 is

$$(\boldsymbol{TW})' = \boldsymbol{W}'\boldsymbol{T}' = (-1 \;\; 3)\begin{bmatrix} 1 & 4 & 3 \\ 2 & 5 & 0 \end{bmatrix}$$

$$= (5 \;\; 11 \;\; -3)$$

Scalar multiplication is the multiplication of a matrix by a single number. Every element in the matrix is multiplied by the scalar. Thus,

$$3\begin{bmatrix} 2 & 1 & 7 \\ 3 & 5 & 9 \end{bmatrix} = \begin{bmatrix} 6 & 3 & 21 \\ 9 & 15 & 27 \end{bmatrix}$$

Orthogonal Vectors

Two vectors $\boldsymbol{v}_1$ and $\boldsymbol{v}_2$ of the same order are **orthogonal** if the vector product

$$\boldsymbol{v}_1'\boldsymbol{v}_2 = 0 \qquad [2.4]$$

If

$$\boldsymbol{v}_1 = \begin{pmatrix} 1 \\ 0 \\ -1 \\ 4 \end{pmatrix} \quad \text{and} \quad \boldsymbol{v}_2 = \begin{pmatrix} 3 \\ 4 \\ -1 \\ -1 \end{pmatrix}$$

then $\boldsymbol{v}_1$ and $\boldsymbol{v}_2$ are orthogonal because

$$\boldsymbol{v}_1'\boldsymbol{v}_2 = (1)(3) + (0)(4) + (-1)(-1) + (4)(-1) = 0$$

Idempotent Matrices

A square matrix is called **idempotent** if it remains unchanged when multiplied by itself. That is, the matrix $\boldsymbol{A}$ is idempotent if $\boldsymbol{AA} = \boldsymbol{A}$. It can be verified that the rank of an idempotent matrix is equal to the sum of the elements on the diagonal (Searle, 1982; Searle and Hausman, 1970). This sum of elements on the diagonal of a square matrix is called the **trace** of the matrix and is denoted by $\mathrm{tr}(\boldsymbol{A})$.

Example 2.2 The matrix

$$\boldsymbol{A} = \frac{1}{6}\begin{bmatrix} 5 & 2 & -1 \\ 2 & 2 & 2 \\ -1 & 2 & 5 \end{bmatrix}$$

is idempotent because

$$\boldsymbol{AA} = \boldsymbol{A}^2 = \frac{1}{6}\begin{bmatrix} 5 & 2 & -1 \\ 2 & 2 & 2 \\ -1 & 2 & 5 \end{bmatrix}\frac{1}{6}\begin{bmatrix} 5 & 2 & -1 \\ 2 & 2 & 2 \\ -1 & 2 & 5 \end{bmatrix}$$

$$= \frac{1}{6}\begin{bmatrix} 5 & 2 & -1 \\ 2 & 2 & 2 \\ -1 & 2 & 5 \end{bmatrix} = \boldsymbol{A}$$

The rank of $\boldsymbol{A}$ is given by

$$r(\boldsymbol{A}) = \mathrm{tr}(\boldsymbol{A}) = \tfrac{1}{6}(5 + 2 + 5) = 2$$ ■

Symmetry is not required for a matrix to be idempotent. However, all idempotent matrices of concern in this text will be symmetric.

Determinants

The **determinant** of a matrix is a scalar computed from the elements of the matrix according to well-defined rules. Determinants are defined only for square matrices and are denoted by $|\boldsymbol{A}|$, where $\boldsymbol{A}$ is a square matrix. The determinant of a 1×1 matrix is the scalar itself. The determinant of a 2×2 matrix,

$$\boldsymbol{A} = \begin{bmatrix} a_{11} & a_{12} \\ a_{21} & a_{22} \end{bmatrix}$$

is defined as

$$|\boldsymbol{A}| = a_{11}a_{22} - a_{12}a_{21} \qquad [2.5]$$

For example, if

$$A = \begin{bmatrix} 1 & 6 \\ -2 & 10 \end{bmatrix}$$

the determinant of A is

$$|A| = (1)(10) - (6)(-2) = 22$$

The determinants of higher-order matrices are obtained by expanding the determinants as linear functions of determinants of 2×2 submatrices. First, it is convenient to define the **minor** and the **cofactor** of an element in a matrix. Let A be a square matrix of order n. For any element a_{rs} in A, a square matrix of order $(n - 1)$ is formed by eliminating the row and column containing the element a_{rs}. Label this matrix A_{rs}, with the subscripts designating the row and column eliminated from A. Then $|A_{rs}|$, the determinant of A_{rs}, is called the **minor** of the element a_{rs}. The product $\theta_{rs} = (-1)^{r+s}|A_{rs}|$ is called the **cofactor** of a_{rs}. Each element in a square matrix has its own minor and cofactor.

The determinant of a matrix of order n is expressed in terms of the elements of any row or column and their cofactors. Using row i for illustration, we can express the determinant of A as

$$|A| = \sum_{j=1}^{n} a_{ij}\theta_{ij} \qquad [2.6]$$

where each θ_{ij} contains a determinant of order $(n - 1)$. Thus, the determinant of order n is expanded as a function of determinants of one less order. Each of these determinants, in turn, is expanded as a linear function of determinants of order $(n - 2)$. This substitution of determinants of one less order continues until $|A|$ is expressed in terms of determinants of 2×2 submatrices of A.

The first step of the expansion will be illustrated for a 3×3 matrix A. To compute the determinant of A, choose any row or column of the matrix. For each element of the row or column chosen, compute the cofactor of the element. Then, if the ith row of A is used for the expansion,

$$|A| = a_{i1}\theta_{i1} + a_{i2}\theta_{i2} + a_{i3}\theta_{i3} \qquad [2.7]$$

Example 2.3 For illustration, let

$$A = \begin{bmatrix} 2 & 4 & 6 \\ 1 & 2 & 3 \\ 5 & 7 & 9 \end{bmatrix}$$

and use the first row for the expansion of $|A|$. The cofactors of the elements in the first row are

$$\theta_{11} = (-1)^{(1+1)}\begin{vmatrix} 2 & 3 \\ 7 & 9 \end{vmatrix} = (18 - 21) = -3$$

$$\theta_{12} = (-1)^{(1+2)}\begin{vmatrix} 1 & 3 \\ 5 & 9 \end{vmatrix} = -(9 - 15) = 6$$

$$\theta_{13} = (-1)^{(1+3)}\begin{vmatrix} 1 & 2 \\ 5 & 7 \end{vmatrix} = (7 - 10) = -3$$

Then, the determinant of A is

$$|A| = 2(-3) + 4(6) + 6(-3) = 0$$

■

If the determinant of a matrix is zero, the matrix is **singular**, or it is not of full rank. Otherwise, the matrix is **nonsingular**. Thus, the matrix A in Example 2.3 is singular. The linear dependency is seen by noting that row 1 is equal to twice row 2. The determinants of larger matrices rapidly become difficult to compute and will be obtained with the help of a computer.

Inverse of a Matrix

Division in the usual sense does not exist in matrix algebra. The concept is replaced by multiplication of the matrix by its **inverse**. The inverse of a matrix A, designated by A^{-1}, is defined as the matrix that gives the identity matrix when multiplied by A. That is,

$$A^{-1}A = AA^{-1} = I \qquad [2.8]$$

The inverse of a matrix may not exist. A matrix has a *unique inverse* if and only if the matrix is square and nonsingular. A matrix is nonsingular if and only if its determinant is not zero.

The inverse of a 2×2 matrix is easily computed. If

$$A = \begin{bmatrix} a_{11} & a_{12} \\ a_{21} & a_{22} \end{bmatrix}$$

then

$$A^{-1} = \frac{1}{|A|}\begin{bmatrix} a_{22} & -a_{12} \\ -a_{21} & a_{11} \end{bmatrix} \qquad [2.9]$$

Note the rearrangement of the elements and the use of the determinant of A as the scalar divisor. For example, if

$$A = \begin{bmatrix} 4 & 3 \\ 1 & 2 \end{bmatrix} \quad \text{then} \quad A^{-1} = \begin{bmatrix} \frac{2}{5} & -\frac{3}{5} \\ -\frac{1}{5} & \frac{4}{5} \end{bmatrix}$$

That this is the inverse of A is verified by multiplication of A and A^{-1}:

$$AA^{-1} = \begin{bmatrix} 4 & 3 \\ 1 & 2 \end{bmatrix} \begin{bmatrix} \frac{2}{5} & -\frac{3}{5} \\ -\frac{1}{5} & \frac{4}{5} \end{bmatrix} = \begin{bmatrix} 1 & 0 \\ 0 & 1 \end{bmatrix}$$

The inverse of a matrix is obtained in general by (1) replacing every element of the matrix with its cofactor, (2) transposing the resulting matrix, and (3) dividing by the determinant of the original matrix, as illustrated in the next example.

Example 2.4 Consider the following matrix:

$$B = \begin{bmatrix} 1 & 3 & 2 \\ 4 & 5 & 6 \\ 8 & 7 & 9 \end{bmatrix}$$

The determinant of B is

$$\begin{aligned} |B| &= 1\begin{vmatrix} 5 & 6 \\ 7 & 9 \end{vmatrix} - 3\begin{vmatrix} 4 & 6 \\ 8 & 9 \end{vmatrix} + 2\begin{vmatrix} 4 & 5 \\ 8 & 7 \end{vmatrix} \\ &= (45 - 42) - 3(36 - 48) + 2(28 - 40) \\ &= 15 \end{aligned}$$

The unique inverse of B exists because $|B| \neq 0$. The cofactors for the elements of the first row of B were used in obtaining $|B|$: $\theta_{11} = 3$, $\theta_{12} = 12$, $\theta_{13} = -12$. The remaining cofactors are

$$\theta_{21} = -\begin{vmatrix} 3 & 2 \\ 7 & 9 \end{vmatrix} = -13 \quad \theta_{22} = \begin{vmatrix} 1 & 2 \\ 8 & 9 \end{vmatrix} = -7 \quad \theta_{23} = -\begin{vmatrix} 1 & 3 \\ 8 & 7 \end{vmatrix} = 17$$

$$\theta_{31} = \begin{vmatrix} 3 & 2 \\ 5 & 6 \end{vmatrix} = 8 \quad \theta_{32} = -\begin{vmatrix} 1 & 2 \\ 4 & 6 \end{vmatrix} = 2 \quad \theta_{33} = \begin{vmatrix} 1 & 3 \\ 4 & 5 \end{vmatrix} = -7$$

Thus, the matrix of cofactors is

$$\begin{bmatrix} 3 & 12 & -12 \\ -13 & -7 & 17 \\ 8 & 2 & -7 \end{bmatrix}$$

and the inverse of B is

$$B^{-1} = \frac{1}{15}\begin{bmatrix} 3 & -13 & 8 \\ 12 & -7 & 2 \\ -12 & 17 & -7 \end{bmatrix}$$

Notice that the matrix of cofactors has been transposed and divided by $|\boldsymbol{B}|$ to obtain $\boldsymbol{B}^{-1}$. It is left as an exercise to verify that this is the inverse of $\boldsymbol{B}$. As with the determinants, computers will be used to find the inverses of larger matrices. ■

2.4 Geometric Interpretation of Vectors

The elements of an $n \times 1$ vector can be thought of as the coordinates of a point in an n-dimensional coordinate system. The vector is represented in this n-space as the directional line connecting the origin of the coordinate system to the point specified by the elements. The direction of the vector is *from* the origin *to* the point; an arrowhead at the terminus indicates direction.

Vector Length

To illustrate, let $\boldsymbol{x}' = (3\ 2)$. This vector is of order 2 and is plotted in two-dimensional space as the line vector going from the origin (0, 0) to the point (3, 2) (see Figure 2.1). This can be viewed as the hypotenuse of a right triangle whose sides are of length 3 and 2, the elements of the vector $\boldsymbol{x}$. The length of $\boldsymbol{x}$ is then given by the Pythagorean theorem as the square root of the *sum of squares* of the elements of $\boldsymbol{x}$. Thus,

$$\text{length}(\boldsymbol{x}) = \sqrt{3^2 + 2^2} = \sqrt{13} = 3.61$$

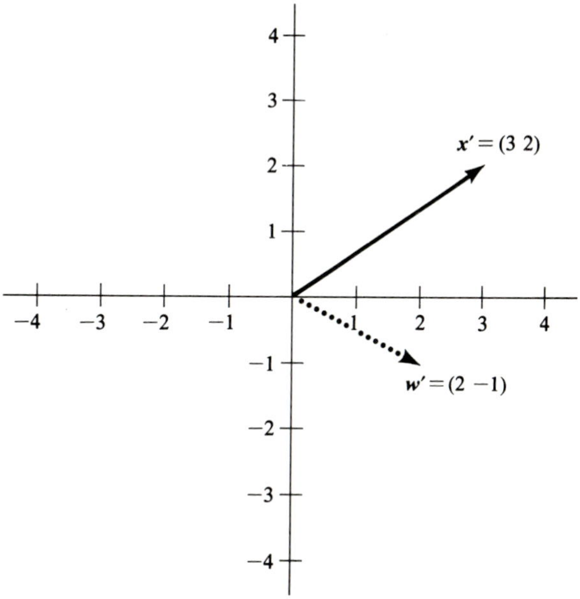

Figure 2.1 The geometric representation of the vectors $\boldsymbol{x} = (3\ 2)$ and $\boldsymbol{w} = (2\ -1)$ in two-dimensional space.

This result extends to the length of any vector regardless of its order. The sum of squares of the elements in a column vector $\boldsymbol{x}$ is given by (the matrix multiplication) $\boldsymbol{x}'\boldsymbol{x}$. Thus, the length of *any* vector $\boldsymbol{x}$ is

$$\text{length}(\boldsymbol{x}) = \sqrt{\boldsymbol{x}'\boldsymbol{x}} \qquad [2.10]$$

Space Defined by x

Multiplication of $\boldsymbol{x}$ by a scalar defines another vector which falls precisely on the line formed by extending the vector $\boldsymbol{x}$ indefinitely in both directions. For example,

$$\boldsymbol{u}' = (-1)\boldsymbol{x} = (-3 \ \ -2)$$

falls on the extension of $\boldsymbol{x}$ in the negative direction. Any point on this indefinite extension of $\boldsymbol{x}$ in both directions can be "reached" by multiplication of $\boldsymbol{x}$ with an appropriate scalar. This set of points constitutes the **space** defined by $\boldsymbol{x}$, or the space **spanned** by $\boldsymbol{x}$. It is a one-dimensional **subspace** of the two-dimensional space in which the vectors are plotted. A single vector of order n defines a one-dimensional subspace of the n-dimensional space in which the vector falls.

Linear Independence

The second vector $\boldsymbol{w}' = (2 \ \ -1)$, shown in Figure 2.1 with a dotted line, defines another one-dimensional subspace. The two subspaces defined by $\boldsymbol{x}$ and $\boldsymbol{w}$ are disjoint subspaces (except for the common origin). The two vectors are said to be **linearly independent** since neither falls in the subspace defined by the other. This implies that one vector cannot be obtained by multiplication of the other vector by a scalar.

Two-Dimensional Subspace

If the two vectors are considered jointly, any point in the plane can be "reached" by an appropriate linear combination of the two vectors. For example, the sum of the two vectors gives the vector $\boldsymbol{y}'$ (see Figure 2.2),

$$\boldsymbol{y}' = \boldsymbol{x}' + \boldsymbol{w}' = (3 \ \ 2) + (2 \ \ -1) = (5 \ \ 1)$$

The two vectors $\boldsymbol{x}$ and $\boldsymbol{w}$ define, or span, the two-dimensional subspace represented by the plane in Figure 2.2. Any third vector of order 2 in this two-dimensional space *must* be a linear combination of $\boldsymbol{x}$ and $\boldsymbol{w}$. That is, there *must* be a linear dependency among any three vectors that fall on this plane.

Vector Addition

Geometrically, the vector $\boldsymbol{x}$ is added to $\boldsymbol{w}$ by moving $\boldsymbol{x}$, while maintaining its direction, until the base of $\boldsymbol{x}$ rests on the terminus of $\boldsymbol{w}$. The resultant vector $\boldsymbol{y}$ is the vector from the origin (0, 0) to the new terminus of $\boldsymbol{x}$. The same result is obtained by moving $\boldsymbol{w}$ along the vector $\boldsymbol{x}$. This process is equivalent to completing the parallelogram using the two original vectors as adjacent sides. The sum $\boldsymbol{y}$ is the diagonal of the parallelogram running from the origin to the opposite corner (see Figure 2.2). Subtraction of two vectors, say $\boldsymbol{w}' - \boldsymbol{x}'$, is most easily viewed as addition of $\boldsymbol{w}'$ and $(-\boldsymbol{x}')$.

Three-Dimensional Subspace

Vectors of order 3 will be considered briefly to show the more general behavior. Each vector of order 3 can be plotted in three-dimensional space; the elements of the vector define the endpoint of the vector. Each vector individually defines a one-dimensional subspace of the three-dimensional space. This sub-

Figure 2.2 Geometric representation of the sum of two vectors.

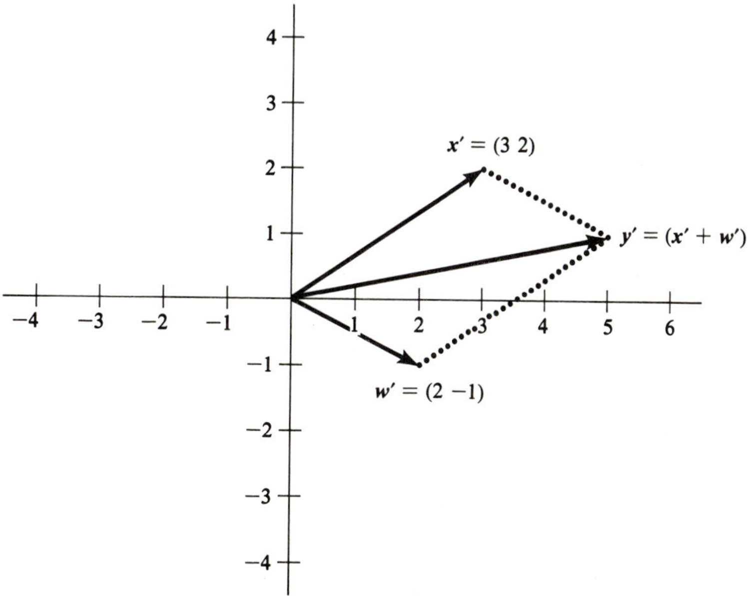

space is formed by extending the vector indefinitely in both directions. Any *two* vectors define a two-dimensional subspace if the two vectors are **linearly independent**—that is, as long as the two vectors do not define the same subspace. The two-dimensional subspace defined by two vectors is the set of points in the *plane* defined by the origin and the endpoints of the two vectors. The two vectors defining the subspace and any linear combination of them lie in this plane.

A three-dimensional space contains an infinity of two-dimensional subspaces. These can be visualized by rotating the plane around the origin. Any third vector that does not fall in the original plane will, in conjunction with either of the first two vectors, define another plane. Any three linearly independent vectors, or any two planes, completely define, or span, the three-dimensional space. Any fourth vector in that three-dimensional subspace must be a linear function of the first three vectors. That is, any four vectors in a three-dimensional subspace *must* contain a linear dependency.

The general results are stated in the box.

> 1. Any vector of order n can be plotted in n-dimensional space and defines a one-dimensional subspace of the n-dimensional space.
> 2. Any p linearly independent vectors of order n, $p \leq n$, define a p-dimensional subspace.
> 3. Any $p + 1$ vectors in a p-dimensional subspace *must* contain a linear dependency.

Orthogonal Vectors

Two vectors, $\boldsymbol{x}$ and $\boldsymbol{w}$, are said to be **orthogonal** to each other if the product $\boldsymbol{x}'\boldsymbol{w}$ or $\boldsymbol{w}'\boldsymbol{x}$ is zero. Geometrically, two orthogonal vectors are perpendicular to each other or they form a right angle at the origin. Two **linearly dependent** vectors form angles of zero or 180 degrees at the origin. All other angles reflect vectors that are neither orthogonal nor linearly dependent. In general, the cosine of the angle between two (column) vectors, $\boldsymbol{x}$ and $\boldsymbol{w}$, is

$$\cos(\alpha) = \frac{\boldsymbol{x}'\boldsymbol{w}}{\sqrt{\boldsymbol{x}'\boldsymbol{x}}\sqrt{\boldsymbol{w}'\boldsymbol{w}}} \qquad [2.11]$$

If the elements of each vector have mean zero, the *cosine* of the angle formed by two vectors is the **product moment correlation** between the two columns of data in the vectors. Thus, orthogonality of two such vectors corresponds to a zero correlation between the elements in the two vectors. If two such vectors are linearly dependent, the correlation coefficient between the elements of the two vectors will be either +1.0 or −1.0, depending on whether the vectors have the same or opposite directions.

2.5 Linear Equations and Solutions

A set of r linear equations in s unknowns is represented in matrix notation as $\boldsymbol{Ax} = \boldsymbol{y}$, where $\boldsymbol{x}$ is a vector of the s unknowns, $\boldsymbol{A}$ is the $r \times s$ matrix of known coefficients on the s unknowns, and $\boldsymbol{y}$ is the $r \times 1$ vector of known constants on the right-hand side of the equations.

A set of equations may have (1) no solution, (2) a unique solution, or (3) an infinite number of solutions. In order to have at least one solution, the equations must be **consistent**. This means that any linear dependencies among the rows of $\boldsymbol{A}$ must also exist among the corresponding elements of $\boldsymbol{y}$ (Searle and Hausman, 1970). For example, the equations

$$\begin{bmatrix} 1 & 2 & 3 \\ 2 & 4 & 6 \\ 3 & 3 & 3 \end{bmatrix} \begin{pmatrix} x_1 \\ x_2 \\ x_3 \end{pmatrix} = \begin{pmatrix} 6 \\ 10 \\ 9 \end{pmatrix}$$

are **inconsistent** because the second row of $\boldsymbol{A}$ is twice the first row, but the second element of $\boldsymbol{y}$ is *not* twice the first element. Since they are *not* consistent, there is no solution to this set of equations. Note that $\boldsymbol{x}' = (1\ \ 1\ \ 1)$ satisfies the first and third equations but not the second. If the second element of $\boldsymbol{y}$ were 12 instead of 10, the equations would be consistent and the solution $\boldsymbol{x}' = (1\ \ 1\ \ 1)$ would satisfy all three equations.

Consistent Equations

One method of determining if a set of equations is consistent is to compare the rank of $\boldsymbol{A}$ to the rank of the augmented matrix $[\boldsymbol{A}\ \ \boldsymbol{y}]$. The equations are consistent if and only if

$$r(\boldsymbol{A}) = r([\boldsymbol{A}\ \ \boldsymbol{y}]) \qquad [2.12]$$

Rank can be determined by using elementary (row and column) operations to reduce the elements below the diagonal to zero. (In a rectangular matrix, the diagonal is defined as the elements $a_{11}, a_{22}, \ldots, a_{dd}$, where d is the lesser of number of rows and number of columns.) The number of nonzero elements remaining *on* the diagonal is the rank of the matrix.

Example 2.5 Elementary operations on

$$A = \begin{bmatrix} 1 & 2 & 3 \\ 2 & 4 & 6 \\ 3 & 3 & 3 \end{bmatrix}$$

give

$$A^* = \begin{bmatrix} 1 & 2 & 3 \\ 0 & -3 & -6 \\ 0 & 0 & 0 \end{bmatrix}$$

so that $r(A) = 2$. [The elementary operations to obtain A^* are (1) subtract 2 times row 1 from row 2, (2) subtract 3 times row 1 from row 3, and (3) interchange rows 2 and 3.] The same elementary operations, plus interchanging columns 3 and 4, on the augmented matrix

$$[A \;\; y] = \begin{bmatrix} 1 & 2 & 3 & 6 \\ 2 & 4 & 6 & 10 \\ 3 & 3 & 3 & 9 \end{bmatrix}$$

yield

$$[A \;\; y]^* = \begin{bmatrix} 1 & 2 & 6 & 3 \\ 0 & -3 & -9 & -6 \\ 0 & 0 & -2 & 0 \end{bmatrix}$$

Thus, $r([A \;\; y]) = 3$. Since $r([A \;\; y]) \neq r(A)$, the equations are *not* consistent and, therefore, they have no solution. ■

Unique Solution

Consistent equations either have a unique solution or an infinity of solutions. If $r(A)$ equals the number of unknowns, the solution is unique and is given by

(1) $\boldsymbol{x} = A^{-1}\boldsymbol{y}$, when A is square; or

(2) $\boldsymbol{x} = A_1^{-1}\boldsymbol{y}$, where A_1 is a full-rank submatrix of A, when A is rectangular.

Example 2.6 The equations $A\boldsymbol{x} = \boldsymbol{y}$ with

$$A = \begin{bmatrix} 1 & 2 \\ 3 & 3 \\ 5 & 7 \end{bmatrix} \quad \text{and} \quad \boldsymbol{y} = \begin{pmatrix} 6 \\ 9 \\ 21 \end{pmatrix}$$

are consistent. (Proof of consistency is left as an exercise.) The rank of A equals the number of unknowns $[r(A) = 2]$, so that the solution is unique. Any two linearly independent equations in the system of equations can be used to obtain the solution. Using the first two rows gives the full-rank equations

$$\begin{bmatrix} 1 & 2 \\ 3 & 3 \end{bmatrix} \begin{pmatrix} x_1 \\ x_2 \end{pmatrix} = \begin{pmatrix} 6 \\ 9 \end{pmatrix}$$

with the solution

$$\begin{pmatrix} x_1 \\ x_2 \end{pmatrix} = \begin{bmatrix} 1 & 2 \\ 3 & 3 \end{bmatrix}^{-1} \begin{pmatrix} 6 \\ 9 \end{pmatrix} = \frac{1}{3} \begin{bmatrix} -3 & 2 \\ 3 & -1 \end{bmatrix} \begin{pmatrix} 6 \\ 9 \end{pmatrix} = \begin{pmatrix} 0 \\ 3 \end{pmatrix}$$

Notice that the solution $x' = (0 \;\; 3)$ satisfies the third equation also. ■

Infinite Solutions

When $r(A)$ in a consistent set of equations is less than the number of unknowns, there is an infinity of solutions.

Example 2.7

Consider the equations $Ax = y$ with

$$A = \begin{bmatrix} 1 & 2 & 3 \\ 2 & 4 & 6 \\ 3 & 3 & 3 \end{bmatrix} \quad \text{and} \quad y = \begin{pmatrix} 6 \\ 12 \\ 9 \end{pmatrix}$$

The rank of A is $r(A) = 2$ and elementary operations on the augmented matrix $[A \;\; y]$ give

$$[A \;\; y]^* = \begin{bmatrix} 1 & 2 & 3 & 6 \\ 0 & -3 & -6 & -18 \\ 0 & 0 & 0 & 0 \end{bmatrix}$$

Thus, $r([A \;\; y]) = 2$, which equals $r(A)$, and the equations are consistent. However, $r(A)$ is less than the number of unknowns, so there is an infinity of solutions. This follows from the fact that one element of x can be chosen arbitrarily and the remaining two chosen to satisfy the set of equations. For example, if x_1 is chosen to be 1, the solution is $x' = (1 \;\; 1 \;\; 1)$, whereas if x_1 is chosen to be 2, the solution is $x' = (2 \;\; -1 \;\; 2)$. ■

Solutions Using Generalized Inverses

A more general method of finding a solution to a set of consistent equations involves the use of **generalized inverses**. There are several definitions of generalized inverses (see Searle, 1971; Searle and Hausman, 1970; Rao, 1973). An adequate definition for our purposes is the following (Searle and Hausman, 1970):

> A generalized inverse of A is any matrix A^- that satisfies the condition $AA^-A = A$.

(A^- will be used to denote a generalized inverse.) The generalized inverse is not unique (unless A is square and of full rank, in which case, $A^- = A^{-1}$). A generalized inverse can be used to express a solution to a set of consistent equations $Ax = y$ as $x = A^-y$. This solution is unique only when $r(A)$ equals the number of unknowns in the set of equations. (The computer will be used to obtain generalized inverses when needed.)

Example 2.8 For illustration, consider the set of consistent equations $Ax = y$, where

$$A = \begin{bmatrix} 1 & 2 \\ 3 & 3 \\ 5 & 7 \end{bmatrix} \quad \text{and} \quad y = \begin{pmatrix} 6 \\ 9 \\ 21 \end{pmatrix}$$

It has been shown that $r(A) = 2$, which equals the number of unknowns, so the solution is unique. A generalized inverse of A is

$$A^- = \frac{1}{18}\begin{bmatrix} -10 & 16 & -4 \\ 8 & -11 & 5 \end{bmatrix}$$

and the unique solution is given by

$$x = A^-y = \begin{pmatrix} 0 \\ 3 \end{pmatrix}$$

It is left as an exercise to verify the matrix multiplication of A^-y and that $AA^-A = A$. ■

Example 2.9 For another illustration of the use of generalized inverses, consider again the consistent equations $Ax = y$ from Example 2.7, where

$$A = \begin{bmatrix} 1 & 2 & 3 \\ 2 & 4 & 6 \\ 3 & 3 & 3 \end{bmatrix} \quad \text{and} \quad y = \begin{pmatrix} 6 \\ 12 \\ 9 \end{pmatrix}$$

This system has been shown to have an infinity of solutions. A generalized inverse of A is

$$A^- = \begin{bmatrix} -\frac{1}{10} & -\frac{2}{10} & \frac{4}{9} \\ 0 & 0 & \frac{1}{9} \\ \frac{1}{10} & \frac{2}{10} & -\frac{2}{9} \end{bmatrix}$$

which gives the solution

$$x' = A^{-}y = (1 \;\; 1 \;\; 1)$$

This happens to agree with the first solution obtained in Example 2.7. Again, it is left as an exercise to verify that $x = A^{-}y$ and $AA^{-}A = A$. ■

2.6 Orthogonal Transformations and Projections

The linear transformation of vector x to vector y, both of order n, is written as $y = Ax$, where A is the $n \times n$ matrix of coefficients effecting the transformation. The transformation is a one-to-one transformation only if A is nonsingular. Then, the inverse transformation of y to x is $x = A^{-1}y$.

Orthogonal Transformations

A linear transformation is an **orthogonal** transformation if $AA' = I$. This condition implies that the row vectors of A are orthogonal and of unit length. Orthogonal transformations maintain distances and angles between vectors. That is, the spatial relationships among the vectors are not changed with orthogonal transformations.

Example 2.10 For illustration, let $y_1' = (3 \;\; 10 \;\; 20)$, $y_2' = (6 \;\; 14 \;\; 21)$, and

$$A = \begin{bmatrix} 1 & 1 & 1 \\ -1 & 0 & 1 \\ -1 & 2 & -1 \end{bmatrix}$$

Then

$$x_1 = Ay_1 = \begin{bmatrix} 1 & 1 & 1 \\ -1 & 0 & 1 \\ -1 & 2 & -1 \end{bmatrix} \begin{pmatrix} 3 \\ 10 \\ 20 \end{pmatrix} = \begin{pmatrix} 33 \\ 17 \\ -3 \end{pmatrix}$$

and

$$x_2 = Ay_2 = \begin{bmatrix} 1 & 1 & 1 \\ -1 & 0 & 1 \\ -1 & 2 & -1 \end{bmatrix} \begin{pmatrix} 6 \\ 14 \\ 21 \end{pmatrix} = \begin{pmatrix} 41 \\ 15 \\ 1 \end{pmatrix}$$

are linear transformations of y_1 to x_1 and y_2 to x_2. These are not orthogonal transformations because

$$AA' = \begin{bmatrix} 3 & 0 & 0 \\ 0 & 2 & 0 \\ 0 & 0 & 6 \end{bmatrix} \neq I$$

The rows of A are mutually orthogonal (the off-diagonal elements are zero) but they do not have unit length. This can be made into an orthogonal transformation by scaling each row vector of A to have unit length by dividing each vector by its length. Thus,

$$x_1^* = A^* y_1 = \begin{bmatrix} \frac{1}{\sqrt{3}} & \frac{1}{\sqrt{3}} & \frac{1}{\sqrt{3}} \\ -\frac{1}{\sqrt{2}} & 0 & \frac{1}{\sqrt{2}} \\ -\frac{1}{\sqrt{6}} & \frac{2}{\sqrt{6}} & -\frac{1}{\sqrt{6}} \end{bmatrix} y_1 = \begin{bmatrix} \frac{33}{\sqrt{3}} \\ \frac{17}{\sqrt{2}} \\ -\frac{3}{\sqrt{6}} \end{bmatrix}$$

and

$$x_2^* = A^* y_2 = \begin{bmatrix} \frac{41}{\sqrt{3}} \\ \frac{15}{\sqrt{2}} \\ \frac{1}{\sqrt{6}} \end{bmatrix}$$

are orthogonal transformations. It is left as an exercise to verify that the orthogonal transformation has maintained the distance between the two vectors; that is, verify that

$$(y_1 - y_2)'(y_1 - y_2) = (x_1^* - x_2^*)'(x_1^* - x_2^*) = 26$$

[The squared distance between two vectors u and v is $(u - v)'(u - v)$.] ■

Projections

Projection of a vector onto a subspace is a special case of a transformation. (Projection is a key step in least squares.) The objective of a projection is to transform y in n-dimensional space to that vector $\hat{y}$ in a subspace such that $\hat{y}$ is as close to y as possible. A linear transformation of y to $\hat{y}$, $\hat{y} = Py$, is a **projection** if and only if P is idempotent and symmetric (Rao, 1973), in which case P is referred to as a projection matrix. The subspace of the projection is defined, or spanned, by the columns or rows of P. If P is a projection matrix, $(I - P)$ is also a projection matrix. However, since P and $(I - P)$ are orthogonal matrices, the projection by $(I - P)$ is onto the subspace *orthogonal* to that defined by P. The rank of a projection matrix is the dimension of the subspace onto which it projects and, since the projection matrix is idempotent, the rank is equal to its trace.

Example 2.11 The matrix

$$A = \frac{1}{6}\begin{bmatrix} 5 & 2 & -1 \\ 2 & 2 & 2 \\ -1 & 2 & 5 \end{bmatrix}$$

is an idempotent matrix. Since A is symmetric, it is also a projection matrix. Thus, the linear transformation

$$\hat{y} = Ay_1 = \frac{1}{6}\begin{bmatrix} 5 & 2 & -1 \\ 2 & 2 & 2 \\ -1 & 2 & 5 \end{bmatrix}\begin{pmatrix} 3 \\ 10 \\ 20 \end{pmatrix} = \begin{pmatrix} 2.5 \\ 11 \\ 19.5 \end{pmatrix}$$

is a projection of $y_1' = (3 \;\; 10 \;\; 20)$ onto the subspace defined by the columns of A. The vector $\hat{y}$ is the unique vector in this subspace that is closest to y_1. That is, $(y_1 - \hat{y})'(y_1 - \hat{y})$ is a minimum. Since A is a projection matrix, so is

$$I - A = \begin{bmatrix} 1 & 0 & 0 \\ 0 & 1 & 0 \\ 0 & 0 & 1 \end{bmatrix} - \frac{1}{6}\begin{bmatrix} 5 & 2 & -1 \\ 2 & 2 & 2 \\ -1 & 2 & 5 \end{bmatrix} = \frac{1}{6}\begin{bmatrix} 1 & -2 & 1 \\ -2 & 4 & -2 \\ 1 & -2 & 1 \end{bmatrix}$$

Then,

$$e = (I - A)y_1 = \frac{1}{6}\begin{bmatrix} 1 & -2 & 1 \\ -2 & 4 & -2 \\ 1 & -2 & 1 \end{bmatrix}\begin{pmatrix} 3 \\ 10 \\ 20 \end{pmatrix} = \begin{pmatrix} \frac{1}{2} \\ -1 \\ \frac{1}{2} \end{pmatrix}$$

is a projection onto the subspace orthogonal to the subspace defined by A. Note that $\hat{y}'e = 0$ and $\hat{y} + e = y_1$. ■

2.7 Eigenvalues and Eigenvectors

Eigenvalues and eigenvectors of matrices are needed for some of the methods to be discussed, including principal component analysis, principal component regression, and assessment of the impact of collinearity. Determining the eigenvalues and eigenvectors of a matrix is a difficult computational problem and computers will be used for all but the very simplest cases. However, the reader needs to develop an understanding of the eigenanalysis of a matrix.

The discussion of eigenanalysis will be limited to real, symmetric, nonnegative definite matrices and only key results will be given. The reader is referred to other texts [such as Searle and Hausman (1970)] for more general discussions. In particular, Searle and Hausman (1970) show several important applications of eigenanalysis of asymmetric matrices. **Real matrices** do not contain any complex numbers as elements. **Symmetric, nonnegative definite matrices** are obtained from products of the type $\boldsymbol{B}'\boldsymbol{B}$ and, if used as the defining matrix in a quadratic form (see Chapter 4), yield only zero or positive scalars.

Definitions

It can be shown that for a real, symmetric matrix $\boldsymbol{A}$ ($n \times n$) there exists a set of n nonnegative scalars, λ_i^2, and n nonzero vectors, z_i, $i = 1, \ldots, n$, such that

$$\boldsymbol{A}z_i = \lambda_i^2 z_i$$

$$\text{or} \quad \boldsymbol{A}z_i - \lambda_i^2 z_i = \boldsymbol{0}$$

$$\text{or} \quad (\boldsymbol{A} - \lambda_i^2 \boldsymbol{I})z_i = \boldsymbol{0}, \quad i = 1, \ldots, n \qquad [2.13]$$

The λ_i^2 are the n **eigenvalues** (characteristic roots or latent roots) of the matrix $\boldsymbol{A}$ and the z_i are the corresponding (column) **eigenvectors** (characteristic vectors or latent vectors). (Notice that λ_i^2 is used to denote the eigenvalues. The positive square roots of the eigenvalues, which are called the **singular values**, will be denoted with λ_i.)

Solution

There are nonzero solutions to equation 2.13 only if the matrix $(\boldsymbol{A} - \lambda_i^2\boldsymbol{I})$ is less than full rank—that is, only if the determinant of $(\boldsymbol{A} - \lambda_i^2\boldsymbol{I})$ is zero. The λ_i^2 and z_i are obtained by solving the general determinantal equation

$$|\boldsymbol{A} - \lambda^2\boldsymbol{I}| = 0 \qquad [2.14]$$

Since $\boldsymbol{A}$ is of order $n \times n$, the determinant of $(\boldsymbol{A} - \lambda_i^2\boldsymbol{I})$ is an nth-degree polynomial in λ^2. Solving this equation gives the n values of λ^2, which are not necessarily distinct. Each value of λ^2 is then used in turn in equation 2.13 to find the companion eigenvector z_i.

When the eigenvalues are distinct, the vector solution to equation 2.13 is unique except for an arbitrary scale factor and sign. By convention, each eigenvector is defined to be the solution vector scaled to have unit length; that is, $z_i'z_i = 1$. Further, the eigenvectors are mutually orthogonal: $z_i'z_j = 0$ when $i \neq j$. When the eigenvalues are not distinct, there is an additional degree of arbitrariness in defining the subsets of vectors corresponding to each subset of nondistinct eigenvalues. Nevertheless, the eigenvectors for each subset can be chosen so that they are mutually orthogonal as well as orthogonal to the eigenvectors of all other eigenvalues. Thus, if $\boldsymbol{Z} = (z_1 \; z_2 \; \ldots \; z_n)$ is the matrix of eigenvectors, then $\boldsymbol{Z}'\boldsymbol{Z} = \boldsymbol{I}$. This implies that $\boldsymbol{Z}'$ is the inverse of $\boldsymbol{Z}$ so that $\boldsymbol{Z}\boldsymbol{Z}' = \boldsymbol{I}$ as well.

Decomposition of a Matrix

Using $\boldsymbol{Z}$ and $\boldsymbol{L}$, defined as the diagonal matrix of the λ_i^2, we can write the initial equations $\boldsymbol{A}z_i = \lambda_i^2 z_i$ as

$$\boldsymbol{A}\boldsymbol{Z} = \boldsymbol{Z}\boldsymbol{L} \qquad [2.15]$$

$$\text{or} \quad \boldsymbol{Z}'\boldsymbol{A}\boldsymbol{Z} = \boldsymbol{L} \qquad [2.16]$$

$$\text{or} \quad \boldsymbol{A} = \boldsymbol{Z}\boldsymbol{L}\boldsymbol{Z}' \qquad [2.17]$$

Equation 2.16 shows that A can be transformed to a diagonal matrix by pre- and postmultiplying by Z' and Z, respectively. Since L is a diagonal matrix, equation 2.17 shows that A can be expressed as the sum of matrices:

$$A = ZLZ' = \sum \lambda_i^2 (z_i z_i') \qquad [2.18]$$

where the summation is over the n eigenvalues and eigenvectors. Each term is an $n \times n$ matrix of rank 1 so that the sum can be viewed as a decomposition of the matrix A into n matrices that are mutually orthogonal. Some of these may be zero matrices if the corresponding λ_i^2 are zero. The rank of A is revealed by the number of nonzero eigenvalues λ_i^2.

Example 2.12 For illustration, consider the matrix

$$A = \begin{bmatrix} 10 & 3 \\ 3 & 8 \end{bmatrix}$$

The eigenvalues of A are found by solving the determinantal equation (equation 2.14),

$$|(A - \lambda^2 I)| = \left| \begin{bmatrix} 10 - \lambda^2 & 3 \\ 3 & 8 - \lambda^2 \end{bmatrix} \right| = 0$$

or

$$(10 - \lambda^2)(8 - \lambda^2) - 9 = (\lambda^2)^2 - 18\lambda^2 + 71 = 0$$

The solutions to this quadratic (in λ^2) equation are

$$\lambda_1^2 = 12.16228 \qquad \text{and} \qquad \lambda_2^2 = 5.83772$$

arbitrarily ordered from largest to smallest. Thus, the matrix of eigenvalues of A is

$$L = \begin{bmatrix} 12.16228 & 0 \\ 0 & 5.83772 \end{bmatrix}$$

The eigenvector corresponding to $\lambda_1^2 = 12.16228$ is obtained by solving equation 2.13 for the elements of z_1:

$$(A - 12.16228I)\begin{pmatrix} z_{11} \\ z_{21} \end{pmatrix} = 0$$

or

$$\begin{bmatrix} -2.1622776 & 3 \\ 3 & -4.162276 \end{bmatrix}\begin{pmatrix} z_{11} \\ z_{21} \end{pmatrix} = 0$$

Arbitrarily setting $z_{11} = 1$ and solving for z_{21}, using the first equation, gives $z_{21} = .720759$. Thus, the vector $z_1' = (1 \;\; .720759)$ satisfies the first equation (and it can be verified that it also satisfies the second equation). Rescaling this vector so it has unit length by dividing by

$$\text{length}(z_1) = \sqrt{z_1' z_1} = \sqrt{1.5194935} = 1.232677$$

gives the first eigenvector

$$z_1' = (.81124 \;\; .58471)$$

The elements of z_2 are found in the same manner to be

$$z_2' = (-.58471 \;\; .81124)$$

Thus, the matrix of eigenvectors for $\boldsymbol{A}$ is

$$\boldsymbol{Z} = \begin{bmatrix} .81124 & -.58471 \\ .58471 & .81124 \end{bmatrix}$$

Notice that the first *column* of $\boldsymbol{Z}$ is the first eigenvector, and the second column is the second eigenvector. ■

Example 2.13 Continuing with Example 2.12, notice that the matrix $\boldsymbol{A}$ is of rank 2 because both eigenvalues are nonzero. The decomposition of $\boldsymbol{A}$ into two rank-1 matrices, $\boldsymbol{A} = \boldsymbol{A}_1 + \boldsymbol{A}_2$, equation 2.18, is given by

$$\boldsymbol{A}_1 = \lambda_1^2 z_1 z_1' = 12.16228 \begin{pmatrix} .81124 \\ .58471 \end{pmatrix} (.81124 \;\; .58471)$$

$$= \begin{bmatrix} 8.0042 & 5.7691 \\ 5.7691 & 4.1581 \end{bmatrix}$$

and

$$\boldsymbol{A}_2 = \lambda_2^2 z_2 z_2' = \begin{bmatrix} 1.9958 & -2.7691 \\ -2.7691 & 3.8419 \end{bmatrix}$$

Since the two columns of $\boldsymbol{A}_1$ are multiples of the same vector $\boldsymbol{u}_1$, they are linearly dependent and, therefore, $r(\boldsymbol{A}_1) = 1$. Similarly, $r(\boldsymbol{A}_2) = 1$. Multiplication of $\boldsymbol{A}_1$ by $\boldsymbol{A}_2$ shows that the two matrices are orthogonal to each other: $\boldsymbol{A}_1 \boldsymbol{A}_2 = \boldsymbol{0}$, where $\boldsymbol{0}$ is a 2×2 matrix of zeros. Thus, the eigenanalysis has decomposed the rank-2 matrix $\boldsymbol{A}$ into two rank-1 matrices. It is left as an exercise to verify the multiplication and that $\boldsymbol{A}_1 + \boldsymbol{A}_2 = \boldsymbol{A}$. ■

Notice that the sum of the eigenvalues in Example 2.12, $\lambda_1^2 + \lambda_2^2 = 18$, is equal to $\text{tr}(\boldsymbol{A})$. This is a general result: The sum of the eigenvalues for any square symmetric matrix is equal to the trace of the matrix. Further, the trace of each of the component rank-1 matrices is equal to its eigenvalue:

$$\text{tr}(A_1) = \lambda_1^2 \quad \text{and} \quad \text{tr}(A_2) = \lambda_2^2$$

2.8 Singular Value Decomposition of a Rectangular Matrix

The eigenanalysis presented in Section 2.7 applies to a square matrix. In this section, the eigenanalysis is used to develop a similar decomposition, called the **singular value decomposition**, for a rectangular matrix. The singular value decomposition is then used to give the **principal component analysis.**

Singular Value Decomposition

Let X be an $n \times p$ matrix with $n > p$. Then $X'X$ is a square symmetric matrix of order $p \times p$. From Section 2.7, $X'X$ can be expressed in terms of its eigenvalues L and eigenvectors Z as

$$X'X = ZLZ' \qquad [2.19]$$

Similarly, XX' is a square symmetric matrix but of order $n \times n$. The rank of XX' will be at most p so there will be at most p nonzero eigenvalues; they are in fact the same p eigenvalues obtained from $X'X$. In addition, $X'X$ will have $n - p$ eigenvalues that are zero. These $n - p$ eigenvalues and their vectors are dropped in the following discussion. Denote with U the matrix of eigenvectors of XX' that correspond to the p eigenvalues common to $X'X$. Each eigenvector u_i will be of order $n \times 1$. Then,

$$XX' = ULU' \qquad [2.20]$$

Equations 2.19 and 2.20 jointly imply that the rectangular matrix X can be written as

$$X = UL^{1/2}Z' \qquad [2.21]$$

where $L^{1/2}$ is the diagonal matrix of the positive square roots of the p eigenvalues of $X'X$. Thus, $L^{1/2}L^{1/2} = L$. Equation 2.21 is the **singular value decomposition** of the rectangular matrix X. The elements of $L^{1/2}$, λ_i, are called the **singular values** and the *column* vectors in U and Z are the left and right singular vectors, respectively.

Since $L^{1/2}$ is a diagonal matrix, the singular value decomposition expresses X as a sum of p rank-1 matrices,

$$X = \sum \lambda_i u_i z_i' \qquad [2.22]$$

where summation is over $i = 1, \ldots, p$. Furthermore, if the eigenvalues have been ranked from largest to smallest, the first of these matrices is the "best" rank-1 approximation to X, the *sum* of the first two matrices is the "best" rank-2 approximation of X, and so forth. These are "best" approximations in the least squares sense; that is, no other matrix (of the same rank) will give a better agreement with the original matrix X as measured by the sum of squared differences between the corresponding elements of X and the approximating matrix (Householder and Young, 1938). The goodness of fit of the approximation in each case is given by the ratio of the sum of the eigenvalues (squares of the singular values) used in the approximation to the sum of all eigenvalues. Thus, the rank-1 approximation has a goodness of fit of $\lambda_1^2/\sum \lambda_i^2$, the rank-2 approximation has a goodness of fit of $(\lambda_1^2 + \lambda_2^2)/\sum \lambda_i^2$, and so forth.

Example 2.14 Singular value decomposition is illustrated using data on average minimum daily temperature, X_1, average maximum daily temperature, X_2, total rainfall, X_3, and total growing degree days, X_4, for six locations. The data were reported by Saeed and Francis (1984) to relate environmental conditions to cultivar-by-environment interactions in sorghum and are used with their kind permission. Each variable has been centered to have zero mean, and standardized to have unit sum of squares. (The centering and standardization are not necessary for a singular value decomposition. The centering removes the mean effect of each variable so that the dispersion about the mean is being analyzed. The standardization puts all variables on an equal basis and is desirable in most cases, particularly when the variables have different units of measure.) The X matrix is

$$
\begin{array}{c} \\ X = \end{array}
\begin{array}{c}
\begin{array}{cccc} \quad X_1 \quad & \quad X_2 \quad & \quad X_3 \quad & \quad X_4 \quad \end{array} \\
\begin{bmatrix}
.178146 & -.523245 & .059117 & -.060996 \\
.449895 & -.209298 & .777976 & .301186 \\
-.147952 & .300866 & -.210455 & -.053411 \\
-.057369 & .065406 & .120598 & -.057203 \\
-.782003 & -.327028 & -.210455 & -.732264 \\
.359312 & .693299 & -.536780 & .602687
\end{bmatrix}
\end{array}
$$

The singular value decomposition of X into $UL^{1/2}Z'$ gives

$$
U = \begin{bmatrix}
-.113995 & .308905 & -.810678 & .260088 \\
.251977 & .707512 & .339701 & -.319261 \\
.007580 & -.303203 & .277432 & .568364 \\
-.028067 & .027767 & .326626 & .357124 \\
-.735417 & -.234888 & .065551 & -.481125 \\
.617923 & -.506093 & -.198632 & -.385189
\end{bmatrix}
$$

$$
L^{1/2} = \begin{bmatrix}
1.496896 & 0 & 0 & 0 \\
0 & 1.244892 & 0 & 0 \\
0 & 0 & .454086 & 0 \\
0 & 0 & 0 & .057893
\end{bmatrix}
$$

$$Z = \begin{bmatrix} .595025 & .336131 & -.383204 & .621382 \\ .451776 & -.540753 & .657957 & .265663 \\ .004942 & .768694 & .639051 & -.026450 \\ .664695 & .060922 & -.108909 & -.736619 \end{bmatrix}$$

The columns of U and Z are the left and right singular vectors, respectively. The first column of U, u_1, the first column of Z, z_1, and the first singular value, $\lambda_1 = 1.496896$, give the best rank-1 approximation of X,

$$X_1 = \lambda_1 u_1 z_1'$$

$$= (1.4969)\begin{pmatrix} -.1140 \\ .2520 \\ .0076 \\ -.0281 \\ -.7354 \\ .6179 \end{pmatrix}(.5950 \;\; .4518 \;\; .0049 \;\; .6647)$$

$$= \begin{bmatrix} -.101535 & -.077091 & -.000843 & -.113423 \\ .224434 & .170403 & .001864 & .250712 \\ .006752 & .005126 & .000056 & .007542 \\ -.024999 & -.018981 & -.000208 & -.027927 \\ -.655029 & -.497335 & -.005440 & -.731725 \\ .550378 & .417877 & .004571 & .614820 \end{bmatrix}$$

The goodness of fit of X_1 to X is measured by

$$\frac{\lambda_1^2}{\sum \lambda_i^2} = \frac{(1.4969)^2}{4} = .56$$

or the sum of squares of the differences between the elements of X and X_1, the lack of fit, is 44% of the total sum of squares of the elements in X. This is not a very good approximation.

The rank-2 approximation to X is obtained by *adding* to X_1 the matrix $X_2 = \lambda_2 u_2 z_2'$. This gives

$$X_1 + X_2 = \begin{bmatrix} .027725 & -.285040 & .295197 & -.089995 \\ .520490 & -.305880 & .678911 & .304370 \\ -.120122 & .209236 & -.290091 & -.015453 \\ -.013380 & -.037673 & .026363 & -.025821 \\ -.753317 & -.339213 & -.230214 & -.749539 \\ .338605 & .758568 & -.479730 & .576438 \end{bmatrix}$$

which has goodness of fit

$$\frac{\lambda_1^2 + \lambda_2^2}{\sum \lambda_i^2} = \frac{(1.4969)^2 + (1.2449)^2}{4} = .95$$

In terms of approximating X with the rank-2 matrix $X_1 + X_2$, the goodness of fit of .95 means that the sum of squares of the discrepancies between X and $(X_1 + X_2)$ is 5% of the total sum of squares of all elements in X. The sum of squares of all elements in X is $\sum \lambda_i^2$, the sum of squares of all elements in $(X_1 + X_2)$ is $(\lambda_1^2 + \lambda_2^2)$, and the sum of squares of all elements in $[X - (X_1 + X_2)]$ is $(\lambda_3^2 + \lambda_4^2)$. In terms of the geometry of the data vectors, the goodness of fit of .95 means that 95% of the dispersion of the "cloud" of points in the original four-dimensional space is, in reality, contained in two dimensions, or the points in four-dimensional space very nearly fall on a plane. Only 5% of the dispersion is lost if the third and fourth dimensions are ignored.

Using all four singular values and their singular vectors gives the complete decomposition of X into four orthogonal rank-1 matrices. The sum of the four matrices equals X, within the limits of rounding error. The analysis has shown, by the relatively small size of the third and fourth singular values, that the last two dimensions contain little of the dispersion and can safely be ignored in interpretation of the data. ■

Principal Component Analysis

The singular value decomposition is the first step in **principal component analysis**. Using the result $X = UL^{1/2}Z'$ and the property that $Z'Z = I$, one can define the $n \times p$ matrix W as

$$W = XZ = UL^{1/2} \tag{2.23}$$

The first column of Z is the first of the right singular vectors of X, or the first eigenvector of $X'X$. Thus, the coefficients in the first eigenvector define the particular linear function of the columns of X (the original variables) that generates the first column of W. The second column of W is obtained using the second eigenvector of $X'X$, and so on. Notice that $W'W = L$. Thus, W is an $n \times p$ matrix which, unlike X, has the property that all its columns are orthogonal. (L is a diagonal matrix so that all off-diagonal elements, the sums of products between columns of W, are zero.) The sum of squares of the ith column of W is λ_i^2, the ith diagonal element of L. Thus, if X is an $n \times p$ matrix of observations on p variables, each column of W is a new variable defined as a linear transformation of the original variables. The ith new variable has sum of squares λ_i^2 and all are pairwise orthogonal. This analysis is called the **principal component analysis** of X, and the columns of W are the **principal components** (sometimes called principal component scores).

Principal component analysis will be used where the columns of X correspond to the observations on different variables. The transformation is to a set of orthogonal variables such that the first principal component accounts for the largest possible amount of the total dispersion, measured by λ_1^2, the second principal component accounts for the largest possible amount of the remaining dispersion, λ_2^2, and so forth. The total dispersion is given by the sum of all eigenvalues, which is equal to the sum of squares of the original variables: $\text{tr}(\boldsymbol{X}'\boldsymbol{X}) = \text{tr}(\boldsymbol{W}'\boldsymbol{W}) = \sum \lambda_i^2$.

Example 2.15 For the Saeed and Francis data used in Example 2.14, each column of $\boldsymbol{Z}$ contains the coefficients that define one of the principal components as a linear function of the original variables. The first vector in $\boldsymbol{Z}$, $z_1' = (.5950\ .4518\ .0049\ .6647)$, has similar first, second, and fourth coefficients with the third coefficient being near zero. Thus, the first principal component is essentially an average of the three temperature variables, X_1, X_2, and X_4. The second column vector in $\boldsymbol{Z}$, $z_2' = (.3361\ -.5408\ .7687\ .0609)$, gives heavy positive weight to X_3, heavy negative weight to X_2, and moderate positive weight to X_1. Thus, the second principal component will be large for those observations that have high rainfall, X_3, and small difference between the maximum and minimum daily temperatures, X_2 and X_1.

The third and fourth principal components account for only 5% of the total dispersion. This small amount of dispersion may be due more to random "noise" than to real patterns in the data. Consequently, the interpretation of these components may not be very meaningful. The third principal component will be large when there is high rainfall and large difference between the maximum and minimum daily temperatures: $z_3' = (-.3832\ .6580\ .6391\ -.1089)$. The variable degree days, X_4, has little involvement in the second and third principal components; the fourth coefficient is relatively small. The fourth principal component is determined primarily by the difference between an average minimum daily temperature and degree days, $z_4' = (.6214\ .2657\ -.0265\ -.7366)$. ■

The principal component vectors are obtained either by the multiplication $\boldsymbol{W} = \boldsymbol{U}\boldsymbol{L}^{1/2}$ or $\boldsymbol{W} = \boldsymbol{X}\boldsymbol{Z}$. The first is easier since it is the simple scalar multiplication of each column of $\boldsymbol{U}$ with the appropriate λ_i.

Example 2.16 The principal component vectors for the Saeed and Francis data of Example 2.14 are (with some rounding)

$$\boldsymbol{W} = \begin{bmatrix} -.1706 & .3846 & -.3681 & .0151 \\ .3772 & .8808 & .1543 & -.0185 \\ .0113 & -.3775 & .1260 & .0329 \\ -.0420 & .0346 & .1483 & .0207 \\ -1.1008 & -.2924 & .0298 & -.0279 \\ .9250 & -.6300 & -.0902 & -.0223 \end{bmatrix}$$

The sum of squares of the first principal component, the first column of $\boldsymbol{W}$, is $\lambda_1^2 = (1.4969)^2 = 2.2407$. Similarly, the sums of squares for the second, third, and fourth principal components are

$$\lambda_2^2 = (1.2449)^2 = 1.5498$$

$$\lambda_3^2 = (.4541)^2 = .2062$$

$$\lambda_4^2 = (.0579)^2 = .0034$$

These sum to 4.0, the total sum of squares of the original three variables after they were standardized. The proportion of the total sum of squares accounted for by the first principal component is $\lambda_1^2/\sum \lambda_i^2 = 2.2407/4 = .56$ or 56%. The first two principal components account for $(\lambda_1^2 + \lambda_2^2)/4 = 3.79/4 = .95$ or 95% of the total sum of squares of the four original variables.

Each of the original data vectors in $\boldsymbol{X}$ was a vector in six-dimensional space and, together, the four vectors defined a four-dimensional subspace. These vectors were not orthogonal. The four vectors in $\boldsymbol{W}$, the principal component vectors, are linear functions of the original vectors and, as such, they fall in the *same* four-dimensional subspace. The principal component vectors, however, are orthogonal and defined such that the first principal component vector has the largest possible sum of squares. This means that the direction of the first principal component axis coincides with the major axis of the ellipsoid of observations, Figure 2.3. Note that the "cloud" of observations, the data points, does *not* change; only the axes are being redefined. The second principal component has the largest possible sum of squares of all vectors orthogonal to the first, and so

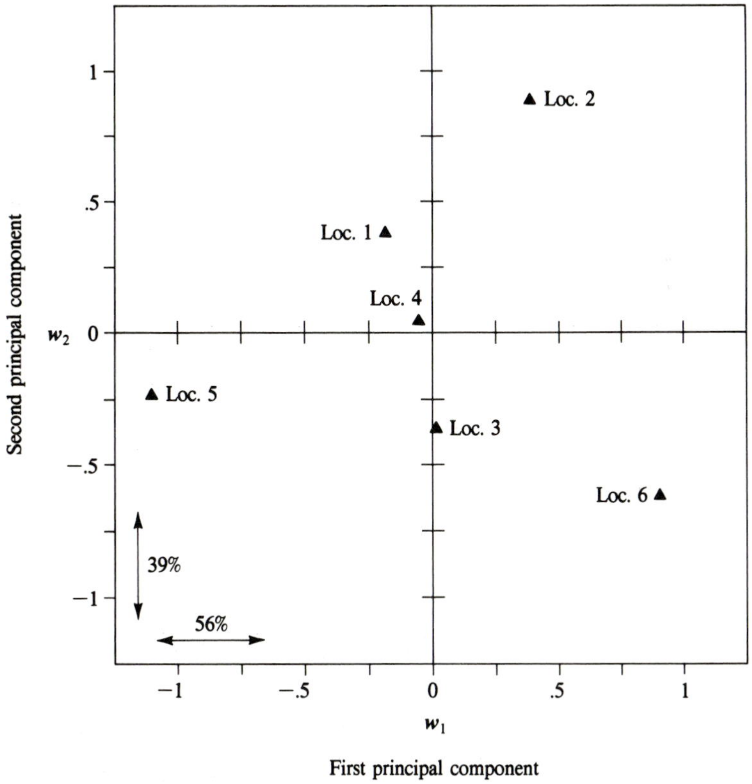

Figure 2.3 The first two principal components of the Saeed and Francis (1984) data on average minimum temperature, average maximum temperature, total rainfall, and growing degree days for six locations. The first principal component primarily reflects average temperature. The second principal component is a measure of rainfall minus the spread between minimum and maximum temperature.

on. The fact that the first two principal components account for 95% of the sum of squares in this example shows that very little of the dispersion among the data points occurs in the third and fourth principal component dimensions. In other words, the variability among these six locations in average minimum and average maximum temperature, total rainfall, and total growing degree days, can be adequately described by considering *only* the two dimensions (or variables) defined by the first two principal components.

The plot of the first two principal components from the Saeed and Francis data, Figure 2.3, shows that locations 5 and 6 differ from each other primarily in the first principal component. This component was noted earlier to be mainly a temperature difference; location 6 is the warmer and has the longer growing season. The other four locations differ primarily in the second principal component, which reflects amount of rainfall and the difference in maximum and minimum temperature. Location 2 has the highest rainfall and tends to have a large difference in maximum and minimum daily temperature. Location 6 is also the lowest in the second principal component, indicating a lower rainfall and small difference between the maximum and minimum temperature. Thus, location 6 appears to be a relatively hot, dry environment with little diurnal temperature variation. ■

2.9 Summary

This chapter has presented the key matrix operations that will be used in this text. The student must be able to use matrix notation and matrix operations. Of particular importance are

- The concepts of rank and the transpose of a matrix
- The special types of matrices: square, symmetric, diagonal, identity, and idempotent
- The elementary matrix operations of addition and multiplication
- The use of the inverse of a square, symmetric matrix to solve a set of equations

The geometry of vectors and projections will be useful in understanding least squares principles. Eigenanalysis and singular value decomposition will be used later in the text.

EXERCISES

* 2.1. Let

$$\boldsymbol{A} = \begin{bmatrix} 1 & 0 \\ 2 & 4 \\ -1 & 2 \end{bmatrix}, \quad \boldsymbol{B} = \begin{bmatrix} 1 & 2 & -1 \\ 0 & 3 & -4 \end{bmatrix},$$

$c' = (1\ \ 2\ \ 0)$, and $d = 2$, a scalar.

Perform the following operations, if possible. If the operation is not possible, explain why.

a. $\boldsymbol{c}'\boldsymbol{A}$
b. $\boldsymbol{A}'\boldsymbol{c}$
c. $\boldsymbol{B}' + \boldsymbol{A}$
d. $\boldsymbol{c}'\boldsymbol{B}$
e. $\boldsymbol{A} - d$
f. $(d\boldsymbol{B}' + \boldsymbol{A})$

2.2. Find the rank of each of the following matrices. Which matrices are full rank?

$$\boldsymbol{A} = \begin{bmatrix} 1 & 1 & 0 & 0 & 0 \\ 1 & 0 & 1 & 0 & 0 \\ 1 & 0 & 0 & 1 & 0 \\ 1 & 0 & 0 & 0 & 1 \end{bmatrix} \qquad \boldsymbol{B} = \begin{bmatrix} 1 & 1 & 0 & 0 \\ 1 & 0 & 1 & 0 \\ 1 & 0 & 0 & 1 \\ 1 & 0 & 0 & 0 \end{bmatrix}$$

$$\boldsymbol{C} = \begin{bmatrix} 1 & 1 & 0 & 0 \\ 1 & 0 & 1 & 0 \\ 1 & 0 & 0 & 1 \\ 1 & -1 & -1 & -1 \end{bmatrix}$$

* 2.3. Use $\boldsymbol{B}$ in exercise 2.2 to compute $\boldsymbol{D} = \boldsymbol{B}(\boldsymbol{B}'\boldsymbol{B})^{-1}\boldsymbol{B}'$. Determine whether $\boldsymbol{D}$ is idempotent. What is the rank of $\boldsymbol{D}$?

2.4. Find the a_{ij} elements to make the following matrix symmetric. Is the matrix idempotent?

$$\boldsymbol{A} = \begin{bmatrix} 1 & 2 & a_{13} & 4 \\ 2 & -1 & 0 & a_{24} \\ 6 & 0 & a_{33} & -2 \\ a_{41} & 8 & -2 & 3 \end{bmatrix}$$

* 2.5. Verify that $\boldsymbol{A}$ and $\boldsymbol{B}$ are inverses of each other.

$$\boldsymbol{A} = \begin{bmatrix} 10 & 5 \\ 3 & 2 \end{bmatrix} \qquad \boldsymbol{B} = \begin{bmatrix} \frac{2}{5} & -1 \\ -\frac{3}{5} & 2 \end{bmatrix}$$

2.6. Find b_{41} such that $\boldsymbol{a}$ and $\boldsymbol{b}$ are orthogonal.

$$\boldsymbol{a} = \begin{pmatrix} 2 \\ 0 \\ -1 \\ 3 \end{pmatrix} \qquad \boldsymbol{b} = \begin{pmatrix} 6 \\ -1 \\ 3 \\ b_{41} \end{pmatrix}$$

* 2.7. Plot the following vectors on a two-dimensional coordinate system:

$$v_1 = \begin{pmatrix} 1 \\ 1 \end{pmatrix} \qquad v_2 = \begin{pmatrix} 4 \\ 1 \end{pmatrix} \qquad v_3 = \begin{pmatrix} 1 \\ -4 \end{pmatrix}$$

By inspection of the plot, which pairs of vectors appear to be orthogonal? Verify numerically that they are orthogonal *and* that all other pairs in this set are *not* orthogonal. Explain from the geometry of the plot how you know there is a linear dependency among the three vectors.

* 2.8. The three vectors in exercise 2.7 are linearly dependent. Find the linear function of v_1 and v_2 that equals v_3. Set up the problem as a system of linear equations to be solved. Let $V = (v_1 \;\; v_2)$, and let $x' = (x_1 \;\; x_2)$ be the vector of unknown coefficients. Then, $Vx = v_3$ is the system of equations to be solved for x.
a. Show that the system of equations is consistent.
b. Show that there is a unique solution.
c. Find the solution.

2.9. Expand the set of vectors in exercise 2.7 to include a fourth vector, $v_4' = (8 \;\; 5)$. Reformulate exercise 2.8 to include the fourth vector by including v_4 in V and an additional coefficient in x. Is this system of equations consistent? Is the solution unique? Find a solution.

2.10. Use the determinant to determine which of the following matrices has a unique inverse.

$$A = \begin{bmatrix} 1 & 1 \\ 4 & 10 \end{bmatrix} \qquad B = \begin{bmatrix} 4 & -1 \\ 0 & 6 \end{bmatrix} \qquad C = \begin{bmatrix} 6 & 3 \\ 4 & 2 \end{bmatrix}$$

*2.11. Given the following matrix:

$$A = \begin{bmatrix} 3 & \sqrt{2} \\ \sqrt{2} & 2 \end{bmatrix}$$

a. Find the eigenvalues and eigenvectors of A.
b. What do your findings tell you about the rank of A?

2.12. Given the following eigenvalues with their corresponding eigenvectors, and the fact that the original matrix was square and symmetric, reconstruct the original matrix.

$$\lambda_1^2 = 6 \qquad z_1 = \begin{pmatrix} 0 \\ 1 \end{pmatrix}$$

$$\lambda_2^2 = 2 \qquad z_2 = \begin{pmatrix} 1 \\ 0 \end{pmatrix}$$

*2.13. Find the inverse of the following matrix:

$$A = \begin{bmatrix} 5 & 0 & 0 \\ 0 & 10 & 2 \\ 0 & 2 & 3 \end{bmatrix}$$

2.14. Let

$$X = \begin{bmatrix} 1 & .2 & 0 \\ 1 & .4 & 0 \\ 1 & .6 & 0 \\ 1 & .8 & 0 \\ 1 & .2 & .1 \\ 1 & .4 & .1 \\ 1 & .6 & .1 \\ 1 & .8 & .1 \end{bmatrix} \qquad Y = \begin{bmatrix} 242 \\ 240 \\ 236 \\ 230 \\ 239 \\ 238 \\ 231 \\ 226 \end{bmatrix}$$

a. Compute $X'X$ and $X'Y$. Verify by separate calculations that the $(i, j) = (2, 2)$ element in $X'X$ is the sum of squares of column 2 in X. Verify that the (2, 3) element is the sum of products between columns 2 and 3 of X. Identify the elements in $X'Y$ in terms of sums of squares or products of the columns of X and Y.
b. Is X of full column rank? What is the rank of $X'X$?
c. Obtain $(X'X)^{-1}$. What is the rank of $(X'X)^{-1}$? Verify by matrix multiplication that $(X'X)^{-1}X'X = I$.
d. Compute $P = X(X'X)^{-1}X'$ and verify by matrix multiplication that P is idempotent. What is $r(P)$?

*2.15. Use X defined in exercise 2.14.
a. Find the singular value decomposition of X. Explain what the singular values tell you about the rank of X.
b. Compute the rank-1 approximation of X; call it X_1. Use the singular values to state the goodness of fit of this rank-1 approximation.
c. Use X_1 to compute a rank-1 approximation of $X'X$; that is, compute $X_1'X_1$. Compare $\text{tr}(X_1'X_1)$ with λ_1^2 and $\text{tr}(X'X)$.

2.16. Use $X'X$ computed in exercise 2.14.
a. Compute the eigenanalysis of $X'X$. What is the relationship between the singular values of X obtained in exercise 2.15 and the eigenvalues obtained for $X'X$?
b. Use the results of the eigenanalysis to compute the rank-1 approximation of $X'X$. Compare this result to the approximation of $X'X$ obtained in exercise 2.15.
c. Show algebraically that they should be identical.

2.17. Verify that

$$A = \frac{1}{15}\begin{bmatrix} 3 & -13 & 8 \\ 12 & -7 & 2 \\ -12 & 17 & -7 \end{bmatrix}$$

is the inverse of

$$B = \begin{bmatrix} 1 & 3 & 2 \\ 4 & 5 & 6 \\ 8 & 7 & 9 \end{bmatrix}$$

*2.18. Show that the equations $Ax = y$ are consistent where

$$A = \begin{bmatrix} 1 & 2 \\ 3 & 3 \\ 5 & 7 \end{bmatrix} \quad \text{and} \quad y = \begin{pmatrix} 6 \\ 9 \\ 21 \end{pmatrix}$$

*2.19. Verify that

$$A^- = \frac{1}{18}\begin{bmatrix} -10 & 16 & -4 \\ 8 & -11 & 5 \end{bmatrix}$$

is a generalized inverse of

$$A = \begin{bmatrix} 1 & 2 \\ 3 & 3 \\ 5 & 7 \end{bmatrix}$$

2.20. Verify that

$$A^- = \begin{bmatrix} -\frac{1}{10} & -\frac{2}{10} & \frac{4}{9} \\ 0 & 0 & \frac{1}{9} \\ \frac{1}{10} & \frac{2}{10} & -\frac{2}{9} \end{bmatrix}$$

is a generalized inverse of

$$A = \begin{bmatrix} 1 & 2 & 3 \\ 2 & 4 & 6 \\ 3 & 3 & 3 \end{bmatrix}$$

*2.21. Use the generalized inverse in exercise 2.20 to obtain a solution to the equations $Ax = y$, where A is defined in exercise 2.20 and $y' = (6 \;\; 12 \;\; 9)$.

2.22. The eigenanalysis of

$$A = \begin{bmatrix} 10 & 3 \\ 3 & 8 \end{bmatrix}$$

in Section 2.7 gave

$$A_1 = \begin{bmatrix} 8.0042 & 5.7691 \\ 5.7691 & 4.1581 \end{bmatrix} \quad \text{and} \quad A_2 = \begin{bmatrix} 1.9958 & -2.7691 \\ -2.7691 & 3.8419 \end{bmatrix}$$

Verify the multiplication of the eigenvectors to obtain A_1 and A_2, verify that $A_1 + A_2 = A$, and that A_1 and A_2 are orthogonal to each other.

*2.23. In Section 2.6, a linear transformation of $\mathbf{y}_1 = (3 \;\; 10 \;\; 20)$ to $\mathbf{x}_1 = (33 \;\; 17 \;\; -3)$ and of $\mathbf{y}_2 = (6 \;\; 14 \;\; 21)$ to $\mathbf{x}_2 = (41 \;\; 15 \;\; 1)$ was made using the matrix

$$\mathbf{A} = \begin{bmatrix} 1 & 1 & 1 \\ -1 & 0 & 1 \\ -1 & 2 & -1 \end{bmatrix}$$

The vectors of $\mathbf{A}$ were then standardized so that $\mathbf{A}'\mathbf{A} = \mathbf{I}$ to produce the *orthogonal* transformation of $\mathbf{y}_1$ and $\mathbf{y}_2$ to $\mathbf{x}_1^* = (33/\sqrt{3} \;\; 17/\sqrt{2} \;\; -3/\sqrt{6})$ and $\mathbf{x}_2^* = (41/\sqrt{3} \;\; 15/\sqrt{2} \;\; 1/\sqrt{6})$, respectively. Show that the squared distance between $\mathbf{y}_1$ and $\mathbf{y}_2$ is unchanged when the orthogonal transformation is made but not when the nonorthogonal transformation is made. That is, show that

$$(\mathbf{y}_1 - \mathbf{y}_2)'(\mathbf{y}_1 - \mathbf{y}_2) = (\mathbf{x}_1^* - \mathbf{x}_2^*)'(\mathbf{x}_1^* - \mathbf{x}_2^*)$$

but that

$$(\mathbf{y}_1 - \mathbf{y}_2)'(\mathbf{y}_1 - \mathbf{y}_2) \neq (\mathbf{x}_1 - \mathbf{x}_2)'(\mathbf{x}_1 - \mathbf{x}_2)$$

CHAPTER
THREE

Multiple Regression in Matrix Notation

We have reviewed linear regression in algebraic notation and have introduced the matrix notation and operations needed to continue with the more complicated models.

This chapter presents the model, and develops the normal equations and solution to the normal equations for a general linear model involving any number of independent variables. The matrix formulation for the variances of linear functions is used to derive the measures of precision of the estimates.

Chapter 1 provided an introduction to multiple regression and suggested that a more convenient notation was needed. Chapter 2 familiarized you with matrix notation and operations of matrices. This chapter states multiple regression results in matrix notation.

3.1 The Model

The linear additive model for relating a dependent variable to p independent variables is

$$Y_i = \beta_0 + \beta_1 X_{i1} + \beta_2 X_{i2} + \cdots + \beta_p X_{ip} + \varepsilon_i \qquad [3.1]$$

The subscript i denotes the observational unit from which the observations on Y and the p independent variables were taken. The second subscript designates the independent variable. The sample size will be denoted with n, $i = 1, \ldots, n$, and p will denote the number of independent variables. There are $(p + 1)$ parameters, β_j, $j = 0, \ldots, p$, to be estimated when the linear model includes the intercept β_0. For convenience, we will use $p' = (p + 1)$.

Matrix Definitions

Four matrices are needed to express the linear model in matrix notation:

$\boldsymbol{Y}$: The $n \times 1$ column vector of observations on the dependent variable, Y_i;

$\boldsymbol{X}$: The $n \times p'$ matrix consisting of a column of ones, which is labeled **1**, followed by the p column vectors of the observations on the independent variables;

$\boldsymbol{\beta}$: The $p' \times 1$ vector of parameters to be estimated; and

$\boldsymbol{\varepsilon}$: The $n \times 1$ vector of random errors.

With these definitions, the linear model can be written as

$$\boldsymbol{Y} = \boldsymbol{X\beta} + \boldsymbol{\varepsilon} \qquad [3.2]$$

or

$$\underset{(n \times 1)}{\begin{pmatrix} Y_1 \\ Y_2 \\ \vdots \\ Y_n \end{pmatrix}} = \underset{(n \times p')}{\begin{pmatrix} 1 & X_{11} & X_{12} & X_{13} & \cdots & X_{1p} \\ 1 & X_{21} & X_{22} & X_{23} & \cdots & X_{2p} \\ \vdots & \vdots & \vdots & \vdots & & \vdots \\ 1 & X_{n1} & X_{n2} & X_{n3} & \cdots & X_{np} \end{pmatrix}} \underset{(p' \times 1)}{\begin{pmatrix} \beta_0 \\ \beta_1 \\ \vdots \\ \beta_p \end{pmatrix}} + \underset{(n \times 1)}{\begin{pmatrix} \varepsilon_1 \\ \varepsilon_2 \\ \vdots \\ \varepsilon_n \end{pmatrix}}$$

The X Matrix

Each column of $\boldsymbol{X}$ contains the values for a particular independent variable. The elements of a particular row of $\boldsymbol{X}$, say row r, are the coefficients on the corresponding parameters in $\boldsymbol{\beta}$ which give $\mathscr{E}(Y_r)$. Notice that β_0 has the constant coefficient 1 for all observations; hence, the column vector **1** is the first column of $\boldsymbol{X}$. Multiplying the first row of $\boldsymbol{X}$ by $\boldsymbol{\beta}$, and adding the first element of $\boldsymbol{\varepsilon}$ confirms that the model for the first observation is

$$Y_1 = \beta_0 + \beta_1 X_{11} + \beta_2 X_{12} + \cdots + \beta_p X_{1p} + \varepsilon_1$$

The vectors $\boldsymbol{Y}$ and $\boldsymbol{\varepsilon}$ are random vectors; the elements of these vectors are random variables. The matrix $\boldsymbol{X}$ is considered to be a matrix of known constants.

The β Vector

The vector $\boldsymbol{\beta}$ is a vector of unknown constants to be estimated from the data. Each element, β_j, is a partial regression coefficient reflecting the change in the dependent variable per unit change in the jth independent variable, **assuming all other independent variables are held constant**. The definition of each partial regression coefficient is dependent on the set of independent variables in the model. Whenever clarity demands, the subscript notation on β_j will be expanded to identify explicitly both the independent variable to which the coefficient applies and the other independent variables in the model. For example, $\beta_{2.13}$ would designate the partial regression coefficient for X_2 in a model that contains X_1, X_2, and X_3.

The Random Vector ε

The usual assumptions about ε_i are now expressed in terms of the random vector $\boldsymbol{\varepsilon}$. $\boldsymbol{\varepsilon}$ is said to have a multivariate normal distribution with mean vector **0** (of order $n \times 1$). The variance of an individual element ε_i is replaced with the variance–covariance matrix for the vector $\boldsymbol{\varepsilon}$. The variance–covariance matrix for any random vector of n elements is defined as an $n \times n$ symmetric matrix with the diagonal elements equal to the variances of the random variables (in order) and the (i, j)th off-diagonal element equal to the covariance between ε_i and ε_j. For example, if $\boldsymbol{Z}$ is a 3×1 vector of random variables z_1, z_2, and z_3, the variance–covariance matrix of $\boldsymbol{Z}$ is the 3×3 matrix

$$\mathbf{Var}(\boldsymbol{Z}) = \begin{bmatrix} \sigma^2(z_1) & \mathrm{Cov}(z_1, z_2) & \mathrm{Cov}(z_1, z_3) \\ \mathrm{Cov}(z_2, z_1) & \sigma^2(z_2) & \mathrm{Cov}(z_2, z_3) \\ \mathrm{Cov}(z_3, z_1) & \mathrm{Cov}(z_3, z_2) & \sigma^2(z_3) \end{bmatrix} \qquad [3.3]$$

The variance–covariance matrix for $\boldsymbol{\varepsilon}$ is $\boldsymbol{I}\sigma^2$ where $\boldsymbol{I}$ is the $n \times n$ identity matrix and σ^2 is the common variance of all ε_i. The distribution of $\boldsymbol{\varepsilon}$ is written in shorthand notation as

$$\boldsymbol{\varepsilon} \sim \mathrm{N}(\mathbf{0}, \boldsymbol{I}\sigma^2) \qquad [3.4]$$

The statement that the variance–covariance matrix of $\boldsymbol{\varepsilon}$, $\mathbf{Var}(\boldsymbol{\varepsilon})$, is $\boldsymbol{I}\sigma^2$ includes the two usual assumptions that

1. The ε_i have common variance σ^2; and
2. They are statistically independent. (Independence is reflected in zero covariances.)

The Y Vector

Since the elements of $\boldsymbol{X}$ and $\boldsymbol{\beta}$ are constants, the $\boldsymbol{X\beta}$ term in the model is a set of constants being added to the vector of random errors, $\boldsymbol{\varepsilon}$. Thus, $\boldsymbol{Y}$ is a random vector with mean vector $\boldsymbol{X\beta}$ and variance–covariance matrix $\boldsymbol{I}\sigma^2$:

$$\mathscr{E}(\boldsymbol{Y}) = \mathscr{E}(\boldsymbol{X\beta} + \boldsymbol{\varepsilon}) = \mathscr{E}(\boldsymbol{X\beta}) + \mathscr{E}(\boldsymbol{\varepsilon}) = \boldsymbol{X\beta} \qquad [3.5]$$

and

$$\mathbf{Var}(\boldsymbol{Y}) = \mathbf{Var}(\boldsymbol{X\beta} + \boldsymbol{\varepsilon}) = \mathbf{Var}(\boldsymbol{\varepsilon}) = \boldsymbol{I}\sigma^2 \qquad [3.6]$$

$\mathbf{Var}(\boldsymbol{Y})$ is the same as $\mathbf{Var}(\boldsymbol{\varepsilon})$ since adding a constant to a random variable does not change the variance. When $\boldsymbol{\varepsilon}$ is normally distributed, $\boldsymbol{Y}$ is also multivariate normally distributed. Thus,

$$\boldsymbol{Y} \sim \mathrm{N}(\boldsymbol{X\beta}, \boldsymbol{I}\sigma^2) \qquad [3.7]$$

This result is based on the assumption that the linear model being used is the correct model. If important independent variables have been omitted or if the functional form of the model is not correct, $\boldsymbol{X\beta}$ will not be the expectation of $\boldsymbol{Y}$.

Example 3.1

For the ozone data used in Example 1.1 (see Table 1.1 on page 4),

$$\boldsymbol{X} = \begin{bmatrix} 1 & .02 \\ 1 & .07 \\ 1 & .11 \\ 1 & .15 \end{bmatrix} \qquad \boldsymbol{Y} = \begin{pmatrix} 242 \\ 237 \\ 231 \\ 201 \end{pmatrix} \qquad \boldsymbol{\beta} = \begin{pmatrix} \beta_0 \\ \beta_1 \end{pmatrix}$$

and $\boldsymbol{\varepsilon}$ is the vector of four (unobservable) random errors. ■

3.2 The Normal Equations and Their Solution

In matrix notation, the normal equations are written as

$$X'X\hat{\beta} = X'Y \qquad [3.8]$$

and the unique solution to the normal equations, if one exists, is

$$\hat{\beta} = (X'X)^{-1}(X'Y) \qquad [3.9]$$

X'X The multiplication $X'X$ generates a $p' \times p'$ matrix where the diagonal elements are the sums of squares of each of the independent variables and the off-diagonal elements are the sums of products between independent variables. The general form is

$$X'X = \begin{bmatrix} n & \sum X_{i1} & \sum X_{i2} & \cdots & \sum X_{ip} \\ \sum X_{i1} & \sum X_{i1}^2 & \sum X_{i1}X_{i2} & \cdots & \sum X_{11}X_{ip} \\ \sum X_{i2} & \sum X_{i1}X_{i2} & \sum X_{i2}^2 & \cdots & \sum X_{i2}X_{ip} \\ \vdots & \vdots & \vdots & & \vdots \\ \sum X_{ip} & \sum X_{i1}X_{ip} & \sum X_{i2}X_{ip} & \cdots & \sum X_{ip}^2 \end{bmatrix} \qquad [3.10]$$

Summation in all cases is over $i = 1$ to n, the n observations in the data. When only one independent variable is involved, $X'X$ consists of only the upper-left 2×2 matrix. Inspection of the normal equations in Chapter 1, equation 1.6, will reveal that the elements in this 2×2 matrix are the coefficients on $\hat{\beta}_0$ and $\hat{\beta}_1$.

X'Y The elements of the matrix product $X'Y$ are the sums of products between each independent variable in turn and the dependent variable:

$$X'Y = \begin{pmatrix} \sum Y_i \\ \sum X_{i1} Y_i \\ \sum X_{i2} Y_i \\ \vdots \\ \sum X_{ip} Y_i \end{pmatrix} \qquad [3.11]$$

The first element, $\sum Y_i$, is the sum of products between the vector of ones (the first column of X) and Y. Again, if only one independent variable is involved, $X'Y$ consists of only the first two elements. The reader can verify that these are the right-hand sides of the two normal equations, equation 1.6.

A Unique Solution The unique solution to the normal equations exists only if the inverse of $X'X$ exists. This, in turn, requires that the matrix X be of full rank; that is, there can be no linear dependencies among the independent variables. The practical implication is that there can be no redundancies in the information contained in X. For example, the amount of nitrogen in a diet is sometimes converted to the amount of protein by multiplication by a constant. Because the same information is reported two ways, a linear dependency is created if both are included in X. Suppose the independent variables in a genetics problem include three variables reporting the observed sample frequencies of three possible alleles (for a particular locus). These three variables, and the **1** vector, create a linear dependency since the sum of the three variables, the sum of the allelic frequencies, must be 1.0. Only two of the allelic frequencies need be reported; the third is redundant since it can be computed from the first two.

Properties of $\hat{\beta}$

Expressing $\hat{\boldsymbol{\beta}}$ as $\hat{\boldsymbol{\beta}} = (\boldsymbol{X}'\boldsymbol{X})^{-1}(\boldsymbol{X}'\boldsymbol{Y}) = [(\boldsymbol{X}'\boldsymbol{X})^{-1}\boldsymbol{X}']\boldsymbol{Y}$ shows that the estimates of the regression coefficients are linear functions of the dependent variable $\boldsymbol{Y}$, with the coefficients being given by $[(\boldsymbol{X}'\boldsymbol{X})^{-1}\boldsymbol{X}']$. Since the X's are constants, the expectations of the estimated regression coefficients involve only the expectation of $\boldsymbol{Y}$. If the model $\boldsymbol{Y} = \boldsymbol{X\beta} + \boldsymbol{\varepsilon}$ is correct, the expectation of $\boldsymbol{Y}$ is $\boldsymbol{X\beta}$ (equation 3.5) and the expectation of $\hat{\boldsymbol{\beta}}$ is

$$\begin{aligned}\mathscr{E}(\hat{\boldsymbol{\beta}}) &= [(\boldsymbol{X}'\boldsymbol{X})^{-1}\boldsymbol{X}']\mathscr{E}(\boldsymbol{Y}) \\ &= [(\boldsymbol{X}'\boldsymbol{X})^{-1}\boldsymbol{X}'](\boldsymbol{X\beta}) \\ &= [(\boldsymbol{X}'\boldsymbol{X})^{-1}\boldsymbol{X}'\boldsymbol{X}]\boldsymbol{\beta} \\ &= \boldsymbol{\beta} \qquad [3.12]\end{aligned}$$

This shows that $\hat{\boldsymbol{\beta}}$ is an unbiased estimate of $\boldsymbol{\beta}$ *if* the chosen model is correct. If the chosen model is *not* correct, say $\mathscr{E}(\boldsymbol{Y}) = \boldsymbol{X}^*\boldsymbol{\beta}$ instead of $\boldsymbol{X\beta}$, then $[(\boldsymbol{X}'\boldsymbol{X})^{-1}\boldsymbol{X}'\boldsymbol{X}^*] \neq \boldsymbol{I}$ and $\mathscr{E}(\hat{\boldsymbol{\beta}})$ does not simplify to $\boldsymbol{\beta}$ as in equation 3.12.

Example 3.2

Matrix operations using $\boldsymbol{X}$ and $\boldsymbol{Y}$ from the ozone example, Example 1.1, give

$$\boldsymbol{X}'\boldsymbol{X} = \begin{bmatrix} 4 & .3500 \\ .3500 & .0399 \end{bmatrix} \qquad \boldsymbol{X}'\boldsymbol{Y} = \begin{pmatrix} 911 \\ 76.99 \end{pmatrix}$$

and

$$(\boldsymbol{X}'\boldsymbol{X})^{-1} = \begin{bmatrix} 1.07547 & -9.43396 \\ -9.43396 & 107.81671 \end{bmatrix}$$

The estimates of the regression coefficients are

$$\hat{\boldsymbol{\beta}} = (\boldsymbol{X}'\boldsymbol{X})^{-1}\boldsymbol{X}'\boldsymbol{Y} = \begin{pmatrix} 253.434 \\ -293.531 \end{pmatrix}$$ ■

3.3 The $\hat{Y}$ Vector and the Residuals Vector

$\hat{Y}$ and P

The vector of estimated means of the dependent variable Y for the values of the independent variables in the data set is computed as

$$\hat{\boldsymbol{Y}} = \boldsymbol{X}\hat{\boldsymbol{\beta}} \qquad [3.13]$$

This is the simplest way to compute $\hat{Y}$. It is useful for later results, however, to express $\hat{Y}$ as a linear function of Y by substituting $(X'X)^{-1}X'Y$ for $\hat{\beta}$. Thus,

$$\begin{aligned} \hat{Y} &= [X(X'X)^{-1}X']Y \\ &= PY \end{aligned} \qquad [3.14]$$

Equation 3.14 defines the matrix P, an $n \times n$ matrix determined entirely by the X's. This matrix plays a particularly important role in regression analysis. It is a symmetric matrix ($P' = P$) that is also idempotent ($PP = P$). Equation 3.14 shows that $\hat{Y}$ is a linear function of Y with the coefficients given by P. (For example, the first row of P contains the coefficients for the linear function of all Y_i that gives $\hat{Y}_1$.)

The expectation of $\hat{Y}$ is

$$\mathscr{E}(\hat{Y}) = P[\mathscr{E}(Y)] = PX\beta = X\beta \qquad [3.15]$$

Thus, $\hat{Y}$ is an unbiased estimate of the means of Y for the particular values of X in the data set, again, *if* the model is correct. The fact that $PX = X$ can be verified using the definition of P:

$$PX = [X(X'X)^{-1}X']X = X[(X'X)^{-1}(X'X)] = X \qquad [3.16]$$

Example 3.3 For the Heagle ozone data used in Example 1.1,

$$P = \begin{bmatrix} 1 & .02 \\ 1 & .07 \\ 1 & .11 \\ 1 & .15 \end{bmatrix} \begin{bmatrix} 1.0755 & -9.4340 \\ -9.4340 & 107.8167 \end{bmatrix} \begin{bmatrix} 1 & 1 & 1 & 1 \\ .02 & .07 & .11 & .15 \end{bmatrix}$$

$$= \begin{bmatrix} .741240 & .377358 & .086253 & -.204852 \\ .377358 & .283019 & .207547 & .132075 \\ .086253 & .207547 & .304582 & .401617 \\ -.204852 & .132075 & .401617 & .671159 \end{bmatrix}$$

Thus, for example,

$$\hat{Y}_1 = .741Y_1 + .377Y_2 + .086Y_3 - .205Y_4 \qquad \blacksquare$$

e The residuals vector, e, reflects the lack of agreement between the observed Y and the estimated $\hat{Y}$:

$$e = Y - \hat{Y} \qquad [3.17]$$

As with $\hat{Y}$, e can be expressed as a linear function of Y by substituting PY for $\hat{Y}$:

$$e = Y - PY = (I - P)Y \qquad [3.18]$$

Recall that least squares estimation minimizes the sum of squares of the residuals; $\hat{\boldsymbol{\beta}}$ has been chosen so that $\boldsymbol{e}'\boldsymbol{e}$ is a minimum. Like $\boldsymbol{P}$, $(\boldsymbol{I} - \boldsymbol{P})$ is symmetric and idempotent. The expectation of the residuals vector is

$$\begin{aligned} \mathscr{E}(\boldsymbol{e}) &= (\boldsymbol{I} - \boldsymbol{P})\mathscr{E}(\boldsymbol{Y}) = (\boldsymbol{I} - \boldsymbol{P})\boldsymbol{X\beta} \\ &= (\boldsymbol{X} - \boldsymbol{PX})\boldsymbol{\beta} = (\boldsymbol{X} - \boldsymbol{X})\boldsymbol{\beta} = \boldsymbol{0} \end{aligned} \qquad [3.19]$$

where $\boldsymbol{0}$ is an $n \times 1$ vector of zeros. Thus, the observed residuals are random variables with mean zero.

$\hat{\boldsymbol{Y}} + \boldsymbol{e}$

This has partitioned $\boldsymbol{Y}$ into two parts, that accounted for by the model, $\hat{\boldsymbol{Y}}$, and the residual, $\boldsymbol{e}$. That the two parts are additive is evident from the fact that $\boldsymbol{e}$ was obtained by difference (equation 3.17), or can be demonstrated as follows:

$$\hat{\boldsymbol{Y}} + \boldsymbol{e} = \boldsymbol{PY} + (\boldsymbol{I} - \boldsymbol{P})\boldsymbol{Y} = (\boldsymbol{P} + \boldsymbol{I} - \boldsymbol{P})\boldsymbol{Y} = \boldsymbol{Y} \qquad [3.20]$$

Example 3.4 Continuing with Example 3.3, we obtain

$$\hat{\boldsymbol{Y}} = \boldsymbol{X}\hat{\boldsymbol{\beta}} = \begin{bmatrix} 1 & .02 \\ 1 & .07 \\ 1 & .11 \\ 1 & .15 \end{bmatrix} \begin{pmatrix} 253.434 \\ -293.531 \end{pmatrix} = \begin{pmatrix} 247.563 \\ 232.887 \\ 221.146 \\ 209.404 \end{pmatrix}$$

The residuals are

$$\boldsymbol{e} = \boldsymbol{Y} - \hat{\boldsymbol{Y}} = \begin{pmatrix} -5.563 \\ 4.113 \\ 9.854 \\ -8.404 \end{pmatrix}$$

The results from the ozone example are summarized in Table 3.1.

Table 3.1 Results for the linear regression of soybean yield on levels of ozone.

X_i	Y_i	$\hat{Y}_i$	e_i
.02	242	247.563	−5.563
.07	237	232.887	4.113
.11	231	221.146	9.854
.15	201	209.404	−8.404

■

3.4 Precision of the Estimates

$\hat{\boldsymbol{\beta}}$, $\hat{\boldsymbol{Y}}$, and $\boldsymbol{e}$ are random vectors because they are functions of the random vector $\boldsymbol{Y}$. The previous section used their linear relationships with $\boldsymbol{Y}$ to derive their expectations: $\mathscr{E}(\hat{\boldsymbol{\beta}}) = \boldsymbol{\beta}$, $\mathscr{E}(\hat{\boldsymbol{Y}}) = \boldsymbol{X\beta}$, and $\mathscr{E}(\boldsymbol{e}) = \boldsymbol{0}$. In this section, the variance–covariance matrix for each of these random vectors is derived using the rules for the variances of linear functions (see Section 1.5).

Variance of a Linear Function

First, the general result for the variance of linear functions is restated in matrix notation. Suppose $\boldsymbol{Y}$ is a random vector with variance–covariance matrix $\mathbf{Var}(\boldsymbol{Y})$. Let $u = \boldsymbol{a}'\boldsymbol{Y}$ be any linear function of the vector $\boldsymbol{Y}$, where $\boldsymbol{a}'$ is the row vector of coefficients defining the linear function. The variance of u is expressed in terms of $\mathbf{Var}(\boldsymbol{Y})$ as

$$\sigma^2(u) = \boldsymbol{a}'[\mathbf{Var}(\boldsymbol{Y})]\boldsymbol{a} \qquad [3.21]$$

If $\mathbf{Var}(\boldsymbol{Y}) = \boldsymbol{I}\sigma^2$, as is assumed in ordinary least squares,

$$\sigma^2(u) = \boldsymbol{a}'(\boldsymbol{I}\sigma^2)\boldsymbol{a} = \boldsymbol{a}'\boldsymbol{a}\sigma^2 \qquad [3.22]$$

Notice that $\boldsymbol{a}'\boldsymbol{a}$ is the sum of squares of the coefficients of the linear function, $\sum a_i^2$, which is the result given in Section 1.5.

Several linear functions of $\boldsymbol{Y}$ can be considered simultaneously by expanding $\boldsymbol{a}'$ into a $k \times n$ matrix of coefficients $\boldsymbol{A}$, where each row of $\boldsymbol{A}$ provides the coefficients for one linear function. Then, $\boldsymbol{U} = \boldsymbol{AY}$ is a column vector containing k new random variables, each of which is a linear function of the random vector $\boldsymbol{Y}$. The $k \times k$ variance–covariance matrix for $\boldsymbol{U}$ is given by

$$\mathbf{Var}(\boldsymbol{U}) = \boldsymbol{A}[\mathbf{Var}(\boldsymbol{Y})]\boldsymbol{A}' \qquad [3.23]$$

Or, when $\mathbf{Var}(\boldsymbol{Y}) = \boldsymbol{I}\sigma^2$,

$$\mathbf{Var}(\boldsymbol{U}) = \boldsymbol{AA}'\sigma^2 \qquad [3.24]$$

The ith diagonal element of $\boldsymbol{AA}'$ is the sum of squares of the coefficients of the ith linear function. This coefficient multiplied by σ^2 gives the variance of the ith linear function. The (i, j)th off-diagonal element is the sum of products of the coefficients of the ith and jth linear functions and, when multiplied by σ^2, gives the covariance between the two linear functions.

The proof of the general form of the variance of linear functions of a random vector is given here primarily as an exercise in matrix algebra. By definition, the variance–covariance matrix of a random vector $\boldsymbol{Y}$ is

$$\mathbf{Var}(\boldsymbol{Y}) = \mathscr{E}\{[\boldsymbol{Y} - \mathscr{E}(\boldsymbol{Y})][\boldsymbol{Y} - \mathscr{E}(\boldsymbol{Y})]'\} \qquad [3.25]$$

where $\mathscr{E}$ denotes taking the expectation of all elements in the matrix that follows. The multiplication in equation 3.25 gives an $n \times n$ matrix with $[Y_i - \mathscr{E}(Y_i)]^2$ terms on the diagonal and $[Y_i - \mathscr{E}(Y_i)][Y_j - \mathscr{E}(Y_j)]$ terms in the off-diagonal positions. By definition, expectations of these elements are variances and co-variances, respectively. This definition of the variance–covariance matrix will be used to derive the variance of $\boldsymbol{U} = \boldsymbol{AY}$. By definition, the variance–covariance matrix for $\boldsymbol{U}$ is

$$\mathbf{Var}(\boldsymbol{U}) = \mathscr{E}\{[\boldsymbol{U} - \mathscr{E}(\boldsymbol{U})][\boldsymbol{U} - \mathscr{E}(\boldsymbol{U})]'\}$$

Substitution of $\boldsymbol{AY}$ for $\boldsymbol{U}$ and factoring gives

$$\begin{aligned}\mathbf{Var}(\boldsymbol{U}) &= \mathscr{E}\{[\boldsymbol{AY} - \mathscr{E}(\boldsymbol{AY})][\boldsymbol{AY} - \mathscr{E}(\boldsymbol{AY})]'\} \\ &= \mathscr{E}\{\boldsymbol{A}[\boldsymbol{Y} - \mathscr{E}(\boldsymbol{Y})][\boldsymbol{Y} - \mathscr{E}(\boldsymbol{Y})]'\boldsymbol{A}'\} \\ &= \boldsymbol{A}\mathscr{E}\{[\boldsymbol{Y} - \mathscr{E}(\boldsymbol{Y})][\boldsymbol{Y} - \mathscr{E}(\boldsymbol{Y})]'\}\boldsymbol{A}' \\ &= \boldsymbol{A}[\mathbf{Var}(\boldsymbol{Y})]\boldsymbol{A}' \end{aligned} \qquad [3.26]$$

The factoring of matrix products must be done carefully; remember that matrix multiplication is not commutative. Therefore, $\boldsymbol{A}$ is factored both to the left (from the first quantity in square brackets) and to the right (from the transpose of the second quantity in square brackets). Remember that transposing a product reverses the order of multiplication. Since $\boldsymbol{A}$ is a matrix of constants it can be factored outside the expectation operator. This leaves an inner matrix which by definition is $\mathbf{Var}(\boldsymbol{Y})$.

Two examples will illustrate the derivation of variances of linear functions using this important result.

Example 3.5 Matrix notation will be used to derive the familiar variance of a sample mean, σ^2/n. The mean of a sample of n observations, $\bar{Y} = \sum Y_i/n$, is written in matrix notation as

$$\bar{Y} = \begin{pmatrix} \frac{1}{n} & \frac{1}{n} & \cdots & \frac{1}{n} \end{pmatrix} \boldsymbol{Y}$$

Thus, $\bar{Y}$ is a linear function of $\boldsymbol{Y}$ with the vector of coefficients being

$$\boldsymbol{a}' = \begin{pmatrix} \frac{1}{n} & \frac{1}{n} & \cdots & \frac{1}{n} \end{pmatrix}$$

Then, if $\mathbf{Var}(\boldsymbol{Y}) = \boldsymbol{I}\sigma^2$,

$$\begin{aligned}\text{Var}(\bar{Y}) &= \boldsymbol{a}'[\mathbf{Var}(\boldsymbol{Y})]\boldsymbol{a} = \boldsymbol{a}'(\boldsymbol{I}\sigma^2)\boldsymbol{a} \\ &= \begin{pmatrix} \frac{1}{n} & \frac{1}{n} & \cdots & \frac{1}{n} \end{pmatrix}(\boldsymbol{I}\sigma^2)\begin{bmatrix} \frac{1}{n} \\ \frac{1}{n} \\ \vdots \\ \frac{1}{n} \end{bmatrix} = n\left(\frac{1}{n}\right)^2\sigma^2 = \frac{\sigma^2}{n} \end{aligned} \qquad [3.27]$$

■

Example 3.6 For the second example, consider two linear contrasts on a set of four treatment means with n observations in each mean. The random vector in this case is the vector of the four treatment means. If the means have been computed from different random samples from the same population, the variance of each mean will be σ^2/n (equation 3.27), and all covariances between the means will be zero. The variance–covariance matrix for the vector of means $\bar{\boldsymbol{Y}}$ is $\mathbf{Var}(\bar{\boldsymbol{Y}}) = \boldsymbol{I}(\sigma^2/n)$. Assume that the two linear contrasts of interest are

$$c_1 = \bar{Y}_1 - \bar{Y}_2 \qquad \text{and} \qquad c_2 = \bar{Y}_1 - 2\bar{Y}_2 + \bar{Y}_3 \qquad [3.28]$$

Notice that $\bar{Y}_4$ is not involved in these contrasts. The contrasts can be written as $\boldsymbol{C} = \boldsymbol{A}\boldsymbol{Y}$, where

$$\boldsymbol{C} = \begin{pmatrix} c_1 \\ c_2 \end{pmatrix} \qquad \text{and} \qquad \boldsymbol{A} = \begin{bmatrix} 1 & -1 & 0 & 0 \\ 1 & -2 & 1 & 0 \end{bmatrix}$$

Thus,

$$\begin{aligned} \mathbf{Var}(\boldsymbol{C}) &= \boldsymbol{A}[\mathbf{Var}(\bar{\boldsymbol{Y}})]\boldsymbol{A}' = \boldsymbol{A}\left[\boldsymbol{I}\left(\frac{\sigma^2}{n}\right)\right]\boldsymbol{A}' \\ &= \boldsymbol{A}\boldsymbol{A}'\left(\frac{\sigma^2}{n}\right) = \begin{bmatrix} 2 & 3 \\ 3 & 6 \end{bmatrix}\frac{\sigma^2}{n} \end{aligned} \qquad [3.29]$$

Thus, the variance of c_1 is $2\sigma^2/n$, the variance of c_2 is $6\sigma^2/n$, and the covariance between the two contrasts is $3\sigma^2/n$. ■

Var($\hat{\beta}$)

The general result for variances of linear functions is now used to obtain the variance–covariance matrices for $\hat{\boldsymbol{\beta}}$, $\hat{\boldsymbol{Y}}$, and $\boldsymbol{e}$. In the following, $\mathbf{Var}(\boldsymbol{Y}) = \boldsymbol{I}\sigma^2$ is assumed. The matrix of coefficients on $\boldsymbol{Y}$ that gives $\hat{\boldsymbol{\beta}}$ is $\boldsymbol{A} = [(\boldsymbol{X}'\boldsymbol{X})^{-1}\boldsymbol{X}']$. Thus,

$$\begin{aligned} \mathbf{Var}(\hat{\boldsymbol{\beta}}) &= [(\boldsymbol{X}'\boldsymbol{X})^{-1}\boldsymbol{X}'][\mathbf{Var}(\boldsymbol{Y})][(\boldsymbol{X}'\boldsymbol{X})^{-1}\boldsymbol{X}']' \\ &= [(\boldsymbol{X}'\boldsymbol{X})^{-1}\boldsymbol{X}'][(\boldsymbol{X}'\boldsymbol{X})^{-1}\boldsymbol{X}']'\sigma^2 \end{aligned}$$

Recalling that the transpose of a product is the product of the transposes and that $\boldsymbol{X}'\boldsymbol{X}$ is symmetric, we obtain

$$\begin{aligned} \mathbf{Var}(\hat{\boldsymbol{\beta}}) &= (\boldsymbol{X}'\boldsymbol{X})^{-1}(\boldsymbol{X}'\boldsymbol{X})(\boldsymbol{X}'\boldsymbol{X})^{-1}\sigma^2 \\ &= (\boldsymbol{X}'\boldsymbol{X})^{-1}\sigma^2 \end{aligned} \qquad [3.30]$$

Thus, the variances and the covariances of the estimated regression coefficients are given by the elements of $(\boldsymbol{X}'\boldsymbol{X})^{-1}$ multiplied by σ^2. The diagonal elements give the variances in the order in which the regression coefficients are listed in $\boldsymbol{\beta}$ and the off-diagonal elements give their covariances.

Example 3.7 In the ozone example, Example 3.3,

$$(X'X)^{-1} = \begin{bmatrix} 1.0755 & -9.4340 \\ -9.4340 & 107.8167 \end{bmatrix}$$

Thus, $\text{Var}(\hat{\beta}_0) = 1.0755\sigma^2$ and $\text{Var}(\hat{\beta}_1) = 107.8167\sigma^2$. The covariance between $\hat{\beta}_0$ and $\hat{\beta}_1$ is $\text{Cov}(\hat{\beta}_0, \hat{\beta}_1) = -9.4340\sigma^2$. ■

Var($\hat{Y}$)

The variance–covariance matrix of $\hat{Y}$ can be derived using either the relationship $\hat{Y} = X\hat{\beta}$ or $\hat{Y} = PY$. Recall that $P = X(X'X)^{-1}X'$. Applying the rules for variances of linear functions to the first relationship gives

$$\begin{aligned} \textbf{Var}(\hat{Y}) &= X[\textbf{Var}(\hat{\beta})]X' \\ &= X(X'X)^{-1}X'\sigma^2 \\ &= P\sigma^2 \end{aligned} \qquad [3.31]$$

The derivation using the second relationship gives

$$\begin{aligned} \textbf{Var}(\hat{Y}) &= P[\textbf{Var}(Y)]P' \\ &= PP'\sigma^2 \\ &= P\sigma^2 \end{aligned} \qquad [3.32]$$

since P is symmetric and idempotent. Therefore, the matrix P multiplied by σ^2 gives the variances and covariances for all $\hat{Y}_i$. P is a large $n \times n$ matrix and often is more than is of interest. The variances of any subset of the $\hat{Y}_i$ can be determined by using only the rows of X, say X_r, that correspond to the data points of interest and applying the first derivation. This gives

$$\textbf{Var}(\hat{Y}_r) = X_r[\textbf{Var}(\hat{\beta})]X_r' = X_r(X'X)^{-1}X_r'\sigma^2 \qquad [3.33]$$

Var($\hat{Y}_{pred}$)

The variances given by $P\sigma^2$ are the appropriate variances for the $\hat{Y}_i$ when they are used to estimate the means of Y for given levels of the independent variables. For prediction of future random observations at the given levels of the independent variables, each of the variances must be increased by σ^2 to account for the variance of the quantity being predicted. Thus, the variance–covariance matrix for prediction is

$$\textbf{Var}(\hat{Y}_{\text{pred}}) = (I + P)\sigma^2 \qquad [3.34]$$

Var(e)

The variance–covariance matrix of the residuals vector $e = (I - P)Y$ is

$$\textbf{Var}(e) = (I - P)\sigma^2 \qquad [3.35]$$

again using the result that $(I - P)$ is a symmetric, idempotent matrix.

Example 3.8

The matrix $\boldsymbol{P} = \boldsymbol{X}(\boldsymbol{X}'\boldsymbol{X})^{-1}\boldsymbol{X}'$ was computed for the ozone example in Example 3.3. Thus, with some rounding of the elements in $\boldsymbol{P}$,

$$\mathbf{Var}(\hat{\boldsymbol{Y}}) = \boldsymbol{P}\sigma^2$$

$$= \begin{bmatrix} .741 & .377 & .086 & -.205 \\ .377 & .283 & .208 & .132 \\ .086 & .208 & .305 & .402 \\ -.205 & .132 & .402 & .671 \end{bmatrix} \sigma^2$$

The variance of the estimated mean of Y when the ozone level is .02 ppm is $\text{Var}(\hat{Y}_1) = .741\sigma^2$. For the ozone level of .11 ppm, the variance of the estimated mean is $\text{Var}(\hat{Y}_3) = .305\sigma^2$. The covariance between the two estimated means is $\text{Cov}(\hat{Y}_1, \hat{Y}_3) = .086\sigma^2$.

If the fitted values are being used for prediction of future observations, the diagonal elements of the matrix of coefficients for $\mathbf{Var}(\hat{\boldsymbol{Y}})$ are increased by unity. Thus, the variances for prediction at .02 ppm and .11 ppm would be $\text{Var}(\hat{Y}_{\text{pred}_1}) = 1.741\sigma^2$ and $\text{Var}(\hat{Y}_{\text{pred}_3}) = 1.305\sigma^2$, respectively. The covariance is unchanged.

The variance–covariance matrix of the residuals is obtained by $\mathbf{Var}(\boldsymbol{e}) = (\boldsymbol{I} - \boldsymbol{P})\sigma^2$. Thus,

$$\text{Var}(e_1) = (1 - .741)\sigma^2 = .259\sigma^2$$

$$\text{Var}(e_3) = (1 - .305)\sigma^2 = .695\sigma^2$$

$$\text{Cov}(e_1, e_3) = -\text{Cov}(\hat{Y}_1, \hat{Y}_3) = -.086\sigma^2$$

It is important to note that the variances of the *observed* residuals are not equal to σ^2 and the covariances are not zero. The assumption of equal variances and zero covariances applies to the ε_i, not the e_i. ■

$Var(\hat{Y}_i) \leq Var(Y_i)$

The variance of any particular $\hat{Y}_i$ and the variance of the corresponding e_i will always add to σ^2 because

$$\boldsymbol{P}\sigma^2 + (\boldsymbol{I} - \boldsymbol{P})\sigma^2 = \boldsymbol{I}\sigma^2 \qquad [3.36]$$

Since variances cannot be negative, each diagonal element of $\boldsymbol{P}$ must be between zero and one: $0 < v_{ii} < 1.0$, where v_{ii} is the ith diagonal element of $\boldsymbol{P}$. Thus, the variance of any $\hat{Y}_i$ is always less than σ^2, the variance of the individual observations. This shows the advantage of fitting a continuous response model, assuming the model is correct, over simply using the individual observed data points as estimates of the mean of Y for the given values of the X's. The greater precision from fitting a response model comes from the fact that each $\hat{Y}_i$ uses information from the surrounding data points. The gain in precision can be quite striking. In Example 3.8, the precision obtained on the estimates of the means for the two intermediate levels of ozone using the linear response equation were

$.283\sigma^2$ and $.305\sigma^2$. To attain the same degree of precision without using the response model would have required more than three observations at each level of ozone.

Var($\hat{Y}_i$) Determined by X's

Equation 3.36 implies that data points having high precision (low variance) on $\hat{Y}_i$ will have low precision (high variance) on e_i and vice versa. Belsley, Kuh, and Welsch (1980) show that the diagonal elements of $\boldsymbol{P}$, v_{ii}, can be interpreted as measures of distance of the corresponding data points from the center of the X-space (from $\bar{X}$ in the case of one independent variable). Points that are far from the center of the X-space have relatively large v_{ii} and, therefore, relatively low precision on $\hat{Y}_i$ and high precision on e_i. The smaller variance of the residuals for the points far from the "center of the data" indicates that the fitted regression line or response surface tends to come closer to the observed values for these points. This aspect of $\boldsymbol{P}$ will be used later to detect the more influential data points.

Precision in Experimental Designs

The variances (and covariances) have been expressed as multiples of σ^2. The coefficients are determined entirely by the $\boldsymbol{X}$ matrix, a matrix of constants that depends on the model being fit and the levels of the independent variables in the study. In designed experiments, the levels of the independent variables are subject to the control of the researcher. Thus, except for the magnitude of σ^2, the precision of the experiment is under the control of the researcher and can be known before the experiment is run. The efficiencies of alternative experimental designs can be compared by computing $(\boldsymbol{X}'\boldsymbol{X})^{-1}$ and $\boldsymbol{P}$ for each design. The design giving the smallest variances for the quantities of interest would be preferred.

3.5 Distribution of Linear Functions of Normal Variables

The assumption that the random errors, the ε_i, are normally distributed implies that the Y_i also are normally distributed. It is a general result that any *linear* function of normally distributed random variables is itself normally distributed (Searle, 1971). The estimated regression coefficients, the fitted values, and the observed residuals are all linear functions of the original observations on the dependent variable. Consequently, the assumption of normality of ε also implies that the random vectors $\boldsymbol{Y}$, $\hat{\boldsymbol{\beta}}$, $\hat{\boldsymbol{Y}}$, and $\boldsymbol{e}$ are each multivariate normally distributed. Their mean vectors and variance–covariance matrices were given in Sections 3.3 and 3.4, respectively.

Distributions of Random Vectors

Thus, with the assumption that $\varepsilon \sim \mathrm{N}(\mathbf{0}, \boldsymbol{I}\sigma^2)$, the distributions of the key random vectors in least squares regression can be summarized as follows:

$$
\begin{aligned}
\boldsymbol{Y} &\sim \mathrm{N}(\boldsymbol{X\beta}, \boldsymbol{I}\sigma^2) \\
\hat{\boldsymbol{\beta}} &\sim \mathrm{N}(\boldsymbol{\beta}, (\boldsymbol{X}'\boldsymbol{X})^{-1}\sigma^2) \\
\hat{\boldsymbol{Y}} &\sim \mathrm{N}(\boldsymbol{X\beta}, \boldsymbol{P}\sigma^2) \\
\boldsymbol{e} &\sim \mathrm{N}(\mathbf{0}, (\mathbf{I} - \mathbf{P})\sigma^2) \\
\hat{\boldsymbol{Y}}_{\text{pred}} &\sim \mathrm{N}(\boldsymbol{X\beta}, (\boldsymbol{I} + \boldsymbol{P})\sigma^2)
\end{aligned}
\qquad [3.37]
$$

This notation indicates that the random vector on the left is distributed as a multivariate normal random variable with mean or expectation vector shown by the first quantity in parentheses and variance–covariance matrix shown by the second quantity. In general, none of the elements in the random vectors computed in least squares regression will be independent of each other; the off-diagonal elements in the variance–covariance matrices are not zero.

Importance of Normality Assumption

The conventional tests of hypotheses and confidence interval estimates of the parameters are based on the assumption that the estimates are normally distributed. Thus, the assumption of normality of the ε_i is critical for these purposes. However, normality is not required for least squares estimation. Even in the absence of normality, the least squares estimates are the best linear unbiased estimates (b.l.u.e.). They are best in the sense of having minimum variance among all linear unbiased estimators. If normality does hold, the least squares estimators are also the maximum likelihood estimators. Maximum likelihood estimators are derived using the criterion of finding those values of the parameters that would have maximized the probability of obtaining the particular sample. The reader is referred to statistical theory texts such as Searle (1971), Graybill (1961), and Cramér (1946) for further discussion of maximum likelihood estimation.

3.6 Summary of Matrix Formulas

Model:	$\boldsymbol{Y} = \boldsymbol{X\beta} + \boldsymbol{\varepsilon}$
Normal equations:	$(\boldsymbol{X}'\boldsymbol{X})\boldsymbol{\beta} = \boldsymbol{X}'\boldsymbol{Y}$
Parameter estimates:	$\hat{\boldsymbol{\beta}} = (\boldsymbol{X}'\boldsymbol{X})^{-1}\boldsymbol{X}'\boldsymbol{Y}$
Fitted values:	$\hat{\boldsymbol{Y}} = \boldsymbol{X}\hat{\boldsymbol{\beta}}$
	$= \boldsymbol{PY}$, where $\boldsymbol{P} = \boldsymbol{X}(\boldsymbol{X}'\boldsymbol{X})^{-1}\boldsymbol{X}'$
Residuals:	$\boldsymbol{e} = \boldsymbol{Y} - \hat{\boldsymbol{Y}}$
	$= (\boldsymbol{I} - \boldsymbol{P})\boldsymbol{Y}$
Variance of $\hat{\boldsymbol{\beta}}$:	$\mathbf{Var}(\hat{\boldsymbol{\beta}}) = (\boldsymbol{X}'\boldsymbol{X})^{-1}\sigma^2$
Variance of $\hat{\boldsymbol{Y}}$:	$\mathbf{Var}(\hat{\boldsymbol{Y}}) = \boldsymbol{P}\sigma^2$
Variance of $\boldsymbol{e}$:	$\mathbf{Var}(\boldsymbol{e}) = (\boldsymbol{I} - \boldsymbol{P})\sigma^2$

EXERCISES

3.1. The linear model in ordinary least squares is $\boldsymbol{Y} = \boldsymbol{X\beta} + \boldsymbol{\varepsilon}$. Assume there are 30 observations and five independent variables. Give the order and rank of:
a. $\boldsymbol{Y}$
b. $\boldsymbol{X}$ (without an intercept in the model)
c. $\boldsymbol{X}$ (with an intercept in the model)

d. $\boldsymbol{\beta}$ (without an intercept in the model)
e. $\boldsymbol{\beta}$ (with an intercept in the model)
f. $\boldsymbol{\varepsilon}$
g. $(\boldsymbol{X}'\boldsymbol{X})$ (with an intercept in the model)
h. $\boldsymbol{P}$ (with an intercept in the model)

* 3.2. For each of the following matrices, indicate whether there will be a unique solution to the normal equations. Show how you arrived at your answer.

$$\boldsymbol{X}_1 = \begin{bmatrix} 1 & 2 & 4 \\ 1 & 3 & 8 \\ 1 & 0 & 6 \\ 1 & -1 & 2 \end{bmatrix} \qquad \boldsymbol{X}_2 = \begin{bmatrix} 1 & 1 & 0 \\ 1 & 1 & 0 \\ 1 & 0 & 1 \\ 1 & 0 & 1 \end{bmatrix} \qquad \boldsymbol{X}_3 = \begin{bmatrix} 1 & 2 & 4 \\ 1 & 1 & 2 \\ 1 & -3 & -6 \\ 1 & -1 & -2 \end{bmatrix}$$

3.3. You have a data set with four independent variables and $n = 42$ observations. If the model is to include an intercept, what would be the order of $\boldsymbol{X}'\boldsymbol{X}$? Of $(\boldsymbol{X}'\boldsymbol{X})^{-1}$? Of $\boldsymbol{X}'\boldsymbol{Y}$? Of $\boldsymbol{P}$?

* 3.4. A data set with one independent variable gave the following $(\boldsymbol{X}'\boldsymbol{X})^{-1}$:

$$(\boldsymbol{X}'\boldsymbol{X})^{-1} = \begin{bmatrix} \frac{31}{177} & \frac{-3}{177} \\ \frac{-3}{177} & \frac{6}{177} \end{bmatrix}$$

How many observations were there in the data set? Find $\sum X_i^2$. Find the corrected sum of squares for the independent variable.

* 3.5. The data in the accompanying table relate grams plant dry weight, Y, to percent soil organic matter, X_1, and kilograms of supplemental soil nitrogen added per 1,000 square meters, X_2.

	Y	X_1	X_2
	78.5	7	2.6
	74.3	1	2.9
	104.3	11	5.6
	87.6	11	3.1
	95.9	7	5.2
	109.2	11	5.5
	102.7	3	7.1
Sums:	652.5	51	32.0
Means:	93.21	7.29	4.57

a. Define $\boldsymbol{Y}$, $\boldsymbol{X}$, $\boldsymbol{\beta}$, and $\boldsymbol{\varepsilon}$ for a model involving both independent variables and an intercept.
b. Compute $\boldsymbol{X}'\boldsymbol{X}$ and $\boldsymbol{X}'\boldsymbol{Y}$.
c. $(\boldsymbol{X}'\boldsymbol{X})^{-1}$ for this problem is

$$(\boldsymbol{X}'\boldsymbol{X})^{-1} = \begin{bmatrix} 1.7995972 & -.0685472 & -.2531648 \\ -.0685472 & .0100774 & -.0010661 \\ -.2531648 & -.0010661 & .0570789 \end{bmatrix}$$

Verify that this is the inverse of $X'X$. Compute $\hat{\beta}$ and write the regression equation. Interpret each estimated regression coefficient. What are the units of measure attached to each regression coefficient?

d. Compute $\hat{Y}$ and e.

e. The P matrix in this case will be a 7×7 matrix. Illustrate the computation of P by computing v_{11}, the first diagonal element, and v_{12}, the second element in the first row. Use the above results and these two elements of P to give the appropriate coefficient on σ^2 for each of the following variances:
   (i) $\text{Var}(\hat{\beta}_1)$
   (ii) $\text{Var}(\hat{Y}_1)$
   (iii) $\text{Var}(\hat{Y}_{\text{pred}_1})$
   (iv) $\text{Var}(e_1)$

3.6. Use the data in exercise 3.5. Center each independent variable by subtracting the mean from each observation. Compute $X'X$, $X'Y$, and $\hat{\beta}$ using the centered data. Were the computations simplified by using centered data? Show that the regression equation obtained using centered data is equivalent to that obtained with the original uncentered data. Compute P using the centered data and compare it to that obtained using the uncentered data.

* 3.7. The matrix P for the Heagle ozone data is given in Example 3.3. Verify that P is symmetric and idempotent. What is the linear function of Y_i that gives $\hat{Y}_3$?

3.8. Compute $(I - P)$ for the Heagle ozone data. Verify that $(I - P)$ is idempotent and that P and $(I - P)$ are orthogonal to each other. What does the orthogonality imply about the vectors $\hat{Y}$ and e?

* 3.9. This exercise uses the Lesser–Unsworth data from exercise 1.19, in which seed weight is related to cumulative solar radiation for two levels of exposure to ozone. Assume that "low ozone" is an exposure of .025 ppm and that "high ozone" is an exposure of .07 ppm.

a. Set up X and β for the regression of seed weight on cumulative solar radiation *and* ozone level. Center the independent variables and include an intercept in the model. Estimate the regression equation and interpret the result.

b. Extend the model to include an independent variable that is the product term between centered cumulative solar radiation and centered ozone level. Estimate the regression equation for this model and interpret the result. What does the presence of the product term contribute to the regression equation?

3.10. This exercise uses the data from exercise 1.21 (number of hospital days for smokers, number of cigarettes smoked, and number of hospital days for control groups of nonsmokers). Exercise 1.21 used the information from the nonsmoker control groups by defining the dependent variable as $Y = \ln$(number of hospital days for smokers/number of hospital days for nonsmokers). Another method of taking into account the experience of the nonsmokers is to use $X_2 = \ln$(number of hospital days for nonsmokers) as an independent variable.

a. Set up X and β for the regression of $Y = \ln$(number of hospital days for smokers) on $X_1 =$ (number of cigarettes)^2 and $X_2 = \ln$(number of hospital days for nonsmokers).

b. Estimate the regression equation and interpret the results. What value of β_2 would correspond to using the nonsmoker experience as was done in exercise 1.21?

3.11. The data in the table relate the annual catch of Gulf Menhaden, *Brevoortia patronus*, to fishing pressure for 1964 to 1979 (Nelson and Ahrenholz, 1986).

Year	*Catch* (*met. ton* × 10^{-3})	*Number Vessels*	*Pressure* (*Vessel-ton-weeks* × 10^{-3})
1964	409.4	76	282.9
1965	463.1	82	335.6
1966	359.1	80	381.3
1967	317.3	76	404.7
1968	373.5	69	382.3
1969	523.7	72	411.0
1970	548.1	73	400.0
1971	728.2	82	472.9
1972	501.7	75	447.5
1973	486.1	65	426.2
1974	578.6	71	485.5
1975	542.6	78	536.9
1976	561.2	81	575.9
1977	447.1	80	532.7
1978	820.0	80	574.3
1979	777.9	77	533.9

Run a linear regression of catch (Y) on fishing pressure (X_1) and number of vessels (X_2). Include an intercept in the model. Interpret the regression equation.

*3.12. A simulation model for peak water flow from watersheds was tested by comparing measured peak flow (cfs) from ten storms with predictions of peak flow obtained from the simulation model. Q_o and Q_p are the observed and predicted peak flows, respectively. Four independent variables were recorded:

X_1 = Area of watershed (mi^2)

X_2 = Average slope of watershed (in percent)

X_3 = Surface absorbency index (0 = complete absorbency, 100 = no absorbency)

X_4 = Peak intensity of rainfall (in/hr) computed on half-hour time intervals

Q_o	Q_p	X_1	X_2	X_3	X_4
28	32	.03	3.0	70	.6
112	142	.03	3.0	80	1.8
398	502	.13	6.5	65	2.0
772	790	1.00	15.0	60	.4
2,294	3,075	1.00	15.0	65	2.3
2,484	3,230	3.00	7.0	67	1.0
2,586	3,535	5.00	6.0	62	.9
3,024	4,265	7.00	6.5	56	1.1
4,279	6,529	7.00	6.5	56	1.4
710	935	7.00	6.5	56	.7

a. Use $Y = \ln(Q_o/Q_p)$ as the dependent variable. The dependent variable will have the value zero if the observed and predicted peak flows agree. Set up the regression problem to determine whether the discrepancy, Y, is related to any of the four independent variables. Use an intercept in the model. Estimate the regression equation.
b. Further consideration of the problem suggested that the discrepancy between observed and predicted peak flow, Y, might go to zero as the values of the four independent variables approach zero. Redefine the regression problem to eliminate the intercept (force $\beta_0 = 0$), and estimate the regression equation.
c. Rerun the regression (without the intercept) using only X_1 and X_4; that is, omit X_2 and X_3 from the model. Do the regression coefficients for X_1 and X_4 change? Explain why they do or do not change.
d. Describe the change in the standard errors of the estimated regression coefficients as the intercept was dropped [part (a) versus part (b)] and as X_2 and X_3 were dropped from the model [part (b) versus part (c)].

*3.13. You have fit a linear model using $\boldsymbol{Y} = \boldsymbol{X\beta} + \boldsymbol{\varepsilon}$ where $\boldsymbol{X}$ involves r independent variables. Now assume that the true model involves an *additional* s independent variables contained in $\boldsymbol{Z}$. That is, the true model is

$$\boldsymbol{Y} = \boldsymbol{X\beta} + \boldsymbol{Z\gamma} + \boldsymbol{\varepsilon}$$

where $\boldsymbol{\gamma}$ are the regression coefficients for the independent variables contained in $\boldsymbol{Z}$.

a. Find $\mathscr{E}(\hat{\boldsymbol{\beta}})$ and show that, in general, $\hat{\boldsymbol{\beta}}$ is a biased estimate of $\boldsymbol{\beta}$.
b. Under what conditions would $\hat{\boldsymbol{\beta}}$ be unbiased?

3.14. The accompanying table shows the part of the data reported by Cameron and Pauling (1978) related to the effects of supplemental ascorbate, vitamin C, in the treatment of colon cancer. The data are taken from Andrews and Herzberg (1985) and are used with permission.

Sex	*Age*	*Days*[a]	*Control*[b]
F	76	135	18
F	58	50	30
M	49	189	65
M	69	1,267	17
F	70	155	57
F	68	534	16
M	50	502	25
F	74	126	21
M	66	90	17
F	76	365	42
F	56	911	40
M	65	743	14
F	74	366	28
M	58	156	31
F	60	99	28
M	77	20	33
M	38	274	80

[a] Days = Number of days survival after date of untreatability.
[b] Control = Average number of days survival of ten control patients for each case.

Use $Y = \ln(\text{days})$ as the dependent variable and $X_1 = \text{sex}$ (coded -1 for males and $+1$ for females), $X_2 = \text{age}$, and $X_3 = \ln(\text{control})$ in a multiple regression to determine if there is any relationship between days survival and sex and age. Define $\boldsymbol{X}$ and $\boldsymbol{\beta}$, and estimate the regression equation. Explain why X_3 is in the model if the purpose is to relate survival to X_1 and X_2.

CHAPTER
FOUR

Analysis of Variance and Quadratic Forms

The previous chapter developed the regression results involving linear functions of the dependent variable, $\hat{\beta}$, $\hat{Y}$, and e. All were shown to be normally distributed random variables if Y was normally distributed.

This chapter develops the regression results for all quadratic functions of Y. The distribution of quadratic forms is used to develop tests of hypotheses and confidence interval estimates and joint confidence regions for β.

The estimates of the regression coefficients, the estimated means, and the residuals have been presented in matrix notation; all were shown to be **linear functions** of the original observations, Y. In this chapter it will be shown that all sums of squares and products are **quadratic functions** of Y. This means that each sum of squares can be written as $Y'AY$, where A is a matrix of coefficients called the **defining matrix**. $Y'AY$ is referred to as a **quadratic form** in Y.

The aim of model fitting is to explain as much of the variation in the dependent variable as possible from information contained in the independent variables. The contributions of the independent variables to the model are measured by partitions of the total sum of squares of Y attributable to, or "explained" by, the independent variables. Each component of the partitioning of the sums of squares is a quadratic form in Y. The degrees of freedom associated with a particular sum of squares and the orthogonality between different sums of squares are determined by the defining matrices in the quadratic forms. The matrix form for a sum of squares makes the computations simple if one has access to a computer package for matrix algebra. Also, the expectations and variances of the sums of squares are easily determined in this form. We give a brief introduction to the properties of quadratic forms before proceeding with the analysis of variance of the dependent variable Y.

4.1 Introduction to Quadratic Forms

Quadratic Form for a Single Linear Function

Consider first a sum of squares with which you are familiar from your earlier statistical methods courses—the sum of squares attributable to a linear contrast. Suppose you are interested in the linear contrast

$$C_1^* = Y_1 + Y_2 - 2Y_3 \qquad [4.1]$$

The sum of squares due to this contrast is

$$\text{SS}(C_1^*) = \frac{(C_1^*)^2}{6} \qquad [4.2]$$

The divisor of 6 is the sum of squares of the coefficients of the contrast. This divisor has been chosen to make the coefficient of σ^2 in the expectation of the sum of squares equal to 1. If we reexpress C_1^* so that the coefficients on the Y_i include $1/\sqrt{6}$, the sum of squares due to the contrast is the square of the contrast. Thus, $C_1 = C_1^*/\sqrt{6}$ can be written in matrix notation as

$$C_1 = \boldsymbol{a}'\boldsymbol{Y} \qquad [4.3]$$

by defining $\boldsymbol{a}' = (1/\sqrt{6}\ \ 1/\sqrt{6}\ \ -2/\sqrt{6})$ and $\boldsymbol{Y}' = (Y_1\ \ Y_2\ \ Y_3)$. The sum of squares for C_1 is then

$$\begin{aligned} \text{SS}(C_1) &= C_1^2 = (\boldsymbol{a}'\boldsymbol{Y})'(\boldsymbol{a}'\boldsymbol{Y}) \\ &= \boldsymbol{Y}'(\boldsymbol{aa}')\boldsymbol{Y} \\ &= \boldsymbol{Y}'\boldsymbol{A}\boldsymbol{Y} \qquad [4.4] \end{aligned}$$

The Defining Matrix

Thus, $\text{SS}(C_1)$ has been written as a **quadratic form in** $\boldsymbol{Y}$ where $\boldsymbol{A}$, the **defining matrix,** is the 3×3 matrix $\boldsymbol{A} = \boldsymbol{aa}'$. The multiplication $\boldsymbol{aa}'$ for this contrast gives

$$\boldsymbol{A} = \boldsymbol{aa}' = \begin{bmatrix} \dfrac{1}{\sqrt{6}} \\ \dfrac{1}{\sqrt{6}} \\ -\dfrac{2}{\sqrt{6}} \end{bmatrix} \begin{pmatrix} \dfrac{1}{\sqrt{6}} & \dfrac{1}{\sqrt{6}} & -\dfrac{2}{\sqrt{6}} \end{pmatrix}$$

$$= \begin{bmatrix} \frac{1}{6} & \frac{1}{6} & -\frac{2}{6} \\ \frac{1}{6} & \frac{1}{6} & -\frac{2}{6} \\ -\frac{2}{6} & -\frac{2}{6} & \frac{4}{6} \end{bmatrix} \qquad [4.5]$$

Completing the multiplication of the quadratic form gives

$$Y'AY = (Y_1 \ \ Y_2 \ \ Y_3)\begin{bmatrix} \frac{1}{6} & \frac{1}{6} & -\frac{2}{6} \\ \frac{1}{6} & \frac{1}{6} & -\frac{2}{6} \\ -\frac{2}{6} & -\frac{2}{6} & \frac{4}{6} \end{bmatrix}\begin{pmatrix} Y_1 \\ Y_2 \\ Y_3 \end{pmatrix}$$

$$= \tfrac{1}{6}[Y_1(Y_1 + Y_2 - 2Y_3) + Y_2(Y_1 + Y_2 - 2Y_3) + Y_3(-2Y_1 - 2Y_2 + 4Y_3)]$$

$$= \tfrac{1}{6}Y_1^2 + \tfrac{1}{6}Y_2^2 + \tfrac{4}{6}Y_3^2 + \tfrac{2}{6}Y_1Y_2 - \tfrac{4}{6}Y_1Y_3 - \tfrac{4}{6}Y_2Y_3 \qquad [4.6]$$

This result is verified by expanding the square of C_1, equation 4.1, in terms of Y_i and dividing by 6.

Comparison of the elements of A, equation 4.5, with the expansion given in equation 4.6 shows that the diagonal elements of the defining matrix are the coefficients on the squared terms and the *sums* of the symmetric off-diagonal elements are the coefficients on the product terms. The defining matrix for a quadratic form will always be written in this symmetric form.

Consider a second linear contrast on $\boldsymbol{Y}$ which is orthogonal to C_1. Let $C_2 = (Y_1 - Y_2)/\sqrt{2} = \boldsymbol{d}'\boldsymbol{Y}$, where $\boldsymbol{d}' = (1/\sqrt{2} \ \ -1/\sqrt{2} \ \ 0)$. The sum of squares for this contrast is

$$\text{SS}(C_2) = \boldsymbol{Y}'\boldsymbol{D}\boldsymbol{Y} \qquad [4.7]$$

where the defining matrix is

$$\boldsymbol{D} = \boldsymbol{dd}' = \begin{bmatrix} \frac{1}{2} & -\frac{1}{2} & 0 \\ -\frac{1}{2} & \frac{1}{2} & 0 \\ 0 & 0 & 0 \end{bmatrix} \qquad [4.8]$$

Degrees of Freedom

Each of these sums of squares has 1 degree of freedom because a single linear contrast is involved in each case. The **degrees of freedom** for a quadratic form is equal to the rank of the defining matrix which, in turn, is equal to the trace of the defining matrix if the defining matrix is **idempotent**. (The defining matrix for a quadratic form does not have to be idempotent. However, the quadratic forms with which we will be concerned will have idempotent defining matrices.) The defining matrices $\boldsymbol{A}$ and $\boldsymbol{D}$ in the two examples are idempotent. It is left to the reader to verify that $\boldsymbol{AA} = \boldsymbol{A}$ and $\boldsymbol{DD} = \boldsymbol{D}$. $\boldsymbol{A}$ and $\boldsymbol{D}$ would not have been idempotent if, for example, the $1/\sqrt{6}$ and $1/\sqrt{2}$ had not been incorporated into the coefficient vectors. Notice that $\text{tr}(\boldsymbol{A}) = \text{tr}(\boldsymbol{D}) = 1$, the degrees of freedom for each contrast.

Quadratic Form for Joint Linear Functions

The quadratic forms defined by $\boldsymbol{A}$ and $\boldsymbol{D}$ treated each linear function separately. That is, each quadratic form was a sum of squares with 1 degree of freedom. The two linear functions can be considered jointly by defining the coefficient matrix $\boldsymbol{K}'$ to be a 2×3 matrix containing the coefficients for both contrasts:

$$\boldsymbol{K}'\boldsymbol{Y} = \begin{bmatrix} \frac{1}{\sqrt{6}} & \frac{1}{\sqrt{6}} & -\frac{2}{\sqrt{6}} \\ \frac{1}{\sqrt{2}} & -\frac{1}{\sqrt{2}} & 0 \end{bmatrix} \begin{pmatrix} Y_1 \\ Y_2 \\ Y_3 \end{pmatrix} \quad [4.9]$$

The defining matrix for the quadratic form $\boldsymbol{Y}'\boldsymbol{K}\boldsymbol{K}'\boldsymbol{Y}$ is

$$\boldsymbol{F} = \boldsymbol{K}\boldsymbol{K}' = \begin{bmatrix} \frac{2}{3} & -\frac{1}{3} & -\frac{1}{3} \\ -\frac{1}{3} & \frac{2}{3} & -\frac{1}{3} \\ -\frac{1}{3} & -\frac{1}{3} & \frac{2}{3} \end{bmatrix} \quad [4.10]$$

In this example, the defining matrix $\boldsymbol{F}$ is idempotent and its trace indicates that there are 2 degrees of freedom for this sum of squares. (The quadratic form defined in this way is idempotent only because the two original contrasts were orthogonal to each other: $\boldsymbol{a}'\boldsymbol{d} = 0$. The general method of defining quadratic forms, sums of squares, for specific hypotheses will be discussed in Section 4.5.1.)

Orthogonal Quadratic Forms

Two quadratic forms (of the same vector $\boldsymbol{Y}$) are orthogonal if the product of the defining matrices is $\boldsymbol{0}$. Orthogonality of the two quadratic forms in the example is verified by the multiplication of $\boldsymbol{A}$ and $\boldsymbol{D}$:

$$\boldsymbol{D}\boldsymbol{A} = \begin{bmatrix} \frac{1}{2} & -\frac{1}{2} & 0 \\ -\frac{1}{2} & \frac{1}{2} & 0 \\ 0 & 0 & 0 \end{bmatrix} \begin{bmatrix} \frac{1}{6} & \frac{1}{6} & -\frac{2}{6} \\ \frac{1}{6} & \frac{1}{6} & -\frac{2}{6} \\ -\frac{2}{6} & -\frac{2}{6} & \frac{4}{6} \end{bmatrix} = \begin{bmatrix} 0 & 0 & 0 \\ 0 & 0 & 0 \\ 0 & 0 & 0 \end{bmatrix} \quad [4.11]$$

which equals $\boldsymbol{A}\boldsymbol{D}$ since both $\boldsymbol{A}$ and $\boldsymbol{D}$ are symmetric. Note that $\boldsymbol{D}\boldsymbol{A} = \boldsymbol{d}\boldsymbol{d}'\boldsymbol{a}\boldsymbol{a}'$ and will be zero if $\boldsymbol{d}'\boldsymbol{a} = 0$. Thus, the quadratic forms associated with two linear functions will be orthogonal if the two vectors of coefficients are orthogonal—that is, if the sum of products of the coefficient vectors, $\boldsymbol{d}'\boldsymbol{a}$, is zero. When the two linear functions are orthogonal, the sum of sums of squares (and degrees of freedom) for the two contrasts considered individually will equal the sum of squares (and degrees of freedom) of the two contrasts considered jointly. For this additivity to hold when more than two linear functions are considered, all must be pairwise orthogonal. Orthogonality of quadratic forms implies that the two pieces of information contained in the individual sums of squares are independent.

The quadratic forms of primary interest in this text are the sums of squares associated with analyses of variance, regression analyses, and tests of hypotheses. All will have idempotent defining matrices.

The following facts about quadratic forms are important [see Searle (1971) for more complete discussions on quadratic forms]:

1. Any sum of squares can be written as $\boldsymbol{Y'AY}$, where $\boldsymbol{A}$ is a square symmetric matrix.
2. The degrees of freedom associated with any quadratic form equal the rank of the defining matrix, which equals its trace when the matrix is idempotent.
3. Two quadratic forms are orthogonal if and only if the product of their defining matrices is the null matrix, $\mathbf{0}$.

Example 4.1 For illustration of quadratic forms, let

$$\boldsymbol{Y}' = (3.55 \quad 3.49 \quad 3.67 \quad 2.76 \quad 1.195)$$

be the vector of mean fungus disease scores on alfalfa. The five treatments were five equally spaced day/night temperature regimes under which the plants were growing at the time of inoculation with the fungus. The total uncorrected sum of squares is

$$\boldsymbol{Y'Y} = 3.55^2 + 3.49^2 + \cdots + 1.195^2 = 47.2971$$

The defining matrix for this quadratic form is the identity matrix of order 5. Since $\boldsymbol{I}$ is an idempotent matrix and $\text{tr}(\boldsymbol{I}) = 5$, this sum of squares has 5 degrees of freedom.

The linear function of $\boldsymbol{Y}$ that gives the total disease score over all treatments is given by $\sum Y_i = \boldsymbol{a}_1' \boldsymbol{Y}$, where

$$\boldsymbol{a}_1' = (1 \quad 1 \quad 1 \quad 1 \quad 1)$$

The sum of squares due to correction for the mean, the correction factor, is $(\sum Y_i)^2/5 = 43.0124$. This is written as a quadratic form as

$$\boldsymbol{Y}'(\boldsymbol{J}/5)\boldsymbol{Y}$$

where $\boldsymbol{J} = \boldsymbol{a}_1 \boldsymbol{a}_1'$ is a 5×5 matrix of ones. The defining matrix, $\boldsymbol{J}/5$, is an idempotent matrix with $\text{tr}(\boldsymbol{J}/5) = 1$. Therefore, the sum of squares due to correction for the mean has 1 degree of freedom.

Based on orthogonal polynomial coefficients for five equally spaced treatments, the linear contrast for temperature effects is given by

$$C_2^* = a_2^{*\prime}Y = (-2 \quad -1 \quad 0 \quad 1 \quad 2)Y$$

Incorporating the divisor, $\sqrt{a_2^* a_2^{*\prime}} = \sqrt{10}$, into the vector of coefficients gives

$$a_2' = \left(-\frac{2}{\sqrt{10}} \quad -\frac{1}{\sqrt{10}} \quad 0 \quad \frac{1}{\sqrt{10}} \quad \frac{2}{\sqrt{10}}\right)$$

The sum of squares due to the linear regression on temperature is given by the quadratic form

$$Y'A_2Y = 2.9594$$

where

$$A_2 = a_2 a_2' = \begin{bmatrix} .4 & .2 & 0 & -.2 & -.4 \\ .2 & .1 & 0 & -.1 & -.2 \\ 0 & 0 & 0 & 0 & 0 \\ -.2 & -.1 & 0 & .1 & .2 \\ -.4 & -.2 & 0 & .2 & .4 \end{bmatrix}$$

The defining matrix A_2 is idempotent with $\text{tr}(A_2) = 1$ and, therefore, the sum of squares has 1 degree of freedom.

The orthogonal polynomial coefficients for the quadratic term, including division by the square root of the sum of squares of the coefficients, is

$$a_3' = \frac{1}{\sqrt{14}}(2 \quad -1 \quad -2 \quad -1 \quad 2)$$

The sum of squares due to quadratic regression is given by the quadratic form

$$Y'A_3Y = 1.2007$$

where

$$A_3 = a_3 a_3' = \begin{bmatrix} .2857 & -.1429 & -.2857 & -.1429 & .2857 \\ -.1429 & .0714 & .1429 & .0714 & -.1429 \\ -.2857 & .1429 & .2857 & .1429 & -.2857 \\ -.1429 & .0714 & .1429 & .0714 & -.1429 \\ .2857 & -.1429 & -.2857 & -.1429 & .2857 \end{bmatrix}$$

The defining matrix A_3 is idempotent and $\text{tr}(A) = 1$ so this sum of squares also has 1 degree of freedom.

It is left to the reader to verify that each of the defining matrices, $\boldsymbol{J}/5$, $\boldsymbol{A}_2$, and $\boldsymbol{A}_3$, is idempotent and that they are pairwise orthogonal to each other. Since they are orthogonal to each other, these three sums of squares represent independent pieces of information. However, they are *not* orthogonal to the uncorrected sum of squares; the defining matrix $\boldsymbol{I}$ is not orthogonal to $\boldsymbol{J}/5$, $\boldsymbol{A}_2$, or $\boldsymbol{A}_3$. In fact, as you know from previous experience, the sums of squares defined by $\boldsymbol{J}/5$, $\boldsymbol{A}_2$, and $\boldsymbol{A}_3$ are partitions of the total uncorrected sum of squares. ■

4.2 Analysis of Variance

The vector of observations on the dependent variable, $\boldsymbol{Y}$, was partitioned in Chapter 3 into the vector of estimated means of Y, $\hat{Y}$, and the residuals vector, $\boldsymbol{e}$. That is,

$$\boldsymbol{Y} = \hat{\boldsymbol{Y}} + \boldsymbol{e} \tag{4.12}$$

This partitioning of $\boldsymbol{Y}$ will be used to provide a similar partitioning of the total sum of squares of the dependent variable.

Partitioning of Y′Y

It has been previously noted that the product

$$\boldsymbol{Y}'\boldsymbol{Y} = \sum Y_i^2 \tag{4.13}$$

gives the total sum of squares of the elements in the column vector $\boldsymbol{Y}$. This is a quadratic form whose defining matrix is the identity matrix: $\boldsymbol{Y}'\boldsymbol{Y} = \boldsymbol{Y}'\boldsymbol{I}\boldsymbol{Y}$. The matrix $\boldsymbol{I}$ is idempotent and its trace is equal to its order, indicating that the total (uncorrected) sum of squares has degrees of freedom equal to the number of elements in the vector. The identity matrix is the *only* full-rank idempotent matrix.

Since $\boldsymbol{Y} = \hat{\boldsymbol{Y}} + \boldsymbol{e}$,

$$\boldsymbol{Y}'\boldsymbol{Y} = (\hat{\boldsymbol{Y}} + \boldsymbol{e})'(\hat{\boldsymbol{Y}} + \boldsymbol{e}) = \hat{\boldsymbol{Y}}'\hat{\boldsymbol{Y}} + \hat{\boldsymbol{Y}}'\boldsymbol{e} + \boldsymbol{e}'\hat{\boldsymbol{Y}} + \boldsymbol{e}'\boldsymbol{e}$$

Substituting $\hat{\boldsymbol{Y}} = \boldsymbol{PY}$ and $\boldsymbol{e} = (\boldsymbol{I} - \boldsymbol{P})\boldsymbol{Y}$ gives

$$\begin{aligned}\boldsymbol{Y}'\boldsymbol{Y} &= (\boldsymbol{PY})'(\boldsymbol{PY}) + (\boldsymbol{PY})'[(\boldsymbol{I} - \boldsymbol{P})\boldsymbol{Y}] + [(\boldsymbol{I} - \boldsymbol{P})\boldsymbol{Y}]'(\boldsymbol{PY}) \\ &\quad + [(\boldsymbol{I} - \boldsymbol{P})\boldsymbol{Y}]'[(\boldsymbol{I} - \boldsymbol{P})\boldsymbol{Y}] \\ &= \boldsymbol{Y}'\boldsymbol{P}'\boldsymbol{PY} + \boldsymbol{Y}'\boldsymbol{P}'(\boldsymbol{I} - \boldsymbol{P})\boldsymbol{Y} + \boldsymbol{Y}'(\boldsymbol{I} - \boldsymbol{P})'\boldsymbol{PY} \\ &\quad + \boldsymbol{Y}'(\boldsymbol{I} - \boldsymbol{P})'(\boldsymbol{I} - \boldsymbol{P})\boldsymbol{Y}\end{aligned} \tag{4.14}$$

Both $\boldsymbol{P}$ and $(\boldsymbol{I} - \boldsymbol{P})$ are symmetric and idempotent so that $\boldsymbol{P}'\boldsymbol{P} = \boldsymbol{P}$ and $(\boldsymbol{I} - \boldsymbol{P})'(\boldsymbol{I} - \boldsymbol{P}) = (\boldsymbol{I} - \boldsymbol{P})$. The two middle terms in equation 4.14 are zero because the two quadratic forms are orthogonal to each other:

$$P'(I - P) = P - P = 0$$

Thus,

$$Y'Y = Y'PY + Y'(I - P)Y = \hat{Y}'\hat{Y} + e'e \qquad [4.15]$$

The total uncorrected sum of squares has been partitioned into two quadratic forms with defining matrices P and $(I - P)$, respectively. $\hat{Y}'\hat{Y}$ is that part of $Y'Y$ that can be attributed to the model being fit and will be labeled SS(Model). The second term, $e'e$, is that part of $Y'Y$ not explained by the model. It is the residual sum of squares after fitting the model and will be labeled SS(Res).

Degrees of Freedom

The orthogonality of the quadratic forms ensures that SS(Model) and SS(Res) are additive partitions. The degrees of freedom associated with each will depend on the rank of the defining matrices. The rank of $P = [X(X'X)^{-1}X']$ is determined by the rank of X. For full-rank models, the rank of X is equal to the number of columns of X, which is also the number of parameters in β. Thus, the degrees of freedom for SS(Model) is p' when the model is of full rank.

The $r(P)$ is also given by $\text{tr}(P)$ because P is idempotent. A result from matrix algebra states that $\text{tr}(ABC) = \text{tr}(BCA)$. Note the rotation of the matrices in the product. Using this property, we have

$$\begin{aligned} \text{tr}(P) &= \text{tr}[X(X'X)^{-1}X'] = \text{tr}[(X'X)^{-1}X'X] \\ &= \text{tr}(I_{p'}) = p' \end{aligned} \qquad [4.16]$$

The subscript on I indicates the order of the identity matrix. The degrees of freedom of SS(Res), $n - p'$, is obtained by noting the additivity of the two partitions or by observing that $\text{tr}(I - P) = \text{tr}(I_n) - \text{tr}(P) = (n - p')$. The order of this identity matrix is n.

Computational Forms

The expressions for the quadratic forms, equation 4.15, are the definitional forms; they show the nature of the sums of squares being computed. There are, however, more convenient computational forms. The computational form for SS(Model) $= \hat{Y}'\hat{Y}$ is

$$\text{SS(Model)} = \hat{\beta}'X'Y \qquad [4.17]$$

The equality is shown by substituting $X\hat{\beta}$ for the first $\hat{Y}$ and $X(X'X)^{-1}X'Y$ for the second. Thus, the sum of squares due to the model can be computed without computing the vector of fitted values or the $n \times n$ matrix P. The $\hat{\beta}$ vector is much smaller than $\hat{Y}$, and $X'Y$ will have already been computed. Since the two partitions are additive, the simplest computational form for SS(Res) $= e'e$ is by subtraction:

$$\text{SS(Res)} = Y'Y - \text{SS(Model)} \qquad [4.18]$$

The definitional and computational forms for this partitioning of the total sum of squares are summarized in Table 4.1.

Table 4.1 Analysis of variance summary for regression analysis.

Source of Variation	Degrees of Freedom	Sum of Squares: Definitional Formula	Sum of Squares: Computational Formula
$\text{Total}_{\text{uncorr}}$	$r(\boldsymbol{I}) = n$	$\boldsymbol{Y}'\boldsymbol{Y}$	
Due to model	$r(\boldsymbol{P}) = p'$	$\hat{\boldsymbol{Y}}'\hat{\boldsymbol{Y}} = \boldsymbol{Y}'\boldsymbol{P}\boldsymbol{Y}$	$\hat{\boldsymbol{\beta}}'\boldsymbol{X}'\boldsymbol{Y}$
Residual	$r(\boldsymbol{I} - \boldsymbol{P}) = (n - p')$	$\boldsymbol{e}'\boldsymbol{e} = \boldsymbol{Y}'(\boldsymbol{I} - \boldsymbol{P})\boldsymbol{Y}$	$\boldsymbol{Y}'\boldsymbol{Y} - \hat{\boldsymbol{\beta}}'\boldsymbol{X}'\boldsymbol{Y}$

Example 4.2 (*Continuation of Example* 3.8) The partitioning of the sums of squares is illustrated using the Heagle ozone example (Table 3.1, page 72). The total uncorrected sum of squares with 4 degrees of freedom is

$$\boldsymbol{Y}'\boldsymbol{Y} = (242 \;\; 237 \;\; 231 \;\; 201)\begin{pmatrix} 242 \\ 237 \\ 231 \\ 201 \end{pmatrix}$$

$$= 242^2 + 237^2 + 231^2 + 201^2 = 208{,}495$$

The sum of squares attributable to the model, SS(Model), can be obtained from the definitional formula, using $\hat{\boldsymbol{Y}}$ from Table 3.1, as

$$\hat{\boldsymbol{Y}}'\hat{\boldsymbol{Y}} = (247.563 \;\; 232.887 \;\; 221.146 \;\; 209.404)\begin{pmatrix} 247.563 \\ 232.887 \\ 221.146 \\ 209.404 \end{pmatrix}$$

$$= 247.563^2 + 232.887^2 + 221.146^2 + 209.404^2$$

$$= 208{,}279.39$$

The more convenient computational formula gives

$$\hat{\boldsymbol{\beta}}'\boldsymbol{X}'\boldsymbol{Y} = (253.434 \;\; -293.531)\begin{pmatrix} 911 \\ 76.99 \end{pmatrix} = 208{,}279.39$$

(See the text following equation 3.12 for $\hat{\boldsymbol{\beta}}$ and $\boldsymbol{X}'\boldsymbol{Y}$.)

The definitional formula for the residual sum of squares (see Table 3.1 for $\boldsymbol{e}$) gives

$$\boldsymbol{e}'\boldsymbol{e} = (-5.563 \;\; 4.113 \;\; 9.854 \;\; -8.404)\begin{pmatrix} -5.563 \\ 4.113 \\ 9.854 \\ -8.404 \end{pmatrix}$$

$$= 215.61$$

The simpler computational formula gives

$$\text{SS(Res)} = Y'Y - \text{SS(Model)} = 208{,}495 - 208{,}279.39$$
$$= 215.61 \qquad \blacksquare$$

Meaning of SS(Regr)

The total *uncorrected* sum of squares has been partitioned into that due to the entire model and a residual sum of squares. Usually, however, one is interested in explaining the variation of Y about its mean, rather than about zero, and in how much the information from the independent variables contributes to this explanation. If no information is available from independent variables, the best predictor of Y is the best available estimate of the population mean. When independent variables are available, the question of interest is how much information the independent variables contribute to the prediction of Y *beyond* that provided by the overall mean of Y.

The measure of the additional information provided by the independent variables is the difference between SS(Model) when the independent variables are included and SS(Model) when no independent variables are included. The model with no independent variables contains only one parameter, the overall mean μ. When μ is the only parameter in the model, SS(Model) will be labeled SS(μ). [SS(μ) is commonly called the **correction factor**.] The *additional* sum of squares accounted for by the independent variable(s) will be called the regression sum of squares and labeled SS(Regr). Thus,

$$\text{SS(Regr)} = \text{SS(Model)} - \text{SS}(\mu) \qquad [4.19]$$

where SS(Model) is understood to be the sum of squares due to the model containing the independent variables.

SS(μ)

The sum of squares due to μ alone, SS(μ), will be determined using matrix notation in order to show the development of the defining matrices for the quadratic forms. The model when μ is the only parameter is still written in the form $Y = X\beta + \varepsilon$, but now X is only a column vector of ones and $\beta = \mu$, a single element. The column vector of ones will be labeled $\mathbf{1}$. Then,

$$\hat{\beta} = (\mathbf{1}'\mathbf{1})^{-1}\mathbf{1}'Y = \left(\frac{1}{n}\right)\mathbf{1}'Y = \bar{Y} \qquad [4.20]$$

and

$$\text{SS}(\mu) = \hat{\beta}'(\mathbf{1}'Y) = \left(\frac{1}{n}\right)(\mathbf{1}'Y)'(\mathbf{1}'Y)$$
$$= \left(\frac{1}{n}\right)Y'(\mathbf{1}\mathbf{1}')Y \qquad [4.21]$$

Notice that $\mathbf{1}'Y = \sum Y_i$ so that SS(μ) is $(\sum Y_i)^2/n$, the familiar result for the sum of squares due to correcting for the mean. Multiplication of $\mathbf{11}'$ gives an $n \times n$ matrix of ones. Convention labels this the $\boldsymbol{J}$ matrix. Thus, the defining matrix for the quadratic form giving the correction factor is

$$\frac{1}{n}(\mathbf{11}') = \frac{1}{n}\begin{bmatrix} 1 & 1 & 1 & \cdots & 1 \\ 1 & 1 & 1 & \cdots & 1 \\ 1 & 1 & 1 & \cdots & 1 \\ \vdots & \vdots & \vdots & & \vdots \\ 1 & 1 & 1 & \cdots & 1 \end{bmatrix} = \frac{1}{n}\boldsymbol{J} \quad [4.22]$$

The matrix $(1/n)\boldsymbol{J}$ is idempotent with rank equal to $\text{tr}[(1/n)\boldsymbol{J}] = 1$ and, hence, the correction factor has 1 degree of freedom.

Quadratic Form for **SS(Regr)**

The *additional* sum of squares attributable to the independent variable(s) in a model is then

$$\begin{aligned} \text{SS(Regr)} &= \text{SS(Model)} - \text{SS}(\mu) \\ &= \boldsymbol{Y'PY} - \boldsymbol{Y}'(\boldsymbol{J}/n)\boldsymbol{Y} \\ &= \boldsymbol{Y}'(\boldsymbol{P} - \boldsymbol{J}/n)\boldsymbol{Y} \end{aligned} \quad [4.23]$$

Thus, the defining matrix for SS(Regr) is $(\boldsymbol{P} - \boldsymbol{J}/n)$. The defining matrix $\boldsymbol{J}/n$ is orthogonal to $(\boldsymbol{P} - \boldsymbol{J}/n)$ and $(\boldsymbol{I} - \boldsymbol{P})$ so that the total sum of squares is now partitioned into three orthogonal components:

$$\begin{aligned} \boldsymbol{Y'Y} &= \boldsymbol{Y}'(\boldsymbol{J}/n)\boldsymbol{Y} + \boldsymbol{Y}'(\boldsymbol{P} - \boldsymbol{J}/n)\boldsymbol{Y} + \boldsymbol{Y}'(\boldsymbol{I} - \boldsymbol{P})\boldsymbol{Y} \\ &= \text{SS}(\mu) + \text{SS(Regr)} + \text{SS(Res)} \end{aligned} \quad [4.24]$$

with 1, $(p' - 1) = p$, and $(n - p')$ degrees of freedom, respectively. Usually $\text{SS}(\mu)$ is subtracted from $\boldsymbol{Y'Y}$ and only the *corrected* sum of squares partitioned into SS(Regr) and SS(Res) is reported.

Example 4.3

For the Heagle ozone example (Example 4.2),

$$\text{SS}(\mu) = \frac{(911)^2}{4} = 207{,}480.25$$

so that

$$\text{SS(Regr)} = 208{,}279.39 - 207{,}480.25 = 799.14$$

The analysis of variance for the ozone example is summarized in Table 4.2 (page 96). ■

The key points to remember are summarized in the next box.

Table 4.2 Summary analysis of variance for the regression of soybean yield on ozone exposure. (Data courtesy A. S. Heagle, North Carolina State University.)

Source of Variation	d.f.	*Sum of Squares*	*Mean Square*
Total_{uncorr}	4	$Y'Y = 208,495$	
Mean	1	$n\overline{Y}^2 = 207{,}480.25$	
Total_{corr}	3	$Y'Y - n\overline{Y}^2 = 1{,}014.75$	
Regression	1	$\hat{\beta}'X'Y - n\overline{Y}^2 = 799.14$	799.14
Residuals	2	$Y'Y - \hat{\beta}'X'Y = 215.61$	107.81

1. The rank of $\boldsymbol{X}$ is determined by the number of linearly independent columns in $\boldsymbol{X}$.
2. The model is a full-rank model if the rank of $\boldsymbol{X}$ equals the number of columns of $\boldsymbol{X}$.
3. The *unique* ordinary least squares solution exists only if the model is of full rank.
4. The defining matrices for the quadratic forms in regression are all idempotent. Examples are $\boldsymbol{I}$, $\boldsymbol{P}$, $(\boldsymbol{I} - \boldsymbol{P})$, and $\boldsymbol{J}/n$.
5. The defining matrices $\boldsymbol{J}/n$, $(\boldsymbol{P} - \boldsymbol{J}/n)$, and $(\boldsymbol{I} - \boldsymbol{P})$ are pairwise orthogonal to each other and, consequently, partition the total uncorrected sum of squares into orthogonal sums of squares.
6. The degrees of freedom for a quadratic form is determined by the rank of the defining matrix which, when it is idempotent, equals its trace.

$$r(\boldsymbol{I}) = n, \quad \text{the only full-rank idempotent matrix}$$
$$r(\boldsymbol{P}) = p'$$
$$r(\boldsymbol{J}/n) = 1$$
$$r(\boldsymbol{P} - \boldsymbol{J}/n) = p$$
$$r(\boldsymbol{I} - \boldsymbol{P}) = n - p'$$

4.3 Expectations of Quadratic Forms

Each of the quadratic forms computed in the analysis of variance of Y is estimating some function of the parameters of the model. The expectations of these quadratic forms must be known if proper use is to be made of the sums of squares and their mean squares. The following results are stated without proofs. The reader is referred to Searle (1971) for more complete development.

General Results

Let $\mathcal{E}(\boldsymbol{Y}) = \boldsymbol{\mu}$, a general vector of expectations, and let $\mathbf{Var}(\boldsymbol{Y}) = \boldsymbol{V}$, a general variance–covariance matrix. Then the general result for the expectation of the quadratic form $\boldsymbol{Y}'\boldsymbol{A}\boldsymbol{Y}$ is

$$\mathcal{E}(\boldsymbol{Y}'\boldsymbol{A}\boldsymbol{Y}) = \text{tr}(\boldsymbol{A}\boldsymbol{V}) + \boldsymbol{\mu}'\boldsymbol{A}\boldsymbol{\mu} \qquad [4.25]$$

Under ordinary least squares assumptions, $\mathcal{E}(\boldsymbol{Y}) = \boldsymbol{X\beta}$ and $\mathbf{Var}(\boldsymbol{Y}) = \boldsymbol{I}\sigma^2$ and the expectation of the quadratic form becomes

$$\mathcal{E}(\boldsymbol{Y}'\boldsymbol{A}\boldsymbol{Y}) = \sigma^2\,\text{tr}(\boldsymbol{A}) + \boldsymbol{\beta}'\boldsymbol{X}'\boldsymbol{A}\boldsymbol{X}\boldsymbol{\beta} \qquad [4.26]$$

The expectations of the quadratic forms in the analysis of variance are obtained from this general result by replacing $\boldsymbol{A}$ with the appropriate defining matrix. When $\boldsymbol{A}$ is idempotent, the coefficient on σ^2 is the degrees of freedom for the quadratic form.

$\mathcal{E}$[SS(Model)]

$$\begin{aligned}\mathcal{E}[\text{SS(Model)}] &= \mathcal{E}(\boldsymbol{Y}'\boldsymbol{P}\boldsymbol{Y}) = \sigma^2\,\text{tr}(\boldsymbol{P}) + \boldsymbol{\beta}'\boldsymbol{X}'\boldsymbol{P}\boldsymbol{X}\boldsymbol{\beta}\\ &= p'\sigma^2 + \boldsymbol{\beta}'\boldsymbol{X}'\boldsymbol{X}\boldsymbol{\beta} \qquad [4.27]\end{aligned}$$

since $\text{tr}(\boldsymbol{P}) = p'$ and $\boldsymbol{PX} = \boldsymbol{X}$. Notice that the second term in equation 4.27 is a quadratic form in $\boldsymbol{\beta}$, including β_0, the intercept.

$\mathcal{E}$[SS(Regr)]

$$\begin{aligned}\mathcal{E}[\text{SS(Regr)}] &= \mathcal{E}[\boldsymbol{Y}'(\boldsymbol{P} - \boldsymbol{J}/n)\boldsymbol{Y}]\\ &= \sigma^2\,\text{tr}(\boldsymbol{P} - \boldsymbol{J}/n) + \boldsymbol{\beta}'\boldsymbol{X}'(\boldsymbol{P} - \boldsymbol{J}/n)\boldsymbol{X}\boldsymbol{\beta}\\ &= p\sigma^2 + \boldsymbol{\beta}'\boldsymbol{X}'(\boldsymbol{I} - \boldsymbol{J}/n)\boldsymbol{X}\boldsymbol{\beta} \qquad [4.28]\end{aligned}$$

since $\boldsymbol{X}'\boldsymbol{P} = \boldsymbol{X}'$. This quadratic form in $\boldsymbol{\beta}$ differs from that for $\mathcal{E}$[SS(Model)] in that $\boldsymbol{X}'(\boldsymbol{I} - \boldsymbol{J}/n)\boldsymbol{X}$ is a matrix of *corrected* sums of squares and products of the X_j. Since the first column of $\boldsymbol{X}$ is a constant, the sums of squares and products involving the first column are zero. Thus, the first row and column of $\boldsymbol{X}'(\boldsymbol{I} - \boldsymbol{J}/n)\boldsymbol{X}$ contain only zeros, which removes β_0 from the quadratic expression. Only the regression coefficients for the independent variables are involved in the expectation of the regression sum of squares.

$\mathcal{E}$[SS(Res)]

$$\begin{aligned}\mathcal{E}[\text{SS(Res)}] &= \mathcal{E}[\boldsymbol{Y}'(\boldsymbol{I} - \boldsymbol{P})\boldsymbol{Y}]\\ &= \sigma^2\,\text{tr}(\boldsymbol{I} - \boldsymbol{P}) + \boldsymbol{\beta}'\boldsymbol{X}'(\boldsymbol{I} - \boldsymbol{P})\boldsymbol{X}\boldsymbol{\beta}\\ &= (n - p')\sigma^2 + \boldsymbol{\beta}'\boldsymbol{X}'(\boldsymbol{X} - \boldsymbol{X})\boldsymbol{\beta}\\ &= (n - p')\sigma^2 \qquad [4.29]\end{aligned}$$

Expectations of the Mean Squares

The coefficient on σ^2 in each expectation is the degrees of freedom for the sum of squares. After division of each expectation by the appropriate degrees of freedom to convert sums of squares to mean squares, the coefficient on σ^2 will be 1 in each case:

$$\mathscr{E}[\text{MS(Regr)}] = \sigma^2 + [\boldsymbol{\beta}'\boldsymbol{X}'(\boldsymbol{I} - \boldsymbol{J}/n)\boldsymbol{X}\boldsymbol{\beta}]/p \qquad [4.30]$$

$$\mathscr{E}[\text{MS(Res)}] = \sigma^2 \qquad [4.31]$$

This shows that the residual mean square, MS(Res), is an unbiased estimate of σ^2. The regression mean square, MS(Regr), is an estimate of σ^2 plus a quadratic function of all β_j except β_0. Comparison of MS(Regr) and MS(Res), therefore, provides the basis for judging the importance of the regression coefficients or, equivalently, of the independent variables. Since the second term in $\mathscr{E}$[MS(Regr)] is a *quadratic* function of $\boldsymbol{\beta}$, which cannot be negative, any contribution from the independent variables to the predictability of Y_i makes MS(Regr) larger *in expectation* than MS(Res). The ratio of the observed MS(Regr) to the observed MS(Res) provides the test of significance of the composite hypothesis that all β_j, except β_0, are zero. Tests of significance are discussed more fully in the following sections.

The expectations assume that the model used in the analysis of variance is indeed the correct model. This is imposed in the preceding derivations when $\boldsymbol{X\beta}$ is substituted for $\mathscr{E}(\boldsymbol{Y}) = \boldsymbol{\mu}$. If the incorrect model has been used, $\mathscr{E}(\boldsymbol{Y}) \neq \boldsymbol{X\beta}$ and the second term in equation 4.29 does not go to zero. Instead, there will remain a quadratic function of regression coefficients of any important independent variables that were mistakenly omitted from the model. In such case, MS(Res) will be a positively biased estimate of σ^2.

Example 4.4

From Example 4.3 using the ozone data, the estimate of σ^2 obtained from MS(Res) is $s^2 = 107.81$ (Table 4.2). This is a very poor estimate of σ^2 because it has only 2 degrees of freedom. Nevertheless, this estimate of σ^2 will be used for the time being. (A better estimate is obtained in Section 4.7.) ■

Estimated Variances

In Chapter 3, the variance–covariance matrices for $\hat{\boldsymbol{\beta}}$, $\hat{\boldsymbol{Y}}$, and $\boldsymbol{e}$ were expressed in terms of the true variance σ^2. **Estimates** of the variance–covariance matrices are obtained by substituting an estimate of σ^2 in each $\mathbf{Var}(\cdot)$ formula; $\boldsymbol{s}^2(\cdot)$ will be used to denote an *estimated* variance–covariance matrix. (Note that boldface type is used to distinguish the matrix of estimates from individual variances.)

Example 4.5

In the ozone example (Example 4.3),

$$\boldsymbol{s}^2(\hat{\boldsymbol{\beta}}) = (\boldsymbol{X}'\boldsymbol{X})^{-1}s^2$$

$$= \begin{bmatrix} 1.0755 & -9.4340 \\ -9.4340 & 107.8167 \end{bmatrix} 107.81$$

$$= \begin{bmatrix} 115.94 & -1{,}017.0 \\ -1{,}017.0 & 11{,}623 \end{bmatrix}$$

Thus,

$$s^2(\hat{\beta}_0) = (1.0755)(107.81) = 115.94$$

$$s^2(\hat{\beta}_1) = (107.8167)(107.81) = 11{,}623$$

$$\text{Cov}(\hat{\beta}_0, \hat{\beta}_1) = (-9.4340)(107.81) = -1{,}017.0$$

In each case, the first number in the product is the appropriate coefficient from the $(\boldsymbol{X}'\boldsymbol{X})^{-1}$ matrix; the second number is s^2. [It is only coincidence that the lower-right diagonal element of $(\boldsymbol{X}'\boldsymbol{X})^{-1}$ is almost identical to s^2.] ■

The estimated variance–covariance matrices for $\hat{\boldsymbol{Y}}$ and $\boldsymbol{e}$ are found similarly by replacing σ^2 with s^2 in the corresponding variance–covariance matrices.

4.4 Distribution of Quadratic Forms

General Results

The probability distributions of the quadratic forms provide the basis for parametric tests of significance. It is at this point (and in making confidence interval statements about the parameters) that the normality assumption on the ε_i comes into play. The results are summarized assuming that normality of $\boldsymbol{\varepsilon}$, and therefore normality of $\boldsymbol{Y}$, is satisfied. When normality is not satisfied, the parametric tests of significance must be regarded as approximations.

A general result from statistical theory [see, for example, Searle (1971)] states:

> If $\boldsymbol{Y}$ is normally distributed, with $\mathscr{E}(\boldsymbol{Y}) = \boldsymbol{\mu}$ and $\textbf{Var}(\boldsymbol{Y}) = \boldsymbol{V}$ ($\boldsymbol{\mu}$ may be $\boldsymbol{X\beta}$ and $\boldsymbol{V}$ may be $\boldsymbol{I}\sigma^2$), then a quadratic function $\boldsymbol{Y}'\boldsymbol{A}\boldsymbol{Y}$ is distributed as a **noncentral chi-square** with
>
> a. degrees of freedom equal to the rank of $\boldsymbol{A}$: df $= r(\boldsymbol{A})$, and
> b. noncentrality parameter $\Omega = \frac{1}{2}(\boldsymbol{\mu}'\boldsymbol{A}\boldsymbol{\mu})$
>
> if and only if $\boldsymbol{AV}$ is idempotent. If $\boldsymbol{V} = \boldsymbol{I}\sigma^2$, the condition reduces to $\boldsymbol{A}\sigma^2$ being idempotent.

t- and F-Statistics

All quadratic forms defined herein, when divided by σ^2, meet this requirement. Thus, the normality assumption on $\boldsymbol{\varepsilon}$ implies that the sums of squares, divided by σ^2, are chi-square random variables. The chi-square distribution and the orthogonality between the quadratic forms provide the basis for the usual tests of significance. For example, when the null hypothesis is true, the t-statistic is the ratio of a normal deviate to the square root of a scaled independent central chi-square random variable. The F-statistic is the ratio of a scaled noncentral chi-square random variable (central chi-square random variable if the null hypothesis is true) to a scaled independent central chi-square random variable. The scaling in each case is division of the chi-square random variable by its degrees of freedom. A **central chi-square distribution** has noncentrality parameter equal to zero.

The Noncentrality Parameter and the F-Test

The noncentrality parameter, $\Omega = \frac{1}{2}(\boldsymbol{\mu}'\boldsymbol{A}\boldsymbol{\mu})$, is important for two reasons. First, the condition that makes the noncentrality parameter of the numerator of the F-ratio equal to zero is an explicit statement of the null hypothesis. Second, the power of the test to detect a false null hypothesis is determined by the magnitude of the noncentrality parameter. The noncentrality parameter of the chi-square distribution is the second term of the expectation of the quadratic form divided by $2\sigma^2$ (see equation 4.26). $\text{SS(Res)}/\sigma^2$ is a central chi-square since the second term is zero (equation 4.29). The noncentrality parameter for $\text{SS(Regr)}/\sigma^2$ (see equation 4.28) is

$$\Omega = \frac{\boldsymbol{\beta}'\boldsymbol{X}'(\boldsymbol{I} - \boldsymbol{J}/n)\boldsymbol{X}\boldsymbol{\beta}}{2\sigma^2} \qquad [4.32]$$

which is a quadratic form of all β_j except β_0. Thus, $\text{SS(Regr)}/\sigma^2$ is a *central* chi-square only if $\Omega = 0$, which requires $\beta_1 = \beta_2 = \cdots = \beta_p = 0$. Therefore, the F-ratio using

$$F = \frac{\text{MS(Regr)}}{\text{MS(Res)}} \qquad [4.33]$$

is a test of the composite hypothesis that all β_j, except β_0, equal zero. This hypothesis will be stated as

$$H_0\colon \ \boldsymbol{\beta}^* = \mathbf{0}$$

$$H_a\colon \ \boldsymbol{\beta}^* \neq \mathbf{0}$$

where $\boldsymbol{\beta}^*$ is the $p \times 1$ vector of regression coefficients *excluding* β_0.

An observed F-ratio, equation 4.33, sufficiently greater than 1 suggests that the noncentrality parameter is not zero. The larger the noncentrality parameter for the numerator chi-square, the larger will be the F-ratio, on the average, and the greater will be the probability of detecting a false null hypothesis. This probability, by definition, is the **power of the test**. (The power of an F-test is also increased by increasing the degrees of freedom for each chi-square, particularly the denominator chi-square.) All of the quantities except $\boldsymbol{\beta}$ in the noncentrality parameter are known *before* the experiment is run (in those cases where the X's are subject to the control of the researcher). Therefore, the relative powers of different experimental designs can be evaluated before the final design is adopted.

Example 4.6 In the Heagle ozone example (Example 4.3),

$$F = \frac{\text{MS(Regr)}}{\text{MS(Res)}} = \frac{799.14}{107.81} = 7.41$$

The critical value for $\alpha = .05$ with 1 and 2 degrees of freedom is $F_{(.05,1,2)} = 18.51$. The conclusion is that these data do not provide sufficient evidence to reject the null hypothesis that β_1 equals zero. Even though MS(Regr) is considerably larger than MS(Res), the difference is not sufficient to be confident that it is not due to random sampling variation from the underlying chi-square distributions. The large critical value of F, 18.51, is a direct reflection of the very limited degrees of freedom for MS(Res) and, consequently, large sampling variation in the F-distribution. A later analysis, which uses a more precise estimate of σ^2 (more degrees of freedom) but the same MS(Regr), will show that β_1 clearly is not zero. ■

The key points from this section are summarized in the box.

1. The expectations of the quadratic forms are model-dependent. If the incorrect model has been used, the expectations are incorrect. This is particularly critical for MS(Res) since it is used repeatedly as the estimate of σ^2. For this reason it is desirable to obtain an estimate of σ^2 that is not model-dependent. This will be discussed in Section 4.7.
2. The expectations of the mean squares provide the basis for choosing the appropriate mean squares for tests of hypothesis with the F-test; the numerator and denominator mean squares must have the same expectations if the null hypothesis is true and the expectation of the numerator mean square must be larger if the alternative hypothesis is true.
3. The assumption of a normal probability distribution for the random errors is necessary for the conventional tests of significance and confidence interval estimates of the parameters to be correct. While tests of significance appear to be reasonably robust against nonnormality, they must be regarded as approximations when the normality assumption is not satisfied.

4.5 General Form for Hypothesis Testing

The ratio of MS(Regr) to MS(Res) provides a test of the null hypothesis that all β_j, except β_0, are simultaneously equal to zero. More flexibility is needed in constructing tests of hypotheses than is allowed by this procedure. This section presents a general method of constructing tests for any hypothesis involving linear functions of $\boldsymbol{\beta}$. The null hypothesis may involve a single linear function, a **simple hypothesis**, or it may involve several linear functions simultaneously, a **composite hypothesis**.

4.5.1 The General Linear Hypothesis

General Hypothesis

The **general linear hypothesis** is defined as

$$H_0: \ \boldsymbol{K}'\boldsymbol{\beta} = \boldsymbol{m}$$
$$H_a: \ \boldsymbol{K}'\boldsymbol{\beta} \neq \boldsymbol{m} \qquad [4.34]$$

where $\boldsymbol{K}'$ is a $k \times p'$ matrix of coefficients defining k linear functions of the β_j to be tested. Each row of $\boldsymbol{K}'$ contains the coefficients for one linear function; $\boldsymbol{m}$ is a $k \times 1$ vector of constants, frequently zeros. The k linear equations in H_0 must be linearly independent (but they need not be orthogonal). Linear independence implies that $\boldsymbol{K}'$ is of full rank, $r(\boldsymbol{K}') = k$, and ensures that the equations in H_0 are consistent for every choice of $\boldsymbol{m}$ (see Section 2.5). The number of linear functions in H_0 cannot exceed the number of parameters in $\boldsymbol{\beta}$; otherwise, $\boldsymbol{K}'$ would not be of full rank.

Example 4.7 Suppose $\boldsymbol{\beta}' = (\beta_0\ \beta_1\ \beta_2\ \beta_3)$ and you wish to test the composite null hypothesis that $\beta_1 = \beta_2$, $\beta_1 + \beta_2 = 2\beta_3$, and $\beta_0 = 20$ or, equivalently,

$$\begin{aligned} H_0:\ \beta_1 - \beta_2 \qquad &= 0 \\ \beta_1 + \beta_2 - 2\beta_3 &= 0 \\ \beta_0 \qquad\qquad &= 20 \end{aligned} \qquad [4.35]$$

These three linear functions can be written in the form $\boldsymbol{K}'\boldsymbol{\beta} = \boldsymbol{m}$ by defining

$$\boldsymbol{K}' = \begin{bmatrix} 0 & 1 & -1 & 0 \\ 0 & 1 & 1 & -2 \\ 1 & 0 & 0 & 0 \end{bmatrix} \quad \text{and} \quad \boldsymbol{m} = \begin{pmatrix} 0 \\ 0 \\ 20 \end{pmatrix} \qquad [4.36]$$

The alternative hypothesis is H_a: $\boldsymbol{K}'\boldsymbol{\beta} \neq \boldsymbol{m}$. The null hypothesis is violated if any one or more of the equalities in H_0 is not true. ■

Estimate and Variance

The least squares estimate of $\boldsymbol{K}'\boldsymbol{\beta} - \boldsymbol{m}$ is obtained by substituting the least squares estimate $\hat{\boldsymbol{\beta}}$ for $\boldsymbol{\beta}$ to obtain $\boldsymbol{K}'\hat{\boldsymbol{\beta}} - \boldsymbol{m}$. Under the ordinary least squares assumptions, including normality, $\boldsymbol{K}'\hat{\boldsymbol{\beta}} - \boldsymbol{m}$ is normally distributed with mean $\mathscr{E}(\boldsymbol{K}'\hat{\boldsymbol{\beta}} - \boldsymbol{m}) = \boldsymbol{K}'\boldsymbol{\beta} - \boldsymbol{m}$, which is zero if the null hypothesis is true, and variance–covariance matrix $\mathbf{Var}(\boldsymbol{K}'\boldsymbol{\beta} - \boldsymbol{m}) = \boldsymbol{K}'(\boldsymbol{X}'\boldsymbol{X})^{-1}\boldsymbol{K}\sigma^2$. The variance is obtained by applying the rules for variances of linear functions (see Section 3.4).

Sum of Squares

The sum of squares for the linear hypothesis H_0: $\boldsymbol{K}'\boldsymbol{\beta} = \boldsymbol{m}$ is computed by (see Searle, 1971)

$$Q = (\boldsymbol{K}'\hat{\boldsymbol{\beta}} - \boldsymbol{m})'[\boldsymbol{K}'(\boldsymbol{X}'\boldsymbol{X})^{-1}\boldsymbol{K}]^{-1}(\boldsymbol{K}'\hat{\boldsymbol{\beta}} - \boldsymbol{m}) \qquad [4.37]$$

This is a quadratic form in $\boldsymbol{K}'\hat{\boldsymbol{\beta}} - \boldsymbol{m}$ with defining matrix

$$\boldsymbol{A} = [\boldsymbol{K}'(\boldsymbol{X}'\boldsymbol{X})^{-1}\boldsymbol{K}]^{-1} \qquad [4.38]$$

The defining matrix, except for division by σ^2, is the inverse of the variance–covariance matrix of the linear functions $\boldsymbol{K}'\hat{\boldsymbol{\beta}} - \boldsymbol{m}$. Thus, $\text{tr}(\boldsymbol{AV}) = \text{tr}(\boldsymbol{I}_k)\sigma^2 = k\sigma^2$ and the expectation of Q (see equation 4.25) is

$$\mathscr{E}(Q) = k\sigma^2 + (\boldsymbol{K}'\boldsymbol{\beta} - \boldsymbol{m})'[\boldsymbol{K}'(\boldsymbol{X}'\boldsymbol{X})^{-1}\boldsymbol{K}]^{-1}(\boldsymbol{K}'\boldsymbol{\beta} - \boldsymbol{m}) \qquad [4.39]$$

With the assumption of normality, Q/σ^2 is distributed as a noncentral chi-square random variable with k degrees of freedom. This is verified by noting that $\boldsymbol{AV}$, properly scaled by $1/\sigma^2$, gives $\boldsymbol{I}_k$, which is idempotent (see Section 4.4). The degrees of freedom are determined from $r(\boldsymbol{A}) = r(\boldsymbol{K}) = k$. The noncentrality parameter is

$$\Omega = \frac{(\boldsymbol{K}'\boldsymbol{\beta} - \boldsymbol{m})'[\boldsymbol{K}'(\boldsymbol{X}'\boldsymbol{X})^{-1}\boldsymbol{K}]^{-1}(\boldsymbol{K}'\boldsymbol{\beta} - \boldsymbol{m})}{2\sigma^2}$$

which is zero when the null hypothesis is true. Thus, Q/k is an appropriate numerator mean square for an F-test of the stated hypothesis.

***F*-Test**

The appropriate denominator of the F-test is any unbiased and independent estimate of σ^2; usually MS(Res) is used. Thus,

$$F = \frac{Q/r(\boldsymbol{K})}{s^2} \qquad [4.40]$$

is a proper F-test of H_0: $\boldsymbol{K}'\boldsymbol{\beta} - \boldsymbol{m} = \boldsymbol{0}$ with numerator degrees of freedom equal to $r(\boldsymbol{K})$ and denominator degrees of freedom equal to the degrees of freedom in s^2.

This general formulation provides a convenient method for testing any hypotheses of interest and is particularly useful when the computations are being done with a matrix algebra computer program. It is important to note, however, that all sums of squares for hypotheses are dependent on the particular model being used. In general, deleting an independent variable or adding an independent variable to the model will change the sum of squares for every hypothesis.

4.5.2 Some Special Cases of the General Form

Three special cases of the general linear hypothesis are of interest.

Case 1. A Simple Hypothesis

When a **simple hypothesis** on $\boldsymbol{\beta}$ is being tested, $\boldsymbol{K}'$ is a single row vector so that $[\boldsymbol{K}'(\boldsymbol{X}'\boldsymbol{X})^{-1}\boldsymbol{K}]$ is a scalar. Its inverse is $1/[\boldsymbol{K}'(\boldsymbol{X}'\boldsymbol{X})^{-1}\boldsymbol{K}]$. The sum of squares for the hypothesis can be written as

$$Q = \frac{(\boldsymbol{K}'\hat{\boldsymbol{\beta}} - \boldsymbol{m})^2}{\boldsymbol{K}'(\boldsymbol{X}'\boldsymbol{X})^{-1}\boldsymbol{K}} \qquad [4.41]$$

and has 1 degree of freedom. The numerator of Q is the square of the linear function of $\hat{\boldsymbol{\beta}}$ and the denominator is its variance, except for σ^2. Thus, the F-ratio is

$$F = \frac{(\boldsymbol{K}'\hat{\boldsymbol{\beta}} - \boldsymbol{m})^2}{[\boldsymbol{K}'(\boldsymbol{X}'\boldsymbol{X})^{-1}\boldsymbol{K}]s^2} \qquad [4.42]$$

The F-test of a *simple* hypothesis can be written as a two-tailed t-test:

$$t = \frac{\boldsymbol{K}'\hat{\boldsymbol{\beta}} - \boldsymbol{m}}{\{[\boldsymbol{K}'(\boldsymbol{X}'\boldsymbol{X})^{-1}\boldsymbol{K}]s^2\}^{1/2}} \qquad [4.43]$$

The denominator is the standard error of the linear function in the numerator.

Case 2. k β_j All Equal Zero

The null hypothesis of interest is that each of k specific regression coefficients is zero. For this case $\boldsymbol{K}'$ is a $k \times p'$ matrix consisting of zeros except for a single one in each row to identify the β_j being tested; $\boldsymbol{m} = \boldsymbol{0}$. With this $\boldsymbol{K}'$, the matrix multiplication $[\boldsymbol{K}'(\boldsymbol{X}'\boldsymbol{X})^{-1}\boldsymbol{K}]$ extracts from $(\boldsymbol{X}'\boldsymbol{X})^{-1}$ the $k \times k$ submatrix consisting of the coefficients for the variances and covariances of the k $\hat{\beta}_j$ being tested. Suppose the null hypothesis to be tested is that β_1, β_3, and β_5 are each equal to zero. The sum of squares Q has the form

$$Q = (\hat{\beta}_1 \ \hat{\beta}_3 \ \hat{\beta}_5)\begin{bmatrix} c_{11} & c_{13} & c_{15} \\ c_{31} & c_{33} & c_{35} \\ c_{51} & c_{53} & c_{55} \end{bmatrix}^{-1}\begin{pmatrix} \hat{\beta}_1 \\ \hat{\beta}_3 \\ \hat{\beta}_5 \end{pmatrix} \qquad [4.44]$$

where c_{ij} is the element from row $(i + 1)$ and column $(j + 1)$ of $(\boldsymbol{X}'\boldsymbol{X})^{-1}$.

The sum of squares for this hypothesis measures the contribution of this subset of k independent variables to a model which already contains the other independent variables. This sum of squares is described as the sum of squares for these k variables **adjusted for** the other independent variables in the model.

Case 3. One β_j Equals Zero; the Partial Sum of Squares

Partial Sum of Squares

The third case is a further simplification of the first two. The hypothesis is that a single β_j is zero; H_0: $\beta_j = 0$. For this hypothesis, $\boldsymbol{K}'$ is a row vector of zeros except for a one in the column corresponding to the β_j being tested. As described in case 2, the sum of squares for this hypothesis is the contribution of X_j *adjusted for* all other independent variables in the model. This sum of squares is called the **partial sum of squares** for the jth independent variable.

The matrix multiplication $[\boldsymbol{K}'(\boldsymbol{X}'\boldsymbol{X})^{-1}\boldsymbol{K}]$ in Q extracts only the $(j + 1)$st diagonal element c_{jj} from $(\boldsymbol{X}'\boldsymbol{X})^{-1}$. This is the coefficient for the variance of $\hat{\beta}_j$. The sum of squares, with 1 degree of freedom, is

$$Q = \frac{\hat{\beta}_j^2}{c_{jj}} \qquad [4.45]$$

This provides an easy method of computing the *partial* sum of squares for any independent variable. For this case, the two-tailed t-test is

$$t = \frac{\hat{\beta}_j}{(c_{jj}s^2)^{1/2}} \qquad [4.46]$$

4.5.3 A Numerical Example

Example 4.8

For illustration of the use of the general linear hypothesis, data from a physical fitness program at North Carolina State University will be used. (The data were provided by A. C. Linnerud and are used with his permission.)

Measurements were taken on $n = 31$ men. In addition to age and weight, oxygen uptake (Y), run time (X_1), heart rate while resting (X_2), heart rate while running (X_3), and maximum heart rate (X_4) while running 1.5 miles were recorded for each subject. The data are given in Table 4.3. The results we discuss

Table 4.3 Physical fitness measurements on 31 men involved in a physical fitness program at North Carolina State University. The variables measured were age (years), weight (kg), oxygen uptake rate (ml per kg body weight per minute), time to run 1.5 miles (minutes), heart rate while resting, heart rate while running (at the same time oxygen uptake was measured), and maximum heart rate while running. (Data courtesy A. C. Linnerud, North Carolina State University.)

Age (yrs)	*Weight (kg)*	O_2 *Uptake (ml/kg/min)*	*Time (min)*	*Heart Rate*		
				Resting	*Running*	*Maximum*
44	89.47	44.609	11.37	62	178	182
40	75.07	45.313	10.07	62	185	185
44	85.84	54.297	8.65	45	156	184
42	68.15	59.571	8.17	40	166	172
38	89.02	49.874	9.22	55	178	180
47	77.45	44.811	11.63	58	176	176
40	75.98	45.681	11.95	70	176	180
43	81.19	49.091	10.85	64	162	170
44	81.42	39.442	13.08	63	174	176
38	81.87	60.055	8.63	48	170	186
44	73.03	50.541	10.13	45	168	168
45	87.66	37.388	14.03	56	186	192
45	66.45	44.754	11.12	51	176	176
47	79.15	47.273	10.60	47	162	164
54	83.12	51.855	10.33	50	166	170
49	81.42	49.156	8.95	44	180	185
51	69.63	40.836	10.95	57	168	172
51	77.91	46.672	10.00	48	162	168
48	91.63	46.774	10.25	48	162	164
49	73.37	50.388	10.08	67	168	168
57	73.37	39.407	12.63	58	174	176
54	79.38	46.080	11.17	62	156	176
52	76.32	45.441	9.63	48	164	166
50	70.87	54.625	8.92	48	146	186
51	67.25	45.118	11.08	48	172	172
54	91.63	39.203	12.88	44	168	172
51	73.71	45.790	10.47	59	186	188
57	59.08	50.545	9.93	49	148	160
49	76.32	48.673	9.40	56	186	188
48	61.24	47.920	11.50	52	170	176
52	82.78	47.467	10.50	53	170	172

Table 4.4 Summary analysis of variance for the regression of oxygen uptake on run time, heart rate while resting, heart rate while running, and maximum heart rate.

Source	d.f.	SS	MS
$\text{Total}_{\text{corr}}$	30	851.3815	
Regression	4	658.2638	164.5659
Residual	26	193.1178	$7.4276 = s^2$

are from the regression of oxygen uptake, Y, on the four variables X_1, X_2, X_3, and X_4.

The model is $\boldsymbol{Y} = \boldsymbol{X\beta} + \boldsymbol{\varepsilon}$, where $\boldsymbol{\beta}' = (\beta_0\ \beta_1\ \beta_2\ \beta_3\ \beta_4)$ with the subscripts matching the identification of the independent variables named above. The estimated regression equation is

$$\hat{Y}_i = 84.26902 - 3.06981X_{i1} + .00799X_{i2} - .11671X_{i3} + .08518X_{i4}$$

The analysis of variance for this model is summarized in Table 4.4. The residual mean square $s^2 = 7.4276$ is the estimate of σ^2 and has 26 degrees of freedom. The tests of hypotheses on $\boldsymbol{\beta}$ require $(\boldsymbol{X}'\boldsymbol{X})^{-1}$:

$$(\boldsymbol{X}'\boldsymbol{X})^{-1} = \begin{bmatrix} 17.423095 & -.1596195 & .0072675 & -.0140450 & -.0779657 \\ -.1596195 & .0236857 & -.0016967 & -.0009852 & .0009482 \\ .0072675 & -.0016967 & .0007776 & -.0000935 & -.0000854 \\ -.0140450 & -.0009852 & -.0000935 & .0005428 & -.0003562 \\ -.0779657 & .0009482 & -.0000854 & -.0003562 & .0007560 \end{bmatrix}$$

The first example will test the composite null hypothesis that the two regression coefficients β_2 and β_4 are zero, H_0: $\beta_2 = \beta_4 = 0$. The alternative hypothesis is that either one or both are not zero. This null hypothesis is written in the general form as

$$\boldsymbol{K}'\boldsymbol{\beta} = \begin{bmatrix} 0 & 0 & 1 & 0 & 0 \\ 0 & 0 & 0 & 0 & 1 \end{bmatrix} \begin{pmatrix} \beta_0 \\ \beta_1 \\ \beta_2 \\ \beta_3 \\ \beta_4 \end{pmatrix} = \begin{pmatrix} 0 \\ 0 \end{pmatrix}$$

Multiplication of the first row vector of $\boldsymbol{K}'$ with $\boldsymbol{\beta}$ gives $\beta_2 = 0$; the second row gives $\beta_4 = 0$.

There are 2 degrees of freedom associated with the sum of squares for this hypothesis, because $r(\boldsymbol{K}) = 2$. The sum of squares is

$$Q = (K'\hat{\beta} - m)'[K'(X'X)^{-1}K]^{-1}(K'\hat{\beta} - m)$$

$$= \begin{bmatrix} .00799 \\ .08518 \end{bmatrix}' \begin{bmatrix} .0007776 & -.0000854 \\ -.0000854 & .0007560 \end{bmatrix}^{-1} \begin{bmatrix} .00799 \\ .08518 \end{bmatrix}$$

$$= 10.0016$$

Notice that the product $K'(X'X)^{-1}K$ extracts the c_{22}, c_{24}, c_{42}, and c_{44} elements from $(X'X)^{-1}$. The F-test of the null hypothesis is

$$F = \frac{Q/2}{s^2} = \frac{(10.0016)/2}{7.4276} = .673$$

The computed F is much smaller than the critical value, $F_{(.05,2,26)} = 3.37$ and, therefore, there is no reason to reject the null hypothesis that β_2 and β_4 are both zero.

The second hypothesis illustrates a case where $m \neq 0$. Suppose prior information suggested that the intercept β_0 for a group of men of this age and weight should be 90. Then the null hypothesis of interest is $\beta_0 = 90$ and, for illustration, we will construct a composite hypothesis by adding this constraint to the two conditions in the first null hypothesis. The null hypothesis is

$$H_0\colon\ K'\beta - m = 0$$

where

$$K'\beta - m = \begin{bmatrix} 1 & 0 & 0 & 0 & 0 \\ 0 & 0 & 1 & 0 & 0 \\ 0 & 0 & 0 & 0 & 1 \end{bmatrix} \begin{pmatrix} \beta_0 \\ \beta_1 \\ \beta_2 \\ \beta_3 \\ \beta_4 \end{pmatrix} - \begin{pmatrix} 90 \\ 0 \\ 0 \end{pmatrix}$$

For this hypothesis

$$(K'\hat{\beta} - m) = \begin{pmatrix} 84.26902 - 90 \\ .00799 \\ .08518 \end{pmatrix} = \begin{pmatrix} -5.73098 \\ .00799 \\ .08518 \end{pmatrix}$$

and

$$[K'(X'X)^{-1}K]^{-1} = \begin{bmatrix} 17.423095 & .0072675 & -.0779657 \\ .0072675 & .0007776 & -.0000854 \\ -.0779657 & -.0000854 & .0007560 \end{bmatrix}^{-1}$$

Notice that $(\boldsymbol{K}'\hat{\boldsymbol{\beta}} - \boldsymbol{m})$ causes the hypothesized $\beta_0 = 90$ to be subtracted from the estimated $\hat{\beta}_0 = 84.26902$. The sum of squares for this composite hypothesis is

$$Q = (\boldsymbol{K}'\boldsymbol{\beta} - \boldsymbol{m})'[\boldsymbol{K}'(\boldsymbol{X}'\boldsymbol{X})^{-1}\boldsymbol{K}]^{-1}(\boldsymbol{K}'\boldsymbol{\beta} - \boldsymbol{m}) = 11.0187$$

and has 3 degrees of freedom. The computed F-statistic is

$$F = \frac{Q/3}{s^2} = \frac{11.0187/3}{7.4276} = .494$$

which, again, is much less than the critical value of F for $\alpha = .05$ and 3 and 26 degrees of freedom, $F_{(.05,3,26)} = 2.98$. There is no reason to reject the null hypothesis that $\beta_0 = 90$ and $\beta_2 = \beta_4 = 0$. ■

4.5.4 Computing Q from Differences in Sums of Squares

Full and Reduced Models

As an alternative to the general formula for Q, equation 4.37, the sum of squares for any hypothesis can be determined from the difference between the residual sums of squares of two models. The current model, in the context of which the null hypothesis is to be tested, is called the **full model**. This model must include all parameters involved in the null hypothesis and will usually include additional parameters. The second model is obtained from the full model by assuming that the null hypothesis is true and imposing its constraints on the full model. The model obtained in this way is called the **reduced model** because it will always have fewer parameters than the full model. For example, the null hypothesis H_0: $\beta_2 = c$, where c is some known constant, gives a reduced model in which β_2 has been replaced with the constant c. Consequently, β_2 is no longer a parameter to be estimated.

Difference in Residual Sums of Squares

The reduced model is a special case of the full model and, hence, its residual sum of squares must always be at least as large as the residual sum of squares for the full model. It can be shown that, for any general hypothesis, the sum of squares for the hypothesis can be computed as

$$Q = \text{SS}(\text{Res}_{\text{reduced}}) - \text{SS}(\text{Res}_{\text{full}}) \qquad [4.47]$$

where "reduced" and "full" identify the two models.

Degrees of Freedom

There are $(n - p')$ degrees of freedom associated with $\text{SS}(\text{Res}_{\text{full}})$. Generating the reduced model by imposing the k linearly independent constraints of the null hypothesis on the full model reduces the number of parameters from p' to $(p' - k)$. Thus, $\text{SS}(\text{Res}_{\text{reduced}})$ has $[n - (p' - k)]$ degrees of freedom. Therefore, Q will have $[(n - p' + k) - (n - p')] = k$ degrees of freedom.

Illustration

Assume that $\boldsymbol{X}$ is a full-rank matrix of order $n \times 4$ and $\boldsymbol{\beta}' = (\beta_0 \;\; \beta_1 \;\; \beta_2 \;\; \beta_3)$. Suppose the null hypothesis to be tested is

$$H_0: \;\; \boldsymbol{K}'\boldsymbol{\beta} = \boldsymbol{m}$$

where

$$K' = \begin{bmatrix} 0 & 1 & -1 & 0 \\ 1 & 0 & 0 & 0 \end{bmatrix} \quad \text{and} \quad m = \begin{pmatrix} 0 \\ 20 \end{pmatrix}$$

The full model was

$$Y_i = \beta_0 + \beta_1 X_{i1} + \beta_2 X_{i2} + \beta_3 X_{i3} + \varepsilon_i$$

If there were n observations, the residual sum of squares from this model would have $(n - 4)$ degrees of freedom. The null hypothesis states that (1) $\beta_1 = \beta_2$ and (2) $\beta_0 = 20$. The reduced model is generated by imposing on the full model the conditions stated in the null hypothesis. Since the null hypothesis states that β_1 and β_2 are equal, one of these two parameters, say β_2, can be eliminated by substitution of β_1 for β_2. Similarly, β_0 is replaced with the constant 20. These substitutions give the reduced model:

$$Y_i = 20 + \beta_1 X_{i1} + \beta_1 X_{i2} + \beta_3 X_{i3} + \varepsilon_i$$

Moving the constant 20 to the left side of the equality and collecting the two terms that involve β_1 gives

$$Y_i - 20 = \beta_1(X_{i1} + X_{i2}) + \beta_3 X_{i3} + \varepsilon_i$$

or

$$Y_i^* = \beta_1 X_{i1}^* + \beta_3 X_{i3} + \varepsilon_i$$

where $Y_i^* = Y_i - 20$ and $X_i^* = X_{i1} + X_{i2}$.

In matrix notation, the reduced model is

$$\boldsymbol{Y}^* = \boldsymbol{X}^* \boldsymbol{\beta}^* + \boldsymbol{\varepsilon}$$

where

$$\boldsymbol{Y}^* = \begin{pmatrix} Y_1 - 20 \\ Y_2 - 20 \\ \vdots \\ Y_n - 20 \end{pmatrix}$$

$$\boldsymbol{X}^* = \begin{bmatrix} (X_{11} + X_{12}) & X_{13} \\ (X_{21} + X_{22}) & X_{23} \\ \vdots & \vdots \\ (X_{n1} + X_{n2}) & X_{n3} \end{bmatrix} = \begin{bmatrix} X_{11}^* & X_{13} \\ X_{21}^* & X_{23} \\ \vdots & \vdots \\ X_{n1}^* & X_{n3} \end{bmatrix}$$

and

$$\boldsymbol{\beta}^* = \begin{pmatrix} \beta_1 \\ \beta_3 \end{pmatrix}$$

The rank of $\boldsymbol{X}^*$ is 2 so that SS(Res$_{\text{reduced}}$) will have $(n - 2)$ degrees of freedom. Consequently,

$$Q = \text{SS}(\text{Res}_{\text{reduced}}) - \text{SS}(\text{Res}_{\text{full}})$$

will have $[(n - 2) - (n - 4)] = 2$ degrees of freedom. Note that this agrees with $r(\boldsymbol{K}') = 2$.

The F-test of the null hypothesis is

$$F = \frac{Q/2}{s^2}$$

with 2 and v degrees of freedom, where v is the degrees of freedom in s^2. The denominator of F must be an unbiased estimate of σ^2 and must be statistically independent of the numerator sum of squares. This means that the defining matrices for the numerator and denominator sums of squares must be orthogonal. This condition is satisfied if σ^2 is estimated from a model that contains at least all of the terms in the full model or is estimated from independent information such as provided by true replication (see Section 4.7).

Example 4.9 The oxygen consumption example (Example 4.8) will be used to illustrate computation of Q using the difference between the residual sums of squares for full and reduced models. The reduced model for the first hypothesis tested, H_0: $\beta_2 = \beta_4 = 0$, is obtained from the full model by setting β_2 and β_4 equal to zero. This leaves a bivariate model containing only X_1 and X_3. Fitting this reduced model gives a residual sum of squares of

$$\text{SS}(\text{Res}_{\text{reduced}}) = 203.1194$$

with $[n - (p' - k)] = (31 - 3) = 28$ degrees of freedom. The residual sum of squares from the full model was

$$\text{SS}(\text{Res}_{\text{full}}) = 193.1178$$

with $(n - p') = 31 - 5 = 26$ degrees of freedom. The difference gives

$$\begin{aligned} Q &= \text{SS}(\text{Res}_{\text{reduced}}) - \text{SS}(\text{Res}_{\text{full}}) \\ &= 203.1194 - 193.1178 = 10.0016 \end{aligned}$$

with $(28 - 26) = 2$ degrees of freedom. This agrees, as it should, with the earlier result for Q obtained in Example 4.8.

The second hypothesis tested in the previous example included the statement that $\beta_0 = 90$ in addition to $\beta_2 = \beta_4 = 0$. The reduced model for this null hypothesis is

$$Y_i = 90 + \beta_1 X_{i1} + \beta_3 X_{i3} + \varepsilon_i$$

or

$$(Y_i - 90) = \beta_1 X_{i1} + \beta_3 X_{i3} + \varepsilon_i$$

The reduced model has a new dependent variable formed by subtracting 90 from every Y_i, has only X_1 and X_3 as independent variables, and has no intercept. The residual sum of squares from this model is

$$\text{SS}(\text{Res}_{\text{reduced}}) = 204.1365$$

with $(31 - 2) = 29$ degrees of freedom. The $\text{SS}(\text{Res}_{\text{full}})$ is the same as before and the difference gives

$$Q = 204.1365 - 193.1178 = 11.0187$$

with 3 degrees of freedom. ■

Using Model or Regression Sums of Squares

The sum of squares Q for any null hypothesis can *always* be computed as a difference in *residual* sums of squares. For null hypotheses where $\boldsymbol{m} = \boldsymbol{0}$, the same result can be obtained, sometimes more conveniently, by taking the difference in the *model* sums of squares; that is,

$$Q = \text{SS}(\text{Model}_{\text{full}}) - \text{SS}(\text{Model}_{\text{reduced}})$$

If β_0 is in the model and *not* involved in the null hypothesis, the difference in regression sums of squares, $\text{SS}(\text{Regr}_{\text{full}}) - \text{SS}(\text{Regr}_{\text{reduced}})$, will also give Q. The first hypothesis in Example 4.9 involved only β_2 and β_4 and had $\boldsymbol{m} = \boldsymbol{0}$. The sum of squares due to regression for the reduced model was $\text{SS}(\text{Regr}_{\text{reduced}}) = 648.2622$. Comparison of this to $\text{SS}(\text{Regr}_{\text{full}}) = 658.2638$ verifies that the difference again gives $Q = 10.0016$.

The difference in regression sums of squares, however, cannot be used to compute Q in the second hypothesis of Example 4.9, where $\beta_0 = 20$ is included in the null hypothesis. Consequently, it is important to develop the habit of either always using the residual sums of squares, since that procedure always gives the correct answer, or being very cautious in the use of differences in regression sums of squares to compute Q.

4.5.5 Sequential Sums of Squares

Definitions

The *partial* sums of squares measure the contributions of the individual variables with each adjusted for *all* other independent variables in the model (Section 4.5.2).

The **sequential sums of squares** measure the contributions of the variables as they are *added* to the model in a particular sequence. The sequential sum of squares for X_j is the difference in SS(Regr), or SS(Res), for the models just before and just after X_j is included. This sum of squares measures the contribution of X_j adjusted *only* for those independent variables that preceded X_j in the model building sequence.

For illustration, suppose a model is to be built by adding variables in the sequence X_3, X_1, X_2. The first model to be fit will contain only X_3 (and the intercept). SS(Regr) from this model is the sequential sum of squares for X_3. The second model to be fit will contain X_3 and X_1. The sequential sum of squares for X_1 is SS(Regr) for this model minus SS(Regr) for the first model. The third model to be fit will contain all three variables and the sequential sum of squares for X_2 is SS(Regr) for this three-variable model minus SS(Regr) for the two-variable model.

Properties of Sequential Sums of Squares

The sequential sums of squares are the successive differences in SS(Regr) for the sequence of models and, consequently, must *add* to SS(Regr) for the full model; that is, the sequential sums of squares are an additive partition of SS(Regr) for the full model. They are dependent on the order in which the variables are added to the model. The *partial* sums of squares, on the other hand, are dependent only on the set of independent variables in the full model, not on the order in which they are added to the model.

Each sequential sum of squares is the appropriate sum of squares for testing the jth partial regression coefficient, H_0: $\beta_j = 0$, for a model that contains X_j and *only* those independent variables that preceded X_j in the sequence. Adjacent sequential sums of squares are additive. Thus, the sum of k adjacent sequential sums of squares has k degrees of freedom and measures the joint contribution of all X_j in the sequence *adjusted* for all variables that preceded them in the model. This sum of squares is appropriate for testing the composite null hypothesis that all partial regression coefficients for the k variables in the sequence are zero, where the model contains these k variables plus those that preceded them in the model.

There are some models (for example, purely nested models and polynomial response models) where there is a logical order in which terms should be added to the model. In such cases, the sequential sums of squares provide the appropriate tests for determining which terms are to be retained in the model. In other cases, prior knowledge of the behavior of the system will suggest a logical ordering of the variables according to their relative importance. Use of this prior information and sequential sums of squares should simplify the process of determining an appropriate model.

4.5.6 The *R*-Notation to Label Sums of Squares

R-Notation

The sum of squares for the null hypothesis that each of a subset of the partial regression coefficients is zero is dependent on both the specific subset of parameters in the null hypothesis and the set of all parameters in the model. To clearly specify both in each case, a more convenient notation for sums of squares is needed. For this purpose, the commonly used ***R*-notation** is introduced.

Let $R(\beta_0\ \beta_1\ \beta_2\ \ldots\ \beta_p) = \text{SS(Model)}$ denote the sum of squares due to the model containing the parameters listed in parentheses. The sum of squares for the hypothesis that each of a subset of β_j, not including β_0, is zero can be obtained by subtraction of SS(Model) for the reduced model from that for the full model. Assume that the subset of β_j being tested against zero consists of the last k β_j. Then

$$\text{SS(Model}_{\text{full}}) = R(\beta_0\ \beta_1\ \ldots\ \beta_{p-k}\ \beta_{p-k+1}\ \ldots\ \beta_p),$$

$$\text{SS(Model}_{\text{reduced}}) = R(\beta_0\ \beta_1\ \ldots\ \beta_{p-k})$$

and

$$\begin{aligned} Q &= \text{SS(Model}_{\text{full}}) - \text{SS(Model}_{\text{reduced}}) \\ &= R(\beta_0\ \beta_1\ \ldots\ \beta_{p-k}\ \beta_{p-k+1}\ \ldots\ \beta_p) - R(\beta_0\ \beta_1\ \ldots\ \beta_{p-k}) \end{aligned} \qquad [4.48]$$

The final R-notation expresses this difference in sums of squares as

$$R(\beta_{p-k+1}\ \beta_{p-k+2}\ \ldots\ \beta_p|\beta_0\ \beta_1\ \ldots\ \beta_{p-k}) \qquad [4.49]$$

The β_j appearing before the vertical bar are those specified to be zero by the null hypothesis, whereas the β_j appearing after the bar are those for which the former are adjusted. Alternatively, the full model consists of all parameters in parentheses, whereas the reduced model contains only those parameters appearing *after* the bar. In this notation,

$$\begin{aligned} \text{SS(Regr)} &= \text{SS(Model)} - \text{SS}(\mu) \\ &= R(\beta_1\ \beta_2\ \ldots\ \beta_p|\beta_0) \end{aligned} \qquad [4.50]$$

While this notation appears cumbersome with the general subscripts on the β_j, it is simple and unambiguous.

Sequential Sums of Squares

To illustrate the R-notation, suppose the linear model contains three independent variables plus an intercept. If the variables are added to the model in the order X_1, X_2, and X_3, the *sequential* sums of squares would be

$R(\beta_1|\beta_0)$: The sum of squares attributable to X_1 in a model with no other independent variables, only X_0;

$R(\beta_2|\beta_0\ \beta_1)$: The sum of squares attributable to X_2 after taking into account the effects of X_1 and X_0;

$R(\beta_3|\beta_0\ \beta_1\ \beta_2)$: The sum of squares attributable to X_3 after taking into account the effects of X_1, X_2, and X_0.

Other sets of sequential sums of squares are obtained if the order in which the independent variables are added is changed.

Partial Sums of Squares

The partial sums of squares for this example would be

$R(\beta_1 | \beta_0 \ \beta_2 \ \beta_3)$,

$R(\beta_2 | \beta_0 \ \beta_1 \ \beta_3)$, and

$R(\beta_3 | \beta_0 \ \beta_1 \ \beta_2)$.

Each is the additional sum of squares accounted for by the parameter (or its corresponding variable) appearing before the vertical bar when added to a model that already contains the parameters appearing after the bar. Each is the appropriate sum of squares for testing the simple hypothesis H_0: $\beta_j = 0$.

4.5.7 Example: Sequential and Partial Sums of Squares

Example 4.10

The oxygen uptake example (Example 4.8) is used to illustrate the R-notation and the sequential and partial sums of squares. The sum of squares due to regression for the full model was

$$\text{SS(Regr)} = R(\beta_1 \ \beta_2 \ \beta_3 \ \beta_4 | \beta_0) = 658.2638$$

with 4 degrees of freedom (Table 4.4). The *sequential* sums of squares, from fitting the model in the order X_1, X_3, X_2, and X_4, and the null hypothesis for which each sum of squares is appropriate is shown in Table 4.5. The full subscript notation on the partial regression coefficients (see Section 3.1) in the null hypotheses is needed to avoid confusion. The F-test of the null hypothesis and the residual mean square and degrees of freedom for each model are also shown in Table 4.5.

Each sequential sum of squares measures the stepwise improvement in the model realized from adding one independent variable. Likewise, the corresponding partial regression coefficients in the null hypotheses are each adjusted for one additional independent variable. The sequential sums of squares add to the total regression sum of squares, $\text{SS(Regr)} = R(\beta_1 \ \beta_2 \ \beta_3 \ \beta_4 | \beta_0) = 658.2638$;

Table 4.5 Sequential sums of squares, residual mean squares, the null hypothesis being tested by each sequential sum of squares, and the F-test of the null hypothesis for the oxygen uptake example.

Sequential Sum of Squares		*Residual Mean Square*	*Null Hypothesis*	F^a
$R(\beta_1 \| \beta_0)$	= 632.9001	7.5338	$\beta_1 = 0$	84.01
$R(\beta_3 \| \beta_0 \ \beta_1)$	= 15.3621	7.2543	$\beta_{3.1} = 0$	2.12
$R(\beta_2 \| \beta_0 \ \beta_1 \ \beta_3)$	= .4041	7.5080	$\beta_{2.13} = 0$	.05
$R(\beta_4 \| \beta_0 \ \beta_1 \ \beta_3 \ \beta_2)$	= 9.5975	7.4276	$\beta_{4.123} = 0$	1.29
Total = SS(Regr)	= 658.2638			

[a]The F-tests have been computed using the residual mean square at each step. The denominator degrees of freedom for the F-tests are 29, 28, 27, and 26, respectively.

this is an orthogonal partitioning. Adjacent sequential sums of squares can be added to generate the appropriate sum of squares for a composite hypothesis. For example, the sequential sums of squares for X_3 and X_2, the second and third lines, respectively, in Table 4.5, can be added to give the additional sum of squares one would obtain from adding *both* X_2 and X_3 in one step to the model containing X_1. Thus,

$$
\begin{aligned}
R(\beta_3|\beta_0\ \beta_1) + R(\beta_2|\beta_0\ \beta_1\ \beta_3) &= 15.3621 + .4041 \\
&= 15.7662 \\
&= R(\beta_2\ \beta_3|\beta_0\ \beta_1)
\end{aligned}
$$

in the R-notation. This is the appropriate sum of squares for testing the composite hypothesis that both $\beta_{2.13}$ and $\beta_{3.12}$ in the three-variable model are zero.

If this particular ordering of the variables was chosen because it was expected that X_1 (run time) would likely be the most important variable with the others being of secondary importance, it is logical to test the composite null hypothesis H_0: $\beta_{2.134} = \beta_{3.124} = \beta_{4.123} = 0$. The sum of the sequential sums of squares for X_2, X_3, and X_4 is the appropriate sum of squares and gives $R(\beta_2\ \beta_3\ \beta_4|\beta_0\ \beta_1) = 25.3637$ with 3 degrees of freedom. This gives $F = 1.14$ which, with 3 and 26 degrees of freedom, does not approach significance. This single test supports the contention that X_1 alone is sufficient to account for oxygen consumption differences among the runners. (Since the variables are not orthogonal, this does *not* rule out the possibility that a model based on the other three variables might do better.)

More specific information is provided by the F-tests of the individual sequential sums of squares (Table 4.5). These show that H_0: $\beta_{4.123} = 0$, H_0: $\beta_{2.13} = 0$, and H_0: $\beta_{3.1} = 0$ cannot be rejected (using $\alpha = .05$). That is, X_4 is not making a significant contribution to a model that already contains X_1, X_2, and X_3; X_2 is not making a significant contribution to a model already containing X_1 and X_3; and X_3 is not making a significant contribution to a model containing only X_1. Of course, these three tests are not independent of the previous test of the composite hypothesis, and with the multiple testing the interpretation of the significance level α is becoming clouded.

The *partial* sums of squares, their null hypotheses, and the F-tests are shown in Table 4.6 (page 116). This is not an orthogonal partitioning; the partial sums of squares will not add to SS(Regr). Each partial sum of squares reflects the contribution of the particular variable as if it were the last to be considered for the model. Hence, it is the appropriate sum of squares for deciding whether the variable might be omitted. The null hypotheses in Table 4.6 reflect the adjustment of each partial regression coefficient for all other independent variables in the model.

The partial sum of squares for X_2, $R(\beta_2|\beta_0\ \beta_1\ \beta_3\ \beta_4) = .0822$ is much smaller than $s^2 = 7.4276$ and provides a clear indication that this variable does not make a significant contribution to a model that already contains X_1, X_3, and X_4. The next logical step in building the model based on tests of the partial sums of squares would be to omit X_2. Even though the tests for $\beta_{3.124}$ and $\beta_{4.123}$

Table 4.6 Partial sums of squares, the null hypothesis being tested by each, and the F-test of the null hypothesis for the oxygen uptake example.

Partial Sum of Squares	*Null Hypothesis*	F^{a}
$R(\beta_1 \vert \beta_0\ \beta_2\ \beta_3\ \beta_4) = 397.8664$	$\beta_{1.234} = 0$	53.57
$R(\beta_3 \vert \beta_0\ \beta_1\ \beta_2\ \beta_4) = 25.0917$	$\beta_{3.124} = 0$	3.38
$R(\beta_2 \vert \beta_0\ \beta_1\ \beta_3\ \beta_4) = .0822$	$\beta_{2.134} = 0$	.01
$R(\beta_4 \vert \beta_0\ \beta_1\ \beta_2\ \beta_3) = 9.5975$	$\beta_{4.123} = 0$	1.29

[a] All F-tests were computed using the residual mean square from the full model, $s^2 = 7.4276$ with 26 degrees of freedom.

are also nonsignificant, one must be cautious in omitting more than one variable at a time on the basis of the partial sums of squares. The partial sums of squares are dependent on which variables are in the model; it will almost always be the case that all partial sums of squares will change when a variable is dropped. (In this case, we know from the sequential sums of squares that all three variables can be dropped. A complete discussion on choice of variables is presented in Chapter 7.) ■

4.6 Confidence Intervals and Joint Confidence Regions

Confidence interval estimates of parameters convey more information to the reader than do simple point estimates. Univariate confidence intervals for several parameters, however, do not take into account correlations among the estimators of the parameters. Furthermore, the individual confidence coefficients do not reflect the overall degree of confidence in the joint statements. Joint confidence regions address these two issues. Univariate confidence interval estimates will be discussed briefly before we proceed to a discussion of joint confidence regions.

4.6.1 Univariate Confidence Intervals

Confidence Intervals for β_j

If $\varepsilon \sim N(\mathbf{0}, \boldsymbol{I}\sigma^2)$, $\hat{\boldsymbol{\beta}}$ and $\hat{\boldsymbol{Y}}$ have multivariate normal distributions (see equation 3.37). With normality, the classical $(1-\alpha)100\%$ confidence interval estimate of each β_j is

$$\hat{\beta}_j \pm t_{(\alpha,\nu)} s(\hat{\beta}_j), \quad j = 0, \ldots, p, \qquad [4.51]$$

where $t_{(\alpha,\nu)}$ is the value of the Student's t-distribution, with ν degrees of freedom, that puts $\alpha/2$ probability in each tail. [In the usual multiple regression problem, $\nu = (n - p')$.] The standard error of $\hat{\beta}_j$ is $s(\hat{\beta}_j) = \sqrt{c_{jj}s^2}$, where s^2 is estimated with ν degrees of freedom and c_{jj} is the jth diagonal element from $(\boldsymbol{X}'\boldsymbol{X})^{-1}$.

Confidence Intervals for $\mathscr{E}(Y_i)$

Similarly, the $(1-\alpha)100\%$ confidence interval estimate of the mean of Y for a particular choice of values for the independent variables, say $\boldsymbol{x}_1' = (1\ X_{i1}\ \ldots\ X_{ip})$, is

$$\hat{Y}_i \pm t_{(\alpha, \nu)} s(\hat{Y}_i) \qquad [4.52]$$

where $s(\hat{Y}_i) = \sqrt{x_i'(X'X)^{-1}x_i s^2}$, in general, or $s(\hat{Y}_i) = \sqrt{v_{ii}s^2}$ if x_i' corresponds to the ith row of X; v_{ii} is the ith diagonal element in P; $t_{(\alpha, \nu)}$ is as defined for equation 4.51.

Example 4.11 The univariate confidence intervals are illustrated with the oxygen uptake example (see Example 4.8). $s^2 = 7.4276$ was estimated with 26 degrees of freedom. The value of Student's t for $\alpha = .05$ and 26 degrees of freedom is $t_{(.05, 26)} = 2.056$. The point estimates of the parameters and the estimated variance–covariance matrix of $\hat{\beta}$ were

$$\hat{\beta}' = (84.2690 \quad -3.0698 \quad .0080 \quad -.1167 \quad .0852)$$

and

$$s^2(\hat{\beta}) = (X'X)^{-1}s^2$$

$$= \begin{bmatrix} 129.4119 & -1.185591 & .053980 & -.104321 & -.579099 \\ -1.185591 & .175928 & -.012602 & -.007318 & .007043 \\ .053980 & -.012602 & .005775 & -.000694 & -.000634 \\ -.104321 & -.007318 & -.000694 & .004032 & -.002646 \\ -.579099 & .007043 & -.000634 & -.002646 & .005616 \end{bmatrix}$$

$s(\hat{\beta}_j)$ is the square root of the $(j + 1)$st diagonal element. If $\boldsymbol{s}$ is defined as the column vector of $s(\hat{\beta}_j)$, the univariate 95% confidence interval estimates can be computed as

$$\mathbf{CL}(\hat{\beta}) = [(\hat{\beta} - t_{(\alpha, \nu)}\boldsymbol{s}) \quad (\hat{\beta} + t_{(\alpha, \nu)}\boldsymbol{s})]$$

$$= \begin{bmatrix} 60.880 & 107.658 \\ -3.932 & -2.207 \\ -.148 & .164 \\ -.247 & .014 \\ -.069 & .239 \end{bmatrix}$$

where the two columns give the lower and upper limits, respectively, for the β_j in the same order as listed in β. ■

4.6.2 Simultaneous Confidence Statements

For the classical univariate confidence intervals, the confidence coefficient, $(1 - \alpha) = .95$, applies to each confidence statement. The level of confidence associated with the statement that all five intervals simultaneously contain their respective parameters, $(1 - \alpha^*)$, is much lower. If the five intervals were statistically independent, which they are not, the overall or joint confidence coefficient would be only $(1 - \alpha^*) = (1 - \alpha)^5 = .77$.

Bonferroni Method

There are two procedures which keep the joint confidence coefficient for several simultaneous statements near a prechosen level, $(1 - \alpha^*)$. The oldest and simplest procedure, commonly called the **Bonferroni method**, constructs the individual confidence intervals as given in equations 4.51 and 4.52, but uses $\alpha = \alpha^*/k$ where k is the number of simultaneous intervals or statements. That is, $t_{(\alpha, \nu)}$ is replaced with $t_{(\alpha^*/k, \nu)}$. This procedure ensures that the true joint confidence coefficient for the k simultaneous statements is *at least* $(1 - \alpha^*)$.

Scheffé's Method

The second procedure applies the general approach developed by Scheffé (1953). **Scheffé's method** provides simultaneous confidence statements for *all* linear combinations of a set of parameters in a d-dimensional subspace of the p'-dimensional parameter space. The Scheffé joint confidence intervals for the p' parameters in $\boldsymbol{\beta}$ and the means of Y, $\mathscr{E}(Y_i)$, are obtained from equations 4.51 and 4.52 by replacing $t_{(\alpha, \nu)}$ with $[p'F_{(\alpha^*, p', \nu)}]^{1/2}$. (If only a subset of d linearly independent parameters β_j are of interest, $t_{(\alpha, \nu)}$ is replaced with $[dF_{(\alpha^*, d, \nu)}]^{1/2}$.) That is,

$$\hat{\beta}_j \pm (p'F_{(\alpha^*, p', \nu)})^{1/2} s(\hat{\beta}_j)$$
$$\hat{Y}_i \pm (p'F_{(\alpha^*, p', \nu)})^{1/2} s(\hat{Y}_i) \qquad [4.53]$$

This method provides simultaneous statements for all linear combinations of the set of parameters. As with the Bonferroni intervals, the joint confidence coefficient for the Scheffé intervals is at least $(1 - \alpha^*)$. That is, the confidence coefficient of $(1 - \alpha^*)$ applies to all confidence statements on the β_j, the $\mathscr{E}(Y_i)$, plus all other linear functions of β_j of interest. Thus, equation 4.53 can be used to establish a confidence *band* on the entire regression surface by computing Scheffé confidence intervals for $\mathscr{E}(Y_i)$ for all values of the independent variables in the region of interest. The confidence band for the simple linear regression case was originally developed by Working and Hotelling (1929) and frequently carries their names.

The reader is referred to Miller (1981) for more complete presentations on Bonferroni and Scheffé methods. Since the Scheffé method provides simultaneous confidence statements on *all* linear functions of a set of parameters, the Scheffé intervals will tend to be longer than Bonferroni intervals, particularly when a small number of simultaneous statements is involved (Miller, 1981). One would choose the method that gave the shorter intervals for the particular application.

Example 4.12

The oxygen uptake data of Example 4.8 has $p = 5$ parameters and $\nu = 26$ degrees of freedom for s^2. In order to attain an overall confidence coefficient no smaller than $(1 - \alpha^*) = .95$ with the Bonferroni method, $\alpha = .05/5 = .01$ would be used, for which $t_{(.01, 26)} = 2.779$. Using this value of t in equation 4.51 gives the Bonferroni simultaneous confidence intervals with an *overall* confidence coefficient at least as large as $(1 - \alpha^*) = .95$:

$$\mathbf{CL}_B(\hat{\boldsymbol{\beta}}) = \begin{bmatrix} 52.655 & 115.883 \\ -4.235 & -1.904 \\ -.203 & .219 \\ -.293 & .060 \\ -.123 & .293 \end{bmatrix}$$

The Scheffé simultaneous intervals for the $p' = 5$ parameters in $\boldsymbol{\beta}$ are obtained by using $[p'F_{(.05,5,26)}]^{1/2} = [5(2.59)]^{1/2} = 3.599$ in place of $t_{(\alpha,v)}$ in equation 4.51. The results are

$$\mathbf{CL}_S(\hat{\boldsymbol{\beta}}) = \begin{bmatrix} 43.331 & 125.207 \\ -4.579 & -1.560 \\ -.265 & .281 \\ -.345 & .112 \\ -.184 & .355 \end{bmatrix}$$

The Bonferroni and Scheffé simultaneous confidence intervals will always be wider than the classical univariate confidence intervals, in which the confidence coefficient applies to each interval. In this example, the Scheffé intervals are wider than the Bonferroni intervals. ■

Simultaneous Versus Joint Confidence Statements

It is tempting, but incorrect, to think of the p'-dimensional volume formed by the intersection of the p' simultaneous confidence intervals, either Bonferroni or Scheffé, as a joint confidence region for all parameters. Each of the simultaneous statements is a confidence statement on the likely values of one parameter *averaged* over all values of the other parameters. The simultaneous statements do not take into account the *joint* distribution of the parameter estimates and, as a result, can be misleading if their intersection is interpreted as a joint confidence region. This distinction will be illustrated after joint confidence regions are defined in the next section.

4.6.3 Joint Confidence Regions

Computation

A joint confidence region for all p' parameters in $\boldsymbol{\beta}$ is obtained from the inequality

$$(\boldsymbol{\beta} - \hat{\boldsymbol{\beta}})'(X'X)(\boldsymbol{\beta} - \hat{\boldsymbol{\beta}}) \leq p's^2 F_{(\alpha, p', v)} \qquad [4.54]$$

where $F_{(\alpha, p', v)}$ is the value of the F-distribution with p' and v degrees of freedom that leaves probability α in the upper tail; v is the degrees of freedom associated with s^2. The left-hand side of this inequality is a quadratic form in $\boldsymbol{\beta}$, because $\hat{\boldsymbol{\beta}}$ and $X'X$ are known quantities computed from the data. The right-hand side is also known from the data. Solving this quadratic form for the boundary of the inequality establishes a p'-dimensional ellipsoid which is the $(1 - \alpha)100\%$ joint confidence region for all the parameters in the model. The slope of the axes and eccentricity of the ellipsoid show the direction and strength, respectively, of correlations between the estimates of the parameters.

Interpretation

An ellipsoidal confidence region with more than two or three dimensions is difficult to interpret. Specific choices of $\boldsymbol{\beta}$ can be checked, with a computer program, to determine whether they fall inside or outside the confidence region. The multidimensional region, however, must be viewed in two or at most three dimensions at a time. One approach to visualizing the joint confidence region is to evaluate the p'-dimensional joint confidence region for specific values of all but two of the parameters. Each set of specified values produces an ellipse which is a two-dimensional "slice" of the multidimensional region. To develop a picture of the entire region, two-dimensional "slices" can be plotted for several choices of values for the other parameters.

An alternative to using the p'-dimensional joint confidence region for all parameters is to construct joint confidence regions for two parameters at a time *ignoring* the other $(p' - 2)$ parameters. The quadratic form for the joint confidence region for a subset of two parameters is obtained from that for all parameters, equation 4.54, by

1. Replacing $(\hat{\boldsymbol{\beta}} - \boldsymbol{\beta})$ with the corresponding vectors involving *only* the two parameters of interest;
2. Replacing $(\boldsymbol{X}'\boldsymbol{X})$ with the *inverse* of the 2×2 variance–covariance matrix for the two parameters; and
3. Replacing $p's^2F_{(\alpha, p', \nu)}$ with $2F_{(\alpha, 2, \nu)}$. Notice that s^2 is not in the second quantity because it has been included in the variance–covariance matrix in step 2.

Thus, if β_j and β_k are the two parameters of interest, the joint confidence region is given by

$$\left[\begin{pmatrix}\hat{\beta}_j\\ \hat{\beta}_k\end{pmatrix} - \begin{pmatrix}\beta_j\\ \beta_k\end{pmatrix}\right]'(s^2(\hat{\beta}_j, \hat{\beta}_k))^{-1}\left[\begin{pmatrix}\hat{\beta}_j\\ \hat{\beta}_k\end{pmatrix} - \begin{pmatrix}\beta_j\\ \beta_k\end{pmatrix}\right] \leq 2F_{(\alpha, 2, \nu)} \qquad [4.55]$$

The confidence coefficient $(1 - \alpha)$ applies to the joint statement on the two parameters being considered at the time. This procedure takes into account the joint distribution of $\hat{\beta}_j$ and $\hat{\beta}_k$ but averages over all possible values of the other parameters. Since this bivariate joint confidence region ignores the joint distribution of $\hat{\beta}_j$ and $\hat{\beta}_k$ with the other $(p' - 2)$ parameter estimates, it suffers from the same conceptual problem as the univariate confidence intervals.

Example 4.13

The oxygen uptake data, given in Example 4.8, are used to illustrate joint confidence regions, but the model is simplified to include only an intercept and two independent variables, time to run 1.5 miles (X_1) and heart rate while running (X_3). The estimate of $\boldsymbol{\beta}$, $\boldsymbol{X}'\boldsymbol{X}$, and the variance–covariance matrix for $\hat{\boldsymbol{\beta}}$

for this reduced model are

$$\hat{\boldsymbol{\beta}}' = (93.0888 \quad -3.14019 \quad -.073510)$$

$$\boldsymbol{X}'\boldsymbol{X} = \begin{bmatrix} 31 & 328.17 & 5{,}259 \\ 328.17 & 3{,}531.797 & 55{,}806.29 \\ 5{,}259 & 55{,}806.29 & 895{,}317 \end{bmatrix}$$

and

$$s^2(\hat{\boldsymbol{\beta}}) = \begin{bmatrix} 68.04308 & -.47166 & -.37028 \\ -.47166 & .13933 & -.00591 \\ -.37028 & -.00591 & .00255 \end{bmatrix}$$

The residual mean square from this model is $s^2 = 7.25426$ with 28 degrees of freedom.

The joint confidence region for all three parameters is obtained from equation 4.54 and is a three-dimensional ellipsoid. The right-hand side of equation 4.54 is

$$p's^2F_{(\alpha, 3, 28)} = 3(7.25426)(2.95)$$

if $\alpha = .05$. This choice of α gives a confidence coefficient of .95 that applies to the joint statement involving all three parameters. The three-dimensional ellipsoid is portrayed in Figure 4.1 with three two-dimensional "slices" (dashed lines) from the ellipsoid at $\beta_0 = 76.59$, 93.09, and 109.59. These choices of β_0 correspond to

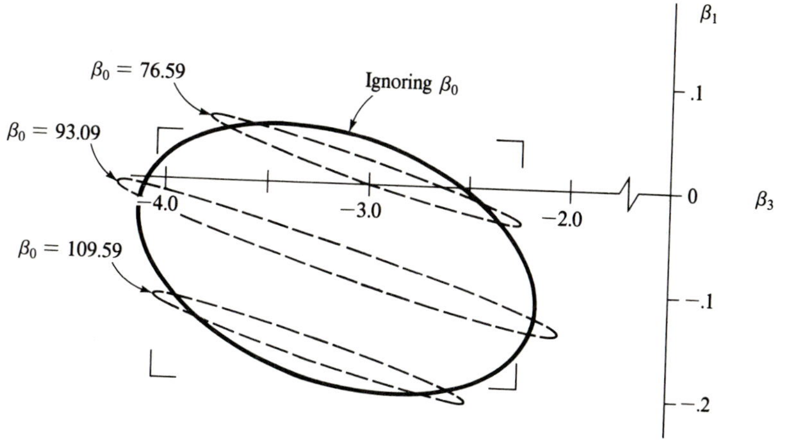

Figure 4.1 Two-dimensional "slices" of the joint confidence region for the regression of oxygen uptake on time to run 1.5 miles (X_1) and heart rate while running (X_3) (dashed ellipses) and the two-dimensional joint confidence region for β_1 and β_3 ignoring β_0 (solid ellipse). The intersection of the Bonferroni univariate confidence intervals is shown as the corners of the rectangle formed by the intersection.

$\hat{\beta}_0$ and $\hat{\beta}_0 \pm 2s(\hat{\beta}_0)$. The "slices" indicate that the ellipsoid is extremely thin in one plane but only slightly elliptical in the other, much like a slightly oval pancake. This is reflecting the high correlation between $\hat{\beta}_0$ and $\hat{\beta}_3$ of $-.89$ and the more moderate correlations of $-.15$ and $-.31$ between $\hat{\beta}_0$ and $\hat{\beta}_1$ and between $\hat{\beta}_1$ and $\hat{\beta}_3$, respectively.

The bivariate joint confidence region for β_1 and β_3 *ignoring* β_0, obtained from equation 4.55, is shown in Figure 4.1 as the ellipse drawn with the solid line. The variance–covariance matrix to be inverted in equation 4.55 is the lower-right 2×2 matrix in $s^2(\hat{\boldsymbol{\beta}})$. The right-hand side of the inequality is $2F_{(\alpha, 2, 28)} = 2(3.34)$ if $\alpha = .05$. The confidence coefficient of .95 applies to the joint statement involving *only* β_1 and β_3. The negative slope in this ellipse reflects the moderate negative correlation between $\hat{\beta}_1$ and $\hat{\beta}_3$. For reference, the Bonferroni confidence intervals for β_1 and β_3, ignoring β_0, using a joint confidence coefficient of .95, are shown by the corners of the rectangle enclosing the intersection region.

The implications as to what are "acceptable" combinations of values for the parameters are very different for the two joint confidence regions. The joint confidence region for all parameters is much more restrictive than the bivariate joint confidence region, or the univariate confidence intervals, would indicate. Allowable combinations of β_1 and β_3 are highly dependent on choice of β_0. Clearly, univariate confidence intervals and joint confidence regions that do not involve all parameters can be misleading. ■

4.7 Estimation of Pure Error

The residual mean square has been used, until now, as the estimate of σ^2. One of the problems with this procedure is the dependence of the residual mean square on the model being fit. Any inadequacies in the model, important independent variables omitted or an incorrect form of the model, will cause the residual mean square to overestimate σ^2. An estimate of σ^2 is needed which is not as dependent on the choice of model being fit at the time.

Definition of Pure Error

The variance σ^2 is the variance of the ε_i about zero or, equivalently, the variance of Y_i about their true means, $\mathscr{E}(Y_i)$. The concept of modeling Y_i assumes that $\mathscr{E}(Y_i)$ is determined by some unknown function of the relevant independent variables. Let $\boldsymbol{x}_i'$ be the row vector of values of all relevant independent variables for the ith observation. Then, all Y_i that have the same $\boldsymbol{x}_i'$ also will have the same true mean regardless of whether the correct model is known. Hence, σ^2 is by definition the variance among statistically independent observations that have the same $\boldsymbol{x}_i'$. Such repeated observations are called **true replicates**. The sample variance of the Y_i among true replicates provides a direct estimate of σ^2 that is independent of the choice of model. (It is, however, dependent on having identified and taken data on all relevant independent variables.) The estimate of σ^2 obtained from true replication is called **pure error**. When several sets of replicate observations are available, the best estimate of σ^2 is obtained by pooling all estimates.

True replication is almost always included in the design of controlled experiments. For example, the estimate of experimental error from the completely random design or the randomized complete block design when there is no block-by-treatment interaction is the estimate of pure error. **Observational studies**, on the other hand, seldom have true replication because they impose no control over the independent variables. Then, true replication occurs only by chance and is very unlikely if several independent variables are involved. In addition, apparent replicates in the observational data may not, in fact, be true replicates due to important variables having been overlooked. Pseudoreplication or **near replication** is sometimes used with observational data to estimate σ^2. These are sets of observations in which the values of the independent variables fall within a relatively narrow range.

Example 4.14

To illustrate the estimation of pure error, the ozone example used in Example 1.1 will be used. The four observations used in that section were the means of five replicate experimental units *at each level of ozone* from a completely random experimental design. The full data set, the treatment means, and the estimates of pure error within each ozone level are given in Table 4.7.

Each s_i^2 is estimated from the variance among the five observations for each ozone level, with 4 degrees of freedom, and is an unbiased estimate of σ^2. Since each is the variation of Y_i about $\overline{Y}_i$ for a given level of ozone, the estimates are in no way affected by the form of the response model that might be chosen to represent the response of yield to ozone. Figure 4.2 (page 124) illustrates that the variation among the replicate observations for a given level of ozone is unaffected by the form of the regression line fit to the data. The best estimate of σ^2 is the pooled estimate

$$s^2 = \frac{\sum(n_i - 1)s_i^2}{\sum(n_i - 1)} = \frac{4(476.61) + \cdots + 4(325.86)}{16}$$

$$= 274.32 \quad \text{with 16 degrees of freedom}$$

The analysis of variance for the completely random design is given (Table 4.8) to emphasize that s^2 is the experimental error from that analysis. The previous regression analysis (Section 1.4, Tables 1.3 and 1.4) used the treatment means

Table 4.7 Replicate yield data for soybeans exposed to chronic levels of ozone and estimates of pure error. (Data courtesy A. S. Heagle, North Carolina State University.)

	Ozone Level (*ppm*)			
	.02	.07	.11	.15
	238.3	235.1	236.2	178.7
	270.7	228.9	208.0	186.0
	210.0	236.2	243.5	206.9
	248.7	255.0	233.0	215.3
	242.4	228.9	233.0	219.5
$\overline{Y}_i$	242.02	236.82	230.74	201.28
s_i^2	476.61	114.83	179.99	325.86

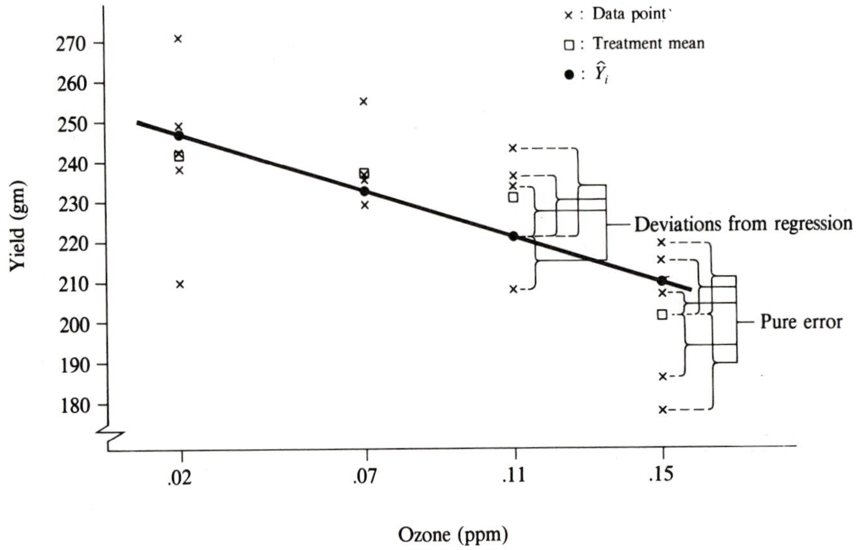

Figure 4.2 Comparison of "pure error" and "deviations from regression" using the data on soybean response to ozone.

Table 4.8 The analysis of variance for the completely random experimental design for the yield response of soybean to ozone.

Source	d.f.	SS	MS
$\text{Total}_{\text{corr}}$	19	9,366.61	
Treatments	3	4,977.47	1,659.16
Regression	1	3,956.31	3,956.31
Deviations	2	1,021.16	510.58
Exp. error	16	4,389.14	274.32

(of $r = 5$ observations). Thus, the sums of squares from that analysis have to be multiplied by $r = 5$ to put them on a "per observation" basis. The analysis of Table 1.4 partitioned the sum of squares among the four treatment means into 1 degree of freedom for the linear regression of Y on ozone level and 2 degrees of freedom for deviations from regression. The middle three lines of Table 4.8 contain the results from the original analysis multiplied by $r = 5$. The numbers differ slightly due to rounding the original means to whole numbers. ■

Mean Square Expectations

The expectations of the mean squares in the analysis of variance show what function of the parameters each mean square is estimating. The mean square expectations for the critical lines in Table 4.8 are

$$\mathscr{E}[\text{MS(Regr)}] = \sigma^2 + \beta_1^2 \sum x^2$$

$$\mathscr{E}[\text{MS(Deviations)}] = \sigma^2 + (\text{Model bias})^2 \qquad [4.56]$$

$$\mathscr{E}[\text{MS(Exp. error)}] = \sigma^2$$

Recall that $\sum x^2$ is used to indicate the *corrected* sum of squares of the independent variable.

The square on "Model bias" emphasizes that any inadequacies in the model cause this mean square to be larger, in expectation, than σ^2. Thus, the deviations mean square is an unbiased estimate of σ^2 only if the linear model is correct. Otherwise, it is biased upward. On the other hand, the "pure error" estimate of σ^2 obtained from the replication in the experiment is unbiased regardless of whether the assumed linear relationship is correct.

The mean square expectation of MS(Regr) is shown as if the linear model relating yield to ozone level is correct. If the model is not correct (for example, if the treatment differences are not due solely to ozone differences), the second term in $\mathscr{E}$[MS(Regr)] will include contributions from all variables that are correlated with ozone levels. This is the case even if the variables have not been identified. The advantage of controlled experiments such as this ozone study is that amount of ozone is, presumably, the only variable changing consistently over the ozone treatments. Random assignment of treatments to the experimental units should destroy any correlation between ozone level and any incidental environmental variable. Thus, treatment differences in this controlled study can be attributed to the effects of ozone and $\mathscr{E}$[MS(Regr)] should not be biased by the effects of any uncontrolled variables. One should not overlook, however, this potential for bias in the regression sum of squares, particularly when observational data are being analyzed.

Testing Adequacy of the Model

The independent estimate of pure error, experimental error, provides the basis for two important tests of significance. The **adequacy of the model** can be checked by testing the null hypothesis that "Model bias" is zero. Any inadequacies in the linear model will make this mean square larger than σ^2 on the average.

Example 4.15 In the ozone example (Example 4.14), the test of the adequacy of the linear model is

$$F = \frac{\text{MS(Deviations)}}{\text{MS(Exp. error)}} = \frac{510.58}{274.32} = 1.86$$

which, if the model is correct, is distributed as F with 2 and 16 degrees of freedom. Comparison against the critical value $F_{(.05,2,16)} = 3.63$ shows this to be nonsignificant, indicating that there is no evidence in these data that the linear model is inadequate for representing the response of soybean to ozone. ■

Testing H_0: $\beta_1 = 0$

The second hypothesis of interest is H_0: $\beta_1 = 0$ against the alternative hypothesis H_a: $\beta_1 \neq 0$. The ratio of the regression mean square to an estimate of σ^2 provides a test of this hypothesis. The mean square expectations show that both mean squares estimate σ^2 when the null hypothesis is true and that the numerator becomes increasingly larger as β_1 deviates from zero. The best estimate of σ^2 is, again, the pure error estimate or experimental error.

Example 4.16 For the ozone example,

$$F = \frac{\text{MS(Regr)}}{\text{MS(Exp. error)}} = \frac{3{,}956.31}{274.32} = 14.42$$

Comparing this to the critical value for $\alpha = .01$, $F_{(.01,1,16)} = 8.53$, indicates that the null hypothesis that $\beta_1 = 0$ should be rejected. This conclusion differs from that from the analysis in Chapter 1 because σ^2 is now estimated with many more degrees of freedom. As a result, the test has more power for detecting departures from the null hypothesis. ■

In summary, multiple, statistically independent observations on the dependent variable for given values of all relevant independent variables is called true replication. True replication provides for an unbiased estimate of σ^2 that is not dependent on the model being used. The estimate of pure error provides a basis for testing the adequacy of the model. True replication should be designed into all studies where possible and the pure error estimate of σ^2, rather than a residual mean square estimate, used for tests of significance and standard errors.

EXERCISES

* 4.1. A dependent variable Y (20 × 1) was regressed onto three independent variables plus an intercept (so that X was of dimension 20 × 4). The following matrices were computed:

$$X'X = \begin{bmatrix} 20 & 0 & 0 & 0 \\ 0 & 250 & 401 & 0 \\ 0 & 401 & 1{,}013 & 0 \\ 0 & 0 & 0 & 128 \end{bmatrix} \qquad X'Y = \begin{pmatrix} 1{,}900.00 \\ 970.45 \\ 1{,}674.41 \\ -396.80 \end{pmatrix}$$

$$Y'Y = 185{,}883$$

a. Compute $\hat{\beta}$ and write the regression equation.

b. Compute the analysis of variance of Y. Partition the sum of squares due to the model into a part due to the mean and a part due to regression on the X's *after adjustment* for the mean. Summarize the results, including degrees of freedom and mean squares, in an analysis of variance table.

c. Compute the estimate of σ^2 and the standard error for *each* regression coefficient. Compute the covariance between $\hat{\beta}_1$ and $\hat{\beta}_2$, $\text{Cov}(\hat{\beta}_1, \hat{\beta}_2)$. Compute the covariance between $\hat{\beta}_1$ and $\hat{\beta}_3$, $\text{Cov}(\hat{\beta}_1, \hat{\beta}_3)$.

d. Drop the last X, X_3, from the model. Reconstruct $X'X$ and $X'Y$ for this model without X_3 and repeat questions (a) and (b). Put X_3 back in the model but drop the next to the last X, X_2, and repeat (a) and (b).

(i) Which of the two independent variables, X_2 or X_3, made the greater contribution to Y in the presence of the remaining X's? That is, compare $R(\beta_2|\beta_0\ \beta_1\ \beta_3)$ and $R(\beta_3|\beta_0\ \beta_1\ \beta_2)$.

(ii) Explain why $\hat{\beta}_1$ changed in value when X_2 was dropped but not when X_3 was dropped.

(iii) Explain the differences in meaning of $\hat{\beta}_1$ in the three models.

e. From inspection of $X'X$, how can you tell that X_1, X_2, and X_3 were expressed as deviations from their respective means? Would $(X'X)^{-1}$ have been easier or harder to obtain if the original X's (without subtraction of their means) had been used? Explain.

4.2. A regression analysis led to the following $\boldsymbol{P} = \boldsymbol{X}(\boldsymbol{X}'\boldsymbol{X})^{-1}\boldsymbol{X}'$ matrix and estimate of σ^2:

$$\boldsymbol{P} = \frac{1}{70}\begin{bmatrix} 62 & 18 & -6 & -10 & 6 \\ 18 & 26 & 24 & 12 & -10 \\ -6 & 24 & 34 & 24 & -6 \\ -10 & 12 & 24 & 26 & 18 \\ 6 & -10 & -6 & 18 & 62 \end{bmatrix} \qquad s^2 = .06$$

a. How many observations were in the data set?
b. How many linearly independent columns are in $\boldsymbol{X}$—that is, what is the rank of $\boldsymbol{X}$? How many degrees of freedom are associated with the *model* sum of squares? Assuming the model contained an intercept, how many degrees of freedom are associated with the *regression* sum of squares?
c. Suppose $\boldsymbol{Y}' = (82 \;\; 80 \;\; 75 \;\; 67 \;\; 55)$. Compute the estimated mean of Y corresponding to the first observation, $\hat{Y}_1$. Compute $s^2(\hat{Y}_1)$. Find the residual, e_1, for the first observation and compute its variance. For which data point will $\hat{Y}_i$ have the smallest variance? For which data point will e_i have the largest variance?

* 4.3. The following $(\boldsymbol{X}'\boldsymbol{X})^{-1}$, $\hat{\boldsymbol{\beta}}$, and residual sum of squares were obtained from the regression of plant dry weight (grams) from $n = 7$ experimental fields on percent soil organic matter (X_1) and kilograms of supplemental nitrogen per 1,000 m^2 (X_2). The regression model included an intercept.

$$(\boldsymbol{X}'\boldsymbol{X})^{-1} = \begin{bmatrix} 1.7995972 & -.0685472 & -.2531648 \\ -.0685472 & .0100774 & -.0010661 \\ -.2531648 & -.0010661 & .0570789 \end{bmatrix}$$

$$\hat{\boldsymbol{\beta}} = \begin{pmatrix} 51.5697 \\ 1.4974 \\ 6.7233 \end{pmatrix}, \qquad \text{SS(Res)} = 27.5808$$

a. Give the regression equation and interpret each regression coefficient. Give the units of measure of each regression coefficient.
b. How many degrees of freedom does SS(Res) have? Compute s^2, the variance of $\hat{\beta}_1$, and the covariance of $\hat{\beta}_1$ and $\hat{\beta}_2$.
c. Determine the 95% univariate confidence interval estimates of β_1 and β_2. Compute the Bonferroni and the Scheffé confidence intervals for β_1 and β_2 using a *joint* confidence coefficient of .95.
d. Suppose previous experience has led you to believe that 1 percentage point increase in organic matter is equivalent to .5 kilogram/1,000 m^2 of supplemental nitrogen in dry matter production. Translate this statement into a null hypothesis on the regression coefficients. Use a t-test to test this null hypothesis against the alternative hypothesis that supplemental nitrogen is *more* effective than this statement would imply.
e. Define $\boldsymbol{K}'$ and $\boldsymbol{m}$ for the general linear hypothesis $H_0\colon \boldsymbol{K}'\boldsymbol{\beta} - \boldsymbol{m} = \boldsymbol{0}$ for testing $H_0\colon 2\beta_1 = \beta_2$. Compute Q and complete the test of significance using the F-test. What is the alternative hypothesis for this test?

f. Give the reduced model you obtain if you impose the null hypothesis in part (e) on the model. Suppose this reduced model gave SS(Res) = 164.3325. Use this result to complete the test of the hypothesis.

4.4. The following analysis of variance summarizes the regression of Y on two independent variables plus an intercept:

Source	d.f.	SS	MS
$\text{Total}_{\text{corr}}$	26	1,211	
Regression	2	1,055	527.5
Residual	24	156	6.5

Variable	*Sequential* SS	*Partial* SS
X_1	263	223
X_2	792	792

a. Your estimate of β_1 is $\hat{\beta}_1 = 2.996$. A friend of yours regressed Y on X_1 and found $\hat{\beta}_1 = 3.24$. Explain the difference in these two estimates.
b. Label each sequential and partial sum of squares using the R-notation. Explain what $R(\beta_1|\beta_0)$ measures.
c. Compute $R(\beta_2|\beta_0)$ and explain what it measures.
d. What is the regression sum of squares due to X_1 after adjustment for X_2?
e. Make a test of significance (use $\alpha = .05$) to determine if X_1 should be retained in the model with X_2.
f. The original data contained several sets of observations having the same values of X_1 and X_2. The *pooled* variance from these replicate observations was $s^2 = 3.8$ with 8 degrees of freedom. With this information, rewrite the analysis of variance to show the partitions of the "Residual" sum of squares into "pure error" and "lack of fit." Make a test of significance to determine whether the model using X_1 and X_2 is adequate.

* 4.5. The accompanying table presents data on one dependent variable and five independent variables.

Y	X_1	X_2	X_3	X_4	X_5
6.68	32.6	4.78	1,092	293.09	17.1
6.31	33.4	4.62	1,279	252.18	14.0
7.13	33.2	3.72	511	109.31	12.7
5.81	31.2	3.29	518	131.63	25.7
5.68	31.0	3.25	582	124.50	24.3
7.66	31.8	7.35	509	95.19	.3
7.30	26.4	4.92	942	173.25	21.1
6.19	26.2	4.02	952	172.21	26.1
7.31	26.6	5.47	792	142.34	19.8

a. Give the linear model in matrix form for regressing Y on the five independent variables. Completely define each matrix and give its order and rank.
b. The following quadratic forms were computed:

$$Y'PY = 404.532 \qquad Y'Y = 405.012 \qquad Y'(I - P)Y = .480$$

$$Y'(I - J/n)Y = 4.078 \qquad Y'(P - J/n)Y = 3.598$$

$$Y'(J/n)Y = 400.934$$

Use a matrix algebra computer program to reproduce each of these sums of squares. Use these results to give the complete analysis of variance summary.

c. The partial sums of squares for X_1, X_2, X_3, X_4, and X_5 are .895, .238, .270, .337, and .922, respectively. Give the R-notation that describes the partial sum of squares for X_2. Use a matrix algebra program to verify the partial sum of squares for X_2.

d. Assume that none of the partial sums of squares for X_2, X_3, and X_4 is significant and that the partial sums of squares for X_1 and X_5 are significant (at $\alpha = .05$). Indicate whether each of the following statements is valid based on these results. If it is not a valid statement, explain why.
   (i) X_1 and X_5 are important causal variables while X_2, X_3, and X_4 are not.
   (ii) X_2, X_3, and X_4 can be dropped from the model with no meaningful loss in predictability of Y.
   (iii) There is no need for all five independent variables to be retained in the model.

4.6. This exercise continues with the analysis of the peak water flow data used in exercise 3.12. In that exercise, several regressions were run to relate $Y = \ln(Q_o/Q_p)$ to three characteristics of the watersheds and a measure of storm intensity. Y measures the discrepancy between peak water flow predicted from a simulation model (Q_p) and observed peak water flow (Q_o). The four independent variables are described in exercise 3.12.

a. The first model used an intercept and all four independent variables.
   (i) Compute SS(Model), SS(Regr), and SS(Res) for this model and summarize the results in an analysis of variance table. Show degrees of freedom and mean squares.
   (ii) Obtain the partial sum of squares for each independent variable and the sequential sums of squares for the variables added to the model in the order X_1, X_4, X_2, X_3.
   (iii) Use tests of significance ($\alpha = .05$) to determine which partial regression coefficients are different from zero. What do these tests suggest as to which variables might be dropped from the model?
   (iv) Construct a test of the null hypothesis H_0: $\beta_0 = 0$ using the general linear hypothesis. What do you conclude from this test?

b. The second model used the four independent variables but forced the intercept to be zero.
   (i) Compute SS(Model), SS(Res), and the partial and sequential sums of squares for this model. Summarize the results in an analysis of variance table.
   (ii) Use the difference in SS(Res) between this model with no intercept and the previous model with an intercept to test H_0: $\beta_0 = 0$. Compare the result with that obtained under (iv) in part (a).
   (iii) Use tests of significance to determine which partial regression coefficients in this model are different from zero. What do these tests tell you in terms of simplifying the model?

c. The third model used the zero-intercept model and only X_1 and X_4.
   (i) Use the results from this model and the zero-intercept model in part (b) to test the composite null hypothesis that β_2 and β_3 are both zero.
   (ii) Use the general linear hypothesis to construct the test of the composite null hypothesis that β_2 and β_3 in the model in part (b) are both zero. Define $\boldsymbol{K}'$ and $\boldsymbol{m}$ for this hypothesis, compute Q, and complete the test of significance. Compare these two tests.

* 4.7. Use the data on annual catch of Gulf Menhaden, number of fishing vessels, and fishing pressure given in exercise 3.11.

a. Complete the analysis of variance for the regression of catch (Y) on fishing pressure (X_1) and number of vessels (X_2) with an intercept in the model. Determine the partial sums of squares for each independent variable. Estimate the standard errors for the regression coefficients and construct the Bonferroni confidence intervals for each using a joint confidence coefficient of 95%. Use the regression equation to *predict* the "catch" if number of vessels is limited to $X_2 = 70$ and fishing pressure is restricted to $X_1 = 400$. Compute the variance of this prediction and the 95% confidence interval estimate of the prediction.
b. Test the hypothesis that the variable "number of vessels" does not add significantly to the explanation of variation in "catch" provided by "fishing pressure" alone (use $\alpha = .05$). Test the hypothesis that "fishing pressure" does not add significantly to the explanation provided by "number of vessels" alone.
c. On the basis of the tests in part (b) would you keep both X_1 and X_2 in the model, or would you eliminate one from the model? If one should be eliminated, which would it be? Does the remaining variable make a significant contribution to explaining the variation in "catch"?
d. Suppose consideration is being given to controlling the annual catch by limiting either the number of fishing vessels or the total fishing pressure. What is your recommendation and why?

4.8. This exercise uses the data in exercise 3.14 relating $Y = \ln(\text{days survival})$ for colon cancer patients receiving supplemental ascorbate to sex (X_1) and age (X_2) of patient and ln(average survival of control group) (X_3).

a. Complete the analysis of variance for the model using all three variables plus an intercept. Compute the partial sum of squares for each independent variable using the formula $\hat{\beta}_j^2/c_{jj}$. Demonstrate that each is the same as the sum of squares obtained by computing Q for the general linear hypothesis that the corresponding β_j is zero. Compute the standard error for each regression coefficient and the 95% confidence interval estimates.
b. Does information on the length of survival of the control group (X_3) help explain the variation in Y? Support your answer with an appropriate test of significance.
c. Test the null hypothesis that "sex of patient" has no effect on survival beyond that accounted for by "age" and survival of the control group. Interpret the results.
d. Test the null hypothesis that "age of patient" has no effect on survival beyond that accounted for by "sex" and survival time of the control group. Interpret the results.
e. Test the composite hypothesis that $\beta_1 = \beta_2 = \beta_3 = 0$. From these results, what do you conclude about the effect of sex and age of patient on the mean survival time of patients in this study receiving supplemental ascorbate? With

the information available in these data, what would you use as the best estimate of the mean ln(days survival)?

* 4.9. The Lesser–Unsworth data (exercise 1.19) was used in exercise 3.9 to estimate a bivariate regression equation relating seed weight to cumulative solar radiation and level of ozone pollution. This exercise continues with the analysis of that model using the centered independent variables.

a. The more complex model used in exercise 3.9 included the independent variables cumulative solar radiation, ozone level, and the product of cumulative solar radiation and ozone level (plus an intercept).

(i) Construct the analysis of variance for this model showing sums of squares, degrees of freedom, and mean squares. What is the estimate of σ^2?

(ii) Compute the standard errors for each regression coefficient. Use a *joint* confidence coefficient of 90% and construct the Bonferroni confidence intervals for the four regression coefficients. Use the confidence intervals to draw conclusions about which regression coefficients are clearly different from zero.

(iii) Construct a test of the null hypothesis that the regression coefficient for the product term is zero (use $\alpha = .05$). Does your conclusion from this test agree with your conclusion based on the Bonferroni confidence intervals? Explain why they need not agree.

b. The simpler model in exercise 3.9 did not use the product term. Construct the analysis of variance for the model using only the two independent variables cumulative solar radiation and ozone level.

(i) Use the residual sums of squares from the two analyses to test the null hypothesis that the regression coefficient on the product term is zero (use $\alpha = .05$). Does your conclusion agree with that obtained in part (a)?

(ii) Compute the standard errors of the regression coefficients for this reduced model. Explain why they differ from those computed in part (a).

(iii) Compute the estimated mean seed weight for the mean level of cumulative solar radiation and .025 ppm ozone. Compute the estimated mean seed weight for the mean level of radiation and .07 ppm ozone. Use these two results to compute the estimated mean loss in seed weight if ozone changes from .025 to .07 ppm. Define a matrix of coefficients $\boldsymbol{K}'$ such that these three linear functions of $\hat{\boldsymbol{\beta}}$ can be written as $\boldsymbol{K}'\hat{\boldsymbol{\beta}}$. Use this matrix form to compute their variances and covariances.

(iv) Compute and plot the 90% *joint* confidence region for β_1 and β_2 *ignoring* β_0. (This joint confidence region will be an ellipse in the two dimensions β_1 and β_2.)

4.10. This is a continuation of exercise 3.10 using the data on number of hospital days for smokers from exercise 1.21. The dependent variable is $Y = \ln$(number of hospital days for smokers). The independent variables are $X_1 =$ (number of cigarettes)^2 and $X_2 = \ln$(number of hospital days for nonsmokers). Note that X_1 is the square of number of cigarettes.

a. Plot Y against number of cigarettes and against the square of number of cigarettes. Do the plots provide any indication of why the square of number of cigarettes was chosen as the independent variable?

b. Complete the analysis of variance for the regression of Y on X_1 and X_2. Does the information on number of hospital days for nonsmokers help explain the variation in number of hospital days for smokers? Make an appropriate test of significance to support your statement. Is Y, after adjustment for number of hospital days for nonsmokers, related to X_1? Make a test of significance to support your statement. Are you willing to conclude from these data that number of cigarettes smoked has a direct effect on the average number of hospital days?

c. It is logical in this problem to expect the number of hospital days for smokers to approach that of nonsmokers as the number of cigarettes smoked goes to zero. This implies that the intercept in this model might be expected to be zero. One might also expect β_2 to be equal to one. (Explain why.) Set up the general linear hypothesis for testing the composite null hypothesis that $\beta_0 = 0$ and $\beta_2 = 1.0$. Complete the test of significance and state your conclusions.

d. Construct the *reduced* model implied by the composite null hypothesis under part (c). Compute the regression for this reduced model, obtain the residual sum of squares, and use the difference in residual sums of squares for the full and reduced models to test the composite null hypothesis. Do you obtain the same result as in part (c)?

e. Based on the preceding tests of significance, decide which model you think is appropriate. State the regression equation for your adopted model. Include standard errors on the regression coefficients.

4.11. You are given the following matrices computed for a regression analysis:

$$\boldsymbol{X'X} = \begin{bmatrix} 9 & 136 & 269 & 260 \\ 136 & 2{,}114 & 4{,}176 & 3{,}583 \\ 269 & 4{,}176 & 8{,}257 & 7{,}104 \\ 260 & 3{,}583 & 7{,}104 & 12{,}276 \end{bmatrix} \qquad \boldsymbol{X'Y} = \begin{bmatrix} 45 \\ 648 \\ 1{,}283 \\ 1{,}821 \end{bmatrix}$$

$$(\boldsymbol{X'X})^{-1} = \begin{bmatrix} 9.610932 & .0085878 & -.2791475 & -.0445217 \\ .0085878 & .5099641 & -.2588636 & .0007765 \\ -.2791475 & -.2588636 & .1395 & .0007396 \\ -.0445217 & .0007765 & .0007396 & .0003698 \end{bmatrix}$$

$$(\boldsymbol{X'X})^{-1}(\boldsymbol{X'Y}) = \begin{bmatrix} -1.163461 \\ .135270 \\ .019950 \\ .121954 \end{bmatrix} \qquad \boldsymbol{Y'Y} = 285$$

a. Use these results to complete the analysis of variance table.

b. Give the computed regression equation and the standard errors of the regression coefficients.

c. Compare each estimated regression coefficient to its standard error and use the t-test to test the simple hypotheses that each regression coefficient is equal to zero. State your conclusions (use $\alpha = .05$).

d. Define $\boldsymbol{K'}$ and $\boldsymbol{m}$ for the composite hypothesis that $\beta_0 = 0$, $\beta_1 = \beta_3$, and $\beta_2 = 0$. Give the rank of $\boldsymbol{K'}$ and the degrees of freedom associated with this test.

e. Give the reduced model for the composite hypothesis in part (d).

*4.12. You are given the following *sequential* and *partial* sums of squares from a regression analysis:

$$R(\beta_3|\beta_0) = 56.9669 \qquad R(\beta_3|\beta_0\ \beta_1\ \beta_2) = 40.2204$$

$$R(\beta_1|\beta_0\ \beta_3) = 1.0027 \qquad R(\beta_1|\beta_0\ \beta_2\ \beta_3) = .0359$$

$$R(\beta_2|\beta_0\ \beta_1\ \beta_3) = .0029 \qquad R(\beta_2|\beta_0\ \beta_1\ \beta_3) = .0029$$

Each sequential and partial sum of squares can be used for the numerator of an F-test. Clearly state the null hypothesis being tested in each case.

4.13. A regression analysis using an intercept and one independent variable gave

$$\hat{Y}_i = 1.841246 + .10934X_{i1}$$

The variance–covariance matrix for $\hat{\boldsymbol{\beta}}$ was

$$s^2(\hat{\boldsymbol{\beta}}) = \begin{bmatrix} .1240363 & -.002627 \\ -.002627 & .0000909 \end{bmatrix}$$

a. Compute the 95% confidence interval estimate of $\hat{\beta}_1$. The estimate of σ^2 used to compute $s^2(\hat{\boldsymbol{\beta}})$ was $s^2 = 1.6360$, the residual mean square from the model using only X_0 and X_1. The data set had $n = 34$ observations.

b. Compute $\hat{Y}$ for $X_1 = 4$. Compute the variance of $\hat{Y}$ if it is being used to estimate the mean of Y when $X_1 = 4$. Compute the variance of $\hat{Y}$ if it is being used to *predict* the mean of a future observation at $X_1 = 4$.

*4.14. You are given the following matrix of simple (product moment) correlations among a dependent variable Y (first variable) and three independent variables:

$$\begin{bmatrix} 1.0 & -.538 & -.543 & .974 \\ -.538 & 1.0 & .983 & -.653 \\ -.543 & .983 & 1.0 & -.656 \\ .974 & -.653 & -.656 & 1.0 \end{bmatrix}$$

a. From inspection of the correlation matrix, which independent variable would account for the greatest variability in Y? What proportion of the corrected sum of squares in Y would be accounted for by this variable? If Y were regressed on all three independent variables (plus an intercept), would the coefficient of determination for the multiple regression be smaller or larger than this proportion?

b. Inspection of the three pairwise correlations among the X variables suggests that at least one of the independent variables will not be useful for the regression of Y on the X's. Explain exactly the basis for this statement and why it has this implication.

CHAPTER
FIVE

Case Study: Five Independent Variables

The last two chapters completed the presentation of the basic regression results for linear models with any number of variables.

This chapter demonstrates the application of least squares regression to a problem involving five independent variables. The full model is fit and then the model is simplified to a two-variable model that conveys most of the information on Y.

The basic steps in ordinary regression analysis have now been covered. This chapter illustrates the application of these methods. Computations and interpretations of the regression results will be emphasized.

5.1 Spartina Biomass Production in the Cape Fear Estuary

The data to be considered are part of a larger study conducted by Dr. Rick Linthurst at North Carolina State University as his Ph.D. thesis research.* The purpose of his research was to identify the important soil characteristics influencing aerial biomass production of the marsh grass *Spartina alterniflora* in the Cape Fear Estuary of North Carolina.

Design

One phase of Linthurst's research consisted of sampling three types of Spartina vegetation (revegetated "dead" areas, "short" Spartina areas, and "tall" Spartina areas) in each of three locations (Oak Island, Smith Island, and Snows Marsh). Samples of the soil substrate from five random sites within each location–vegetation type (giving 45 total samples) were analyzed for 14 soil physico-chemical characteristics each month for several months. In addition, above-ground biomass at each sample site was measured each month. The data used in this case study involve only the September sampling and the following five substrate measurements:

* Rick A. Linthurst (1979). Aeration, nitrogen, pH and salinity as factors affecting *Spartina alterniflora* growth and dieback. Ph.D. thesis. North Carolina State University.

X_1 = Salinity o/oo ($SALINITY$)

X_2 = Acidity as measured in water (pH)

X_3 = Potassium ppm (K)

X_4 = Sodium ppm (Na)

X_5 = Zinc ppm (Zn)

The dependent variable Y is aerial biomass, gm^{-2}. The data from the September sampling for these six variables are given in Table 5.1 (page 136).

Objective

The objective of this phase of the Linthurst research was to identify the substrate variables showing the stronger relationships to biomass. These variables would then be used in controlled studies to investigate causal relationships. The purpose of this case study is to use multiple linear regression to relate *total* variability in Spartina biomass production to total variability in the five substrate variables. For this analysis, total variation among vegetation types, locations, and samples within vegetation types and locations is being used. It is left as an exercise for the student to study separately the relationships shown by the variation *among* vegetation types and locations (using the location–vegetation type means) and the relationships shown by the variation among samples within location–vegetation type combinations.

5.2 Regression Analysis for the Full Model

Model

The initial model will assume that $BIOMASS$, Y, can be adequately characterized by linear relationships with the five independent variables plus an intercept. Thus, the linear model

$$Y = X\beta + \varepsilon \qquad [5.1]$$

is completely specified by defining Y, X, and β and stating the appropriate assumptions about the distribution of the random errors ε. Y is the vector of $BIOMASS$ measurements

$$Y' = (676 \;\; 516 \;\; \ldots \;\; 1{,}560)$$

X (45 × 6) consists of the column vector $\mathbf{1}$, the 45 × 1 column vector of ones, and the five column vectors of data for the substrate variables, $X_1 = SALINITY$, $X_2 = pH$, $X_3 = K$, $X_4 = Na$, and $X_5 = Zn$:

$$X = [\mathbf{1} \; X_1 \; X_2 \; X_3 \; X_4 \; X_5]$$

$$= \begin{bmatrix} 1 & 33 & 5.00 & 1{,}441.67 & 35{,}184.5 & 16.4524 \\ 1 & 35 & 4.75 & 1{,}299.19 & 28{,}170.4 & 13.9852 \\ \vdots & \vdots & \vdots & \vdots & \vdots & \vdots \\ 1 & 28 & 5.40 & 856.96 & 16{,}892.2 & 19.2420 \end{bmatrix} \qquad [5.2]$$

Table 5.1 Aerial biomass (*BIO*) and five physico-chemical properties of the substrate (salinity (*SAL*), *pH*, *K*, *Na*, and *Zn*) in the Cape Fear Estuary of North Carolina. (Data used with permission of Dr. R. A. Linthurst.)

Obs.	*Loc.*	*Type*	*BIO*	*SAL*	*pH*	*K*	*Na*	*Zn*
1	OI	DVEG	676	33	5.00	1,441.67	35,184.5	16.4524
2	OI	DVEG	516	35	4.75	1,299.19	28,170.4	13.9852
3	OI	DVEG	1,052	32	4.20	1,154.27	26,455.0	15.3276
4	OI	DVEG	868	30	4.40	1,045.15	25,072.9	17.3128
5	OI	DVEG	1,008	33	5.55	521.62	31,664.2	22.3312
6	OI	SHRT	436	33	5.05	1,273.02	25,491.7	12.2778
7	OI	SHRT	544	36	4.25	1,346.35	20,877.3	17.8225
8	OI	SHRT	680	30	4.45	1,253.88	25,621.3	14.3516
9	OI	SHRT	640	38	4.75	1,242.65	27,587.3	13.6826
10	OI	SHRT	492	30	4.60	1,282.95	26,511.7	11.7566
11	OI	TALL	984	30	4.10	553.69	7,886.5	9.8820
12	OI	TALL	1,400	37	3.45	494.74	14,596.0	16.6752
13	OI	TALL	1,276	33	3.45	526.97	9,826.8	12.3730
14	OI	TALL	1,736	36	4.10	571.14	11,978.4	9.4058
15	OI	TALL	1,004	30	3.50	408.64	10,368.6	14.9302
16	SI	DVEG	396	30	3.25	646.65	17,307.4	31.2865
17	SI	DVEG	352	27	3.35	514.03	12,822.0	30.1652
18	SI	DVEG	328	29	3.20	350.73	8,582.6	28.5901
19	SI	DVEG	392	34	3.35	496.29	12,369.5	19.8795
20	SI	DVEG	236	36	3.30	580.92	14,731.9	18.5056
21	SI	SHRT	392	30	3.25	535.82	15,060.6	22.1344
22	SI	SHRT	268	28	3.25	490.34	11,056.3	28.6101
23	SI	SHRT	252	31	3.20	552.39	8,118.9	23.1908
24	SI	SHRT	236	31	3.20	661.32	13,009.5	24.6917
25	SI	SHRT	340	35	3.35	672.15	15,003.7	22.6758
26	SI	TALL	2,436	29	7.10	525.65	10,225.0	.3729
27	SI	TALL	2,216	35	7.35	563.13	8,024.2	.2703
28	SI	TALL	2,096	35	7.45	497.96	10,393.0	.3205
29	SI	TALL	1,660	30	7.45	458.38	8,711.6	.2648
30	SI	TALL	2,272	30	7.40	498.25	10,239.6	.2105
31	SM	DVEG	824	26	4.85	936.26	20,436.0	18.9875
32	SM	DVEG	1,196	29	4.60	894.79	12,519.9	20.9687
33	SM	DVEG	1,960	25	5.20	941.36	18,979.0	23.9841
34	SM	DVEG	2,080	26	4.75	1,038.79	22,986.1	19.9727
35	SM	DVEG	1,764	26	5.20	898.05	11,704.5	21.3864
36	SM	SHRT	412	25	4.55	989.87	17,721.0	23.7063
37	SM	SHRT	416	26	3.95	951.28	16,485.2	30.5589
38	SM	SHRT	504	26	3.70	939.83	17,101.3	26.8415
39	SM	SHRT	492	27	3.75	925.42	17,849.0	27.7292
40	SM	SHRT	636	27	4.15	954.11	16,949.6	21.5699
41	SM	TALL	1,756	24	5.60	720.72	11,344.6	19.6531
42	SM	TALL	1,232	27	5.35	782.09	14,752.4	20.3295
43	SM	TALL	1,400	26	5.50	773.30	13,649.8	19.5880
44	SM	TALL	1,620	28	5.50	829.26	14,533.0	20.1328
45	SM	TALL	1,560	28	5.40	856.96	16,892.2	19.2420

The vector of parameters is

$$\boldsymbol{\beta}' = (\beta_0\ \beta_1\ \beta_2\ \beta_3\ \beta_4\ \beta_5) \qquad [5.3]$$

The random errors $\boldsymbol{\varepsilon}$ are assumed to be normally distributed, $\boldsymbol{\varepsilon} \sim N(\mathbf{0}, \boldsymbol{I}\sigma^2)$. The assumption that the variance–covariance matrix for $\boldsymbol{\varepsilon}$ is $\boldsymbol{I}\sigma^2$ contains the two assumptions of independence of the errors and common variance σ^2.

5.2.1 The Correlation Matrix

A useful starting point in any multiple regression analysis is to compute the matrix of correlations among all variables including the dependent variable. This provides a "first look" at the simple linear relationships among the variables. The correlation matrix is obtained by

$$\hat{\boldsymbol{\rho}} = \boldsymbol{S}[\boldsymbol{W}'(\boldsymbol{I} - \boldsymbol{J}/n)\boldsymbol{W}]\boldsymbol{S} \qquad [5.4]$$

where $\boldsymbol{W}$ is the (45×6) matrix of *BIOMASS* (Y) and the five independent variables, and $\boldsymbol{S}$ is a diagonal matrix of the reciprocals of the square roots of the corrected sums of squares of each variable. The corrected sums of squares are given by the diagonal elements of $\boldsymbol{W}'(\boldsymbol{I} - \boldsymbol{J}/n)\boldsymbol{W}$. For the Linthurst data,

$$\hat{\boldsymbol{\rho}} = \begin{array}{c} \begin{matrix} Y & SAL & pH & K & Na & Zn \end{matrix} \\ \begin{bmatrix} 1 & -.103 & .774 & -.205 & -.272 & -.624 \\ & 1 & -.051 & -.021 & .162 & -.421 \\ & & 1 & .019 & -.038 & -.722 \\ & & & 1 & .792 & .074 \\ & & & & 1 & .117 \\ & & \text{Symmetric} & & & 1 \end{bmatrix} \end{array}$$

The first row of $\hat{\boldsymbol{\rho}}$ contains the simple correlations of the dependent variable with each of the independent variables. The two variables *pH* and *Zn* have reasonably high correlations with *BIOMASS*. They would "account for" 60% ($r^2 = .774^2$) and 39%, respectively, of the variation in *BIOMASS* if used separately as the only independent variable in the regressions. *Na* and *K* are about equally correlated with *BIOMASS* but at a much lower level than *pH* and *Zn*. There appears to be almost no correlation between *SALINITY* and *BIOMASS*.

There are two high correlations among the independent variables, *K* and *Na* with $r = .79$ and *pH* and *Zn* at $r = -.72$. The impact of these correlations on the regression results will be noted as the analysis proceeds. With the exception of a moderate correlation between *SALINITY* and *Zn*, all other correlations are quite small.

5.2.2 Multiple Regression Results: The Full Model

Summary of Results

The results of the multiple regression analysis using all five independent variables are summarized in Table 5.2. There is a strong relationship between *BIOMASS* and the independent variables. The coefficient of determination, R^2, is .677, or 68% of the sum of squares in *BIOMASS* can be associated with the variation in these five independent variables. The test of the composite hypothesis that all five regression coefficients are zero is highly significant; $F = 16.37$ compared to $F_{(.01,5,39)} = 3.53$.

Computations

The computations for this analysis were done using a matrix algebra computer program (SAS/IML, 1985a) operating on the X and Y matrices only. The steps in the language of SAS/IML and an explanation of each step are given in Table 5.3. The simplicity of matrix arithmetic can be appreciated only if one attempts to do the analysis with, say, a hand calculator. Obtaining $(\boldsymbol{X}'\boldsymbol{X})^{-1}$ is the most difficult and requires the use of a computer for all but the simplest problems. Most of the other computations are relatively easy. Notice that the large 45×45 $\boldsymbol{P}$ matrix is not computed and generally is not needed in its entirety. The $\hat{\boldsymbol{Y}}$ vector is more easily computed as $\hat{\boldsymbol{Y}} = \boldsymbol{X}\hat{\boldsymbol{\beta}}$, rather than as $\hat{\boldsymbol{Y}} = \boldsymbol{PY}$. The only need for $\boldsymbol{P}$ is for $\mathbf{Var}(\hat{\boldsymbol{Y}}) = \boldsymbol{P}\sigma^2$ and $\mathbf{Var}(\boldsymbol{e}) = (\boldsymbol{I} - \boldsymbol{P})\sigma^2$. Even then, the variance of an individual $\hat{Y}_i$ or e_i of interest can be computed using only the ith row of X, rather than the entire X matrix.

Residual Mean Square

The residual mean square, $s^2 = 158{,}622$ with 39 degrees of freedom, is an unbiased estimate of σ^2 if this five-variable model is the correct model. Of course, this is almost certainly not the correct model because (1) important variables may have been excluded, or (2) the mathematical form of the model may not be correct. (Including *unimportant* variables generally will not bias the estimate of σ^2.) Therefore, s^2 must be regarded as the tentative "best" estimate of σ^2 and will be used for tests of significance and for computing the standard errors of the estimates.

Inconsistencies in the Results

The regression of *BIOMASS* on these five independent variables is highly significant. Yet, only one partial regression coefficient, $\hat{\beta}_2$ for *pH*, is significantly different from zero, with $t = 3.48$. Recall that the simple correlation between

Table 5.2 Results of the regression of *BIOMASS* on the five independent variables *SALINITY, pH, K, Na,* and *Zn* (Linthurst September data).

Variable	$\hat{\beta}_j$	$s(\hat{\beta}_j)$	t	*Partial* SS
SAL	−30.285	24.031	−1.26	251,921
pH	305.525	87.879	3.48	1,917,306
K	−.2851	.3484	−.82	106,211
Na	−.0087	.0159	−.54	47,011
Zn	−20.676	15.055	−1.37	299,209

Analysis of variance for *BIOMASS*.

Source	d.f.	*Sum of Squares*	*Mean Square*	
Total	44	19,170,963		
Regression	5	12,984,700	2,596,940	$F = 16.37$
Residual	39	6,186,263	158,622	

Table 5.3 The matrix algebra steps for the regression analysis results. It is assumed that Y and X have been properly defined in the matrix program.

Program Step[a]	*Matrix Being Computed*
INVX=INV(X`*X);	$(X'X)^{-1}$
B=INVX*X`*Y;	$\hat{\beta}$
CF=SUM(Y)# #2/NROW(X);	$Y'(J/n)Y = (\sum Y)^2/n$, the correction factor. The "# #2" squares SUM(Y).
SST=Y`*Y−CF;	$Y'(I - J/n)Y$, corrected sum of squares for *BIOMASS*.
SSR=B`*X`*Y−CF;	$Y'(P - J/n)Y =$ SS(Regr). Notice that P need not be computed.
SSE=SST−SSR;	$Y'(I - P)Y =$ SS(Res)
S2=SSE/(NROW(X)−NCOL(X));	s^2, the estimate of σ^2, with degrees of freedom $= n - r(X)$
SEB=SQRT(VECDIAG(INVX)*S2);	Vector of standard errors of $\hat{\beta}$. "VECDIAG" creates a vector from the diagonal elements of **INVX**.
T=B/SEB;	Vector of Student's t for $H_0\colon \beta_j = 0$. "/" indicates elementwise division of **B** by **SEB**.
PART=B# #2/VECDIAG(INVX);	Vector of partial sums of squares
YHAT=X*B;	$\hat{Y}$, estimated means for Y
E=Y−YHAT;	e, estimated residuals

[a] Program steps for SAS/IML (1985a), an interactive matrix language program developed by SAS Institute, Inc., Cary, North Carolina.

BIOMASS and *pH* showed that *pH alone* would account for 60%, or 11.5 million, of the total corrected sum of squares for *BIOMASS*. When *pH* is used in a model with the other four variables, however, its partial sum of squares, 1,917,306, is only 10% of the total sum of squares and less than 15% of the regression sum of squares for all five variables. On the other hand, the partial sum of squares for *SALINITY* is larger than the simple correlation between *BIOMASS* and *SALINITY* would suggest.

These apparent inconsistencies are typical of regression results when the independent variables are not orthogonal. They are not inconsistencies if the meaning of the word "partial" in partial regression coefficients and partial sums of squares is remembered. "Partial" indicates that the regression coefficient or the sum of squares is the contribution of that particular independent variable *after* taking into account the effects of all other independent variables. Only when an independent variable is orthogonal to *all* other independent variables are its simple and partial regression coefficients and its simple and partial sums of squares equal.

5.3 Simplifying the Model

Removing Variables

The t-tests of the partial regression coefficients, $H_0\colon \beta_j = 0$, would seem to suggest that four of the five independent variables are unimportant and could be dropped from the model. The dependence of the partial regression coefficients and sums of squares on the other variables in the model, however, means that one must be cautious in removing more than one variable at a time from the regression model. Removing one variable from the model will cause the regression coefficients and the partial sums of squares for the remaining variables to change (unless they are orthogonal to the variable dropped). The results thus far do indicate that not all five independent variables are needed in the model. It would appear that any one of the four, *SALINITY*, *pH*, *Na*, or *K*, could be dropped without causing a significant decline in predictability of *BIOMASS*. It is not clear at this stage of the analysis, however, that more than one can be dropped.

There are several approaches for deciding which variables to include in the final model. These will be studied in Chapter 7. For this example, one variable at a time will be eliminated—the one whose elimination will cause the smallest increase in the residual sum of squares. The process will stop when the partial sums of squares for all variables remaining in the model are significant ($\alpha = .05$). As will be discussed in Chapter 7, data-driven variable selection and multiple testing to arrive at the final model alter the true significance levels; probability levels and confidence intervals should be used with caution.

A 4-Variable Model

The variable *Na* has the smallest partial sum of squares in the five-variable model. This means that *Na* is the least important of the five variables in accounting for the variability in *BIOMASS* *after* the contributions of the other four variables have been taken into account. As a result, *Na* is the logical variable to eliminate first. And, since the partial sum of squares for *Na*, $R(\beta_4|\beta_1\ \beta_2\ \beta_3\ \beta_5\ \beta_0) = 47{,}011$, is not significant, there is no reason $X_4 = Na$ should not be eliminated.

Dropping *Na* means that $\boldsymbol{X}$ must be redefined to be the 45×5 matrix consisting of $\mathbf{1}$, $\boldsymbol{X}_1 = SALINITY$, $\boldsymbol{X}_2 = pH$, $\boldsymbol{X}_3 = K$, and $\boldsymbol{X}_5 = Zn$; the column vector of *Na* observations, $\boldsymbol{X}_4$, is removed from $\boldsymbol{X}$. Similarly, $\boldsymbol{\beta}$ must be redefined by removing β_4. The regression analysis using these four variables (Table 5.4) shows the decrease in the regression sum of squares, now with 4 degrees of freedom, and the increase in the residual sum of squares to be exactly equal to the partial sum of squares for *Na* in the previous stage. This demonstrates the

Table 5.4 Results of the regression of *BIOMASS* on the four variables *SALINITY*, *pH*, *K*, and *Zn* (Linthurst data).

Variable	$\hat{\beta}_j$	$s(\hat{\beta}_j)$	t	*Partial* SS
SAL	−35.94	21.48	−1.67	436,496
pH	293.9	84.5	3.48	1,885,805
K	−.439	.202	−2.17	732,606
Zn	−23.45	14.04	−1.67	434,796

Analysis of variance.

Source	d.f.	SS	MS
Total	44	19,170,963	
Regression	4	12,937,689	3,234,422
Residual	40	6,233,274	155,832

meaning of "partial sum of squares." In the absence of independent information on σ^2, the residual mean square from this reduced model will now be used (tentatively) as the point estimate of σ^2, $s^2 = 155{,}832$. (Notice that the increase in the residual sum of squares does not necessarily imply an increase in the residual mean square.)

A 3-Variable Model

The partial sums of squares at the four-variable stage (Table 5.4) show *SALINITY* and *Zn* to be equally unimportant to the model; the partial sum of squares for *Zn* is slightly smaller and both are nonsignificant. The next step in the search for the final model is to eliminate one of these two variables. Again, it is not safe to assume that both variables can be dropped because they are not orthogonal.

Since *Zn* has the slightly smaller partial sum of squares, *Zn* will be eliminated and *pH*, *SALINITY*, and *K* retained as the three-variable model. One could have used the much higher simple correlation between *Zn* and *BIOMASS*, $r = -.62$ versus $r = -.10$, to argue that *SALINITY* is the variable to eliminate at this stage. This is a somewhat arbitrary choice with the information at hand, and illustrates one of the problems of this sequential method of searching for the appropriate model. There is no assurance that choosing to eliminate *Zn* first will lead to the best model by whatever criterion is used to measure "goodness" of the model.

Again, $\mathbf{X}$ and $\boldsymbol{\beta}$ are redefined, so that *Zn* is eliminated, and the computations repeated. This analysis gives the results in Table 5.5. The partial sum of squares for *pH* increases dramatically when *Zn* is dropped from the model, from 1.9 million to 11.5 million. This is due to the strong correlation between *pH* and *Zn* ($r = -.72$). When two independent variables are highly correlated, either positively or negatively, much of the predictive information contained in either can be usurped by the other. Thus, a very important variable may appear as insignificant if the model contains a correlated variable and, conversely, an otherwise unimportant variable may take on false significance.

A 2-Variable Model

The contribution of *SALINITY* in the three-variable model is even smaller than it was before *Zn* was dropped and is far from being significant. The next step is to drop *SALINITY* from the model. In this particular example, one would

Table 5.5 Results of the regression of *BIOMASS* on the three variables *SALINITY*, *pH*, and *K* (Linthurst data).

Variable	$\hat{\beta}_j$	$s(\hat{\beta}_j)$	t	*Partial* SS
SAL	−12.06	16.37	−.74	88,239
pH	410.21	48.83	8.40	11,478,835
K	−.490	.204	−2.40	935,178

Analysis of variance.

Source	d.f.	SS	MS
Total	44	19,170,963	
Regression	3	12,502,893	4,167,631
Residual	41	6,668,070	162,636

Table 5.6 Results of the regression of *BIOMASS* on the two variables *pH* and *K* (Linthurst data).

Variable	$\hat{\beta}_j$	$s(\hat{\beta}_j)$	t	*Partial* SS
pH	412.04	48.50	8.50	11,611,782
K	−.487	.203	−2.40	924,266

Analysis of variance.

Source	d.f.	SS	MS
Total	44	19,170,963	
Regression	2	12,414,654	6,207,327
Residual	42	6,756,309	160,865

not have been misled by eliminating both *SALINITY* and *Zn* at the previous step. This is not true in general.

The two-variable model containing *pH* and *K* gives the results in Table 5.6. Since the partial sums of squares for both *pH* and *K* are significant, the simplification of the model will stop with this two-variable model. The degree to which the linear model consisting of the two variables *pH* and *K* accounts for the variability in *BIOMASS* is $R^2 = .65$, only slightly smaller than the $R^2 = .68$ obtained with the original five-variable model.

5.4 Results of the Final Model

The Equation

This particular method of searching for an appropriate model led to the two-variable model consisting of *pH* and *K*. The regression equation is

$$\hat{Y}_i = -507.0 + 412.0X_{i2} - .4871X_{i3} \qquad [5.5]$$

or, expressed in terms of the centered variables,

$$\hat{Y}_i = 1{,}000.8 + 412.0(X_{i2} - 4.60) - .4871(X_{i3} - 797.62)$$

where $X_2 = pH$ and $X_3 = K$. This equation accounts for 65% of the variation in the observed values of aerial *BIOMASS*. That is, the predicted values computed from $\hat{\boldsymbol{Y}} = \boldsymbol{X}\hat{\boldsymbol{\beta}}$ account for 65% of the variation of *BIOMASS* or, conversely, the sum of squares of the residuals, $\boldsymbol{e}'\boldsymbol{e}$, is 35% of the original corrected sum of squares of *BIOMASS*. The square root of R^2 is the simple correlation between *BIOMASS* and $\hat{Y}$:

$$r(Y, \hat{Y}) = \sqrt{.65} = .80$$

$s^2(\hat{\beta})$

The estimate of σ^2 from this final model is $s^2 = 160{,}865$ with $(n - p') = 42$ degrees of freedom. The variance–covariance matrix for the regression coefficients is

$$s^2(\hat{\boldsymbol{\beta}}) = (\boldsymbol{X}'\boldsymbol{X})^{-1}s^2$$

$$= \begin{bmatrix} .4865711 & -.0663498 & -.0001993 \\ & .0146211 & -.0000012 \\ \text{Symmetric} & & .00000026 \end{bmatrix} (160{,}865)$$

$$= \begin{bmatrix} 78{,}272 & -10{,}673 & -32.0656 \\ & 2{,}352.0 & -.18950 \\ \text{Symmetric} & & .04129 \end{bmatrix}$$

The square roots of the diagonal elements give the standard errors of the estimated regression coefficients in the order in which they are listed in $\boldsymbol{\beta}$. In this model,

$$\boldsymbol{\beta}' = (\beta_0 \;\; \beta_2 \;\; \beta_3)$$

Thus, the standard errors of the estimated regression coefficients are

$$\begin{aligned} s(\hat{\beta}_0) &= \sqrt{78{,}272} = 280 \\ s(\hat{\beta}_2) &= \sqrt{2{,}352.0} = 48.50 \\ s(\hat{\beta}_3) &= \sqrt{.04129} = .2032 \end{aligned} \qquad [5.6]$$

The regression coefficients for pH and K are significantly different from zero as shown by the t-test (Table 5.6). The critical value of Student's t is $t_{(.05,42)} = 2.018$. (The intercept $\hat{\beta}_0 = -507.0$ is not significantly different from zero, $t = -1.81$, and if one had reason to believe that β_0 should be zero the intercept could be dropped from the model.)

Univariate Confidence Intervals

The **univariate 95% confidence interval estimates** of the regression coefficients (Section 4.6.1),

$$\hat{\beta}_j \pm t_{(.05,42)}s(\hat{\beta}_j),$$

are

$$\begin{aligned} -1{,}072 < \beta_0 &< 58 \\ 314 < \beta_2 &< 510 \\ -.898 < \beta_3 &< -.077 \end{aligned}$$

The value of Student's t for these intervals is $t_{(.05,42)} = 2.018$. The confidence coefficient of .95 applies to each interval statement.

Bonferroni Confidence Intervals

The **Bonferroni confidence intervals** (Section 4.6.2), using a joint confidence coefficient of .95, are

$$-1{,}206 < \beta_0 < 192$$

$$291 < \beta_2 < 533$$

$$-.995 < \beta_3 < .021$$

The joint confidence coefficient of $1 - \alpha^*$ is obtained by using the value of Student's t for $\alpha = \alpha^*/p'$: $t_{(.05/3,42)} = 2.50$. The Bonferroni intervals are necessarily wider than the univariate confidence intervals to allow for the fact that the confidence coefficient of .95 applies to the statement that all three intervals contain their true regression coefficients. In this example, the Bonferroni interval for β_3 overlaps zero, whereas the univariate 95% confidence interval did not.

Joint Confidence Region

The **95% joint confidence region** for the three regression coefficients is determined from the quadratic inequality shown in equation 4.54 (Section 4.6.3). The resulting three-dimensional ellipsoid is illustrated schematically in Figure 5.1. The intersection of the Bonferroni confidence intervals is shown as the box in Figure 5.1. The Bonferroni region includes large areas that are excluded by the joint confidence region. Many of the parameter points in the box are far outside

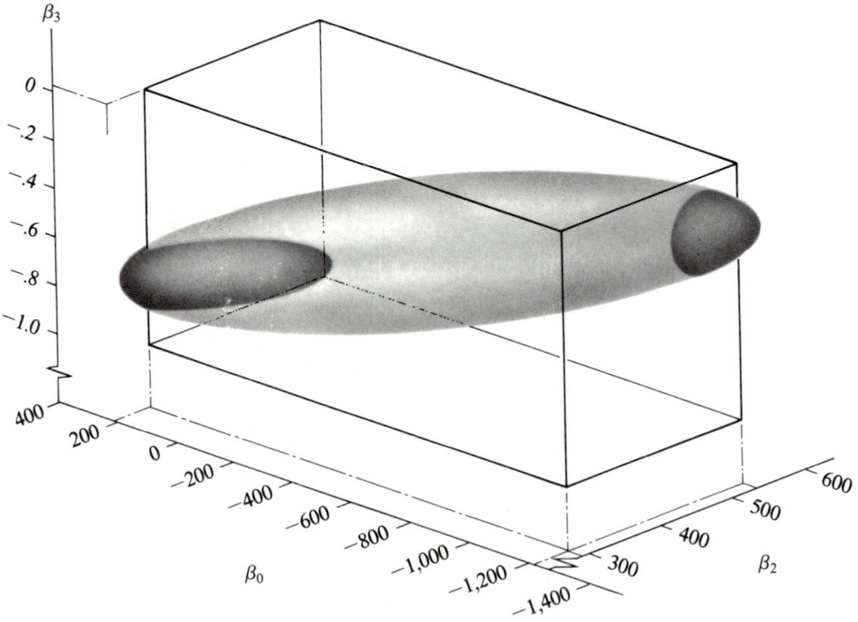

Figure 5.1 Three-dimensional 95% joint confidence region for β_0, β_2, and β_3. The intersection of the Bonferroni confidence intervals is shown as the box.

the ellipsoid and, therefore, would be very unlikely parameter combinations. Clearly, any of the univariate confidence intervals can be misleading if interpreted in terms of combinations of parameter values that might be acceptable.

The general shape of the confidence region can be seen from the three-dimensional figure. However, it is very difficult to read the parameter values corresponding to any particular point in the figure. Furthermore, the joint confidence ellipsoid for more than three parameters cannot be pictured.

A more useful representation of the joint confidence region is obtained by plotting two-dimensional "slices" through the ellipsoid for pairs of parameters of particular interest. This is done by evaluating the joint confidence equation at specific values of the other parameters. Three two-dimensional ellipses for β_2 and β_3 are shown in Figure 5.2. These are obtained from the ellipsoid in Figure 5.1 by determining the boundaries for β_2 and β_3 at β_0 equal to $-1{,}066$, -507, and 53, respectively. These choices for β_0 correspond to $\hat{\beta}_0$ and $\hat{\beta}_0 \pm 2s(\hat{\beta}_0)$. These slices help picture the three-dimensional ellipsoid but they are not to be interpreted individually as joint confidence regions for β_2 and β_3.

Alternatively, one can determine the two-dimensional 95% joint confidence region for β_2 and β_3 *ignoring* β_0. This region is also shown in Figure 5.2 as the ellipse with the dashed line. In this case, $\hat{\beta}_2$ and $\hat{\beta}_3$ are only slightly negatively correlated so that the two-dimensional joint confidence region is only slightly

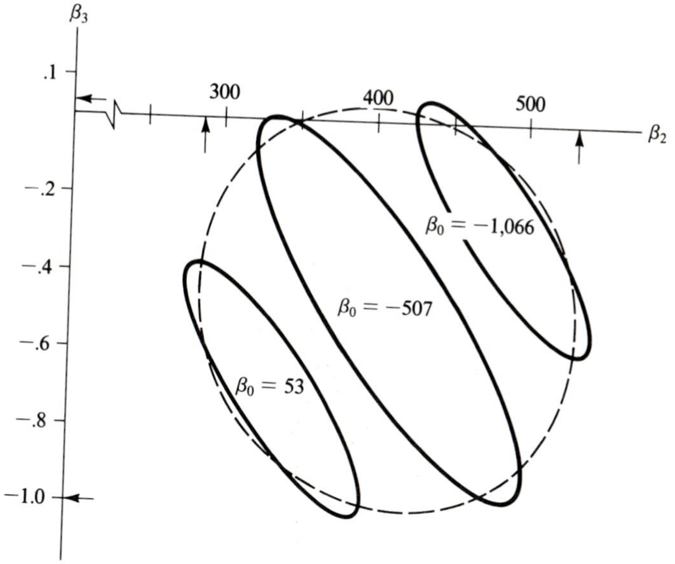

Figure 5.2 Two-dimensional slices, corresponding to three values of β_0, of the 95% three-dimensional joint confidence region and the joint confidence region for β_2 and β_3 *ignoring* β_0 (shown in dashed line). The arrows indicate limits of the Bonferroni joint confidence intervals for β_2 and β_3, obtained using a joint confidence coefficient of .95 for the three parameters in the model.

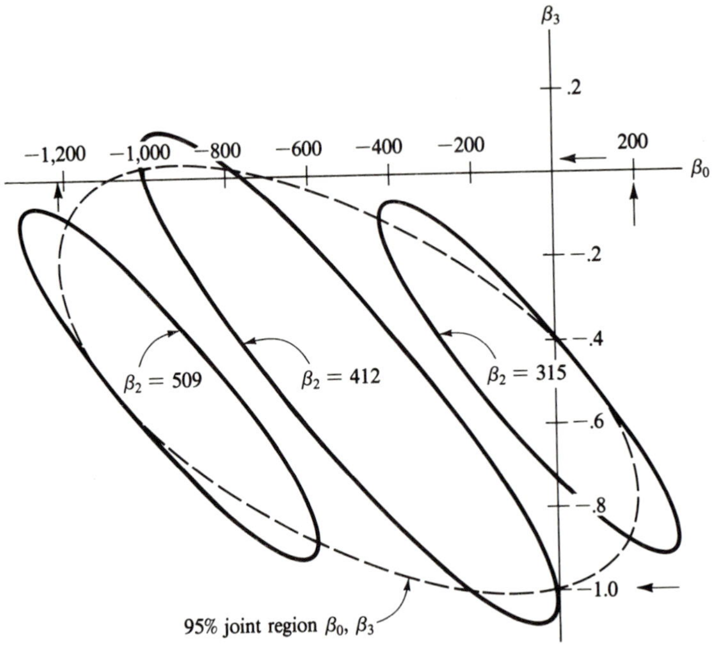

Figure 5.3 Two-dimensional slices of the joint confidence region for three values of β_2 and the joint confidence region for β_0 and β_3 ignoring β_2 (shown in dashed line). The arrows indicate the limits of the intersection of the Bonferroni confidence intervals for β_0 and β_3.

elliptical. However, the very elliptical slices from the original joint confidence region show that the choice of β_2 and β_3 for a given value of β_0 are more restricted than the two-dimensional joint confidence region would lead one to believe. This illustrates the information obscured by confidence intervals or regions that do not take into account the joint distribution of the parameter estimates. The limits on the intersection of the Bonferroni confidence intervals for β_2 and β_3, computed using a joint confidence coefficient of .95 for the three parameters in the model, are shown by the arrows on the axes.

Two-dimensional slices through the joint confidence region in another direction, for given values of β_2, and the two-dimensional confidence region for β_0 and β_3 ignoring β_2 are shown in Figure 5.3. The strong negative correlation between $\hat{\beta}_0$ and $\hat{\beta}_3$ is evident in the two-dimensional joint confidence region and the slices from the three-dimensional region. Again, it is clear that reasonable combinations of β_0 and β_3 are dependent on the assumed value of β_2, a result that is not evident from the two-dimensional joint confidence region ignoring β_2. The limits on the intersection of the Bonferroni confidence intervals for β_0 and β_3 are shown by the arrows on the axes.

$\hat{Y}_1$ and $s^2(\hat{Y}_1)$

$\hat{\boldsymbol{Y}}$ and $\boldsymbol{e}$ for this example are not given. They are easily computed as shown in Table 5.2. Likewise, $s^2(\hat{\boldsymbol{Y}}) = \boldsymbol{P}s^2$ and $s^2(\boldsymbol{e}) = (\boldsymbol{I} - \boldsymbol{P})s^2$ are not given; each is a 45 × 45 matrix. Computation of $\hat{Y}_i$ and its estimation and prediction variances will be illustrated using the first data point. Each $\hat{Y}_i$ is computed using the corresponding row vector from $\boldsymbol{X}$, which is designated $\boldsymbol{x}_i'$. For the first observation,

$$\boldsymbol{x}_1' = (1 \quad 5.00 \quad 1{,}441.67)$$

Thus,

$$\hat{Y}_1 = \boldsymbol{x}_1'\hat{\boldsymbol{\beta}}$$

$$= (1 \quad 5.00 \quad 1{,}441.67)\begin{pmatrix} -506.9774 \\ 412.0392 \\ -.4871 \end{pmatrix} = 850.99$$

The variance of $\hat{Y}_1$ depends on whether it is to be used as an estimate of the mean aerial *BIOMASS* at this specific level of pH (X_2) and K (X_3) or as a prediction of a future observation. As an estimate of the mean, $s^2(\hat{Y}_1) = v_{11}s^2$, where v_{11} is the first diagonal element from $\boldsymbol{P}$. The ith diagonal element of $\boldsymbol{P}$ can be obtained individually as $v_{ii} = \boldsymbol{x}_i'(\boldsymbol{X}'\boldsymbol{X})^{-1}\boldsymbol{x}_i$. Or, the variance for any one $\hat{Y}_i$ is obtained as the variance of a linear function of $\hat{\boldsymbol{\beta}}$. Thus,

$$s^2(\hat{Y}_1) = \boldsymbol{x}_1'[s^2(\hat{\boldsymbol{\beta}})]\boldsymbol{x}_1$$

$$= (1 \quad 5.00 \quad 1{,}441.67)\begin{bmatrix} 78{,}272 & -10{,}673 & -32.0656 \\ & 2{,}352.0 & -.18950 \\ \text{Symmetric} & & .04129 \end{bmatrix}\begin{pmatrix} 1 \\ 5.00 \\ 1{,}441.67 \end{pmatrix}$$

$$= 20{,}978.78$$

Its standard error is

$$s(\hat{Y}_1) = \sqrt{20{,}978.78} = 144.8$$

If used as a prediction, the variance of $\hat{Y}_1$ is increased by $s^2 = 160{,}865$ to account for the variability of the random variable being predicted. This gives

$$s^2(\hat{Y}_{\text{pred}_1}) = 20{,}979 + 160{,}865 = 181{,}843$$

or the standard error of prediction is

$$s(\hat{Y}_{\text{pred}_1}) = \sqrt{181{,}843} = 426.4$$

e_1 and $s^2(e_1)$

The residual for the first observation is

$$e_1 = Y_1 - \hat{Y}_1 = 676 - 850.99 = -174.99$$

The estimated variance of e_1 is

$$s^2(e_1) = (1 - v_{11})s^2$$

or, since $s^2(\hat{Y}_1) = v_{11}s^2$ has already been computed, $s^2(e_1)$ is easily obtained as

$$s^2(e_1) = s^2 - s^2(\hat{Y}_1)$$

$$= 160{,}865 - 20{,}979 = 139{,}886$$

The standard error is

$$s(e_1) = \sqrt{139{,}886} = 374.0$$

Confidence Intervals on $\mathscr{E}(\hat{Y}_i)$

These variances are used to compute confidence interval estimates of each of the corresponding parameters. Student's t has 42 degrees of freedom, the degrees of freedom in the estimate of σ^2. For illustration, the 95% confidence interval estimate of the mean *BIOMASS* production when $pH = 5.00$ and $K = 1{,}441.67$ ppm, $\mathscr{E}(Y_1)$, is

$$\hat{Y}_1 \pm t_{(.05,42)}s(\hat{Y}_1)$$

or

$$850.99 \pm (2.018)(144.8)$$

which gives

$$558.7 < \mathscr{E}(\hat{Y}_1) < 1{,}143.3$$

as the limits. Thus, these results indicate that, with 95% confidence, the true mean *BIOMASS* for $pH = 5.00$ and $K = 1{,}441.67$ is between 559 and 1,143 gm^{-2}.

The example will stop at this point. A complete analysis would include plots of regression results to verify that the regression equation gives a reasonable characterization of the observed data and that the residuals are behaving as they should. Such an extended analysis, however, would require topics not yet discussed.

5.5 General Comments

The original objective of the Linthurst research was to identify important soil variables that were influencing the amount of *BIOMASS* production in the marshes. The wording of this objective implies that the desire is to establish causal links.

Cannot Establish Causality

Observational data *cannot* be used to establish causal relationships. Any analysis of observational data must build on the observed relationships, or the correlational structure, in the sample data. There are many reasons why correlations might exist in any set of data, only one of which is a causal pathway involving the variables. Some of the correlations observed will be fortuitous, accidents of the sampling of random variables. This is particularly likely if small numbers of observations are taken or if the sample points are not random. Some of the correlations will result from accidents of nature or from the variables being causally related to other unmeasured variables which, in turn, are causally related to the dependent variable. Even if the linkage between an independent and dependent variable is causal in origin, the direction of the causal pathway cannot be established from the observational data alone. The only way causality can be established is in controlled experiments where the causal variable is changed and the impact on the response variable is observed.

Thus, it is incorrect in this case study to conclude that pH and K are important *causal* variables in $BIOMASS$ production. The least squares analysis has established only that variation in $BIOMASS$ is associated with variation in pH and K. The *reason* for the association is not established. Furthermore, there is no assurance that this analysis has identified all of the variables which show significant association with $BIOMASS$. The reasonably high correlation between pH and Zn, for example, has caused the regression analysis to eliminate Zn from the model; the partial sum of squares for Zn is nonsignificant after adjustment for pH. This sequential method of building the model may have eliminated an important causal variable.

Interpreting the Regression Equation

Another common purpose of least squares is to develop prediction equations for the behavior of the dependent variable. Observational data are frequently the source of information for this purpose. Even here, care must be used in interpreting the results. The results from this case study predict that, on the average, $BIOMASS$ production changes by 412 gm^{-2} for each unit change in pH and $-.5$ gm^{-2} for each ppm change in K. This prediction is appropriate for the population being sampled by this set of data, the marshes in the Cape Fear Estuary of North Carolina. It is not appropriate if the population has been changed by some event nor is it appropriate for points outside the population represented by the sample.

The regression coefficient for pH gives the expected change in $BIOMASS$ per unit change in pH. This statement treats the other variables in the system two different ways, depending on whether they are included in the prediction equation. The predicted change in $BIOMASS$ per unit change in pH ignores all variables *not included* in the final prediction equation. This means that any change in pH, for which a prediction is being made, will be accompanied by simultaneous changes in these ignored variables. The nature of these changes will be controlled by the correlational structure of the data. For example, Zn would be expected to decrease on the average as pH is increased due to the negative correlation between the two variables. Thus, this predicted change in $BIOMASS$ is really associated with the *simultaneous* increase in pH *and* decrease in *Zn*. It is incorrect to think the prediction is for a situation where, somehow, *Zn* is not allowed to change.

On the other hand, the predicted change of 412 gm^{-2} *BIOMASS* associated with a unit change in *pH* assumes that the other variables *included* in the prediction equation, in this case *K*, are being held constant. Again, this is unrealistic when the variables in the regression equation are correlated.

The appropriate view of the regression equation obtained from observational data is as a description of the response surface of the dependent variable, where the independent variables in the equation are serving as surrogates for the many variables that have been omitted from the equation. The partial regression coefficients are the slopes of the response surface in the directions represented by the corresponding independent variables. Any attempt to ascribe these slopes, or changes, to the particular independent variables in the model implicitly assumes a causal relationship of the independent variable to the dependent variable and that all other variables in the system, for which the variables in the equation serve as surrogates, are unimportant in the process.

The response surface equation obtained from observational data can serve as a useful prediction equation as long as care is taken to ensure that the points for which predictions are to be made are valid points in the sampled population. This requires that the values of the independent variables for the prediction points must be in the sample space. It is easy, for example when one variable at a time is being changed, to create prediction points that are outside the sample space. Predictions for these points can be very much in error.

EXERCISES

The accompanying table provides simulated data on peak rate of flow Q (cfs) of water from six watersheds following storm episodes. The storm episodes have been chosen from a larger data set to give a range of storm intensities. The independent variables are

X_1 = Area of watershed (mi^2)

X_2 = Area impervious to water (mi^2)

X_3 = Average slope of watershed (in percent)

X_4 = Longest stream flow in watershed (in thousands of feet)

X_5 = Surface absorbency index; 0 = complete absorbency, 100 = no absorbency

X_6 = Estimated soil storage capacity (inches of water)

X_7 = Infiltration rate of water into soil (inches/hour)

X_8 = Rainfall (inches)

X_9 = Time period during which rainfall exceeded $\frac{1}{4}$ inch/hour.

Computations with this set of data will require a computer.

X_1	X_2	X_3	X_4	X_5	X_6	X_7	X_8	X_9	Q
.03	.006	3.0	1	70	1.5	.25	1.75	2.0	46
.03	.006	3.0	1	70	1.5	.25	2.25	3.7	28
.03	.006	3.0	1	70	1.5	.25	4.00	4.2	54
.03	.021	3.0	1	80	1.0	.25	1.60	1.5	70
.03	.021	3.0	1	80	1.0	.25	3.10	4.0	47
.03	.021	3.0	1	80	1.0	.25	3.60	2.4	112
.13	.005	6.5	2	65	2.0	.35	1.25	.7	398
.13	.005	6.5	2	65	2.0	.35	2.30	3.5	98
.13	.005	6.5	2	65	2.0	.35	4.25	4.0	191
.13	.008	6.5	2	68	.5	.15	1.45	2.0	171
.13	.008	6.5	2	68	.5	.15	2.60	4.0	150
.13	.008	6.5	2	68	.5	.15	3.90	3.0	331
1.00	.023	15.0	10	60	1.0	.20	.75	1.0	772
1.00	.023	15.0	10	60	1.0	.20	1.75	1.5	1,268
1.00	.023	15.0	10	60	1.0	.20	3.25	4.0	849
1.00	.023	15.0	10	65	2.0	.20	1.80	1.0	2,294
1.00	.023	15.0	10	65	2.0	.20	3.10	2.0	1,984
1.00	.023	15.0	10	65	2.0	.20	4.75	6.0	900
3.00	.039	7.0	15	67	.5	.50	1.75	2.0	2,181
3.00	.039	7.0	15	67	.5	.50	3.25	4.0	2,484
3.00	.039	7.0	15	67	.5	.50	5.00	6.5	2,450
5.00	.109	6.0	15	62	1.5	.60	1.50	1.5	1,794
5.00	.109	6.0	15	62	1.5	.60	2.75	3.0	2,067
5.00	.109	6.0	15	62	1.5	.60	4.20	5.0	2,586
7.00	.055	6.5	19	56	2.0	.50	1.80	2.0	2,410
7.00	.055	6.5	19	56	2.0	.50	3.25	4.0	1,808
7.00	.055	6.5	19	56	2.0	.50	5.25	6.0	3,024
7.00	.063	6.5	19	56	1.0	.50	1.25	2.0	710
7.00	.063	6.5	19	56	1.0	.50	2.90	3.4	3,181
7.00	.063	6.5	19	56	1.0	.50	4.76	5.0	4,279

*5.1. Compute the correlation matrix for all variables including the dependent variable Q. By inspection of the correlations determine which variables are most likely to contribute significantly to variation in Q. If you could use only one independent variable in your model, which would it be?

5.2. Compute the correlation matrix using $LQ = \ln(Q)$ and the logarithms of all independent variables. How does this change the correlations and your conclusions about which variables are most likely to contribute significantly to variation in LQ?

*5.3. Use $LQ = \ln(Q)$ as the dependent variable and the logarithms of all nine independent variables plus an intercept as the "full" model. Compute the least squares regression equation and test the composite null hypothesis that all partial regression coefficients for the independent variables are zero. Compare the estimated partial regression coefficients to their standard errors. Which partial regression coefficients are significantly different from zero? Which independent variable would you eliminate first to simplify the model?

5.4. Eliminate the least important variable from the model in exercise 5.3 and recompute the regression. Are all partial sums of squares for the remaining variables significant ($\alpha = .05$)? If not, continue to eliminate the least important independent variable at each stage and recompute the regression. Stop when all independent variables in the model are significant (use $\alpha = .05$). What do the results indicate about the need for the intercept? Does it make sense to have $\beta_0 = 0$ in this exercise? Summarize the results of your final model in an analysis of variance table. Discuss your conclusions about what factors are important in peak flow rates.

*5.5. Determine the 95% univariate confidence interval estimates of the regression coefficients for your final model in exercise 5.4. Determine the 95% Bonferroni confidence interval estimates.

5.6. Construct the 95% joint confidence region for the partial regression coefficients for X_8 and X_9 *ignoring* the parameters for the other variables in your final model in exercise 5.4.

CHAPTER SIX

Geometric Interpretation of Least Squares

Matrix notation has been used to present least squares regression and the application of least squares has been demonstrated.

This chapter presents the geometry of least squares. The data vectors are represented by vectors plotted in n-space and the basic concepts of least squares are illustrated using relationships among the vectors. The intent of this chapter is to give insight into the basic principles of least squares. This chapter is not essential for an understanding of the remaining topics.

All concepts of ordinary least squares can be visualized by applying a few principles of geometry. Many find the geometric interpretation more helpful than the cumbersome algebraic equations in understanding the concepts of least squares. Partial regression coefficients, sums of squares, degrees of freedom, and most of the properties and problems of ordinary least squares have direct *visual* analogs in the geometry of vectors.

This chapter is presented solely to enhance your understanding. Although the first exposure to the geometric interpretation may seem somewhat confusing, the geometry usually enhances understanding of the least squares concepts. You are encouraged to study this chapter in the spirit in which it is presented. It is not an essential chapter for the use and understanding of regression. Review of Section 2.4 before reading this chapter may prove helpful.

6.1 The Linear Model and the Solution

X-space

In the geometric interpretation of least squares, X is viewed as a collection of p' column vectors. It is assumed for this discussion that the column vectors of X are linearly independent (or, any linear dependencies that might have existed in X have been eliminated). Each column vector of X can be plotted as a vector in n-dimensional space (see Section 2.4). That is, the n elements in each column vector provide the coordinates for identifying the endpoint of the vector plotted in n-space. The p' vectors *jointly* define a p'-dimensional subspace of the n-

dimensional space in which they are plotted ($p' < n$). This p'-dimensional subspace consists of the set of points that can be reached by *linear* functions of the p' vectors of X. This subspace is called the **X-space**. (When the vectors of X are not linearly independent, the dimensionality of the X-space is determined by the rank of X.)

$\mathscr{E}(Y)$ Vector

The Y vector is also a vector in n-dimensional space. Its expectation

$$\mathscr{E}(Y) = X\beta = \beta_0 \mathbf{1} + \beta_1 X_1 + \cdots + \beta_p X_p \quad [6.1]$$

is a linear function of the column vectors of X with the elements of β being the coefficients. Thus, the linear model

$$Y = X\beta + \varepsilon \quad [6.2]$$

says that the mean vector $\mathscr{E}(Y) = X\beta$ falls *exactly* in the X-space. The specific point at which $\mathscr{E}(Y)$ falls is determined by the true, and unknown, partial regression coefficients in β.

Y Vector

The vector of observations on the dependent variable, Y, will fall somewhere in n-dimensional space around its mean $\mathscr{E}(Y)$, with its exact position being determined by the random elements in ε. The model (equation 6.2) states that Y is the sum of the two vectors $\mathscr{E}(Y)$ and ε. While $\mathscr{E}(Y)$ is in the X-space, ε and, consequently, Y are random vectors in n-dimensional space. Neither ε nor Y will fall in the X-space (unless an extremely unlikely sample has been drawn).

Example 6.1

To illustrate these relationships, we must limit ourselves to three-dimensional space. The concepts illustrated in two and three dimensions extend to n-dimensional geometry. Assume that X consists of two vectors X_1 and X_2, each of order 3, so that they can be plotted in three-dimensional space (Figure 6.1). The plane in Figure 6.1 represents the two-dimensional subspace defined by X_1

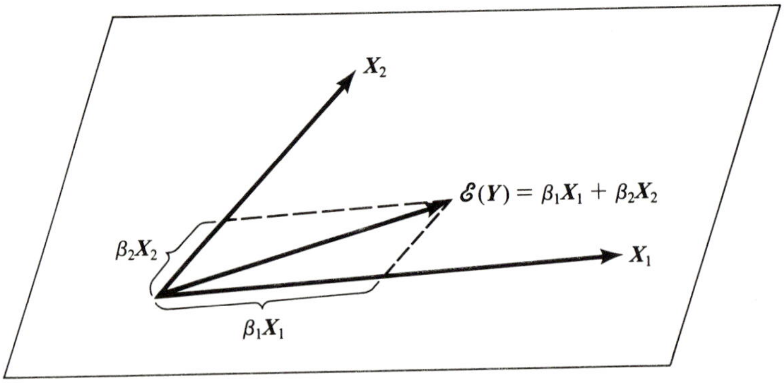

Figure 6.1 The geometric interpretation of $\mathscr{E}(Y)$ as a linear function of X_1 and X_2. The plane represents the space defined by the two independent vectors. The vector $\mathscr{E}(Y)$ is shown as the sum of $\beta_1 X_1$ and $\beta_2 X_2$.

and X_2. The vector $\mathscr{E}(Y)$ lies in this plane and represents the true mean vector of Y, as the linear function of X_1 and X_2 expressed in the model. The dashed lines in Figure 6.1 show the addition of the vectors $\beta_1 X_1$ and $\beta_2 X_2$ to give the vector $\mathscr{E}(Y)$. This, of course, assumes that the model is correct. In practice, $\mathscr{E}(Y)$ is not known because $\boldsymbol{\beta}$ is not known. The purpose of the regression analysis is to find "best" estimates of β_1 and β_2. ■

The Partial Regression Coefficients

The position of $\mathscr{E}(Y)$ in Figure 6.1 represents a case where both β_1 and β_2 are positive; the vectors to be added to give $\mathscr{E}(Y)$, $\beta_1 X_1$ and $\beta_2 X_2$, have the same direction as the original vectors X_1 and X_2. When $\mathscr{E}(Y)$ falls outside the angle formed by X_1 and X_2, one or both of the regression coefficients must be negative. Multiplication of a vector by a negative coefficient reverses the direction of the vector. For example, $-.1X_1$ defines a vector that is $\frac{1}{10}$ the length of X_1 and has opposite direction to X_1. Figure 6.2 partitions the two-dimensional X-space according to the signs β_1 and β_2 take when $\mathscr{E}(Y)$ falls in the particular region.

Figure 6.3 (page 156) uses the same X-space and $\mathscr{E}(Y)$ as Figure 6.1, but includes Y, at some distance from $\mathscr{E}(Y)$ and *not* in the X-space (because of ε), and $\hat{Y}$. Since $\hat{Y}$ is a linear function of the columns of X, $\hat{Y} = X\hat{\boldsymbol{\beta}}$, it must fall in the X-space. The *estimated* regression coefficients, $\hat{\beta}_1$ and $\hat{\beta}_2$, are shown as the multiples of X_1 and X_2 that give $\hat{Y}$ when summed. The estimated regression coefficients serve the same role in determining $\hat{Y}$ that the true regression coefficients β_1 and β_2 do in determining $\mathscr{E}(Y)$. Of course, $\hat{Y}$ will almost certainly never coincide with $\mathscr{E}(Y)$. Figure 6.3 is drawn so that both $\hat{\beta}_1$ and $\hat{\beta}_2$ are positive. The signs of $\hat{\beta}_1$ and $\hat{\beta}_2$ are determined by the region of the X-space in which $\hat{Y}$ falls, as illustrated in Figure 6.2 for β_1 and β_2.

The e Vector

The short vector connecting $\hat{Y}$ to Y in Figure 6.3 is the vector of residuals $\boldsymbol{e}$. The least squares principle requires that $\hat{\boldsymbol{\beta}}$, and hence $\hat{Y}$, be chosen such that $\sum(Y_i - \hat{Y}_i)^2 = \boldsymbol{e}'\boldsymbol{e}$ is minimized. But $\boldsymbol{e}'\boldsymbol{e}$ is the squared length of $\boldsymbol{e}$. Geometrically, it is the squared distance from the end of the Y vector to the end of the $\hat{Y}$ vector. Thus, $\hat{Y}$ must be that unique vector in the X-space that is closest to Y in n-space. The closest point on the plane to Y (in Figure 6.3) is the point that would be

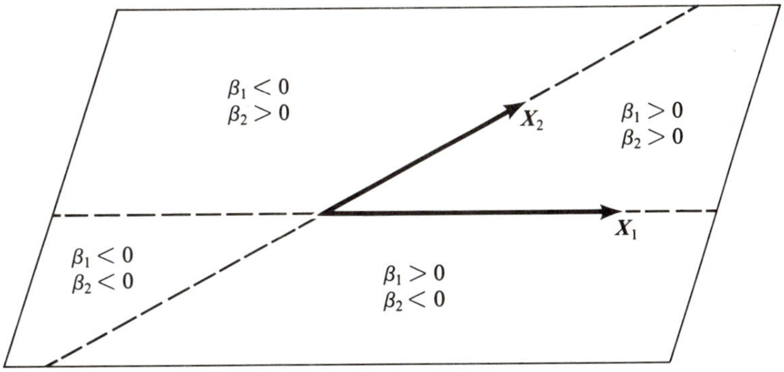

Figure 6.2 Partitions of the two-dimensional X-space according to the signs β_1 and β_2 take when $\mathscr{E}(Y)$ falls in the indicated region.

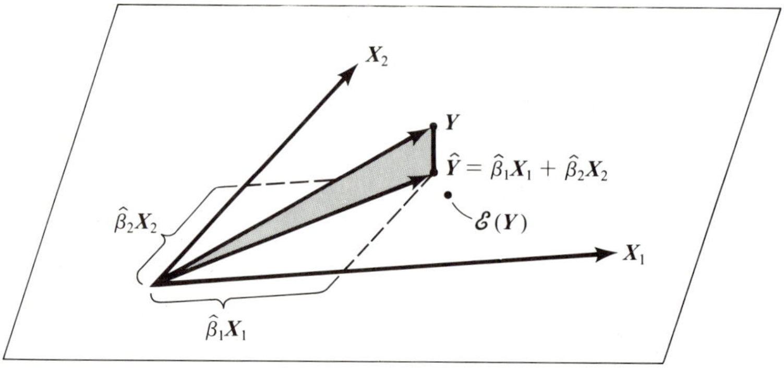

Figure 6.3 The geometric relationship of $\boldsymbol{Y}$ and $\hat{\boldsymbol{Y}}$ to the X-space. $\boldsymbol{Y}$ is not in the plane defined by $\boldsymbol{X}_1$ and $\boldsymbol{X}_2$. The perpendicular projection from $\boldsymbol{Y}$ to the plane defines the vector $\hat{\boldsymbol{Y}}$, which is in the plane. The estimated regression coefficients are the proportions of $\boldsymbol{X}_1$ and $\boldsymbol{X}_2$ that, when added, give $\hat{\boldsymbol{Y}}$. The short vector connecting $\hat{\boldsymbol{Y}}$ to $\boldsymbol{Y}$ is the vector of residuals $\boldsymbol{e}$.

reached with a perpendicular projection from $\boldsymbol{Y}$ to the plane. That is, $\boldsymbol{e}$ must be perpendicular to the X-space. $\hat{\boldsymbol{Y}}$ is shown as the shadow on the plane cast by $\boldsymbol{Y}$ with a light directly "overhead."

Visualize the floor of a room being the plane defined by the X-space. Let one corner of the room at the floor be the origin of the three-dimensional coordinate system, the line running along one baseboard be the $\boldsymbol{X}_1$ vector, and the line running along the adjoining baseboard be the $\boldsymbol{X}_2$ vector. Thus, the floor of the room is the X-space. Let the $\boldsymbol{Y}$ vector run from the origin to a point in the ceiling. It is obvious that the point on the floor *closest* to the point in the ceiling is the point directly beneath. That is, the "projection" of $\boldsymbol{Y}$ onto the X-space must be a perpendicular projection onto the floor. A line from the end of $\hat{\boldsymbol{Y}}$ to $\boldsymbol{Y}$ must form a right angle with the floor. This "vertical" line from $\hat{\boldsymbol{Y}}$ to $\boldsymbol{Y}$ is the vector of observed residuals, $\boldsymbol{e} = \boldsymbol{Y} - \hat{\boldsymbol{Y}}$. The two vectors $\hat{\boldsymbol{Y}}$ and $\boldsymbol{e}$ clearly add to $\boldsymbol{Y}$.

Common sense told us that $\boldsymbol{e}$ must be perpendicular to the plane for $\hat{\boldsymbol{Y}}$ to be the closest possible vector to $\boldsymbol{Y}$. The least squares procedure requires this to be the case. The normal equations

$$\boldsymbol{X}'\boldsymbol{X}\hat{\boldsymbol{\beta}} = \boldsymbol{X}'\boldsymbol{Y}$$

imply that

$$\boldsymbol{X}'\boldsymbol{Y} - \boldsymbol{X}'\boldsymbol{X}\hat{\boldsymbol{\beta}} = \boldsymbol{0}$$

or

$$\boldsymbol{X}'(\boldsymbol{Y} - \boldsymbol{X}\hat{\boldsymbol{\beta}}) = \boldsymbol{X}'(\boldsymbol{Y} - \hat{\boldsymbol{Y}}) = \boldsymbol{X}'\boldsymbol{e} = \boldsymbol{0} \qquad [6.3]$$

The statement $\boldsymbol{X}'\boldsymbol{e} = \boldsymbol{0}$ shows that $\boldsymbol{e}$ must be orthogonal (or perpendicular) to each of the column vectors in $\boldsymbol{X}$. (The sum of products of the elements of $\boldsymbol{e}$ with those of each vector in $\boldsymbol{X}$ is zero.) Hence, $\boldsymbol{e}$ must be perpendicular to any linear function of these vectors in order for the result to be a least squares result.

P Matrix

$\hat{\boldsymbol{Y}}$ may also be written as $\hat{\boldsymbol{Y}} = \boldsymbol{PY}$. The matrix $\boldsymbol{P} = \boldsymbol{X}(\boldsymbol{X}'\boldsymbol{X})^{-1}\boldsymbol{X}'$ is the matrix that *projects* $\boldsymbol{Y}$ onto the p'-dimensional subspace defined by the columns of $\boldsymbol{X}$. In other words, premultiplying $\boldsymbol{Y}$ by $\boldsymbol{P}$ gives $\hat{\boldsymbol{Y}}$ such that the vector $\boldsymbol{e}$ is perpendicular to the X-space and as short as possible. $\boldsymbol{P}$ is called a **projection matrix**; hence its label $\boldsymbol{P}$.

SS(Res)

The "room" analogy can be used to show another property of least squares regression. The regression of $\boldsymbol{Y}$ on one independent variable, say $\boldsymbol{X}_1$, cannot give a smaller residual sum of squares $\boldsymbol{e}'\boldsymbol{e}$ than the regression on $\boldsymbol{X}_1$ and $\boldsymbol{X}_2$ jointly. The X-space defined by $\boldsymbol{X}_1$ alone is the set of points along the baseboard representing $\boldsymbol{X}_1$. Therefore, the projection of $\boldsymbol{Y}$ onto the space defined only by $\boldsymbol{X}_1$ (as if $\boldsymbol{X}_1$ were the only variable in the regression) must be to a point along this baseboard. The subspace defined by $\boldsymbol{X}_1$ alone is part of the subspace defined jointly by $\boldsymbol{X}_1$ and $\boldsymbol{X}_2$. Therefore, no point along this baseboard can be closer to the end of the $\boldsymbol{Y}$ vector than the closest point on the entire floor (the X-space defined by $\boldsymbol{X}_1$ and $\boldsymbol{X}_2$ jointly). The two vectors of residuals, that from the regression of $\boldsymbol{Y}$ on $\boldsymbol{X}_1$ along and that from the regression of $\boldsymbol{Y}$ on $\boldsymbol{X}_1$ and $\boldsymbol{X}_2$ jointly, would be the same length only if the projection onto the floor happened to fall exactly at the baseboard. In this case, $\hat{\beta}_{2.1}$ must be zero. This illustrates a general result that the residual sum of squares from the regression of $\boldsymbol{Y}$ on a subset of independent variables *cannot* be smaller than the residual sum of squares from the regression on the full set of independent variables.

The results of this section are summarized in the box.

1. $\boldsymbol{Y}$ is a vector in n-space.
2. Each column vector of $\boldsymbol{X}$ is a vector in n-space.
3. The p' linearly independent vectors of $\boldsymbol{X}$ define a p'-dimensional subspace.
4. The linear model specifies that $\mathscr{E}(\boldsymbol{Y}) = \boldsymbol{X\beta}$ is in the X-space; the vector $\boldsymbol{Y}$ is (almost certainly) not in the X-space.
5. The least squares solution, $\hat{\boldsymbol{Y}} = \boldsymbol{X}\hat{\boldsymbol{\beta}} = \boldsymbol{PY}$, is that point in the X-space that is closest to $\boldsymbol{Y}$.
6. The residuals vector $\boldsymbol{e}$ is orthogonal to the X-space.
7. The right triangle formed by $\boldsymbol{Y}$, $\hat{\boldsymbol{Y}}$, and $\boldsymbol{e}$ expresses $\boldsymbol{Y}$ as the sum of the other two vectors, $\boldsymbol{Y} = \hat{\boldsymbol{Y}} + \boldsymbol{e}$.

6.2 Sums of Squares and Degrees of Freedom

Length of Y

The Pythagorean theorem in two-dimensional space states that the length of the hypotenuse of a right triangle is the square root of the sum of the squares of the sides of the triangle. In Section 2.4 it was explained that this extends into n dimensions to state that the length of any vector is the square root of the sum of the squares of *all* its elements. Thus, $\mathbf{Y}'\mathbf{Y}$, the uncorrected sum of squares of the dependent variable, is the squared length of the vector $\mathbf{Y}$.

Partitioning the Total Sum of Squares

The vectors $\mathbf{Y}$, $\hat{\mathbf{Y}}$, and $\mathbf{e}$ form a right triangle with $\mathbf{Y}$ being the hypotenuse (Figure 6.3). One side of the triangle, $\hat{\mathbf{Y}}$, lies in the X-space; the other side, $\mathbf{e}$, is perpendicular to the X-space. The Pythagorean theorem can be used to express the length of $\mathbf{Y}$ in terms of the lengths of $\hat{\mathbf{Y}}$ and $\mathbf{e}$:

$$\text{length}(\mathbf{Y}) = \sqrt{[\text{length}(\hat{\mathbf{Y}})]^2 + [\text{length}(\mathbf{e})]^2}$$

Squaring both sides yields

$$\mathbf{Y}'\mathbf{Y} = \hat{\mathbf{Y}}'\hat{\mathbf{Y}} + \mathbf{e}'\mathbf{e} \qquad [6.4]$$

Thus, the partitioning of the total sum of squares of $\mathbf{Y}$, $\mathbf{Y}'\mathbf{Y}$, into SS(Model) $= \hat{\mathbf{Y}}'\hat{\mathbf{Y}}$ and SS(Res) $= \mathbf{e}'\mathbf{e}$ corresponds to expressing the squared length of the vector $\mathbf{Y}$ in terms of the squared lengths of the sides of the right triangle.

Degrees of Freedom

The "degrees of freedom" associated with each sum of squares is the number of dimensions in which that vector is "free to move." $\mathbf{Y}$ is free to fall anywhere in n-dimensional space and, hence, has n degrees of freedom. $\hat{\mathbf{Y}}$, on the other hand, must fall in the X-space and, hence, has degrees of freedom equal to the dimension of the X-space—two in Figure 6.3 or p' in general. The residual vector $\mathbf{e}$ can fall anywhere in the subspace of the n-dimensional space that is *orthogonal* to the X-space. This subspace has dimensionality $(n - p')$ and, hence, $\mathbf{e}$ has $(n - p')$ degrees of freedom. In Figure 6.3, $\mathbf{e}$ has $(3 - 2) = 1$ degree of freedom. In general, the degrees of freedom associated with $\hat{\mathbf{Y}}$ and $\mathbf{e}$ will be $r(\mathbf{X})$ and $[n - r(\mathbf{X})]$, respectively.

Figures 6.1–6.3 have been described as if all vectors were of order 3 so that they could be fully represented in the three-dimensional figures. This is being more restrictive than needed. Three vectors of *any* order define a three-dimensional subspace and, if one forgoes plotting the individual vectors in n-space, the relationships among the three vectors can be illustrated in three dimensions as in Figures 6.1–6.3.

Example 6.2

This example uses the data from exercise 1.4, which relate heart rate at rest to kilograms of body weight. The model to be fit includes an intercept so that the two vectors defining the X-space are $\mathbf{1}$, the vector of ones, and $\mathbf{X}_1$, the vector of body weights. The $\mathbf{Y}$ and $\mathbf{X}_1$ vectors in the original data are an order of magnitude longer than $\mathbf{1}$, so both $\mathbf{Y}$ and $\mathbf{X}_1$ have been scaled by $\frac{1}{20}$ for purposes of this illustration. The rescaled data are

$$\mathbf{X}_1' = (4.50 \;\; 4.30 \;\; 3.35 \;\; 4.45 \;\; 4.05 \;\; 3.75)$$

$$\mathbf{Y}' = (3.10 \;\; 2.25 \;\; 2.00 \;\; 2.75 \;\; 3.20 \;\; 2.65)$$

The X-space is defined by $\mathbf{1}$ and $\mathbf{X}_1$. The lengths of the vectors are

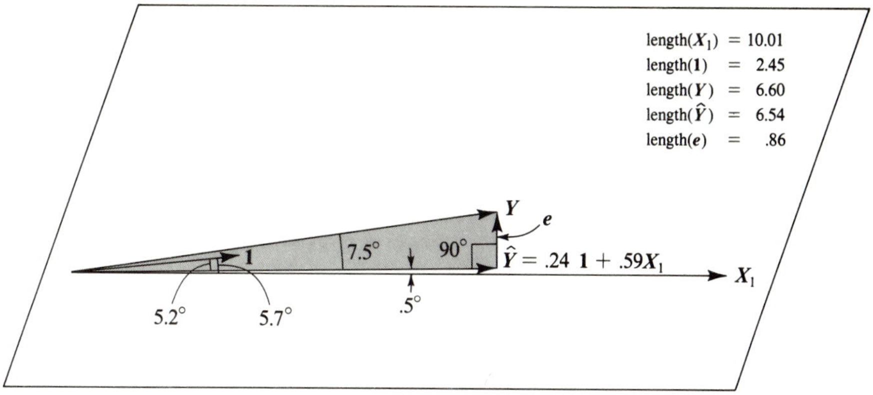

Figure 6.4 Geometric interpretation of the regression of heart rate at rest (Y) on kilograms body weight (X_1). The model was $\boldsymbol{Y} = \boldsymbol{X\beta} + \boldsymbol{\varepsilon}$, where $\boldsymbol{X} = (\mathbf{1} \ \ \boldsymbol{X}_1)$. The plane in the figure is the X-space defined by **1** and $\boldsymbol{X}_1$. The data are from exercise 1.4 with both $\boldsymbol{X}_1$ and $\boldsymbol{Y}$ scaled by $\frac{1}{20}$. Angles between vectors are shown in degrees. $\boldsymbol{Y}$ protrudes away from the plane at an angle of 7.5°. Perpendicular projection of $\boldsymbol{Y}$ onto the plane defines $\hat{\boldsymbol{Y}}$, which forms an angle of 5.2° with **1** and .5° with $\boldsymbol{X}_1$.

$$\text{length}(\mathbf{1}) = \sqrt{\mathbf{1}'\mathbf{1}} = \sqrt{6} = 2.45$$

$$\text{length}(\boldsymbol{X}_1) = \sqrt{\boldsymbol{X}_1'\boldsymbol{X}_1} = \sqrt{100.23} = 10.01$$

and the angle between the two vectors, $\theta(\mathbf{1}, \boldsymbol{X}_1)$, is

$$\theta(\mathbf{1}, \boldsymbol{X}_1) = \text{arc-cosine}\left(\frac{\mathbf{1}'\boldsymbol{X}_1}{\sqrt{\mathbf{1}'\mathbf{1}}\sqrt{\boldsymbol{X}_1'\boldsymbol{X}_1}}\right)$$

$$= \text{arc-cosine}\left(\frac{24.4}{\sqrt{6}\sqrt{100.23}}\right) = 5.7°$$

The vectors **1** and $\boldsymbol{X}_1$ are plotted in Figure 6.4 using their relative lengths and the angle between them. The X-space defined by **1** and $\boldsymbol{X}_1$ is the plane represented by the parallelogram.

The $\boldsymbol{Y}$ vector is drawn as protruding above the surface of the plane at an angle of $\theta(\boldsymbol{Y}, \hat{\boldsymbol{Y}}) = 7.5°$, the angle between $\boldsymbol{Y}$ and $\hat{\boldsymbol{Y}}$. [All angles are computed as illustrated for $\theta(\mathbf{1}, \boldsymbol{X}_1)$. The length of $\boldsymbol{Y}$ is

$$\text{length}(\boldsymbol{Y}) = \sqrt{\boldsymbol{Y}'\boldsymbol{Y}} = \sqrt{43.4975} = 6.60$$

This is the square root of the uncorrected sum of squares of Y which, since $\boldsymbol{Y}$ can fall anywhere in six-dimensional space, has 6 degrees of freedom. The projection of $\boldsymbol{Y}$ onto the plane defines $\hat{\boldsymbol{Y}}$ as the sum

$$\hat{\boldsymbol{Y}} = .24\,\mathbf{1} + .59\boldsymbol{X}_1$$

The angles between $\hat{\boldsymbol{Y}}$ and the two X-vectors are $\theta(\hat{\boldsymbol{Y}}, \mathbf{1}) = 5.2°$ and $\theta(\hat{\boldsymbol{Y}}, \boldsymbol{X}_1) = .5°$. The length of $\hat{\boldsymbol{Y}}$ is the square root of SS(Model):

$$\text{length}(\hat{\boldsymbol{Y}}) = \sqrt{\hat{\boldsymbol{Y}}'\hat{\boldsymbol{Y}}} = \sqrt{42.7552} = 6.54$$

Since $\hat{\boldsymbol{Y}}$ must fall in the two-dimensional X-space, SS(Model) has 2 degrees of freedom. The residuals vector $\boldsymbol{e}$ connecting $\hat{\boldsymbol{Y}}$ to $\boldsymbol{Y}$ is perpendicular to the plane and its length is the square root of SS(Res):

$$\text{length}(\boldsymbol{e}) = \sqrt{\boldsymbol{e}'\boldsymbol{e}} = \sqrt{.7423} = .86$$

Since $\boldsymbol{e}$ must be orthogonal to the X-space, SS(Res) has 4 degrees of freedom. Thus, the squared lengths of $\boldsymbol{Y}$, $\hat{\boldsymbol{Y}}$, and $\boldsymbol{e}$, and the dimensions in which each is free to move reflect the analysis of variance of the regression results.

In this example, $\hat{\boldsymbol{Y}}$ falls very close to $\boldsymbol{X}_1$; the angle between the two vectors is only .5°. This suggests that very nearly the same predictability of $\boldsymbol{Y}$ would be obtained from the regression of $\boldsymbol{Y}$ on $\boldsymbol{X}_1$ alone—that is, if the model forced the regression line to pass through the origin. If the no-intercept model is adopted, the X-space becomes the one-dimensional space defined by $\boldsymbol{X}_1$. The projection of $\boldsymbol{Y}$ onto this X-space gives $\hat{\boldsymbol{Y}} = .65\boldsymbol{X}_1$. That is, $\hat{\boldsymbol{Y}}$ falls *on* $\boldsymbol{X}_1$. The length of $\hat{\boldsymbol{Y}}$ is

$$\text{length}(\hat{\boldsymbol{Y}}) = \sqrt{42.7518} = 6.54$$

which is trivially shorter than that obtained with the intercept model, $\sqrt{42.7518}$ versus $\sqrt{42.7552}$. The residuals vector is, correspondingly, only slightly longer:

$$\text{length}(\boldsymbol{e}) = \sqrt{.7457} = .86$$

■

Centered Independent Variables

It is common in least squares regression to express the model in terms of **centered independent variables**. That is, each independent variable is coded to have zero mean by subtracting the mean of the variable from each observation. The only effect, geometrically, of centering the independent variable is to shift the position, in the original X-space, of the vector representing the independent variable so that it is orthogonal to $\mathbf{1}$. In general, when more than one independent variable is involved, each centered variable will be orthogonal to $\mathbf{1}$. The centering will change the angles between the vectors of the independent variables but the X-space remains as defined by the original variables.

Example 6.3

The geometric interpretation of the effect of centering the independent variable is illustrated in Figure 6.5 for the heart rate/body weight data from Example 6.2. Let $\boldsymbol{X}_c$ be the centered vector. $\boldsymbol{X}_c$ is obtained by the subtraction

$$\boldsymbol{X}_c = \boldsymbol{X}_1 - (4.0667)\mathbf{1}$$

where 4.0667 is the mean of the elements in $\boldsymbol{X}_1$. Since $\boldsymbol{X}_c$ is a linear function of $\mathbf{1}$ and $\boldsymbol{X}_1$, it is by definition in the space defined by $\mathbf{1}$ and $\boldsymbol{X}_1$. Thus, the X-space

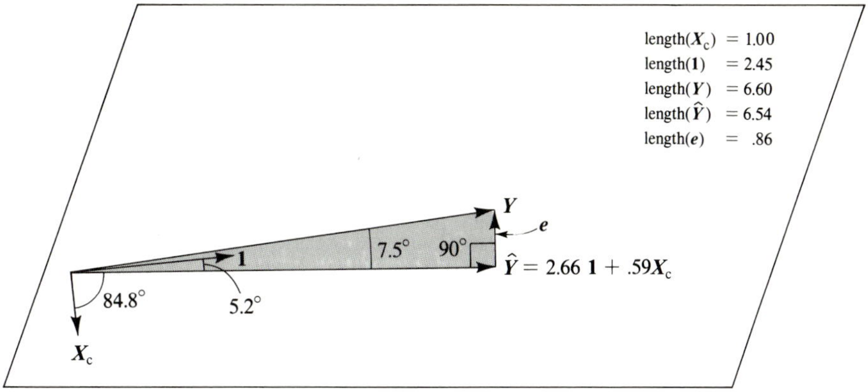

Figure 6.5 Geometric interpretation of the regression of heart rate at rest (Y) on kilograms body weight using the centered variable (X_c). The plane in the figure is defined by **1** and X_c and is identical to the plane defined by **1** and X_1 in Figure 6.4. All vectors are the same as in Figure 6.4 except X_c replaces X_1.

defined by **1** and X_c in Figure 6.5 is identical to the X-space defined by **1** and X_1 in Figure 6.4. Centering the independent variable does not alter the definition of the X-space. The centered vector X_c is orthogonal to **1**, because $\mathbf{1}'X_c = 0$, and has length 1.002. Y is the same as in Figure 6.4 and, because the X-space is the same, the projection of Y onto the X-space must give the same $\hat{Y}$. The regression equation, however, is now expressed in terms of a linear function of **1** and X_c rather than in terms of **1** and X_1. ■

6.3 Correction for the Mean and Sequential Regressions

Correction for the Mean

Equation 6.4 gave the partitioning of the total *uncorrected* sum of squares for Y. Interest is usually in partitioning the total *corrected* sum of squares. The partitioning of the corrected sum of squares is obtained by subtracting the sum of squares attributable to the mean, or the correction factor, from both $Y'Y$ and SS(Model):

$$\begin{aligned} Y'Y - \text{SS}(\mu) &= [\text{SS(Model)} - \text{SS}(\mu)] + e'e \\ &= \text{SS(Regr)} - e'e \end{aligned} \qquad [6.5]$$

The correction for the mean, SS(μ), is the sum of squares attributable to a model that contains only the constant term, β_0. Geometrically, this is equivalent to projecting Y onto the one-dimensional space defined by **1**. The least squares estimate of β_0 is $\bar{Y}$, and the residuals vector from this projection is the vector of deviations of Y_i from $\bar{Y}$, $y_i = Y_i - \bar{Y}$. The squared length of this residuals vector is the *corrected* sum of squares for Y. Since the space defined by **1** is a one-dimensional space, this residuals vector lies in $(n - 1)$-dimensional space and has $(n - 1)$ degrees of freedom.

SS(Regr) and the partial regression coefficients are the results obtained when this residuals vector is, in turn, projected onto the p-dimensional subspace ($p = p' - 1$) defined by the independent variables where each independent variable has also been "corrected for" its mean. Thus, obtaining SS(Regr) can be viewed as a two-stage process. First, Y and the independent variables are each projected onto the space defined by **1**. Then, the *residuals* vector for Y is projected onto the space defined by the *residuals* vectors for the independent variables. The squared length of $\hat{Y}$ for this second projection is SS(Regr).

Sequential Sums of Squares

The **sequential sum of squares** for an independent variable is an extension of this process. Now, however, Y and the independent variable of current interest are first projected onto the space defined by *all* independent variables that precede the current X in the model, not just **1**. Then, the *residuals* vector for Y (call it e_y) is projected onto the space defined by the *residuals* vector for the current X (call it e_x). The sequential sum of squares for the current independent variable is the squared length of $\hat{Y}$ for this projection of e_y onto e_x. Note that both the dependent variable and the current independent variable have been "adjusted" for all preceding independent variables. At each step in the sequential analysis, the new X-space is a one-dimensional space and, therefore, the sequential sum of squares at each stage has 1 degree of freedom.

Since the residuals vector in least squares is always orthogonal to the X-space onto which Y is projected, e_y and e_x are both orthogonal to *all* independent variables previously included in the model. Because of this orthogonality to the previous X-space, the sequential sums of squares and degrees of freedom are additive. That is, the sum of the sequential sums of squares and the sum of the degrees of freedom for each step are equal to what would have been obtained if a single model containing all independent variables had been used.

6.4 The Collinearity Problem

Definition of Collinearity

The partial regression coefficient and partial sum of squares for any independent variable are, in general, dependent on which other independent variables are in the model. In the case study in Chapter 5, it was observed that the changes in regression coefficients and sums of squares as other variables were added to or removed from the model could be large. This dependence of the regression results for each variable on what other variables are in the model derives from the independent variables *not* being mutually orthogonal. Lack of orthogonality of the independent variables is to be expected in observational studies, those in which the researcher is restricted to making observations on nature as it exists. In such studies, the researcher

> ... cannot impose on a subject, or withhold from the subject, a procedure or treatment whose effects he desires to discover, or cannot assign subjects at random to different procedures (Cochran, 1983).

On the other hand, orthogonality or near orthogonality is usually designed into controlled experiments.

The extreme case of nonorthogonality, where two or more independent variables are very nearly linearly dependent, creates severe problems in least squares regression. This is referred to as the **collinearity problem**. The regression coefficients become extremely unstable; they are very sensitive to small random errors in Y and may fluctuate wildly as independent variables are added to or removed from the model. The instability in the regression results is reflected in very large standard errors for the partial regression coefficients. Frequently, none of the individual partial regression coefficients will be significantly different from zero even though their combined effect is highly significant.

Geometry of Collinearity

The impact of collinearity can be illustrated geometrically. Figure 6.6 shows a two-dimensional subspace defined by $\boldsymbol{X}_1$ and $\boldsymbol{X}_2$. The plane is the X-space and the heavy dot, with the shaded area around it, represents $\mathscr{E}(\boldsymbol{Y})$. The shaded area represents the distribution of projections of $\boldsymbol{Y}$, $\hat{\boldsymbol{Y}}$, onto the X-space one might obtain from repeated samplings of the dependent variable. Panels (a) and (b) represent the case where $\boldsymbol{X}_1$ and $\boldsymbol{X}_2$ are orthogonal. Panels (c) and (d) represent the case where $\boldsymbol{X}_1$ and $\boldsymbol{X}_2$ are nearly collinear; the angle between the vectors is small. The position of $\boldsymbol{X}_2$ relative to $\mathscr{E}(\boldsymbol{Y})$ remains the same in all cases; the position of $\boldsymbol{X}_1$ has been shifted to create the collinearity. (This diagram does not really show a severe case of collinearity. For clarity in the diagram, the angle between $\boldsymbol{X}_1$ and $\boldsymbol{X}_2$ was kept reasonably large.)

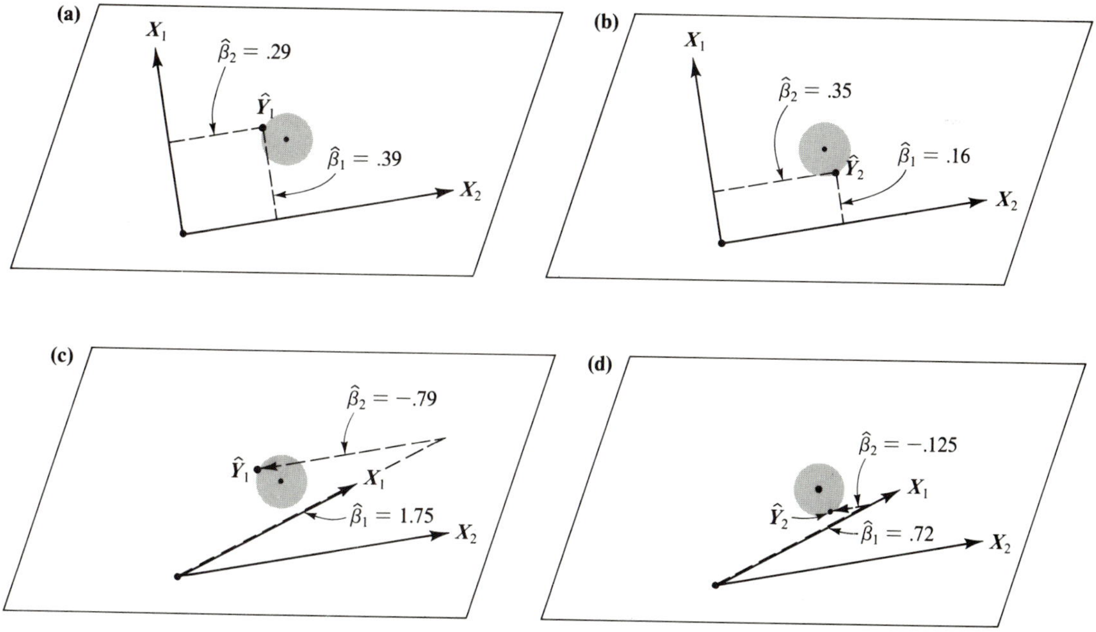

Figure 6.6 Illustration of the effect of collinearity on the stability of the partial regression coefficients. Panels (a) and (b) show the partial regression coefficients for two hypothetical $\hat{\boldsymbol{Y}}$ vectors [for the same $\mathscr{E}(\boldsymbol{Y})$] when $\boldsymbol{X}_1$ and $\boldsymbol{X}_2$ are orthogonal. Panels (c) and (d) show the results when $\boldsymbol{X}_1$ is shifted to make the two vectors more nearly collinear.

Two possible projections, $\hat{Y}_1$ and $\hat{Y}_2$, representing two independent sets of observations, are used to illustrate the relative sensitivity of the partial regression coefficients to variation in Y in the collinear case compared to the orthogonal case. It is assumed that the true model, $\mathscr{E}(Y)$, is known. The model also assumes that the X's are fixed and measured without error so that the planes are well defined (fixed) in both cases. Thus, only the effect of variation in Y, different samples of ε, is being illustrated by the difference between $\hat{Y}_1$ and $\hat{Y}_2$ in Figure 6.6.

The partial regression coefficients under orthogonality [panels (a) and (b) of Figure 6.6] changed from $\hat{\beta}_1 = .39$ and $\hat{\beta}_2 = .29$ for $\hat{Y}_1$ to $\hat{\beta}_1 = .16$ and $\hat{\beta}_2 = .35$ for $\hat{Y}_2$. With collinearity, however, the partial regression coefficients changed from $\hat{\beta}_1 = 1.75$ and $\hat{\beta}_2 = -.79$ for $\hat{Y}_1$ to $\hat{\beta}_1 = .72$ and $\hat{\beta}_2 = -.125$ for $\hat{Y}_2$ [panels (c) and (d)]. Comparable shifts in Y caused larger changes in the partial regression coefficients when the X-vectors were not orthogonal. As the two vectors approach collinearity (the angle between the vectors approaches 0° or 180°), the sensitivity of the regression coefficients to random changes in Y increases dramatically. In the limit, when the angle is 0° or 180°, the two vectors are linearly dependent and no longer define a two-dimensional subspace. In such cases, it is not possible to estimate β_1 and β_2 separately; only the joint effect of X_1 and X_2 on Y is estimable.

Variation in the X-Vectors

Figure 6.6 illustrates the relative impact of variation in ε on the partial regression coefficients in the orthogonal and nonorthogonal cases. In most cases, and particularly when the data are observational, the X-vectors are also subject to random variation in the population being sampled. Consequently, even if the independent variables are measured without error, repeated samples of the population will yield different X-vectors. Measurement error on the independent variables adds another component of variation to the X-vectors. Geometrically, this means that the X-space defined by the observed X's, the plane in Figure 6.6, will vary from sample to sample; the amount of variation in the plane will depend on the amount of sampling variation and measurement error in the independent variables.

The impact of sampling variation and measurement error in the independent variables is magnified with increasing collinearity of the X-vectors. Imagine balancing a cardboard (the plane) on two pencils (the vectors). If the pencils are at right angles, the plane is relatively insensitive to small movements in the tips of the pencils. On the other hand, if the pencils form a very small angle with each other (the vectors are nearly collinear), the plane becomes very unstable and its orientation changes drastically as the pencils are shifted even slightly. In the limit as the angle goes to 0° (the two vectors are linearly dependent), the pencils merge into one and all support for the plane disappears.

In summary, collinearity causes the partial regression coefficients to be sensitive to small changes in Y; the solution to the normal equations becomes unstable. In addition, sampling variation and measurement error in the independent variables cause the X-space to be poorly defined, which magnifies the sensitivity of the partial regression coefficients to collinearity. The instability in the least squares solution due to variation in Y is reflected in larger standard errors on the partial regression coefficients. The instability due to sampling variation and measurement error in the independent variables, however, is ignored in the usual regression analysis, because the independent variables are assumed to be fixed constants.

6.5 Summary

The following regression results are obtained from the geometric interpretation of least squares:

1. The data vectors, $\boldsymbol{Y}$ and $\boldsymbol{X}_j$, are vectors in n-dimensional space.
2. The linear model states that the true mean of $\boldsymbol{Y}$, $\mathcal{E}(\boldsymbol{Y})$, is in the X-space, a p'-dimensional subspace of the n-dimensional space.
3. $\hat{\boldsymbol{Y}}$ is the point in the X-space closest to $\boldsymbol{Y}$; $\boldsymbol{e}$ is orthogonal to the X-space.
4. The partial regression coefficients multiplied by their respective X-vectors define the set of vectors that must to added to "reach" $\hat{\boldsymbol{Y}}$.
5. The vectors $\hat{\boldsymbol{Y}}$ and $\boldsymbol{e}$ are the two sides of a right triangle whose hypotenuse is $\boldsymbol{Y}$. Thus, $\boldsymbol{Y} = \hat{\boldsymbol{Y}} + \boldsymbol{e}$.
6. The squared lengths of the sides of the right triangle give the partitioning of the sums of squares of $\boldsymbol{Y}$: $\boldsymbol{Y}'\boldsymbol{Y} = \hat{\boldsymbol{Y}}'\hat{\boldsymbol{Y}} + \boldsymbol{e}'\boldsymbol{e}$.
7. The correlation structure among the X's influences the regression results. Only if $\boldsymbol{X}_1$ and $\boldsymbol{X}_2$ are orthogonal will $\hat{\beta}_1 = \hat{\beta}_{1.2}$.
8. Regression of $\boldsymbol{Y}$ on one independent variable, say $\boldsymbol{X}_1$, cannot give smaller $\boldsymbol{e}'\boldsymbol{e}$ than regression on $\boldsymbol{X}_1$ and $\boldsymbol{X}_2$ jointly. More generally, regression on a subset of independent variables cannot give a better fit (smaller $\boldsymbol{e}'\boldsymbol{e}$) than regression on all variables.
9. If $\boldsymbol{X}_1$ and $\boldsymbol{X}_2$ are nearly collinear, small variations in $\boldsymbol{Y}$ cause large shifts in the partial regression coefficients. The regression results become unstable.

EXERCISES

*6.1. Use Figure 6.3 as plotted to approximate the values of $\hat{\beta}_1$ and $\hat{\beta}_2$. Where would $\hat{\boldsymbol{Y}}$ have to have fallen for $\hat{\beta}_1$ to be negative? For $\hat{\beta}_2$ to be negative? For both to be negative?

6.2. Construct a figure similar to Figure 6.3 except draw the projection of $\boldsymbol{Y}$ onto the space defined by $\boldsymbol{X}_1$. Similarly, draw the projection of $\boldsymbol{Y}$ onto the space defined by $\boldsymbol{X}_2$.
   a. Approximate the values of the simple regression coefficients in each case and compare them to the partial regression coefficients in Figure 6.3.
   b. Identify the residuals vector in both cases and in Figure 6.3.
   c. Convince yourself that the shortest residuals vector is the one in Figure 6.3.

6.3. Construct a diagram similar to Figure 6.3 except make $\boldsymbol{X}_1$ and $\boldsymbol{X}_2$ orthogonal to each other. Convince yourself that, when the independent variables are orthogonal, the simple regression coefficients from the projection of $\boldsymbol{Y}$ onto $\boldsymbol{X}_1$ and $\boldsymbol{X}_2$ separately equal the partial regression coefficients from the projection of $\boldsymbol{Y}$ onto the space defined by $\boldsymbol{X}_1$ and $\boldsymbol{X}_2$ jointly.

*6.4. Plot a plane defined by two vectors of order 10. Represent the two vectors *in this plane* by $X_1 = (5\ \ 2)$ and $X_2 = (0\ \ -4)$. Suppose the projection of Y onto this plane plots at $(-1\ \ 3)$.

a. Approximate the regression coefficients for this projection.
b. Compute the sum of squares due to the regression on X_1 and X_2 jointly. How many degrees of freedom does this sum of squares have?
c. Do you have enough information to compute the residual sum of squares? How many degrees of freedom would the residual sum of squares have?
d. Suppose someone told you that the original vector Y had length 3. Would there be any reason to doubt their statement?

6.5. Plot the two vectors $X_1 = (5\ \ 0)$ and $X_2 = (-4\ \ .25)$. Suppose two different samples of Y give projections onto this X-space at $\hat{Y}_1 = (4\ \ .5)$ and $\hat{Y} = (4\ \ -.5)$. Approximate the partial regression coefficients for the two cases. Note the shift in the partial regression coefficients for the two cases. Compare this shift to what would have been realized if $X_2 = (0\ \ 4)$, orthogonal to X_1.

*6.6. Data from exercise 1.9 relating plant biomass, Y, to total accumulated solar radiation, X, was used to fit a no-intercept model. $\hat{Y}$ and e were determined from the regression equation. The matrix $W\,(8 \times 4)$ was defined as

$$W = [X\ \ Y\ \ \hat{Y}\ \ e]$$

and the following matrix of sums of squares and products was computed:

$$W'W = \begin{bmatrix} 1{,}039{,}943.1 & 1{,}255{,}267.1 & 1{,}255{,}267.1 & 0 \\ 1{,}255{,}267.1 & 1{,}523{,}628.9 & 1{,}515{,}174.7 & 8{,}454.2 \\ 1{,}255{,}267.1 & 1{,}515{,}174.7 & 1{,}515{,}174.7 & 0 \\ 0 & 8{,}454.2 & 0 & 8{,}454.2 \end{bmatrix}$$

a. Determine the length of each (column) vector in W.
b. Compute the angles between all pairs of vectors.
c. Use the lengths of the vectors and the angles between the vectors to show graphically the regression results. What is the dimension of the X-space? Why is the angle between X and $\hat{Y}$ zero? Estimate the regression coefficient from the figure you construct.

6.7. This exercise uses the data given in exercise 1.19 relating seed weight of soybeans, Y, to cumulative seasonal solar radiation, X, for two levels of ozone exposure. For simplicity in plotting , rescale X by dividing by 2 and Y by dividing by 100 for this exercise.

a. Use the "Low Ozone" data to compute the linear regression of Y on X (with an intercept). Compute $\hat{Y}$ and e, the lengths of all vectors, and the angle between each pair of vectors. Use the vector lengths and angles to display graphically the regression results (similar to Figure 6.4). Use your figure to "estimate" the regression coefficients. From the relative positions of the vectors, what is your judgment as to whether the intercept is needed in the model?
b. Repeat part (a) using the "High Ozone" data.
c. Compare the graphical representations of the two regressions. What is your judgment as to whether the regressions are homogeneous—that is, are the same basic relationships, within the limits of random error, illustrated in both figures?

*6.8. The angle, θ, between the intercept vector, **1**, and an independent variable vector, $\boldsymbol{X}$, depends on the coefficient of variation of the independent variable. Use the relationship

$$\cos(\theta) = \frac{\mathbf{1}'\boldsymbol{X}}{\sqrt{\mathbf{1}'\mathbf{1}}\sqrt{\boldsymbol{X}'\boldsymbol{X}}}$$

to show the relationship to the coefficient of variation. What does this relationship imply about the effect on the angle of scaling the $\boldsymbol{X}$ by a constant? What does it imply about the effect of adding a constant to or subtracting a constant from $\boldsymbol{X}$?

CHAPTER
SEVEN

Model Development: Selection of Variables

The discussion of least squares regression thus far has presumed that the model was known with respect to which variables were to be included and the form these variables should take.

This chapter discusses methods of deciding which variables should be included in the model. It is still assumed that the variables are in the appropriate form. The effect of variable selection on least squares, the use of automated methods of selecting variables, and criteria for choice of subset model are discussed.

The previous chapters dealt with computation and interpretation of least squares regression. With the exception of the case study in Chapter 5, it has been assumed that the independent variables to be used in the model, and the form in which they would be expressed, were known. The properties of the least squares estimators were based on the assumption that the model was correct.

Most regression problems, however, require decisions on which variables to include in the model, the form the variables should take (for example, X, X^2, $1/X$, etc.), and the functional form of the model. This chapter discusses the choice of variables to include in the model. It is assumed that there is a set of t candidate variables, which presumably includes all relevant variables, from which a subset of r variables is to be chosen for the regression equation. The candidate variables may include different forms of the same basic variable, such as X and X^2, and the selection process may include constraints on which variables are to be included. For example, X may be forced into the model if X^2 is in the selected subset; this is a common constraint in building polynomial models (see Chapter 14).

There are three distinct problem areas related to this general topic:

1. The theoretical effects of variable selection on the least squares regression results
2. The computational methods for finding the "best" subset of variables for each subset size
3. The choice of subset size (for the final model), or the "stopping rule"

An excellent review of these topics is provided by Hocking (1976). This chapter gives some of the key results on the effects of variable selection, discusses the conceptual operation of automated variable selection procedures (without getting involved in the computational algorithms), and presents several of the commonly used criteria for choice of subset size.

7.1 Uses of the Regression Equation

The purpose of the least squares analysis—how the regression equation is to be used—will influence the manner in which the model is constructed. Hocking (1976) relates six potential uses of regression equations given by Mallows (1973b):

1. Providing a good description of the behavior of the response variable
2. Prediction of future responses and estimation of mean responses
3. Extrapolation, or prediction of responses outside the range of the data
4. Estimation of parameters
5. Control of a process by varying levels of input
6. Developing realistic models of the process

Each objective has different implications on how much emphasis is placed on eliminating variables from the model, on how important it is that the retained variables be causally related to the response variable, and on the amount of effort devoted to making the model realistic. The concern in this chapter is the selection of variables. Decisions on causality and realism must depend on information from outside the specific data set—for example, on details of how the data were obtained (the experimental design), and on fundamental knowledge of how the particular system operates.

Describing the Behavior of Y

When the object is **simple description** of the behavior of the response variable in a particular data set, there is little reason to be concerned about elimination of variables from the model, about causal relationships, or about the realism of the model. The best description of the response variable, in terms of minimum residual sum of squares, will be provided by the full model, and it is unimportant whether the variables are causally related or the model is realistic.

Why Eliminate Variables?

Elimination of variables becomes more important for the other purposes of least squares regression. Regression equations with fewer variables have the appeal of simplicity, as well as an economic advantage in terms of obtaining the necessary information to use the equations. In addition, there is a theoretical advantage of eliminating irrelevant variables and, in some cases, even variables that contain some predictive information about the response variable; this will be discussed in Section 7.2. The motivation to eliminate variables is tempered by the biases and loss of predictability that are introduced when relevant variables are eliminated. The objective is to reach a compromise where the final equation satisfies the purpose of the study.

Prediction and Estimation

Of the uses of regression, **prediction** and **estimation** of mean responses are the most tolerant toward eliminating variables. At the same time, it is relatively unimportant whether the variables are causally related or the model is realistic. It is tacitly assumed that prediction and estimation are to be within the X-space of the data and that the system continues to operate as it did when the data were collected. Thus, any variables that contain predictive information on the dependent variable, and for which information can be obtained at a reasonable cost, are useful variables. Of course, more faith could be placed in predictions and estimates based on established causal relationships, because of the protection such models provide against inadvertent extrapolations and unrecognized changes in the correlational structure of the system.

Extrapolation

Extrapolation requires more care in choice of variables. There should be more concern that all relevant variables are retained so that the behavior of the system is described as fully as possible. Extrapolations (beyond the X-space of the data) are always dangerous but can become disastrous if the equation is not a reasonably correct representation of the true model. Any extrapolation carries with it the assumption that the correlational structure observed in the sample continues outside the sample space. Validation and continual updating are essential for equations that are intended to be used for extrapolations (such as forecasts).

Estimation of Parameters

One should also be conservative in eliminating variables when **estimation of parameters** is the objective. This is to avoid the bias introduced when a relevant variable is dropped (see Section 7.2). There is an advantage in terms of reduced variance of the estimates if variables truly unrelated to the dependent variable are dropped.

Control of a System

Control of a system also implies that good estimates of the parameters are needed, but it further implies that the independent variables must have a causal effect on the response variable. Otherwise, one cannot intervene in a system and effect a change by altering the value of independent variables.

Developing Realistic Models

The objective of basic research is often related to **building realistic models**, usually the most preliminary stages of model building. Understanding the process is the ultimate goal. Whether explicitly stated or not, there will be the desire to identify the variables that are important, through some causal link, in the expression of the dependent variable. For this purpose, variable selection procedures based on the observed correlational structure in a particular set of data become relatively unimportant. At best, they can serve as tools for identifying classes of variables that warrant further study of the causal relationships, usually in controlled experiments. As the objective of the research becomes more oriented toward understanding the process, there will be increasing emphasis on developing models whose functional forms realistically reflect the behavior of the system.

Model Building Depends on Purpose

The purpose of introducing these differing objectives is to emphasize that the approach to the selection of variables will depend on the objectives of the analysis. Furthermore, how far a researcher can move in the direction of establishing the importance of variables or causality depends on the source and nature of the data. Least squares regression results reflect only the correlational structure of the data being analyzed. Of itself, least squares analysis cannot establish causal relationships. Causality can be established only from controlled experiments in which the value of the suspected causal variable is changed and the impact on the dependent variable measured. The results from any variable selection procedure, and particularly those that are automated, need to be studied carefully to make sure the models suggested are consistent with the state of knowledge of the process being modeled. No variable selection procedure can substitute for the insight of the researcher.

7.2 Effects of Variable Selection on Least Squares

The effects of variable selection on the least squares results are explicitly developed only for the case where selection is *not* based on information from the current data. This often is not the case, as in the variable selection techniques to be discussed in this chapter, but the theoretical results for this situation provide motivation for variable selection.

Theoretical Effects of Eliminating Variables

Assume that the correct model involves t independent variables but that a subset of p variables (chosen randomly or on the basis of external information) is used in the regression equation. Let $\boldsymbol{X}_p$ and $\boldsymbol{\beta}_p$ denote submatrices of $\boldsymbol{X}$ and $\boldsymbol{\beta}$ that relate to the p selected variables. $\hat{\boldsymbol{\beta}}_p$ will denote the least squares estimate of $\boldsymbol{\beta}_p$ obtained from the p-variate subset model. Similarly, $\hat{Y}_{pi}$, $\hat{Y}_{\text{pred}_{pi}}$, and MS($\text{Res}_p$) will denote the estimated mean for the ith observation, the prediction for the ith observation, and the mean squared residual, respectively, obtained from the p-variate subset model. Hocking (1976) summarizes the following properties:

1. MS(Res_p) is a *positively* biased estimate of σ^2 unless the true regression coefficients for all deleted variables are zero.
2. $\hat{\boldsymbol{\beta}}_p$ is a biased estimate of $\boldsymbol{\beta}_p$ and $\hat{Y}_{pi}$ is a biased estimate of $\mathscr{E}(Y_i)$ unless the true regression coefficient for each deleted variable is zero or, in the case of $\hat{\boldsymbol{\beta}}_p$, each deleted variable is orthogonal to the p retained variables.
3. $\hat{\boldsymbol{\beta}}_p$, $\hat{Y}_{pi}$, and $\hat{Y}_{\text{pred}_{pi}}$ are generally *less* variable than the corresponding statistics obtained from the t-variate model.
4. There are conditions under which the mean squared errors (variance plus squared bias) of $\hat{\boldsymbol{\beta}}_p$, $\hat{Y}_{pi}$, and $\hat{Y}_{\text{pred}_{pi}}$ are smaller than the variances of the estimates obtained from the t-variate model.

Thus, a bias penalty is paid whenever relevant variables, those with $\beta_j \neq 0$, are omitted from the model (statements 1 and 2). On the other hand, there is an advantage in terms of decreased variance for both estimation and prediction if variables are deleted from the model (statement 3). Furthermore, there may be cases in which there is a gain in terms of mean squared error of estimation and prediction from omitting variables whose true regression coefficients are *not* zero (statement 4).

Sampled-Based Selection of Variables

These results provide motivation for selecting subsets of variables, but they do not apply directly to the usual case where variable selection is based on analyses of the current data. The general nature of these effects may be expected to persist, but selection of variables based on their performance in the sample data introduces another class of biases that confound these results. The process of searching through a large number of potential subset models for the one that best fits the data capitalizes on the random variation in the sample to "overfit" the data. That is to say, the chosen subset model can be expected to show a higher degree of agreement with the sample data than the true equation would show with the population data. Another problem of sample-based selection is that relative importance of variables *as manifested in the sample* will not necessarily reflect relative importance in the population. The best subset in the sample, by whatever criterion, need not be the best subset in the population. Important variables in the population may appear unimportant in the sample and consequently be omitted from the model, and vice versa.

Bias in Residual Mean Squared Error

Simulation studies of the effects of subset selection (Berk, 1978) gave sample mean squared errors that were biased downward as much as 25% below the population residual variance when the sample size was less than 50. The bias decreased, as sample size increased, to 2 or 3% when there were several hundred observations in the sample. The percentage bias tended to be largest when the number of variables in the subset was relatively small, $\frac{1}{5}$ to $\frac{1}{2}$ of the number of variables in the full model. This bias in the residual mean squared error translated into bias in the F-ratios for "testing" the inclusion of a variable. The bias in F tended to be largest (positive) for inclusion of the first or second predictor, dropped to near zero before half the variables were added, and became a negative bias as more variables were added.

7.3 All Possible Regressions

Nonorthogonality Among the Independent Variables

When the independent variables in the data set are orthogonal, as they might be in a designed experiment, the least squares results for each variable remain the same regardless of which other variables are in the model. In these cases, the results from a single least squares analysis can be used to choose those independent variables to keep in the model. Usually, however, the independent variables will not be orthogonal. Nonorthogonality is to be expected with observational data and will frequently occur in designed experiments due to unforeseen mishaps. Lack of orthogonality among the independent variables causes the least squares results for each independent variable to be dependent on which other variables are in the model. The full subscript notation for the partial regression coefficients and the R-notation for sums of squares explicitly identify the variables in the model for this reason.

Computing All Possible Regressions

Conceptually, the only way of ensuring that the best model for each subset size has been found is to compute all possible subset regressions. This is feasible when the total number of variables is relatively small, but rapidly becomes a major computing problem even for moderate numbers of independent variables. For example, if there are 10 independent variables from which to choose, there are $2^{10} - 1 = 1{,}023$ possible models to be evaluated. Much effort has been devoted to finding computing algorithms that capitalize on the computations already done for previous subsets in order to reduce the total amount of computing for all possible subsets (e.g., Furnival, 1971). Furnival (1971) also pointed out that much less computing is required if one is satisfied with only the residual sum of squares from each subset model.

Finding Best Subsets

More recently, attention has focused on identifying the best subsets within each subset size without computing all possible subsets. These methods utilize the basic least squares property that the residual sum of squares cannot decrease when a variable is dropped from a model. Thus, comparison of residual sums of squares from different subset models is used to eliminate the need to compute other subsets. For example, if a two-variable subset has already been found that gives a residual sum of squares less than some three-variable model, then none of the two-variable subsets of the three-variable model need be computed; they will all give residual sums of squares larger than that from the three-variable model and, hence, larger than for the two-variable model already found. The **leaps-and-bounds algorithm** of Furnival and Wilson (1974) combines comparisons of residual sums of squares for different subset models with clever control over the sequence in which subset regressions are computed. This algorithm guarantees finding the best m subset regressions within each subset size with considerably less computing than is required for all possible subsets. The RSQUARE method in PROC REG (SAS Institute, Inc., 1985d) uses the leaps-and-bounds algorithm. These computing advances have made all possible regressions a viable option in most cases.

Example 7.1

The Linthurst data used in the case study in Chapter 5 will be used to illustrate the model selection methods of this chapter. First, the regressions for all possible models are computed to find the "best" model for this data set and to serve as references for the stepwise methods to follow. The five independent variables used in Chapter 5 will also be used here as potential variables for the model. Thus, there are $2^5 - 1 = 31$ possible regression models: five one-variable, ten two-variable, ten three-variable, five four-variable, and one five-variable model.

The RSQUARE method in PROC REG (SAS Institute, Inc., 1985d) was used to compute all possible regressions. In Table 7.1 (page 174), the subset models are ranked within each subset size (p') from the best to the worst fitting model. (Table 7.1 includes the results from four criteria to be discussed later. For the present discussion, only the coefficient of determination, R^2, will be used.) The full model, $p' = 6$, accounts for $100R^2 = 67.7\%$ of the variation in the dependent variable *BIOMASS*. No subset of the independent variables can give a larger R^2.

Of the univariate subsets, the best, pH, accounted for 59.9% of the variation in *BIOMASS*, 8% below the maximum. The second best univariate subset, Zn, accounted for only 39% of the variation in Y. The best two-variable model, pH and Na, accounted for 65.8%, only 2% below the maximum. The second best two-variable subset, pH and K, is very nearly as good, with $100R^2 = 64.8\%$. Note that the second best single variable is not contained in either of the two best two-variable subsets.

There are three three-variable models that are equally effective for all practical purposes, with $100R^2$ ranging from 65.9% to 66.3%. All three of these

Table 7.1 Summary statistics R^2, MS(Res), R^2_{adj}, and C_p from all possible regressions for Linthurst data using the five independent variables *SALINITY, pH, K, Na,* and *Zn.* All models included an intercept. Data used with permission.

p'	*Variables*	R^2	MS(Res)	R^2_{adj}	C_p
2	*pH*	.5994	178,618	.5901	7.42
	Zn	.3899	272,011	.3757	32.74
	Na	.0740	412,835	.0525	70.91
	K	.0419	427,165	.0196	74.80
	SAL	.0106	441,091	−.0124	78.57
3	*pH Na*	.6584	155,909	.6422	2.28
	pH K	.6476	160,865	.6308	3.59
	pH Zn	.6083	178,801	.5896	8.34
	SAL pH	.6034	181,030	.5845	8.93
	SAL Zn	.5526	204,209	.5313	15.07
	Na Zn	.4300	260,164	.4029	29.89
	K Zn	.4152	266,932	.3874	31.67
	SAL Na	.0776	421,031	.0337	72.48
	K Na	.0743	422,520	.0303	72.87
	SAL K	.0534	432,069	.0083	75.40
4	*pH Na Zn*	.6625	157,833	.6378	3.80
	pH K Na	.6604	158,811	.6355	4.05
	SAL pH Na	.6590	159,424	.6341	4.21
	SAL pH K	.6522	162,636	.6267	5.04
	pH K Zn	.6521	162,677	.6266	5.05
	SAL pH Zn	.6366	169,900	.6101	6.91
	SAL K Zn	.5765	198,026	.5455	14.19
	SAL Na Zn	.5644	203,666	.5326	15.64
	K Na Zn	.4300	266,509	.3883	31.89
	SAL K Na	.0776	431,296	.0101	74.48
5	*SAL pH K Zn*	.6749	155,832	.6423	4.30
	SAL pH Na Zn	.6718	157,312	.6390	4.67
	pH K Na Zn	.6642	160,955	.6306	5.59
	SAL pH K Na	.6617	162,137	.6279	5.89
	SAL K Na Zn	.5773	202,589	.5350	16.09
6	*SAL pH K Na Zn*	.6773	158,622	.6359	6

subsets include *pH* and *Na*. Thus, it makes little difference which of the three variables, *SAL*, *Zn*, or *K*, is added to the best two-variable subset. The two best four-variable subsets are also equally effective; the best in this subset does not include the best two-variable or three-variable subsets.

A key point to note from the all-possible-regressions analysis is that more than one model is in contention for nearly every subset size. With only minor differences in R^2 for the best two or three subsets in each case, it is very likely that other considerations, such as behavior of the residuals, cost of obtaining information, or prior knowledge on the importance of the variables, could shift the final choice of model away from the "best" subset.

For this example, adding a second independent variable to the model increased R^2 by 6%. However, the third, fourth, and fifth variables increased R^2

by only .4%, 1.2%, and .2%, respectively. The improvement obtained from the second variable would appear worthwhile, but the value of adding the third, fourth, and fifth variables is questionable. Further discussion of choice of subset size will be delayed until the different criteria for the choice of subset size have been discussed. ■

7.4 Stepwise Regression Methods

Alternative variable selection methods have been developed which identify good (although not necessarily the best) subset models, with considerably less computing than is required for all possible regressions. These methods are referred to as **stepwise regression methods**. The subset models are identified sequentially by adding or deleting, depending on the method, the one variable that has the greatest impact on the residual sum of squares. These stepwise methods are not guaranteed to find the "best" subset for each subset size, and the results produced by different methods may not agree with each other.

Forward Selection

Forward stepwise selection of variables chooses the subset models by adding one variable at a time to the previously chosen subset. Forward selection starts by choosing as the one-variable subset the independent variable that accounts for the largest amount of variation in the dependent variable. This will be the variable having the highest simple correlation with Y. At each successive step, the variable in the subset of variables *not* already in the model that causes the largest decrease in the residual sum of squares is added to the subset. This will be the variable that has the highest correlation with the *residuals* from the current model. Without a termination rule, forward selection continues until all variables are in the model.

Backward Elimination

Backward elimination of variables chooses the subset models by starting with the full model and then eliminating at each step the one variable whose deletion will cause the residual sum of squares to increase the least. This will be the variable in the current subset model that has the smallest partial sum of squares. Without a termination rule, backward elimination continues until the subset model contains one variable.

Stepwise Selection

Neither forward selection nor backward elimination takes into account the effect that the addition or deletion of a variable can have on the contributions of other variables to the model. A variable added early to the model in forward selection can become unimportant after other variables are added, or variables previously dropped in backward elimination can become important after other variables are dropped from the model. The variable selection method commonly labeled **stepwise regression** is a forward selection process that rechecks at each step the importance of all previously included variables. If the partial sums of squares for any previously included variables do not meet a minimum criterion to stay in the model, the selection procedure changes to backward elimination and variables are dropped one at a time until all remaining variables meet the minimum criterion. Then, forward selection resumes.

Stepwise selection of variables requires more computing than forward or backward selection but has an advantage in terms of the number of potential subset models checked before the model for each subset size is decided. It is reasonable to expect stepwise selection to have a greater chance of choosing the best subsets in the sample data, but selection of the best subset for each subset size is not guaranteed.

Stopping Rules

The computer programs for the stepwise selection methods generally include criteria for terminating the selection process. In forward selection, the common criterion is the ratio of the reduction in residual sum of squares caused by the next candidate variable to be considered to the residual mean square from the model including that variable. This criterion can be expressed in terms of a critical "F-to-enter" or in terms of a critical "significance level to enter" (SLE), where F is the "F-test" of the partial sum of squares of the variable being considered. The forward selection terminates when no variable outside the model meets the criterion to enter. This "F-test," and the ones to follow, should be viewed only as *stopping rules* rather than as classical tests of significance. The use of the data to select the most favorable variables creates biases that invalidate these ratios as tests of significance (Berk, 1978).

The stopping rule for backward elimination is the "F-test" of the largest partial sum of squares of the variables remaining in the model. Again, this criterion can be stated in terms of an "F-to-stay" or as a "significance level to stay" (SLS). Backward elimination terminates when all variables remaining in the model meet the criterion to stay.

The stopping rule for stepwise selection of variables uses both the forward selection and backward elimination criteria. The variable selection process terminates when all variables in the model meet the criterion to stay *and* no variables outside the model meet the criterion to enter (except, perhaps, for the variable that was just eliminated). The criterion for a variable to enter the model need not be the same as the criterion for the variable to stay. There is some advantage in using a more relaxed criterion for entry to force the selection process to consider a larger number of subsets of variables.

Example 7.2

(*Continuation of Example* 7.1) The FORWARD, BACKWARD, and STEPWISE methods of variable selection in PROC REG (SAS Institute, Inc., 1985d) are illustrated with the Linthurst data. In this program, the termination rules are expressed in terms of "significance level to enter" (SLE) and "significance level to stay" (SLS). For this example, the criteria were set at SLE = .50 in forward selection, SLS = .10 in backward elimination, and SLE = .50 and SLS = .15 in stepwise selection. The values were chosen for forward and backward selection to allow the procedures to continue through most of the subset sizes. One can then tell by inspection of the results where the selection would have terminated with more stringent criteria.

The results from the *forward selection* method applied to the Linthurst data are summarized in Table 7.2. The F-ratio given is the ratio of the partial sum of squares for the variable to the mean square residual for the model containing all previously admitted variables plus the one being considered.

The best single variable is pH, which gives $(100)R^2 = 59.9\%$ (see Table 7.1)

Table 7.2 Summary statistics for forward selection of variables for the Linthurst data using significance level for variable to enter the model of SLE = .50. All models included an intercept.

Step	*Variable*	*Partial* SS	MS(Res)	R^2	F[a]	Prob > F[b]
1.	Determine best single variable and test for entry:					
	SAL	204,048	441,091	.0106	.46	.5001
	pH	11,490,388	178,618	.5994	64.33	.0001
	K	802,872	427,165	.0419	1.88	.1775
	Na	1,419,069	412,835	.0740	3.44	.0706
	Zn	7,474,474	272,011	.3899	27.48	.0001
	Best 1-variable model: *pH*				$C_p = 7.42$	
2.	Determine best second variable and test for entry:					
	SAL	77,327	181,030	.6034	.43	.5170
	K	924,266	160,865	.6476	5.75	.0211
	Na	1,132,401	155,909	.6584	7.26	.0101
	Zn	170,933	178,801	.6083	.96	.3338
	Best 2-variable model: *pH Na*				$C_p = 2.28$	
3.	Determine best third variable and test for entry:					
	SAL	11,778	159,424	.6590	.07	.7871
	K	36,938	158,811	.6604	.23	.6322
	Zn	77,026	157,833	.6625	.49	.4888
	Best 3-variable model: *pH Na Zn*				$C_p = 3.80$	
4.	Determine best fourth variable and test for entry:					
	SAL	178,674	157,312	.6718	1.136	.2929
	K	32,964	160,955	.6642	.205	.6533
	Best 4-variable model: *pH Na Zn SAL*				$C_p = 4.67$	
5.	Test last variable for entry:					
	K	106,211	158,622	.6773	.670	.4182
	Last variable is added with SLE = .50				$C_p = 6.00$	

[a] F is the ratio of the partial sum of squares for the variable divided by the residual mean square for the model including that variable.

[b] Prob > F is the probability of a larger F-value assuming the ratio is a valid F-statistic. The variable selection process biases these statistics; Prob > F is used as a variable selection criterion and not as a test of significance.

and $F = 64.3$. The corresponding significance level is far beyond the significance level needed to enter, SLE = .50. The second step of the forward selection computes the partial sums of squares for each of the remaining variables, *SALINITY*, *K*, *Na* and *Zn*, in a model that contains *pH* plus that particular variable. The partial sum of squares for *Na* is the largest and gives $F = 7.26$, or Prob > F = .0101, which satisfies the criterion for entry. Thus, *Na* is added to the model and the selection process goes to step 3. At the third step, the partial sum of squares for *Zn* is the largest and Prob > F = .4888 just meets the criterion for entry. *SALINITY* meets the criterion for entry at the fourth step, and *K* at the fifth step.

In this case, the choice of SLE = .50 allowed all variables to be included in the model. The selection would have stopped at the two-variable model with *pH* and *Na* had SLE been chosen anywhere between .4888 and .0101. Any choice of

Table 7.3 Summary statistics for the backward elimination of variables for the Linthurst data using significance level for staying of SLS = .10. All models included an intercept.

Step	*Variable*	*Partial* SS	$R^{2\,a}$	F^b	Prob > F^c
0	Model: All variables; $R^2 = .6773$, $C_p = 6$, $s^2 = 158{,}622$ with 39 d.f.				
	SAL	251,921	.6642	1.59	.2151
	pH	1,917,306	.5773	12.09	.0013
	K	106,211	.6718	.67	.4182
	Na	47,011	.6749	.30	.5893
	Zn	299,209	.6617	1.89	.1775
1	Model: *Na* removed; $R^2 = .6749$, $C_p = 4.30$, $s^2 = 155{,}832$ with 40 d.f.				
	SAL	436,496	.6521	2.80	.1020
	pH	1,885,805	.5765	12.10	.0012
	K	732,606	.6366	4.70	.0361
	Zn	434,796	.6522	2.79	.1027
2	Model: *Zn* removed; $R^2 = .6522$, $C_p = 5.04$, $s^2 = 162{,}636$ with 41 d.f.				
	SAL	88,239	.6476	.54	.4656
	pH	11,478,835	.0534	70.58	.0001
	K	935,178	.6034	5.75	.0211
	Delete *SAL*				
3	Model: *SAL* removed; $R^2 = .6476$, $C_p = 3.59$, $s^2 = 160{,}865$ with 42 d.f.				
	pH	11,611,782	.0419	72.18	.0001
	K	924,266	.5994	5.75	.0211

Stop. Prob > F for both remaining variables exceeds SLS = .10. Final model contains *pH* and *K* (plus an intercept).

[a] R^2 is for the model with the indicated variable removed.
[b] F is the ratio of the partial sum of squares for the indicated variable to s^2 for that step.
[c] Prob > F is the probability of a larger F-value assuming the ratio is a valid F-statistic. The variable selection process biases these statistics; Prob > F is used as a variable selection criterion and not as a test of significance.

SLE less than .0101 would have stopped the selection process with the one-variable model.

Forward selection chose the best subset models for $p = 1$, 2, and 3, but the second best model for $p = 4$ (see Table 7.1). This illustrates the fact that the stepwise methods are not guaranteed to find the best subset model for each subset size. In addition, the stepwise methods do not alert the user to the fact that other subsets at each stage may be as good. For example, one is not aware from the forward selection results that two other three-variable subsets [(*pH*, *K*, *Na*) and (*SAL*, *pH*, *Na*)] are essentially equivalent to the one chosen.

The stepwise regression results using *backward elimination* are summarized in Table 7.3. Starting with the full model, the procedure eliminates the variable with the smallest partial sum of squares if its sum of squares does not meet the criterion to stay in the model. In this example, the significance level to stay is set at SLS = .10. *Na* has the smallest partial sum of squares and is eliminated from the model because Prob > F = .5893 is larger than SLS = .10. This leaves (*SAL*, *pH*, *K*, *Zn*) as the chosen four-variable subset. Of these four variables, *Zn* has the smallest partial sum of squares (by a very small margin over *SALINITY*) and Prob > F = .1027, slightly larger than the criterion to stay SLS = .10.

Therefore, Zn is dropped from the model leaving (SAL, pH, K) as the chosen three-variable model. At the next step, SAL is dropped, giving (pH, K) as the chosen two-variable model. Both pH and K meet the criterion to stay (Prob $> F$ is less than SLS), and the backward selection process stops with that model.

Backward elimination identifies the best four-variable subset whereas forward selection did not. On the other hand, backward elimination chose the fourth best three-variable subset and the second best two-variable subset, whereas forward selection identified the best subset at these stages. If SLS had been set low enough, say at .02, backward elimination would have gone one step further and correctly identified pH as the best one-variable subset.

The *stepwise* method of variable selection applied to the Linthurst data starts the same as forward selection (Table 7.2). After the second step when pH and Na are both in the model, the stepwise method rechecks the contribution of each variable to determine if each should stay in the model. The partial sums of squares are

$$R(\beta_{pH}|\beta_{Na}) = 11{,}203{,}720$$

$$R(\beta_{Na}|\beta_{pH}) = 1{,}132{,}401$$

The mean square residual for this model is MS(Res) $= 155{,}909$ with 42 degrees of freedom. Both give large F-ratios with Prob $> F$ much smaller than SLS $=$ .15 so that both pH and Na are retained.

The forward selection phase of stepwise selection resumes with the choice of Zn as the third variable to be added (Table 7.2). Again, the contribution of each variable in the model is rechecked to determine if each should stay. The partial sums of squares are

$$R(\beta_{pH}|\beta_{Na}\ \beta_{Zn}) = 4{,}455{,}726$$

$$R(\beta_{Na}|\beta_{pH}\ \beta_{Zn}) = 1{,}038{,}493$$

$$R(\beta_{Zn}|\beta_{pH}\ \beta_{Na}) = 77{,}026$$

The mean square residual for this model is MS(Res) $= 157{,}833$ with 41 degrees of freedom. Both pH and Na meet the criterion to stay, but the F-ratio for Zn is less than 1.0 with Prob $> F = .4888$, which does not meet the criterion of SLS $= .15$. Therefore, Zn, which has just been added, is immediately dropped from the model.

The stepwise procedure then checks to see if any variables other than Zn meet the criterion to enter the model. The two remaining variables to be checked are $SALINITY$ and K. The partial sum of squares for each, adjusted for pH and Na, are given in step 3 of the forward selection, Table 7.2. The Prob $> F$ for both variables is larger than SLE $= .50$. Therefore, no other variables meet the criterion to enter the model and all variables in the model meet the criterion to stay so the selection terminates with the two-variable subset (pH, Na).

In general, the rechecking of previous decisions in stepwise selection should improve the chances of identifying the best subsets at each subset size. In this particular example, the choice of SLS = .15 caused the stepwise selection to terminate early. If SLS had been chosen equal to SLE = .50, stepwise regression would have followed the same path as forward selection until the fifth variable, K, had been added to the model. Then, rechecking the variables in the model would have caused Na to be dropped from the model, leaving (SAL, pH, K, Zn) as the selected four-variable subset. This is the best four-variable subset (Table 7.1), which forward selection failed to identify. ■

Warnings on Using Stepwise Methods

Even in the small example just discussed, there are several close contenders within most subset sizes, as shown by all possible regressions (Table 7.1). Each stepwise regression method reveals only one subset at each step and, if the stopping criteria are set to select a "best" subset size, only part of the subset models are identified. (Choice of criteria for this purpose will be discussed in Section 7.5.) In general, it is not recommended that the automated stepwise regression methods be used blindly to identify a "best" model. It is imperative that any model obtained in this manner be thoroughly checked for any inadequacies (see Chapter 9) and validated against an independent data set before being adopted (see Section 7.6).

If stepwise variable selection methods are to be used, they are best used as screening tools to identify contender models. For this purpose, forward selection and backward elimination methods alone provide very narrow views of the possible models; stepwise selection would be somewhat better. An even better option would be the joint use of all three methods. If forward selection and backward elimination identify the same subsets, then it is known that they will have identified the best subset in each subset size (Berk, 1978). One still would not have information on close contenders within each subset size. For screening purposes, the choice of the termination criteria could be such as to provide the greatest exposure to alternative models. For forward selection, this means that SLE should be large, say SLE = .5 or larger. For backward elimination, SLS should be small. For the stepwise method of selection, SLE should be large but the choice of SLS is not so easily specified. It may be worthwhile to try more than one choice of each.

For the purpose of identifying several contender models, one should not overlook the possible use of a program that utilizes the leaps-and-bounds algorithm, such as the RSQUARE option in PROC REG (SAS Institute, Inc., 1985d). This algorithm guarantees that the best m subset models within each subset size will be identified. Changing m from 1 to 10 approximately doubles the computing time (Furnival and Wilson, 1974). While the computing cost will be higher than for one of the stepwise methods, the cost may not be excessive, and considerably more information is obtained.

7.5 Criteria for Choice of Subset Size

Many criteria for choice of subset size have been proposed. Hocking (1976) reviews eight stopping rules, Bendel and Afifi (1977) compare eight (not all the same as Hocking's) in forward selection, and the RSQUARE method in PROC REG (SAS, 1985d) provides the option of computing twelve. Most of the criteria are monotone functions of the residual sum of squares for a given subset size and, consequently, give identical rankings of the subset models within each subset size. However, the choice of criteria may lead to different choices of subset size, and they may give different impressions of the magnitude of the differences among subset models. The latter may be particularly relevant when the purpose is to identify several competing models for further study.

Four commonly used criteria will be discussed briefly. In addition, the choice of F-to-enter and F-to-stay, or the corresponding "significance levels," SLE and SLS, will be reviewed. The four commonly used criteria to be discussed are

1. Coefficient of determination, R^2
2. Residual mean square, MS(Res)
3. Adjusted coefficient of determination, R^2_{adj}
4. Mallows' C_p statistic

The values for these criteria are given in Table 7.1 for all possible subsets from the Linthurst data.

7.5.1 Coefficient of Determination

Behavior of R^2

The **coefficient of determination**, R^2, is the proportion of the total (corrected) sum of squares of the dependent variable "explained" by the independent variables in the model:

$$R^2 = \frac{\text{SS(Regr)}}{\text{SS(Total)}} \qquad [7.1]$$

The objective is to select a model that accounts for as much of the variation in Y as is practical. Since R^2 cannot decrease as independent variables are added to the model, the model that gives the maximum R^2 will necessarily be the model that contains all independent variables. The typical plot of R^2 against the number of variables in the model is a steeply upward sloping curve that levels off near the maximum R^2 once the more important variables have been included. Thus, the use of the R^2 criterion for model building requires a judgment as to whether the increase in R^2 from additional variables justifies the increased complexity of the model. The subset size is chosen near the bend where the curve tends to flatten.

Example 7.3

(*Continuation of Example* 7.1) The best one-variable subset accounted for $100R^2 = 59.9\%$ of the variation in *BIOMASS*, the best two-variable subset accounted for $100R^2 = 65.8\%$, and the best three-variable subset accounted for $100R^2 = 66.3\%$ (see Figure 7.1 on page 182). The increase in R^2 from two to three variables was small and R^2 is close to the maximum of $100R^2 = 67.7\%$. Thus, the R^2 criterion leads to the choice of the two-variable subset containing *pH* and *Na* as the "best." ■

Figure 7.1 R^2, R^2_{adj}, and MS(Res) plotted against p' for the best model from each subset size for the Linthurst data.

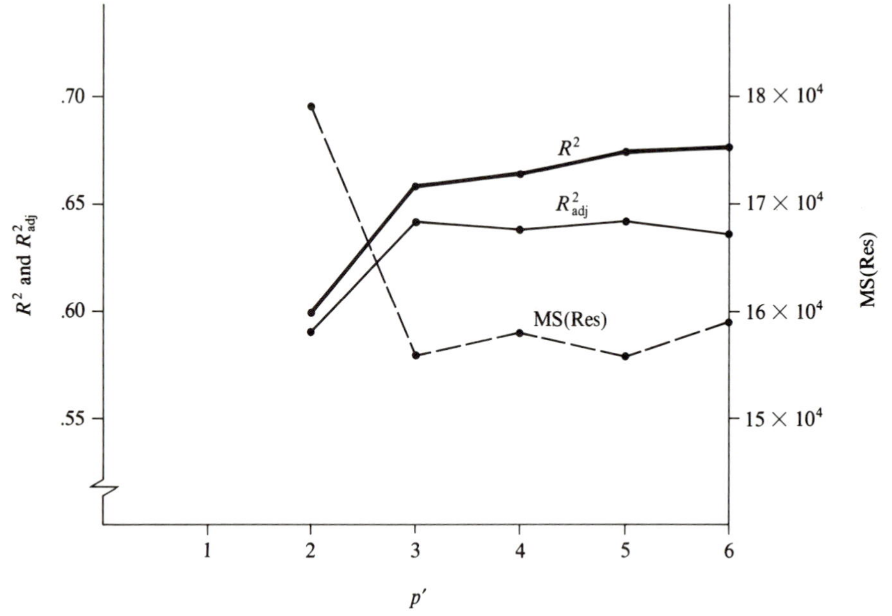

7.5.2 Residual Mean Square

***Expected Behavior of* MS(Res)**

The **residual mean square**, MS(Res), is an estimate of σ^2 if the model contains all relevant independent variables. If relevant independent variables have been omitted, the residual mean square is biased upward. Including an unimportant independent variable will have little impact on the residual mean square. Thus, the expected behavior of the residual mean square, as variables are added to the model, is for it to decrease toward σ^2 as important independent variables are added to the model and to fluctuate around σ^2 once all relevant variables have been included.

Behavior of* MS(Res) *with Variable Selection

The previous paragraph describes the expected behavior of MS(Res) when the selection of variables is not based on sample data. Berk (1978) demonstrated with simulation that selection of variables based on the sample data causes MS(Res) to be biased downward. In his studies, the bias was as much as 25% when sample sizes were less than 50. The bias tended to reach its peak in the early stages of forward selection, when one-third to one-half of the total number of variables had been admitted to the model. In backward elimination, the bias tended to peak when slightly more than half of the variables had been eliminated. These results suggest that the pattern of MS(Res) as variables are added in a variable selection procedure will be to drop slightly below σ^2 in the intermediate stages of the selection and then to return to near σ^2 as the full model is approached. It is unlikely that a bias of this magnitude would be detectable in plots of MS(Res) against number of variables, particularly in small samples where the bias is most serious.

The pattern of the residual mean squares, as variables are added to the model, is used to judge when the residual mean square is estimating σ^2 and, by inference, when the model contains the important independent variables. In larger regression problems, with many independent variables and several times as many observations, a plot of the residual mean square against the number of parameters in the model will show when the plateau has been reached. The plateau may not be clearly defined in smaller problems.

Example 7.4

For the Linthurst data (Example 7.1), MS(Res) drops from MS(Res) = 178,618 for the best one-variable subset to MS(Res) = 155,909 for the best two-variable subset, and then changes little beyond that point (see Table 7.1 and Figure 7.1). The two-variable subset would be chosen by this criterion. ■

7.5.3 Adjusted Coefficient of Determination

Behavior of R^2_{adj}

The **adjusted R^2**, which is labeled as R^2_{adj}, is a rescaling of R^2 by degrees of freedom so that it involves a ratio of mean squares rather than sums of squares:

$$R^2_{adj} = 1 - \frac{\text{MS(Res)}}{\text{MS(Total)}} = 1 - \frac{(1 - R^2)(n - 1)}{(n - p')} \qquad [7.2]$$

This expression removes the impact of degrees of freedom and gives a quantity that is more comparable than R^2 over models involving different numbers of parameters. Unlike R^2, R^2_{adj} need not always increase as variables are added to the model. The value of R^2_{adj} will tend to stabilize around some upper limit as variables are added. The simplest model with R^2_{adj} near this upper limit is chosen as the "best" model. R^2_{adj} is closely related to MS(Res) (see equation 7.2), and will lead to the same conclusions.

Example 7.5

For the Linthurst data, the maximum R^2_{adj} for the one-variable subset is R^2_{adj} = .590 (see Table 7.1 and Figure 7.1). This increases to .642 for the two-variable subset, and then shows no further increase; R^2_{adj} = .638, .642, and .636 for p' = 4, 5, and 6, respectively. ■

7.5.4 Mallows' C_p Statistic

Behavior of C_p

The C_p statistic is an estimate of the standardized total mean squared error of estimation for the current set of data (Hocking, 1976). The C_p statistic and the C_p plot were initially described by Mallows [see Mallows (1973a) for earlier references]. The C_p statistic is computed as

$$C_p = \frac{\text{SS(Res)}_p}{s^2} + 2p' - n \qquad [7.3]$$

where SS(Res)_p is the residual sum of squares from the p-variable subset model being considered and s^2 is an estimate of σ^2, either from independent information or, more commonly, from the model containing all independent variables. When the model is correct, the residual sum of squares is an unbiased estimate of $(n - p')\sigma^2$; in this case, C_p is (approximately) equal to p'. When important independent variables have been omitted from the model, the residual sum of squares is an estimate of $(n - p')\sigma^2$ *plus* a positive quantity reflecting the contribution of the omitted variables; in this case, C_p is expected to be greater than p'.

C_p Plot

The C_p plot presents C_p as a function of p' for the better subset models and provides a convenient method of selecting the subset size and judging the competitor subsets. The usual pattern is for the minimum C_p statistic for each subset size, $C_{p\,\min}$, to be much larger than p' when p' is small, to decrease toward p' as the important variables are added to the model, and then to fall below or fluctuate around p'. When the residual mean square from the full model has been used as s^2, C_p will equal p' for the full model. A value of C_p near p' indicates little bias in MS(Res) as an estimate of σ^2. (This interpretation assumes that s^2 in the denominator of C_p is an unbiased estimate of σ^2. If s^2 has been obtained from the full model, s^2 is an unbiased estimate of σ^2 only if the full model contains all relevant variables.)

C_p Criterion

Different criteria have been advanced for the use of C_p. Mallows (1973a) suggested that all subset models with small C_p and with C_p close to p' be considered for further study. Hocking (1976) defined two criteria depending on whether the model is intended primarily for prediction or for parameter estimation. He used the criterion $C_p \le p'$ for prediction. For parameter estimation, Hocking argued that fewer variables should be eliminated from the model, to avoid excessive bias in the estimates, and provided the selection criterion $C_p \le 2p' - t$, where t is the number of variables in the full model.

Example 7.6

The C_p plot for the Linthurst example is given in Figure 7.2. Only the smaller C_p statistics, the dots, are shown for each value of p', with the $C_{p\,\min}$ values connected by the dashed line. The figure includes two reference lines corresponding to Hocking's two criteria, $C_p = p'$ and $C_p = 2p' - t$. The C_p statistics for all subsets are given in Table 7.1. For the one-variable subsets, $C_{p\,\min} = 7.42$, well above $p' = 2$. For the two-variable subsets, $C_{p\,\min} = 2.28$, just below $p' = 3$. The next best two-variable subset has $C_p = 3.59$, somewhat above $p' = 3$. Three three-variable subsets give C_p close to $p' = 4$ with $C_{p\,\min} = 3.80$. The C_p statistics for the four-variable subsets identify two subsets with $C_p \le p'$. Two other subsets have C_p slightly greater than p'.

Mallows' C_p criterion (which requires C_p small and near p') identifies the two-variable subsets (pH, Na) and (pH, K), and possibly the three-variable subsets (pH, Na, Zn), (pH, K, Na), and $(SALINITY, pH, Na)$, as contending subsets for further study. Preference would be given to (pH, Na) if this model appears to be adequate when subjected to further study. Hocking's criterion for selection of the best subset model *for prediction* leads to the two-variable model (pH, Na); $C_p = 2.28$ is less than $p' = 3$. The more restrictive criterion for subset selection *for parameter estimation* leads to the best four-variable subset $(SALINITY, pH, K, Zn)$; $C_p = 4.30$ is less than $2p' - t = 5$. ■

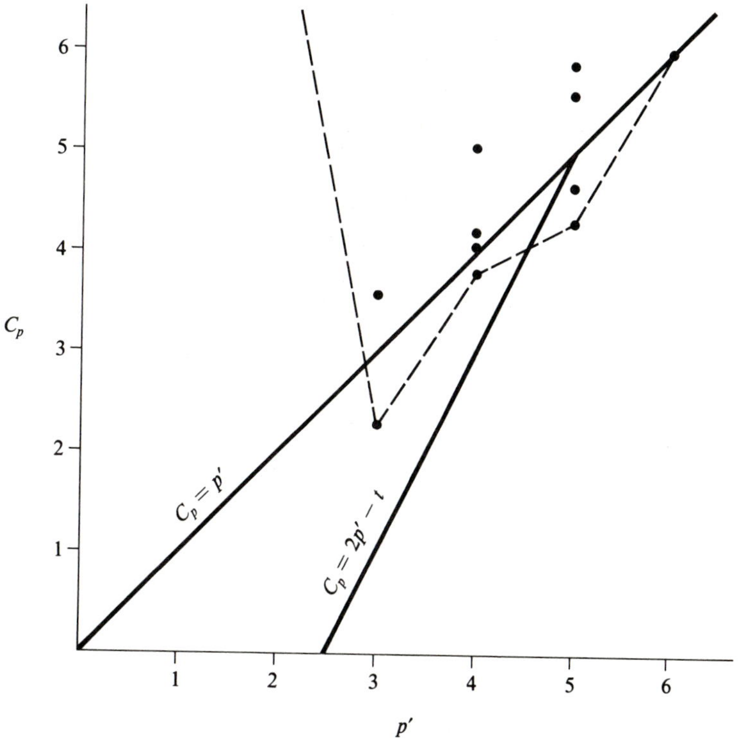

Figure 7.2 The C_p plot for the Linthurst data. The dashed line connects $C_{p\,\min}$ for each subset size. The two solid lines are the reference lines for subset selection according to Hocking's criteria.

7.5.5 "Significance Levels" for Choice of Subset Size

Use of "Significance Levels"

F-to-enter and F-to-stay, or the equivalent "significance levels," in the stepwise variable selection methods serve as subset size selection criteria when they are chosen so as to terminate the selection process before all subset sizes have been considered. Bendel and Afifi (1977) compared several stopping rules for forward selection and showed that the sequential F-test based on a constant "significance level" compared very favorably. The optimum "significance level to enter," SLE, varied between SLE = .15 and .25. Although not the best of the criteria they studied, the sequential F-test with SLE = .15 allowed one to do "almost best" when $n - p \leq 20$. When $n - p \geq 40$, the C_p statistic was preferred over the sequential F-test but by a very slight margin if SLE = .20 were used.

This is similar to the conclusion reached by Kennedy and Bancroft (1971) for the sequential F-test but where the order of importance of the variables was known a priori. They concluded that the significance level should be .25 for forward selection and .10 for backward elimination. Bendel and Afifi did not speculate on the choice of "significance level to stay," SLS, in backward elimination. For stepwise selection, they recommended the same levels of SLE as for forward selection and half that level for SLS.

Example 7.7 For the Linthurst data of Example 7.1, the Bendel and Afifi level of SLE = .15 would have terminated forward selection with the two-variable subset (*pH*, *Na*) (see Table 7.2). The Kennedy and Bancroft suggestion of using SLS = .10 for backward elimination gives the results shown in Table 7.3 terminating with the two-variable subset (*pH*, *K*). In this case, the backward elimination barely bypassed the second step where the least significant of the four variables had Prob > F = .1027. The recommended significance levels of SLE = 2(SLS) = .15 for the stepwise selection method terminates at the same point as forward selection. ■

Conclusions

In summary of the choice of subset size, some of the other conclusions of Bendel and Afifi (1977) regarding stopping rules are of importance. First, the use of all independent variables is a very poor rule unless $n - p$ is very large. For their studies, the use of all variables was always inferior to the best stopping rule. This is consistent with the theoretical results (Section 7.2) that showed larger variances for $\hat{\beta}$, $\hat{Y}$, and $\hat{Y}_{\text{pred}}$ for the full models. Second, most of the stopping rules do poorly if $n - p \leq 10$. The C_p statistic does poorly when $n - p \leq 10$ (but is recommended for $n - p \geq 40$). Third, the lack-of-fit test of the $(t - p)$ variables that have been dropped (an intuitively logical procedure but not discussed in this text) is generally very poor as a stopping rule regardless of the significance level used. Finally, an unbiased version of the coefficient of determination generally did poorly unless $n - p$ was large. This suggests that R^2, and perhaps R^2_{adj} and MS(Res), may not serve as good stopping rules for subset size selection.

Mallows' C_p statistic and significance levels appear to be the most favored criteria for subset size selection. The C_p statistic was not the optimum choice of Bendel and Afifi in the intermediate-sized data sets and it did poorly for very small samples. Significance level as a criterion did slightly better than C_p in the intermediate-sized studies. The poor performance of C_p in the small samples should not be taken as an indictment. First, none of the criteria did well in such studies and, second, no variable selection routine or model building exercise should be taken seriously when the sample sizes are as small as $n - p \leq 10$.

7.6 Model Validation

Importance of Validation

Validation of a fitted regression equation is the demonstration or confirmation that the model is sound and effective for the purpose for which it was intended. This is *not* equivalent to demonstrating that the fitted equation agrees well with the data from which it was computed. Validation of the model requires assessing the effectiveness of the fitted equation against an *independent* set of data, and is essential if confidence in the model is to be expected.

Results from the regression analysis—R^2, MS(Res), and so forth—do not necessarily reflect the degree of agreement one might obtain from future applications of the equation. The model building exercise has searched through many possible combinations of variables and mathematical forms for the model. In addition, least squares estimation has given the best possible agreement of the chosen model with the observed data. As a result, the fitted equation is expected to fit the data from which it was computed better than it will fit an independent set of data. In fact, the fitted equation quite likely will fit the sample data better than the *true* model would if it were known.

A fitted model should be validated for the specific objective for which it was planned. An equation that is good for predicting Y_i in a given region of the X-space might be a poor predictor in another region of the X-space, or for estimation of a mean change in Y for a given change in X even in the same region of the X-space. Two criteria are of interest:

1. Does the fitted regression equation provide unbiased predictions of the quantities of interest?
2. Is the precision of the prediction good enough (the variance small enough) to accomplish the objective of the study?

Mean Squared Error of Prediction

Both quantities, bias and variance, are sometimes incorporated into a single measure called the **mean squared error of prediction** (MSEP). Mean squared error of prediction is defined as the average squared difference between independent observations and predictions from the fitted equation for the corresponding values of the independent variables. The mean squared error of prediction incorporates both the variance of prediction and the square of the bias of the prediction.

Example 7.8

For illustration, suppose a model has been developed to predict maximum rate of runoff from watersheds following rain storms. The independent variables are rate of rainfall (inches per hour), acreage of watershed, average slope of land in the watershed, soil moisture levels, soil type, amount of urban development, and amount and type of vegetative cover. The dependent variable is maximum rate of runoff ($\text{ft}^3\text{sec}^{-1}$), or peak flow. Assume the immediate interest in the model is prediction of peak flow for a particular watershed. The model is to be validated for this watershed by comparing observed rates of peak flow with the model predictions for ten episodes of rain. The observed peak flow, the predicted peak flow, and the error of prediction are given in Table 7.4 (page 188) for each episode.

The average prediction bias is $\bar{\delta} = -289$ $\text{ft}^3\text{sec}^{-1}$; the peak flow in these data is underestimated by approximately 6%. The variance of the prediction error is $s^2(\delta) = 255{,}477$, or $s(\delta) = 505$ $\text{ft}^3\text{sec}^{-1}$. The standard error of the estimated mean bias is $s(\bar{\delta}) = 505/\sqrt{10} = 160$. A t-test of the hypothesis that the bias is zero gives $t = -1.81$, which, with 9 degrees of freedom and $\alpha = .05$, is not significant.

The mean squared error of prediction is

$$\text{MSEP} = \frac{\delta'\delta}{n} = 313{,}450$$

or

Table 7.4 Observed rate of runoff, predicted rate of runoff, and prediction error for validation of water runoff model. Results are listed in increasing order of runoff (ft^3 sec^{-1}).

	Predicted P	*Observed* Y	*Prediction Error* $\delta = P - Y$
	2,320	2,380	−60
	3,300	3,190	110
	3,290	3,270	20
	3,460	3,530	−70
	3,770	3,980	−210
	4,210	4,390	−180
	5,470	5,400	70
	5,510	5,770	−260
	6,120	6,890	−770
	6,780	8,320	−1,540
Mean	4,423	4,712	−289

$$\text{MSEP} = \frac{(n-1)s^2(\delta)}{n} + (\bar{\delta})^2$$

$$= \frac{9(255{,}477)}{10} + (-289)^2 = 313{,}450$$

The bias term contributes 27% of MSEP. The square root of MSEP gives 560 ft^{3}sec^{-1}, an approximate 12% error in prediction.

Even though the average bias is not significantly different from zero, the very large prediction error on the largest peak flow (Table 7.4) suggests that the regression equation is not adequate for heavy rainfalls. Review of the data from which the equation was developed shows very few episodes of rainfall as heavy as the last in the validation data set. If the last rainfall episode is omitted from the computations, the average bias drops to $\bar{\delta} = -150$ ft^{3}sec^{-1} with a standard deviation of $s(\delta) = 265$, or a standard error of the mean of $s(\bar{\delta}) = 88.2$. Again, the average bias is not significantly different from zero using these nine episodes. However, the error of prediction on the largest rainfall differs from zero by $-1{,}540/265 = -5.8$ standard deviations. This is a clear indication that the regression equation is seriously biased for the more intense rainfalls and must be modified before it can be used with confidence. ■

Choosing the Data Set for Validation

In Example 7.8, the peak flow model was being validated for a particular watershed. If the intended use of the model had been prediction of peak flow from several watersheds over a large geographical area, this sample of data would have been inadequate for validation of the model. Validation on one watershed would not provide assurance that the equation would function well over a wide range of watersheds. The data to be used for validation of a model must represent the population for which the predictions are to be made.

Splitting the Data Set

It often is impractical to obtain an adequate independent data set with which to validate a model. If the existing data set is sufficiently large, an alternative is to use that data for both estimation and validation. One approach is to divide the data set into two representative halves; one half is then used to develop the regression model and the other half is used for validation of the model. Snee (1977) suggests that the total sample size should be greater than $2p' + 25$ before splitting the sample is considered. Of course, one could reverse the roles of the two data sets and have double estimation and validation. Presumably, after the validation, and assuming satisfactory results, one would prefer to combine the information from the two halves to obtain one model that would be better than either alone.

***Estimating* MSEP**

Methods have been devised for estimating the mean squared error of prediction, MSEP, when it is not practical to obtain new independent data. The C_p statistic can be considered an estimator of MSEP. Weisberg (1981) presents a method of allocating C_p to the individual observations which facilitates detecting inadequacies in the model. Another approach is to measure the discrepancy between each observation and its prediction but where that observation was not used in the development of the prediction equation. The sum of squares of these discrepancies is the *PRESS* statistic given by Allen (1971b). Let $\hat{Y}_{\text{pred}_{i(i)}}$ be the prediction of observation i where the (i) indicates that the ith observation was not used in the development of the regression equation. Then,

$$PRESS = \sum_{i}^{n} (Y_i - \hat{Y}_{\text{pred}_{i(i)}})^2 \qquad [7.4]$$

The individual discrepancies are of particular interest for model validation. Unusually large discrepancies or patterns to the discrepancies can indicate inadequacies in the model. Bunke and Droge (1984) derive a best unbiased estimator and a minimum mean squared error estimator of MSEP where there is replication and all variables are assumed to have a multivariate normal distribution.

Validation of the model based on an independent sampling of the population is to be preferred to the use of estimates of mean squared error of prediction based on the original sample data. Part of the error of prediction may arise because the original data do not adequately represent the original population. Or, the population may have changed in some respects since the original sample was taken. Estimates of MSEP computed from the original data cannot detect these sources of inadequacies in the model.

EXERCISES

* 7.1. Show algebraically the relationship between R^2 and MS(Res).

7.2. Show algebraically the relationship between R^2 and C_p. Between MS(Res) and C_p.

7.3. Show that C_p is approximately an estimate of p' when the model is correct.

7.4. Use the relationship between R^2 and MS(Res), exercise 7.1, to show equality between the two forms of R^2_{adj} in equation 7.2.

* 7.5. The following approach was used to determine the effect of acid rain on agricultural production. U.S. Department of Agriculture statistics on crop production, fertilizer practices, insect control, fuel costs, land costs, equipment costs, labor costs, and so forth for each county in the geographical area of interest were paired with county

level estimates of average pH of rainfall for the year. A multiple regression analysis was run in which "production ($)" was used as the dependent variable and all input costs plus pH of rainfall were used as independent variables. A stepwise regression analysis was used with pH forced to be in all regressions. The partial regression coefficient on pH from the model chosen by stepwise regression was taken as the measure of the impact of acid rain on crop production.

a. Discuss the validity of these data for establishing a causal relationship between acid rain and crop production.
b. Suppose a causal effect of acid rain on crop production had already been established from other research. Discuss the use of the partial regression coefficient for pH from these data to predict the change in crop production that would result if rain acidity were to be decreased. Do you see any reason the prediction might not be valid?
c. Suppose the regression coefficient for pH were significantly negative (higher pH predicts lower crop production). Do you see any problem with inferring that stricter government air pollution standards on industry would result in an increase in crop production?
d. Do you see any potential for bias in the estimate of the partial regression coefficient for pH resulting from the fact that pH was forced to be in every model?

7.6. The final model in the Linthurst example in this chapter used pH and Na content of the marsh substrate as the independent variables for predicting biomass (in the forward selection and stepwise methods). The regression equation was

$$\hat{Y}_i = -476 + 407X_{\text{pH}} - .0233X_{\text{Na}}$$

What inference are you willing to make about the relative importance of pH and Na versus salinity, K, and Zn as biologically important variables in determining biomass? When all five variables were in the model, the partial regression coefficient for pH was a nonsignificant $-.009$ ($\pm .016$). Does this result modify your inference?

The remaining questions use the simulated data on peak flow of water used in the exercises in Chapter 5. Use $LQ = \ln(Q)$ as the dependent variable with the logarithms of the nine independent variables.

* 7.7. Determine the total number of possible models when there are nine independent variables, as in the peak water flow problem. Your computing resources may not permit computing all possible regressions. Use a program such as METHOD = RSQUARE option in PROC REG (SAS Institute, Inc., 1985d) to find the $n = 6$ "best" subsets in each stage. This will require using the SELECT = n option. Plot the behavior of the C_p statistic and determine the "best" model.

7.8. Use a forward selection variable selection method to search for an acceptable model for the peak flow data. Use SLE = .50 for entry of a variable into the model. What subset model is selected? What subset model is selected if you use SLE = .15? Compute and plot the C_p statistic for the models from SLE = .50. What subset model do you select for prediction using C_p?

* 7.9. Repeat exercise 7.8 using backward elimination. Use SLS = .10 for elimination of a variable. What subset model is selected? Compute and plot the C_p statistic for the models used and decide on the "best" model. Does backward elimination give the same model as forward selection in exercise 7.8?

7.10. Repeat exercise 7.8 using the stepwise method of variable selection. Use SLE = .50 and SLS = .20 for elimination of a variable from the model. What subset model is selected? Plot the C_p statistic for the models used to decide which model to adopt. Do you arrive at the same model as with forward selection? As with backward elimination?

*7.11. Give a complete summary of the results for the model you adopted from the backward elimination method in exercise 7.9. Give the analysis of variance, the partial regression coefficients, their standard errors, and R^2.

7.12. Your analysis of the peak flow data has been done on $\ln(Q)$. Reexpress your final model on the original scale (by taking the antilogarithm of your equation). Does this equation make sense; that is, are the variables the ones you would expect to be important and do they enter the equation the way common sense would suggest? Are there omitted variables you would have thought important?

CHAPTER
EIGHT

Class Variables in Regression

In all previous discussions, the independent variables were continuous or quantitative variables. There are many situations in which this is too restrictive.

This chapter introduces the use of categorical or class variables in regression models. The use of class variables broadens the scope of regression to include the classical analysis of variance models and models containing both continuous and class variables, such as analysis of covariance models and models to test homogeneity of regressions over groups.

To this point, only quantitative variables have been used as independent variables in regression models. This chapter extends the models to include qualitative (or categorical) variables. Quantitative variables are the result of some measurement such as length, weight, temperature, area, or volume. There is always a logical ordering attached to the measurements of such variables. Qualitative variables, on the other hand, identify the state, category, or class to which the observation belongs, such as hair color, sex, breed, or country of origin. There may or may not be a logical ordering to the classes. Such variables will be called **class variables**.

Class variables greatly increase the flexibility of regression models. This chapter shows how class variables are included in regression models with the use of **indicator variables** or **dummy variables**. The classical analyses of variance for the standard experimental designs are then shown to be special cases of ordinary least squares regression using class variables. This forms the basis for the more general linear model analysis of unbalanced data where conventional analyses of variance are no longer valid (Chapter 16). Then class variables and continuous variables are used jointly to discuss the test of homogeneity of regressions (Section 8.6) and the analysis of covariance (Section 8.7).

Some of the material in the analysis of variance sections of this chapter (Sections 8.2–8.5) will not be used again until Chapter 16. This material is placed here, rather than immediately preceding Chapter 16, in order to provide the reader with an early appreciation of the generality of regression analyses, and to provide the tools for tests of homogeneity that will be used from time to time throughout the text.

8.1 Description of Class Variables

Class Variables

A **class variable** identifies, by an appropriate code, the distinct classes or *levels* of the variable. For example, a code that identifies the different genetic lines, or cultivars, in a field experiment is a class variable. The classes or levels of the variable are the code names or numbers that have been assigned to represent the cultivars. The variation in the dependent variable attributable to this class variable is the total variation among the cultivar classes. It does not make sense to think of a continuous response curve relating a dependent variable to a class variable. There frequently is no logical ordering of the class variable or, if there is a logical ordering, the relative spacing of the classes on a quantitative scale is often not well defined.

Quantitative Variables as Class Variables

There are situations in which a quantitative variable is treated (temporarily) as a class variable. That is, the quantitative information contained in the variable is ignored and only the distinct categories or classes are considered. For example, assume the treatments in an experiment are the amounts of fertilizer applied to each experimental unit. The independent variable "amount of fertilizer" is, of course, quantitative. However, as part of the total analysis of the effects of the fertilizer, the total variation among the treatment categories is of interest. The sum of squares "among levels of fertilizer" is the treatment sum of squares and is obtained by using the variable "amount of fertilizer" as a class variable. For this purpose, the quantitative information contained in the variable "amount of fertilizer" is ignored; the variable is used only to identify the grouping or class identification of the observations. Subsequent analyses to determine the nature of the response curve would use the quantitative information in the variable.

The completely random and the randomized complete block experimental designs will be used to illustrate the use of class variables in the least squares regression model. Then, a class variable will be introduced to test homogeneity of regression coefficients (for a continuous variable) over the levels of the class variable. Finally, continuous and class variables will be combined to give the analysis of covariance in the regression context.

8.2 The Model for One-Way Structured Data

The Model

The model for one-way structured data, of which the completely random design (CRD) is the most common example, can be written either as

$$Y_{ij} = \mu_i + \varepsilon_{ij}$$

or

$$Y_{ij} = \mu + \tau_i + \varepsilon_{ij} \qquad [8.1]$$

where $\mu_i = \mu + \tau_i$ is the mean of the ith group or treatment and ε_{ij} is the random error associated with the jth observation in the ith group, $j = 1, \ldots, r$. The group mean μ_i in the first form is expressed in the second form in terms of an overall constant μ and the effect of the ith group or treatment τ_i, $i = 1, \ldots, t$. The first form is called the **means model**; the second is the classical **effects model** (equation 8.1).

The model assumes that the members of each group are randomly selected from the population of individuals in that group or, in the case of the completely random experimental design, that each treatment has been randomly assigned to r experimental units. (The number of observations in each group or treatment need not be constant but is assumed to be constant for this discussion.)

Class Variable Defined

The data set consists of two columns of information, one containing the response for the dependent variable Y_{ij} and one designating the group or treatment from which the observation came. The code used to designate the group is the class variable. In the case of the CRD, the class variable is the treatment code. For convenience, the class variable will be called *treatment* and $i = 1, 2, \ldots, t$ will designate the *level* of the class variable.

Model in Matrix Notation

It is easier to see the transition of this model to matrix form if the observations are listed:

$$
\begin{aligned}
Y_{11} &= \mu + \tau_1 + \varepsilon_{11} \\
Y_{12} &= \mu + \tau_1 + \varepsilon_{12} \\
&\vdots \\
Y_{1r} &= \mu + \tau_1 + \varepsilon_{1r} \\
Y_{21} &= \mu + \tau_2 + \varepsilon_{21} \\
&\vdots \\
Y_{2r} &= \mu + \tau_2 + \varepsilon_{2r} \\
&\vdots \\
Y_{tr} &= \mu + \tau_t + \varepsilon_{tr}
\end{aligned}
\qquad [8.2]
$$

The observations are ordered so that the first r observations are from the first treatment, the second r observations are from the second treatment, and so forth. The particular order is not critical as long as the same order is consistently used. The total number of observations is $n = rt$ so that the vector of observations on the dependent variable $\boldsymbol{Y}$ is of order $n \times 1$. The total number of parameters is $t + 1$: μ and t τ's. The vector of parameters is written

$$\boldsymbol{\beta}' = (\mu \;\; \tau_1 \;\; \tau_2 \;\; \cdots \;\; \tau_t) \qquad [8.3]$$

Dummy Variables

In order to express the algebraic model (equation 8.1) in matrix form, we must define $\boldsymbol{X}$ such that the product $\boldsymbol{X\beta}$ associates μ with every observation but each τ_i with only the observations from the ith group. Including μ with every observation is the same as including the common intercept in the usual regression equation. Therefore, the first column of $\boldsymbol{X}$ is $\mathbf{1}$, a column of ones. The remaining columns of $\boldsymbol{X}$ assign the treatment effects to the appropriate observations. This is done by defining a series of **indicator variables** or **dummy variables**, variables that take only the values zero or one. A dummy variable is defined for each level of the class variable. The ith dummy variable is an $n \times 1$ column vector with ones in the rows corresponding to the observations receiving the ith treatment and zeros elsewhere. Thus, $\boldsymbol{X}$ is of order $n \times (t + 1)$.

Example 8.1

To illustrate the pattern, assume there are four treatments ($t = 4$) with two replications per treatment ($r = 2$). Then $\boldsymbol{X}$ is an 8×5 matrix and $\boldsymbol{\beta}$ is 5×1:

$$\boldsymbol{X} = \begin{bmatrix} 1 & 1 & 0 & 0 & 0 \\ 1 & 1 & 0 & 0 & 0 \\ 1 & 0 & 1 & 0 & 0 \\ 1 & 0 & 1 & 0 & 0 \\ 1 & 0 & 0 & 1 & 0 \\ 1 & 0 & 0 & 1 & 0 \\ 1 & 0 & 0 & 0 & 1 \\ 1 & 0 & 0 & 0 & 1 \end{bmatrix} \qquad \boldsymbol{\beta} = \begin{pmatrix} \mu \\ \tau_1 \\ \tau_2 \\ \tau_3 \\ \tau_4 \end{pmatrix} \tag{8.4}$$

The second column of $\boldsymbol{X}$ is the dummy variable identifying the observations from treatment 1, the third column identifies the observations from treatment 2, and so on. For this reason, the dummy variables are sometimes called **indicator variables** and $\boldsymbol{X}$ the **indicator matrix**. The reader should verify that multiplication of $\boldsymbol{X}$ by $\boldsymbol{\beta}$ generates the same pattern of model effects shown in equation 8.2.

■

***X* Is Singular**

With these definitions of $\boldsymbol{X}$ and $\boldsymbol{\beta}$, the model for the completely random design can be written as

$$\boldsymbol{Y} = \boldsymbol{X\beta} + \boldsymbol{\varepsilon} \tag{8.5}$$

which is the usual matrix form of the least squares model. The difference now is that $\boldsymbol{X}$ is not a full-rank matrix; $r(\boldsymbol{X})$ is less than the number of columns of $\boldsymbol{X}$. The singularity in $\boldsymbol{X}$ is evident from the fact that the sum of the last four columns is equal to the first column. This singularity indicates that the model as defined has too many parameters; it is overparameterized.

Since $\boldsymbol{X}$ is not of full rank, the unique $(\boldsymbol{X}'\boldsymbol{X})^{-1}$ does not exist. Therefore, there is no unique solution to the normal equations as there is with the full-rank models. The absence of a unique solution indicates that at least some of the parameters in the model cannot be estimated; they are said to be **nonestimable**. (Estimability will be discussed more fully later.)

SS(Regr)

Recall that the degrees of freedom associated with the model sum of squares is determined by the rank of X. In the full-rank models, $r(X)$ always equals the number of columns of X. Here, however, there is one linear dependency among the columns of X, so the rank of X is t rather than $t + 1$. There will be only t degrees of freedom associated with SS(Model). Adjusting the sum of squares for μ uses one degree of freedom, leaving $(t - 1)$ degrees of freedom for SS(Regr). This SS(Regr) is the partial sum of squares for the t dummy variables defined from the class variable. For convenience, we will refer to SS(Regr) more simply as the sum of squares for the class variable. This sum of squares, with $(t - 1)$ degrees of freedom, is the treatment sum of squares in the analysis of variance for the completely random experimental design.

Approaches When X Is Singular

There are two approaches to handling linear models that are not of full rank:

1. Redefine, or reparameterize, the model so that it is a full-rank model; or
2. Use one of the nonunique solutions to the normal equations to obtain the regression results.

Reparameterization of the model was the standard approach before computers and is still used in many instances. Understanding reparameterization is helpful in understanding the results of the second approach, which is used in most computer programs for the analysis of general linear models.

8.3 Reparameterizing to Remove Singularities

Purpose

The purpose of reparameterization is to redefine the model so that it is of full rank. This is accomplished by imposing linear constraints on the parameters so as to reduce the number of unspecified parameters to equal the rank of X. Then, with X of full rank, ordinary least squares can be used to obtain a solution. If there is one singularity in X, one constraint must be imposed, or the number of parameters must be reduced by 1. Two singularities require the number of parameters to be reduced by 2, and so on. There are several alternative reparameterizations for each case. Three common ones will be illustrated, each of which gives a full-rank model.

Notation

Each reparameterization carries with it a redefinition of the parameters remaining in the model and corresponding modifications in X. To distinguish the reparameterized model from the original model, an asterisk will be appended to β and X, and to the individual parameters when the same symbols are used for both sets. Thus, the reparameterized models will be written as $Y = X^*\beta^* + \varepsilon$ with X^* and β^* appropriately defined.

8.3.1 Reparameterizing with the Means Model

Defining the Model

The means model, letting $\mu_i = \mu + \tau_i$, is presented here as a reparameterization of the classical effects model. The $(t + 1)$ parameters in the effects model are replaced with the t parameters μ_i. The model becomes

$$Y_{ij} = \mu_i + \varepsilon_{ij} \qquad [8.6]$$

(This redefinition of the model is equivalent to imposing the constraint that $\mu = 0$ in the original model, leaving only τ_1 to τ_4 to be estimated. Because of the obvious link of the new parameters to the group means, the usual notation for a population mean, μ, is used in place of τ.)

While the means model is used here as a reparameterization of the effects model, it is a valid model in its own right and is often proposed as the more direct approach to the analysis of data (Hocking, 1985). The essential difference between the two models is that the algebraic form of the classical effects model conveys the structure of the data, which in turn generates logical hypotheses and sums of squares in the analysis. The means model, on the other hand, conveys the structure of the data in constraints imposed on the μ_i and in hypotheses specified by the analyst. This text emphasizes the use of the classical effects model. The reader is referred to Hocking (1985) for discussions on the use of the means model.

Defining the Matrices

The reparameterized model is written as

$$\boldsymbol{Y} = \boldsymbol{X}^*\boldsymbol{\beta}^* + \boldsymbol{\varepsilon}$$

where

$$\boldsymbol{X}^* = \begin{bmatrix} 1 & 0 & 0 & 0 \\ 1 & 0 & 0 & 0 \\ 0 & 1 & 0 & 0 \\ 0 & 1 & 0 & 0 \\ 0 & 0 & 1 & 0 \\ 0 & 0 & 1 & 0 \\ 0 & 0 & 0 & 1 \\ 0 & 0 & 0 & 1 \end{bmatrix} \qquad \boldsymbol{\beta}^* = \begin{pmatrix} \mu_1 \\ \mu_2 \\ \mu_3 \\ \mu_4 \end{pmatrix} \qquad [8.7]$$

The columns of $\boldsymbol{X}^*$ are the dummy variables defined for the original matrix, equation 8.4. For the general case, $\boldsymbol{X}^*$ will be a matrix of order $(n \times t)$, where $n = rt$ is the total number of observations. In this form, $\boldsymbol{X}^*$ is of full rank and ordinary least squares regression can be used to estimate the parameters β^*.

Solution

The form of $\boldsymbol{X}^*$ in this reparameterization makes the least squares arithmetic particularly simple. $\boldsymbol{X}^{*\prime}\boldsymbol{X}^*$ is a diagonal matrix of order $(t \times t)$ with the diagonal elements being the number of replications r of each treatment. Thus, $(\boldsymbol{X}^{*\prime}\boldsymbol{X}^*)^{-1}$ is diagonal with diagonal elements $1/r$. $\boldsymbol{X}^{*\prime}\boldsymbol{Y}$ is the vector of treatment sums. The least squares solution is

$$\hat{\boldsymbol{\beta}}^{*\prime} = (\overline{Y}_{1.} \;\; \overline{Y}_{2.} \;\; \ldots \;\; \overline{Y}_{t.}) \qquad [8.8]$$

which is the vector of treatment means.

Meaning of $\hat{\beta}^*$

Since this is the least squares solution to a full-rank model, $\hat{\boldsymbol{\beta}}^*$ is the best linear unbiased estimate of $\boldsymbol{\beta}^*$, but not of $\boldsymbol{\beta}$. (The parameters $\boldsymbol{\beta}$ in the original model are not estimable.) It is helpful in understanding the results of the reparameterized model to know what function of the original parameters is being estimated by $\hat{\boldsymbol{\beta}}^*$. This is determined by finding the expectation of $\hat{\boldsymbol{\beta}}^*$ in terms of the expectation of $\boldsymbol{Y}$ from the original model, $\mathscr{E}(\boldsymbol{Y}) = \boldsymbol{X\beta}$:

$$\begin{aligned}\mathscr{E}(\hat{\boldsymbol{\beta}}^*) &= [(\boldsymbol{X}^{*\prime}\boldsymbol{X}^*)^{-1}\boldsymbol{X}^{*\prime}]\mathscr{E}(\boldsymbol{Y}) \\ &= [(\boldsymbol{X}^{*\prime}\boldsymbol{X}^*)^{-1}\boldsymbol{X}^{*\prime}\boldsymbol{X}]\boldsymbol{\beta}\end{aligned} \qquad [8.9]$$

Notice that the last $\boldsymbol{X}$ is the original matrix. Evaluating this expectation for the current reparameterization gives

$$\mathscr{E}(\hat{\boldsymbol{\beta}}^*) = \begin{bmatrix} 1 & 1 & 0 & 0 & 0 \\ 1 & 0 & 1 & 0 & 0 \\ 1 & 0 & 0 & 1 & 0 \\ 1 & 0 & 0 & 0 & 1 \end{bmatrix}\begin{pmatrix} \mu \\ \tau_1 \\ \tau_2 \\ \tau_3 \\ \tau_4 \end{pmatrix} = \begin{pmatrix} \mu + \tau_1 \\ \mu + \tau_2 \\ \mu + \tau_3 \\ \mu + \tau_4 \end{pmatrix} \qquad [8.10]$$

Thus, each element of $\hat{\boldsymbol{\beta}}^*$, $\hat{\mu}_i = \overline{Y}_{i.}$, is an estimate of $\mu + \tau_i$. This is the expectation of the ith group mean under the original model.

Estimable Functions of β

Unbiased estimates of other estimable functions of the original parameters are obtained by using appropriate linear functions of $\hat{\boldsymbol{\beta}}^*$. For example, $(\tau_1 - \tau_2)$ is estimated by $\hat{\mu}_1 - \hat{\mu}_2 = \overline{Y}_{1.} - \overline{Y}_{2.}$. Notice, however, that there is no linear function of $\hat{\boldsymbol{\beta}}^*$ that provides an unbiased estimate of μ, or of one of the τ_i. These are *nonestimable functions* of the original parameters, and no reparameterization of the model will provide estimates of such nonestimable quantities.

SS(Model) *and* SS(Res)

The sum of squares due to this model is the uncorrected treatment sum of squares,

$$\begin{aligned}\text{SS(Model)} &= \hat{\boldsymbol{\beta}}^{*\prime}\boldsymbol{X}^{*\prime}\boldsymbol{Y} \\ &= \frac{\sum_{i=1}^{t}(Y_{i.})^2}{r}\end{aligned} \qquad [8.11]$$

because the elements of $\hat{\boldsymbol{\beta}}^*$ are the treatment means and the elements of $\boldsymbol{X}^{*\prime}\boldsymbol{Y}$ are the treatment sums. The residual sum of squares is the pooled sum of squares from among the replicate observations within each group,

$$\begin{aligned}\text{SS(Res)} &= \boldsymbol{Y}'\boldsymbol{Y} - \text{SS(Model)} \\ &= \sum_{i=1}^{t}\sum_{j=1}^{r} Y_{ij}^2 - \frac{\sum_{i=1}^{t}(Y_{i.})^2}{r} \\ &= \sum_{i=1}^{t}\sum_{j=1}^{r}(Y_{ij} - \overline{Y}_{i.})^2\end{aligned} \qquad [8.12]$$

Table 8.1 Relationship between the conventional analysis of variance and ordinary least squares regression computations for the completely random experimental design.

Source of Variation	d.f.	*Traditional* AOV SS	*Regression* SS
$\text{Total}_{\text{uncorr}}$	nt	$\sum\sum Y_{ij}^2$	$Y'Y$
Model	t	$\sum(Y_{i.})^2/r$	$\hat{\beta}'X'Y$
C.F.	1	$n\bar{Y}^2$	$n\bar{Y}^2$
Treatments	$t-1$	$\sum(Y_{i.})^2/r - n\bar{Y}^2$	$\hat{\beta}'X'Y - n\bar{Y}^2$
Residual	$t(n-1)$	$\sum\sum Y_{ij}^2 - \sum(Y_{i.})^2/r$	$Y'Y - \text{SS(Model)}$

and has $(n - t)$ degrees of freedom. A dot in a subscript indicates that the observations have been summed over that subscript; thus, $Y_{i.}$ is the ith treatment sum and $\bar{Y}_{i.}$ is the ith treatment mean.

Treatment Sum of Squares

SS(Model) measures the squared deviations of the treatment means from zero. Comparisons among the treatment means are of greater interest. Sums of squares for these comparisons are generated using the general linear hypothesis (discussed in Section 4.5). For example, the sum of squares for the null hypothesis that all μ_i are equal, which implies that all τ_i are zero, is obtained by constructing a $\boldsymbol{K}'$ matrix of rank $(t - 1)$ to account for all differences among the t treatment parameters. One such $\boldsymbol{K}'$ is

$$\boldsymbol{K}' = \begin{bmatrix} 1 & -1 & 0 & 0 \\ 0 & 1 & -1 & 0 \\ 0 & 0 & 1 & -1 \end{bmatrix} \qquad [8.13]$$

This matrix defines the three nonorthogonal but linearly independent contrasts of treatment 1 versus treatment 2, treatment 2 versus treatment 3, and treatment 3 versus treatment 4. Any set of three linearly independent contrasts would produce the sum of squares for the hypothesis that all μ_i are equal. The sum of squares for this hypothesis is the treatment sum of squares. In general, the treatment sum of squares can be obtained by defining $\boldsymbol{K}'$ so that $r(\boldsymbol{K}') = (t - 1)$.

Alternatively, the treatment sum of squares can be obtained by using the difference in sums of squares between full and reduced models. The reduced model for the null hypothesis that all μ_i are equal contains only one parameter, a constant mean μ. The sum of squares for such a model is $\text{SS}(\mu) = n\bar{Y}_{..}^2$, or the sum of squares due to correction for the mean, commonly called the **correction factor** (C.F.). Thus, the treatment sum of squares for the completely random experimental design can be obtained as $\text{SS(Model)} - \text{SS}(\mu)$. The relationship between the conventional analysis of variance and the regression analysis for the completely random design is summarized in Table 8.1.

8.3.2 Reparameterizing Using the Constraint $\sum \tau_i = 0$

Redefining the Model

The original model defined the τ_i as deviations from μ. If μ is thought of as the overall true mean of the t treatments, it is reasonable to impose the condition that the sum of the treatment deviations about the true mean is zero; that is, $\sum \tau_i = 0$. This implies that one τ_i can be expressed as the negative of the sum of the other τ_i. The number of parameters to be estimated is thus reduced by 1.

The constraint $\sum \tau_i = 0$ is used to express the last treatment effect τ_t in terms of the first $(t - 1)$ treatment effects. Thus,

$$\tau_t = -(\tau_1 + \tau_2 + \cdots + \tau_{t-1})$$

is substituted for τ_t everywhere in the original model. In the example, $\tau_4 = -(\tau_1 + \tau_2 + \tau_3)$ so that the model for each observation in the fourth group changes from

$$Y_{4j} = \mu + \tau_4 + \varepsilon_{4j}$$

to

$$Y_{4j} = \mu + (-\tau_1 - \tau_2 - \tau_3) + \varepsilon_{4j}$$

This substitution eliminates τ_4, reducing the number of parameters from 5 to 4 or, in general, from $(t + 1)$ to t. The vector of redefined parameters is

$$\boldsymbol{\beta}^{*\prime} = (\mu^* \ \ \tau_1^* \ \ \tau_2^* \ \ \tau_3^*) \qquad [8.14]$$

$\boldsymbol{X}^*$

The design matrix $\boldsymbol{X}^*$ for this reparameterization is obtained from the original $\boldsymbol{X}$ as follows. The dummy variable for treatment 4, the last column of $\boldsymbol{X}$, equation 8.4, identifies the observations that contain τ_4 in the model. For each such observation, the substitution of $-(\tau_1 + \tau_2 + \tau_3)$ for τ_4 is effected by replacing the "0" coefficients on τ_1, τ_2, and τ_3 with "-1" and dropping the dummy variable for τ_4. Thus, $\boldsymbol{X}^*$ for this reparameterization is

$$\boldsymbol{X}^* = \begin{bmatrix} 1 & 1 & 0 & 0 \\ 1 & 1 & 0 & 0 \\ 1 & 0 & 1 & 0 \\ 1 & 0 & 1 & 0 \\ 1 & 0 & 0 & 1 \\ 1 & 0 & 0 & 1 \\ 1 & -1 & -1 & -1 \\ 1 & -1 & -1 & -1 \end{bmatrix} \qquad [8.15]$$

$\hat{\boldsymbol{\beta}}^*$

Again, the reparameterized model is of full rank and ordinary least squares gives an unbiased estimate of the new parameters defined in $\boldsymbol{\beta}^*$. The expectation of $\hat{\boldsymbol{\beta}}^*$ in terms of the parameters in the original model is found from equation 8.9 using $\boldsymbol{X}^*$ from the current reparameterization. This gives

$$\mathscr{E}(\hat{\boldsymbol{\beta}}^*) = \begin{bmatrix} 1 & \frac{1}{4} & \frac{1}{4} & \frac{1}{4} & \frac{1}{4} \\ 0 & \frac{3}{4} & -\frac{1}{4} & -\frac{1}{4} & -\frac{1}{4} \\ 0 & -\frac{1}{4} & \frac{3}{4} & -\frac{1}{4} & -\frac{1}{4} \\ 0 & -\frac{1}{4} & -\frac{1}{4} & \frac{3}{4} & -\frac{1}{4} \end{bmatrix} \begin{pmatrix} \mu \\ \tau_1 \\ \tau_2 \\ \tau_3 \\ \tau_4 \end{pmatrix} = \begin{pmatrix} \mu + \bar{\tau} \\ \tau_1 - \bar{\tau} \\ \tau_2 - \bar{\tau} \\ \tau_3 - \bar{\tau} \end{pmatrix} \qquad [8.16]$$

where $\bar{\tau}$ is the average of the four τ_i. Thus, $\hat{\mu}^*$ is an estimate of $(\mu + \bar{\tau})$, $\hat{\tau}_1^*$ is an estimate of $(\tau_1 - \bar{\tau})$, and so forth. While $\hat{\tau}_4^*$ is not given explicitly in $\hat{\boldsymbol{\beta}}^*$, it can be obtained by imposing the original constraint. Thus,

$$\hat{\tau}_4^* = -(\hat{\tau}_1^* + \hat{\tau}_2^* + \hat{\tau}_3^*)$$

and is an unbiased estimate of $(\tau_4 - \bar{\tau})$.

Functions of the β_i

Other estimable functions of the original parameters are obtained from appropriate linear functions of $\hat{\boldsymbol{\beta}}^*$. For example, the least squares estimate of the ith treatment mean, $(\mu + \tau_i)$, is given by $(\hat{\mu}^* + \hat{\tau}_i^*)$. The estimate of the difference between two treatment effects, say $(\tau_2 - \tau_3)$, is given by $(\hat{\tau}_2^* - \hat{\tau}_3^*)$.

Treatment Sum of Squares

The analysis of variance for the completely random design is obtained from this reparameterization in much the same way as with the means reparameterization. The sum of squares for treatments is obtained as the sum of squares for the null hypothesis

$$H_0\colon \ \tau_1^* = \tau_2^* = \tau_3^* = 0$$

or as

$$\text{SS(Model)} - \text{SS}(\mu)$$

In terms of the original parameters, this null hypothesis is satisfied only if all τ_i are equal.

8.3.3 Reparameterizing Using the Constraint $\tau_t = 0$

Redefining the Model

Another method of reducing the number of parameters in an overparameterized model is to arbitrarily set the required number of parameters equal to zero. In the model for the completely random experimental design, one constraint is needed so that one parameter—usually the last τ_i—is set equal to zero. In the example with four treatments, setting $\tau_4 = 0$ gives

$$\boldsymbol{\beta}^{*\prime} = (\mu^* \ \ \tau_1^* \ \ \tau_2^* \ \ \tau_3^*)$$

and an $\boldsymbol{X}^*$ that contains only the first four columns of the original $\boldsymbol{X}$. As with the other reparameterizations, this model is of full rank and ordinary least squares can be used to obtain the solution $\hat{\boldsymbol{\beta}}^*$. The constraint implies that $\hat{\tau}_4^* = 0$.

$\hat{\boldsymbol{\beta}}^*$

The expectation of $\hat{\boldsymbol{\beta}}^*$ in terms of the parameters in the original model (from equation 8.9 using the current $\boldsymbol{X}^*$) is

$$\mathscr{E}(\hat{\boldsymbol{\beta}}^*) = \begin{bmatrix} 1 & 0 & 0 & 0 & 1 \\ 0 & 1 & 0 & 0 & -1 \\ 0 & 0 & 1 & 0 & -1 \\ 0 & 0 & 0 & 1 & -1 \end{bmatrix} \begin{pmatrix} \mu \\ \tau_1 \\ \tau_2 \\ \tau_3 \\ \tau_4 \end{pmatrix} = \begin{pmatrix} \mu + \tau_4 \\ \tau_1 - \tau_4 \\ \tau_2 - \tau_4 \\ \tau_3 - \tau_4 \end{pmatrix} \quad [8.17]$$

With this parameterization, $\hat{\mu}^*$ is an estimate of the mean of the fourth treatment, $\mu + \tau_4$, and each $\hat{\tau}_i^*$ estimates the difference between the true means of the ith treatment and the fourth treatment. The ith treatment mean $\mu + \tau_i$ is estimated by $\hat{\mu}^* + \hat{\tau}_i^*$. The difference between two means $(\tau_i - \tau_{i'})$ is estimated by $(\hat{\tau}_i^* - \hat{\tau}_{i'}^*)$.

Treatment Sum of Squares

The treatment sum of squares for this parameterization is given as the sum of squares for the composite null hypothesis

$$H_0\colon\ \tau_i^* = 0 \quad \text{for } i = 1, 2, 3$$

or as

$$\text{SS(Model)} - \text{SS}(\mu)$$

In terms of the original parameters, this hypothesis implies that the first three τ_i are each equal to τ_4 (equation 8.17), or that $\tau_1 = \tau_2 = \tau_3 = \tau_4$.

Estimable Functions

Each of the three reparameterizations introduced in this section has provided estimates of the meaningful functions of the original parameters, the true means of the treatments, and all contrasts among the true treatment means. These are estimable functions of the original parameters. As a general result, if a function of the original parameters is estimable, it can be estimated from $\hat{\boldsymbol{\beta}}^*$ obtained from any reparameterization. Furthermore, the same numerical estimate for any estimable function of the original parameters will be obtained from every reparameterization. Estimability is discussed more fully in Chapter 16 and the reader is referred to Searle (1971) for the theoretical developments.

8.3.4 Reparameterization: A Numerical Example

Example 8.2

A small numerical example illustrates the three reparameterizations. An artificial data set was generated to simulate an experiment with $t = 4$ and $r = 2$. The conventional one-way model was used with the parameters chosen to be $\mu = 12$, $\tau_1 = -3$, $\tau_2 = 0$, $\tau_3 = 2$, and $\tau_4 = 4$. A random observation from a normal distribution with mean zero and unit variance was added to each expectation to simulate random error. (The τ_i are chosen so they do not add to zero for this illustration.) The vector of observations generated in this manner was

$$\boldsymbol{Y} = \begin{bmatrix} Y_{11} \\ Y_{12} \\ Y_{21} \\ Y_{22} \\ Y_{31} \\ Y_{32} \\ Y_{41} \\ Y_{42} \end{bmatrix} = \begin{bmatrix} \mu + \tau_1 + \varepsilon_{11} \\ \mu + \tau_1 + \varepsilon_{12} \\ \mu + \tau_2 + \varepsilon_{21} \\ \mu + \tau_2 + \varepsilon_{22} \\ \mu + \tau_3 + \varepsilon_{31} \\ \mu + \tau_3 + \varepsilon_{32} \\ \mu + \tau_4 + \varepsilon_{41} \\ \mu + \tau_4 + \varepsilon_{42} \end{bmatrix} = \begin{bmatrix} 8.90 \\ 8.76 \\ 11.78 \\ 12.07 \\ 14.50 \\ 12.48 \\ 16.79 \\ 16.57 \end{bmatrix} \qquad [8.18]$$

The parameter estimates from these data for each of the three reparameterizations and their expectations in terms of the original parameters are shown in Table 8.2. Most notable are the numerical differences in $\hat{\boldsymbol{\beta}}^*$ for the different parameterizations. All convey the same information but in very different packages. The results from the means model are the most directly useful; each regression coefficient estimates the corresponding group mean. Contrasts among the τ_i are estimated by the same contrasts among the estimated regression coefficients. For example,

$$\hat{\mu}_1^* - \hat{\mu}_2^* = 8.8300 - 11.9250 = -3.0950$$

is an estimate of $(\tau_1 - \tau_2)$, which is known to be -3 from the simulation model.

Table 8.2 Estimates obtained from simulated data for three reparameterizations of the one-way model, $t = 4$ and $r = 2$. Expectations of the estimates are in terms of the parameters of the original singular model.

Reparameterization: *Means Model*		$\sum \tau_i = 0$		$\tau_4 = 0$[a]	
$\hat{\boldsymbol{\beta}}^*$	$\mathscr{E}(\hat{\boldsymbol{\beta}}^*)$	$\hat{\boldsymbol{\beta}}^*$	$\mathscr{E}(\hat{\boldsymbol{\beta}}^*)$	$\hat{\boldsymbol{\beta}}^*$	$\mathscr{E}(\hat{\boldsymbol{\beta}}^*)$
8.830	$\mu + \tau_1$	12.731	$\mu + \bar{\tau}$	16.680	$\mu + \tau_4$
11.925	$\mu + \tau_2$	-3.901	$\tau_1 - \bar{\tau}$	-7.850	$\tau_1 - \tau_4$
13.490	$\mu + \tau_3$	$-.806$	$\tau_2 - \bar{\tau}$	-4.755	$\tau_2 - \tau_4$
16.680	$\mu + \tau_4$	.759	$\tau_3 - \bar{\tau}$	-3.190	$\tau_3 - \tau_4$

[a] The solution obtained from the general linear models solution in PROC GLM corresponds to that for $\tau_4 = 0$.

The reparameterization using the "sum" constraint gives $\hat{\mu}^* = 12.73125$, which is an estimate of the overall mean plus the average of the treatment effects. [From the simulation model, $(\mu + \bar{\tau})$ is known to be 12.75.] Each of the other computed regression coefficients is estimating the deviation of a τ_i from $\bar{\tau}$. The estimate of $(\tau_4 - \bar{\tau})$ is obtained by forcing the $\hat{\tau}_i^*$ to satisfy the constraint. This gives

$$\hat{\tau}_4^* = -(-3.90125 - .80625 + .75875) = 3.94875$$

The sum of the first two estimates,

$$\hat{\mu}^* + \hat{\tau}_1^* = 12.73125 + (-3.90125) = 8.8300$$

is an estimate of $(\mu + \tau_1)$. This estimate is identical to that obtained for $(\mu + \tau_1)$ from the means model. Similarly, the estimate of $(\tau_1 - \tau_2)$,

$$\hat{\tau}_1^* - \hat{\tau}_2^* = -3.90125 - (-.80625) = 3.095$$

is the same as that obtained from the means model.

The third reparameterization using $\tau_4 = 0$ gives $\hat{\mu}^* = 16.6800$, which is an estimate of $(\mu + \tau_4)$, the true mean of the fourth group. The sum of the first two regression coefficients again estimates $(\mu + \tau_1)$ as

$$\hat{\mu}^* + \hat{\tau}_1^* = 16.6800 + (-7.8500) = 8.8300$$

Each $\hat{\tau}_i^*$ estimates the difference in effects between the ith group and the fourth group. The numerical values obtained for these estimates are identical to those obtained from the other models. ■

Unique Results from Reparameterizations

The results from these three reparameterizations illustrate general results. Least squares estimates of $\boldsymbol{\beta}^*$ obtained from different reparameterizations estimate different functions of the original parameters. The relationship of the redefined parameters to those in the original model must be known in order to properly interpret these estimates. Even though the solution appears to change with the different reparameterizations, all give identical numerical estimates of every estimable function of the original parameters. This includes $\hat{\boldsymbol{Y}} = \boldsymbol{X}^*\boldsymbol{\beta}^*$ and $\boldsymbol{e} = \boldsymbol{Y} - \hat{\boldsymbol{Y}}$. Further, sums of squares associated with any estimable contrast on $\boldsymbol{\beta}$ are identical, which implies that all parameterizations give the same analysis of variance. In Example 8.2, all models gave

$$\text{SS(Regr)} = 64.076238 \quad \text{and} \quad \text{SS(Res)} = 2.116250$$

8.4 Generalized Inverse Approach to Singular Models

Nonunique Solutions

When $\boldsymbol{X}$ is not of full rank there is no unique solution to the normal equations, $(\boldsymbol{X}'\boldsymbol{X})\boldsymbol{\beta} = \boldsymbol{X}'\boldsymbol{Y}$. A general approach to models of less than full rank is to use one of the nonunique solutions to the normal equations. This is accomplished by using a **generalized inverse** of $\boldsymbol{X}'\boldsymbol{X}$. (The generalized inverse of a matrix $\boldsymbol{A}$ will be denoted by $\boldsymbol{A}^-$.) There are many different kinds of generalized inverses which, to some extent, have different properties. The reader is referred to Searle (1971) for complete discussions on generalized inverses. It is sufficient for now to know that a generalized inverse provides one of the infinity of solutions that satisfies the normal equations. Such a solution will be denoted with $\boldsymbol{\beta}^0$ to emphasize the fact that it is not a unique solution. $\hat{\boldsymbol{\beta}}$ will be reserved as the label for the unique least squares solution when it exists. Thus,

$$\beta^0 = (X'X)^- X'Y \qquad [8.19]$$

Computers will be used to obtain the generalized inverse solutions.

Estimable Functions

Since β^0 is not unique, its elements per se are meaningless. Another generalized inverse would give another set of numbers from the same data. However, many of the regression results obtained from using a nonunique solution are unique; the same numerical results are obtained regardless of which solution is used. It was observed in Section 8.3 that all reparameterizations gave identical estimates of estimable functions of the parameters. This important result applies to all generalized inverse solutions to the normal equations. Any estimable function of the original parameters is *uniquely* estimated by the same linear function of one of the nonunique solutions β^0. That is, if $K'\beta$ is estimable, then $K'\beta^0$ is the least squares estimate of $K'\beta$ *and* the estimate is unique with respect to choice of solution. Such estimates of estimable linear functions of the original parameters have all the desirable properties of least squares estimators.

Unique Results

Results concerning other unique quantities follow from this statement. For example, $X\beta$ is an estimable function of β and, hence, $\hat{Y} = X\beta^0$ is the unique unbiased estimate of $X\beta$. Then, $e = Y - \hat{Y}$ must be unique. Since SS(Model) = $\hat{Y}'\hat{Y}$ and SS(Res) = $e'e$, these sums of squares are also unique with respect to choice of solution. The uniqueness extends to the partitions of the sums of squares, as long as the sums of squares relate to hypotheses that are estimable linear functions of the parameters.

Thus, the generalized inverse approach to models of less than full rank provides all the results of interest. The only quantities not estimated uniquely are those quantities for which the data contain no information—the nonestimable functions of β.

PROC GLM

The generalized inverse approach is used for the least squares analysis of models of less than full rank by many computer programs, including PROC GLM (SAS Institute, Inc., 1985d, 1985f). (GLM stands for General Linear Models.) In this procedure, any variable that is to be regarded as a class variable must be identified in a CLASS statement in the program. Each class variable will generate singularities that make the model less than full rank. (Singularities can also result from linear dependencies among continuous variables, but this chapter is concerned with the use of class variables in regression models.) Since the estimates of the regression coefficients in the singular model are not unique, PROC GLM does not print the solution β^0 unless it is specifically requested. The user is given the message that the model is not of full rank. The unique results from the analysis are obtained by requesting estimation of specific estimable functions and tests of **testable hypotheses**. (A testable hypothesis is one in which the linear functions of parameters in the null hypothesis are estimable functions.)

When a class variable is specified, PROC GLM creates β and the set of dummy variables for the X matrix as was done in Section 8.2. No reparameterization is done so that X remains singular. The particular generalized inverse used by PROC GLM gives the same solution as that obtained with reparameterization using the constraint $\tau_t = 0$. The solution vector in PROC GLM contains an estimate for every parameter including τ_t^*. But, because each $\hat{\tau}_i^*$ is estimating $\tau_i - \tau_t$, the numerical value of $\hat{\tau}_t^*$ is always zero. Thus, the PROC GLM solution for the simulated data from the completely random design is the same as that given in the last column of Table 8.2, except the vector of estimates includes $\hat{\tau}_4^* = 0$ in the fifth position. The estimates obtained for all estimable functions and sums of squares are identical to those obtained from the reparameterizations.

8.5 The Model for Two-Way Classified Data

Defining the Model

The conventional model for two-way classified data, of which the randomized complete block design (RCB) is the most common example, is

$$Y_{ij} = \mu + \gamma_i + \tau_j + \varepsilon_{ij} \qquad [8.20]$$

where μ is an overall mean, γ_i is the effect of the ith block, τ_j is the effect of the jth treatment, and ε_{ij} is the random error. In this model there are two class variables—"block" and "treatment"—which identify the particular block and treatment associated with the ijth experimental unit. There are b levels $(i = 1, \ldots, b)$ of the block class variable and t levels $(j = 1, \ldots, t)$ of the treatment class variable.

X and β

Defining the $\boldsymbol{X}$ matrix for this model requires b dummy variables for blocks and t dummy variables for treatments. The vector of observations will be assumed to be ordered with all of the treatments occurring in order for the first block followed by the treatments in order for the second block, and so forth. The parameter vector $\boldsymbol{\beta}$ will be defined with the block effects occurring before the treatment effects. For illustration, assume that $b = 2$ and $t = 4$ for a total of $bt = 8$ observations. Then,

$$\boldsymbol{X} = \begin{bmatrix} 1 & 1 & 0 & 1 & 0 & 0 & 0 \\ 1 & 1 & 0 & 0 & 1 & 0 & 0 \\ 1 & 1 & 0 & 0 & 0 & 1 & 0 \\ 1 & 1 & 0 & 0 & 0 & 0 & 1 \\ 1 & 0 & 1 & 1 & 0 & 0 & 0 \\ 1 & 0 & 1 & 0 & 1 & 0 & 0 \\ 1 & 0 & 1 & 0 & 0 & 1 & 0 \\ 1 & 0 & 1 & 0 & 0 & 0 & 1 \end{bmatrix} \qquad \boldsymbol{\beta} = \begin{bmatrix} \mu \\ \gamma_1 \\ \gamma_2 \\ \tau_1 \\ \tau_2 \\ \tau_3 \\ \tau_4 \end{bmatrix} \qquad [8.21]$$

The second and third columns of $\boldsymbol{X}$ are the dummy variables for blocks; the last four columns are the dummy variables for treatments.

Degrees of Freedom

There are two linear dependencies in $\boldsymbol{X}$. The sum of the block dummy variables (columns 2 and 3) and the sum of the treatment dummy variables (the last four columns) both equal column 1. Thus, the rank of $\boldsymbol{X}$ is $r(\boldsymbol{X}) = 7 - 2 = 5$, which is the degrees of freedom for SS(Model). In the conventional RCB analysis of variance these degrees of freedom are partitioned into 1 for the correction factor, $(b - 1) = 1$ for SS(Blocks), and $(t - 1) = 3$ for SS(Treatments).

Reparameterizing Using $\gamma_2 = 0$ and $\tau_4 = 0$

Reparameterizing this model to make it full rank requires two constraints. The effective number of parameters must be reduced to 5, the rank of $\boldsymbol{X}$. The simplest constraints to impose would be $\gamma_2 = 0$ and $\tau_4 = 0$. These constraints have the effect of eliminating γ_2 and τ_4 from $\boldsymbol{\beta}$ and columns 3 and 7 from $\boldsymbol{X}$. Thus, $\boldsymbol{X}^*$ would be an 8×5 matrix consisting of columns 1, 2, 4, 5, and 6 from $\boldsymbol{X}$ and $\boldsymbol{\beta}^*$ would be

$$\boldsymbol{\beta}^{*\prime} = (\mu^* \;\; \gamma_1^* \;\; \tau_1^* \;\; \tau_2^* \;\; \tau_3^*) \qquad [8.22]$$

Reparameterizing Using Sum Constraint

The constraints requiring the sum of the effects to be zero would be $\gamma_1 + \gamma_2 = 0$ and $\sum \tau_i = 0$. These constraints are imposed by substituting $-\gamma_1$ for γ_2 and $-(\tau_1 + \tau_2 + \tau_3)$ for τ_4 in the original model. This reduces the number of parameters by 2 and gives

$$\boldsymbol{X}^* = \begin{bmatrix} 1 & 1 & 1 & 0 & 0 \\ 1 & 1 & 0 & 1 & 0 \\ 1 & 1 & 0 & 0 & 1 \\ 1 & 1 & -1 & -1 & -1 \\ 1 & -1 & 1 & 0 & 0 \\ 1 & -1 & 0 & 1 & 0 \\ 1 & -1 & 0 & 0 & 1 \\ 1 & -1 & -1 & -1 & -1 \end{bmatrix} \qquad [8.23]$$

Either of these reparameterizations will generate the conventional analysis of variance of two-way classified data when the least squares regression concepts are applied. The full model consists of μ^*, the γ_i^*, and the τ_j^*. The residual mean square from this model estimates σ^2. The general linear hypothesis can be used to generate the sum of squares for testing the null hypothesis that γ_1^* is zero. In the more general case, this would be a composite hypothesis that all γ_i^* are zero. The sum of squares, Q, generated for this hypothesis will have 1 degree of freedom [or, in general, $(b - 1)$ degrees of freedom] and is algebraically identical to SS(Blocks) in the conventional analysis of variance. Similarly, the sum of squares associated with the composite hypothesis that all τ_j^* are zero is identical to SS(Treatments) in the conventional analysis of variance. These sums of squares can also be computed from the procedure based on $[\text{SS}(\text{Res}_{\text{reduced}}) - \text{SS}(\text{Res}_{\text{full}})]$.

Reparameterizing Using the Means Model

The model could also be made full rank by using the means model reparameterization. Each cell of the two-way table would be assigned its own mean. Thus,

$$Y_{ij} = \mu_{ij} + \varepsilon_{ij} \qquad [8.24]$$

where $\mu_{ij} = \mu + \gamma_i + \tau_j$ in terms of the parameters of the original model. This model is different from the original, however. The original model specified a column (or treatment) effect and a row (or block) effect which added to give the "cell" effect; the same column effect was imposed on all rows and the same row effects applied to all columns. Deviations from the sum of the block and treatment effects were assumed to be random error. The means model as given, on the other hand, imposes no restrictions on the relationships among the μ_{ij}. The means model is made analogous to the classical RCB effects model by imposing constraints on the μ_{ij} so as to satisfy the conditions of no interaction in every 2×2 subtable of the $b \times t$ table of μ_{ij}. The reader is referred to Hocking (1985) for complete discussions on analyses using means models.

Generalized Inverse Approach

The generalized inverse approach can also be used for two-way classified data. The two class variables would be used to generate the singular $\boldsymbol{X}$ (equation 8.21) and a generalized inverse would be used to obtain a (nonunique) solution. SS(Res) from that analysis would be the interaction sum of squares for the two-way table, which in the RCB design is the estimate of experimental error. Appropriate hypotheses on the subsets of parameters generate the usual analysis of variance for two-way data.

Two-Way Model with Interaction Effects

A more general model for two-way classified data includes interaction effects in the model. Suppose the γ_i and τ_j are the effects of two treatment factors, A and B, with a levels of factor A and b levels of factor B. Let the interaction effects between the two factors be represented by $(\gamma\tau)_{ij}$ and assume there are r observations in each cell, $k = 1, \ldots, r$. The linear model is

$$Y_{ijk} = \mu + \gamma_i + \tau_j + (\gamma\tau)_{ij} + \varepsilon_{ijk} \qquad [8.25]$$

where $i = 1, \ldots, a$ and $j = 1, \ldots, b$. In matrix notation, $\boldsymbol{\beta}$ contains $(1 + a + b + ab) = (a + 1)(b + 1)$ parameters and $\boldsymbol{X}$ contains an equal number of columns. The number of rows of $\boldsymbol{X}$ will equal the number of observations, $n = abr$. The r observations from the same treatment combination have the same expectation (equation 8.25), so that there will be ab distinct rows in $\boldsymbol{X}$ with r repeats of each.

Example 8.3

For illustration, assume $a = 2$ and $b = 4$. Then $\boldsymbol{X}$ contains 15 columns and 8 distinct rows. Each of the 8 rows will be repeated r times. Then,

$$\boldsymbol{X} = \begin{bmatrix} 1&1&0&1&0&0&0&1&0&0&0&0&0&0&0 \\ 1&1&0&0&1&0&0&0&1&0&0&0&0&0&0 \\ 1&1&0&0&0&1&0&0&0&1&0&0&0&0&0 \\ 1&1&0&0&0&0&1&0&0&0&1&0&0&0&0 \\ 1&0&1&1&0&0&0&0&0&0&0&1&0&0&0 \\ 1&0&1&0&1&0&0&0&0&0&0&0&1&0&0 \\ 1&0&1&0&0&1&0&0&0&0&0&0&0&1&0 \\ 1&0&1&0&0&0&1&0&0&0&0&0&0&0&1 \end{bmatrix} \qquad [8.26]$$

where only the 8 distinct rows of $\boldsymbol{X}$ are shown.

The first 7 columns of $\boldsymbol{X}$ are as defined in equation 8.21. The last 8 columns are the dummy variables for the interaction effects. The dummy variable for $(\gamma\tau)_{ij}$ takes the value 1 if the observation is from the ijth treatment combination, and 0 otherwise. The dummy variable for $(\gamma\tau)_{ij}$ can also be obtained as the element-by-element product of the dummy variables for the corresponding γ_i and τ_j effects. (This is a general result that extends to higher-order interaction effects.) While X contains 15 columns, its rank is only 8. (The rank of X cannot be greater than the number of linearly independent rows.) Thus, there must be 7 linear dependencies among the columns of X. These dependencies would have to be identified if the model were to be reparameterized. The generalized inverse approach, however, uses X as defined.

■

Computing Load

The size of X increases very rapidly as additional factors and particularly their interactions are added to the model. The number of columns of X required for each set of interaction effects is the product of the number of levels of all the factors in the interaction. The total number of parameters in a model with class variables and their interactions is 1 plus the product of the number of levels of all class variables in the model. It is not uncommon for the full X matrix of a reasonably sized experiment to have more than 100 columns. The computational load of finding the generalized inverse and operating on this very large X matrix can become exorbitant.

On the other hand, the conventional analysis of variance formulas, which result from the least squares analysis of balanced data, are computationally very efficient. Very large models can be easily analyzed. Consequently, analyses of variance programs should always be used in preference to the generalized inverse (the general linear models) approach when data are balanced. The more general approach has been introduced to demonstrate the link between least squares regression analysis and the conventional analyses of variance, and to set the stage for the analysis of unbalanced data (Chapter 16).

8.6 Class Variables to Test Homogeneity of Regressions

Consider the situation where two or more subsets of data are available, each of which provides information on the dependent variable of interest and the potential predictor variables. The subsets of data originate from different levels of one or more class variables. For example, data relating yield in corn to levels of nitrogen and phosphorus fertilization may be available for several corn hybrids grown in several environments. Yield is the dependent variable, amount of nitrogen fertilizer and amount of phosphorus fertilizer are independent variables, and "hybrid" and "environment" are two class variables.

The objective is to model the response of yield to changing rates of nitrogen and phosphorus fertilization. The question is whether a single regression equation will adequately describe the relationship for all hybrids and environments or will different regressions be required for each hybrid–environment combination. The most complete description of the response (the best fit to the data) would be obtained by allowing each combination to have its own regression equation. This would be inefficient, however, if the responses were similar over all groups; the researcher would be estimating more parameters than necessary. On the other hand, a single regression equation to represent the response for all groups will not characterize any one group as well and could be very misleading if the relationships differed among groups. The simplicity of the single regression equation is to be preferred if it can be justified. Intermediate models may allow a common regression for some independent variables but require others to have different regression coefficients for different subsets of data.

Illustration

The decision to use a regression coefficient for each subset or a common regression coefficient for all subsets is based on the test of homogeneity of regression coefficients over levels of the class variable. The test of homogeneity will be illustrated assuming a linear relationship between a dependent variable and one independent variable. The general method extends to any number of independent variables and any functional relationship.

Defining the Model

Suppose the data consist of t groups with n_i observations in each group. There will be $\sum n_i = n$ data points, each consisting of an observation on the Y, X, and the class variable identifying the group from which the observation came. The most general model for this situation allows each group to have its own intercept and slope coefficient. The separate models can be written as

$$\begin{aligned}
\text{Group 1:}\quad & Y_{1j} = \beta_{10} + \beta_{11}X_{1j} + \varepsilon_{1j} \\
\text{Group 2:}\quad & Y_{2j} = \beta_{20} + \beta_{21}X_{2j} + \varepsilon_{2j} \\
& \vdots \\
\text{Group } t\text{:}\quad & Y_{tj} = \beta_{t0} + \beta_{t1}X_{tj} + \varepsilon_{tj}
\end{aligned} \qquad [8.27]$$

If the subscript i designates the group code, or the level of the class variable, the models can be written as

$$Y_{ij} = \beta_{i0} + \beta_{i1}X_{ij} + \varepsilon_{ij} \qquad [8.28]$$

where $i = 1, \ldots, t$ and $j = 1, \ldots, n_i$. This model contains $2t$ parameters: t β_0-parameters and t β_1-parameters. The random errors ε_{ij} for all groups are assumed to be normally and independently distributed with zero mean and common variance σ^2.

Model in Matrix Notation

The model encompassing all t groups is written in matrix notation by using t dummy variables to identify the levels of the class variable "group." Let

$$W_1 = \begin{cases} 1 & \text{if the observation is from group 1} \\ 0 & \text{otherwise} \end{cases}$$

$$W_2 = \begin{cases} 1 & \text{if the observation is from group 2} \\ 0 & \text{otherwise} \end{cases}$$

$$\vdots$$

$$W_t = \begin{cases} 1 & \text{if the observation is from group } t \\ 0 & \text{otherwise} \end{cases}$$

Then

$$\begin{aligned} Y_{ij} &= W_1(\beta_{10} + \beta_{11}X_{1j}) + W_2(\beta_{20} + \beta_{21}X_{2j}) \\ &\quad + \cdots + W_t(\beta_{t0} + \beta_{t1}X_{tj}) + \varepsilon_{ij} \\ &= \beta_{10}W_1 + \beta_{11}(W_1X_{1j}) + \beta_{20}W_2 + \beta_{21}(W_2X_{2j}) \\ &\quad + \cdots + \beta_{t0}W_t + \beta_{t1}(W_tX_{tj}) + \varepsilon_{ij} \end{aligned} \qquad [8.29]$$

or

$$\boldsymbol{Y} = \boldsymbol{X\beta} + \boldsymbol{\varepsilon} \qquad [8.30]$$

where

$$\boldsymbol{X} = \begin{bmatrix} 1 & X_{11} & 0 & 0 & \cdots & 0 & 0 \\ \vdots & \vdots & \vdots & \vdots & & \vdots & \vdots \\ 1 & X_{1n_1} & 0 & 0 & \cdots & 0 & 0 \\ 0 & 0 & 1 & X_{21} & \cdots & 0 & 0 \\ \vdots & \vdots & \vdots & \vdots & & \vdots & \vdots \\ 0 & 0 & 1 & X_{2n_2} & \cdots & 0 & 0 \\ \vdots & \vdots & \vdots & \vdots & & \vdots & \vdots \\ 0 & 0 & 0 & 0 & \cdots & 1 & X_{t1} \\ \vdots & \vdots & \vdots & \vdots & & \vdots & \vdots \\ 0 & 0 & 0 & 0 & \cdots & 1 & X_{tn_t} \end{bmatrix} \qquad \boldsymbol{\beta} = \begin{bmatrix} \beta_{10} \\ \beta_{11} \\ \beta_{20} \\ \beta_{21} \\ \vdots \\ \beta_{t0} \\ \beta_{t1} \end{bmatrix}$$

The odd-numbered columns of X are the dummy variables and provide for the t β_0's in the model. The even-numbered columns are the elementwise products of the dummy variables and the independent variable. These bring in the level of the X variable times the appropriate β_{i1} only when the observations are from the ith group. This is a full-rank model; $r(X) = 2t$ and there are $2t$ parameters to be estimated.

The two columns associated with any particular group are orthogonal to all other columns. Therefore, the results of the least squares regression using this large model to encompass all groups are identical to the results that would be obtained if each group were analyzed separately. The SS(Model) will have $2t$ degrees of freedom and will be the sum of the SS(Model) quantities from the separate analyses. The residual mean square from this full analysis will be identical to the pooled residual mean squares from the separate analyses. The pooled residual mean square is the best estimate of σ^2 unless a pure error estimate is available.

Testing Homogeneity of Slopes

There are several tests of homogeneity of interest. The test of homogeneity of slopes of regression lines is most common in the context of allowing the intercepts to be different. Thus, the different groups are allowed to have different mean levels of Y but are required to have the same response to changes in the independent variable. The null hypothesis is the composite hypothesis

$$H_0: \beta_{11} = \beta_{21} = \cdots = \beta_{t1} \qquad [8.31]$$

The difference in SS(Res) for full and reduced models will be used to test this hypothesis of common β_1. The reduced model is obtained from equation 8.29 by replacing the t different slopes, β_{i1}, with a common slope β_1:

$$Y_{ij} = \beta_{10} W_1 + \beta_{20} W_2 + \cdots + \beta_{t0} W_t + \beta_1 X_{ij} + \varepsilon_{ij} \qquad [8.32]$$

The independent variable is no longer multiplied by the dummy variables W_i. The $\boldsymbol{X}$ matrix for the reduced model consists of t columns for the dummy variables plus one column of the observations on the independent variable; the X_{ij} are no longer separated by groups. The rank of $\boldsymbol{X}$ in the reduced model is $(t + 1)$, t degrees of freedom for estimating the t intercepts and 1 degree of freedom for estimating the common slope.

The difference between the residual sum of squares for the full model and the residual sum of squares for the reduced model,

$$Q = \text{SS}(\text{Res}_{\text{reduced}}) - \text{SS}(\text{Res}_{\text{full}}) \qquad [8.33]$$

has $(t - 1)$ degrees of freedom, $(\sum n_i - t - 1) - (\sum n_i - 2t)$. This is the appropriate sum of squares for testing the composite null hypothesis given in equation 8.31. The test statistic is an F-ratio with $Q/(t - 1)$ as the numerator and the residual mean square from the full model as the denominator. A nonsignificant F-ratio leads to the conclusion that the regressions of Y on X for the several groups are adequately represented by a series of parallel lines. The differences in the "heights" of the lines, controlled by the t intercepts, reflect the mean differences among the groups.

Testing Homogeneity of Intercepts

The same general procedure can be used to test other hypotheses. The composite null hypothesis of common intercepts, β_{i0}, in the presence of heterogeneous slopes is not a meaningful hypothesis unless there is some logic in expecting the regressions for all groups to converge to a common value of Y at $X = 0$. (The intercept is usually defined as the value of Y at $X = 0$ or, if the X's are centered, the value of Y at $X = \bar{X}$. The origin of the independent variable can be shifted by adding a constant to or subtracting a constant from each value of X so that it is possible to test convergence of the regression lines to any chosen value of X.) It is quite common, however, to test homogeneity of intercepts after having decided that the groups have common slope. For this test, the reduced model with t β_{i0}-parameters and common β_1 (equation 8.32) becomes the full model. The new reduced model for $H_0: \beta_{10} = \beta_{20} = \cdots = \beta_{t0}$ is the simple regression model

$$Y_{ij} = \beta_0 + \beta_1 X_{ij} + \varepsilon_{ij} \qquad [8.34]$$

The $\boldsymbol{X}$ matrix for this reduced model has only two columns, the column of ones for the intercept and the column of X_{ij}. The difference in residual sums of squares for this model and the full model will have $(t - 1)$ degrees of freedom and is appropriate for testing the null hypothesis of equal intercepts in the presence of equal slopes.

A numerical example showing the tests of homogeneity of regression coefficients is presented in Section 8.8.

8.7 Analysis of Covariance

Use of Covariate to Improve Precision

The classical purpose of the analysis of covariance is to improve the precision of the experiment by statistical control of variation among experimental units. A useful covariate identifies variation among the experimental units that is also associated with variation in the dependent variable. For example, variation in density of plants in the experimental units causes variation in yield of most plant species, or variation in age or body weight of animals often causes variation in rate of gain in feeding trials. The covariance analysis removes this source of variation from experimental error and adjusts the treatment means for differences attributable to the covariate. For this purpose, the covariate should not be affected by the treatments. Otherwise, adjustment for the covariate will bias the estimates of treatment effects and possibly lead to incorrect inferences.

As an illustration, consider a study to measure the effects of nutrient levels on the growth rate of a species of bacteria. It is well known that temperature has an effect on growth rate. Therefore, any differences in temperature of the experimental units can be expected to cause differences in growth rates even if the experimental units receive the same nutrient treatment. Such differences will inflate experimental error and, to the extent that the nutrient groups differ in mean temperature, cause biases in the observed treatment effects. Suppose the available resources do not permit sufficient control of temperature to rule out these effects. Covariance analysis, with the measured temperature of each experimental unit as the covariate, could be used to adjust the observed growth rates to a common temperature.

Use of Covariate to Interpret Treatment Effects

A second use of the analysis of covariance is as an aid in the interpretation of treatment effects on a primary response variable. In this case, the covariate is another response variable that may be involved in the response of the primary response variable. The questions to be addressed by the covariance analysis are whether the treatment effects on the primary response variable are essentially independent of those on the secondary variable (the covariate) and, if not, how much of the effect on the primary response variable might be attributed to the indirect effects of the treatments on the covariate. For this purpose, it is quite likely that the covariate will be affected by the treatments.

Analysis of covariance is a special case of regression analysis where both continuous and class variables are used. The class variables take into account the experimental design features as discussed earlier in this chapter. The covariate will (almost) always be a continuous variable for which the experimental results are to be "adjusted."

Two-Way Model with Covariate

The usual linear model for the analysis of covariance for a randomized complete block design is

$$Y_{ij} = \mu + \gamma_i + \tau_j + \beta(X_{ij} - \bar{X}_{..}) + \varepsilon_{ij} \qquad [8.35]$$

where the term $\beta(X_{ij} - \bar{X}_{..})$ has been added to the RCB model, equation 8.20, to incorporate the effect of the covariate, X_{ij}, on the dependent variable. The covariate is expressed in terms of the deviations about its sample mean $\bar{X}_{..}$. This emphasizes that it is the variation in the covariate that is of interest, and simplifies the subsequent adjustment of the treatment means. Equation 8.35 is the simplest form in which a covariate effect can be included in a model—one covariate acting in a linear manner. The covariate model can be extended to include more than one covariate and more complicated relationships.

Model in Matrix Notation

The covariance model is written in matrix form by augmenting the design matrix $\boldsymbol{X}$ and parameter vector $\boldsymbol{\beta}$ for the appropriate experimental design. $\boldsymbol{X}$ is expanded to include a column vector of $(X_{ij} - \bar{X}_{..})$. $\boldsymbol{\beta}$ is expanded to include the regression coefficient for the covariate, β. The ordering of the observations for the covariate must be identical to the ordering of observations in $\boldsymbol{Y}$. The numerical example in Section 8.8 will illustrate $\boldsymbol{X}$ and $\boldsymbol{\beta}$.

Quantities of Interest

The covariance model is of less than full rank, because the design matrix to which the covariate vector was appended is singular. None of the singularities, however, involves the covariate vector. Reparameterization or the generalized inverse approach is used to obtain the relevant sums of squares and to estimate the estimable functions of the parameters. The quantities of primary interest are:

1. Partial sums of squares attributable to the covariate and to differences among the treatments
2. Estimate of experimental error after removal of the variation attributable to the covariate
3. Estimated treatment means and mean contrasts after adjustment to a common level of the covariate

The covariance analysis will first be discussed as if the purpose of the analysis were to increase precision of the experiment. Then, the key changes in interpretation will be noted for the case when covariance analysis is being used to help interpret the treatment effects.

Analysis of Variance

The partial sums of squares for the class variables, "blocks" and "treatments" in the RCB, and the covariate are shown in Table 8.3. These are *not* additive partitions of the total sum of squares even when the data are balanced. The covariate destroys the orthogonality that might have been present in the basic experimental design. The error variance is estimated from the residual mean square, the "block by treatment" interaction mean square after adjustment for the covariate. The degrees of freedom for residual reflect the loss of 1 degree of freedom for estimating β for the covariate.

This model and analysis assume that the basic datum is one observation on the ijth experimental unit, so that the residual mean square from the regression

Table 8.3 Partial sums of squares and mean squares from the analysis of covariance for a randomized complete block design with b blocks and t treatments.

Source	d.f.	*Partial* SS[a]	MS
Total	$bt - 1$	$Y'Y - \text{C.F.}$	
Blocks	$b - 1$	$R(\gamma'\|\tau'\ \beta\ \mu)$	
Treatments	$t - 1$	$R(\tau'\|\gamma'\ \beta\ \mu)$	
Covariate	1	$R(\beta\|\gamma'\ \tau'\ \mu)$	
Residual	$(b - 1)(t - 1) - 1$	$Y'Y - R(\gamma'\ \tau'\ \beta\ \mu)$	s^2

[a] γ' and τ' designate the row vectors of effects for the class variables "blocks" and "treatments," respectively.

analysis is also the error variance. If the data involve multiple samples from each experimental unit, the residual mean square in Table 8.3 will contain both experimental error and sampling error. For the covariance analysis to proceed correctly in the presence of sampling, the experimental unit means must be used in the analysis. Otherwise, the estimate of β will be based on pooled sampling and experimental errors rather than on experimental error alone.

Testing the Effect of the Covariate

The presence of the covariate reduces the residual sum of squares by the amount $R(\beta|\gamma'\ \tau'\ \mu)$, the partial sum of squares attributable to the covariate. This reflects the direct impact of the covariate on the magnitude of σ^2 and, hence, on the precision of the experiment. The null hypothesis that the covariate has no effect, H_0: $\beta = 0$, is tested with

$$F = \frac{R(\beta|\gamma'\ \tau'\ \mu)}{s^2} \qquad [8.36]$$

which has 1 and $[(b - 1)(t - 1) - 1]$ degrees of freedom. If F is not significant at the chosen α, it is concluded that the covariate is not important in controlling precision and the covariance analysis is abandoned. Interpretations are based on the conventional analysis of variance. If the null hypothesis is rejected, it is concluded that the covariate is effective in increasing precision and the covariance analysis is continued to obtain estimates of treatment means and contrasts adjusted for the effects of the covariate. The residual mean square is the estimate of σ^2 for all subsequent computations.

Testing Treatment Effects

The partial sum of squares for each class variable, $R(\tau'|\gamma'\ \beta\ \mu)$ or $R(\gamma'|\tau'\ \beta\ \mu)$, is the appropriate sum of squares for testing the composite null hypothesis that all effects for that class variable are zero. As always, these sums of squares can be computed either by defining an appropriate $\boldsymbol{K}'$ for the general linear hypothesis or by the difference between residual sums of squares for full and reduced models. The partial sum of squares for a class variable adjusted for the covariate measures the variability among the levels of the class variable as if all observations had occurred at the mean level of the covariate. The null hypothesis that all treatment effects are zero is tested by

$$F = \frac{R(\tau'|\gamma'\ \beta\ \mu)/(t - 1)}{s^2} \qquad [8.37]$$

Unadjusted Treatment Means

The conventional, unadjusted treatment means are computed as simple averages of the observations in each treatment. The vector of unadjusted treatment means can be written as

$$\bar{\boldsymbol{Y}} = \boldsymbol{T}'\boldsymbol{Y} \tag{8.38}$$

where $\boldsymbol{T}$ is defined as the matrix of the t treatment dummy variables with each divided by the number of observations in the treatment. Thus, $\boldsymbol{T}$ is

$$\boldsymbol{T} = \frac{1}{b}\begin{bmatrix} 1 & 0 & \cdots & 0 \\ \vdots & \vdots & & \vdots \\ 1 & 0 & & 0 \\ 0 & 1 & & 0 \\ \vdots & \vdots & & \vdots \\ 0 & 1 & & 0 \\ & & \ddots & \\ 0 & 0 & & 1 \\ \vdots & \vdots & & \vdots \\ 0 & 0 & \cdots & 1 \end{bmatrix} \tag{8.39}$$

when there are b observations per treatment. The expectation of $\bar{\boldsymbol{Y}}$ is

$$\mathscr{E}(\bar{\boldsymbol{Y}}) = \boldsymbol{T}'\boldsymbol{X}\boldsymbol{\beta} \tag{8.40}$$

If the model includes a covariate, the expectation of the ith mean contains the term $\beta(\bar{X}_{i.} - \bar{X}_{..})$ in addition to the appropriate linear function of the other model effects. Because of this term, comparisons among the treatment means include differences due to the covariate unless $\beta = 0$ or $\bar{X}_{i.}$ is the same for all treatments.

Adjusted Treatment Means

The **adjusted treatment means** are designed to remove this confounding. Adjustment is accomplished either by estimating directly from $\boldsymbol{\beta}^0$ the linear function of the parameters of interest, or by subtracting an estimate of the bias term from each unadjusted treatment mean. The linear functions of the parameters that need to be estimated are appropriately defined by equation 8.40 if $\boldsymbol{X}$ is redefined by replacing the column of covariate values with a column of zeros. If this redefined $\boldsymbol{X}$ is labeled $\boldsymbol{X}_c$, the linear functions to be estimated by the adjusted treatment means are

$$\mathscr{E}(\bar{\boldsymbol{Y}}_{\text{adj}}) = \boldsymbol{T}'\boldsymbol{X}_c\boldsymbol{\beta} \tag{8.41}$$

where $\bar{\boldsymbol{Y}}_{\text{adj}}$ denotes the vector of adjusted treatment means. The least squares estimate of the *adjusted* treatment means is given by the same linear function of the least squares solution $\boldsymbol{\beta}^0$,

$$\bar{\boldsymbol{Y}}_{\text{adj}} = \boldsymbol{T}'\boldsymbol{X}_c\boldsymbol{\beta}^0 \tag{8.42}$$

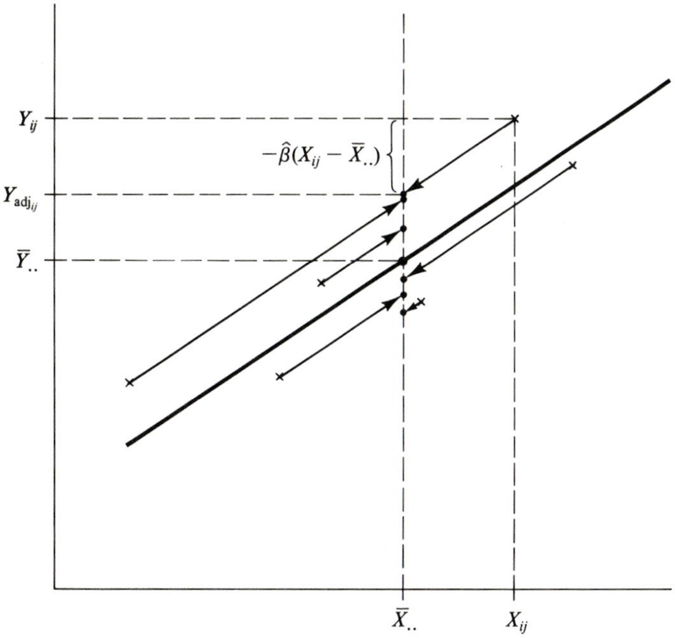

Figure 8.1 Illustration of the adjustment of the response variable Y for differences in the covariate X.

The adjusted treatment means are estimates of the treatment means for the case where all treatments have the mean level of the covariate, $X = \bar{X}_{..}$. The adjustment can be made to any level of the covariate, say C, by defining $\mathbf{X}_c$ to be the matrix with the column vector of covariate values replaced with $(C - \bar{X}_{..})$ rather than with zeros.

Alternatively, each adjusted treatment mean can be obtained by removing the bias $\beta(\bar{X}_{i.} - \bar{X}_{..})$ from the corresponding unadjusted treatment mean. This leads to the more traditional method of computing the adjusted treatment means:

$$\bar{Y}_{\text{adj}_{i.}} = \bar{Y}_{i.} - \hat{\beta}(\bar{X}_{i.} - \bar{X}_{..}) \qquad [8.43]$$

The covariance adjustment is illustrated in Figure 8.1. The diagonal line passing through the point $(\bar{X}_{..}, \bar{Y}_{..})$ is the regression line with slope $\hat{\beta}$ relating the dependent variable to the covariate. The original observations are represented with $\times$'s. The adjustment can be viewed as moving each observation along a path parallel to the fitted regression line from the observed value of the covariate, $X = X_{ij}$, to the common value $X = \bar{X}_{..}$. The dots on the vertical line at $X = \bar{X}_{..}$ represent the adjusted observations. The amount each Y_{ij} is adjusted during this shift is determined by the slope of the regression line and the change in X,

$$Y_{\text{adj}_{ij}} = Y_{ij} - \hat{\beta}(X_{ij} - \bar{X}_{..})$$

Averaging the adjusted observations within each treatment gives the adjusted treatment means, equation 8.43.

Variances of Adjusted Treatment Means

The variance–covariance matrix of the adjusted treatment means follows directly from the matrix equation for the variance of a linear function. Thus,

$$\mathbf{Var}(\bar{Y}_{\text{adj}}) = (\boldsymbol{T}'\boldsymbol{X}_c)(\boldsymbol{X}'\boldsymbol{X})^-(\boldsymbol{T}'\boldsymbol{X}_c)'\sigma^2 \qquad [8.44]$$

The variances of the adjusted treatment means, the diagonal elements of equation 8.44, simplify to the classical formula for the variance:

$$s^2(\bar{Y}_{\text{adj}i.}) = \left[\frac{1}{n} + \frac{(\bar{X}_{i.} - \bar{X}_{..})}{E_{xx}}\right]\sigma^2 \qquad [8.45]$$

where E_{xx} is the residual sum of squares from the RCB analysis of variance *of the covariate.*

Covariance to Help Interpret Treatment Effects

When the covariance analysis is being used to aid interpretation of the treatment effects, the primary interest is in comparison of the treatment means and sums of squares before and after adjustment for the covariate. The adjustment of the means and sums of squares is *not* viewed as a method of obtaining *unbiased* estimates of treatment effects. Rather, the changes in the means and sums of squares provide some indication of the proportion of the treatment effects that can be viewed as direct effects on Y versus possible indirect effects on Y through X, or through some other variable which in turn affects both X and Y. For example, highly significant treatment effects that remain about the same after adjustment for X would suggest that most of the treatment effects on Y are essentially independent of any treatment effects on X. On the other hand, dramatic changes in the treatment effects with adjustment would suggest that X and Y are closely linked in the system being studied so that the responses of both variables to the treatments are highly correlated.

The test of the null hypothesis $H_0\colon \beta = 0$ is a test of the hypothesis that the correlation between the residuals for X and the residuals for Y is zero, after both have been adjusted for block and treatment effects. If the covariate was chosen because it was expected to have a direct impact on Y, then β would be expected to be nonzero and this test would serve only as a confirmation of some link between the two variables. A nonsignificant test would suggest that the link between the two variables is very weak, or the power of the test is not adequate to detect the link. In either case, any effort devoted to interpretation of the adjusted treatment means and sums of squares would not be very productive.

8.8 Numerical Examples

Two examples will be used. The first example will combine several concepts covered in this chapter:

1. Analysis of variance as a regression problem including reparameterization

2. Use of dummy variables to test homogeneity of regressions
3. Analysis of covariance to aid in the interpretation of treatment effects

The second example will illustrate the more classical use of covariance and will use a generalized inverse solution to the normal equations.

Example 8.4 The purpose of the first study was to compare ascorbic acid content in cabbage from two genetic lines (cultivars) planted on three different dates (Table 8.4). The experimental design was a completely random design with $r = 10$ experimental units for each combination of planting date and genetic line, for a total of 60 observations. It was anticipated that ascorbic acid content might be dependent on the size of the cabbage head; hence, head weight was recorded for possible use as a covariate. (The data are from the files of the late Dr. Gertrude M. Cox.)

Ascorbic acid content is the dependent variable of interest and head weight will be used as a covariate. The variables "date" and "line" will be treated as class variables. The first analysis will be the conventional analysis of variance for the factorial experiment. Then, in anticipation of the analysis of covariance, the homogeneity of regression coefficients, relating ascorbic acid content to head size, over the six date–line treatment combinations will be tested. Finally, the analysis of covariance will be run.

Table 8.4 Head weight and ascorbic acid content for two cabbage varieties on three planting dates.

	Planting Date					
	16		20		21	
Line Number	*Head Wt.*	*Ascorbic Content*	*Head Wt.*	*Ascorbic Content*	*Head Wt.*	*Ascorbic Content*
39	2.5	51	3.0	65	2.2	54
	2.2	55	2.8	52	1.8	59
	3.1	45	2.8	41	1.6	66
	4.3	42	2.7	51	2.1	54
	2.5	53	2.6	41	3.3	45
	4.3	50	2.8	45	3.8	49
	3.8	50	2.6	51	3.2	49
	4.3	52	2.6	45	3.6	55
	1.7	56	2.6	61	4.2	49
	3.1	49	3.5	42	1.6	68
52	2.0	58	4.0	52	1.5	78
	2.4	55	2.8	70	1.4	75
	1.9	67	3.1	57	1.7	70
	2.8	61	4.2	58	1.3	84
	1.7	67	3.7	47	1.7	71
	3.2	68	3.0	56	1.6	72
	2.0	58	2.2	72	1.4	62
	2.2	63	2.3	63	1.0	68
	2.2	56	3.8	54	1.5	66
	2.2	72	2.0	60	1.6	72

The purpose of the covariance analysis in this example is as an aid in interpreting the effects of planting date and genetic line on ascorbic acid content, rather than for control of random variation among the experimental units. It is expected that the covariable head weight will be affected by the date and line treatment factors. Hence, adjustment of ascorbic acid content to a common head weight would redefine treatment effects.

8.8.1 Analysis of Variance

The conventional model for a factorial set of treatments in a completely random design is

$$Y_{ijk} = \mu + \gamma_i + \tau_j + (\gamma\tau)_{ij} + \varepsilon_{ijk} \qquad [8.46]$$

where γ_i are the "date" effects ($i = 1, 2, 3$), τ_j are the "line" effects ($j = 1, 2$), and $(\gamma\tau)_{ij}$ are the "date by line" interaction effects. This model contains 12 parameters to define only six group means. Thus, there are six linear dependencies in the model and reparameterization requires six constraints. There must be one constraint on the γ_i, one on the τ_j, and four on the $(\gamma\tau)_{ij}$.

For this illustration, the means model will be used as the reparameterized model and then the general linear hypothesis will be used to partition the variation among the six treatments into "date," "line," and "date by line" sums of squares. Thus, the (full-rank) model for the analysis of variance is

$$Y_{ijk} = \mu_{ij} + \varepsilon_{ijk} \qquad [8.47]$$

where μ_{ij} is the true mean of the ijth date–line group. In this model $\boldsymbol{X}$ is of order (60×6) where each column is a dummy variable showing the incidence of the observations for one of the date–line groups. That is, the ijth dummy variable takes the value one if the observation is from the ijth date–line group; otherwise the dummy variable takes the value zero. It is assumed that the elements of $\boldsymbol{\beta}^*$ are in the order $\boldsymbol{\beta}^{*\prime} = (\mu_{11}\ \mu_{12}\ \mu_{21}\ \mu_{22}\ \mu_{31}\ \mu_{32})$.

The least squares analysis using this model gives SS(Model) = 205,041.9 with 6 degrees of freedom and SS(Residual) = 2,491.1 with 54 degrees of freedom. The least squares estimates of μ_{ij} are the group means:

$$\hat{\boldsymbol{\beta}}^{*\prime} = (50.3\ \ 62.5\ \ 49.4\ \ 58.9\ \ 54.8\ \ 71.8)$$

Each $\hat{\mu}_{ij}$ is estimating $\mu + \gamma_i + \tau_j + (\gamma\tau)_{ij}$, the mean of the treatment group in terms of the original parameters. These are the estimated group means for ascorbic acid ignoring any differences in head weight since the model does not include the covariate.

The partitions of SS(Model) are obtained by appropriate definition of $\boldsymbol{K}'$ for general linear hypotheses on the μ_{ij}. For this purpose, it is helpful to view the μ_{ij} as a 3×2 "date by line" table of means. The marginal means for this table, $\bar{\mu}_{i.}$ and $\bar{\mu}_{.j}$, represent the "date" means and the "line" means, respectively. For each sum of squares to be computed, the appropriate null hypothesis will be stated in terms of the μ_{ij}, the appropriate $\boldsymbol{K}'$ will be defined for the null hypothesis, and the sum of squares Q computed using the general linear hypothesis, equation 4.37, will be given. In all hypotheses $\boldsymbol{m} = 0$ and Q is computed as

$$Q = (\boldsymbol{K}'\hat{\boldsymbol{\beta}}^*)'[\boldsymbol{K}'(\boldsymbol{X}^{*\prime}\boldsymbol{X}^*)^{-1}\boldsymbol{K}]^{-1}(\boldsymbol{K}'\hat{\boldsymbol{\beta}}^*)$$

1. *Correction factor:* The sum of squares due to the correction for the mean, the correction factor, measures the deviation of the overall mean $\bar{\mu}_{..}$ from zero. The overall mean is zero only if the sum of the μ_{ij} is zero. Therefore,

$$H_0: \ \bar{\mu}_{..} = 0 \quad \text{or} \quad \sum\sum \mu_{ij} = 0$$

$$\boldsymbol{K}_1' = (1 \ 1 \ 1 \ 1 \ 1 \ 1) \qquad r(\boldsymbol{K}_1) = 1 \qquad [8.48]$$

$$Q_1 = 201{,}492.1 \quad \text{with 1 degree of freedom}$$

2. *Sum of squares for "dates":* The hypothesis of no date effects is equivalent to the hypothesis that the three marginal means $\bar{\mu}_{i.}$ are equal. The equality of the three means can be expressed in terms of two linearly independent differences being zero:

$$H_0: \ \bar{\mu}_{1.} = \bar{\mu}_{2.} = \bar{\mu}_{3.}$$

or

$$H_0: \ (\mu_{11} + \mu_{12}) - (\mu_{21} + \mu_{22}) = 0$$

$$\text{and} \quad (\mu_{11} + \mu_{12}) + (\mu_{21} + \mu_{22}) - 2(\mu_{31} + \mu_{32}) = 0$$

$$\boldsymbol{K}_2' = \begin{bmatrix} 1 & 1 & -1 & -1 & 0 & 0 \\ 1 & 1 & 1 & 1 & -2 & -2 \end{bmatrix} \qquad r(\boldsymbol{K}_2) = 2 \qquad [8.49]$$

$$Q_2 = 909.3 \quad \text{with 2 degrees of freedom}$$

3. *Sum of squares for "lines":* The hypothesis of no "line" effects is equivalent to the hypothesis that the two marginal means for "lines" $\bar{\mu}_{.j}$ are equal or that the difference is zero:

$$H_0: \ \bar{\mu}_{.1} = \bar{\mu}_{.2}$$

or

$$H_0: \ \mu_{11} + \mu_{21} + \mu_{31} - \mu_{12} - \mu_{22} - \mu_{32} = 0$$

$$\boldsymbol{K}_3' = (1 \ -1 \ 1 \ -1 \ 1 \ -1) \qquad r(\boldsymbol{K}_3) = 1 \qquad [8.50]$$

$$Q_3 = 2{,}496.15 \quad \text{with 1 degree of freedom}$$

4. *Sum of squares for "dates by lines":* The null hypothesis of no interaction effects between "dates" and "lines" is equivalent to the hypothesis that the difference between lines is the same for all dates, or that the differences among dates are the same for all lines. The former is easier to visualize because there are only two lines and one difference between lines for each date. There are three such differences which, again, require two linearly independent statements:

$$H_0\colon\ \mu_{11} - \mu_{12} = \mu_{21} - \mu_{22} = \mu_{31} - \mu_{32}$$

or

$$H_0\colon\ (\mu_{11} - \mu_{12}) - (\mu_{21} - \mu_{22}) = 0$$

$$\text{and}\quad (\mu_{11} - \mu_{12}) + (\mu_{21} - \mu_{22}) - 2(\mu_{31} - \mu_{32}) = 0$$

$$\boldsymbol{K}_4' = \begin{bmatrix} 1 & -1 & -1 & 1 & 0 & 0 \\ 1 & -1 & 1 & -1 & -2 & 2 \end{bmatrix} \qquad r(\boldsymbol{K}_4) = 2 \qquad [8.51]$$

$Q_4 = 144.25$ with 2 degrees of freedom

The $\boldsymbol{K}'$ matrix appropriate for the hypothesis of no interaction is the most difficult matrix to define. The statements were generated using the fact that interaction measures the failure of the simple effects to be consistent over all levels of the other factor. It should be observed, however, that $\boldsymbol{K}_4'$ is easily generated as the elementwise product of each row vector in $\boldsymbol{K}_2'$ with the row vector in $\boldsymbol{K}_3'$. Interaction contrasts can always be generated in this manner.

This analysis of variance is summarized in Table 8.5. The results are identical to those from the conventional analysis of variance for a two-factor factorial in a completely random experimental design. The residual mean square serves as the denominator for F-tests of the treatment effects (if treatment effects are fixed effects). There are significant differences among the planting dates and between the two genetic lines for ascorbic acid content. The interaction between dates and lines is not significant, indicating that the difference between the lines is reasonably constant over all planting dates.

Table 8.5 Factorial analysis of variance of ascorbic acid content of cabbage.

Source	d.f.	*Sum of Squares*	*Mean Square*
$\text{Total}_{\text{uncorr}}$	60	207,533.0	
Model	6	205,041.9	
C.F.	1	201,492.1	
Dates	2	909.3	454.7
Lines	1	2,496.2	2,496.2
Dates × Lines	2	144.3	72.2
Residual	54	2,491.1	46.1

8.8.2 Test of Homogeneity of Regression Coefficients

The analysis of covariance assumes that all treatments have the same relationship between the dependent variable and the covariate. In preparation for the covariance analysis of the cabbage data (Section 8.8.3), this section gives the test of homogeneity of the regression coefficients.

The full model for the test of homogeneity allows each treatment group to have its own regression coefficient relating ascorbic acid content to head size. The means model used in the analysis of variance (equation 8.47) is expanded to give

$$Y_{ijk} = \mu_{ij} + \beta_{ij}(X_{ijk} - \bar{X}_{...}) + \varepsilon_{ijk} \qquad [8.52]$$

where the ij subscripts on β allow for a different regression coefficient for each of the six treatment groups. There are now 12 parameters and $\boldsymbol{X}^*$ must be of order (60×12). Each of the additional six columns in $\boldsymbol{X}^*$ consists of the covariate values for one of the treatment groups. The elements in the column for the ijth group take the values $(X_{ijk} - \bar{X}_{...})$ if the observation is from that group and zero otherwise. These six columns can be generated by elementwise multiplication of the dummy variable for each treatment by the original vector of $(X_{ijk} - \bar{X}_{...})$. The $\boldsymbol{X}^*$ matrix has the form

$$\boldsymbol{X}^* = \begin{bmatrix} \mathbf{1} & \mathbf{0} & \mathbf{0} & \mathbf{0} & \mathbf{0} & \mathbf{0} & \boldsymbol{x}_{11} & \mathbf{0} & \mathbf{0} & \mathbf{0} & \mathbf{0} & \mathbf{0} \\ \mathbf{0} & \mathbf{1} & \mathbf{0} & \mathbf{0} & \mathbf{0} & \mathbf{0} & \mathbf{0} & \boldsymbol{x}_{12} & \mathbf{0} & \mathbf{0} & \mathbf{0} & \mathbf{0} \\ \mathbf{0} & \mathbf{0} & \mathbf{1} & \mathbf{0} & \mathbf{0} & \mathbf{0} & \mathbf{0} & \mathbf{0} & \boldsymbol{x}_{21} & \mathbf{0} & \mathbf{0} & \mathbf{0} \\ \mathbf{0} & \mathbf{0} & \mathbf{0} & \mathbf{1} & \mathbf{0} & \mathbf{0} & \mathbf{0} & \mathbf{0} & \mathbf{0} & \boldsymbol{x}_{22} & \mathbf{0} & \mathbf{0} \\ \mathbf{0} & \mathbf{0} & \mathbf{0} & \mathbf{0} & \mathbf{1} & \mathbf{0} & \mathbf{0} & \mathbf{0} & \mathbf{0} & \mathbf{0} & \boldsymbol{x}_{31} & \mathbf{0} \\ \mathbf{0} & \mathbf{0} & \mathbf{0} & \mathbf{0} & \mathbf{0} & \mathbf{1} & \mathbf{0} & \mathbf{0} & \mathbf{0} & \mathbf{0} & \mathbf{0} & \boldsymbol{x}_{32} \end{bmatrix}$$

where each *element* in $\boldsymbol{X}^*$ is a column vector of order 10×1; $\boldsymbol{x}_{ij}$ is the 10×1 column vector of the deviations of head weight from the overall mean head weight for the ijth treatment group. The least squares analysis using this model gives $\text{SS(Res}_{\text{full}}) = 1{,}847.2$ with $60 - 12 = 48$ degrees of freedom.

The reduced model for the null hypothesis of homogeneity of regression coefficients

$$H_0\colon \quad \beta_{ij} = \beta \text{ for all } ij \text{ combinations}$$

is

$$Y_{ijk} = \mu_{ij} + \beta(X_{ijk} - \bar{X}_{...}) + \varepsilon_{ijk} \qquad [8.53]$$

There are seven parameters in this reduced model—the six μ_{ij} plus the common β. (This is the covariance model that will be used in the next section.) The least squares analysis of this reduced model gives $\text{SS(Res}_{\text{reduced}}) = 1{,}975.1$ with 53 degrees of freedom.

The difference in residual sums of squares for the full and reduced models is

$$Q = SS(Res_{reduced}) - SS(Res_{full})$$
$$= 1{,}975.1 - 1{,}847.2 = 127.9$$

with $53 - 48 = 5$ degrees of freedom. This is the appropriate numerator sum of squares for F-test of the null hypothesis. The appropriate denominator for the F-test is the residual mean square from the full model,

$$s^2 = \frac{1{,}847.24}{48} = 38.48$$

Thus,

$$F = \frac{127.9/5}{38.48} = .66$$

which is nonsignificant. A common regression coefficient for all treatments is sufficient for describing the relationship between ascorbic acid content and the head weight of cabbage in these data.

If the regression coefficients are heterogeneous, the covariance analysis for whatever purpose must be used with caution. The meaning of "adjusted treatment means" is not clear when the responses to the covariate differ. The choice of the common level of the covariate to which adjustment is made becomes critical. The treatment differences and even the ranking of the treatments can depend on this choice.

8.8.3 Analysis of Covariance

The analysis of covariance is used on the ascorbic acid content of cabbage as an aid in interpreting the treatment effects. The differences among adjusted treatment means are *not* to be interpreted as treatment effects. The changes in the sums of squares and treatment means as they are adjusted provide insight into the degree of relationship between the treatment effects on the two response variables, ascorbic acid content and head weight.

The model for the analysis of covariance, using the means parameterization and a common regression of ascorbic acid on head size for all groups, was given as the reduced model in the test of homogeneity, equation 8.53. The least squares analysis of this model gives the analysis of covariance. The $\boldsymbol{X}^*$ matrix from the analysis of variance is augmented with the column of observations on the covariate, expressed as deviations from the mean of the covariate. The vector of parameters is expanded to include β, the regression coefficient for the covariate.

Least squares analysis for this model gives SS(Model) = 205,557.9 with 7 degrees of freedom and SS(Residual) = 1,975.1 with 53 degrees of freedom. The decrease in the residual sum of squares from the analysis of variance model to the covariance model is due to the linear regression on the covariate. This difference in SS(Res) for the two models is the partial sum of squares for β, $R(\beta|\boldsymbol{\mu}') = 2{,}491.1 - 1{,}975.1 = 516.0$ with 1 degree of freedom, and is the appropriate numerator sum of squares for the F-test of the null hypothesis $H_0: \beta = 0$. The denominator is the residual mean square from the covariance model, $s^2 = 1{,}975.1/53 = 37.3$.

The F-test of $H_0: \beta = 0$ is

$$F = \frac{516.0}{37.3} = 13.8$$

with 1 and 53 degrees of freedom, which is significant beyond $\alpha = .001$. This confirms that there is a significant correlation between the variation in ascorbic acid content and head size after both have been adjusted for other effects in the model. This can be interpreted as a test of the hypothesis that the correlation between the random plot-to-plot errors of the two traits is zero.

General linear hypotheses are used to compute the partial sum of squares attributable to each of the original class variables. These sums of squares will differ from the analysis of variance sums of squares because they will now be adjusted for the covariate. The $\boldsymbol{K}'$ matrices defined in the analysis of variance, equations 8.48–8.51, need to be augmented on the right with a column of zeros as coefficients for β so that $\boldsymbol{K}'$ and $\hat{\boldsymbol{\beta}}^*$ conform for multiplication. These sums of squares are no longer additive partitions of the model sum of squares because the adjustment for the covariate has destroyed the orthogonality. An additional $\boldsymbol{K}'$ could be defined for the hypothesis that $\beta = 0$, but the appropriate F-test based on the difference in residual sums of squares has already been performed in the previous paragraph. The analysis of variance summary for the covariance model is given in Table 8.6.

A comparison of Tables 8.5 and 8.6 shows major decreases in the sums of squares for "dates" and "lines" after adjustment for differences in head weight. The test for "date by line" effects is nonsignificant both before and after adjustment. The sum of squares for "dates" was reduced from a highly significant 909 to a just-significant 240 ($\alpha = .05$). The sum of squares for "lines" was reduced by

Table 8.6 Partial sums of squares for the analysis of covariance of ascorbic acid content for the cabbage data. The covariate is head weight.

Source	d.f.	*Sum of Squares*	*Mean Square*
$\text{Total}_{\text{uncorr}}$	60	207,533.0	
Model	7	205,557.9	
C.F.	1	201,492.1	
Dates	2	239.8	119.9
Lines	1	1,237.3	1,237.3
Dates × Lines	2	30.7	15.4
Covariate	1	516.0	516.0
Residual	53	1,975.1	37.27

Table 8.7 Adjustment of treatment means for ascorbic acid content in cabbage for differences in the covariable head weight.

Group	*Mean Head Weight*	*Mean Ascorbic Acid (unadjusted)*	*Adjustment* $-\hat{\beta}(\bar{X}_{ij.} - \bar{X}_{...})$	*Mean Ascorbic Acid (adjusted)*[a]
11	3.18	50.3	2.64	52.94 (2.06)
12	2.26	62.5	−1.50	61.00 (1.97)
21	2.80	49.4	.93	50.33 (1.95)
22	3.11	58.9	2.33	61.23 (2.03)
31	2.74	54.8	.66	55.46 (1.94)
32	1.47	71.8	−5.06	66.74 (2.36)
Mean	2.593	57.95		57.95

[a] Standard errors of adjusted treatment means are shown in parentheses. The standard error on each unadjusted treatment mean is 2.15.

half but is still highly significant. These results suggest that a significant part of the variation in ascorbic acid content among dates of planting and between lines is associated with variation in head size. However, not all of the variation in ascorbic acid content can be explained by variation in head size.

The estimate of the parameters is

$$\hat{\beta}^{*\prime} = (52.94 \quad 61.00 \quad 50.33 \quad 61.23 \quad 55.46 \quad 66.74 \quad -4.503)$$

The $\hat{\mu}_{ij}$ from the means reparameterization are estimates of the treatment means for ascorbic acid content, which are now adjusted for differences in head weight. (The estimate of the parameters contains the adjusted treatment means only because the means reparameterization was used and the covariate was centered. Otherwise, linear functions of the parameter estimates would have to be used to compute the adjusted means.) The estimate of the regression coefficient for the covariate is $\hat{\beta} = -4.5026$. Each increase of 1 unit in head weight is associated with a *decrease* in ascorbic acid content of 4.5 units on the average.

The adjustments to mean ascorbic acid content for differences in mean head weight are shown in Table 8.7. The biggest adjustment is for the third planting date for line 2, which had a very small head weight and high ascorbic acid content. Adjustment for head size reduced the average difference in ascorbic acid content between the two lines from about 12 units to 10 units. The first two planting dates differ very little for either line, but the third planting date gives appreciably higher ascorbic acid content even after adjustment for smaller head size in that planting date.

The analysis shows that there is considerable genetic and environmental correlation between ascorbic acid content and head size in cabbage. Some of the higher ascorbic acid content in line 2 in the third planting date may be attributable to the smaller head size produced by that treatment combination. This does not mean, however, that this adjusted mean is a better estimate of ascorbic acid content of line 2 when planted late. The smaller head size may be an innate trait of line 2 when grown under the environmental conditions of the late planting. If so, the adjustment to a common head size underestimates the ascorbic acid content for line 2 grown under those conditions. ■

The next example illustrates the classical use of covariance to control experimental error.

Example 8.5 The data for the example are from a study to compare seven sources of phosphorus each applied at two rates (40 and 80 lbs/A). The experimental design is a randomized complete block experimental design with $b = 3$ blocks. The dependent variable is 3-year dry weight forage production (lbs/A). The covariate is soil phosphorus content (ppm P_2O_5) measured at the beginning of the study. The data are given in Table 8.8. (The data are from the files of the late Dr. Gertrude M. Cox.)

The linear model for a factorial set of treatments in a randomized complete block design is

$$Y_{ijk} = \mu + \rho_i + \gamma_j + \tau_k + (\gamma\tau)_{jk} + \varepsilon_{ijk} \quad [8.54]$$

where

ρ_i = effect of ith block ($i = 1, 2, 3$)

γ_j = effect of jth source of phosphorus ($j = 1, \ldots, 7$)

τ_k = effect of kth rate of application ($k = 1, 2$)

$(\gamma\tau)_{jk}$ = interaction effect of jth source and kth rate

Table 8.8 Average dry forage yields (lbs/A) from a study of sources and rates of phosphorus fertilization. The experimental design was a randomized complete block design with seven sources of phosphorus, each applied at two rates (lbs/A). The phosphorus content of the soil (ppm of P_2O_5) at the beginning of the study was recorded for use as a possible covariate. (Data are from the files of the late Dr. Gertrude M. Cox.)

Treatment		*Block* I		*Block* II		*Block* III	
Source	*Rate*	*Phos.*	*Forage*	*Phos.*	*Forage*	*Phos.*	*Forage*
SUPER	40	32.0	2,475	43.2	3,400	51.2	3,436
SUPER	80	44.8	3,926	56.0	4,145	75.2	3,706
TSUPER	40	43.2	2,937	52.8	2,826	27.2	3,288
TSUPER	80	41.6	3,979	64.0	4,065	36.8	4,344
BSLAG	40	49.6	3,411	62.4	3,418	46.4	2,915
BSLAG	80	51.2	4,420	62.4	4,141	48.0	4,297
FROCK	40	48.0	3,122	75.2	3,372	22.4	1,576
FROCK	80	48.0	4,420	76.8	3,926	24.0	1,666
RROCK	40	54.4	2,334	60.8	2,530	49.6	1,275
RROCK	80	60.8	3,197	59.2	3,444	46.4	2,414
COLOID	40	72.0	3,045	59.2	2,206	19.2	540
COLOID	80	76.8	3,333	32.0	410	70.4	4,294
CAMETA	40	64.0	3,594	62.4	3,787	44.8	3,312
CAMETA	80	62.4	3,611	76.8	4,211	48.0	4,379

Table 8.9 Analysis of variance of dry forage from the phosphorus fertilization data.

Source	d.f.	*Sum of Squares*	*Mean Square*	*F-Value*	Prob > *F*
Corrected total	41	41,719,241			
BLOCK	2	1,520,897	760,449	1.03	.3700
SOURCE	6	13,312,957	2,218,826	3.01	.0226
RATE	1	7,315,853	7,315,853	9.94	.0040
SOURCE*RATE	6	435,267	72,544	.10	.9959
Error	26	19,134,266	735,933		

The covariate is included in the model by adding the term $\beta(X_{ijk} - \bar{X}_{...})$ to equation 8.54. In this example, the covariate was measured before the treatments were applied to the experimental units, so there is no chance the covariate could have been affected by the treatments.

The analysis of variance model contains 27 parameters but the rank of $\boldsymbol{X}$ is $r(\boldsymbol{X}) = 17$; reparameterization would therefore require 10 constraints. Analysis of these data will use the generalized inverse approach, rather than reparameterization, to obtain the solution to the normal equations. PROC ANOVA and PROC GLM, the general linear models procedure (SAS Institute, Inc., 1985d, 1985f) are used for the analyses.

The analysis of variance is obtained from PROC ANOVA using the following statements:

```
PROC ANOVA;
  CLASS BLOCK SOURCE RATE;
  MODEL FORAGE = BLOCK SOURCE RATE SOURCE*RATE;
```

The CLASS statement identifies the variables that are to be regarded as class variables. Whenever a class variable is encountered in the MODEL statement, the program constructs a dummy variable for each level of the class variable. Thus, $\boldsymbol{X}$ will contain three dummy variables for BLOCK, seven dummy variables for SOURCE, and two dummy variables for RATE. An interaction between two (or more) class variables in the MODEL statement instructs the program to construct a dummy variable for each unique joint level of the two factors; there will be 14 dummy variables for SOURCE*RATE.

The summary of the analysis of variance for the experiment is given in Table 8.9. There are significant differences among the sources of phosphorus ($\alpha = .05$) and highly significant differences between the rates of application ($\alpha = .01$). Block effects and source-by-rate interaction effects are not significant. The residual mean square is $s^2 = 735,933$ and the coefficient of variation is 26.7%.

The purpose of the covariance analysis is to use the information on soil phosphorus content to "standardize" the experimental results to a common level of soil phosphorus and, thereby, improve the precision of the comparisons. The analysis of covariance is obtained from PROC GLM (PROC ANOVA cannot handle a continuous variable) by expanding the model statement to include the covariate, PHOSDEV:

MODEL FORAGE=BLOCK SOURCE RATE SOURCE*RATE
PHOSDEV/SOLUTION;

The variable PHOSDEV has been previously defined in the program as the centered covariate. The "/SOLUTION" portion of the statement requests PROC GLM to print the solution $\boldsymbol{\beta}^0$ to the normal equations.

The analysis of covariance is summarized in Table 8.10. The lower two sections of Table 8.10 present the sequential sums of squares (TYPE I in SAS) and the partial sums of squares (TYPE III in SAS). Since the covariate was placed last in the model statement and the experimental design was balanced, the first four lines of the sequential sums of squares reproduce the analysis of variance sums of squares (Table 8.9).

The first question to ask of the analysis is whether the covariate has improved the precision of the comparisons. The residual mean square after adjustment for the covariate is $s^2 = 384{,}776$. (PROC ANOVA and PROC GLM always label the residual mean square as ERROR. In many cases this is a misnomer.) This is a reduction of 48% from $s^2 = 735{,}933$ in the analysis of variance (Table 8.9), or the coefficient of variation has been reduced from 26.7% to 19.3%. The reduction in the residual sum of squares is the partial sum of squares for the covariate and provides a test of the hypothesis $H_0\colon \beta = 0$, where β is the regression coefficient on PHOSDEV. This test gives $F = 24.73$ with 1 and 25 degrees of freedom, which is significant beyond $\alpha = .0001$ [$\hat{\beta} = 39.7801$ with $s(\hat{\beta}) = 7.9996$]. The use of the covariate, initial soil phosphorus content, has greatly improved the precision of the experiment.

Table 8.10 Covariance analysis for dry forage yield from a randomized complete block design with seven sources of phosphorus applied at two rates. The covariate is amount of soil phosphorus in the plot at the beginning of the 3-year study.

Source	d.f.	*Sum of Squares*	*Mean Square*	*F-Value*	Prob > *F*
Model	16	32,099,838	2,006,240	5.21	.0001
Error	25	9,619,403	384,776		
Corrected total	41	41,719,241			
Sequential Sums of Squares:					
Source	d.f.	SS	MS	*F-Value*	Prob > *F*
BLOCK	2	1,520,897	760,449	1.98	.1596
SOURCE	6	13,312,957	2,218,826	5.77	.0007
RATE	1	7,315,853	7,315,853	19.01	.0002
SOURCE*RATE	6	435,267	72,544	.19	.9773
PHOSDEV	1	9,514,863	9,514,863	24.73	.0001
Partial Sums of Squares:					
Source	d.f.	SS	MS	*F-Value*	Prob > *F*
BLOCK	2	1,173,100	586,550	1.52	.2373
SOURCE	6	15,417,193	2,569,532	6.68	.0003
RATE	1	3,623,100	3,623,100	9.42	.0051
SOURCE*RATE	6	999,892	166,649	.43	.8497
PHOSDEV	1	9,514,863	9,514,863	24.73	.0001

Adjustment of the treatment effects for differences in the covariate changed the treatment sums of squares (compare the sequential and partial sums of squares in Table 8.10) but did not change any of the conclusions from the F-tests of the treatment effects. Sources of phosphorus and rates of application remain significant, both beyond $\alpha = .01$, and the source-by-rate interaction remains nonsignificant. The absence of any interaction between sources and rates of fertilization means that differences in forage production among the 14 phosphorus fertilization treatments can be summarized in the marginal means for the two treatment factors, sources and rates. However, both sets of means need to be adjusted to remove biases due to differences in initial levels of soil phosphorus.

The adjusted SOURCE marginal means are obtained as

$$\overline{Y}_{\text{adj}.j.} = \overline{Y}_{.j.} - \hat{\beta}(\overline{X}_{.j.} - \overline{X}_{...})$$

where $\overline{X}_{.j.}$ is the marginal mean for the covariate for those experimental plots receiving the jth source of phosphorus, and $\overline{X}_{...}$ is the overall mean for the covariate; $\hat{\beta} = 39.7801$. This adjusts the SOURCE means to the common level of initial soil phosphorus $\overline{X}_{...} = 52.4$ ppm. Similarly, the adjusted RATE marginal means are obtained as

$$\overline{Y}_{\text{adj}..k} = \overline{Y}_{..k} - \hat{\beta}(\overline{X}_{..k} - \overline{X}_{...})$$

The unadjusted marginal means and the steps in the adjustment to obtain the adjusted means are shown in Table 8.11. The standard errors of the adjusted

Table 8.11 Unadjusted and adjusted treatment means for "Source" and "Rate" of phosphorus fertilization. There was no "Rate-by-Source" interaction so the experimental results are summarized in terms of the marginal means.[a]

Treatment	*Forage Mean (unadj.)*[b]	*Phosphorus Mean Deviation*	*Covariance Adjustment*	*Forage Mean (adj.)*	*Std. Error*
"Source" means:					
BSLAG	3,767.0	.914	−36.4	3,730.6	253.3
CAMETA	3,815.7	7.314	−291.0	3,524.7	259.9
COLOID	2,304.7	2.514	−100.1	2,204.6	254.0
FROCK	3,013.7	−3.352	133.3	3,147.0	254.7
RROCK	2,532.3	2.781	−110.6	2,421.7	254.2
SUPER	3,514.7	−2.019	80.3	3,595.0	253.8
TSUPER	3,573.2	−8.152	324.3	3,897.5	261.5
"Rate" means:					
40	2,800.0	−2.895	115.2	2,915.1	137.3
80	3,634.7	2.895	−115.2	3,519.5	137.3

[a] "Phosphorus mean deviation" is $(\overline{X}_{.j.} - \overline{X}_{...})$ for "Source" means and $(\overline{X}_{..k} - \overline{X}_{...})$ for "Rate" means. "Covariance adjustment" is $-\hat{\beta}$(Phosphorus mean deviation), where $\hat{\beta} = 39.7801$.

[b] The standard errors for the unadjusted treatment means are $s(\overline{Y}_{.j.}) = 350.2$ for the "Source" means and $s(\overline{Y}_{..k}) = 229.3$ for the "Rate" means.

treatment means are also shown. The standard errors on the unadjusted treatment means were $s(\bar{Y}_{.j.}) = 350.2$ and $s(\bar{Y}_{..k}) = 229.3$. The differences between standard errors for the unadjusted and adjusted means show a marked increase in precision from the use of the covariate.

PROC GLM computes the adjusted means as linear functions of the solution $\boldsymbol{\beta}^0$. The appropriate linear functions to be estimated for each mean are determined by the expectations of means in balanced data with the covariate set equal to $\bar{X}_{...}$. For example, the expectation of the marginal mean for the first source, BSLAG, is

$$\mathscr{E}(\bar{Y}_{.1.}) = \mu + \frac{\rho_1 + \rho_2 + \rho_3}{3} + \gamma_1 + \frac{\tau_1 + \tau_2}{2} + \frac{(\gamma\tau)_{11} + (\gamma\tau)_{12}}{2}$$

The expectation contains, in addition to $\mu + \gamma_1$, the average of the block effects, ρ_i, the average of the rate effects, τ_k, and the average of the interaction effects in which source 1 is involved. The covariate is not involved in this expectation because adjusting to the mean level of the covariate is equivalent to adjusting to PHOSDEV = 0 when the centered covariate is used. This is the particular linear function of $\boldsymbol{\beta}$ that is to be estimated as the marginal FORAGE mean for SOURCE = BSLAG. The estimate is obtained by computing the *same* linear function of $\boldsymbol{\beta}^0$. The adjusted means are obtained from PROC GLM with the statement

```
LSMEANS SOURCE RATE/STDERR;
```

The "/STDERR" asks for the standard errors on the adjusted means to be printed.

Interpretations of the treatment effects are based on the adjusted treatment means. In this example, adjustment for differences in the covariate changed the ranking of the four best sources of phosphorus, which did not differ significantly, and decreased the difference between the two rates of application. The adjusted means suggest an average rate of change in forage of 15 lbs/A for each 1b/A of phosphorus compared to 21 lbs/A suggested by the unadjusted means. ∎

EXERCISES

8.1. Use matrix multiplication to verify that the linear model in equation 8.5, where $\boldsymbol{X}$ and $\boldsymbol{\beta}$ are as defined in equation 8.4, generates the combinations of effects shown in equation 8.2.

8.2. Determine the number of rows and columns in $\boldsymbol{X}$ before reparameterization for one-way structured data with t groups (or treatments) and n observations in each group. How does the order of $\boldsymbol{X}$ change if there are n_i observations in each group?

* 8.3. Suppose you have one-way structured data with $t = 3$ groups. Define the linear model such that μ is the mean of the first group and the second and third groups are measured as deviations from the first. Is $\boldsymbol{X}$ for this model of full rank? Does this form of the model relate to any of the three reparameterizations?

8.4. The accompanying table gives survival data for tropical corn borer under field conditions in Thailand (1974). Researchers inoculated thirty experimental plots with egg masses of the corn borer on the same date by placing egg masses on each corn plant in the plot. After each of 3, 6, 9, 12, and 21 days, the plants in six random plots were dissected and the surviving larvae were counted. This gives a completely random experimental design with the treatments being "days after inoculation." (Data are used with permission of Dr. L. A. Nelson, North Carolina State University.)

Days After Inoculation	*Numbers of Larvae Surviving in Six Plots*					
3	17	22	26	20	11	14
6	37	26	24	11	11	16
9	8	5	12	3	5	4
12	14	8	4	6	3	3
21	10	13	5	7	3	4

a. Do the classical analysis of variance for the completely random design. Include in your analysis a partitioning of the sum of squares for treatments to show the linear regression on "number of days" and deviations from linearity.

b. Regard "days after inoculation" as a class variable. Define $\boldsymbol{Y}$, $\boldsymbol{X}$, and $\boldsymbol{\beta}$ so that the model for the completely random design $Y_{ij} = \mu + \tau_i + \varepsilon_{ij}$ can be represented in matrix form. Show enough of each matrix to make evident the order in which the observations are listed. Identify the singularity that makes $\boldsymbol{X}$ not of full rank.

c. Show the form of $\boldsymbol{X}$ and $\boldsymbol{\beta}$ for each of the three reparameterizations—the means model, the $\sum \tau_i = 0$ constraint, and the $\tau_5 = 0$ constraint.

d. Choose one of the reparameterizations to compute $R(\boldsymbol{\tau}'|\mu)$ and SS(Res). Summarize the results in an analysis of variance table and compare with the analysis of variance obtained in part (a).

e. Use SAS PROC GLM, or a similar program for the analysis of less than full-rank models, to compute the analysis of variance. Ask for the solution to the normal equations so that "estimates" of $\boldsymbol{\beta}$ are obtained. Compare these sums of squares and estimates of $\boldsymbol{\beta}$ with the results from your reparameterization in part (d). Show that the unbiased estimates of $\mu + \tau_1$ and $\tau_1 - \tau_2$ are the same from both analyses.

f. Now regard X as a quantitative variable and redefine $\boldsymbol{X}$ and $\boldsymbol{\beta}$ so that $\boldsymbol{Y} = \boldsymbol{X\beta} + \boldsymbol{\varepsilon}$ expresses Y as a linear function of "number of days." Compute SS(Regr) and compare the result with that obtained in part (a). Test the null hypothesis that the linear regression coefficient is zero. Test the null hypothesis that the linear function adequately represents the relationship.

* 8.5. Use $\boldsymbol{X}$ and $\boldsymbol{\beta}$ as defined for the completely random design, equation 8.4. Define $\boldsymbol{K}'$ for the null hypothesis H_0: $\tau_1 = \tau_2$. Define $\boldsymbol{K}'$ for the null hypothesis H_0: $\tau_3 = \tau_4$. Define $\boldsymbol{K}'$ for the *composite* null hypothesis H_0: $\tau_1 = \tau_2$ and $\tau_3 = \tau_4$ and $\tau_1 + \tau_2 = \tau_3 + \tau_4$. Is each of these hypotheses testable? How does the sum of squares generated by the composite hypothesis relate to the analysis of variance?

8.6. Show that the means model reparameterization for the completely random design is equivalent to imposing the constraint that $\mu = 0$.

* 8.7. Use the means model reparameterization on a randomized complete block design with $b = 2$ and $t = 4$. As discussed in the text, this reparameterization leaves zero degrees of freedom for the estimate of error. However, experimental error can be estimated as the block-by-treatment interaction sum of squares. Define $\boldsymbol{K}'$ for the means reparameterization so that the sum of squares obtained from Q is the error sum of squares.

8.8. Show $\boldsymbol{X}^*$ and $\boldsymbol{\beta}^*$ for the model for the randomized complete block design (equation 8.20) with $b = 2$ and $t = 4$ using the constraint $\gamma_2 = 0$ and $\mu_j = \mu + \tau_j$. Determine the expectation of $\hat{\boldsymbol{\beta}}^*$ in terms of the original parameters.

8.9. Use matrix multiplication of $\boldsymbol{X}$ and $\boldsymbol{\beta}$ in equation 8.21 to verify that the linear model in equation 8.20 is obtained.

8.10. Determine the general result for the number of columns in $\boldsymbol{X}$ for two-way classified data when there are b levels of one factor and t levels of the other factor if the model does not contain interaction effects. How many additional columns are needed if the model does contain interaction effects?

*8.11. A randomized complete block experimental design was used to determine the joint effects of temperature and concentration of herbicide on absorption of two herbicides on a commercial charcoal material. There were 2 blocks and a total of 20 treatment combinations—2 temperatures by 5 concentrations by 2 herbicides. (The data shown in the table are used with permission of Dr. J. B. Weber, North Carolina State University.)

Block	*Temp.* (°C)	*Herb.*	*Concentration* × 10^5 20	40	60	80	100
1	10	A	.280	.380	.444	.480	.510
		B	.353	.485	.530	.564	.620
	55	A	.266	.332	.400	.436	.450
		B	.352	.474	.556	.590	.625
2	10	A	.278	.392	.440	.470	.500
		B	.360	.484	.530	.566	.611
	55	A	.258	.334	.390	.436	.446
		B	.358	.490	.560	.570	.600

The usual linear model for a randomized complete block experiment, $Y_{ij} = \mu + \gamma_i + \tau_j + \varepsilon_{ij}$, where γ_i is the effect of the ith block and τ_j is the effect of the jth treatment, can be expanded to include the main and interaction effects of the three factors:

$$Y_{ijkl} = \mu + \gamma_i + T_j + H_k + C_l + (TH)_{jk} + (TC)_{jl}$$
$$+ (HC)_{kl} + (THC)_{jkl} + \varepsilon_{ijkl}$$

where T_j, H_k, and C_l refer to the effects of temperature, herbicide, and concentration, respectively. The combinations of letters refer to the corresponding interaction effects.

a. Show the form of $\boldsymbol{X}$ and $\boldsymbol{\beta}$ for the usual RCB model, the model containing γ_i and τ_j. Assume the data in $\boldsymbol{Y}$ are listed in the order that would be obtained if successive rows of data in the table were appended into one vector. What is the order of $\boldsymbol{X}$ and how many singularities does it have? Use $\gamma_2 = 0$ and $\tau_{20} = 0$ to reparameterize the model and compute the sums of squares for blocks and treatments.

b. Define $\boldsymbol{K}'$ for the singular model in part (a) for the composite null hypothesis that there is no temperature effect at any of the combinations of herbicide and concentration. [*Note:* τ_1 is the effect for the treatment having temperature 10°, herbicide A, and concentration 20×10^{-5}. τ_{11} is the effect for the similar treatment except with 55° temperature. The null hypothesis states that these two effects must be equal or their difference must be zero, and similarly for all other combinations of herbicide and concentration.] How many degrees of freedom does this sum of squares have? Relate these degrees of freedom to degrees of freedom in the conventional factorial analysis of variance. Define $\boldsymbol{K}'$ for the null hypothesis that the *average* effects of temperature are zero. How many degrees of freedom does this sum of squares have and how does it relate to the analysis of variance?

c. Show the form of $\boldsymbol{X}$ and $\boldsymbol{\beta}$ if the factorial model with *only* the main effects, T_j, H_k, and C_l, is used. How many singularities does this $\boldsymbol{X}$ matrix contain? Show the form of $\boldsymbol{X}^*$ if the "sum" constraints are used. Use this reparameterized form to compute the sums of squares due to temperature, due to herbicides, and due to concentration.

d. Demonstrate how $\boldsymbol{X}$ in part (c) is augmented to include the $(TH)_{jk}$ effects. How many columns are added to $\boldsymbol{X}$? How many additional singularities does this introduce? How many columns would be added to $\boldsymbol{X}$ to accommodate the $(TC)_{jl}$ effects? The $(HC)_{kl}$ effects? The $(THC)_{jkl}$ effects? How many singularities does each introduce?

e. Use PROC ANOVA in SAS, or a similar computer package, to compute the full factorial analysis of variance. Regard blocks, temperature, herbicide, and concentration as class variables.

8.12. The effect of supplemental ascorbate, vitamin C, on survival time of terminal cancer patients was studied. [Data given here are from Cameron and Pauling (1978) as reported in Andrews and Herzberg (1985).] The survival time (Days) of each treated patient was compared to the mean survival time of a control group (Cont.) of 10 similar patients. Age of patient was also recorded. For this exercise, the results are used from three cancer types—stomach, bronchus, and colon. There were 13, 17, and 17 patients in the three groups, respectively. For this question use the logarithm of the ratio of days survival of the treated patient to the mean days survival of his or her control group as the dependent variable.

Stomach Cancer				*Bronchus Cancer*				*Colon Cancer*			
Sex	*Age*	*Days*	*Cont.*	*Sex*	*Age*	*Days*	*Cont.*	*Sex*	*Age*	*Days*	*Cont.*
F	61	124	38	M	74	74	33	F	76	135	18
M	69	12	18	M	74	423	18	F	58	50	30
F	62	19	36	M	66	16	20	M	49	189	65
F	66	45	12	M	52	450	58	M	69	1,267	17
M	63	257	64	F	48	87	13	F	70	155	57
M	79	23	20	F	64	115	49	F	68	534	16
M	76	128	13	M	70	50	38	M	50	502	25
M	54	46	51	M	77	50	24	F	74	126	21
M	62	90	10	M	71	113	18	M	66	90	17
F	69	876	19	M	70	857	18	F	76	365	42
M	46	123	52	M	39	38	34	F	56	911	40
M	57	310	28	M	70	156	20	M	65	743	14
F	59	359	55	M	70	27	27	F	74	366	28
				M	55	218	32	M	58	156	31
				M	74	138	27	F	60	99	28
				M	69	39	39	M	77	20	33
				M	73	231	65	M	38	274	80

a. Use the means model reparameterization to compute the analysis of variance for ln(survival ratio). Determine $X^{*\prime}X^*$, $X^{*\prime}Y$, $\hat{\beta}^*$, SS(Model), SS(Res), and s^2. What is the least squares estimate of the mean ln(survival ratio) for each cancer group and what is the standard error of each mean? Two different kinds of hypotheses are of interest: (1) Does the treatment increase survival time; that is, is ln(survival ratio) significantly greater than zero for each type cancer? (2) Are there significant differences among the cancer types in the effect of the treatment? Use a t-test to test the null hypothesis that the true mean ln(survival ratio) for each group is zero. Use an F-test to test the significance of differences among cancer types.

b. The ages of the patients in the study varied from 38 to 79; the mean age was 64.3191 years. Augment the X^* matrix in part (a) with the vector of centered ages. Compute the residual sum of squares and the estimate of σ^2 for this model. Compute the standard error of each estimated regression coefficient. Use a t-test to test the null hypothesis that the partial regression coefficient for the regression of ln(survival ratio) on age is zero. Use the difference in residual sums of squares between this model and the previous model to test the same null hypothesis. How are these two tests related? What is your conclusion about the importance of adjusting for age differences?

c. Since the means model was used in part (b) and ages were expressed as deviations from the mean age, the first three regression coefficients in $\hat{\beta}$ are the estimates of the cancer group means *adjusted* to the mean age of 64.3191. Construct K' for the hypothesis that the true means, adjusted for age differences, of the stomach and bronchus cancer groups, the first and second groups, are the same as for colon cancer, the third group. Complete the test and state your conclusion.

d. Describe how X_c^* would be defined to adjust all observations to age 60 for all patients. Show the form of T for averaging the adjusted observations to obtain the adjusted group means. The adjusted group means are obtained as $T'X_c^*\hat{\beta}^*$, equation 8.42. Compute $T'X_c^*$ and $s^2(\bar{Y}_{\text{adj}})$ for this example.

e. Even though the average regression on age did not appear important, it was decided that each cancer group should be allowed to have its own regression on age to verify that age was not important in any of the three groups. Illustrate how X^* would be expanded to accommodate this model and complete the test of the null hypothesis that the regressions on age are the same for all three cancer groups. State your conclusion.

*8.13. The means reparameterization was used on the cabbage data example (Example 8.4) in the text. Define β^* and X^* for this model (equation 8.46) using the reparameterization constraints $\gamma_3 = \tau_2 = (\gamma\tau)_{31} = (\gamma\tau)_{32} = (\gamma\tau)_{12} = (\gamma\tau)_{22} = 0$. Define K' for the reparameterized model so as to obtain the sum of squares for the "dates-by-lines" interaction.

8.14. Equation 8.53 defines the reduced model for H_0: $\beta_{ij} = \beta$ for all ij. Define the reduced model for the test of homogeneity of regressions *within* lines:

$$H_0\colon\ \beta_{11} = \beta_{21} = \beta_{31} \quad \text{and} \quad \beta_{12} = \beta_{22} = \beta_{32}$$

Find SS(Res) for this reduced model and complete the test of homogeneity.

*8.15. The means model was used in the cabbage data example (equation 8.47) and K' was defined to partition the sums of squares. Develop a reduced model that reflects H_0: $\bar{\mu}_{1.} = \bar{\mu}_{2.} = \bar{\mu}_{3.}$. Use the full and reduced models to obtain the sum of squares for this hypothesis and verify that it is equivalent to that obtained using K' (equation 8.49) in the text.

*8.16. The covariance analysis of the phosphorus study in Section 8.8.3 assumed a common regression of forage yield on soil phosphorus. Use a general linear analysis program (such as PROC GLM in SAS) to test the homogeneity of regressions over the 14 treatment groups.

8.17. The Linthurst data used in Chapters 5 and 7 came from nine sites classified according to location (*LOC*) and type of vegetation (*TYPE*). (The data are given in Table 5.1 on page 136.) Do the analysis of variance on *BIOMASS*, partitioning the sum of squares into that due to *LOC*, *TYPE*, and *LOC*-by-*TYPE* interaction. The regression models in Chapter 7 indicated that *pH* and *Na* were important variables in accounting for the variation in *BIOMASS*. Add these two variables to your analysis of variance model as covariates (center each) and compute the analysis of covariance. Obtain the adjusted *LOC*, *TYPE*, and *LOC*-by-*TYPE* treatment means. Interpret the results of the covariance analysis. For what purpose is the analysis of covariance being used in this case?

CHAPTER
NINE

Problem Areas in Least Squares

All discussions to this point have assumed that the least squares assumptions of normality, common variance, and independence are valid, and that the data are correct and representative of the intended populations.

In reality, the least squares assumptions hold only approximately and one can expect the data to contain either errors or observations that are somewhat unusual compared to the rest of the data. This chapter presents a synopsis of the problem areas that commonly arise in least squares analysis.

The least squares regression method discussed in the previous chapters was based on the assumptions that the errors are additive (to the fixed-effects part of the model) and are normally distributed independent random variables with common variance σ^2. Least squares estimation based on these assumptions is referred to as **ordinary least squares**. When the assumptions of independence and common variance hold, least squares estimators have the desirable property of being the best (minimum variance) among all possible linear unbiased estimators. When the normality assumption is satisfied, the least squares estimators are also maximum likelihood estimators.

Three of the major problem areas in least squares analysis relate to failures of the basic assumptions—normality, common variance, and independence of the errors. Other problem areas are overly influential data points, outliers, inadequate specification of the functional form of the model, near-linear dependencies among the independent variables (collinearity), and independent variables being subject to error. This chapter is a synopsis of these problem areas with brief discussions on how they might be detected, their impact on least squares, and what might be done to remedy or at least reduce the problem. Subsequent chapters discuss in greater detail techniques for detecting the problems, transformations of variables as a means of alleviating some of the problems, and analysis of the correlational structure of the data to understand the nature of the collinearity problem. This process of checking the validity of the assumptions, the behavior of the data, and the adequacy of the model is an important step in every regression analysis. It should not, however, be regarded as a substitute for a proper validation of the regression equation against an independent set of data.

The emphasis here is on making the user aware of problem areas in the data or the model and insofar as possible removing the problems. An alternative to least squares regression when the assumptions are not satisfied is **robust regression**. Robust regression refers to a general class of statistical procedures designed to reduce the sensitivity of the estimates to failures in the assumptions of the parametric model. For example, the least squares approach is known to be sensitive to gross errors, or outliers, in the data because the solution minimizes the *squared* deviations. A robust regression procedure would reduce the impact of such errors by reducing the weight given to large residuals. This can be done by minimizing the sum of absolute residuals, for example, rather than the sum of squared residuals. In the general sense, procedures for detecting outliers and influential observations can be considered part of robust regression. Except for this connection, robust regression will not be discussed in this text. The reader is referred to Huber (1981) and Hampel, Ronchetti, Rousseeuw, and Stahel (1986) for discussions on robust statistics.

9.1 Nonnormality

Importance of Normality

The assumption that the residuals, ε, are normally distributed is not necessary for estimation of the regression parameters and partitioning of the total variation. Normality is needed only for tests of significance and construction of confidence interval estimates of the parameters. The t-test, F-test, and chi-square test require the underlying random variables to be normally distributed. Likewise, the conventional confidence interval estimates depend on the normal distribution, either directly or through Student's t-distribution.

"Nonnormal" Data

Experience has shown that normality is a reasonable assumption in many cases. However, in some situations it is not appropriate to assume normality. Count data will frequently behave more like Poisson-distributed random variables. The proportion of subjects that show a response to the agent in toxicity studies is a binomially distributed random variable if the responses are independent. Time to failure in reliability studies and time to death in toxicity studies will tend to have asymmetric distributions and, hence, not be normally distributed.

The impact of nonnormality on least squares depends on the degree of departure from normality and the specific application. Nonnormality does not affect the estimation of the parameters; the least squares estimates are still best linear unbiased estimates if the other assumptions are met. The tests of significance and confidence intervals, however, are affected by nonnormality. In general, the probability levels associated with the tests of significance or the confidence coefficients will not be correct. The F-test is generally regarded as being reasonably robust against nonnormality.

Effect on Confidence Intervals

Confidence interval estimates can be more seriously affected by nonnormality, particularly when the underlying distribution is highly skewed or has fixed boundaries. The two-tailed symmetric confidence interval estimates based on normality will not, in fact, be allocating equal probability to each tail if the distribution is asymmetric and may even violate natural boundaries for the parameter. The confidence interval estimate for proportion of affected individuals in a toxicity study, for example, may be less than 0 or greater than 1 if the estimates ignore the nonnormality in the problem.

Detecting Nonnormality

Plots of the observed residuals, e, and skewness and kurtosis coefficients are helpful in detecting nonnormality. The **skewness coefficient** measures the asymmetry of the distribution, whereas **kurtosis** measures the tendency of the distribution to be too flat or too peaked. The skewness coefficient for the normal distribution is 0; the kurtosis coefficient is 3.0. Some statistical computing packages provide these coefficients in the univariate statistics analysis. (Often, the kurtosis coefficient is expressed as a deviation from the value for the normal distribution.) When the sample size is sufficiently large, a frequency distribution of the residuals can be used to judge symmetry and kurtosis. A full-normal or half-normal plot, which gives a straight line under normality, is probably easier to use. These plots compare the ordered residuals from the data to the expected values of ordered observations from a normal distribution (with mean zero and unit variance). The full-normal plot uses the signed residuals; the half-normal plot uses the absolute values of the residuals. Different shapes of the normal plots reveal different kinds of departure from normality. More details on these plots will be given in Section 10.1.

Improving Normality

Transformation of the dependent variable to a form that is more nearly normally distributed is the usual recourse to nonnormality. Statistical theory says that such a transformation exists if the distribution of the original dependent variable is known. Many of the common transformations (such as the arcsin, the square root, the logarithmic, and the logistic transformations) were developed for situations in which the random variables were expected a priori to have specific nonnormal distributions.

In many cases, the sample data provide the only information available for determining the appropriate normalizing transformation. The plots of the residuals may suggest transformations, or several transformations might be tried and the one adopted which most nearly satisfies the normality criteria. Alternatively, an empirical method of estimating the appropriate power transformation might be used (Box and Cox, 1964). Chapter 11 is devoted to transformations of variables.

9.2 Heterogeneous Variances

Importance of Homogeneous Variance

The assumption of common variance plays a key role in ordinary least squares. The assumption implies that every observation on the dependent variable contains the same amount of information. Consequently, all observations in ordinary least squares receive the same weight. On the other hand, heterogeneous variances imply that some observations contain more information than others. Rational use of the data would require that more weight be given to those observations that contain the most information.

The minimum variance property of ordinary least squares estimators is directly dependent on this assumption. Equal weighting, as in ordinary least squares, does not give the minimum variance estimates of the parameters if the variances are not equal. Therefore, the direct impact of heterogeneous variances in ordinary least squares is a loss of precision in the estimates compared to the precision that would have been realized if the heterogeneous variances had been taken into account.

Data Having Heterogeneous Variances

Heterogeneous variance, as with nonnormality, is expected a priori with certain kinds of data. The same situations that give nonnormal distributions will invariably give heterogeneous variances because the variance in most nonnormal distributions is related to the mean of the distribution. Even in situations where the underlying distributions are normal within groups, the variances of the underlying distributions may change from group to group. Most commonly, larger variances will be associated with groups having the larger means. Various plots of the residuals are useful for revealing heterogeneous variances.

Decreasing Heterogeneity

Two approaches to handling heterogeneous variances are transformation of the dependent variable and use of weighted least squares; the former is probably the more common. The transformation is chosen to make the variance homogeneous (or more nearly so) on the transformed scale. Prior information on the probability distribution of the dependent variable or empirical information on the relationship of the variance to the mean may suggest a transformation. For example, the arcsin transformation is designed to stabilize the variance when the dependent variable is binomially distributed. Weighted least squares uses the original metric of the dependent variable but gives each observation weight according to the relative amount of information it contains. Weighted least squares is discussed in Section 11.5.1.

9.3 Correlated Errors

Origin of Correlated Errors

Correlations among the residuals may arise from many sources. It is common for data collected in a time sequence to have correlated errors; the error associated with an observation at one point in time will tend to be correlated with the errors of the immediately preceding observations. Almost any physical process that is continuous over time will show serial correlations. Hourly measurements on the pollutant emissions from a coal smokestack, for example, have very high serial correlations. Biological studies in which repeated measurements are made over time on the same individuals, such as plant and animal growth studies or clinical trials, will usually have correlated residuals.

Many of the experimental designs, including the randomized complete block design and the split-plot design, allow us to capitalize on the correlated errors among the observations within a block or within a whole plot to improve the precision of certain comparisons. The observations among samples within experimental units will have correlated errors, and the conventional analyses take these correlations into account. In some cases, however, correlations may be introduced inadvertently by the way the experiment is managed. For example, the grouping of experimental units for convenience in exposing them to a treatment, applying nutrient solution, taking measurements, and so forth, will tend to introduce positively correlated errors among the observations within the groups. These correlations are frequently overlooked and are not taken into account in the conventional analyses.

Impact of Correlated Errors

The impact of correlated errors on the ordinary least squares results is loss in precision in the estimates, similar to the effect of heterogeneous variances. Correlated errors that are not recognized appropriately in the analysis will seriously bias the estimates of variances with the direction and magnitude of the bias depending on the nature of the correlations. This, in turn, causes all measures of precision of the estimates to be biased and invalidates tests of significance.

Detecting Correlated Errors

The nature of the data will frequently suggest the presence of correlated errors. Any data set collected in a time sequence should be considered suspect and treated as time series data unless the correlation can be shown to be negligible. There are many texts devoted to the analysis of time series data (e.g., Fuller, 1976; Bloomfield, 1976). A clear understanding of the design and conduct of the experiment will reveal many potential sources of correlated errors. The more troublesome to detect are the inadvertent correlated errors arising from inadequate randomization of the experiment or failure to adhere to the randomization plan. In such cases, inordinately small error variances may provide the clue. In other cases, plotting of the residuals according to the order in which the data were collected or the grouping used in the laboratory may reveal patterns of residuals that suggest correlated errors.

Handling Correlated Errors

The remedy to the problem of correlated errors is to utilize a model that takes into account the correlation structure in the data. Various time series models and analyses have been constructed to accommodate specific correlated error structures. **Generalized least squares** is a general approach to the analysis of data having correlated errors. This is an extension of weighted least squares where the entire variance–covariance matrix of the residuals is used. The difficulty with generalized least squares is that the covariances are usually not known and must be estimated from the data. This is a difficult estimation problem, unless the correlation structure is simple, and poor estimation of the correlation matrix can cause a loss in precision, rather than a gain, compared to ordinary least squares. Generalized least squares is discussed in Section 11.5.2.

9.4 Influential Data Points and Outliers

Influential Data Points

The method of ordinary least squares gives equal weight to every observation. However, every observation does not have equal impact on the various least squares results. For example, the slope in a simple linear regression problem is influenced most by the observations having values of the independent variable farthest from the mean. A single point far removed from the other data points can have almost as much influence on the regression results as all other points combined. Such observations are called **influential points** or **high leverage points**.

The potential influence of a data point on the least squares results is determined by its position in the X-space relative to the other points. In general, the more "distant" the point is from the center of the data points in the X-space, the greater is its potential for influencing the regression results.

Outliers

The term *outlier* refers to an observation which in some sense is inconsistent with the rest of the observations in the data set. An observation can be an outlier due to the dependent variable or any one or more of the independent variables having values outside expected limits. The term *potentially influential observation* will be used to refer to an observation that is an outlier in one or more of the independent variables. The term *outlier* will be restricted to a data point for which the value of the dependent variable is inconsistent with the rest of the sample. The phrase *outlier in the residuals* will refer to a data point for which the observed residual is larger than might reasonably be expected from random variation alone. The context of the usage will make clear whether *outlier* refers to the value of the dependent variable or of the residual.

Origin of Outliers and Influential Points

A data point may be an outlier or a potentially influential point because of errors in the conduct of the study (machine malfunction; recording, coding, or data entry errors; failure to follow the experimental protocol) or because the data point is from a different population. The latter could result, for example, from management changes that take the system out of the realm of interest or the occurrence of atypical environmental conditions. A valid data point may appear to be an outlier—have an outlier residual—because the model being used is not adequately representing the process. On the other hand, a data point that is truly an outlier may not have an outlier residual, and almost certainly will not if it happens also to be an influential point. The influential data points tend to force the regression so that such points have small residuals.

Handling Outliers and Influential Points

Influential points and outliers need to be identified. Little confidence can be placed in regression results that have been dominated by a few observations, regardless of the total size of the study. The first concern should be to verify that these data points are correct. Clearly identifiable errors should be corrected if possible or else eliminated from the data set. Data points that are not clearly identified as errors or that are found to be correct should be studied carefully for the information they might contain about the system being studied. Do they reflect inadequacies in the model or inadequacies in the design of the study? Outliers and overly influential data points should not be discarded indiscriminately. The outlier might be the most informative observation in the study.

Detection of Influential Points

Detection of the potentially more influential points is by inspection of the diagonal elements of $\boldsymbol{P}$, the projection matrix. The diagonal elements of $\boldsymbol{P}$ are measures of the Euclidean distances between the corresponding sample points and the centroid of the sample X-space. Whether a potentially influential point has, in fact, been influential is determined by measuring directly the impact of each data point on various regression results. Appropriate influence statistics will be discussed in Section 10.2.

Detection of Outliers

Outliers are detected by analysis of the observed residuals. It is usually recommended that the residuals first be standardized to have a common variance. Some suggest the use of recursive residuals (Hedayat and Robson, 1970). A residual that is several standard deviations from zero identifies a data point that needs careful review. Plots of residuals for detecting nonnormality and heterogeneous variances are also effective in identifying outliers. The detection of outliers will be discussed in Section 10.1.

9.5 Model Inadequacies

Missing Independent Variables

The ordinary least squares estimators are unbiased if the model is correct. They will not be unbiased if the model is incorrect in any of several different ways. If, for example, an important independent variable has been omitted from the model, the residual mean square is a (positively) biased estimate of σ^2 and the regression coefficients for all independent variables are biased (unless the omitted variable is orthogonal to all variables in the model). The common linear model that uses only the first power of the independent variables assumes that the relationship of Y to each of the independent variables is linear and that the effect of each independent variable is independent of the other variables. Omitting any important higher-order polynomial terms, including product terms, has the same effect as omitting an independent variable.

Approximating the "True" Model

One does not expect a complicated physical, chemical, or biological process to be linear in the parameters. In this sense, the ordinary linear least squares model (including higher-degree polynomial terms) must be considered an approximation of the true process model. The rationale for using a linear model, in cases where the true relationship is almost certainly nonlinear, is that any nonlinear function can be approximated to any degree of accuracy desired with an appropriate number of terms of a linear function. Thus, the linear model is used to provide what is believed to be a satisfactory approximation in some limited region of interest. To the extent that the approximation is not adequate, the least squares results will contain biases similar to those created by omitting a variable.

Detecting Inadequacies

Detection of model inadequacies will depend on the nature of the problem and the amount of information available on the system. Bias in the residual mean square and, hence, indication of an omitted term, can be detected if an independent estimate of σ^2 is available as would be the case in most designed experiments. In other cases, previous experience might provide some idea of the size of σ^2 from which a judgment can be made as to the presence of bias in the residual mean square. Overlooked higher-order polynomial terms are usually easily detected by appropriate residuals plots. Independent variables that are missing altogether are more difficult to detect. Unusual patterns of behavior in the residuals may provide clues.

Nonlinear Models

More realistic nonlinear models might be formulated as alternatives to the linear approximations. Some nonlinear models will be such that they can be linearized by an appropriate transformation of the dependent variable. These are called **intrinsically linear models**. Ordinary least squares can be used on linearized models if the assumptions on the errors are satisfied after the transformation is made. The intrinsically nonlinear models require the use of **nonlinear least squares** for the estimation of the parameters. The nonlinear form of even the intrinsically linear models might be preferred if it is believed that the least squares assumptions are more nearly satisfied in that form. Nonlinear models and nonlinear least squares are discussed in Chapter 14.

9.6 The Collinearity Problem

Near-Singularities in X

Singularity of X results when some linear function of the columns of X is exactly equal to the zero vector. Such cases become obvious when the least squares analysis is attempted because the unique $(X'X)^{-1}$ does not exist. A more troublesome situation arises when the matrix is only close to being singular; a linear function of the vectors is nearly zero. Redundant independent variables—the same information expressed in different forms—will cause X to be nearly singular. Interdependent variables that are closely linked in the system being studied can cause near-singularities in X.

A unique solution to the normal equations exists in these nearly-singular cases but the solution is very unstable. Small changes (random noise) in the variables, Y or X, can cause drastic changes in the estimates of the regression coefficients. The variances of the regression coefficients, for the independent variables involved in the near-singularity, become very large. In effect, the variables involved in the near-singularity can serve as surrogates for each other so that widely different combinations of the independent variables can be used to give nearly the same value of Y. The difficulties that arise from X being nearly singular are referred to collectively as the **collinearity problem**. The collinearity problem was defined geometrically in Section 6.4.

Effects of Collinearity

The impact of collinearity on least squares is very serious if primary interest is in the regression coefficients per se or if the purpose is to identify "important" variables in the process. The estimates of the regression coefficients can differ greatly from the parameters they are estimating, even to the point of having incorrect sign. The collinearity will allow "important" variables to be replaced in the model with incidental variables that are involved in the near-singularity. Hence, the regression analysis provides little indication of the relative importance of the independent variables.

The use of the regression equation for prediction is not seriously affected by collinearity as long as the correlational structure observed in the sample persists in the prediction population and prediction is carefully restricted to the sample X-space. However, prediction to a system where the observed collinearity does not exist or for points outside the sample space can be very misleading. The sample X-space in the presence of near-collinearities becomes very narrow in certain dimensions so that it is easy to choose prediction points that fall outside the sample space and, at the same time, difficult to detect when this has been done. Points well within the limits of each independent variable may be far outside the sample space.

Detecting Collinearity

Most regression computer programs are not designed to warn the user automatically of the presence of near-collinearities. Certain clues are present, however: unreasonable values for regression coefficients, large standard errors, nonsignificant partial regression coefficients when the model provides a reasonable fit, and known important variables appearing as unimportant in the regression results. High correlations between independent variables will identify near-collinearities involving two variables but may miss those involving more than two variables. A more direct approach to detecting the presence of collinearity is with a singular value decomposition of $\boldsymbol{X}$ or an eigenanalysis of $\boldsymbol{X}'\boldsymbol{X}$. These topics were discussed in Sections 2.7 and 2.8. Their use and other collinearity diagnostics are discussed in Section 10.3.

Handling Collinearity

The remedies for the collinearity problem depend on the objective of the model fitting exercise. If the objective is prediction, collinearity causes no serious problem within the sample X-space. The limitations discussed above must be understood, however. When primary interest is in estimation of the regression coefficients, one of the biased regression methods may be useful (Chapter 12). A better solution, when possible, is to obtain new data or additional data such that the sample X-space is expanded to remove the near-singularity. It is not likely that this will be possible when the near-singularity is the result of internal constraints of the system being studied. When the primary interest of the research is to identify the "important" variables in a system or to model the system, the regression results in the presence of severe collinearity will not be very helpful and can be misleading. It is more productive for this purpose to concentrate on understanding the correlational structure of the variables and how the dependent variable fits into this structure. Principal component analysis, Gabriel's (1971) biplot, and principal component regression can be helpful in understanding this structure. These topics are discussed in Chapter 12.

9.7 Errors in the Independent Variables

Model

The original model assumed that the independent variables were measured without error; they were considered to be constants in the regression model. With the errors-in-variables model, the true values of the independent variables are masked by measurement errors. Thus, the observed X_i is

$$X_i = \eta_i + \delta_i \qquad [9.1]$$

where η_i is the unobserved true value of X_i and δ_i is the measurement error. The error δ_i is assumed to have zero mean and variance σ_δ^2. The regression model assumes Y_i is a function of the true value η_i:

$$Y_i = \mu + \beta\eta_i + \varepsilon_i \qquad [9.2]$$

Bias in $\hat{\beta}$

Applying ordinary least squares to this problem, using X_i in place of the nonobservable η_i, gives a biased estimate of β. In general, $\hat{\beta}$ underestimates the absolute value of the true slope, $|\beta|$. If the measurement errors δ_i are independent of η_i, the expectation of $\hat{\beta}$ is approximately

$$\mathscr{E}(\hat{\beta}) \doteq \frac{\beta}{1 + n\sigma_\delta^2/\sum x_i^2} \qquad [9.3]$$

The denominator of equation 9.3 is always greater than 1 whenever there is measurement error, $\sigma_\delta^2 > 0$. The bias is small if σ_δ^2 is small relative to $\sum x_i^2/n$.

Proposals for Estimating β

There have been numerous proposals for estimating β under these conditions, none of which has been entirely satisfactory. Riggs, Guarnieri, and Addelman (1978) used computer simulation to study the behavior of a large number of published estimators and several additional ones they developed. One of the more obvious solutions is to attempt to correct $\hat{\beta}$ for bias using $\sum x_i^2$ and assuming σ_δ^2 is known (Madansky, 1959). Riggs et al. found this method to yield very erratic results whenever errors were large. Another suggestion divides the data into halves (Wald, 1940) or thirds (Bartlett, 1949) and uses the mean values of the extreme groups to compute an average slope. The Wald and Bartlett procedures were nearly identical in performance but gave only slightly less bias than the ordinary least squares estimate and tended to have less precision.

Another approach minimizes the sum of squares of the perpendicular distances to the regression line, properly weighted so that ordinary least squares regression of Y on X is obtained if $\sigma_\delta^2 = 0$ and the regression of X on Y is obtained if $\sigma_\varepsilon^2 = 0$. This estimator is given by

$$\tilde{\beta} = \frac{\sum y^2 - \lambda \sum x^2 + [(\sum y^2 - \lambda \sum x^2)^2 + 4\lambda(\sum xy)^2]^{1/2}}{2(\sum xy)} \qquad [9.4]$$

where $\lambda = \sigma_\varepsilon^2/\sigma_\delta^2$. Riggs et al. (1978) found this to be the most generally satisfactory of the published procedures, but warned that it tended to give highly unreliable estimates when σ_δ^2 was large and n was small. The reader is referred to Riggs, Guarnieri, and Addelman (1978) for more discussion of the problem and the summary of their comparisons.

Instrumental Variables

A different approach to the errors-in-variables problem is to use information from other variables that are correlated with η_i, but not with δ_i, to obtain consistent estimators of β. Such variables are called **instrumental variables**. The reader is referred to Feldstein (1974) and Carter and Fuller (1980) for more discussion on the use of instrumental variables.

Control with Design

The errors-in-variables issue greatly complicates the regression problem. There appears to be no one solution that does well in all situations. It appears best to avoid the problem whenever possible. The bias from ordinary least squares is dependent on the ratio of σ_δ^2 to $\sum x^2/n$. Thus, the problem can be minimized by designing the research so that the dispersion in X is large relative to any measurement errors. In such cases, ordinary least squares should be satisfactory.

9.8 Summary

This chapter is a synopsis of the common problems in least squares regression, emphasizing their importance and encouraging the user to be critical of his or her own results. Because least squares is a powerful and widely used tool, it is important that the user be aware of its pitfalls. Some of the diagnostic techniques (such as the analysis of residuals) are useful for detection of several different problems. Similarly, some of the remedial methods (such as transformations) attack more than one kind of problem. The following three chapters are devoted to discussions of the tools for detecting the problems and some of the remedies.

EXERCISES

*9.1. Several levels of a drug were used to assess its toxic effects on a particular animal species. Twenty-four animals were used and each was administered a particular dose of the drug. After a fixed time interval, each animal was scored as 0 if it showed no ill effects and as 1 if a toxic effect was observed. That is, the dependent variable takes the value of 0 or 1 depending on the absence or presence of a toxic reaction.

a. Which assumptions of ordinary least squares would you expect not to be satisfied with this dependent variable?
b. The dependent variable was used in a linear regression on dose. The resulting regression equation was $\hat{Y} = -.214 + .159X$. Plot this regression line for $X = 1$ to $X = 8$. Superimpose on the plot what you might expect the observed data to look like if 24 approximately equally spaced dose levels were used. What problems do you see now?
c. The researcher anticipated using $\hat{Y}$ to estimate the proportion of affected individuals at the given dose. What is the estimated proportion of individuals that will be affected by a dose of $X = 2$ units? Use the conventional method to compute the 95% confidence interval estimate of the mean at $X = 2$ if $s^2 = .1284$ with 22 degrees of freedom, $\bar{X} = 4.5$, and $\sum(X_i - \bar{X})^2 = 126$. Comment on the nature of this interval estimate.
d. Suppose each observation consisted of the proportion of mosquitoes in a cage of 50 that showed response to the drug rather than the response of a single animal. Would this have helped satisfy some of the least squares assumptions? Which?

9.2. Identify an independent variable in your area of research that you would not expect to be normally distributed. How is this variable usually handled in the analysis of experimental results?

*9.3. Suppose there are three independent observations that are to be averaged. The known variances of the three observations are 4, 9, and 16. Two different averages are proposed, the simple arithmetic average and the weighted average where each observation is weighted by the reciprocal of its variance. Use variances of linear functions, equation 3.21 and following, to demonstrate that the weighted average has the smaller variance. Can you find any other weighting that will give an even smaller variance?

9.4. Find a data set from your area of research in which you do *not* expect the variances to be homogeneous. Explain how you expect the variances to behave. How are these data usually handled in analysis?

*9.5. A plant physiologist was studying the relationship between intercepted solar radiation and plant biomass produced over the growing season. Several experimental plots under different growing conditions were monitored for radiation. Several times during the growing season biomass samples were taken from the plots to measure growth. The resulting data for each experimental plot showed cumulative solar radiation and biomass for the several times the biomass was measured during the season. Would you expect the dependent variable, biomass, to have constant variance over the growing season? Would you expect the several measurements of biomass on each plot to be statistically independent? Would you expect the measurements from different random experimental units to be statistically independent?

*9.6. The relatively greater influence of observations farther from the center of the X-space can be illustrated using simple linear regression. Express the slope of the regression line as $\hat{\beta}_1 = \sum(X_i - \bar{X})Y_i/\sum(X_i - \bar{X})^2$. In this form it is clear that a perturbation of the amount δ on any $Y_{i'}$ changes $\hat{\beta}_1$ by the amount $\delta(X_{i'} - \bar{X})/\sum(X_i - \bar{X})^2$. (Substitute $Y_{i'} + \delta$ for $Y_{i'}$ in $\hat{\beta}_1$ and subtract out $\hat{\beta}_1$.) Assume a perturbation of $\delta = 1$ on each Y_i in turn. Compute the amount $\hat{\beta}_1$ would change if the values of X are 0, 1, 2, and 9. Compute $\boldsymbol{P} = \boldsymbol{X}(\boldsymbol{X}'\boldsymbol{X})^{-1}\boldsymbol{X}'$ for this example. Which observation has the largest diagonal element of $\boldsymbol{P}$?

9.7. Find an example in your field for which you might expect collinearity to be a problem. Explain why you expect there to be collinearity.

CHAPTER
TEN

Regression Diagnostics

Chapter 9 summarized the problems that are encountered in least squares regression and the impact of these problems on the least squares results.

This chapter presents methods for detecting problem areas. Included are graphical methods for detecting failures in the assumptions, unusual observations, and inadequacies in the model, statistics to flag observations that are dominating the regression, and methods of detecting when strong relationships among the independent variables are affecting the results.

Regression diagnostics is the general class of techniques for detecting problems in regression—problems with either the model or the data set. This is an active field of research with many recent publications. It is not clear which of the proposed techniques will eventually prove most useful. Some of the simpler techniques that appear to be gaining favor are presented in this chapter. Belsley, Kuh, and Welsch (1980) and Cook and Weisberg (1982) are recommended for a more thorough coverage of the theory and methods of diagnostic techniques.

10.1 Residuals Analysis

Characteristics of the Residuals

Analysis of the regression residuals, or some transformation of the residuals, is very useful for detecting inadequacies in the model or problems in the data. The true errors in the regression model are assumed to be normally and independently distributed random variables with zero mean and common variance, $\varepsilon \sim N(\mathbf{0}, \boldsymbol{I}\sigma^2)$. The observed residuals, however, are not independent and do not have common variance, even when the $\boldsymbol{I}\sigma^2$ assumption is valid. Under the usual least squares assumptions, $\boldsymbol{e} = (\boldsymbol{I} - \boldsymbol{P})\boldsymbol{Y}$ has a multivariate normal distribution with $\mathscr{E}(\boldsymbol{e}) = \mathbf{0}$ and $\mathbf{Var}(\boldsymbol{e}) = (\boldsymbol{I} - \boldsymbol{P})\sigma^2$. The diagonal elements of $\mathbf{Var}(\boldsymbol{e})$ are not equal, so the observed residuals do not have common variance; the off-diagonal elements are not zero, so they are not independent.

Standardized Residuals

The heterogeneous variances in the observed residuals are easily corrected by standardizing each residual. The variances of the residuals are estimated by the diagonal elements of $(\boldsymbol{I} - \boldsymbol{P})s^2$. Dividing each residual by its standard deviation gives a **standardized residual**, denoted with r_i:

$$r_i = \frac{e_i}{s\sqrt{1 - v_{ii}}} \qquad [10.1]$$

where v_{ii} is the ith diagonal element of $\boldsymbol{P}$. All standardized residuals have unit variance. The standardized residuals behave much like a Student's t random variable except for the fact that the numerator and denominator of r_i are not independent.

Studentized Residuals

Belsley, Kuh, and Welsch (1980) suggest standardizing each residual with an estimate of its standard deviation that is independent of the residual. This is accomplished by using, as the estimate of σ^2 for the ith residual, the residual mean square from an analysis where that observation has been omitted. This variance is labeled $s^2_{(i)}$, where the subscript in parentheses indicates that the ith observation has been omitted from the analysis. The result is the **Studentized residual**, denoted r_i^*:

$$r_i^* = \frac{e_i}{s_{(i)}\sqrt{1 - v_{ii}}} \qquad [10.2]$$

Each Studentized residual is distributed as Student's t with $(n - p' - 1)$ degrees of freedom when normality of ε holds. As with e_i and r_i, the r_i^* are not independent of each other. Belsley, Kuh, and Welsch show that the Studentized residuals can be obtained from the ordinary residuals without rerunning the regression with the observation omitted.

Notation

The standardized residuals, r_i, are called Studentized residuals in many references (e.g., Cook and Weisberg, 1982; Pierce and Gray, 1982; Cook and Prescott, 1981; and SAS Institute, Inc., 1985d, 1985f). Cook and Weisberg refer to r_i as the Studentized residual with *internal* Studentization in contrast to *external* Studentization for r_i^*. The r_i^* are called the cross-validatory or jackknife residuals by Atkinson (1983) and RSTUDENT by Belsley, Kuh, and Welsch (1980) and SAS Institute, Inc. (1985d, 1985f). The terms *standardized* and *Studentized* will be used in this text as labels to distinguish between r_i and r_i^*.

Using Residuals

The observed residuals and the scaled versions of the observed residuals have been used extensively to study validity of the regression model and its assumptions. The heterogeneous variances of the observed residuals and the lack of independence among all three types of residuals complicate interpretation of their behavior. In addition, there is a tendency for inadequacies in the data to be spread over several residuals. For example, an outlier will have the effect of inflating residuals on several other observations and may itself have a relatively small residual. Further, the residuals from least squares regression will tend to be "supernormal." That is, when the normality assumption is *not* met, the observed residuals from a least squares analysis will fit the normal distribution more closely than would the original ε_i (Huang and Bolch, 1974; Quesenberry and Quesenberry, 1982; Cook and Weisberg, 1982). As a result, there will be a tendency for failures in the model to go undetected when residuals are used for judging goodness of fit of the model.

In spite of the problems associated with their use, the observed, standardized, and Studentized residuals have proven useful for detecting model inadequacies and outliers. For most cases, the three types of residuals give very similar patterns and lead to similar conclusions. The heterogeneous variances of e_i can confound the comparisons somewhat, and for that reason use of one of the standardized residuals, r_i or r_i^*, is to be preferred if they are readily available. The primary advantage of the Studentized residuals over the standardized residuals is their closer connection to the t-distribution. This allows the use of Student's t as a convenient criterion for judging whether the residuals are inordinately large.

Exact tests of the behavior of the observed residuals are not available; approximations and subjective judgments must be used. The use of the standardized or Studentized residuals as a check for an outlier is a multiple testing procedure, since the residual to be tested will be the largest out of the sample of n, and appropriate allowances on α must be made. The first-order Bonferroni bound on the probability would suggest using the critical value of t for $\alpha = \alpha^*/n$, as was done for the Bonferroni confidence intervals in Chapter 4. (α^* is the desired overall significance level.) Cook and Prescott (1981), in a study assessing the accuracy of the Bonferroni significance levels for detecting outliers in linear models, conclude that the bounds can be expected to be reasonably accurate if the correlations among the residuals are not excessively large. Cook and Weisberg (1982) suggest using $\alpha = v_{ii}\alpha^*/p'$ for testing the ith Studentized residual. This choice of α maintains the overall significance level but gives greater power to cases with large v_{ii}.

Recursive Residuals

Another class of residuals, **recursive residuals**, are constructed so that they are independent and identically distributed when the model is correct and are recommended by some for residuals analysis (Hedayat and Robson, 1970; Brown, Durbin, and Evans, 1975; Galpin and Hawkins, 1984; Quesenberry, 1986). Recursive residuals are computed from a sequence of regressions starting with a base of p' observations (p' = number of parameters to be estimated) and adding one observation at each step. The regression equation computed at each step is used to compute the residual for the next observation to be added. This sequence continues until the last residual has been computed. There will be $(n - p')$ recursive residuals; the p' observations used as the base will fit the data exactly.

Computation of Recursive Residuals

Assume a particular ordering of the data has been adopted for the purpose of computing the recursive residuals. Let y_r and $\boldsymbol{x}_r'$ be the rth rows from $\boldsymbol{Y}$ and $\boldsymbol{X}$, respectively. Let $\boldsymbol{X}_r$ be the first r rows of $\boldsymbol{X}$ and $\hat{\boldsymbol{\beta}}_r$ be the least squares solution using the first r observations in the chosen ordering. Then the *recursive residual* is defined as

$$w_r = \frac{y_r - \boldsymbol{x}_r'\hat{\boldsymbol{\beta}}_{r-1}}{[1 + \boldsymbol{x}_r'(\boldsymbol{X}_{r-1}'\boldsymbol{X}_{r-1})^{-1}\boldsymbol{x}_r]^{1/2}} \qquad [10.3]$$

for $r = p' + 1, \ldots, n$. The original proposal defined the recursive residuals for time sequence data. Galpin and Hawkins (1984) contend, however, that they are useful for all data sets, but particularly so when there are natural orderings to the data.

Quesenberry's Uniform Random Variables

Quesenberry (1986) uses a more general theoretical development of tests of goodness of fit that generates the use of recursive residuals. His general approach leads to a set of independent and identically distributed *uniform* random variables on the interval (0, 1), which contain all relevant information for testing goodness of fit of the model including all assumptions. In addition, independent and identically distributed *normal* random variables can be obtained from the uniform random variables, which can then be tested by any of the tests for normality. The major obstacle to the use of Quesenberry's method is lack of a generally available computer program. The key steps in Quesenberry's method are outlined in Appendix 10.A, preceding the chapter exercises.

Characteristics of Recursive Residuals

Recursive residuals are independent and have common variance σ^2. Each is explicitly associated with a particular observation and, consequently, recursive residuals seem to avoid some of the "spreading" of model defects that occurs with ordinary residuals. Since the recursive residuals are independently and identically distributed, exact tests for normality and outliers can be used. The major criticisms of recursive residuals are that greater computational effort is required, no residuals are associated with the first p' observations used as the base, and the residuals are not unique because the data can be ordered in different ways. Appropriate computer programs can remove the first problem. The last two are partially overcome by computing recursive residuals for different orderings of the data.

Graphical Techniques

Graphical techniques are very effective for detecting abnormal behavior of residuals. If the model is correct and the assumptions are satisfied, the residuals should appear in any plot as random variation about zero. Any convincing pattern to the residuals would suggest some inadequacy in the model or the assumptions. To emphasize the importance of plotting, Anscombe (1973) presents four (artificial) data sets that give identical least squares regression results [same $\hat{\beta}$, $\hat{Y}$, SS(Total), SS(Regression), SS(Residual), and R^2], but that are strikingly different when plotted. The fitted model appears equally good in all cases if one looks only at the quantitative results. The plots of Y versus X (Figure 10.1), however, show obvious differences [adapted from Anscombe (1973)].

Anscombe Plots

The first data set, Figure 10.1(a), shows a typical linear relationship between Y and X with apparent random scatter of the data points above and below the regression line. This is the expected pattern if the model is adequate and the ordinary least squares assumptions hold.

The data in Figure 10.1(b) show a distinct quadratic relationship and a very patterned set of residuals. It is clear from the plot that the linear model is inadequate and that the fit would be almost perfect if the model were expanded to include a quadratic term.

Figure 10.1(c) illustrates a case where there is a strict linear relationship between Y and X except for one (aberrant?) data point. Removal of this one point would cause the residual sum of squares to go to zero. The residuals pattern is a clear indication of a problem with the data or the model. If this is a valid data

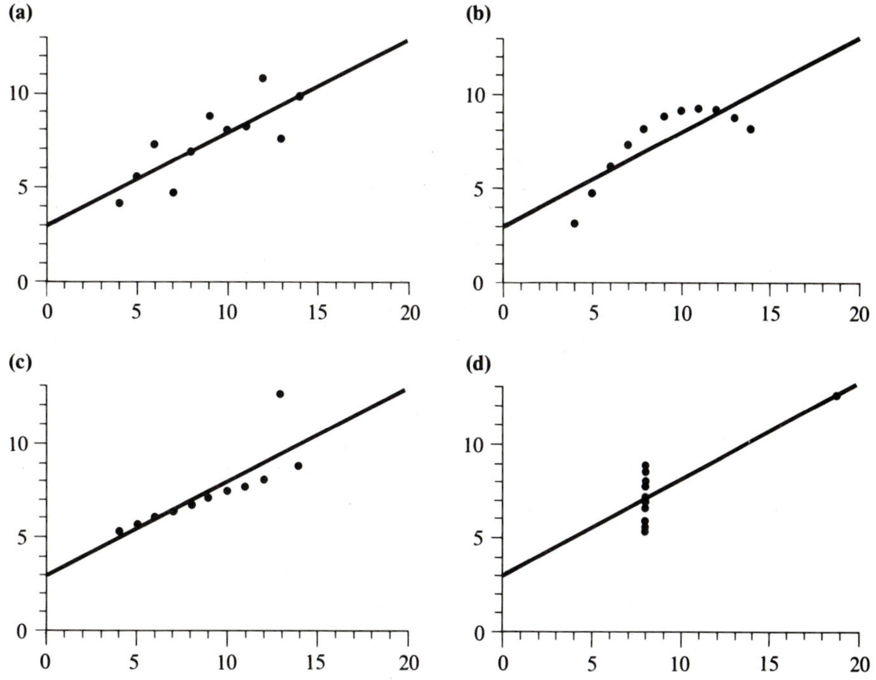

Figure 10.1 Four data sets that give the same quantitative results for the linear regression of Y on X. [Adapted from Anscombe (1973).]

point, the model must be inadequate. It may be that an important independent variable has been omitted.

The data in Figure 10.1(d) represent a case where the entire regression relationship is determined by one observation. This observation is a particularly influential point because it is so far removed (on the X-scale) from the other data points. Even if this is a valid data point, one could place little faith in estimates of regression parameters so heavily dependent on a single observation.

The Anscombe plots emphasize the power of simple graphical techniques for detecting inadequacies in the model. There are several informative plots one might use. No single plot can be expected to detect all types of problems. The following plots are presented as if the ordinary residuals e_i are being used. In all cases, the standardized, Studentized, or recursive residuals could be used.

10.1.1 Plots of e Versus $\hat{Y}$

Expected Behavior

The plot of the residuals against the fitted values of the dependent variable is particularly useful. A random scattering of the points above and below the line $e = 0$ with nearly all the data points being within the band defined by $e = \pm 2s$ (Figure 10.2 on page 254) is expected if the assumptions are satisfied. ($\hat{\mathbf{Y}}$ is used rather than $\mathbf{Y}$ because $\mathbf{e}$ is orthogonal to $\hat{\mathbf{Y}}$ but not to $\mathbf{Y}$. A plot of $\mathbf{e}$ versus $\mathbf{Y}$ will show a pattern due to this lack of orthogonality.)

Detecting Heterogeneous Variances

Any pattern in the *magnitude* of the dispersion about zero associated with changing $\hat{Y}_i$ suggests heterogeneous variances of ε_i. The fan-shaped pattern in Figure 10.3 is the typical pattern when the variance increases with the mean of

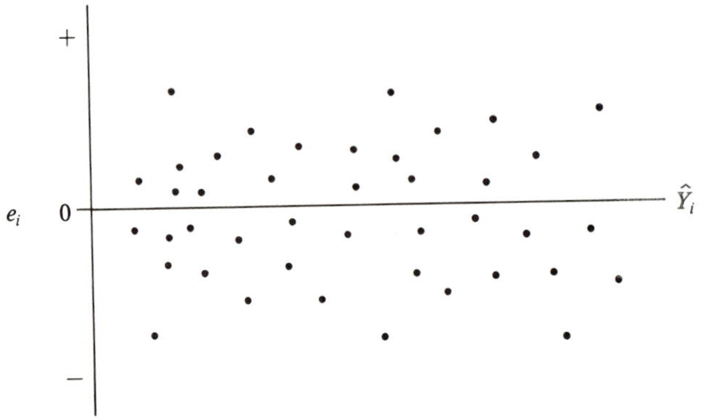

Figure 10.2 Typical pattern expected for a plot of ***e*** versus $\hat{Y}$ when assumptions are met.

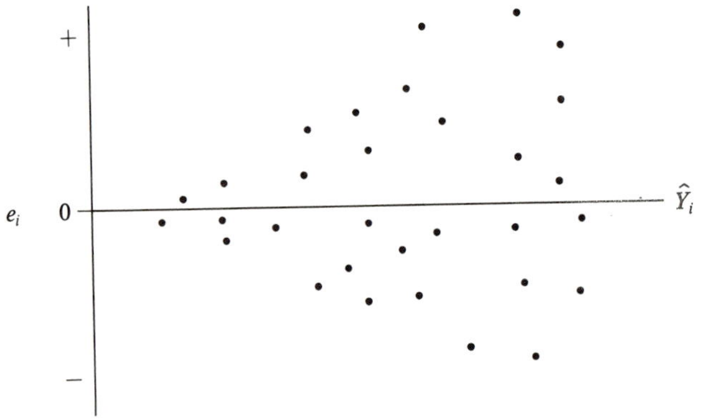

Figure 10.3 Plot of ***e*** versus $\hat{Y}$ showing increasing dispersion (larger variance) with larger $\hat{Y}$.

the dependent variable. This is the pattern to be expected if the dependent variable has a Poisson or a log-normal distribution, for example, or if the errors are multiplicative rather than additive. Binomially distributed data would show greater dispersion when the proportion of "successes" is in the intermediate range.

Detecting Model Inadequacies

Any asymmetry of the distribution of the residuals about zero suggests a problem with the model or the basic assumptions. A majority of relatively small negative residuals and fewer but larger positive residuals would suggest a positively skewed distribution of residuals rather than the assumed symmetric normal distribution. (A skewed distribution would be more evident in either a frequency plot or a normal plot of the residuals.) A preponderance of negative residuals for some regions of $\hat{Y}$ and positive residuals in other regions, such as the curved pattern of residuals in Figure 10.4, suggests a systematic error in the data or an important variable missing from the model. The obvious candidate in this illustration would be the square of one of the present independent variables. A missing independent variable can cause unusual patterns of residuals depending on the scatter of the data with respect to that variable.

Outlier Residuals

An outlier residual would appear in any of the plots of ***e*** as a point well outside the band containing most of the residuals. However, an outlier in ***Y*** will not necessarily have an outlier residual.

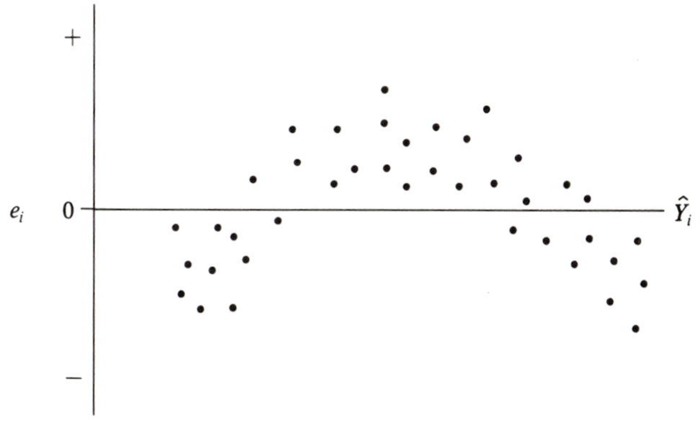

Figure 10.4 An asymmetric (curved) pattern of residuals plotted against $\hat{Y}$ suggests that the model is missing an important independent variable, perhaps a quadratic term.

Example 10.1 The Lesser–Unsworth data in exercise 1.19 related seed weight of soybeans to cumulative solar radiation for plants exposed to two different levels of ozone. The Studentized residuals from the regression of Y_i = (seed weight)$^{1/2}$ on solar radiation and ozone level are plotted against $\hat{Y}_i$ in Figure 10.5. The residuals

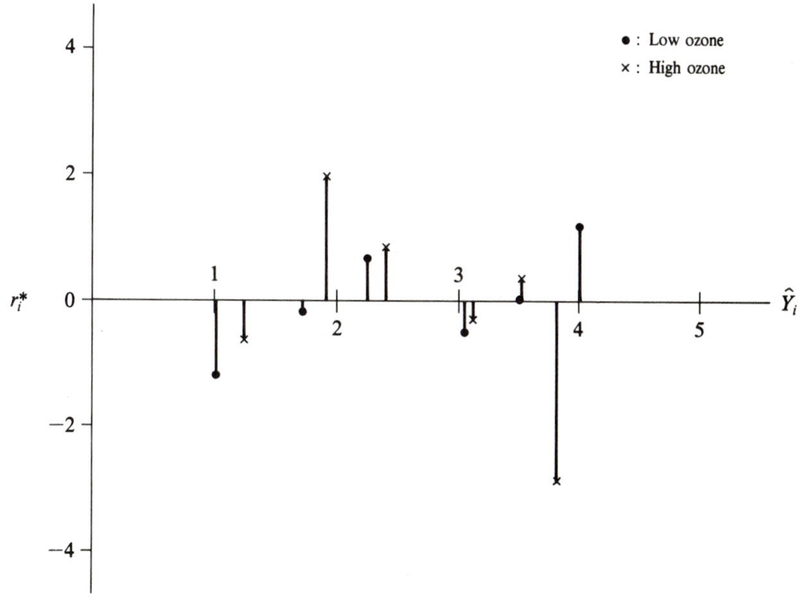

Figure 10.5 Plot of r_i^* versus $\hat{Y}_i$ for the Lesser–Unsworth data (exercise 1.19) relating seed weight of soybeans to cumulative solar radiation for two levels of ozone exposure. The model included linear regression of (seed weight)$^{1/2}$ on ozone level and solar radiation.

for the low and high levels of ozone are shown as dots and $\times$'s, respectively. One observation from the high ozone treatment seems to stand out from the others. Is this residual the result of an error in the data, an incorrect model, or simply random variation in the data?

The value of this Studentized residual is $r_{12}^* = -2.8369$. This is distributed as Student's t with $(n - p' - 1) = 8$ degrees of freedom. The probability of $|t| > 2.8369$ is slightly less than .02. Allowing for the fact that this is the most extreme residual out of a sample of 12, it does not appear to be unusually large. Overall, the remaining residuals tend to show an upward trend, suggesting that this observation is pulling the regression line down. Inspection of the residuals by treatment, however, shows that the high ozone treatment, the $\times$'s, have a slight downward slope. Perhaps the large residual results from an incorrect model that forces both ozone treatments to have a common regression on solar radiation. ■

Example 10.2 The standardized residuals from the regression of oxygen uptake on time to run a fixed course, resting heart rate, heart rate while running, and maximum heart rate while running, Table 4.3, are plotted against $\hat{Y}_i$ in Figure 10.6. Although the pattern is not definitive, there is some semblance of the fan-shaped pattern of residuals suggesting heterogeneous variance. The larger dispersion for the higher levels of oxygen consumption could also result from the model being inadequate in this region. Perhaps the faster runners, who tended to use more oxygen, differed in ways not measured by the four variables. ■

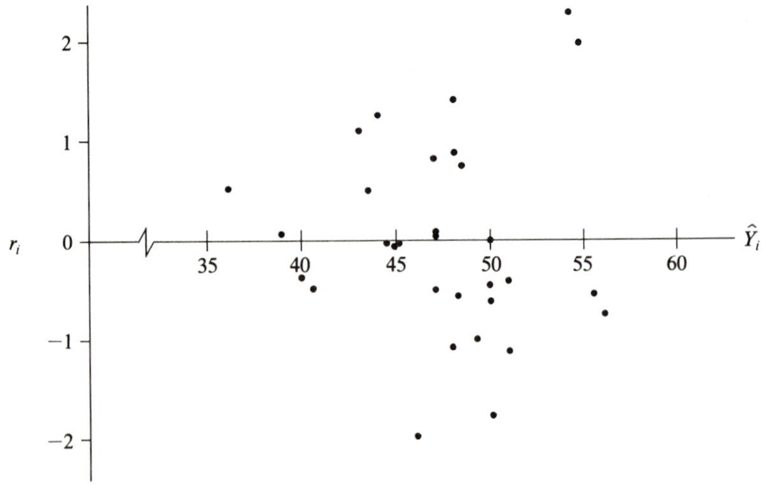

Figure 10.6 Plot of r_i versus $\hat{Y}_i$ for the regression of oxygen uptake on running time, resting heart rate, running heart rate, and maximum heart rate. The original data are given in Table 4.3.

10.1.2 Plots of *e* Versus X_i

Interpretation

Plots of the residuals against the independent variables have interpretations similar to plots of residuals against $\hat{Y}$. Differences in magnitude of dispersion about zero suggest heterogeneous variances. A missing higher-degree polynomial term for the independent variable should be evident in these plots. However, inadequacies in the model associated with one variable, such as a missing higher-degree polynomial term, can be obscured by the effects and distribution of other independent variables. The partial regression leverage plots (to be discussed in Section 10.1.6) may be more revealing when several independent variables are involved.

Outliers and Influential Points

Outlier residuals will be evident. Observations that appear as isolated points at the extremes of the X_i scale are potentially influential because of their extreme values for that particular independent variable. Such points will tend to have small residuals because of their high leverage or influence. However, data points can be far outside the sample X-space without being outside the limits of any one independent variable by having unlikely combinations of values for two or more variables. Such points are potentially influential but will not be easily detected by any univariate plots.

Example 10.3

(*Continuation of Example* 10.1) The plot of the Studentized residuals against radiation from the regression of seed weight on ozone exposure and cumulative solar radiation (Lesser–Unsworth data) is given in Figure 10.7. [Seed weight is

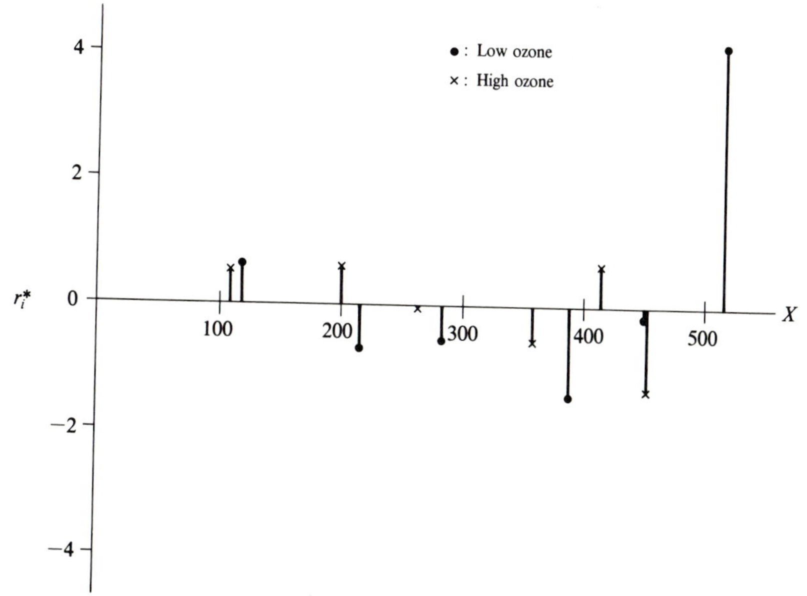

Figure 10.7 Plot of the Studentized residuals versus radiation (X) for the Lesser–Unsworth data. The residuals are from the regression of seed weight on ozone level and cumulative solar radiation.

being used as the dependent variable rather than (seed weight)$^{1/2}$ as in Figure 10.5.] One residual (not the same as in Figure 10.5) stands out as a possible outlier. In this case, $r_6^* = 4.1565$ and is very close to being significant, $\alpha^* = .05$. It is evident from the general negative slope of the other residuals that this point has had a major effect on the regression coefficient. ■

10.1.3 Plots of *e* Versus Time

Data collected over time on individual observational units will often have serially correlated residuals. That is, the residual at one point in time depends to some degree on the previous residuals. Classical time series data, such as the data generated by the continuous monitoring of some process, are readily recognized as such and are expected to have correlated residuals. Time series models and analyses take into account these serial correlations and should be used in such cases (Fuller, 1976; Bloomfield, 1976).

Causes of Correlated Residuals

There are many opportunities, however, for time effects to creep into data which normally would not be thought of as time series data. For example, resource limitations may force the researcher to run the experiment over some period of time to obtain even one observation on each treatment. This is common in industrial experiments where an entire production process may be utilized to produce an observation. The time of day or time of week can have effects on the experimental results even though the process is thought to be well controlled.

Even in biological experiments, where it is usual for all experimental units to be under observation at the same time, some phases of the study may require extended periods of time to complete. For example, autopsies on test animals to determine the incidence of precancerous cell changes may require several days. The simple recording of data in a field experiment may take several days. All such situations provide the opportunity for "time" to have an impact on the differences among the experimental observations.

A plot of the residuals against time may reveal effects not previously thought to be important and, consequently, not taken into account in the design of the study. Serial correlations will appear as a tendency of the residuals to follow each other as they oscillate around zero.

Example 10.4

The standardized residuals from a regression adjusting yearly Menhaden catch from 1964 to 1979 for a linear time trend are shown in Figure 10.8. [Data are taken from Nelson and Ahrenholz (1986) and are given in exercise 3.11.] The serial correlation is relatively weak in this case; the lag-one serial correlation is .114. (The lag-one serial correlation is the correlation between residuals one time unit apart.) Even though the serial correlation is weak, the residuals show the typical pattern of the positive and negative residuals occurring in runs. ■

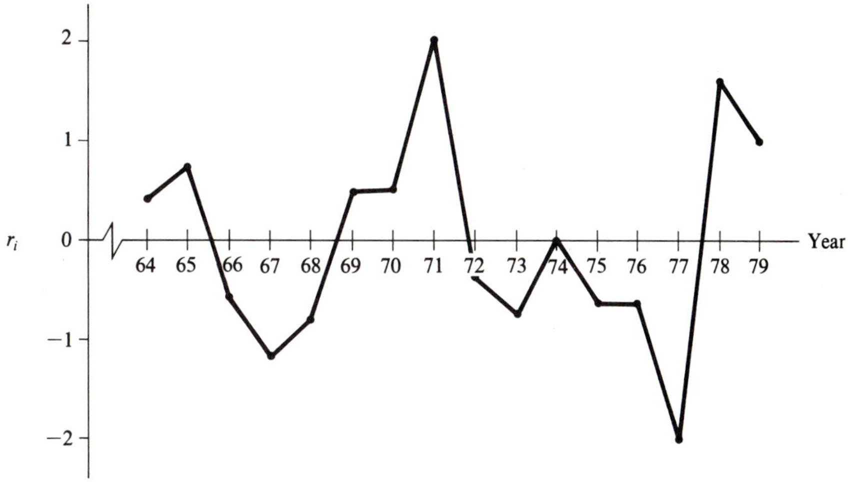

Figure 10.8 Plot of r_i versus year of catch for the regression of yearly Menhaden catch on year. [Data are from Nelson and Ahrenholz (1986).]

Changes in the production process, drifting of monitoring equipment, time-of-day effects, time-of-week effects, and so forth, will show up as shifts in the residuals plot. "Time" in this context can be the sequence in which the treatments are imposed, in which measurements are taken, or in which experimental units are tended. Alternatively, "time" could represent the spatial relationship of the experimental units during the course of the trial. In the last case, plots of $\boldsymbol{e}$ versus "time" might detect environmental gradients within the space of the experiment.

Runs Test

The **runs test** is frequently used to detect serial correlation. The test consists of counting the number of runs, or sequences of positive and negative residuals, and comparing the result to the expected number of runs under the null hypothesis of independence. (The lack of statistical independence among the observed residuals will confound the runs test to some degree. This effect can probably be ignored as long as a reasonable proportion of the total degrees of freedom is devoted to the residual sum of squares.)

Example 10.5

(*Continuation of Example* 10.4) The data of annual catch of Menhaden from 1964 to 1979 when regressed against time show the following sequence of positive and negative residuals (see Figure 10.8):

$$+ \quad + \quad - \quad - \quad - \quad + \quad + \quad + \quad - \quad - \quad - \quad - \quad - \quad - \quad + \quad +$$

There are $u = 5$ runs in a sample consisting of $n_1 = 7$ positives and $n_2 = 9$ negatives. The cumulative probabilities for number u of runs in sample sizes of (n_1, n_2) are given by Swed and Eisenhart (1943) for $n_1 + n_2 \leq 20$. In this example with $(n_1, n_2) = (7, 9)$, the probability of $u \le 5$ is .035, indicating significant departure from independence. This low number of runs suggests the presence of a positive serial correlation. ■

If n_1 and n_2 are greater than 10, a normal approximation for the distribution of runs can be used, where

$$\mu = \frac{2n_1 n_2}{n_1 + n_2} + 1 \qquad [10.4]$$

and

$$\sigma^2 = \frac{2n_1 n_2(2n_1 n_2 - n_1 - n_2)}{(n_1 + n_2)^2(n_1 + n_2 - 1)} \qquad [10.5]$$

Then

$$z = \frac{(u - \mu + \frac{1}{2})}{\sigma} \qquad [10.6]$$

is the standardized normal deviate, where the $\frac{1}{2}$ is the correction for continuity.

Example 10.6

(*Continuation of Example* 10.5) Applying the normal approximation to the Menhaden catch data, even though n_1 and n_2 are less than 10, gives $\mu = 8.875$ and $\sigma^2 = 3.6094$, which yields $z = -1.78$. The probability of z being less than -1.78 is .0375, very close to the probability of .035 taken from Swed and Eisenhart (1943). ■

10.1.4 Plots of e_i Versus e_{i-1}

A serial correlation in time series data is more clearly revealed with a plot of each residual against the immediately preceding residual. A positive serial correlation would produce a scatter of points with a clear positive slope as in Figure 10.9.

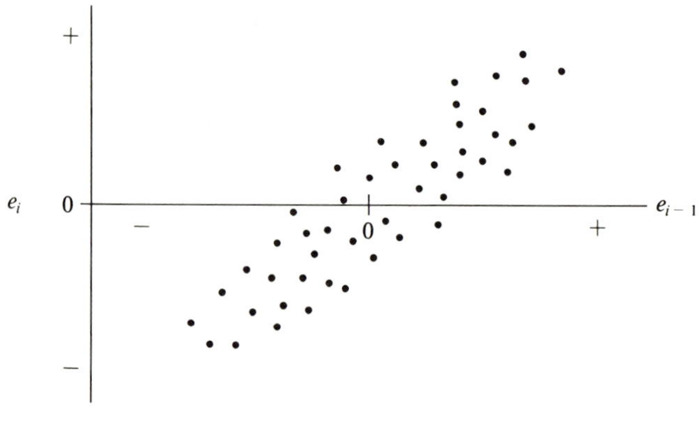

Figure 10.9 Typical plot of e_i versus e_{i-1} showing a positive serial correlation among successive residuals.

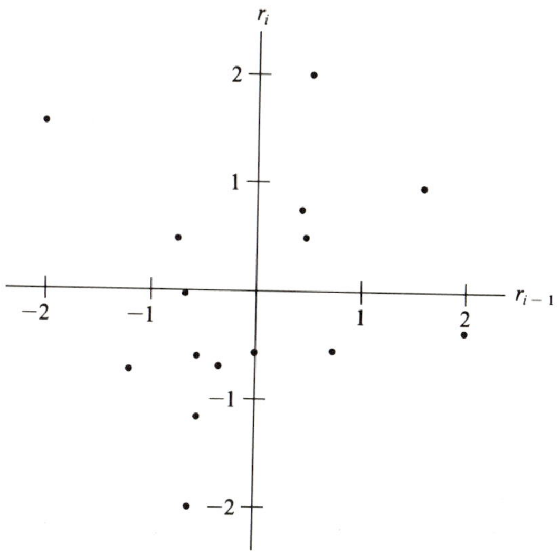

Figure 10.10 Plot of r_i versus r_{i-1} for the Menhaden catch data. The residuals are from the regression of annual catch on year of catch.

Example 10.7 The plot of e_i versus e_{i-1} for the Menhaden catch data is shown in Figure 10.10. The extreme point in the upper left quadrant is the plot of the second largest positive residual (1978) against the largest negative residual (1977). This sudden shift in catch from 1977 to 1978 is largely responsible for the serial correlation being as small as it is. Even so, the positive serial correlation is evident. ■

Durbin–Watson Test

The presence of serial correlation in the residuals is also detected by the Durbin–Watson test for independence (Durbin and Watson, 1951). The Durbin–Watson test statistic is

$$d = \frac{\sum_{i=2}^{n}(e_i - e_{i-1})^2}{\sum_{i=1}^{n} e_i^2} \qquad [10.7]$$

The Durbin–Watson statistic d gets smaller as the serial correlation ρ increases. The one-tailed Durbin–Watson test of the null hypothesis of independence, H_0: $\rho = 0$, against the alternative hypothesis H_a: $\rho > 0$, uses two critical values, d_U and d_L, which depend on n, p, and the choice of α. Critical values for the Durbin–Watson test statistic are given in Appendix Table C. The test procedure rejects the null hypothesis if $d < d_L$, does not reject the null hypothesis if $d > d_U$, and is inconclusive if $d_L < d < d_U$. Tests of significance for the alternative hypothesis H_a: $\rho < 0$ use the same critical values, d_U and d_L, but the test statistic is first subtracted from 4.

Some statistical computing packages routinely provide the Durbin–Watson test of the serial correlation of the residuals. In PROC GLM (SAS Institute, Inc., 1985d, 1985f) for example, the Durbin–Watson statistic is reported as part of the standard results whenever the residuals are requested, even though the data may not be time series data. The statistic is computed on the residuals in the order in which the data are listed in the data set. Care must be taken to ensure that the test is appropriate and that the ordering of the data is meaningful before the Durbin–Watson test is used.

10.1.5 Normal Probability Plots

The normal probability plot is designed to detect nonnormality. It is the plot of the ordered residuals against the normal order statistics for the appropriate sample size. The normal order statistics are the expected values of ordered observations from the normal distribution with zero mean and unit variance.

Defining Normal Order Statistics

Let $z_1, z_2, \ldots, z_n$ be the observations from a random sample of size n. The n observations ordered (and relabeled) so that $z_{(1)} \leq z_{(2)} \leq \cdots \leq z_{(n)}$ give the sample order statistics. The average for each $z_{(i)}$ over repeated samplings gives the ith order statistic for the probability distribution being sampled. These are the normal order statistics if the probability distribution being sampled is the normal distribution with zero mean and unit variance. For example, the normal order statistics for a sample of size 5 are -1.163, $-.495$, 0, .495, and 1.163. The expected value of the smallest observation in a sample of size 5 from a N(0, 1) distribution is -1.163, the second smallest has expectation $-.495$, and so forth.

The normal order statistics were tabled for sample sizes to $n = 20$ by Pearson and Hartley (1966, *Biometrika Tables for Statisticians*), and have been reproduced in many references (e.g., Weisberg, 1985, Table D; Rohlf and Sokal, 1981, Table 27). In some references the indexing of the normal order statistics is in the reverse order so that the first order statistic refers to the largest. The order statistics are easily approximated by any computer program that provides the inverse function of the cumulative normal distribution. Thus, $z_{(i)} \doteq \Phi^{-1}(p)$, where p is chosen as a function of the ranks of the residuals. Several choices of p have been suggested. Blom's (1958) suggestion of using

$$p = \frac{r_i - \frac{3}{8}}{n + \frac{1}{4}} \qquad [10.8]$$

where r_i is the rank and n is the sample size, provides an excellent approximation if $n \geq 5$. Plotting the ordered observed residuals against their normal order statistics provides the normal plot.

Expected Behavior

The expected result from a normal plot when the residuals are a sample from a normal distribution is a straight line passing through zero with the slope of the line determined by the standard deviation of the residuals. There will be random deviations from a straight line due to sampling variation of the sample order statistics. Some practice is needed to develop judgment for the amount of departure one should allow before concluding that nonnormality is a problem. Daniel and Wood (1980) give illustrations of the amount of variation in normal probability plots of samples from normal distributions. Normal probability plots for small samples will not be very informative, because of sampling variation, unless departures from normality are large.

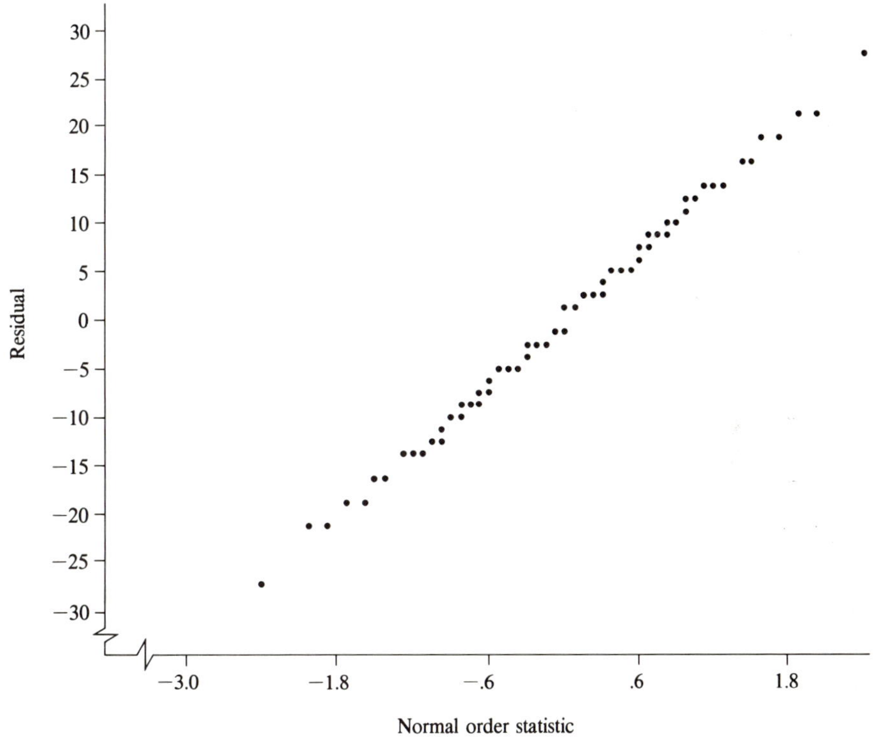

Figure 10.11 Normal plot of residuals from the analysis of variance of final plant heights in a study of blue mold infection on tobacco. (Data courtesy of M. Moss and C. C. Main, North Carolina State University.)

Example 10.8 Figure 10.11 shows a well-behaved normal plot of the residuals from an analysis of variance of final plant heights in a study of blue mold infection in tobacco. (Data provided courtesy of M. Moss and C. C. Main, North Carolina State University.) There is a total of 80 observations and the residual sum of squares has 36 degrees of freedom. The amount of dependence among the residuals is related to the proportion of degrees of freedom used by the model, $\frac{44}{80}$ in this case. This relatively high degree of dependence among the residuals and the "supernormal" tendencies of least squares residuals mentioned earlier may be contributing to the very normal-appearing behavior of this plot. ■

Interpretation of Normal Plots

The pattern of the departure from the expected straight line in a normal plot suggests the nature of the nonnormality. A skewed distribution will show a curved normal plot with the direction of the curve determined by the direction of the skewness. An S-shaped curve suggests heavy-tailed or light-tailed distributions (Figure 10.12, page 264), depending on the direction of the S. (Heavy-tailed distributions have a relatively higher frequency of extreme observations than the normal distribution; light-tailed distributions have relatively fewer.) Other model defects can mimic the effects of nonnormality. For example, hetero-

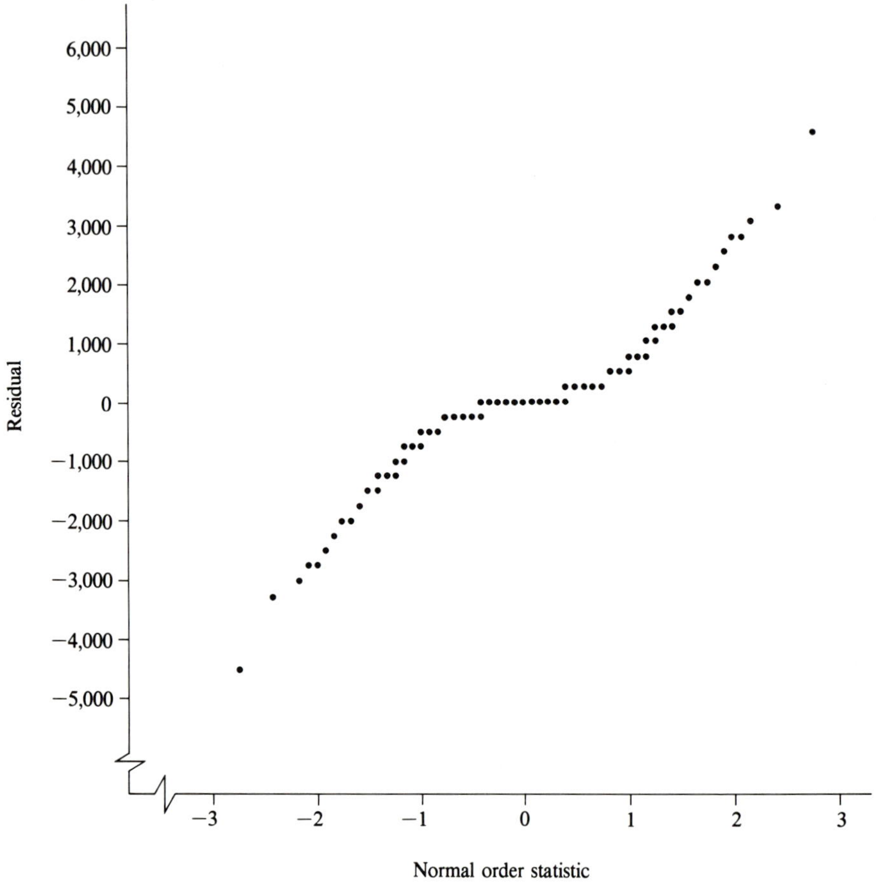

Figure 10.12 A normal probability plot with a pattern typical of a light-tailed distribution. In this case, the S-shape resulted from heterogeneous variances in the data.

geneous variances or outlier residuals will give the appearance of a lighted-tailed distribution. The ordinary least squares residuals are constrained to have zero mean (if the model includes the intercept term). The recursive residuals, on the other hand, are not so constrained and, thus, the normal plot of recursive residuals need not pass through the origin. This is interpreted as an indication of an outlier in the base set of observations or as a model misfit such as an omitted variable (Galpin and Hawkins, 1984).

Tests for Normality

There are many tests for nonnormality under independence. However, these tests must be used with caution when applied to regression residuals, because the residuals are not independent. The limiting distributions of the test statistics show that they are appropriate for regression residuals if the sample size is infinite (Pierce and Kopecky, 1979). For finite samples, however, all are approximations and the question becomes one of how large the sample must be for the approximation to be satisfactory. The required sample size depends on the number of parameters, p', and the nature of $\boldsymbol{P}$, which is determined by the configuration of the X's (Cook and Weisberg, 1982). Simulation studies have suggested that the approximation is adequate, insofar as size of the test is concerned, for samples as small as $n = 20$ when there are four or six independent variables (White and MacDonald, 1980; Pierce and Gray, 1982), or with $n = 40$ when there are eight independent variables (Pierce and Gray, 1982). (The **size** of a test is the probability of rejecting the null hypothesis when it is true.) However, caution must be used; Weisberg (1980) gives an example using an experimental design matrix with $n = 20$ where the observed size of the test is near $\alpha = .30$, rather than the nominal $\alpha = .10$ level.

Shapiro–Francia Statistic

It appears that many tests for normality applied to regression residuals will provide acceptable approximations if the sample size is reasonable, say $n > 40$ or $n > 80$ if p' is large. The size and power of the tests in small samples make them of questionable value. The Shapiro–Francia (1972) W' test statistic for normality, a modification of the Shapiro–Wilk (1965) W, provides a direct quantitative measure of the degree of agreement between the normal plot and the expected straight line. The Shapiro–Francia statistic is the squared correlation between the observed ordered residuals and the normal order statistics. Let $\boldsymbol{u}$ be the vector of *centered* observed ordered residuals (the e_i, r_i, or r_i^*) and let $\boldsymbol{z}$ be the vector of normal order statistics. Then

$$W' = \frac{(\boldsymbol{u}'\boldsymbol{z})^2}{(\boldsymbol{u}'\boldsymbol{u})(\boldsymbol{z}'\boldsymbol{z})} \qquad [10.9]$$

The observed residuals are expressed as deviations from their mean. The e_i will have zero mean, if the model includes an intercept, but this does not apply to r_i or r_i^*. The null hypothesis of normality is rejected for sufficiently small values of W'. Critical values for W' are tabulated by Shapiro and Francia (1972) for $n = 35$, 50, 51(2)99, and are reproduced in Appendix Table D. For $n < 50$, the percentage points provided by Shapiro and Wilk (1965) for W are good approximations of those for W' (Weisberg, 1974).

Additional Tests

Other test statistics are frequently used as tests for nonnormality. For example, PROC UNIVARIATE (SAS Institute, Inc., 1985c) uses the Shapiro–Wilk W statistic if $n < 2{,}000$ and the Kolmogorov D statistic if $n > 2{,}000$. PROC UNIVARIATE also reports skewness and kurtosis coefficients for the sample; these are sometimes used for testing normality.

10.1.6 Partial Regression Leverage Plots

When several independent variables are involved, the relationship of the residuals to one independent variable can be obscured by effects of other variables. **Partial regression leverage plots** are an attempt to remove the confounding effects of the other variables. Let $\boldsymbol{e}(j)$ denote the residuals from the regression of the dependent variable on all independent variables *except* the jth. Similarly, let $\boldsymbol{u}(j)$ denote the residuals from the regression of the jth independent variable on all *other* independent variables. The plot of $\boldsymbol{e}(j)$ versus $\boldsymbol{u}(j)$ is the partial regression leverage plot for the jth variable. Note that both $\boldsymbol{e}(j)$ and $\boldsymbol{u}(j)$ have been adjusted for all other independent variables in the model.

Interpretation

This plot reflects what the least squares regression is "seeing" when the jth variable is being added last to the model. The slope of the linear regression line in the partial regression leverage plot is the partial regression coefficient for that independent variable in the full model. The deviations from the linear regression line correspond to the residuals $\boldsymbol{e}$ from the full model.

Any curvilinear relationships not already taken into account in the model should be evident from the partial regression leverage plots. The plot is useful for detecting outliers and high-leverage points and for showing how several leverage points might be interacting to influence the partial regression coefficients.

Example 10.9

The partial regression leverage plot of catch versus fishing pressure for the Menhaden yearly catch data of Example 10.4 is given in Figure 10.13. In this case, yearly catch was regressed on number of vessels and fishing pressure; the model also included an intercept. Thus, the partial residuals for catch and fishing pressure are adjusted for the intercept and number of vessels. The figure shows a clear linear relationship between catch and pressure and there may be some suggestion of a slight curvilinear relationship. None of the points appears to be an obvious outlier. The two leftmost points and the uppermost point appear to be influential points in terms of the possible curvilinear relationship. ■

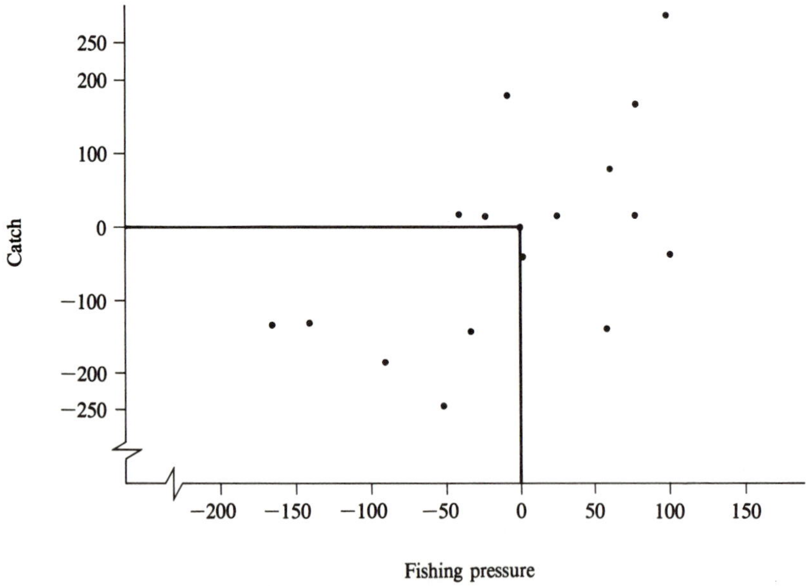

Figure 10.13 Partial regression leverage plot for catch versus fishing pressure from the regression of yearly catch of Menhaden on number of vessels and fishing pressure. The model included an intercept. [Data from Nelson and Ahrenholz (1986).]

10.2 Influence Statistics

Potentially Influential Points

Potentially influential points or points with high leverage are the data points that are far from the center of the X-space. A measure of the distance of the ith data point from the centroid of the X-space is provided by v_{ii}, the ith diagonal element of the projection matrix $\boldsymbol{P}$ (called the Hat matrix in some references). The ith diagonal element of $\boldsymbol{P}$ is given by

$$v_{ii} = \boldsymbol{x}_i'(\boldsymbol{X}'\boldsymbol{X})^{-1}\boldsymbol{x}_i \qquad [10.10]$$

where $\boldsymbol{x}_i'$ is the ith row of $\boldsymbol{X}$. The limits on v_{ii} are $1/n \leq v_{ii} \leq 1/c$, where c is the number of rows of $\boldsymbol{X}$ that have the same values as the ith row. The lower bound $1/n$ is attained only if every element in $\boldsymbol{x}_i$ is equal to the mean for that independent variable—in other words, only if the data point falls on the centroid. The larger values reflect data points that are farther from the centroid. The upper limit of 1 (when $c = 1$) implies that the leverage for the data point is so high as to force the regression line to pass exactly through that point. The variance of $\hat{Y}$ for such a point is σ^2 and the variance of the residual is zero. The average value of v_{ii} is p'/n. [There are n v_{ii}-elements and the sum, tr($\boldsymbol{P}$), is p'.] Belsley, Kuh, and Welsch (1980) suggest using $v_{ii} > 2p'/n$ to identify potentially influential points or leverage points.

Identifying the Influential Points

The diagonal elements of $\boldsymbol{P}$ only identify data points that are far from the centroid of the sample X-space. Such points are potentially but not necessarily influential in determining the results of the regression. The general procedure for assessing the influence of a point in a regression analysis is to determine the changes that occur when that observation is omitted. Several measures of influence have been developed using this concept. They differ in the particular regression result on which the effect of the deletion is measured, and the standardization used to make them comparable over observations. All influence statistics can be computed from the results of the single regression using all data.

Influence Measures

Four influence measures are discussed, each of which measures the effect of deleting the ith observation:

1. Cook's D_i, which measures the effect on $\hat{\boldsymbol{\beta}}$
2. DFFITS$_i$, which measures the effect on $\hat{Y}_i$
3. DFBETAS$_{j(i)}$, which measures the effect on $\hat{\beta}_j$
4. COVRATIO$_i$, which measures the effect on the variance–covariance matrix of the parameter estimates

The first three of these, Cook's D, DFFITS, and DFBETAS, can be thought of as special cases of a general approach for measuring the impact of deleting the ith observation on any set of k linearly independent functions of $\hat{\boldsymbol{\beta}}$ (Cook and Weisberg, 1982). Let $\boldsymbol{U} = \boldsymbol{K}'\boldsymbol{\beta}$ be a set of linear functions of $\boldsymbol{\beta}$ of interest. Then the change in the estimate of $\boldsymbol{U}$ when the ith observation is dropped is given by $\boldsymbol{K}'(\hat{\boldsymbol{\beta}} - \hat{\boldsymbol{\beta}}_{(i)})$, where $\hat{\boldsymbol{\beta}}_{(i)}$ is the vector of regression coefficients estimated with the ith observation omitted. This change can be written in a quadratic form similar to the quadratic form for the general linear hypothesis in Chapter 4:

$$\frac{[\boldsymbol{K}'(\hat{\boldsymbol{\beta}} - \hat{\boldsymbol{\beta}}_{(i)})]'[\boldsymbol{K}'(\boldsymbol{X}'\boldsymbol{X})^{-1}\boldsymbol{K}]^{-1}[\boldsymbol{K}'(\hat{\boldsymbol{\beta}} - \hat{\boldsymbol{\beta}}_{(i)})]}{r(\boldsymbol{K})\hat{\sigma}^2} \qquad [10.11]$$

If $\boldsymbol{K}'$ is chosen as $\boldsymbol{I}_{p'}$ and s^2 is used for $\hat{\sigma}^2$, Cook's D results. If $\boldsymbol{K}'$ is chosen as $\boldsymbol{x}_i'$, the ith row of $\boldsymbol{X}$, and $s^2_{(i)}$ is used for $\hat{\sigma}^2$, the result is $(\text{DFFITS}_i)^2$. Choosing $\boldsymbol{K}' = (0 \ldots 0\ 1\ 0 \ldots 0)$, where the 1 occurs in the $(j + 1)$st position, and using $s^2_{(i)}$ for $\hat{\sigma}^2$ gives $(\text{DFBETAS}_{j(i)})^2$.

10.2.1 Cook's *D*

Computation

Cook's D (Cook, 1977; Cook and Weisberg, 1982) is designed to measure the shift in $\hat{\boldsymbol{\beta}}$ when a particular observation is omitted. It is a combined measure of the impact of that observation on all regression coefficients. Cook's D is defined as

$$D_i = \frac{(\hat{\boldsymbol{\beta}}_{(i)} - \hat{\boldsymbol{\beta}})'(\boldsymbol{X}'\boldsymbol{X})(\hat{\boldsymbol{\beta}}_{(i)} - \hat{\boldsymbol{\beta}})}{p's^2} \qquad [10.12]$$

Computationally, D_i is more easily obtained as

$$D_i = \frac{r_i^2}{p'}\left(\frac{v_{ii}}{1 - v_{ii}}\right) \qquad [10.13]$$

where r_i is the standardized residual and v_{ii} is the ith diagonal element of $\boldsymbol{P}$ computed from the full regression. Notice that D_i is large if the standardized residual is large and if the data point is far from the centroid of the X-space—that is, if v_{ii} is large.

Interpretation

Cook's D measures the distance from $\hat{\boldsymbol{\beta}}$ to $\hat{\boldsymbol{\beta}}_{(i)}$ in terms of the joint confidence ellipsoids about $\hat{\boldsymbol{\beta}}$. Thus, if D_i is approximately equal to $F_{(\alpha, p', n-p')}$, the $\hat{\boldsymbol{\beta}}_{(i)}$ vector is (approximately) on the $100(1 - \alpha)\%$ confidence ellipsoid of $\boldsymbol{\beta}$. This should not be treated as a test of significance. A shift in $\hat{\boldsymbol{\beta}}$ to the ellipsoid corresponding to $\alpha = .50$ from omitting a single data point would be considered a major shift. For reference, the 50th percentile for the F-distribution is 1.0 when the numerator and denominator degrees of freedom are equal and is always less than 1.0 if the denominator degrees of freedom is the larger. The 50th percentile does not get smaller than .8 unless the numerator degrees of freedom is only 1 or 2. Thus, Cook's D_i in the vicinity of .8–1.0 would indicate a shift to near the 50th percentile in most situations.

Cook's D can also be written in the form

$$D_i = \frac{(\hat{\boldsymbol{Y}}_{(i)} - \hat{\boldsymbol{Y}})'(\hat{\boldsymbol{Y}}_{(i)} - \hat{\boldsymbol{Y}})}{p's^2} \qquad [10.14]$$

where $\hat{Y}_{(i)} = X\hat{\beta}_{(i)}$. In this form, Cook's D can be interpreted as the Euclidean distance between $\hat{Y}_{(i)}$ and $\hat{Y}$ and, hence, measures the shift in $\hat{Y}$ caused by deleting the ith observation.

10.2.2 DFFITS

Computation

Equation 10.13 showed that Cook's D provides a measure of the shift in $\hat{Y}$ when the ith observation is not used in the estimation of β. A closely related measure is provided by DFFITS (Belsley, Kuh, and Welsch, 1980), defined as

$$\text{DFFITS}_i = \frac{\hat{Y}_i - \hat{Y}_{(i)i}}{s_{(i)}\sqrt{v_{ii}}}$$

$$= \left(\frac{v_{ii}}{1 - v_{ii}}\right)^{1/2} \frac{e_i}{s_{(i)}(1 - v_{ii})^{1/2}} \qquad [10.15]$$

where $\hat{Y}_{i(i)}$ is the estimated mean for the ith observation but where the ith observation was not used in estimating β. Notice that σ has been estimated with $s_{(i)}$, the estimate of σ obtained without the ith observation. $s_{(i)}$ is obtained without redoing the regression by using the relationship

$$(n - p' - 1)s_{(i)}^2 = (n - p')s^2 - \frac{e_i^2}{1 - v_{ii}} \qquad [10.16]$$

Interpretation

The relationship of DFFITS to Cook's D is

$$D_i = (\text{DFFITS}_i)^2\left(\frac{s_{(i)}^2}{p's^2}\right) \qquad [10.17]$$

Belsley, Kuh, and Welsch (1980) suggest that DFFITS larger in absolute value than $2\sqrt{p'/n}$ be used to flag influential observations. Ignoring the difference between s^2 and $s_{(i)}^2$, this cutoff number for DFFITS suggests a cutoff of $4/n$ for Cook's D.

Atkinson's C_i

A modified version of Cook's D suggested by Atkinson (1983) is even more closely related to DFFITS:

$$C_i = |r_i^*|\left[\left(\frac{n - p'}{p'}\right)\left(\frac{v_{ii}}{1 - v_{ii}}\right)\right]^{1/2}$$

$$= \left(\frac{n - p'}{p'}\right)^{1/2} |\text{DFFITS}_i| \qquad [10.18]$$

The cutoff point for DFFITS for flagging large values translates into a cutoff for C_i of $2[(n - p')/n]^{1/2}$. Atkinson recommends that signed values of C_i be plotted in any of the ways customary for residuals. (This recommendation can be extended to any of the measures of influence.) Very nearly identical interpretations are obtained from DFFITS_i, Cook's D_i, and Atkinson's C_i if these reference numbers are used. There is no need to use more than one.

10.2.3 DFBETAS

Computation

Cook's D_i reveals the impact of the ith observation on the entire vector of the estimated regression coefficients. The influential observations for the individual regression coefficients are identified by $\text{DFBETAS}_{j(i)}$, $j = 1, 2, \ldots, p'$ (Belsley, Kuh, and Welsch, 1980), where each $\text{DFBETAS}_{j(i)}$ is the standardized change in $\hat{\beta}_j$ when the ith observation is deleted from the analysis. Thus,

$$\text{DFBETAS}_{j(i)} = \frac{\hat{\beta}_j - \hat{\beta}_{j(i)}}{s_{(i)}\sqrt{c_{jj}}} \qquad [10.19]$$

where c_{jj} is the $(j + 1)$st diagonal element from $(\boldsymbol{X}'\boldsymbol{X})^{-1}$. Although the formula is not quite as simple as for DFFITS_i, $\text{DFBETAS}_{j(i)}$ can also be computed from the results of the original regression. The reader is referred to Belsley, Kuh, and Welsch (1980) for details.

Interpretation

$\text{DFBETAS}_{j(i)}$ measures the change in $\hat{\beta}_j$ in multiples of its standard error. While this looks like a t-statistic, it should not be interpreted as a test of significance. Values of $\text{DFBETAS}_{j(i)}$ greater than 2 would certainly indicate a major, but very unlikely, impact from a single point. The cutoff point of $2/\sqrt{n}$ is suggested by Belsley, Kuh, and Welsch as the point that will tend to highlight the same proportion of influential points.

10.2.4 COVRATIO

Computation

The impact of the ith observation on the variance–covariance matrix of the estimated regression coefficients is measured by the ratio of the determinants of the two variance–covariance matrices. Belsley, Kuh, and Welsch (1980) formulate this as

$$\text{COVRATIO} = \frac{\det(s^2_{(i)}[\boldsymbol{X}'_{(i)}\boldsymbol{X}_{(i)}]^{-1})}{\det(s^2[\boldsymbol{X}'\boldsymbol{X}]^{-1})}$$

$$= \left[\left(\frac{n - p' - 1}{n - p'} + \frac{r_i^{*2}}{n - p'}\right)^p (1 - v_{ii})\right]^{-1} \qquad [10.20]$$

Interpretation

The determinant of a variance–covariance matrix is a generalized measure of variance. Thus, COVRATIO reflects the impact of the ith observation on the precision of the estimates of the regression coefficients. Values near 1 indicate the ith observation has little effect on the precision of the estimates. A value of

COVRATIO greater than 1 indicates that the presence of the ith observation increases the precision of the estimates; a ratio less than 1 indicates that the presence of the observation impairs the precision of the estimates. Belsley, Kuh, and Welsch (1980) suggest that values of COVRATIO outside the limits $1 \pm 3(p'/n)$ be considered extreme for purposes of identifying influential points.

Using the Influence Statistics

The influence statistics are to be used as diagnostic tools for identifying the observations having the greatest impact on the regression results. While some of the influence measures resemble test statistics, they are not to be interpreted as tests of significance for influential observations. The large number of influence statistics that can be generated can cause confusion. One should concentrate on the diagnostic tool that measures the impact on the quantity of primary interest. The first two statistics, Cook's D_i and DFFITS$_i$, are very similar and provide "overall" measures of the influence of each observation. One of these will be of primary interest in most problems. In those cases where interest is in the estimation of particular regression parameters, DFBETAS$_{j(i)}$ for those j of interest will be most helpful.

Example 10.10

(*Continuation of Example* 10.3) The Studentized residuals and DFFITS$_i$ for the Lesser–Unsworth example are plotted against observation number in Figure 10.14. The close relationship between DFFITS$_i$ and r_i^* is evident; DFFITS$_i$ is the product of r_i^* and $[v_{ii}/(1 - v_{ii})]^{1/2}$ (equation 10.15). The latter is a measure of the potential leverage of the observation which, in this example, varies from .48 for observation 9 to .80 for observation 1. The suggested cutoff value for DFFITS is $2\sqrt{p'/n} = 2\sqrt{3/12} = 1$. Only DFFITS$_6$ exceeds this value and the residual for observation 6 certainly appears to be an outlier: $r_6^* = 4.16$. The

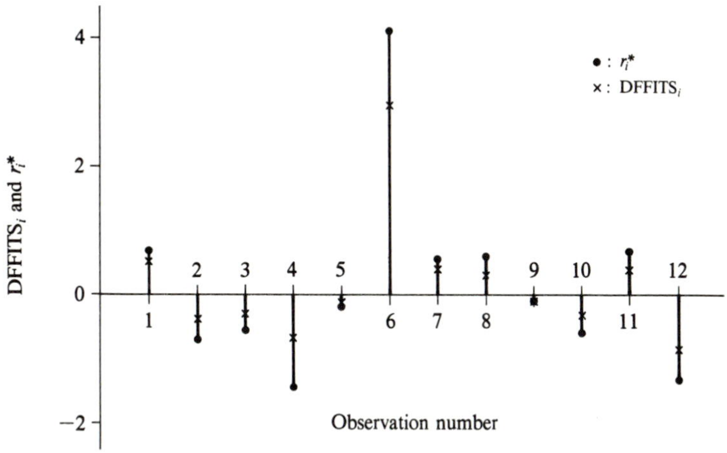

Figure 10.14 Studentized residuals and DFFITS$_i$ plotted against observation number from the regression of seed weight on ozone level and cumulative solar radiation using the Lesser–Unsworth data.

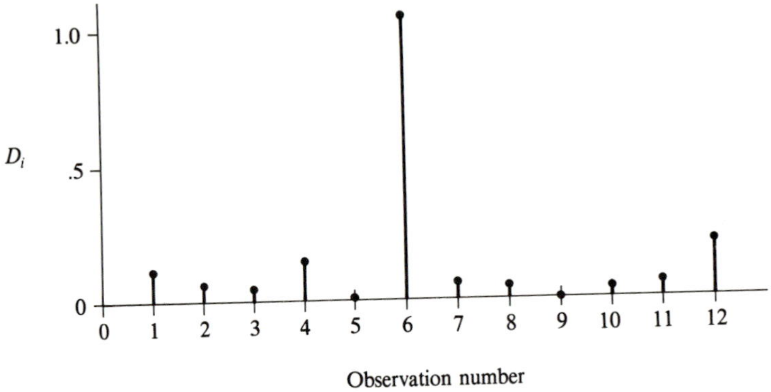

Figure 10.15 Cook's D_i plotted against observation number from the regression of seed weight on ozone level and cumulative solar radiation using the Lesser–Unsworth data.

closely related Cook's D_i are plotted against observation number in Figure 10.15. The most influential point on $\hat{\boldsymbol{\beta}}$ is observation 6 with $D_6 = 1.06$. Thus, deleting observation 6 from the analysis causes $\hat{\boldsymbol{\beta}}$ to shift beyond the .50 confidence ellipsoid of $\boldsymbol{\beta}$, $F_{(.50;3,9)} = .852$. The cutoff point translated from DFFITS to Cook's D is $4/n = .33$; only D_6 exceeds this number.

The impact of each observation on the estimate of β_0 and β_1, where β_1 is the regression of seed weight on total solar radiation, is shown in the plots of DFBETAS_0 and DFBETAS_1 in Figure 10.16. The suggested cutoff point for DFBETAS_j is $2/\sqrt{n} = .58$ in this example. None of the observations exceeds the cutoff point for DFBETAS_0 [Figure 10.16(a)] and only observation 6 exceeds the cutoff for DFBETAS_1 [Figure 10.16(b)]. This illustrates a case where an observation has major impact on the regression (D_6, DFFITS_6, and $\text{DFBETAS}_{1(6)}$ are large) but has very little effect on the estimation of one parameter, in this case β_0.

The suggested cutoff values for COVRATIO_i in the Lesser–Unsworth example are $1 \pm 3p'/n = (.25, 1.75)$. Observations 1, 5, and 7 exceed the upper cutoff point (values not shown), indicating that the presence of these three observations has the greatest impact on *increasing* the precision of the parameter estimates. $\text{COVRATIO}_6 = .07$ is the only one that falls below the lower limit. This indicates that the presence of observation 6 greatly *decreases* the precision of the estimates; the large residual from this observation will cause s^2 to be much larger than $s^2_{(6)}$.

The influence diagnostics on the Lesser–Unsworth example flag observation 6 as a serious problem in this analysis. This could be due to observation 6 being in error in some sense or the model not adequately representing the relationship between seed weight, solar radiation, and ozone exposure. The seed weight and radiation values for observation 6 were both the largest in the sample. There is no obvious error in either. The most logical explanation of the impact of this

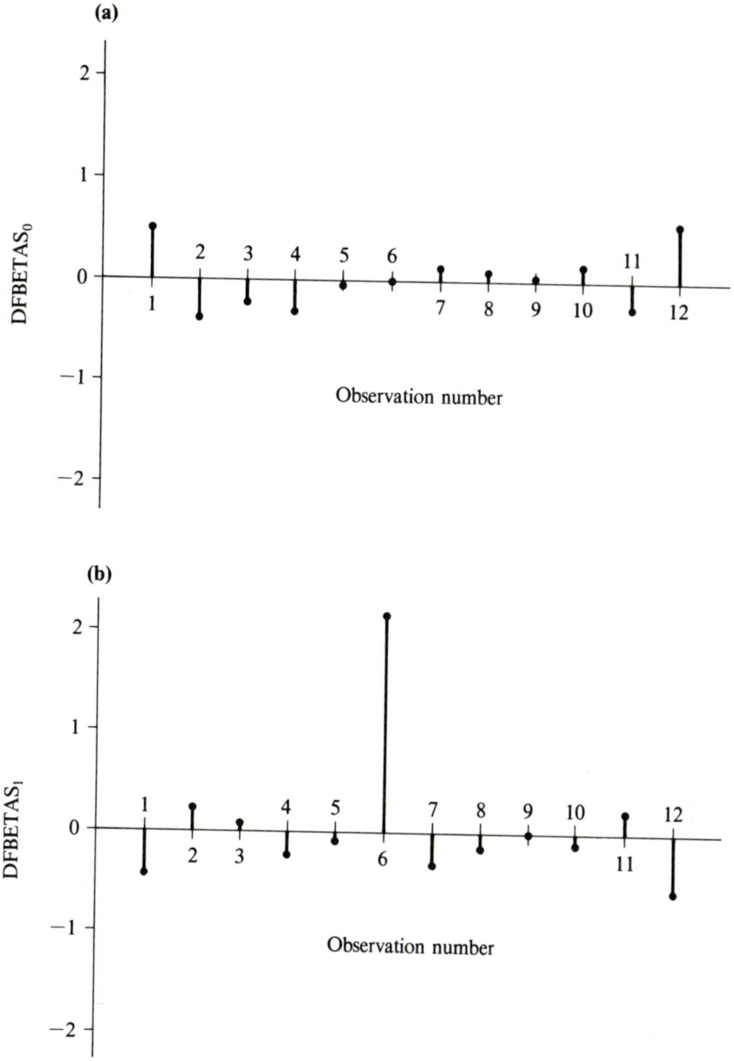

Figure 10.16 DFBETAS$_{0(i)}$ and DFBETAS$_{1(i)}$ plotted against observation number from the regression of seed weight on ozone level and cumulative solar radiation using the Lesser–Unsworth data.

observation is that the linear model does not adequately represent the relationship for these extreme values. ■

10.3 Collinearity Diagnostics

Effects of Collinearity

The collinearity problem in regression refers to the problems created when there are near-singularities among the columns of the $\boldsymbol{X}$ matrix; certain linear combinations of the columns of $\boldsymbol{X}$ are nearly zero. This implies that there are (near) redundancies among the independent variables; essentially the same information is being provided in more than one way. Geometrically, collinearity results when at least one dimension of the X-space is very poorly defined in the sense that there is almost no dispersion among the data points in that dimension.

Limited dispersion in an independent variable results in a very poor (high-variance) estimate of the regression coefficient for that variable. This can be viewed as a result of the near-collinearity between the variable and the column of ones (for the intercept) in $\boldsymbol{X}$. (A variable that has very little dispersion relative to its mean is very nearly a multiple of the vector of ones.) This is an example of collinearity that is easy to detect by simple inspection of the amount of dispersion in the individual independent variables. The more usual, and more difficult to detect, collinearity problem arises when the near-singularity involves several independent variables. The dimension of the X-space in which there is very little dispersion is some linear combination of the independent variables, and may not be detectable from inspection of the dispersion of the individual independent variables.

The result of collinearity involving several variables is high variance for the regression coefficients of all variables involved in the near-singularity. In addition, and perhaps more important, it becomes virtually impossible to separate the influences of the independent variables and very easy to pick points for prediction that are (unknowingly) outside the sample X-space, representing extrapolations.

Detecting Collinearity with Eigenanalysis

The presence of collinearity is detected with the singular value decomposition of $\boldsymbol{X}$ or the eigenanalysis of $\boldsymbol{X}'\boldsymbol{X}$ (Sections 2.7 and 2.8). The eigenvalues, λ_i^2, provide measures of the amount of dispersion in the dimensions corresponding to the principal component axes of the X-space. The elements in the eigenvectors are the coefficients (for the independent variables) defining the principal component axes. All principal components are pairwise orthogonal.

The first principal component axis is defined so as to identify the direction through the X-space that has the maximum dispersion. The second principal component axis identifies the dimension orthogonal to the first that has the second most variation and so forth until the last principal component axis identifies the dimension with the least dispersion. The relative sizes of the eigenvalues reveal the relative amounts of dispersion in the different dimensions of the X-space, and the eigenvectors identify the linear combinations of the independent variables that define those dimensions. The smaller eigenvalues, and their eigenvectors, are of particular interest for the collinearity diagnostics.

Standardizing X

The eigenanalysis for purposes of detecting collinearity typically is done on $\boldsymbol{X}'\boldsymbol{X}$ *after* $\boldsymbol{X}$ has been scaled so that the length of each vector, the sum of squares of each column, is 1. Thus, $\text{tr}(\boldsymbol{X}'\boldsymbol{X}) = p'$. This standardization is necessary to prevent the eigenanalysis from being dominated by one or two of the independent variables. The sum of the eigenvalues equals the trace of the matrix being analyzed, $\sum \lambda_k^2 = \text{tr}(\boldsymbol{X}'\boldsymbol{X})$, which is the sum of the sums of squares of the independent variables including X_0. The independent variables in their original units of measure would contribute unequally to this total sum of squares and, hence, to the eigenvalues. A simple change of scale of a variable, such as from inches to centimeters, would change the contribution of the variable to the principal components if the vectors were not rescaled to have equal length.

The standardization of $\boldsymbol{X}$ is accomplished by dividing the elements of each column vector by the square root of the sum of squares of the elements. In matrix form, define a diagonal $(p' \times p')$ matrix, $\boldsymbol{STD}$, which consists of square roots of the diagonal elements of $\boldsymbol{X}'\boldsymbol{X}$. The standardized $\boldsymbol{X}$ matrix, $\boldsymbol{Z}$, is given by

$$\boldsymbol{Z} = \boldsymbol{X}(\boldsymbol{STD}^{-1}) \qquad [10.21]$$

The eigenanalysis is done on $\boldsymbol{Z}'\boldsymbol{Z}$.

Centering the Independent Variables

Some authors argue that the independent variables should first be centered by subtracting the mean of each independent variable. This centering makes all independent variables orthogonal to the intercept column and, hence, removes any collinearity that involves the intercept. Marquardt (1980) calls this the "nonessential collinearity." Any independent variable that has a very small coefficient of variation, small dispersion relative to its mean, will be highly collinear with the intercept and yet, when centered, be orthogonal to the intercept. Belsley, Kuh, and Welsch (1980) and Belsley (1984) argue that this correction for the mean is part of the multiple regression arithmetic and should be taken into account when assessing the collinearity problem. For further discussion on this topic, the reader is referred to Belsley (1984) and the discussions following his article by Cook (1984), Gunst (1984), Snee and Marquardt (1984), and Wood (1984).

"Nonessential" Collinearity

The seriousness of collinearity and whether it is "nonessential" collinearity depend on the specific objectives of the regression. Even under severe collinearity, certain linear functions of the parameters may be estimated with adequate precision. For example, the estimate of the *change* in Y between two points in X may be very precisely estimated even though the estimates of some of the parameters are highly variable. If these linear functions also happen to be the quantities of primary interest, the collinearity might be termed "nonessential." However, any collinearity, including collinearity with the intercept, that destroys the stability of the quantities of interest cannot be so termed.

This discussion of collinearity diagnostics will assume that the noncentered independent variables (scaled to have unit vector length) are being used. The diagnostics from the centered data can be used when they are more relevant for the problem. In any specific case, it is best to look at the seriousness of the collinearity in terms of the objectives of the study.

10.3.1 Condition Number and Condition Index

Condition Number

The **condition number**, $K(\boldsymbol{X})$, of a matrix $\boldsymbol{X}$ is defined as the ratio of the largest singular value to the smallest singular value (Belsley, Kuh, and Welsch, 1980),

$$K(\boldsymbol{X}) = \frac{\lambda_{\max}}{\lambda_{\min}} \qquad [10.22]$$

The condition number provides a measure of the sensitivity of the solution to the normal equations to small changes in X or Y. A large condition number indicates that a near-singularity is causing the matrix to be poorly conditioned. For reference, the condition number of a matrix is 1 when all the columns are pairwise orthogonal and scaled to have unit length; all λ_k are equal to 1.

Condition Index

The condition number concept is extended to provide the **condition index** for each (principal component) dimension of the X-space. The condition index, δ_k, for the kth principal component dimension of the X-space is

$$\delta_k = \frac{\lambda_{\max}}{\lambda_k} \qquad [10.23]$$

The largest condition index is also the condition number, $K(X)$, of the matrix. Thus, condition indices identify the dimensions of the X-space where dispersion is limited enough to cause problems with the least squares solution.

Interpretation

Belsley, Kuh, and Welsch (1980) suggest that condition indices around 10 indicate weak dependencies that may be starting to affect the regression estimates. Condition indices of 30 to 100 indicate moderate to strong dependencies, and indices larger than 100 indicate serious collinearity problems. The number of condition indices in the critical range indicates the number of near-dependencies contributing to the collinearity problem.

mci

Another measure of collinearity involves the ratios of the squares of the eigenvalues. Thisted (1980) suggested

$$mci = \sum_{j=1}^{p'} \left(\frac{\lambda_{p'}^4}{\lambda_j^4}\right) \qquad [10.24]$$

as a multicollinearity index, where $\lambda_{p'}^2$ is the smallest eigenvalue of $X'X$. Values of mci near 1.0 indicate high collinearity; values greater than 2.0 indicate little or no collinearity.

Example 10.11

A small numerical example is used to illustrate the measures of collinearity. An X matrix, 20×4, consists of the intercept column, and three independent variables constructed in the following way:

X_1 is the sequence of numbers 20 to 29 and repeated.

X_2 is $X_1 - 25$, with the first and 11th observations changed to -4 (from -5) to avoid a complete linear dependency.

X_3 is a periodic sequence running 5, 4, 3, 2, 1, 2, 3, 4, 5, 6 and repeated. X_3 is designed to be nearly orthogonal to the variation in X_1 and X_2.

The singular values and the condition indices using the noncentered, unit-length vectors for this X matrix are given in Table 10.1. The largest condition index $\delta_4 = 1.702410/.006223 = 273.6$ indicates a severe collinearity problem. This is the condition number, $K(X)$, of X as scaled. The condition indices for the other dimensions do not indicate any collinearity problem; they are well below the value

Table 10.1 The singular values and the condition indices for Example 10.11.

Principal Component	*Singular Values*	*Condition Index*
1	1.702410	1.00
2	1.003349	1.70
3	.308303	5.52
4	.006223	273.60

of 10 suggested by Belsley, Kuh, and Welsch as the point at which collinearity may be severe enough to begin having an effect. The multicollinearity index of Thisted is very close to 1 ($mci = 1.061$), which indicates severe collinearity. ■

10.3.2 Variance Inflation Factor

Definition

Another common measure of collinearity is the **variance inflation factor** for the jth regression coefficient, VIF_j. The variance inflation factors are computed from the correlation matrix $\hat{\boldsymbol{\rho}}$ of the independent variables. Thus, the independent variables are centered and standardized to unit length. The diagonal elements of $\hat{\boldsymbol{\rho}}^{-1}$, the inverse of $\hat{\boldsymbol{\rho}}$, are the variance inflation factors. The link between VIF_j and collinearity (of the standardized and centered variables) is through the relationship

$$VIF_j = \frac{1}{1 - R_j^2} \qquad [10.25]$$

where R_j^2 is the coefficient of determination from the regression of X_j on the other independent variables. If there is a near-singularity involving X_j and the other independent variables, R_j^2 will be near 1.0 and VIF_j will be large. If X_j is orthogonal to the other independent variables, R_j^2 will be 0 and VIF_j will be 1.0.

The term *variance inflation factor* comes from the fact that the variance of the jth regression coefficient can be shown to be directly proportional to VIF_j (Theil, 1971; Berk, 1977):

$$s^2(\hat{\beta}_j) = \frac{\sigma^2}{\boldsymbol{x}_j'\boldsymbol{x}_j}(VIF_j) \qquad [10.26]$$

where $\boldsymbol{x}_j$ is the jth *column* of $\boldsymbol{X}$, centered and scaled to have unit length.

Interpretation

The variance inflation factors are simple diagnostics for detecting overall collinearity problems that do not involve the intercept. They will not detect multiple near-singularities nor identify the source of the singularities. The maximum variance inflation factor has been shown to be a lower bound on the condition number (Berk, 1977). Snee and Marquardt (1984) suggest that there is no practical difference between Marquardt's (1970) guideline for serious collinearity, $VIF > 10$, and Belsley, Kuh, and Welsch's (1980) condition number of 30.

Example 10.12 The variance inflation factors computed from the correlation matrix of the independent variables for Example 10.11 are

$$VIF_1 = 169.4$$

$$VIF_2 = 175.7$$

$$VIF_3 = 1.7$$

The variance inflation factors indicate that the estimates of β_1 and β_2 would be seriously affected by the very-near-singularity in $\mathbf{X}$. In this case, the near-singularity is known to be due to the near-redundancy between X_1 and X_2. Notice that the variance inflation factor of $\hat{\beta}_3$ is near 1, the expected result if all variables are orthogonal. The variance inflation factors are computed on the centered and scaled data, and as a result are orthogonal to the intercept column. Thus, the variance inflation factors in this example indicate a collinearity problem that does not involve the intercept. ■

10.3.3 Variance Decomposition Proportions

Partitioning $\mathrm{Var}(\hat{\beta}_j)$

The variance of each estimated regression coefficient can be expressed as a function of the eigenvalues, λ_k^2, of $\mathbf{X}'\mathbf{X}$ and the elements of the eigenvectors. Let u_{jk} be the jth element of the kth eigenvector. Then,

$$\mathrm{Var}(\hat{\beta}_j) = \sigma^2 \sum_k \left(\frac{u_{jk}^2}{\lambda_k^2}\right) \qquad [10.27]$$

The summation is over the $k = 1, \ldots, p'$ principal component dimensions. Thus, the variance of each regression coefficient can be decomposed into the contributions from each of the principal components. The size of each contribution (for the variance of the jth regression coefficient) is determined by the square of the ratio of the jth element from the kth eigenvector, u_{jk}, to the singular value λ_k.

The major contributions to the variance of a regression coefficient occur when the coefficient in the eigenvector is large, in absolute value, and the eigenvalue is small. A large coefficient u_{jk} indicates that the jth independent variable is a major contributor to the kth principal component. The small eigenvalues identify the near-singularities that are the source of the instability in the least squares estimates. Not all regression coefficients need be affected. If the jth variable is not significantly involved in the near-singularity, its coefficient in the kth eigenvector, u_{jk}, will be near zero and its regression coefficient will remain stable even in the presence of the collinearity.

Variance Decomposition Proportions

It is helpful to express each of the contributions as a proportion of the total variance for that particular regression coefficient. These partitions of the variances are called the **variance decomposition proportions**.

Example 10.13 The variance decomposition proportions for the data of Example 10.11 are given in Table 10.2. The entries in any one column show the proportions of the variance for that regression coefficient that come from the principal components indicated on the left. For example, 34% of the variance of $\hat{\beta}_3$ comes from the fourth principal component, 65% from the third, and only slightly over 1% from the first and second principal components. ■

Table 10.2 Variance decomposition proportions for Example 10.11 using all principal components (upper half of table) and with the fourth principal component deleted (lower half).

Principal Component	*Variance Proportion*			
	Intercept	X_1	X_2	X_3
1	.0000[a]	.0000[a]	.0000[a]	.0102
2	.0000[a]	.0000[a]	.0055	.0008
3	.0001	.0001	.0003	.6492
4	.9999	.9998	.9942	.3398
	1.0000	1.0000	1.0000	1.0000
1	.070	.060	.001	.015
2	.002	.002	.942	.001
3	.928	.939	.057	.983
	1.000	1.000	1.000	1.000

[a] Variance proportions are less than 10^{-4}.

Interpretation The critical information in Table 10.2 is how the variances are being affected by the last principal component, the one with the least dispersion and the greatest impact on the collinearity problem. For reference, if the columns of $\boldsymbol{X}$ were orthogonal, the variance decomposition proportions would be all 0 except for a single 1 in each row and column. That is, each principal component would contribute to the variance of only one regression coefficient. Serious collinearity problems are indicated when a principal component with a small eigenvalue contributes heavily—more than 50%—to two or more regression coefficients.

Example 10.14 (*Continuation of Example* 10.13) The fourth principal component is responsible for over 99% of $s^2(\hat{\beta}_0)$, $s^2(\hat{\beta}_1)$, and $s^2(\hat{\beta}_2)$. The fourth principal component had a condition index of $\delta_4 = 274$, well above the critical point for severe collinearity. The fourth principal component identifies a nearly singular dimension of the X-space that is causing severe variance inflation of these three regression coefficients. Notice, however, that the variance of $\hat{\beta}_3$ is not seriously affected by this near-singularity. This implies that X_3 is not a major component of the near-singularity defined by the fourth principal component. ■

The interpretation of the variance decomposition proportions requires two conditions for the result to be an indication of serious collinearity:

1. The condition index for the principal component must be "large."
2. The variance decomposition proportions must show that the principal component is a major contributor (>50%) to at least two regression coefficients.

Multiple Near-Singularities

More than one near-singularity may be causing variance inflation problems. In such a situation, the variance decomposition table will be dominated by the principal component with the smallest eigenvalue so that the effect of other near-singularities may not be apparent. The variance contributions of the other principal components should then be found by rescaling each column so that the proportions add to 100% *without* the dominating principal component. This approximates what would happen to the variance proportions if the dominating principal component were "removed."

Example 10.15

Since the condition index ($\delta_3 = 5.5$) for the third principal component from Example 10.14 is not in the critical range, the analysis of the variance proportions normally would not proceed any further. However, to illustrate the process, we give in the lower portion of Table 10.2 the variance decomposition proportions for the example without the fourth principal component. If the condition index for the third principal component had been sufficiently high, this result would be suggesting that this dimension also was causing variance inflation problems. ■

Linear Functions

The variance decomposition proportions provide useful information when the primary interest is in the regression coefficients per se. When the primary objective of the regression analysis is the use of the estimated regression coefficients in some linear function, such as in a prediction equation, it is more relevant to measure the contributions of the principal components to the variance of the linear function of interest. Let $c = \boldsymbol{K}'\hat{\boldsymbol{\beta}}$ be the linear function of interest. The variance of c is

$$\sigma^2(c) = \boldsymbol{K}'(\boldsymbol{X}'\boldsymbol{X})^{-1}\boldsymbol{K}\sigma^2 \qquad [10.28]$$

which can be decomposed into the contributions from each of the principal components as

$$\sigma^2(c) = \sum_k \left(\frac{(\boldsymbol{K}'\boldsymbol{u}_k)^2}{\lambda_k^2}\right)\sigma^2 \qquad [10.29]$$

Each term reflects the contribution to the variance of the corresponding principal component.

Example 10.16

(*Continuation of Example* 10.15) Suppose the linear function of interest is $c = \boldsymbol{K}'\hat{\boldsymbol{\beta}}$, where

$$\boldsymbol{K}' = (1 \quad 25 \quad 0 \quad 3)$$

Table 10.3 The variance partitions and the variance proportions for the linear function $\boldsymbol{K}'\boldsymbol{\beta}$, where $\boldsymbol{K}' = (1 \;\; 25 \;\; 0 \;\; 3)$.

Principal Component		*Variance Partition*	*Variance Proportion*
1		.0451	.7542
2		.0003	.0050
3		.0142	.2375
4		.0002	.0033
	Total	.0597	1.0000

Then the variance of $c = \boldsymbol{K}'\hat{\boldsymbol{\beta}}$ is $\sigma^2(c) = .0597\sigma^2$. The partitions of this variance into the contributions from the four principal components and the variance proportions are given in Table 10.3. For this (deliberately chosen) linear function, the fourth principal component, which was causing the collinearity problem and the severe variance inflation of the regression coefficients, is having almost no impact. Thus, if this particular linear function were the primary objective of the analysis, the near-singularity identified by the fourth principal component could be termed a "nonessential collinearity." This can be viewed as a generalization of the concept of "nonessential ill-conditioning" used by Marquardt (1980) to refer to near-singularities involving the intercept. ■

10.4 Regression Diagnostics on the Linthurst Data

Example 10.17 The Linthurst data were used in Chapter 5 to illustrate the choice of variables in a model building process. In that exercise, the modeling started with five independent variables, *SALINITY*, *pH*, *K*, *Na*, and *Zn*, and ended with a model that contained two variables. The usual assumptions of ordinary least squares were made and all of the variables were assumed to be related linearly to the dependent variable *BIOMASS*. In this section, the regression diagnostics are presented for the Linthurst data for the five-variable regression model.

The residuals, e_i, standardized residuals, r_i (called STUDENT residual in PROC REG), and Cook's D for the regression of *BIOMASS* on the five independent variables were obtained from the RESIDUAL option in PROC REG (SAS Institute, Inc., 1985d, 1985f) and are given in Table 10.4 (page 282). The Studentized residuals, r_i^* (called RSTUDENT in PROC REG), and several influence statistics were obtained from the INFLUENCE option and are given in Table 10.5 (page 283).

Residuals The standardized residual for observation 34 is the largest, with a value of $r_{34} = 2.834$; this residual is 2.834 standard deviations away from zero. When expressed as the Studentized residual, its value is $r_{34}^* = 3.14$. Four other Studentized residuals are greater than 2.0 in absolute value. This frequency of large residuals (11%) is higher than might be expected from a sample size of 45. An approximate chi-square test, however, does not show a significant departure

Table 10.4 Residuals analysis from the regression of *BIOMASS* on the five independent variables *SAL, pH, K, Na,* and *Zn* (from SAS PROC REG, option R).

Obs.	Y_i	$\hat{Y}_i$	$s(\hat{Y}_i)$	e_i	$s(e_i)$	r_i	*Cook's D*
1	676	724	176	−48	357	−.135	.001
2	516	740	142	−224	372	−.601	.009
3	1,052	691	127	361	378	.956	.017
4	868	815	114	53	382	.140	.000
5	1,008	1,063	321	−55	235	−.236	.017
6	436	958	126	−522	378	−1.381	.035
7	544	527	214	17	336	.050	.000
8	680	827	141	−147	373	−.394	.004
9	640	676	174	−36	358	−.101	.000
10	492	911	165	−419	362	−1.155	.046
11	984	1,166	167	−182	362	−.503	.009
12	1,400	573	147	827	370	2.232	.130*
13	1,276	816	153	460	368	1.252	.045
14	1,736	953	137	783	374	2.093	.099*
15	1,004	898	166	106	362	.293	.003
16	396	355	135	41	375	.109	.000
17	352	577	127	−225	377	−.595	.007
18	328	586	139	−258	373	−.691	.011
19	392	586	118	−194	380	−.511	.004
20	236	494	131	−258	376	−.687	.010
21	392	596	122	−204	379	−.537	.005
22	268	570	120	−302	380	−.795	.010
23	252	584	124	−332	378	−.877	.014
24	236	479	100	−243	386	−.631	.004
25	340	425	131	−85	376	−.227	.001
26	2,436	2,296	170	140	360	.388	.006
27	2,216	2,202	196	14	347	.040	.000
28	2,096	2,230	187	−134	351	−.381	.007
29	1,660	2,408	171	−748	360	−2.080	.163*
30	2,272	2,369	168	−97	361	−.270	.003
31	824	1,110	115	−286	381	−.750	.008
32	1,196	982	118	214	381	.562	.005
33	1,960	1,155	120	805	380	2.120	.075
34	2,080	1,008	124	1,072	378	2.834	.145*
35	1,764	1,254	136	510	374	1.363	.041
36	412	959	111	−547	383	−1.431	.029
37	416	626	133	−210	376	−.558	.006
38	504	624	107	−120	384	−.313	.001
39	492	588	99	−96	386	−.250	.001
40	636	838	95	−202	387	−.521	.003
41	1,756	1,526	129	230	377	.610	.007
42	1,232	1,298	97	−66	386	−.171	.000
43	1,400	1,401	106	−1	384	−.004	.000
44	1,620	1,306	113	314	382	.822	.010
45	1,560	1,265	90	295	388	.759	.005

Table 10.5 Influence statistics from the regression of *BIOMASS* on the five independent variables *SAL, pH, K, Na,* and *Zn* (from SAS PROC REG, option INFLUENCE).

						DFBETAS					
Obs.	e_i	r_i^*	v_{ii}	COVRATIO	DFFITS	*Intercept*	*SAL*	*pH*	*K*	*Na*	*Zn*
1	−48	−.133	.195	1.447*	−.065	.010	−.004	−.004	−.002	−.032	.001
2	−224	−.596	.127	1.266	−.228	.074	−.086	−.014	−.081	−.016	−.007
3	361	.955	.101	1.128	.321	.123	−.094	−.166	−.005	.152	−.171
4	53	.138	.082	1.269	.041	.020	−.020	−.019	−.010	.027	−.021
5	−55	−.233	.651*	3.318*	−.318	.065	−.030	−.108	.245	−.244	−.083
6	−522	−1.398	.100	.961	−.466	.054	−.069	.022	−.220	.007	.078
7	17	.050	.289*	1.642*	.032	−.019	.022	.009	.026	−.021	.013
8	−147	−.390	.125	1.304	−.147	−.075	.069	.081	−.030	−.041	.091
9	−36	−.100	.191	1.443*	−.049	.029	−.034	−.014	−.017	.004	−.014
10	−419	−1.160	.172	1.146	−.529	−.310*	.285	.317*	−.068	−.177	.378*
11	−182	−.498	.175	1.362	−.229	−.174	.116	.172	.004	.022	.180
12	827	2.359	.135	.595*	.934*	−.151	.442*	−.150	−.293	.092	.020
13	460	1.261	.148	1.073	.526	.307*	−.126	−.398*	−.052	−.023	−.351*
14	783	2.193	.119	.649	.806*	.133	.165	−.346*	−.041	−.090	−.331*
15	106	.289	.173	1.395	.132	.107	−.076	−.104	−.062	.042	−.098
16	41	.107	.115	1.317	.039	−.014	.013	.011	−.011	.005	.024
17	−225	−.590	.102	1.232	−.199	−.020	.027	.000	.081	−.028	−.061
18	−258	−.687	.121	1.236	−.255	.013	−.032	−.010	.084	.024	−.093
19	−194	−.506	.088	1.230	−.157	.008	−.056	.036	.041	.006	−.007
20	−258	−.682	.108	1.218	−.238	.043	−.118	.046	.039	.006	−.014
21	−204	−.532	.094	1.234	−.172	−.100	.070	.104	.106	−.084	.069
22	−302	−.791	.090	1.165	−.249	−.022	.012	.017	.074	.008	−.069
23	−332	−.874	.097	1.149	−.287	.010	−.075	.054	−.069	.163	−.044
24	−243	−.626	.063	1.173	−.162	.011	−.043	.030	−.014	.050	−.037
25	−85	−.224	.108	1.300	−.078	.041	−.057	−.012	−.007	.022	−.037
26	140	.384	.181	1.395	.181	.074	−.074	−.006	−.047	.025	−.091
27	14	.039	.243	1.543*	.022	−.011	.012	.013	.005	−.010	.006
28	−134	−.376	.222	1.468*	−.201	.090	−.094	−.118	.011	.037	−.042
29	−748	−2.177	.184	.709	−1.034*	−.130	.154	−.250	.235	−.010	.247
30	−97	−.267	.178	1.406*	−.124	−.023	.026	−.024	.033	−.012	.038
31	−286	−.745	.083	1.168	−.224	−.141	.174	.069	.052	−.108	.097
32	214	.557	.087	1.219	.172	−.066	.060	.059	.126	−.139	.078
33	805	2.225	.091	.617	.704	−.044	−.179	.291	.027	.048	.249
34	1,072	3.140	.098	.325*	1.032*	.584*	−.752*	−.309*	−.183	.533*	−.406*
35	510	1.379	.117	.988	.502	−.125	.041	.213	.307*	−.341*	.211
36	−547	−1.451	.078	.917	−.421	−.119	.206	.015	−.114	.039	−.002
37	−210	−.553	.111	1.253	−.196	.060	−.023	−.069	−.079	.076	−.119
38	−120	−.309	.072	1.241	−.086	−.026	.035	.020	−.023	.011	−.002
39	−96	−.247	.062	1.235	−.064	−.001	.009	−.001	−.015	.009	−.020
40	−202	−.516	.057	1.188	−.127	−.059	.065	.047	−.043	.018	.033
41	230	.605	.106	1.233	.208	.033	−.081	.058	.017	−.044	.026
42	−66	−.168	.060	1.237	−.043	.010	.001	−.024	−.004	.009	−.020
43	−1	−.004	.070	1.257	−.001	.000	.000	−.001	−.000	.000	−.000
44	314	.819	.081	1.144	.242	−.127	.075	.180	.080	−.105	.159
45	295	.755	.051	1.127	.176	−.056	.013	.109	.025	−.024	.083

from an expected 5% frequency of residuals greater than 2.0 in absolute value. (This test has an additional approximation compared with the conventional "goodness-of-fit" test because the residuals are not independent.)

These large residuals must not be interpreted, however, as indicating that these points are in error or that they do not belong to the population sampled. Of course, the data should be carefully checked to verify that there are no errors and that the points represent legitimate observations. But as a general rule, outlier points should not be dropped from the data set unless they are found to be in error and the error cannot be corrected. An excessively high frequency of large residuals on a carefully edited data set is probably an indication of an inadequate model. The model and the system being modeled should be studied carefully. Perhaps an important independent variable has been overlooked or the relationships are not linear as has been assumed.

10.4.1 Plots of Residuals

***e* Versus $\hat{Y}$**

The plot of the ordinary least squares residuals against the predicted values, Figure 10.17(a), shows the presence of five predicted values that are greater than 2,000, much larger than any of the others. Four of the five residuals associated with these points are not particularly notable, but the fifth point is the largest negative residual, -748 or a standardized residual of $r_{29} = -2.080$. A second feature of interest in Figure 10.17(a) is the apparently greater spread among the positive residuals than among the negative residuals. This suggests that the distribution of the residuals might be skewed. The skewness is seen more clearly in a frequency polygon of the residuals, Figure 10.18 (page 286). There are four residuals greater than 2.0 but only one less than -2.0, and there is a high frequency of relatively small negative residuals.

Normal Probability Plot

The normal probability plot of the standardized residuals, Figure 10.19, shows a distinct curvature, rather than the straight line expected of normally distributed data. The shape of this normal plot, except for the additional bend caused by the four most negative residuals, is consistent with the positively skewed distribution suggested by the frequency polygon. The most negative residual, $r_{29} = -2.080$, is sufficiently larger in magnitude than the other negative residuals to raise the possibility that it might be an outlier. (An extreme standardized residual of -2.080 is not large for a normal distribution but seems large in view of the positive skewness and the fact that the next largest negative residual is -1.431. The overall behavior of the residuals suggests that they may not be normally distributed. A transformation of the dependent variable might improve the symmetry of the distribution.

Standardized Residuals Versus $\hat{Y}$

The values for the dependent variable, *BIOMASS*, cover a wide range from 236 to 2,436 (Table 10.4). In such cases it is not uncommon for the variance of the dependent variable to increase with the mean of the dependent variable. The plot of the standardized residuals against $\hat{Y}_i$ does not suggest any increase in dispersion for the larger $\hat{Y}_i$. The five random samples taken at each of the nine sites, however, provide independent estimates of variation for *BIOMASS*. These "within-sampling-site" variances are not direct estimates of σ^2 because the five

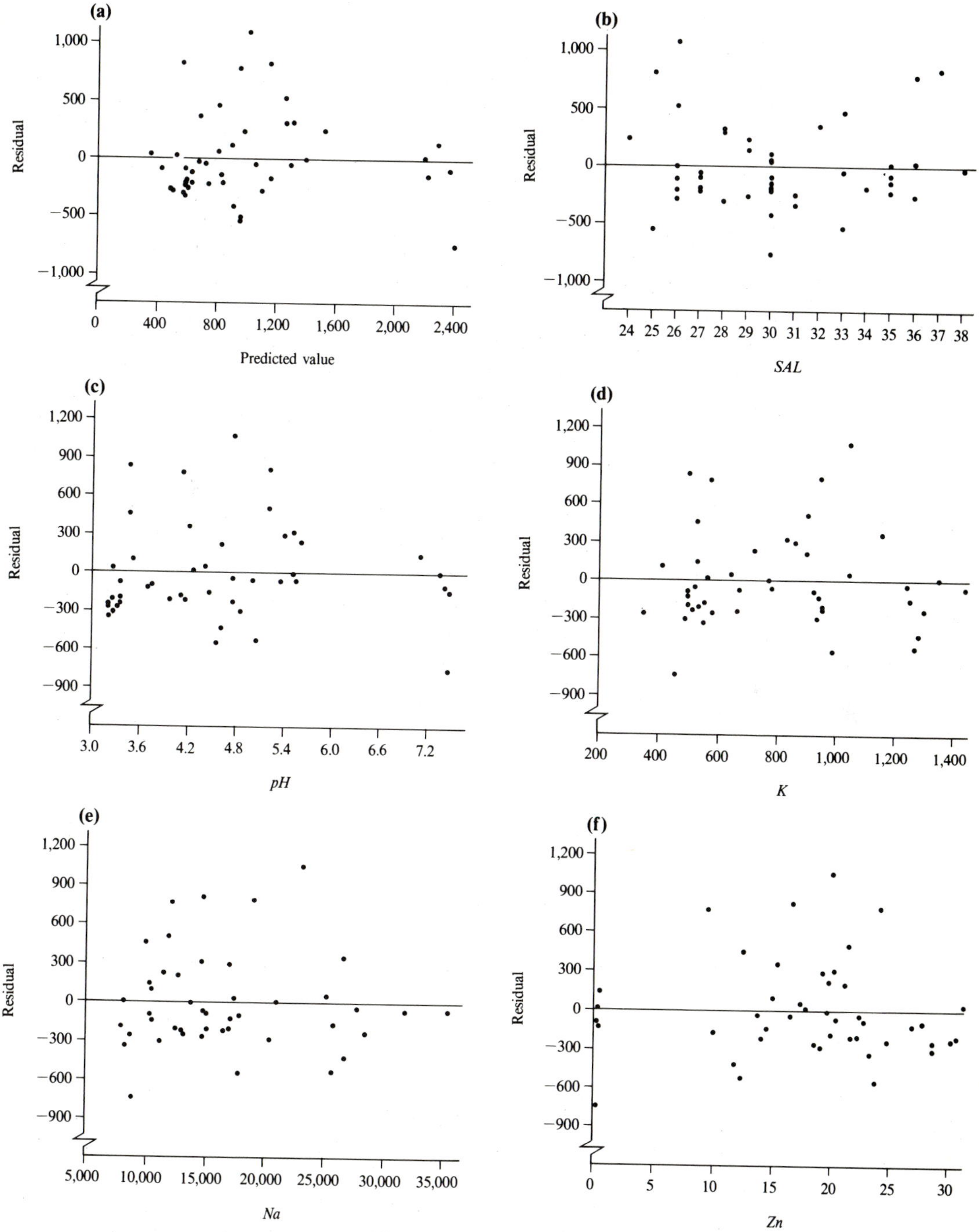

Figure 10.17 Least squares residuals plotted against the predicted values (a) and each of the five independent variables [(b)–(f)] for the Linthurst September data.

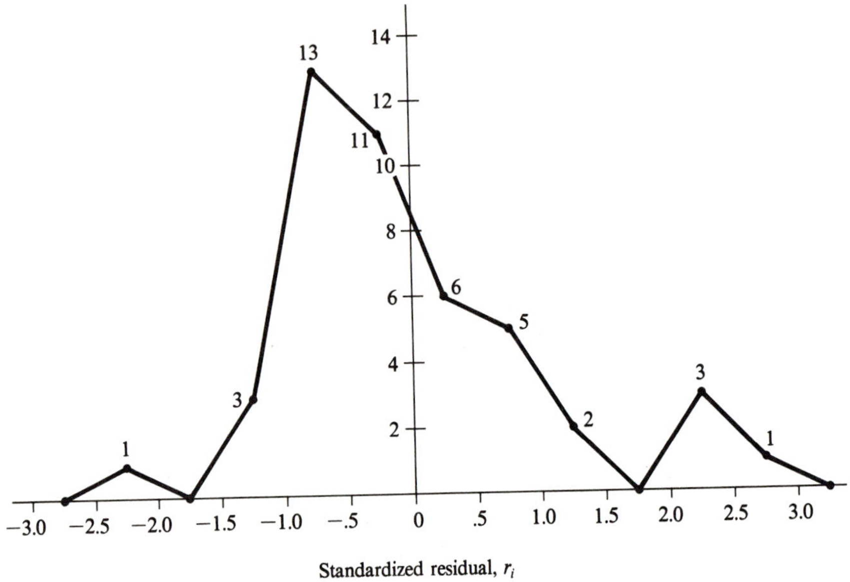

Figure 10.18 Frequency polygon of the standardized residuals from the regression of *BIOMASS* on the five independent variables *SALINITY*, *pH*, *K*, *Na*, and *Zn*, for the Linthurst September data.

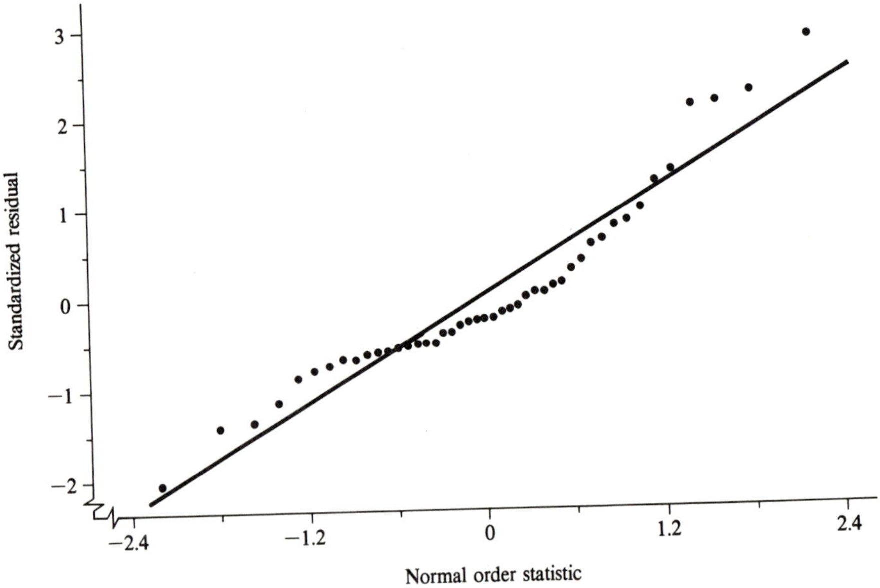

Figure 10.19 Normal plot of the standardized residuals from the regression of *BIOMASS* on the five independent variables for the Linthurst September data.

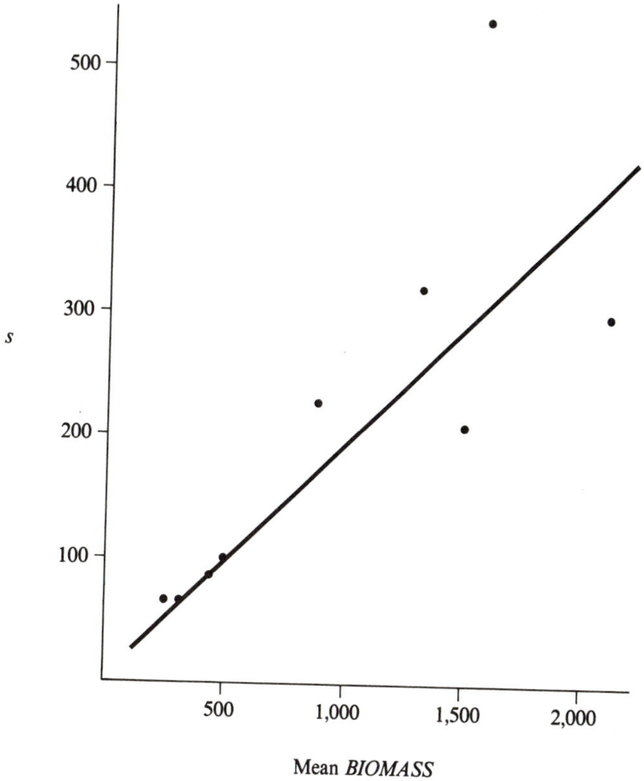

Figure 10.20 The standard deviation among observations within sites plotted against the mean *BIOMASS* from the five observations at each site for the Linthurst September data.

samples at each site are not true replicates; the values of the independent variables are not the same in all samples. They do provide, however, a measure of the differences in variance at very different levels of *BIOMASS*.

Standard Deviation Versus Mean

The plot of the standard deviation from each site versus the mean *BIOMASS* at each site, Figure 10.20, suggests that the standard deviation increases at a rate approximately proportional to the mean. As will be seen in Chapter 11, this suggests the logarithmic transformation of the dependent variable to stabilize the variance. The logarithmic transformation would also reduce the positive skewness noticed earlier. Continued analysis of these data would entail a transformation of *BIOMASS* to ln(*BIOMASS*), or some other similar transformation, and perhaps a change in the model as a result of the transformation. For the present purpose, however, the analysis will be continued on the original scale.

Residuals Versus X_j

Inspection of the remaining plots in Figure 10.17—the residuals versus the independent variables—provides only one suggestion that the relationship of *BIOMASS* with the independent variable is other than linear. The residuals plot for *SALINITY*, Figure 10.17(b), suggests a slight curvilinear relationship between *BIOMASS* and *SALINITY*. A quadratic term for *SALINITY* in the model might be helpful. The five extreme points noticed in Figure 10.17(a) appear again as high values for *pH*, Figure 10.17(c), and as low values for *Zn*, Figure 10.17(f). These points are the five points from one sampling site, observations 26–30, and they are clearly having a major impact on the regression results. This site had very high *BIOMASS*, high *pH*, and low *Zn*.

Partial Regression Leverage Plots

The effects of the other independent variables may obscure relationships in plots of the residuals against any one independent variable. The partial regression leverage plots, Figure 10.21, are intended to avoid this problem. Each partial regression leverage plot shows the relationship between the dependent variable and one of the independent variables (including the intercept as an independent variable) after both have been adjusted for the effects of the other independent variables. The partial regression coefficient for the independent variable is shown by the slope of the relationship in the partial residuals plots, and any highly influential points will stand out as points around the periphery of the plot. Some of the critical observations in the plots have been labeled with their observation numbers for easier reference.

The following points are notable from the partial regression leverage plots:

1. The partial plot for *SALINITY* does not reveal the curvilinear relationship suggested by Figure 10.17(b); that pattern must have been an artifact of the effects of the other variables.
2. Observation 34 repeatedly has a large positive residual for *BIOMASS* and may be having a marked influence on several regression coefficients. This is also the observation with the largest standardized residual, $r_{34} = 2.834$. It is important that the data for this point be verified.
3. The partial plots for *K*, Figure 10.21(d), and *Na*, Figure 10.21(e), show that points 5 and 7 are almost totally responsible for any significant relationship between *BIOMASS* and *K* and between *BIOMASS* and *Na*. Without these two points in the data set, there would be no obvious relationship in either case.
4. Several other data points repeatedly occur on the periphery of the plots but not in such extreme positions. Point 29, the observation with the largest negative residual, always has a small partial residual for the independent variable. That is, point 29 never deviates far from the zero mean for each independent variable after adjustment for the other variables. It is therefore unlikely that this observation has any great impact on any of the partial regression coefficients in this model. Nevertheless, it would be wise to recheck the data for this observation also.
5. One of the inadequacies of the influence statistics for detecting influential observations is illustrated with the points 5, 27, and 28 in the partial plot for *pH*, Figure 10.21(c). These three points have the largest partial residuals for *pH* and would appear to have a major impact on the regression coefficient for *pH*. (Visualize what the slope of the regression would be if all three points were missing.) However, dropping only one of the three points may not appreciably affect the slope because the other

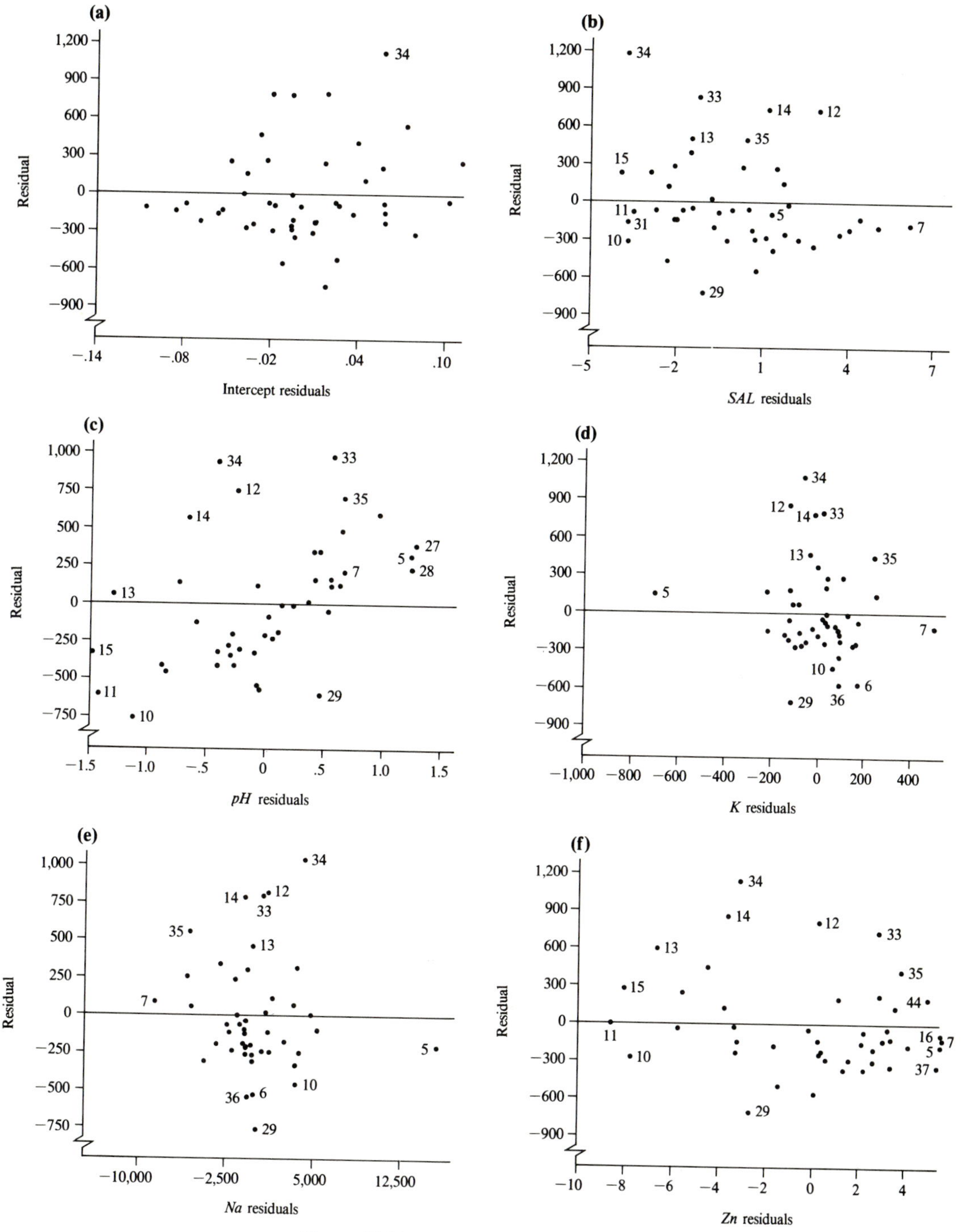

Figure 10.21 The partial regression leverage plots from the regression of *BIOMASS* on the intercept and five independent variables for the Linthurst data. Numbers associated with specific points refer to observation number.

two points are still "pulling" the line in the same direction. This illustrates that the simple influence statistics, where only one observation is dropped at a time, may not detect influential observations when several points are having similar influence. The partial residuals plots will show these jointly influential points.

Except for pH, the partial plots do not show any relationship between Y and the independent variable. This is consistent with the regression results using these five variables; only pH had a partial regression coefficient significantly different from zero (Table 5.2, page 138). In the all-possible-regressions analysis (Table 7.1, page 174), K and Na were about equally effective as the second variable in a two-variable model. The failure to see any association between Y and either of these two variables in the partial plots results from the collinearity in these data. (The collinearity will be shown in Section 10.4.3.) Collinearity among the independent variables will tend to obscure regression relationships in the partial plots.

10.4.2 Influence Statistics

The influence statistics have been presented in Table 10.4 (Cook's D) and Table 10.5. The reference values for the influence statistics for this example, $p' = 6$ and $n = 45$, are as follows:

v_{ii}, elements of $\boldsymbol{P}$ (called HAT DIAG in PROC REG): Average value is $p'/n = 6/45 = .133$. A point is potentially influential if $v_{ii} \geq 2p'/n = .267$.

Cook's D: Cutoff value for Cook's D is $4/n = 4/45 = .09$ if the relationship to DFFITS is used.

DFFITS: Absolute values greater than $2\sqrt{p'/n} = 2\sqrt{6/45} = .73$ indicate influence on $\hat{Y}_i$.

DFBETAS$_j$: Absolute values greater than $2/\sqrt{n} = .298$ indicate influence on $\hat{\beta}_j$.

COVRATIO: Values outside the interval $1 \pm 3p'/n = (.6, 1.4)$ indicate a major effect on the generalized variance.

The points that exceed these limits are marked with an asterisk in Tables 10.4 (Cook's D) and 10.5. Nine observations appear potentially influential, based on values of v_{ii}, or influential by Cook's D, DFFITS, or one or more of the DFBETAS$_j$; COVRATIO is ignored for the moment. These nine points are summarized in Table 10.6.

v_{ii} The influence statistics must be studied in conjunction with the partial regression leverage plots, Figure 10.21. The plots will give insight into why certain observations are influential and others are not. The diagonal elements of $\boldsymbol{P}$, the v_{ii}, reflect the relative distance an observation is from the centroid of the sample X-space and, hence, that point's potential for influencing the regression results. Two observations, 5 and 7, are flagged by v_{ii}. It is difficult to detect from inspection of the data, Table 5.1, that these two points are the most "distant." Although both have values near the extremes for one or more of the variables,

Table 10.6 Nine observations showing potential influence (v_{ii}) or influence in the Linthurst data. The asterisk in the column indicates that the measure exceeded its critical value.

				DFBETAS					
Obs.	v_{ii}	*Cook's D*	DFFITS	*Intercept*	*SAL*	*pH*	*K*	*Na*	*Zn*
5	*								
7	*								
10				*		*			*
12		*	*		*				
13				*		*			*
14		*	*			*			*
29		*	*						
34		*	*	*	*	*		*	*
35							*	*	

neither has the most extreme value for any of the variables. They do appear as extreme points in several of the residuals plots, particularly the plots for *K* and *Na*. Note, however, that neither observation is detected as being influential by any of the measures of influence. This appears to be a contradiction, but the measures of influence show the impact when only that one observation is dropped from the analysis. In the partial plots for *K* and *Na* it is clear that the two observations are operating in concert; eliminating either 5 or 7 has little effect on the regression coefficient because of the influence of the remaining observation. Similarly, several points (16, 7, 5, and 37) are operating together in the partial plot for *Zn* to mask the effect of eliminating one of these points. In other cases, as with point 5 in the partial plot for *SALINITY* or point 7 in the partial plot for *pH*, the potentially influential point is, in fact, not influential for that particular regression coefficient because it is not an extreme point in that dimension.

***Cook's D,* DFFITS, *and* DFBETAS**

Cook's *D* and DFFITS are very similar measures and identify the same four observations as being influential: observations 12, 14, 29, and 34. Dropping any one of these four points causes a relatively large shift in $\hat{\beta}$ or $\hat{Y}$, depending on the interpretation used. They are consistently on the periphery of the partial plots. Point 33 is also on the periphery in all plots but was not flagged by either Cook's *D* or DFFITS. However, its value for both measures is only slightly below the cutoff. Of these four points, only 34 has influence on most of the individual regression coefficients; only DFBETAS for *K* is not flagged. This is consistent with the position of observation 34 in the partial plots.

Finally, there are three observations, 10, 13, and 35, that have been flagged as having influence on one or more regression coefficients but which were not detected by any of the general influence measures, v_{ii}, Cook's *D*, or DFFITS. In these cases, however, the largest $\text{DFBETAS}_{j(i)}$ was .3457, only slightly above the critical value of .298.

COVRATIO

The COVRATIO statistic identifies nine observations as being influential with respect to the variance–covariance matrix of $\hat{\beta}$; all but two of these nine points increase the precision of the estimates. The two points, 12 and 34, whose presence inflates the generalized variance (COVRATIO < 1.0) are two points that were influential for several regression coefficients. These two points have the largest standardized residuals, so when they are eliminated the estimate of σ^2 and the generalized variance decrease. Thus, in this case, the low COVRATIO might be reflecting inadequacies in the model.

Discussion

What is gained from the partial regression leverage plots and the influence measures? They must be viewed as diagnostic techniques, as methods for studying the relationship between the regression equation and the data. These are not tests of significance, and flagging an observation as influential does not imply that the observation is somehow in error. Of course, an error in the data can make an observation very influential and, therefore, careful editing of the data should be standard practice. Detection of a highly influential point suggests that the editing of the data, and perhaps the protocol for collecting the data, be rechecked.

A point may be highly influential because, due to inadequate sampling, it is the only observation representing a particular region of the X-space. Is this the reason points 5 and 7 are so influential? They are the two most "remote" points and are almost totally responsible for the estimates of the regression coefficients for K and Na. More data might "fill in the gaps" in the X-space between these two points and the remaining sample points and, as a result, tend to validate these regression estimates. Alternatively, more data might confirm that these two points are anomalies for the population and, hence, invalidate the present regression estimates. If one is forced to be content with this set of data, it would be prudent to be cautious regarding the importance of K and Na because they are so strongly influenced by these two data points.

The purpose of the diagnostic techniques is to identify weaknesses in the regression model or the data. Remedial measures, correction of errors in the data, elimination of true outliers, collection of better data, or improvement of the model, will allow greater confidence in the final product.

10.4.3 Collinearity Diagnostics

Condition Number and Condition Index

The collinearity diagnostics (Table 10.7) were obtained from the COLLIN option in PROC REG (SAS Institute, Inc., 1985d, 1985f). The collinearity measures are obtained from the eigenanalysis of the standardized $\boldsymbol{X'X}$; the sum of squares for

Table 10.7 Collinearity diagnostics for the regression of *BIOMASS* on the five independent variables *SAL*, *pH*, *K*, *Na*, and *Zn*, Linthurst data (from SAS PROC REG, option COLLIN).

Principal Component Dimension	*Eigenvalues*	*Condition Index*	*Variance Decomposition Proportion*					
			Intercept	*SAL*	*pH*	*K*	*Na*	*Zn*
1	5.57664	1.000	.0001	.0002	.0006	.0012	.0013	.0011
2	.21210	5.128	.0000	.0007	.0265	.0004	.0000	.1313
3	.15262	6.045	.0015	.0032	.0141	.0727	.1096	.0155
4	.03346	12.910	.0006	.0713	.1213	.2731	.2062	.0462
5	.02358	15.380	.0024	.0425	.1655	.5463	.5120	.0497
6	.00160	58.977	.9954	.8822	.6719	.1062	.1709	.7561

each column is unity and the eigenvalues must add to $p' = 6$. The condition number for X is 58.98, an indication of moderate to strong collinearities. The condition indices for the fourth and fifth dimensions are greater than 10, indicating that these two dimensions of the X-space may also be causing some collinearity problems.

Variance Decomposition Proportions

The variance decomposition proportions show that the sixth principal component dimension is accounting for more than 50% of the variance in four of the six regression coefficients. Thus, the intercept, *SALINITY*, *pH*, and *Zn* are the four independent variables primarily responsible for the near-singularity causing the collinearity problem. (The eigenvectors would be required to determine the specific linear function of the X vectors that causes the near-singularity.)

If the sixth principal component dimension is eliminated from consideration and the variance proportions of the remaining dimensions are restandardized to add to 1.0, the variance proportions associated with the fifth principal component dimension account for more than 50% of the remaining variance for four of the six regression coefficients. Similarly, eliminating the fifth principal component dimension leaves the fourth principal component dimension accounting for more than 50% of the variance of four of the six regression coefficients.

Thus, it appears that the last three principal component dimensions may be contributing to instability of the regression coefficients. The course of action to take in the face of this problem will be discussed in Chapter 12. ■

10.A Appendix: Uniform Identically Distributed Residuals

Quesenberry (1986) uses a general theoretical development of tests of goodness of fit which generates the use of recursive residuals. The general approach of Quesenberry uses the concept that for any class of models there is a transformation of the n observations into a set of $n - p^*$ independently and identically distributed uniform random variables on the interval (0, 1) [abbreviated as i.i.d. U(0, 1)], where p^* is the number of parameters to be estimated. These uniform random variables, u_i, contain all of the relevant information for testing the goodness of fit of the model including all assumptions. If the model is not correct or the assumptions are not valid, the u_i will fail to be i.i.d. U(0, 1). Thus, testing goodness of fit of the u_i to the uniform distribution is a general test for the entire model. Further, independently and identically distributed *normal* random variables can be obtained from the u_i by $z_i = \Phi^{-1}(u_i)$, where $\Phi^{-1}(\cdot)$ is the inverse of the N(0, 1) probability distribution function. These can then be tested by any of the tests for normality.

Quesenberry's method reduces to the following in the classical regression model with the usual normality assumptions. Assume there are $p^* = p' + 1$ parameters to be estimated—the p' parameters in $\boldsymbol{\beta}$ plus σ^2. Thus, computation of the recursive residuals starts with a base of $p' + 1$ observations. The index r in the following starts with $p' + 2$.

1. Compute the following least squares results for the base set of $(r - 1)$ observations: $(\boldsymbol{X}'_{r-1}\boldsymbol{X}_{r-1})^{-1}$, $\hat{\boldsymbol{\beta}}_{r-1}$, and $S^2_{r-1} = \text{SS(Res)}$.

2. Compute the recursive residual, labeled w_r, as follows:

$$w_r = \frac{y_r - x_r'\hat{\beta}_{r-1}}{[1 + x_r'(X_{r-1}'X_{r-1})^{-1}x_r]^{1/2}} \qquad \text{[A.1]}$$

3. The rth **Studentized recursive residual** is computed as

$$B_r = \frac{(r - p' - 1)^{1/2}w_r}{S_{r-1}} \qquad \text{[A.2]}$$

4. The **uniform recursive residual** is computed as

$$u_r = G_{r-p'-1}(B_r) \qquad \text{[A.3]}$$

where $G_v(B_r)$ denotes the Student t-distribution function with v degrees of freedom evaluated at B_r.

5. The rth observation is now included in the data base and the following recursive relations are used to update the results:

$$(X_r'X_r)^{-1} = (X_{r-1}'X_{r-1})^{-1} - \frac{(X_{r-1}'X_{r-1})^{-1}x_r x_r'(X_{r-1}'X_{r-1})^{-1}}{1 + x_r'(X_{r-1}'X_{r-1})^{-1}x_r}$$

$$\hat{\beta}_r = \hat{\beta}_{r-1} + (X_r'X_r)^{-1}x_r(y_r - x_r'\hat{\beta}_{r-1}) \qquad \text{[A.4]}$$

$$S_r^2 = S_{r-1}^2 + w_r^2$$

6. Then, w_{r+1}, B_{r+1}, and u_{r+1} for the next cycle are computed as above.

This process is continued for $r = p' + 1$ to n to give a total of $n - p' - 1$ residuals. The B_j are independently and identically distributed Student's t random variables. The u_j are i.i.d. uniform random variables on the (0, 1) interval. The i.i.d. N(0, 1) random variables z_i are obtained by using u_i in the argument of $\Phi^{-1}(u_i)$, the inverse of the N(0, 1) probability distribution function. That is, z_i is the value of the standardized normal deviate for which $\text{Prob}(z \leq z_i) = u_i$.

Testing the goodness of fit of Quesenberry's uniform residuals to a uniform distribution is equivalent to testing the entire model including the normality assumptions. There are many tests for testing a uniform distribution. Quesenberry (1986) recommends the Watson (1961) U^2 statistic if $N = n - p' - 1$ is less than 10; he recommends both U^2 and the smooth test of Neyman (1937), using the p_4^2 statistic, if N is greater than 10. The U^2 statistic is

$$U^2 = \frac{1}{12N} + \sum_{i=1}^{N}\left(\frac{2i - 1}{2N} - u_i\right)^2 - N(\bar{u} - .5) \qquad \text{[A.5]}$$

where $\bar{u}$ is the arithmetic mean of the u_i. A modification of U^2 (Stephens, 1970) has critical values that are nearly constant for $N > 10$ and Quesenberry (1986) provides critical values for $N \leq 10$. The modified U^2 statistic is

$$U^2_{\text{MOD}} = \left(U^2 - \frac{.1}{N} + \frac{.1}{N^2}\right)\left(1 + \frac{.8}{N}\right) \qquad [A.6]$$

The critical values for U^2_{MOD} for N large are .152, .187, and .267 for $\alpha = .1, .05$, and .01, respectively. The null hypothesis of uniformity is rejected if the test statistic exceeds the critical value.

EXERCISES

* 10.1. Plot the following Studentized residuals against $\hat{Y}_i$. Does the pattern suggest any problem with the model or the data?

r_i^*	$\hat{Y}_i$	r_i^*	$\hat{Y}_i$	r_i^*	$\hat{Y}_i$	r_i^*	$\hat{Y}_i$
−.53	10	−.92	11	−1.55	15	−.82	18
.23	19	−.45	23	−1.00	26	.47	32
−.36	38	.75	41	1.27	43	1.85	48
1.16	49	.04	49	.96	51	−1.03	60
−.25	65	−.92	67	−1.84	69	.52	73
−.80	76	−.88	79	.57	85	−.25	90
1.51	93	1.62	99	.65	100		

10.2. Plot the following Studentized residuals against the corresponding $\hat{Y}_i$. What does the pattern in the residuals suggest?

r_i^*	$\hat{Y}_i$	r_i^*	$\hat{Y}_i$	r_i^*	$\hat{Y}_i$	r_i^*	$\hat{Y}_i$
−.53	60	−.92	81	−1.55	83	−.82	78
.23	19	−.45	53	−1.00	63	.47	42
−.36	48	.75	41	1.27	23	1.85	98
1.16	29	.04	49	.96	21	−1.03	80
−.25	65	−.92	57	−1.84	72	.52	33
−.80	76	−.88	69	.57	65	−.25	30
1.51	13	1.62	19	.65	25		

* 10.3. For each of the following questions, choose the *one* diagnostic you would use (for example, a plot or an influence statistic) to answer the question. Describe your choice and what you would expect to see if there were no problem.

a. Do the ε_i have homogeneous variance?
b. Is the regression being unduly influenced by observation 11?
c. Is the regression on X_3 really linear as the model states?
d. Is there an observation that does not seem to fit the model?
e. Has an important independent variable been omitted from the model?

10.4. For each of the following diagnostic tools, indicate which aspects of ordinary least squares are being checked and the types of results that would indicate problems.

a. Normal plot of r_i^*
b. Plot of $\boldsymbol{e}$ versus $\hat{\boldsymbol{Y}}$

c. Cook's D
d. v_{ii}, the diagonal elements of $\boldsymbol{P}$
e. DFBETAS$_j$

* 10.5. The collinearity diagnostics in PROC REG in SAS gave the eigenvalues 2.1, 1.7, .8, .3, and .1 for a set of data.
a. Compute the condition number for the matrix and the condition index for each principal component dimension. Is there cause for concern about collinearity in these data?
b. Compute Thisted's measure of collinearity, *mci*. Does the value of *mci* indicate a collinearity problem?

10.6. A regression problem gave largest and smallest eigenvalues of 3.29 and .02, and the following variance decomposition proportions corresponding to the last principal component:

Parameter:	$\hat{\beta}_0$	$\hat{\beta}_1$	$\hat{\beta}_2$	$\hat{\beta}_3$	$\hat{\beta}_4$	$\hat{\beta}_5$
Variance Proportion:	.72	.43	.18	.85	.71	.02

a. Do these results indicate collinearity problems?
b. Which $\hat{\beta}$'s, if any, are "suffering" from collinearity? Explain the basis for your conclusion.

* 10.7. PROC REG (in SAS) was run on a set of data with $n = 40$ observations on Y and three independent variables. The collinearity diagnostics gave the results in the accompanying table.

			Variance Proportion			
Number	*Eigenvalue*	*Cond. Index*	*Intercept*	X_1	X_2	X_3
1	3.84682	1.00	.0007	.0010	.0043	.0075
2	.09992	6.205	.0032	.0059	.1386	.8647
3	.04679	9.067	.0285	.0942	.7645	.0912
4	.00647	24.379	.9676	.8990	.0926	.0366

a. What is the rank of $\boldsymbol{X}$ in this model?
b. What is the condition number for $\boldsymbol{X}$? What does that say about the potential for collinearity problems?
c. Interpret the variance proportions for the fourth principal component. Is there variance inflation from the collinearity? Which regression coefficients are being affected most?
d. Compute the variance proportions for the *third* principal component after the fourth has been removed. Considering the condition index and the variance proportions for the third principal component, is there variance inflation from the third component?

10.8. An experiment was designed to estimate the response surface relating Y to two quantitative independent variables. A 4×4 factorial set of treatments was used with $X_1 = 1, 2, 3,$ and 4, and $X_2 = 65, 70, 75,$ and 80.
a. Set up $\boldsymbol{X}$ for the linear model,

$$Y_{ij} = \beta_0 + \beta_1 X_{1i} + \beta_2 X_{2i} + \varepsilon_{ij}$$

(You need use only the 16 distinct rows of X.) Do the singular value decomposition on the scaled X. Is there any indication of collinearity problems?

b. Redefine the model so that X_1 and X_2 are both expressed as deviations from their means. Redo the singular value decomposition. Have the collinearity diagnostics changed? Explain the differences, if any.

c. Use the centered X's but include squares of the X's in the model. Redo the singular value decomposition. Have the collinearity diagnostics changed? Explain the changes.

* 10.9. The following are the results of a principal component analysis, on $\boldsymbol{Z}$, of data collected from a fruitfly experiment attempting to relate a measure of fly activity, WBF = wing beat frequency, to the chemical activity of four enzymes, SDH, FUM, GH, and GO. Measurements were made on $n = 21$ strains of fruitfly. (Data courtesy of Dr. Laurie-Alberg, North Carolina State University.)

Eigenvalues: 2.1970 1.0790 .5479 .1762

	Eigenvectors			
Variable	1st	2nd	3rd	4th
SDH	.547	−.465	−.252	−.649
FUM	.618	−.043	−.367	.694
GH	.229	.870	−.306	−.312
GO	.516	.158	.842	−.005

a. Compute the proportion of the dispersion in the X-space accounted for by each principal component.

b. Compute the condition number for $\boldsymbol{Z}$ and the condition index for each principal component. What do the results suggest about possible variance inflation from collinearity?

c. Describe the first principal component in terms of the original centered and standardized variables. Describe the second principal component.

d. The sum of the variances of the estimates of the least squares regression coefficients, $\text{tr}[\textbf{Var}(\hat{\boldsymbol{\beta}})] = \sum(1/\lambda_j^2)\sigma^2$, must be larger than σ^2/λ_4^2. Compute this minimum (in terms of σ^2). How does this value compare to the minimum if the four variables had been orthogonal?

10.10. The following questions relate to the residuals analysis reported in Tables 10.4 and 10.5.

a. Compute $s^2(\hat{Y}_i) + s^2(e_i)$ for several choices of i. How do you explain the fact that you obtain very nearly the same number each time?

b. Find the largest and smallest $s(\hat{Y}_i)$ and the largest and smallest v_{ii}. How do you explain that they are the same observations in each case?

c. A COVRATIO equal to 1.0 implies that the ith point has no real impact on the overall precision of the estimates. A COVRATIO less than 1.0 indicates that the presence of the ith observation has *decreased* the precision of the estimates (e.g., observation 12). How do you explain the *presence* of an additional observation causing less precision?

d. Cook's D provides a measure of the shift in $\hat{\boldsymbol{\beta}}$. The DFBETAS measure shifts in the individual $\hat{\beta}_j$. Explain the fact that observation 29, which has the largest value of Cook's D, has no DFBETAS$_j$ that exceed the cutoff point, whereas observation 34, which has the next largest value of Cook's D, shows major shifts in all but one of the regression coefficients. Conversely, explain why observation 10 has a small Cook's D but shows major shifts in the intercept and the regression coefficients for pH and Zn.

*10.11. The accompanying table reports data on percentages of sand, silt, and clay at 20 sites. [The data are from Nielsen, Biggar, and Erh (1973), as presented by Andrews and Herzberg (1985). The depths 1, 2, and 3 correspond to depths 1, 6, and 12 in Andrews and Herzberg.] Use the sand, silt, and clay percentages at the three depths as nine columns of an X matrix.

Plot	*Depth* 1			*Depth* 2			*Depth* 3		
No.	*Sand*	*Silt*	*Clay*	*Sand*	*Silt*	*Clay*	*Sand*	*Silt*	*Clay*
1	27.3	25.3	47.4	34.9	24.2	40.7	20.7	36.7	42.6
2	40.3	20.4	39.4	42.0	19.8	38.2	45.0	25.3	29.8
3	12.7	30.3	57.0	25.7	25.4	49.0	13.1	37.6	49.3
4	7.9	27.9	64.2	8.0	26.6	64.4	22.1	30.8	47.1
5	16.1	24.2	59.7	14.3	30.4	55.3	5.6	33.4	61.0
6	10.4	27.8	61.8	18.3	27.6	54.1	8.2	34.4	57.4
7	19.0	33.5	47.5	27.5	37.6	34.9	0.0	30.1	69.9
8	15.5	34.4	50.2	11.9	38.8	49.2	4.4	40.8	54.8
9	21.4	27.8	50.8	20.2	30.3	49.3	18.9	36.1	45.0
10	19.4	25.1	55.5	15.4	35.7	48.9	3.2	44.4	52.4
11	39.4	25.5	35.6	42.6	23.6	33.8	38.4	32.5	29.1
12	32.3	32.7	35.0	20.6	28.6	50.8	26.7	37.7	35.6
13	35.7	25.0	39.3	42.5	20.1	37.4	60.7	13.0	26.4
14	35.2	19.0	45.8	32.5	27.0	40.5	20.5	42.5	37.0
15	37.8	21.3	40.9	44.2	19.1	36.7	52.0	21.2	26.8
16	30.4	28.7	40.9	30.2	32.0	37.8	11.1	45.1	43.8
17	40.3	16.1	43.6	34.9	20.8	44.2	5.4	44.0	50.6
18	27.0	28.2	44.8	37.9	30.3	31.8	8.9	57.8	32.8
19	32.8	18.0	49.2	23.2	26.3	50.5	33.2	26.8	40.0
20	26.2	26.1	47.7	29.5	34.9	35.6	13.2	34.8	52.0

a. From the nature of the variables, is there any reason to expect a collinearity problem if these nine variables were to be used as independent variables in a multiple regression analysis?

b. Center and scale the variables and do a singular value decomposition on $\boldsymbol{Z}$. Does the SVD indicate the presence of a collinearity problem? Would you have obtained the same results if the variables had not been centered and the intercept included? Explain.

CHAPTER
ELEVEN

Transformation of Variables

Several methods for detecting problem areas were discussed in Chapter 10 and their applications to real data were demonstrated.

This chapter discusses the use of transformations of variables to simplify relationships, to stabilize variances, and to improve normality. Weighted least squares and generalized least squares are presented as methods of handling the problems of heterogeneous variances and lack of independence.

There are many situations in which transformations of the dependent or independent variables are helpful in least squares regression. Chapter 9 suggested transformation of the *dependent* variable as a possible remedy for some of the problems in least squares. In this chapter, the reasons for making transformations, including transformations on the independent variables, and the methods used to choose the appropriate transformations are discussed more fully. Generalized least squares and weighted least squares are included in this chapter, because they can be viewed as ordinary least squares regression on a transformed dependent variable.

11.1 Reasons for Making Transformations

There are three basic reasons for transforming variables in regression. Transformations of the dependent variable were indicated in Chapter 9 as possible remedies for nonnormality and for heterogeneous variances of the errors. A third reason for making transformations is to simplify the relationship between the dependent variable and the independent variables.

The Simplest Model

A basic rule of science says that, all other things being equal, the simplest model that describes the observed behavior of the system should be adopted. Simple relationships are more easily understood and communicated to others. With statistical models, the model with the fewest parameters is considered the simplest, straight-line relationships are considered simpler than curvilinear relationships, and models linear in the parameters are simpler than nonlinear models.

Curvilinear Relationships

Curvilinear relationships between two variables frequently can be simplified by a transformation on either one or both of the variables. The power family of transformations and a few of the two-bend transformations are discussed for this purpose (Section 11.2).

Nonlinear Models

Many models nonlinear in the parameters can be linearized, reexpressed as a linear function of the parameters, by appropriate transformations. For example, the relationship

$$Y = \alpha X^{\beta}$$

is linearized by taking the logarithm of both sides of the equality, giving

$$\ln(Y) = \ln(\alpha) + \beta[\ln(X)]$$

or

$$Y^* = \alpha^* + \beta X^*$$

The nonlinear relationship between Y and X is represented by the linear relationship between Y^* and X^*.

Heterogeneous Variances and Nonnormality

The effects of heterogeneous variances and nonnormality on least squares regression have already been noted (Chapter 9). Transformation of the dependent variable was indicated as a possible remedy for both. Sections 11.3 and 11.4 discuss the choice of transformations for these two situations. Alternatively, weighted least squares or its more general version, generalized least squares, can be used to account for different degrees of precision in the observations. These methods are discussed in Section 11.5.

Throughout this discussion, it should be remembered that it may not be possible to find a set of transformations that will satisfy all objectives. A transformation on the dependent variable to simplify a nonlinear relationship will destroy both homogeneous variances and normality if these assumptions were satisfied with the original dependent variable. Or, a transformation to stabilize variance may cause nonnormality. Fortunately, transformations for homogeneity of variance and normality tend to go hand-in-hand so that often both assumptions are more nearly satisfied after an appropriate transformation (Bartlett, 1947). If one must make a choice, stabilizing variance is usually given precedence over improving normality. Many recommend that simplifying the relationship should take precedence over all. The latter would seem to depend on the intrinsic value and the general acceptance of the relationship being considered. If a nonlinear model is meaningful and is readily interpreted, a transformation to linearize the model would not seem wise if it creates heterogeneous variance or nonnormality.

11.2 Transformations to Simplify Relationships

It is helpful to differentiate two situations where transformations to simplify relationships might be considered. In the first case, there is *no* prior idea of the form the model should take. The objective is to empirically determine mathematical forms of the dependent and independent variables that allow the observed relationship to be represented in the simplest form, preferably a straight line. The model is to be linear in the parameters; only the form in which the variables are expressed is being considered.

In the second case, *prior* knowledge of the system suggests a nonlinear mathematical function, nonlinear in the parameters, for relating the dependent variable to the independent variable(s). The purpose of the transformation in this case is to reexpress the nonlinear model in a form that is linear in the parameters and for which ordinary least squares can be used. Such linearization of nonlinear models is not always possible but when it is possible the transformation to be used is dictated by the functional form of the model.

"One-Bend" Transformations

The power family of transformations, $X^* = X^k$, provides a useful set of transformations for "straightening" a single bend in the relationship between two variables. These are referred to as the **"one-bend" transformations** (Tukey, 1977; Mosteller and Tukey, 1977). Ordering the transformations according to the exponent k gives a sequence of power transformations, which Mosteller and Tukey (1977) call the *ladder of reexpressions.* The common powers considered are

$$k = -1, -\tfrac{1}{2}, 0, \tfrac{1}{2}, 1, 2,$$

where the power transformation $k = 0$ is to be interpreted as the logarithmic transformation. The power $k = 1$ implies no transformation.

Ladder of Transformations

The rule for straightening a "one-bend" relationship is to move up or down the ladder of transformations according to the direction in which the bulge of the curve points. For example, if the bulge in the curve points toward lower values of Y, as in the exponential decay and growth curves shown in Figure 11.1 (page 302), moving down the ladder of transformations to $\sqrt{Y}$, $\ln(Y)$, and $1/Y$ will tend to straighten the relationship. [In the specific case of the exponential function, it is known that the logarithmic transformation ($k = 0$) will give a linear relationship.] For the exponential decay curve, the bulge also points toward lower values of X. Therefore, moving down the ladder for a power transformation of X will also tend to straighten the relationship. For the exponential growth curve, however, one must move up the ladder to X^2 or X^3 for a power transformation on X to straighten the relationship; the bulge points upward with respect to X. The inverse polynomial curve (Figure 11.2) points upward with respect to Y and downward with respect to X. Therefore, *higher* powers of Y or *lower* powers of X will tend to straighten the relationship.

How far one moves on the ladder of transformations depends on the sharpness of the curvature. This is easily determined when only one independent

Figure 11.1
Examples of the exponential growth curve and the exponential decay curve.

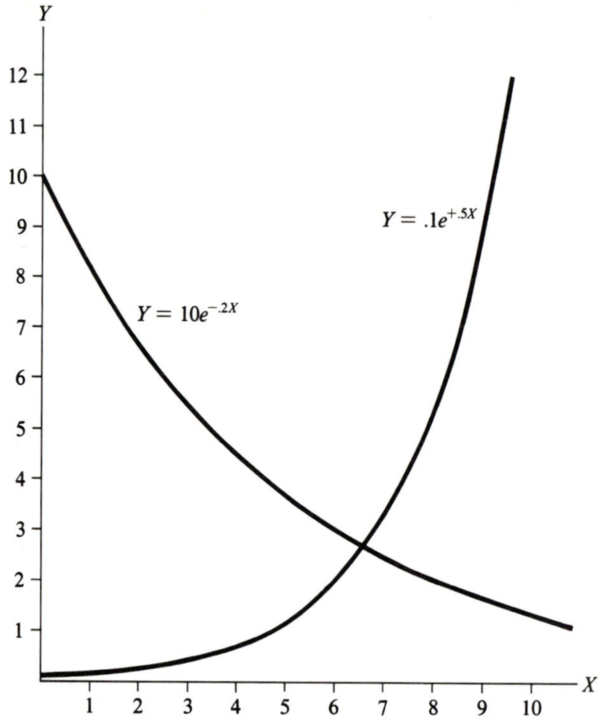

Figure 11.2
Examples of the inverse polynomial model and the logistic model.

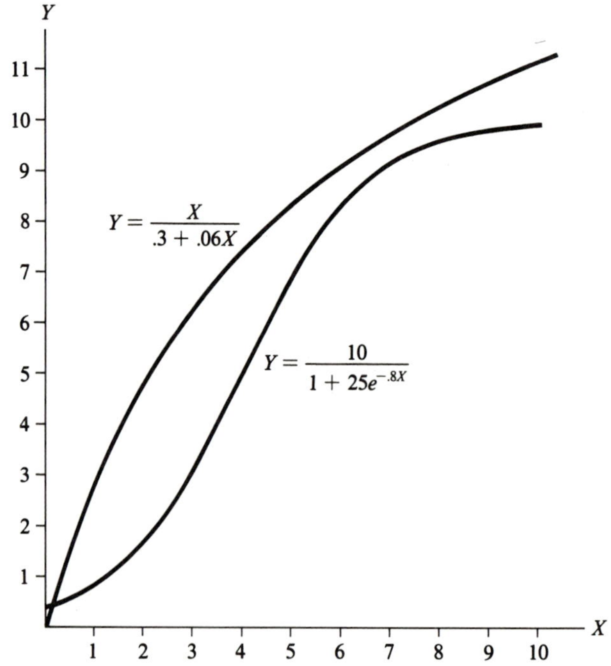

variable is involved by trying several transformations on a few observations covering the range of the data and then choosing the transformation that makes the points most nearly collinear. Several independent variables make the choice more difficult, particularly when the data are not balanced or when there are interactions among the independent variables. The partial regression leverage plots for the first-degree polynomial model will show the relationship between Y and a particular independent variable *after* adjustment for all other independent variables, and should prove helpful in determining the power transformation. Since only one transformation on Y can be used in any one analysis, attention must focus on transformations of the independent variables when several independent variables are involved.

Box–Tidwell Method

Box and Tidwell (1962) give a computational method for determining the power transformations on *independent* variables such that lower-order polynomial models of the transformed variables might be used. They assume that the usual least squares assumptions are well enough satisfied on the present scale of Y (perhaps after some transformation) so that further transformations to simplify relationships must be done on the independent variables. The **Box–Tidwell method** is a general method applicable to any model and any class of transformations. However, its consideration here will be restricted to the polynomial model and power transformations on individual X's. The steps of the Box–Tidwell method will be given for a full second-degree polynomial model in two variables. The simplifications of the procedure and an illustration for the first-degree polynomial model are given.

Procedure

The proposed second-degree model is

$$\begin{aligned} Y_i &= F(\boldsymbol{U}, \boldsymbol{\beta}) + \varepsilon_i \\ &= \beta_0 + \beta_1 U_{i1} + \beta_2 U_{i2} + \beta_{11} U_{i1}^2 + \beta_{22} U_{i2}^2 + \beta_{12} U_{i1} U_{i2} + \varepsilon_i \end{aligned}$$

where $i = 1, \ldots, n$ and $j = 1, 2$. The U_{ij} are power transformations on X_{ij}:

$$U_{ij} = \begin{cases} X_{ij}^{\alpha_j} & \text{if } \alpha_j \neq 0 \\ \ln(X_{ij}) & \text{if } \alpha_j = 0 \end{cases} \qquad [11.1]$$

The objective is to find the α_1 and α_2 for transforming X_{i1} and X_{i2} to U_{i1} and U_{i2}, respectively, that provide the best fit of $F(\boldsymbol{U}, \hat{\boldsymbol{\beta}})$ to $\boldsymbol{Y}$. The steps in the Box–Tidwell method to approximate the α_j are as follows:

1. Fit the polynomial model to $\boldsymbol{Y}$ to obtain the regression equation in the *original* variables, $\hat{\boldsymbol{Y}} = F(\boldsymbol{X}, \hat{\boldsymbol{\beta}})$.
2. Differentiate $\hat{\boldsymbol{Y}}$ with respect to each independent variable and evaluate the partial derivatives for each of the n observations to obtain $W_{ij} = \partial(\hat{\boldsymbol{Y}})/\partial X_j$, $i = 1, \ldots, n$. For the quadratic model,

$$W_{i1} = \hat{\beta}_1 + 2\hat{\beta}_{11} X_{i1} + \hat{\beta}_{12} X_{i2}$$

and

$$W_{i2} = \hat{\beta}_2 + 2\hat{\beta}_{22}X_{i2} + \hat{\beta}_{12}X_{i1}$$

(For the first-degree polynomial model, the partial derivatives are simply the constants $W_{i1} = \hat{\beta}_1$ and $W_{i2} = \hat{\beta}_2$.)

3. Create two new independent variables, Z_{i1} and Z_{i2}, by multiplying each W_{ij} by the corresponding values of $X_{ij}[\ln(X_{ij})], j = 1, 2$.
4. Refit the polynomial model *augmented* with the two new variables Z_1 and Z_2. Let $\hat{\gamma}_j$ be the partial regression coefficient obtained for Z_j.
5. Compute the desired power transformations as $\hat{\alpha}_j = \hat{\gamma}_j + 1, j = 1, 2$.

This is the end of the first round of iteration to approximate the coefficients for the power transformation. The α_j are then used to transform the original X's (according to equation 11.1) and the process is repeated using the power-transformed variables as if they were the original variables. The α_j obtained on the second iteration are used to make a power transformation on the *previously* transformed variables. (This is equivalent to transforming the original variables using the product of α_j from the first and second steps as the power on the jth variable.) The iteration terminates when α_j converges close enough to 1.0 to cause only trivial changes in the power transformation.

Example 11.1 The Box–Tidwell method is illustrated using data from an experiment to test tolerance of certain families of pine to saltwater flooding (Land, 1973). Three seedlings from each of eight families of pine were subjected to 0, 72, or 144 hours of flooding in a completely random experimental design. The data are given in Table 11.1. The response variable is the chloride content (% dry matter) of the pine needles. (The $Y = .00\%$ chloride measurement for family 3 was changed to $Y = .01$ and $X = 0$ hours flooding was changed to $X = 1$ hour. Both changes were made to avoid problems with taking logarithms in the Box–Tidwell method and in the Box–Cox method to be used in exercise 11.1.)

Table 11.1 Chloride content (percent dry weight) of needles of pine seedlings exposed to 0, 72, or 144 hours of flooding with saltwater. Nine seedlings of each of eight genetic families were used in a completely random experimental design. (Data from S. B. Land, Jr., 1973, Ph.D. Thesis, North Carolina State University, and used with permission.)

	Hours of Flooding with Saltwater								
Family	0			72			144		
1	.36	.47	.30	3.54	4.35	4.88	6.13	6.49	7.04
2	.32	.63	.51	4.95	4.45	1.50	6.46	4.35	2.18
3	.00	.43	.72	4.26	3.89	6.54	5.93	6.29	9.62
4	.54	.70	.49	3.69	2.81	4.08	5.68	4.68	5.79
5	.44	.42	.39	3.01	4.08	4.54	6.06	6.05	6.97
6	.55	.57	.45	2.32	3.57	3.59	4.32	6.11	6.49
7	.20	.51	.27	3.16	3.17	3.75	4.79	5.74	5.95
8	.31	.44	.84	2.80	2.96	2.04	10.58	4.44	1.70

The regression of $Y = (\%\ \text{chloride})^{1/2}$ on X = hours of exposure, allowing a different intercept for each family, required a quadratic polynomial to adequately represent the relationship. The Box–Tidwell method will be used to search for a power transformation on X that will allow the relationship to be represented by a straight line. The first step fits the model

$$Y_{ijk} = \beta_{0i} + \beta X_j + \varepsilon_{ijk}$$

where $i = 1, \ldots, 8$ designates the family, X_j is the number of hours of flooding, $j = 1, 2, 3$, and $k = 1, 2, 3$ designates the seedling within each i, j combination. The estimate of the regression coefficient is $\hat{\beta} = .01206$. This is the partial derivative of $\hat{Y}_{ijk}$ with respect to X when the model is linear in X; therefore, $W_i = \hat{\beta}$ in step 2. Thus, the new independent variable is

$$Z_j = .01206X_j[\ln(X_j)]$$

The model is augmented with Z_j to give

$$Y_{ijk} = \beta_{0i} + \beta X_j + \gamma Z_j + \varepsilon_{ijk}$$

Fitting this model gives $\hat{\gamma} = -.66971$; thus, $\hat{\alpha} = \hat{\gamma} + 1 = .33029$ is the estimated power transformation on X from the first iteration. The cycle is repeated using the transformed $X(1)_j = (X_j)^{.33029}$ in place of X_j.

The second iteration gives $\hat{\beta} = .41107$, $\hat{\gamma} = .22405$, and $\hat{\alpha} = 1.22405$. Thus, the estimated power transformation on $X(1)_j$ is $X(2)_j = [X(1)_j]^{1.22405}$. The third iteration uses $X(2)_j$ in place of $X(1)_j$.

The third iteration gives $\hat{\beta} = .26729$, $\hat{\gamma} = -.00332$, and $\hat{\alpha} = .99668$. If the iterations were to continue, the new independent variable would be $X(3)_j = [X(2)_j]^{.99668}$. Since $\hat{\alpha}$ is very close to 1.0, giving only trivial changes in $X(2)_j$, the iterations can stop. The estimated power transformation on X is the product of the three $\hat{\alpha}$'s $(.33029)(1.22405)(.99668) = .4023$, which is close to the square-root transformation on X. In this example, a linear model using the transformed $X^* = X^{.4023}$ provides the same degree of fit as a quadratic model using the original X_j; the residual sums of squares from the two models are very nearly identical. ■

Estimating Power with Nonlinear Regression

An alternative method of determining the power transformations is to include the powers on the independent variables as parameters in the model and use nonlinear least squares to estimate all parameters simultaneously (Chapter 14). In some cases, this may lead to overparameterization of the model and failure of the procedure to find a solution. There is no assurance that appropriate power transformations will exist to make the chosen polynomial fit the data. The usual precautions should be taken to verify that the model is adequate for the purpose.

Transformations and Model Assumptions

The objective to this point has been to find the power transformation of *either* Y or X that most nearly straightens the relationship. However, any transformation on the *dependent* variable will also affect the distributional properties of Y. Hence, the normality and common variance assumptions on ε must be considered at the same time as transformations to simplify relationships. The power family of transformations on the *dependent* variable will be considered in Section 11.4, where the criterion is to have $\mathscr{E}(Y)$ adequately represented by a relatively simple model *and* the assumptions of normality and constant variance approximately satisfied (Box and Cox, 1964).

Two-Bend Transformations

Relationships that show more than one bend, such as the classical S-shaped growth curve (see the logistic curve in Figure 11.2), cannot be straightened with the power family of transformations. A few commonly used two-bend transformations are:

1. Logit: $Y^* = \frac{1}{2}\ln[p/(1 - p)]$
2. Arcsin (or angular): $Y^* = \arcsin(\sqrt{p})]$
3. Probit: $Y^* = \Phi^{-1}(p)$, where $\Phi^{-1}(p)$ is the standard normal deviate that gives a cumulative probability of p.

These transformations are generally applied to situations where the variable p is the proportion of "successes" and consequently bounded by 0 and 1. The effect of the transformation in all three cases is to "stretch" the upper and lower tails, the values of p near 1 and 0, making the relationship more nearly linear (Bartlett, 1947). The logit is sometimes preferred as a means of simplifying a model that involves products of probabilities. The probit transformation arises as the logical transformation when, for example, the chance of survival of an organism to a toxic substance is related to the dose, or ln(dose), of the toxin through a normal probability distribution of sensitivities. That is, individuals in the population vary in their sensitivities to the toxin, and the threshold dose (perhaps on the logarithmic scale) that "kills" individuals has a normal distribution. In such case, the probit transformation translates the proportion affected into a linear relationship with dose, or ln(dose). The logit transformation has a similar interpretation but where the threshold distribution is the logistic distribution.

Intrinsically Linear Models

Nonlinear models that can be linearized are called **intrinsically linear models.** The function $Y = \alpha X^{\beta}$ in Section 11.1 was linearized by taking the logarithm of both Y and X. If a positive multiplicative random error is incorporated to make it a statistical model, the model becomes

$$Y_i = \alpha X_i^{\beta}\varepsilon_i \qquad [11.2]$$

The linearized form of this model is

$$\ln(Y_i) = \ln(\alpha) + \beta[\ln(X_i)] + \ln(\varepsilon_i)$$

or

$$Y_i^* = \alpha^* + \beta X_i^* + \varepsilon_i^* \qquad [11.3]$$

where $\alpha^* = \ln(\alpha)$, $X_i^* = \ln(X_i)$, and $\varepsilon_i^* = \ln(\varepsilon_i)$. This transformation is repeated here to emphasize the impact of the transformation of Y on the random error.

The least squares model assumes that the random errors are *additive*. Thus, in order for the random error to be additive on the logarithmic scale, they must have been *multiplicative* on the original scale. Further, the ordinary least squares assumptions of normality and homogeneous variances apply to the $\varepsilon_i^* = \ln(\varepsilon_i)$, not to the ε_i. The implication is that linearization of models, and transformations in general, must also take into account the least squares assumptions. It may be better in some cases, for example, to forgo linearization of a model if the transformation destroys normality or homogeneous variances. Likewise, it may not be desirable to go to extreme lengths to achieve normality or homogeneous variances if it entails the use of an excessively complicated model.

Exponential Models

Another example of an intrinsically linear model is the **exponential growth model**,

$$Y_i = \alpha e^{\beta X_i} \varepsilon_i \qquad [11.4]$$

This growth function starts at $Y_i = \alpha$ when $X = 0$ and increases exponentially with a *relative* rate of growth equal to β ($\alpha > 0, \beta > 0$). The **exponential decay model** has the same form but with a negative exponential term. The decay model starts at $Y_i = \alpha$ when $X = 0$ and *declines* at a relative rate equal to β. The two exponential functions are illustrated in Figure 11.1. Both are linearized with the logarithmic transformation. Thus, for the growth model,

$$Y_i^* = \alpha^* + \beta X_i + \varepsilon_i^*$$

where Y_i^*, α^*, and ε_i^* are the natural logarithms of the corresponding quantities in the original model.

Inverse Polynomial Model

One version of the **inverse polynomial model** has the form

$$Y_i = \frac{X}{\alpha + \beta X_i + \varepsilon_i} \qquad [11.5]$$

This function, illustrated in Figure 11.2, is a monotonically increasing function of X that very slowly approaches the asymptote $Y = 1/\beta_0$. The reciprocal transformation on Y, $Y^* = 1/Y$, gives

$$Y_i^* = \beta + \alpha\left(\frac{1}{X_i}\right) + \varepsilon_i^*$$

Thus, Y^* is a first-degree polynomial in $1/X$ with intercept β and slope α. Values of X equal to zero must be avoided for this transformation to work.

Logistic Model

The frequently used **logistic growth model** is

$$Y_i = \frac{\alpha}{1 + \gamma e^{-\beta X_i} \varepsilon_i} \qquad [11.6]$$

This function gives the characteristic growth curve starting at $Y = \alpha/(1 + \gamma)$ at

$X = 0$ and approaching the asymptote $Y = \alpha$ as X gets large (Figure 11.2). The function is intrinsically linear only if the value of α is known, as is the case, for example, when the dependent variable is the proportion of individuals showing reaction to a treatment. If α is known, the model is linearized by defining

$$Y^* = \ln\left(\frac{\alpha}{Y} - 1\right)$$

and the model becomes

$$Y_i^* = \gamma^* - \beta X_i + \varepsilon_i^*$$

where $\gamma^* = \ln(\gamma)$ and $\varepsilon_i^* = \ln(\varepsilon_i)$.

In these examples, the placement of the error in the original model was such that the transformed model had an *additive* error. If there were reason to believe that the errors were additive in the original models, all would have become intrinsically nonlinear. The least squares assumptions on the behavior of the errors apply to the errors *after* transformation. Decisions as to how the errors should be incorporated into the models will depend on one's best judgment as to how the system operates and the analysis of the behavior of the residuals before and after transformation.

Approximating Functions with Polynomials

Any mathematical function relating Y to one or more independent variables can be approximated to any degree of precision desired with an appropriate polynomial in the independent variables. This is the fundamental reason polynomial models have proven so useful in regression, although seldom would one expect a polynomial model to be the true model for a physical, chemical, or biological process. Even intrinsically nonlinear models can be simplified, if need be, in the sense that they can be approximated with polynomial models, which are linear in the parameters. (Some caution is needed in using a polynomial to approximate a nonlinear response that has an asymptote. The polynomial will tend to oscillate about the asymptote.) The regression coefficients in the polynomial model will usually be nonlinear functions of the original parameters. This will make it more difficult to extract the physical meaning from the polynomial model than from the original nonlinear model. Nevertheless, polynomial models will continue to serve as very useful approximations, at least over limited regions of the X-space, of the more complicated, and usually unknown, true models.

11.3 Transformations to Stabilize Variances

Links Between Mean and Variance

The variance and the mean are independent in the normal probability distribution. All other common distributions have a direct link between the mean and the variance. For example, the variance is equal to the mean in the Poisson distribution, the distribution frequently associated with count data. A plot of the

Poisson variance against the mean would be a straight line with a slope of 1. The variance of the count of a binomially distributed random variable is $np(1 - p)$ and the mean is np. A plot of the binomial variance against the mean would show zero variance at $p = 0$ and $p = 1$ and maximum variance at $p = \frac{1}{2}$. The variance of a chi-square distributed random variable is equal to twice its mean. As with the Poisson, this is a linear relationship between the variance and the mean but with a steeper slope. A priori, one should expect variances to be heterogeneous when the random variable is not normally distributed.

Even in cases where there is no obvious reason to suspect nonnormality, there often is an association between the mean and the variance. Most commonly, the variance increases as the mean increases. It is prudent to suspect heterogeneous variances if the data for the dependent variable cover a wide range, such as a doubling or more in value between the smallest and largest observations.

General Transformation to Stabilize Variance

If the functional relationship between the variance and the mean is known, a transformation exists that will make the variance (approximately) constant (Bartlett, 1947). Let

$$\sigma^2 = \Omega(\mu)$$

where $\Omega(\mu)$ is the function of the mean μ that gives the variance. Let $f(\mu)$ be the transformation needed to stabilize the variance. Then $f(\mu)$ is the indefinite integral

$$f(\mu) = \int \frac{1}{[\Omega(\mu)]^{1/2}}\, d\mu$$

Example 11.2

For example, if σ^2 is proportional to μ, $\sigma^2 = c\mu$ as in the Poisson,

$$f(\mu) = \int \frac{1}{[(c\mu)]^{1/2}}\, d\mu = \sqrt{\mu}$$

except for a proportionality constant and the constant of integration. Thus, the square-root transformation on the dependent variable would stabilize the variance in this case. ■

Variance Proportional to Power of Mean

In general, if the variance is (approximately) proportional to μ^{2k}, the appropriate transformation to stabilize the variance is $Y^* = Y^{1-k}$. In the Poisson example, $k = \frac{1}{2}$. When $k = 1$, the variance is proportional to the square of the mean and the logarithmic transformation is appropriate; Y^0 is interpreted as the logarithmic transformation. When the relationship between the mean and the variance is not known, empirical results can be used to approximate the relationship and suggest a transformation.

When the variance is proportional to a power of the mean, the transformation required to stabilize the variance is a power transformation on the dependent variable—the same family of transformations used for "straightening" one-bend relationships. Thus, a possible course of action is to use a power transformation on the *dependent* variable to stabilize the variance and another power transformation on the *independent* variable to "straighten" the relationship.

Arcsin Transformation

The variance may not be proportional to a power of the mean. A binomially distributed random variable, for example, has maximum variance at $p = \frac{1}{2}$ with decreasing variance as p goes toward either zero or one, $\sigma^2(\hat{p}) = p(1 - p)/n$. The transformation that approximately stabilizes the variance is the arcsin transformation, $Y^* = \arcsin(\sqrt{\hat{p}}) = \sin^{-1}\sqrt{\hat{p}}$. This assumes that the number of Bernoulli trials in each $\hat{p}_i$ is constant. While the arcsin transformation is designed for binomial data, it seems to stabilize the variance sufficiently in many cases where the variance is not entirely binomial in origin.

The arcsin transformation is the only one of the three two-bend transformations given in Section 11.2 that also stabilizes the variance (if the data are binomially distributed). The other two, the logit and the probit, although they are generally applied to binomial data, will not stabilize the variance.

A word of caution is in order regarding transformation of data in the form of proportions. Not all such data are binomially distributed, and therefore they should not be automatically subjected to the arcsin transformation. For example, chemical proportions that vary over a relatively narrow range, such as the oil content in soybeans, may be very nearly normally distributed with constant variance.

11.4 Transformations to Improve Normality

Transformations to improve normality have generally been given lower priority than transformations to simplify relationships or to stabilize variance. Even though least squares estimation per se does not require normality and moderate departures from normality are known not to be serious (Bartlett, 1947), there are sufficient reasons to be concerned about normality (see Section 9.1).

Fortunately, transformations to stabilize variance often have the effect of also improving normality. The logit, arcsin, and probit transformations that are used to stabilize variance and straighten relationships also make the distribution more normal-like by "stretching" the tails of the distribution, values near 0 or 1, to give a more bell-shaped distribution. Likewise, the power family of transformations, which have been discussed for straightening one-bend relationships and stabilizing variance, are also useful for increasing symmetry (decreasing skewness) of the distribution. The expectation is that the distribution will also be more nearly normal. The different criteria for deciding which transformation to make will not necessarily lead to the same choice, but it often happens that the optimum transformation for one will improve the other.

Box–Cox Method

Box and Cox (1964) present a computational method for determining a power transformation for the *dependent* variable where the objective is to obtain a simple, normal, linear model that satisfies the usual least squares assumptions. The Box–Cox criterion combines the objectives of the previous sections—simple relationship and homogeneous variance—with the objective of improving normality. The method is presented in this section because it is the only approach that directly addresses normality. The Box–Cox method results in estimates of the power transformation (λ), σ^2, and $\boldsymbol{\beta}$ that make the distribution of the transformed data as close to normal as possible [at least in large samples and as measured by the Kullback–Leibler information number (Hernandez and Johnson, 1980)]. However, normality is not guaranteed to result from the Box–Cox transformation and all the usual precautions should be taken to check the validity of the model.

The **Box–Cox method** uses the parametric family of transformations defined, in standardized form, as

$$Y_i^{(\lambda)} = \begin{cases} \dfrac{Y_i^{\lambda} - 1}{\lambda(\dot{Y})^{\lambda-1}} & \text{for } \lambda \neq 0 \\ \dot{Y}\ln(Y_i) & \text{for } \lambda = 0 \end{cases} \qquad [11.7]$$

where $\dot{Y}$ is the *geometric* mean of the original observations,

$$\dot{Y} = \exp\{\textstyle\sum [\ln(Y_i)]/n\}$$

The method assumes that for some λ the $Y_i^{(\lambda)}$ satisfy all the normal-theory assumptions of least squares; that is, they are independently and normally distributed with mean $\boldsymbol{X\beta}$ and common variance σ^2. With these assumptions, the maximum likelihood estimates of λ, $\boldsymbol{\beta}$, and σ^2 are obtained. [Hernandez and Johnson (1980) point out that this is not a valid likelihood because $Y_i^{(\lambda)}$ cannot be normal except in the special case of the original distribution being log-normal. Nevertheless, the Box–Cox method has proven to be useful.]

Estimating λ

The maximum likelihood solution is obtained by doing the least squares analysis on the transformed data for several choices of λ from, say, $\lambda = -1$ to 1. Let SS[Res(λ)] be the residual sum of squares from fitting the model to $Y_i^{(\lambda)}$ for the given choice of λ and let $\hat{\sigma}^2(\lambda) = \{\text{SS}[\text{Res}(\lambda)]\}/n$. The likelihood for each choice of λ is given by

$$L_{\max}(\lambda) = -\tfrac{1}{2}n\ln[\hat{\sigma}^2(\lambda)] \qquad [11.8]$$

Maximizing the likelihood is equivalent to minimizing the residual sum of squares. The maximum likelihood solution for $\hat{\lambda}$, then, is obtained by plotting SS[Res(λ)] against λ and reading off the value where the minimum, $\text{SS}[\text{Res}(\lambda)]_{\min}$, is reached. It is unlikely that the exact power transformation defined by $\hat{\lambda}$ will be used. It is more common to use one of the standard power transformations, $\lambda = \frac{1}{2}, 0, -\frac{1}{2}, -1$, in the vicinity of $\hat{\lambda}$.

Confidence Intervals on λ

Approximate confidence intervals on λ can be determined by drawing a horizontal line on the graph at

$$\{SS[Res(\lambda)]_{min}\}\left(1 + \frac{t^2_{(\alpha, v)}}{v}\right) \qquad [11.9]$$

where v is the degrees of freedom for $SS[Res(\lambda)]_{min}$ and $t_{(\alpha, v)}$ is the critical value of Student's t with $\alpha/2$ probability in each tail. Confidence limits on λ are given as the values of λ where the horizontal line intersects the $SS[Res(\lambda)]$ curve (Box, Hunter, and Hunter, 1978).

Considerations Before Using Box–Cox

The functional relationship between Y and the independent variables is specified in $\boldsymbol{X\beta}$ before the maximum likelihood estimate of λ is obtained. Thus, the solution obtained, $\hat{\lambda}$, is conditional on, and can be sensitive to, the presumed form of the model (Cook and Wang, 1983). The Box–Cox method is attempting to simultaneously satisfy the three objectives, $\mathscr{E}(Y^{(\lambda)}) = \boldsymbol{X\beta}$, constant variance, and normality. The relative weights given to satisfying the three objectives will depend on which will yield the greatest impact on the likelihood function. For example, if $\boldsymbol{X\beta}$ specifies a linear relationship between $Y^{(\lambda)}$ and X when the observed relationship between Y and X is very curvilinear, it is likely that pressure to "straighten" the relationship will dominate the solution. The transformed data can be even more nonnormal and their variances more heterogeneous.

If emphasis is to be placed on improving normality or constancy of variance, the functional form of the model specified by $\boldsymbol{X\beta}$ should be flexible enough to provide a reasonable fit to a range of transformations, including no transformation. For example, suppose the data show a curvilinear relationship that could be straightened with an appropriate power transformation. Specifying $\boldsymbol{X\beta}$ as a linear model would force the Box–Cox transformation to try to straighten the relationship. On the other hand, a quadratic model for $\boldsymbol{X\beta}$ would reduce the pressure to straighten the relationship and allow more pressure on improving normality and constancy of variance. Box and Cox (1964) show how to partition the effects of simple model, constant variance, and normality on the maximum likelihood estimate of λ.

Example 11.3

This example of a Box–Cox transformation is from a combined analysis of residuals from four studies on the effects of ozone and sulfur dioxide on soybean yields.* Each of the studies was subjected to the appropriate analysis of variance for the experimental design for that year. The observed residuals were pooled for checking model assumptions. There were a total of 174 residuals and 80 degrees of freedom for the pooled residual sum of squares.

Plots of the residuals suggested an increase in variance associated with increased yield (Figure 11.3). The normal plot of residuals was slightly S-shaped, consistent with slightly heavy tails, but not sufficiently nonnormal to give concern. The Box–Cox standardized transformation, equation 11.7, was applied for

* Analyses by V. M. Lesser on data courtesy of A. S. Heagle, North Carolina State University.

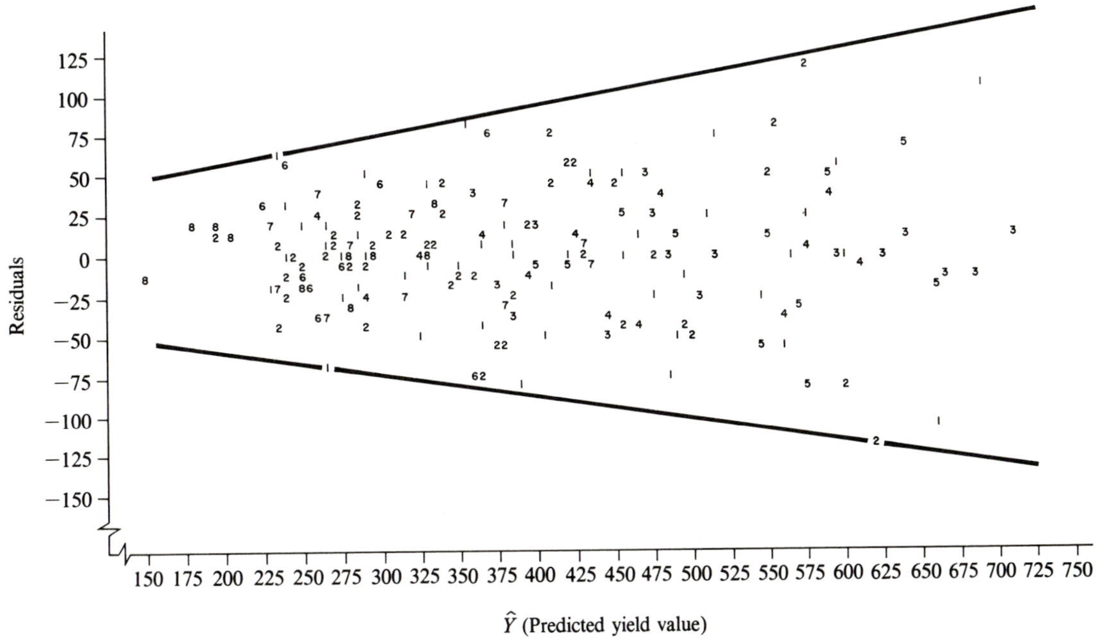

Figure 11.3 Plot of e_i versus $\hat{Y}_i$ (untransformed) from the combined analysis of four experiments on the effects of ozone and sulfur dioxide on soybean yields.

$\lambda = -1, -\frac{1}{2}, 0, \frac{1}{2}, 1$, and the analyses of variance repeated for each λ. The plot of the pooled residual sum of squares against λ, Figure 11.4 (page 314), suggested $\hat{\lambda} = -.05$ with 95% confidence limits of approximately $-.55$ to .40. The confidence limits on λ overlap both $\lambda = 0$ and $\lambda = -.5$ but, since $\hat{\lambda}$ was much nearer 0 than $-.5$, the logarithmic transformation was adopted. The plot of the residuals of the log-transformed data showed no remaining trace of heterogeneous variance or nonnormality (Figure 11.5), and the normal plot of the residuals was noticeably straighter. ■

11.5 Generalized Least Squares

There will be cases where it is necessary, or at least deemed desirable, to use a dependent variable that does not satisfy the assumption of homogeneous variances. The transformation required to stabilize the variances may not be desirable because it destroys a good relationship between Y and X, or it destroys the additivity and normal distribution of the residuals. It may be that no transformation adequately stabilizes the variances, or a transformation made to simplify a relationship leaves heterogeneous variances. The logit and probit transformations, for example, do not stabilize the variances. The arcsin transformation of binomial proportions will stabilize the variances only if the sample

Figure 11.4 Residual sum of squares plotted against λ for the Box–Cox transformation in the soybean experiments. The lower and upper limits of the approximate 95% confidence interval estimate of λ are shown by λ^- and λ^+, respectively.

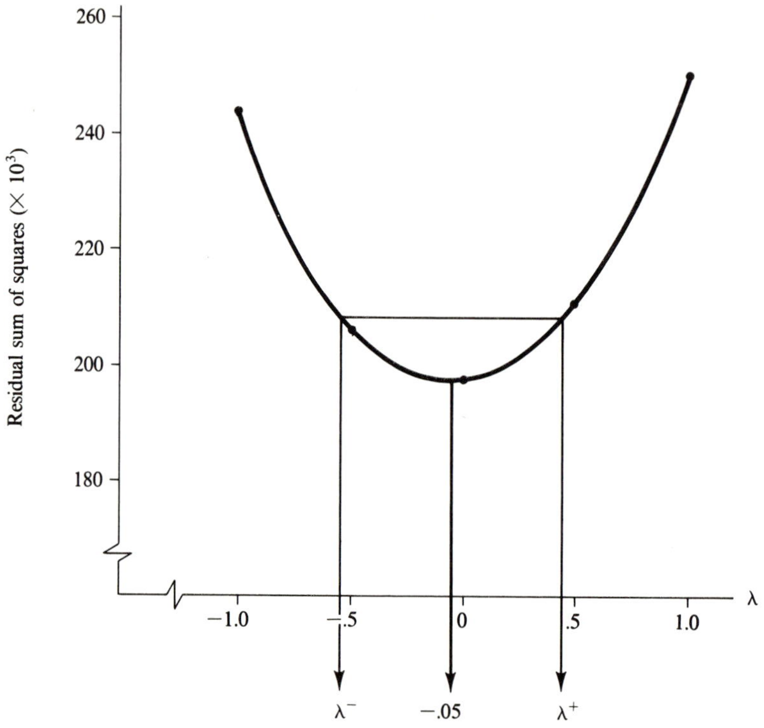

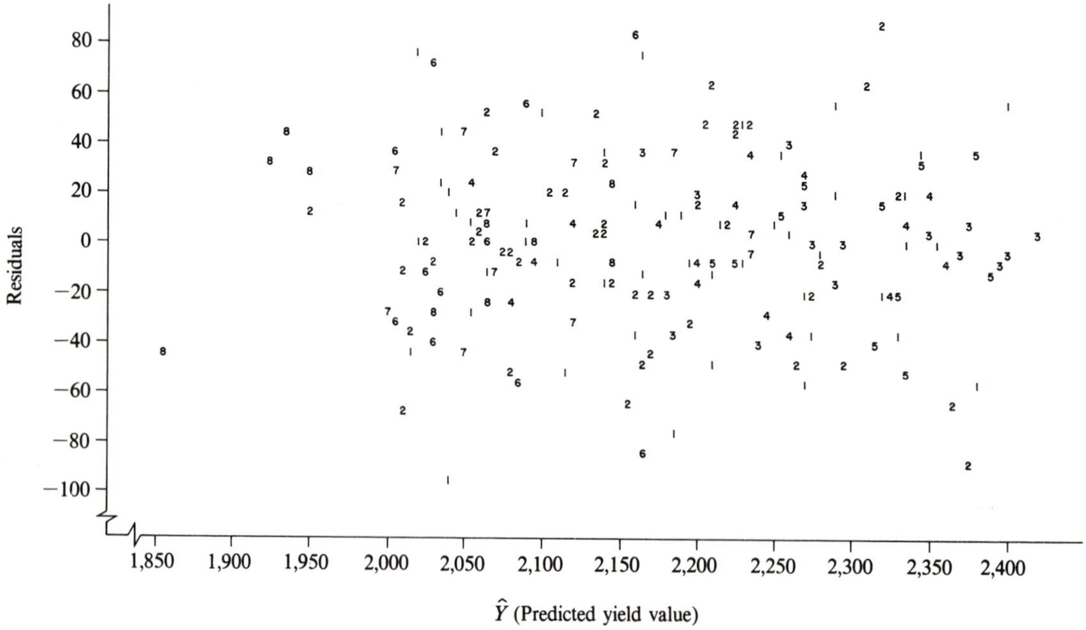

Figure 11.5 Plot of e_i versus $\hat{Y}$, after the logarithmic transformation, from the combined analysis of four experiments on the effects of ozone and sulfur dioxide on soybean yields.

sizes, n_i, are equal. Otherwise, the variances will be proportional to $1/n_i$ and remain unequal after transformation. If treatment means are based on unequal numbers of observations, the variances will differ even if the original observations had homogeneous variances. Analysis on the original scale is preferred in such cases.

Weighted versus Generalized Least Squares

Ordinary least squares estimation does not provide minimum variance estimates of the parameters when $\mathbf{Var}(\boldsymbol{\varepsilon}) \neq \boldsymbol{I}\sigma^2$. This section presents the estimation procedure that does provide minimum variance linear unbiased estimates when the variance–covariance matrix of the errors is an arbitrary symmetric positive definite matrix, $\mathbf{Var}(\boldsymbol{\varepsilon}) = \boldsymbol{V}$. This procedure will be considered in two steps, although the same principle is involved in both. First, the case will be considered where the ε_i have unequal variances but are independent; $\boldsymbol{V}$ is a diagonal matrix of the unequal variances. Second, the general case will be considered where, in addition to heterogeneous variances, the errors are not independent. Convention labels the first case **weighted least squares** and the second (more general) case **generalized least squares.**

11.5.1 Weighted Least Squares

The linear model is assumed to be

$$\boldsymbol{Y} = \boldsymbol{X\beta} + \boldsymbol{\varepsilon} \qquad [11.10]$$

with

$$\begin{aligned}\mathbf{Var}(\boldsymbol{\varepsilon}) &= \boldsymbol{V}\sigma^2 \\ &= \mathrm{Diag}(a_1^2 \ a_2^2 \ \ldots \ a_n^2)\sigma^2\end{aligned}$$

The variance of ε_i, and Y_i, is $a_i^2\sigma^2$, and all covariances are zero.

General Principle

The variance of a random variable is changed when the random variable is multiplied by a constant:

$$\sigma^2(cZ) = c^2[\sigma^2(Z)] \qquad [11.11]$$

where c is a constant. If the constant is chosen to be proportional to the reciprocal of the standard deviation of Z, $c = k/\sigma(Z)$, the variance of the rescaled variable is k^2:

$$\sigma^2(cZ) = \left(\frac{k}{\sigma(Z)}\right)^2 \sigma^2(Z) = k^2 \qquad [11.12]$$

Thus, if each observation in $\boldsymbol{Y}$ is divided by the proportionality factors, a_i, the rescaled independent variables will have equal variances σ^2 and ordinary least squares can be applied.

This is the principle followed in weighted least squares. The dependent variable is rescaled such that $\boldsymbol{V} = \boldsymbol{I}\sigma^2$ after rescaling and then ordinary least squares is applied to the rescaled variables. (The same principle is used in generalized least squares although the weighting is more complicated.) This rescaling gives weight to each observation proportional to the reciprocal of its standard deviation. The observations with the greater precision (smaller standard deviation) receive the greater weight.

Matrix Formulation

The matrix formulation of weighted regression is as follows. Define the matrix $\boldsymbol{V}^{1/2}$ to be the diagonal matrix consisting of the square roots of the diagonal elements of $\boldsymbol{V}$, so that $\boldsymbol{V}^{1/2}\boldsymbol{V}^{1/2} = \boldsymbol{V}$. The weighting matrix $\boldsymbol{W}$ that rescales $\boldsymbol{Y}$ to have common variances is

$$\boldsymbol{W} = (\boldsymbol{V}^{1/2})^{-1}$$

$$= \begin{bmatrix} 1/a_1 & 0 & \cdots & 0 \\ 0 & 1/a_2 & & 0 \\ \vdots & \vdots & \ddots & \vdots \\ 0 & 0 & \cdots & 1/a_n \end{bmatrix} \qquad [11.13]$$

where the a_i are constants that reflect the proportional differences in the variances of ε_i. Notice that $\boldsymbol{WW} = \boldsymbol{V}^{-1}$. Premultiplying both sides of the model by $\boldsymbol{W}$ gives

$$\boldsymbol{WY} = \boldsymbol{WX\beta} + \boldsymbol{W\varepsilon} \qquad [11.14]$$

or

$$\boldsymbol{Y}^* = \boldsymbol{X}^*\boldsymbol{\beta} + \boldsymbol{\varepsilon}^* \qquad [11.15]$$

where $\boldsymbol{Y}^* = \boldsymbol{WY}$, $\boldsymbol{X}^* = \boldsymbol{WX}$, and $\boldsymbol{\varepsilon}^* = \boldsymbol{W\varepsilon}$. The variance of $\boldsymbol{\varepsilon}^*$ is, from the variance of linear functions,

$$\mathbf{Var}(\boldsymbol{\varepsilon}^*) = \boldsymbol{W}[\mathbf{Var}(\boldsymbol{\varepsilon})]\boldsymbol{W}' = \boldsymbol{WVW}\sigma^2 = \boldsymbol{I}\sigma^2 \qquad [11.16]$$

since $\boldsymbol{WVW} = (\boldsymbol{V}^{1/2})^{-1}\boldsymbol{V}^{1/2}\boldsymbol{V}^{1/2}(\boldsymbol{V}^{1/2})^{-1} = \boldsymbol{I}$. The usual assumption of equal variances is met and ordinary least squares can be used on $\boldsymbol{Y}^*$ and $\boldsymbol{X}^*$ to estimate $\boldsymbol{\beta}$.

$\hat{\boldsymbol{\beta}}$

The weighted least squares estimate of $\boldsymbol{\beta}$ is

$$\hat{\boldsymbol{\beta}} = (\boldsymbol{X}^{*\prime}\boldsymbol{X}^*)^{-1}\boldsymbol{X}^{*\prime}\boldsymbol{Y}^* \qquad [11.17]$$

or, expressed in terms of the original $\boldsymbol{X}$ and $\boldsymbol{Y}$,

$$\hat{\boldsymbol{\beta}} = (\boldsymbol{X}'\boldsymbol{WWX})^{-1}(\boldsymbol{X}'\boldsymbol{WWY})$$

$$= (\boldsymbol{X}'\boldsymbol{V}^{-1}\boldsymbol{X})^{-1}(\boldsymbol{X}'\boldsymbol{V}^{-1}\boldsymbol{Y}) \qquad [11.18]$$

The variance of $\hat{\beta}$ is

$$\mathbf{Var}(\hat{\beta}) = (X'V^{-1}X)^{-1}\sigma^2 \qquad [11.19]$$

Weighted least squares, which is equivalent to ordinary least squares applied to the transformed variables, finds the solution $\hat{\beta}$ that minimizes $e^{*\prime}e^* = e'V^{-1}e$, not $e'e$. The analysis of variance of interest is the analysis of Y^*. The fitted values and the residuals on the transformed scale, $\hat{Y}^*$ and e^*, are the appropriate quantities to inspect for behavior of the model. Not all regression programs automatically provide the weighted residuals e^*; BMDP does (Dixon, 1981). Usually, the regression results will be presented on the original scale so that some of the following results are given for both scales. The transformation between scales for the fitted values and for the residuals is the same as the original transformation between Y and Y^*.

$\hat{Y}$ and $\hat{Y}^*$

The fitted values on the transformed scale are obtained by

$$\begin{aligned}\hat{Y}^* &= X^*\hat{\beta} \\ &= X^*(X^{*\prime}X^*)^{-1}X^{*\prime}Y^* = P^*Y^* \end{aligned} \qquad [11.20]$$

where P^* is the projection matrix for projecting Y^* onto the space defined by X^*. The $\hat{Y}^*$ are transformed back to the original scale by

$$\hat{Y} = W^{-1}\hat{Y}^* = X\hat{\beta} \qquad [11.21]$$

Their respective variances are

$$\mathbf{Var}(\hat{Y}^*) = X^*(X'V^{-1}X)^{-1}X^{*\prime}\sigma^2 = P^*\sigma^2 \qquad [11.22]$$

and

$$\mathbf{Var}(\hat{Y}) = X(X'V^{-1}X)^{-1}X'\sigma^2 \qquad [11.23]$$

e and e^*

The observed residuals are $e^* = Y^* - \hat{Y}^*$ on the transformed scale and $e = Y - \hat{Y}$ on the original scale. Their variances are

$$\begin{aligned}\mathbf{Var}(e^*) &= [I - X^*(X'V^{-1}X)^{-1}X^{*\prime}]\sigma^2 \\ &= (I - P^*)\sigma^2 \end{aligned} \qquad [11.24]$$

and

$$\mathbf{Var}(e) = [V - X(X'V^{-1}X)^{-1}X']\sigma^2 \qquad [11.25]$$

Note that the usual properties of ordinary least squares apply to the transformed variables Y^*, e^*, and X^*.

Example 11.4

For illustration, suppose the dependent variable is a vector of treatment means with unequal numbers, r_i, of observations per mean. If the original observations have equal variances, the means will have variances σ^2/r_i. Thus,

$$V\sigma^2 = \begin{bmatrix} 1/r_1 & 0 & \dots & 0 \\ 0 & 1/r_2 & \dots & 0 \\ \vdots & & & \vdots \\ 0 & 0 & \dots & 1/r_n \end{bmatrix} \sigma^2 \qquad [11.26]$$

The weighting matrix that gives $\mathbf{Var}(\varepsilon^*) = I\sigma^2$ is

$$W = \begin{bmatrix} \sqrt{r_1} & 0 & \dots & 0 \\ 0 & \sqrt{r_2} & \dots & 0 \\ \vdots & & \ddots & \vdots \\ 0 & 0 & \dots & \sqrt{r_n} \end{bmatrix} \qquad [11.27]$$

■

Estimating the Weights

In Example 11.4, it is clear that the variances of the dependent variable will not be equal and what the weighting matrix should be. In other cases, the variances may not be known a priori and their relative sizes will have to be determined from the data. If true replicates are available in the data set, the different variances can be estimated from the variance among the replicates for each group. In the absence of true replication, one might estimate the variances by using "near" replicates, groups of observations having nearly the same level of the independent variable(s). The variances of the "near" replicates might be plotted against the means of the "near" replicates, from which the relationship between the variance and the mean might be deduced and used to "estimate" the variance for each Y_i.

Computer Programs

A weighted least squares procedure is available in most least squares computer programs. Care must be used to specify the appropriate "weights" for the specific program. The weights in PROC GLM and PROC REG (SAS Institute, Inc., 1985d, 1985f), for example, must be specified as a column vector of the *squares* of the diagonal elements in W.

11.5.2 Generalized Least Squares

Generalized least squares extends the usual linear model to allow for an arbitrary positive definite variance–covariance matrix of ε, $\mathbf{Var}(\varepsilon) = V\sigma^2$. The diagonal elements need not be equal and the off-diagonal elements need not be 0. The positive definite condition ensures that it is a proper variance matrix; that is, any linear function of the observations will have a positive variance. As with weighted least squares, a linear transformation is made on Y such that the transformed model will satisfy the least squares assumption of $\mathbf{Var}(\varepsilon^*) = I\sigma^2$.

Finding the Weighting Matrix

For any positive definite matrix V it is possible to find a nonsingular symmetric matrix T such that

$$TT = T^2 = V \qquad [11.28]$$

Since T is nonsingular, it has an inverse T^{-1}. Premultiplying the model by T^{-1} gives

$$Y^* = X^*\beta + \varepsilon^* \qquad [11.29]$$

where $Y^* = T^{-1}Y$, $X^* = T^{-1}X$, and $\varepsilon^* = T^{-1}\varepsilon$. With this transformation,

$$\text{Var}(\varepsilon^*) = T^{-1}VT^{-1}\sigma^2 = I\sigma^2 \qquad [11.30]$$

and ordinary least squares is again appropriate for Y^* and X^*. Weighted least squares is a special case of generalized least squares. If V is a diagonal matrix, the appropriate T^{-1} is W as defined in equation 11.13.

Many of the least squares regression computer programs are not designed to handle generalized least squares. It is always possible, however, to make the indicated transformation, equation 11.29, and use ordinary least squares, or to resort to a matrix algebra computer program to do generalized least squares.

Warnings

One must be somewhat cautious in the use of generalized least squares. As with weighted least squares, the sum of squares $e^{*\prime}e^*$ is minimized and $\hat{\beta}$ is the best linear unbiased estimator of β *if* V is known. In most cases, however, V is unknown and must be estimated from the data. When an estimate of V is used, the solution obtained is no longer the minimum variance solution. In the worst cases where there is limited information with which to estimate V, the generalized least squares estimates can have larger variances than the ordinary least squares estimates. (This comment also applies to weighted least squares, but there the estimation problem is much less difficult.) Further, it is possible for the generalized least squares regression line, if plotted on the original scale, to "miss" the data. That is, all of the observed data points can fall on one side of the regression line. The necessary condition for this to occur is sufficiently large positive off-diagonal elements in V. This does not depend on whether V is known or estimated. Estimation of V, however, will likely cause the problem to occur more frequently. Such a result is not a satisfactory solution to a regression problem even though it may be the best linear unbiased estimate (as it is when V is known). Plotting the data and the regression line on the original scale will make the user aware of any such results.

Example 11.5

The example used to illustrate weighted and generalized least squares comes from an effort to develop a prediction equation for tree diameter at 54 inches above the ground (*DBH*) based on data from diameters at various stump heights. The objective was to predict amount of timber illegally removed from a tract of land and *DBH* was one of the measurements needed. Diameter at 54 inches (*DBH*) and stump diameters (*SD*) at stump heights (*SHt*) of 2, 4, 6, 8, 10, and 12 inches above ground were measured on 100 standing trees in an adjacent, similar stand. The trees were grouped into 2-inch *DBH* classes. There were $n = 4, 16, 42, 26, 9$, and 3 trees in *DBH* classes 6, 8, 10, 12, 14, and 16 inches, respectively.

Table 11.2 Averages by DBH class of logarithms of the ratios of stump diameter to diameter at 54 inches of 100 pine trees, $\bar{Y}_{ij.}$, where $Y_{ijk} = \ln(SD_{ijk}) - \ln(DBH_{ik})$. The values for the independent variable $X_k = 54^c - (SHt)^c$ for $c = .1$ are shown in the last row.

		Stump Height (inches above ground)					
DBH (in.)	*No. Trees*	2	4	6	8	10	12
6	4	.3435	.3435	.2715	.1438	.0719	.0719
8	16	.3143	.2687	.2548	.2294	.1674	.1534
10	42	.2998	.2514	.2083	.1733	.1463	.1209
12	26	.3097	.2705	.2409	.1998	.1790	.1466
14	9	.2121	.1859	.1597	.1449	.1039	.1039
16	3	.2549	.2549	.1880	.1529	.1529	.1529
	X_k	.4184	.3415	.2940	.2590	.2313	.2081

It was argued that the ratio of DBH to the stump diameter at a particular height should be a monotonically decreasing function approaching 1 as the stump height approached 54 inches. This relationship has the form of an exponential decay function but with much sharper curvature than the exponential function allows. These considerations led to a model in which the dependent variable was defined as

$$Y_{ijk} = [\ln(SD_{ijk}) - \ln(DBH_{ik})]$$

and the independent variable as

$$X_j = [54^c - (SHt_j)^c],$$

where i is the DBH class ($i = 1, \ldots, 6$); j is the stump height class ($j = 1, \ldots, 6$); k is the tree within each DBH class ($k = 1, \ldots, n_i$); and $\ln(SD_{ijk})$ and $\ln(DBH_{ik})$ are the logarithms of stump diameters and DBH. The averages of Y_{ijk} over k for each DBH–stump height category are given in Table 11.2. The exponent c, applied to the stump heights, was used to straighten the relationship (on the logarithmic scale) and was chosen by finding the value, $c = .1$, which minimized the residual sum of squares for the linear relationship. Thus, the model is

$$\bar{Y}_{ij.} = \beta X_j + \bar{\varepsilon}_{ij.},$$

a no-intercept model, where the $\bar{Y}_{ij.}$ are the DBH–stump height cell means of Y_{ijk} given in Table 11.2. Thus, $\boldsymbol{Y}$ is a 36×1 vector of the six values of $\bar{Y}_{1j.}$ in the first row of Table 11.2 followed by the six values of $\bar{Y}_{2j.}$ in the second row, and so on. The $\boldsymbol{X}$ vector consists of six repeats of the six values of X_j corresponding to the six stump heights.

It is not appropriate to assume $\mathbf{Var}(\boldsymbol{\varepsilon}) = \boldsymbol{I}\sigma^2$ in this example for two reasons. First, the dependent variable consists of averages of differing numbers of trees within each DBH class ranging from $n = 3$ to $n = 42$. Second, all Y_{ijk} from the same tree (same i and k) are correlated due to the fact that DBH_{ik} is involved in the definition of Y_{ijk} in each case. Observations in different DBH classes are independent because different trees are involved. It will be assumed that the variance–covariance matrix of the observations within each DBH class is the same over DBH classes. Thus, the 36×36 variance–covariance matrix $\mathbf{Var}(\boldsymbol{\varepsilon})$ will have the form

$$\mathbf{Var}(\boldsymbol{\varepsilon}) = \begin{bmatrix} \boldsymbol{V}/4 & \mathbf{0} & \mathbf{0} & \mathbf{0} & \mathbf{0} & \mathbf{0} \\ \mathbf{0} & \boldsymbol{V}/16 & \mathbf{0} & \mathbf{0} & \mathbf{0} & \mathbf{0} \\ \mathbf{0} & \mathbf{0} & \boldsymbol{V}/42 & \mathbf{0} & \mathbf{0} & \mathbf{0} \\ \mathbf{0} & \mathbf{0} & \mathbf{0} & \boldsymbol{V}/26 & \mathbf{0} & \mathbf{0} \\ \mathbf{0} & \mathbf{0} & \mathbf{0} & \mathbf{0} & \boldsymbol{V}/9 & \mathbf{0} \\ \mathbf{0} & \mathbf{0} & \mathbf{0} & \mathbf{0} & \mathbf{0} & \boldsymbol{V}/3 \end{bmatrix} \quad [11.31]$$

where $\boldsymbol{V}$ is the 6×6 variance–covariance matrix for Y_{ijk} from the same tree. That is, the diagonal elements of $\boldsymbol{V}$ are variances of Y_{ijk} for a given stump height and the off-diagonal elements are covariances between Y_{ijk} at two different stump heights for the same tree.

The estimate of $\boldsymbol{V}$ was obtained by defining six variables from the Y_{ijk}, one for each stump height (level of j). Thus, the matrix $\boldsymbol{Y}$ of data is 100×6 (there were 100 trees), with each column containing the measurements from one of the six stump heights. The variance–covariance matrix $\boldsymbol{V}$ was estimated as

$$\hat{\boldsymbol{V}} = [\boldsymbol{Y}'(\boldsymbol{I} - \boldsymbol{J}/n)\boldsymbol{Y}]/99$$

$$= \begin{bmatrix} 86.2 & 57.2 & 63.0 & 53.9 & 48.9 & 52.5 \\ 57.2 & 71.4 & 59.5 & 45.2 & 35.0 & 39.3 \\ 63.0 & 59.5 & 100.2 & 73.8 & 51.8 & 50.6 \\ 53.9 & 45.2 & 73.8 & 97.3 & 62.9 & 53.7 \\ 48.9 & 35.0 & 51.8 & 62.9 & 76.5 & 59.3 \\ 52.5 & 39.3 & 50.6 & 53.7 & 59.3 & 78.6 \end{bmatrix} 10^{-4} \quad [11.32]$$

The correlations in $\hat{\boldsymbol{V}}$ range from .47 to .77. It is likely that the form of $\hat{\boldsymbol{V}}$ could be simplified by assuming, for example, a common variance or equality of subsets of the correlations. This would improve the estimates of the weights if the simplications were justified. For this example, the general covariance matrix was used.

Generalized least squares was used to estimate β and its standard error. $\hat{\boldsymbol{V}}$ was multiplied by (99×10^2), rounded to two digits, and then substituted for $\boldsymbol{V}$ in equation 11.31 to give the weighting matrix for generalized least squares. The computations were done with IML (SAS Institute, Inc., 1985a), which is an interactive matrix program. The regression equation obtained was

$$\hat{Y}_{ij} = .7277X_j$$

with $s(\hat{\beta}) = .0270$. The regression coefficient is significantly different from zero.

For comparison, the unweighted regression and the weighted regression using only the numbers of trees in the *DBH* classes as weights were also run. The resulting regression equations differed little from the generalized regression results but the computed variances of the estimates were very different. The computed results from the two regressions were as follows:

Unweighted:
$\hat{Y}_{ij} = .6977X_j \quad \text{with} \quad s(\hat{\beta}) = .0237$

Weighted by n_i:
$\hat{Y}_{ij} = .7147X_j \quad \text{with} \quad s(\hat{\beta}) = .0148$

Comparison of the standard errors appears to indicate a *loss* in precision from using generalized least squares. However, the variances computed by the standard regression formulas assumed that $\mathbf{Var}(\varepsilon) = \boldsymbol{I}\sigma^2$ in the unweighted case and $\mathbf{Var}(\varepsilon) = \text{Diag}(\{1/n_i\})\sigma^2$ in the weighted regression, neither of which is correct in this example.

The correct variance when ordinary least squares is used but where $\mathbf{Var}(\varepsilon) \neq \boldsymbol{I}\sigma^2$ is given by

$$\sigma^2(\hat{\boldsymbol{\beta}}) = (\boldsymbol{X}'\boldsymbol{X})^{-1}\boldsymbol{X}'[\mathbf{Var}(\varepsilon)]\boldsymbol{X}(\boldsymbol{X}'\boldsymbol{X})^{-1} \qquad [11.33]$$

When weighted least squares is used but with an incorrect weight matrix $\boldsymbol{V}$, the correct variance is given by

$$\sigma^2(\hat{\boldsymbol{\beta}}) = (\boldsymbol{X}'\boldsymbol{V}^{-1}\boldsymbol{X})^{-1}\boldsymbol{X}'\boldsymbol{V}^{-1}[\mathbf{Var}(\varepsilon)]\boldsymbol{V}^{-1}\boldsymbol{X}(\boldsymbol{X}'\boldsymbol{V}^{-1}\boldsymbol{X})^{-1} \qquad [11.34]$$

When $\hat{\boldsymbol{V}}$ (equation 11.32) is substituted in equation 11.31 to give an estimate of $\mathbf{Var}(\varepsilon)$, equations 11.33 and 11.34 give estimates of the variances of the regression coefficients for the unweighted and weighted (by n_i) analyses. The resulting standard errors of $\hat{\beta}$ are

Unweighted:
$s(\hat{\beta}) = .04850$

Weighted by n_i:
$s(\hat{\beta}) = .03215$

The efficiency of generalized least squares relative to unweighted least squares and to weighting by n_i is 3.22 and 1.42, respectively, in this example. [The relative efficiency of two estimates, $\hat{\theta}_1$ to $\hat{\theta}_2$, is measured as the ratio of variances, R.E. $= s^2(\hat{\theta}_2)/s^2(\hat{\theta}_1)$.] These relative efficiencies are biased in favor of generalized regression because an estimated variance–covariance matrix has been used in place of the true variance–covariance matrix. Nevertheless, in this example they show major increases in precision which result from accounting for unequal variances and correlation structure in the data. Comparison of the standard errors computed from the unweighted analysis and the weighted analysis with the results of equations 11.33 and 11.34 illustrates the underestimation of variances that commonly occurs when positively correlated errors in the data are ignored. ■

11.6 Summary

The first sections of this chapter discussed transformations of the independent and dependent variables to make the model simpler in some sense, or to make the assumptions of homogeneous variance and normality more nearly satisfied. Transformations on the independent variable affect only the form of the model. Transformations to stabilize variances or to more nearly satisfy normality must be made on the dependent variable. The power family of transformations plays an important role in all three cases.

The ladder of transformations and the rules for determining the transformation are easily applied as long as the model is reasonably simple. In more complex cases, the Box–Tidwell method provides power transformations on the independent variables that give the best fit to a particular model; the result is dependent on the model chosen. The Box–Cox transformation provides a power transformation on the dependent variable with the more general criterion of satisfying all aspects of the distribution assumption on $\mathbf{Y}$: $\mathbf{Y} \sim N(\mathbf{X}\boldsymbol{\beta}, \mathbf{I}\sigma^2)$. The result and the relative emphasis the method gives to simplifying the model, stabilizing variance, and improving normality are dependent on the choice of $\mathbf{X}\boldsymbol{\beta}$. In no case are we assured that the appropriate power transformation exists to satisfy all criteria. All precautions should be taken to verify the adequacy of the model and the least squares results.

The last section covered weighted least squares and generalized least squares. These methods address the specific situation where the scale of the dependent variable has already been decided but where the basic assumption of $\mathbf{Var}(\boldsymbol{\varepsilon}) = \mathbf{I}\sigma^2$ is not satisfied. In such cases, the minimum variance estimators are obtained only if the true $\mathbf{Var}(\boldsymbol{\varepsilon})$ is taken into account by using weighted least squares or generalized least squares, as the situation requires.

EXERCISES

* 11.1. This exercise uses Land's data on tolerance of certain families of pine to saltwater flooding given in Table 11.1. For this exercise, replace Hours = 0 with 1 and $Y = .00$ in family 3 with .01 to avoid problems with taking logarithms.

a. Plot Y = chloride content against X = hours. Summarize what the plot suggest about homogeneous variances, about normality, and about the type of response curve needed if *no* transformations are made.

b. Use the plot of the data and the ladder of transformations to suggest a transformation on Y that might straighten the relationship. Suggest a transformation on X that might straighten the relationship. In view of your answer to part (a), would you prefer the transformation on Y or on X?

c. Assume a common quadratic relationship of $Y^{(\lambda)}$ with X for all families, but allow each family to have its own intercept. Use the Box–Cox transformation for $\lambda = 0, .2, .3, .4, .5, .7$, and 1.0, and plot the residual sum of squares in each case against λ. At what value of λ does the minimum residual sum of squares occur? Graphically determine 95% confidence limits on λ. What power transformation on Y do you choose?

d. Repeat part (c) using a *linear* relationship between $Y^{(\lambda)}$ and X. Show how this changes the Box–Cox results and explain (in words) why the results differ.

e. Use the Box–Cox transformation adopted in part (c) as the dependent variable. If $Y^{(\lambda)}$ is regressed on X using the quadratic model in part (c), the quadratic term is highly significant. Use the Box–Tidwell method to find a power transformation on X that will straighten the relationship. Plot the residuals from the regression of $Y^{(\lambda)}$ on X^{α}, the Box–Tidwell transformation on X, against $\hat{Y}$ and in a normal plot. Do you detect any problems?

11.2. The Land data given in Table 11.1 are percentage data. Are they binomially distributed data? Would you a priori expect the arcsin transformation to work?

* 11.3. A replicated corn yield trial (25 entries in three blocks) grown at five locations gave data in which the response variable varied from 55 bu/acre in a particularly dry location to 190 bu/acre in the most favorable environment. The mean yields and the experimental error variances (each with 48 degrees of freedom) for the five locations were as follows:

Mean Yield	*Error Variance*
55	68
105	139
131	129
148	325
190	375

Consider two options for handling the heterogeneous variances in a combined analysis of variance: (1) an appropriate transformation on Y, and (2) weighted least squares.

a. What transformation would you suggest from inspection of the relationship between the mean and the variance?
b. Explain what your weighting matrix would be if you used weighted least squares. This will be a very large matrix. Explain how you could do the weighting without forming this matrix.
c. A third option would be to ignore the heterogeneous variances and proceed with the combined analysis. Discuss the merits of the three alternatives and how you would decide which to use.

11.4. The monomolecular growth model has the form

$$\mathscr{E}(Y_i) = \alpha(1 - \beta e^{-kt})$$

Is this model nonlinear in the parameters? Can it be linearized with an appropriate transformation on Y? Can it be linearized if α is known?

* 11.5. A dose–response model based on the Weibull function can be written as

$$\mathscr{E}(Y_i) = \alpha\{\exp[-(X/\gamma)^{\delta}]\}$$

Does taking the logarithm of Y linearize this model?

11.6. A nonlinear model for a chemical reaction rate can be formulated as

$$\mathscr{E}(Y_i) = \alpha X/(1 + \beta X_1 + \gamma X_2)$$

Does the reciprocal transformation on Y give a model that is linear in the parameters? Does a redefinition of the parameters make the model linear in the parameters?

* 11.7. The water runoff data in exercise 5.1 were analyzed using $\ln(Q)$, where Q was the

peak rate of flow. Use the Box–Cox method with a linear model containing the logarithms of all nine independent variables to determine the transformation on Q. Is $\lambda = 0$ within the 95% confidence interval estimate of λ?

11.8. The growth data (Y = dry weight in grams) in the accompanying table were taken on four different independent experimental units at each of six different ages (X = age in weeks).

X (weeks of age)					
1	2	3	5	7	9
8	35	57	68	76	85
10	38	63	76	95	98
12	42	68	86	103	105
15	48	74	90	105	110

a. Plot Y versus X. Use the ladder of transformations to determine a power transformation on Y that will straighten the relationship. Determine a power transformation on X that will straighten the relationship.

b. Use the Box–Tidwell method to determine a power transformation on X for the linear model. Does this differ from what you decided using the ladder of transformations? Is there any problem with the behavior of the residuals?

c. Observe the nature of the dispersion of Y for each level of X. Does there appear to be any problem with respect to the least squares assumption of constant variance? Will either of your transformations in part (a) improve the situation? [Do trial transformations on Y for the first, fourth, and sixth levels of X (ages 1, 5, and 9), and observe the change in the dispersion.]

* 11.9. Use the data in exercise 11.8 and the Box–Cox method to arrive at a transformation on Y. Recall that the Box–Cox method assumes a particular model $\mathscr{E}(Y) = X\boldsymbol{\beta}$. For this exercise, use $\mathscr{E}(Y_i) = \beta_0 + \beta_1 X_i$. Plot SS[Res($\lambda$)] versus λ, find the minimum, and determine approximate 95% confidence limits on λ. What choice of λ does the Box–Cox method suggest for this model? Fit the resulting regression equation, plot the transformed data and the regression equation, and observe the nature of the residuals. Does the transformation appear to be satisfactory with respect to the straight-line relationship? With respect to the assumption of constant variance? [*Note:* The purposes of exercises 11.9–11.12 are, in addition to demonstrating the use of the Box–Cox transformation, to show the dependence of the method on the assumed model and to illustrate that obtaining the power transformation via the Box–Cox method does not guarantee either that the model fits or that the usual least squares assumptions are automatically satisfied.]

11.10. Repeat exercise 11.9 using the quadratic polynomial model in X. With this model, to which transformation does the Box–Cox method lead and does it appear satisfactory?

*11.11. Repeat exercise 11.9 using $X^* = \ln(X)$ in the linear model. What transformation do you obtain this time and is it satisfactory?

11.12. Repeat exercise 11.11, but allow a quadratic model in $X^* = \ln(X)$. What transformation do you obtain and does it appear to be more satisfactory?

*11.13. The corn borer survival data, number of larvae surviving 3, 6, 9, 12, and 21 days after inoculation, in exercise 8.4 were analyzed without transformation. "Number of larvae" might be expected not to have homogeneous variance. Plot the residuals from the analysis of variance against $\hat{Y}$. Do they provide any indication of a problem? Use the Box–Cox method to estimate a transformation for "number of larvae" where $X\beta$ is defined for the analysis of variance model. Is a transformation suggested? If so, do the appropriate transformation and summarize the results.

11.14. Show that P^* in $\hat{Y}^* = P^* Y^*$, equation 11.20, is idempotent.

11.15. Use equation 11.21 to obtain the coefficient matrix on Y, the original variable, that gives $\hat{Y}$. Show that this matrix is idempotent.

11.16. Use equation 11.21 to express $\hat{Y}'\hat{Y}$ as a quadratic function of Y. Likewise, obtain $e'e$, where $e = Y - \hat{Y}$, as a quadratic function of Y. Show that:

a. Neither coefficient matrix is idempotent.

b. The two coefficient matrices are not orthogonal.

What are the implications of these results?

11.17. Use the variance of linear functions to derive $\mathbf{Var}(\hat{\beta})$, equation 11.19.

11.18. Use the variance of linear functions to derive $\mathbf{Var}(\hat{Y}^*)$, equation 11.22.

11.19. Derive $\mathbf{Var}(\hat{\beta})$ when ordinary least squares is used to estimate β but where $\mathbf{Var}(\varepsilon) \neq I\sigma^2$, equation 11.33.

11.20. The data used in the generalized least squares analysis in the text to develop a model to relate DBH (diameter at breast height, 54 inches) to diameters at various stump heights (Example 11.5) are given in Table 11.2. The numbers in the table are $\bar{Y}_{ij.}$, where Y_{ijk} and X_j are defined in the text. The estimated variance–covariance matrix is shown in equation 11.32.

a. Use a computer matrix program to do the generalized least squares analysis on these data as outlined in Example 11.5. Notice that the model contains a zero intercept. Give the regression equation, the standard error of the regression coefficient, and the analysis of variance summary. (Your answers may differ slightly from those in Example 11.5 unless the variance–covariance matrix is rounded as described.)

b. It would appear reasonable to simplify the variance–covariance matrix, equation 11.32, by assuming homogeneous variances and common covariances. Average the appropriate elements of $\hat{V}$ to obtain a common variance and a common covariance. Redo the generalized least squares regression with V redefined in this way. Compare the results with the results in part (a) and the unweighted regression results given in Example 11.5.

CHAPTER
TWELVE

Collinearity

Chapters 9–11 have outlined the problem areas, discussed methods of detecting the problems, and discussed the use of transformations to alleviate the problems.

This chapter addresses the collinearity problem, with the emphasis on understanding the relationships among the independent variables rather than on the routine application of biased regression methods. Principal component analysis and Gabriel's biplots are used to explore the correlational structure. Two of the biased regression methods are presented and their limitations are discussed.

Origins of Collinearity

The collinearity problem in regression arises when at least one linear function of the independent variables is very nearly equal to zero. (Technically, a set of vectors is collinear when a linear function is *exactly* equal to zero. In general discussions of the collinearity problem, the term "collinear" is often used to apply to linear functions that are only *approximately* zero. This convention will be followed in this text.) This near-singularity may arise in several ways:

1. An inbuilt mathematical constraint on variables that forces them to add to a constant will generate a collinearity. For example, frequencies of alleles at a locus will add to 1 if the frequencies of all alleles are recorded, or nearly to 1 if a rare allele is not scored. Generating new variables as transformations of other variables can produce a collinearity among the set of variables involved. Ratios of variables or powers of variables frequently will be nearly collinear with the original variables.
2. Component variables of a system may show near-linear dependencies because of the biological, physical, or chemical constraints of the system. Various measures of size of an organism will show dependencies as will amounts of chemicals in the same biological pathway, or measures of rainfall, temperature, and elevation in an environmental system. Such correlational structures are properties of the system and can be expected to be present in all observational data obtained from the system.
3. Inadequate sampling may generate data in which the near-linear depen-

dencies are an artifact of the data collection process. Unusual circumstances also can cause unlikely correlations to exist in the data, correlations that may not be present in later samplings or samplings from other similar populations.

4. A bad experimental design may cause some model effects to be nearly completely confounded with others. This is the result of choosing levels of the experimental factors in such a way that there are near-linear dependencies among the columns of $\boldsymbol{X}$ representing the different factors. Usually, experimental designs are constructed so as to ensure that the different treatment factors are orthogonal, or very nearly orthogonal, to each other.

One may not always be able to identify clearly the origin of the collinearity problem, but it is important to understand its nature as much as possible. Knowing the nature of the collinearity problem will often suggest to the astute researcher its origin and, in turn, appropriate ways of handling the problem and of interpreting the regression results.

The first section of this chapter discusses methods of analyzing the correlational structure of the X-space with a view toward understanding the nature of the collinearity. The second section introduces biased regression as one of the classical methods of handling the collinearity problem. For all discussions in this chapter, the matrix of centered and scaled independent variables, $\boldsymbol{Z}$, is used so that $\boldsymbol{Z}'\boldsymbol{Z}$ is the correlation matrix. The artificial data set used in Section 10.3 to illustrate the measures of collinearity is used here. Chapter 13 is a case study using the methods discussed in this chapter.

12.1 Understanding the Structure of the *X*-Space

Correlation Matrix $Z'Z$

The matrix of sums of squares and products of the centered and scaled independent variables $\boldsymbol{Z}'\boldsymbol{Z}$, scaled so that the sum of squares of each variable is unity, is a useful starting point for understanding the structure of the X-space. (This is the correlation matrix if the independent variables are random variables and, for convenience, will be referred to as the *correlation matrix* even when the X's are fixed constants.) The off-diagonal elements of this matrix are the cosines of the angles between the corresponding centered and scaled vectors in the X-space. Values near 1.0 or -1.0 indicate nearly collinear vectors; values near 0 indicate nearly orthogonal vectors.

Example 12.1

The correlation matrix for the artificial data from Example 10.11 shows a very high correlation between X_1 and X_2 of $r_{12} = .996$ (Table 12.1). This indicates a near-linear dependency, which is known to exist from the manner in which the data were constructed. The relatively low correlations of X_1 and X_2 with X_3 suggest that X_3 is not involved in the collinearity problem. ■

Correlations will reveal linear dependencies involving two variables, but

Table 12.1 Correlation matrix of the independent variables for the artificial data set demonstrating collinearity.

	X_1	X_2	X_3
X_1	1.000	.996	.290
X_2	.996	1.000	.342
X_3	.290	.342	1.000

they frequently will not reveal linear dependencies involving several variables. Individual pairwise correlations can be relatively small when several variables are involved in a linear dependency. Thus, the absence of high correlations cannot be interpreted as an indication of no collinearity problem.

Detecting Near-Singularities

Near-linear dependencies involving any number of variables are revealed with a **singular value decomposition** of the matrix of independent variables, or with an **eigenanalysis** of the sums of squares and products matrix. (See Sections 2.7 and 2.8 for discussions of eigenanalysis and singular value decomposition.) For the purpose of detecting near-singularities, the independent variables should always be scaled so that the vectors are of equal length. In addition, the independent variables are often centered to remove collinearities with the intercept. (Refer to Section 10.3 for discussion on this point.) The discussion here is presented in terms of the centered and scaled independent variables $\boldsymbol{Z}$. The eigenvectors of $\boldsymbol{Z}'\boldsymbol{Z}$ that correspond to the smaller eigenvalues identify the linear functions of the Z's that show least dispersion. It is these specific linear functions that are causing the collinearity problem if one exists.

Example 12.2

The results of the eigenanalysis of the correlation matrix for Example 12.1 are shown in Table 12.2. The eigenvalues reflect a moderate collinearity problem, with the condition number being $(2.16698/.00290)^{1/2} = 27.3$. (This differs from the results in Section 10.3 because collinearities involving the intercept have been eliminated by centering the variables.) The eigenvector corresponding to the smallest eigenvalue defines the third principal component, the dimension causing the collinearity problem, as

$$\boldsymbol{W}_3 = .69998\boldsymbol{Z}_1 - .71299\boldsymbol{Z}_2 + .04100\boldsymbol{Z}_3$$

The variables primarily responsible for the near-singularity are Z_1 and Z_2, as shown by their relatively large coefficients in the third eigenvector. The coefficient

Tabel 12.2 Eigenvalues and eigenvectors of the correlation matrix of independent variables for the artificial data set.

Eigenvalue	*Eigenvector*
$\lambda_1^2 = 2.166984$	$\boldsymbol{v}_1' = (.65594 \quad .66455 \quad .35793)$
$\lambda_2^2 = .830117$	$\boldsymbol{v}_2' = (-.28245 \quad -.22365 \quad .93285)$
$\lambda_3^2 = .002898$	$\boldsymbol{v}_3' = (.69998 \quad -.71299 \quad .04100)$

for Z_3 is relatively close to zero. The coefficients on Z_1 and Z_2 are very similar in magnitude but opposite in sign, suggesting that the near-singularity is due to $(Z_1 - Z_2)$ being nearly zero. This is known to be true from the way the data were constructed; X_2 was defined as $(X_1 - 25)$ with two of the 20 numbers changed by one digit to avoid a complete singularity. After centering and scaling, Z_1 and Z_2 are very nearly identical so that their difference is almost zero.

Inspection of the first eigenvector shows that the major dispersion in the Z-space is in the dimension defined as a weighted average of the three variables

$$\boldsymbol{W}_1 = .65594\boldsymbol{Z}_1 + .66455\boldsymbol{Z}_2 + .35793\boldsymbol{Z}_3$$

with Z_1 and Z_2 receiving nearly twice as much weight as Z_3. $\boldsymbol{W}_1$ is the first principal component. The second dimension, the second principal component, is dominated by Z_3:

$$\boldsymbol{W}_2 = -.28245\boldsymbol{Z}_1 - .22365\boldsymbol{Z}_2 + .93285\boldsymbol{Z}_3$$ ■

Gabriel's Biplot

The correlational structure of the independent variables is displayed with Gabriel's biplot (Gabriel, 1971, 1972, 1978). This is an informative plot that shows (1) the relationships among the independent variables, (2) the relative similarities of the individual data points, and (3) the relative values of the observations for each independent variable. The name "biplot" comes from this simultaneous presentation of both row (observation) and column (variable) information in one plot.

The biplot uses the singular value decomposition of $\boldsymbol{Z}$, $\boldsymbol{Z} = \boldsymbol{U}\boldsymbol{L}^{1/2}\boldsymbol{V}'$. The matrices $\boldsymbol{L}^{1/2}$ and $\boldsymbol{V}$ can be obtained from the results of the eigenanalysis of $\boldsymbol{Z}'\boldsymbol{Z}$ shown in Table 12.2. $\boldsymbol{V}$ is the matrix of eigenvectors, each column being an eigenvector, and $\boldsymbol{L}^{1/2}$ is the diagonal matrix of the positive square roots of the eigenvalues. More computations are required to obtain $\boldsymbol{U}$. If the dispersion in the Z-space can be adequately represented by two dimensions, one biplot using the first and second principal component information will convey most of the information in $\boldsymbol{Z}$. If needed, additional biplots of first and third, and second and third principal components can be used. Each biplot is the projection of the dispersion in Z-space onto the plane defined by the two principal components being used in the biplot.

Example 12.3

(*Continuation of Example* 12.2) The first two principal component dimensions account for

$$\frac{\lambda_1^2 + \lambda_2^2}{\sum \lambda_i^2} = \frac{2.166984 + .830118}{3} = .999 \qquad [12.1]$$

or 99.9% of the total dispersion in the three dimensions. Therefore, a single biplot of the first and second principal components suffices; only .1% of the information in $\boldsymbol{Z}$ is ignored by not using the third principal component.

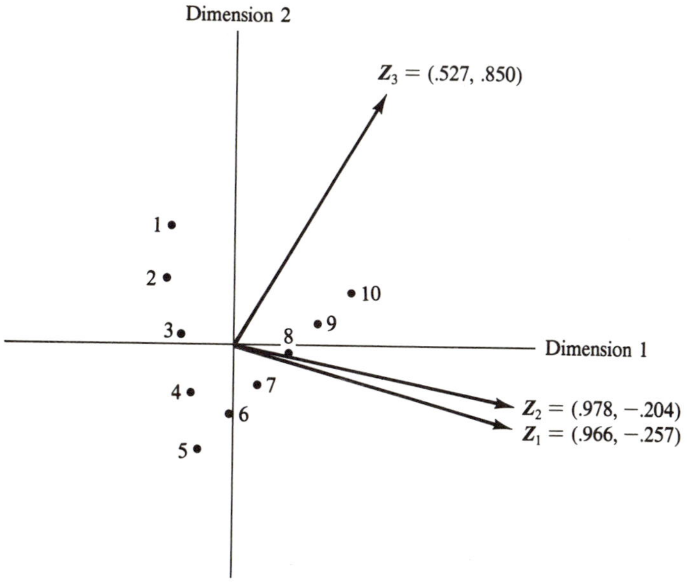

Figure 12.1 Gabriel's biplot of the first two principal component dimensions for Example 12.3.

The biplot using the first two principal component dimensions is shown in Figure 12.1. The vectors in the figure are the vectors of the independent variables *as seen in this two-dimensional projection.* The coordinates for the endpoints of the vectors, which are called *column markers,* are obtained from $\boldsymbol{L}^{1/2}\boldsymbol{V}'$:

$$\boldsymbol{L}^{1/2}\boldsymbol{V}' = \begin{bmatrix} .9656 & .9783 & .5269 \\ -.2573 & -.2038 & .8499 \\ .0377 & -.0384 & .0022 \end{bmatrix} \quad [12.2]$$

The first and second elements in column 1 are the coordinates for the $\boldsymbol{Z}_1$ vector in the biplot using the first and second principal components, the first and second elements in column 2 are the coordinates of the $\boldsymbol{Z}_2$ vector, and so on. The third number in each column of $\boldsymbol{L}^{1/2}\boldsymbol{V}'$ gives the coordinate in the third dimension for each vector, which is being ignored in this biplot. Notice, however, that none of the variable vectors is very far from zero in the third dimension. This reflects the small amount of dispersion in that dimension. ■

Variable Information

Since the $\boldsymbol{Z}_j$ vectors were scaled to have unit length in the original n-dimensional space, the deviation of each vector length from unity in the biplot provides a direct measure of how far the original vector is from the plane being plotted. Thus, plotted vectors that are close to having unit length are well represented by the biplot and relationships among such vectors are accurately displayed. Conversely, plotted vectors that are appreciably shorter than unity are not well represented in that particular biplot; other biplots should be used to study relationships involving these vectors. In Example 12.3, all three plotted vectors are very close to having unit length.

Observation Information

The dots in the biplot represent the observations. The coordinates for the observations, called *row markers*, are the elements of $\boldsymbol{U}$ from the singular value decomposition. Recall that the principal components can be written as $\boldsymbol{W} = \boldsymbol{U}\boldsymbol{L}^{1/2}$. Thus, each column of $\boldsymbol{U}$ is one of the principal components rescaled to remove λ_j.

Example 12.4

(*Continuation of Example* 12.3) The first ten rows of $\boldsymbol{U}$ are

$$\boldsymbol{U} = \begin{bmatrix} -.2350 & .4100 & -.4761 \\ -.2365 & .2332 & .4226 \\ -.2011 & .0363 & .2355 \\ -.1656 & -.1605 & .0485 \\ -.1302 & -.3574 & -.1385 \\ -.0222 & -.2491 & -.0985 \\ .0857 & -.1407 & -.0584 \\ .1937 & -.0323 & -.0184 \\ .3016 & .0761 & .0217 \\ .4096 & .1844 & .0617 \\ \vdots & \vdots & \vdots \end{bmatrix} \qquad [12.3]$$

The second ten rows in $\boldsymbol{U}$ duplicate the first ten in this example. The first and second columns are the first and second principal components, respectively, except for multiplication by λ_1 and λ_2. These two columns are the coordinates for the observations in the biplot (Figure 12.1). The first observation, for example, has coordinates $(-.2350, .4100)$. The horizontal and vertical scales for plotting the row markers need not be the same as the scales for the column markers. Often the scales for the row markers will be shown across the top and across the right side of the plot, as will be illustrated later in Figure 12.2. ■

The key elements for the interpretation of the biplot are listed in the box.

Interpretation of Gabriel's Biplot

1. The length of the variable vector in a biplot, relative to its length in the original n-space, indicates how well the two-dimensional biplot represents that vector. Vectors that do not lie close to the plane defined by the two principal components being used in the biplot will project onto the biplot as much shorter vectors than they are in n-space. For such variables, that particular biplot will be a poor representation of the relationship among the variables and interpretations involving them should be avoided.
2. The angle between two variable vectors reflects their pairwise correlation as seen in this two-dimensional projection. The correlation is the

cosine of the angle. Hence, a 90° angle indicates zero correlation; a 0° or 180° angle indicates a correlation of 1.0 or −1.0, respectively. [The angles between the vectors translate into correlations only because the variables have been centered before the eigenanalysis was done. The biplot is also used for some purposes on uncentered and/or unscaled data. See Bradu and Gabriel (1974, 1978), Gabriel (1971, 1972, 1978), and Corsten and Gabriel (1976) for examples.]

3. The spatial proximity of individual observations reflects their similarities with respect to this set of independent variables and as seen in the two dimensions being plotted. Points close together have similar values and vice versa.
4. The relative values of the observations for a particular variable are seen by projecting the observation points onto the variable vector, extended as required in either the positive or negative direction. The vector points in the direction of the largest values for the variable.

Example 12.5

The biplot of Example 12.4 shows that Z_1 and Z_2 are very highly positively correlated; the angle between the two vectors is close to zero. Z_1 and Z_2 are nearly orthogonal to Z_3 since both angles are close to 90°. One would have to conclude from this biplot that Z_1 and Z_2 are providing essentially the same information.

While the three variables technically define a three-dimensional space, two of the vectors are so nearly collinear that the third dimension is almost nonexistent. No regression will be able to separate the effects of Z_1 and Z_2 on Y from this set of data; the data are inadequate for this purpose. Furthermore, if the collinearity between Z_1 and Z_2 is a reflection of the innate properties of the system, additional data collected in the same way will show the same collinearity, and clear separation of their effects on any dependent variable will not be possible. When that is the case, it is probably best to define a new variable that reflects the (Z_1, Z_2)-axis and avoid the use of Z_1 and Z_2 per se. On the other hand, if the collinearity between Z_1 and Z_2 is a result of inadequate sampling or a bad experimental design, additional data will remove the collinearity and then separation of the effects of Z_1 and Z_2 on the dependent variable might be possible.

The proximity of the observations (points) to each other reflects their similarities for the variables used in the biplot. For example, points 1 and 2 are very much alike but are quite different from points to 10 and 5. Real data will frequently show clusters of points that reflect meaningful groupings of the observations.

The perpendicular projection of the observations (points) onto one of the vectors, extended in either direction as needed, gives the *relative* values of the observations for that variable. If the projection of the observations onto either the Z_1 or Z_2 axis is visualized, the points as numbered monotonically increase in value. Projection of the observations onto the Z_3 vector shows that their values for Z_3 decrease to the fifth point and then increase to the tenth point. (Recall that points 11–20 are repeats of points 1–10.) This pattern is a direct reflection of the original values for the three variables. ■

Example 12.6

A second example of a biplot is taken from Shy-Modjeska, Riviere, and Rawlings (1984). The biplot shown in Figure 12.2 displays the relationships among nephrotoxicity, physiological, and pharmacokinetic variables. The study used 24 adult female beagles that were subtotally nephrectomized ($\frac{3}{4}$ or $\frac{7}{8}$ of the kidneys were surgically removed) and assigned to one of four different treatments. A control group of six dogs was used. Nine variables measuring renal function are used in this biplot. Complete data were obtained on 29 of the 30 animals. Six of the 24 nephrectomized animals developed toxicity. The biplot presents the information from the first two principal component dimensions of the 29 × 9 data matrix. These two dimensions account for 76% of the total dispersion in the Z-space.

The biplot represents most of the vectors reasonably well. The shortest

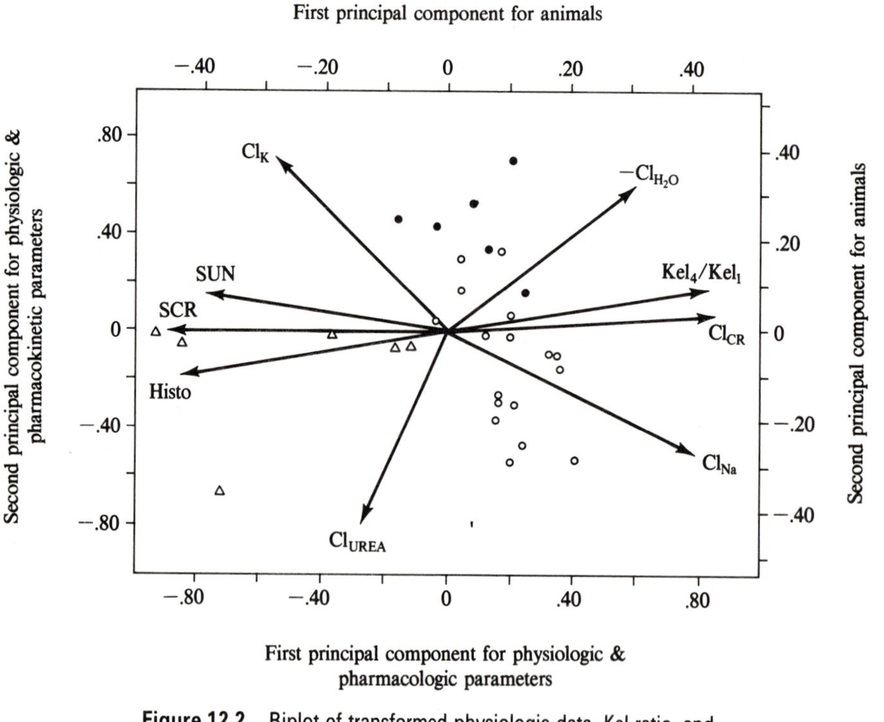

Figure 12.2 Biplot of transformed physiologic data, Kel ratio, and histopathologic index for 23 subtotally nephrectomized and six control animals. The first two dimensions accounted for 76% of the dispersion in the full matrix. Triangles designate dogs that developed toxicity, open circles designate dogs that were nephrectomized but did not develop toxicity, and the closed circles designate control animals. (Used with permission.)

vectors are Cl_{UREA} and Cl_{H_2O}. All other vectors are at least 80% of their original length. The complex of five variables labeled SUN, SCR, Histo, Kel_4/Kel_1, and Cl_{CR} comprises a highly correlated system in these data. The first three are highly positively correlated, as are the last two (the vectors point in the same direction), whereas there are high negative correlations between the two groups (the vectors point in different directions). The variable Cl_{Na}, on the other hand, is reasonably highly negatively correlated with Cl_K. Cl_{UREA} and Cl_{H_2O} also appear to be highly negatively correlated, but these are the two shortest vectors and may not be well represented in this biplot.

The horizontal axis across the bottom and the vertical axis on the left of Figure 12.2 are the scales for the column markers (variables) and row markers (animals), respectively, for the first principal component. The horizontal axis across the top and the vertical axis on the right are the scales for the column and row markers, respectively, for the second principal component. The vectors for the complex of five variables first mentioned are closely aligned with the axis of the first principal component; the first principal component is defined primarily by these five variables. Variation along the second principal component axis is primarily due to the variables Cl_{UREA}, Cl_K, Cl_{H_2O}, and Cl_{Na}, although these four variables are not as closely aligned with the axis.

The observations, the animals, tend to cluster according to their treatment received. Visualizing the projections of these points onto the vectors indicates how the animals differ for these nine variables. The major differences among the animals will be along the first principal component axis and are due to the difference between toxic and nontoxic animals. The toxic animals tend to have high values for SUN, SCR, and Histo and low values for Kel_4/Kel_1 and Cl_{CR}. This suggests that these are the key variables to study as indicators of toxicity. (Which of the five variables caused the toxicity or are a direct result of the toxicity cannot be determined from these data. The biplot is simply showing the association of variables.) One toxic animal, represented by the triangle in the lower left quadrant, is very different from all other animals. It has high values for the toxicity variables and a very high level of Cl_{UREA}. This would suggest a review of the data for this particular animal to ensure correctness of the values. If all appears to be in order, the other characteristics of the animal need to be studied to try to determine why it is responding so differently. The control animals separate from the nontoxic animals in the dimension of the second principal component. They have higher values for Cl_K and lower values for Cl_{UREA} and Cl_{Na} than the nontoxic animals.

This biplot accounts for 76% of the dispersion. Although this is the major part of the variation, a sizable proportion is being ignored. In this case, one would also study the information provided by the third dimension by biplotting the first and third and, perhaps, the second and third dimensions. These plots would reveal whether the negative correlation between Cl_{UREA} and Cl_{H_2O} is as strong as the first biplot suggests. ■

Overview of Gabriel's Biplot

Gabriel's biplot is a graphical technique for revealing relationships in a matrix of data. It is an exploratory tool and is not intended to provide estimates of parameters or tests of significance. Its graphical presentation of (1) the correlational structure among the variables, (2) the similarity of the observations, and (3) the relative values of the data points for the variables measured can be most helpful in understanding a complex set of data.

For the data set used in Examples 12.1–12.5, it is clear from the correlation matrix and the biplot that there is sufficient collinearity to cause a severe problem for ordinary least squares. With the high degree of collinearity between Z_1 and Z_2, it is unreasonable to expect any regression method to identify properly the contributions to Y of these two independent variables. Similarly, the biplot from Shy-Modjeska et al. (1984) showed a highly correlated complex of five variables that appeared to separate toxic from nontoxic animals. However, any regression analysis that attempts to assign relative importance to the five variables can be expected to be very misleading. "Seeing" the nature of the correlational structure in these data enhances the understanding of the problem and should introduce caution into the use of regression results. If it is important that effects of the individual variables be identified, data must be obtained in which the strong dependencies among the independent variables have been sufficiently weakened so that the collinearity problem no longer exists. In cases where the structure in the data is intrinsic to the system, as it may be in the toxicity study of Example 12.6, it will be necessary to obtain data using experimental protocols that will disrupt the natural associations among the variables before reliable estimates of the effects can be obtained.

12.2 Biased Regression Methods

Biased Estimators

The least squares estimators of the regression coefficients are the best linear unbiased estimators. That is, of all possible estimators that are both linear functions of the data and unbiased for the parameters being estimated, the least squares estimators have the smallest variance. In the presence of collinearity, however, this minimum variance may be unacceptably large. Relaxing the least squares condition that estimators be *unbiased* opens for consideration a much larger set of possible estimators from which one with better properties in the presence of collinearity might be found. **Biased regression** refers to this class of regression methods in which *unbiasedness* is no longer required. Such methods have been suggested as a possible solution to the collinearity problem. (See Chapters 9 and 10.) The motivation for biased regression methods rests in the potential for obtaining estimators that are closer, on the average, to the parameter being estimated than are the least squares estimators.

Mean Squared Error of an Estimator

A measure of average "closeness" of an estimator to the parameter being estimated is the **mean squared error (MSE)** of the estimator. If $\tilde{\theta}$ is an estimator of θ, the mean squared error of $\tilde{\theta}$ is defined as

$$\text{MSE}(\tilde{\theta}) = \mathscr{E}(\tilde{\theta} - \theta)^2 \qquad [12.4]$$

Recall that the variance of an estimator $\tilde{\theta}$ is defined as

$$\text{Var}(\tilde{\theta}) = \mathscr{E}[\tilde{\theta} - \mathscr{E}(\tilde{\theta})]^2 \qquad [12.5]$$

Note that MSE is the average squared deviation of the estimator from the *parameter* being estimated, whereas variance is the average squared deviation of the estimator from its *expectation*. If the estimator is unbiased, $\mathscr{E}(\tilde{\theta}) = \theta$ and $\text{MSE}(\tilde{\theta}) = \sigma^2(\tilde{\theta})$. Otherwise, MSE is equal to the variance of the estimator plus the square of its bias, where $\text{Bias}(\tilde{\theta}) = \mathscr{E}(\tilde{\theta}) - \theta$. It is possible for the variance of a biased estimator to be sufficiently smaller than the variance of an unbiased estimator to more than compensate for the bias introduced. In such a case, the biased estimator is closer on the average to the parameter being estimated than is the unbiased estimator. Such is the hope with the biased regression techniques.

Possible Advantage of Biased Estimators

The possible advantage of biased estimators is illustrated in Figure 12.3. The normal curve centered at θ in the diagram represents the probability distribution of an unbiased estimator of θ; its expectation is equal to θ. The spread of this curve reflects the variance of the unbiased estimator. The normal curve centered at $\mathscr{E}(\tilde{\theta})$ represents the probability distribution of a biased estimator; the bias is the difference between $\mathscr{E}(\tilde{\theta})$ and θ. The smaller spread in this distribution reflects its smaller variance. By allowing some bias, it may be possible to find an estimator for which the sum of its variance and squared bias, MSE, is smaller than the variance of the unbiased estimator.

Warnings Concerning Biased Regression Methods

Several biased regression methods have been proposed as solutions to the collinearity problem. These include Stein shrinkage (Stein, 1960), ridge regression (Hoerl and Kennard, 1970a, 1970b), and principal component regression and variations thereof (Lott, 1973; Hawkins, 1973; Hocking, Speed, and Lynn, 1976; Marquardt, 1970; Webster, Gunst, and Mason, 1974). While ridge regression has received the greatest acceptance, all have been used with apparent success in various problems. Nevertheless, biased regression methods have not been universally accepted and should be used with caution. The MSE justification for biased regression methods makes it clear that such methods can provide better *estimates* of the parameters in the sense of mean squared error. It does not necessarily follow that a biased regression solution is acceptable or even "better" than the least squares solution for purposes other than estimation of parameters.

While collinearity does not affect the precision of the estimated *responses* (and predictions) at the observed points in the X-space, it does cause variance

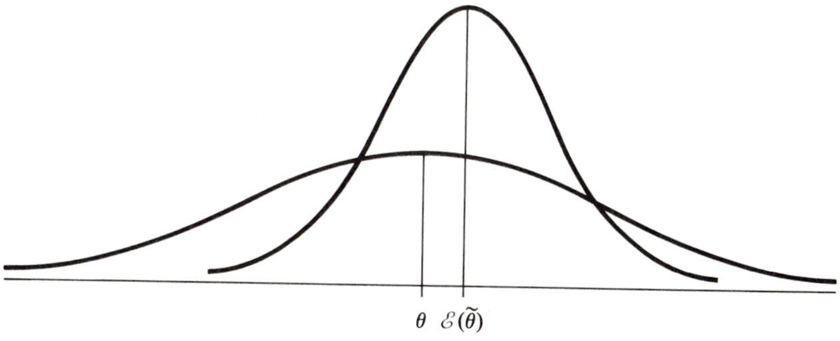

Figure 12.3 Illustration of a biased estimator having smaller mean squared error than an unbiased estimator.

inflation of estimated responses at other points. Park (1981) shows that the restrictions on the parameter estimates implicit in principal component regression are also optimal in the MSE sense for estimation of responses over certain regions of the X-space. This suggests that biased regression methods may be beneficial in certain cases for estimation of responses also. However, caution must be exercised when using collinear data for estimation and prediction of responses for points other than the observed sample points.

The biased regression methods do not seem to have much to offer when the objective is to assign some measure of "relative importance" to the independent variables involved in a collinearity. In essence, the biased estimators of the regression coefficients for the variables involved in the collinearity are weighted averages of the least squares regression coefficients for those variables. Consequently, each is reflecting the joint effects of all variables in the complex. (This will be illustrated later with the data from Examples 12.1–12.5.) The best recourse to the collinearity problem when the objective is to assign relative importance is to recognize that the data are inadequate for the purpose and to obtain better data, perhaps from controlled experiments.

Two of the biased regression methods, ridge regression and principal component regression, will be discussed briefly. The biased regression methods attack the collinearity problem by computationally suppressing the effects of the collinearity. Ridge regression does this by reducing the apparent magnitude of the correlations. Principal component regression attacks the problem by regressing Y on the important principal components and then parceling out the effect of the principal component variables to the original variables.

12.2.1 Ridge Regression

Justification for Ridge Regression

Ridge regression builds on the fact that a singular square matrix can be made nonsingular by adding a constant to the diagonal of the matrix. That is, if $\boldsymbol{X}'\boldsymbol{X}$ is singular, then $(\boldsymbol{X}'\boldsymbol{X} + k\boldsymbol{I})$ is nonsingular, where k is some small positive constant. This concept is used in ridge regression to reduce the impact of collinearity. A small positive constant is added to the diagonal of the nearly singular $\boldsymbol{X}'\boldsymbol{X}$. This makes the off-diagonal elements appear relatively less important and, in effect, suppresses the near-singularities among the independent variables.

Model

Ridge regression works with the centered and scaled independent variables $\boldsymbol{Z}$ so that the sum of squares and products matrix of the independent variables is the correlation matrix. Using $\boldsymbol{Z}$, which does not include the vector of ones for the intercept, we can write the model as

$$\boldsymbol{Y} = \boldsymbol{1}\beta_0 + \boldsymbol{Z\beta} + \boldsymbol{\varepsilon} \qquad [12.6]$$

where $\boldsymbol{1}$ is the column vector of ones and $\boldsymbol{\beta}$ is the vector of all regression coefficients except β_0. Since the independent variables are expressed as deviations from their respective means, $\boldsymbol{Z}$ is orthogonal to $\boldsymbol{1}$, and $\boldsymbol{\beta}$ and β_0 can be estimated separately. The estimate of β_0 is $\overline{Y}$.

Solution

The ridge regression estimate of $\boldsymbol{\beta}$, labeled $\tilde{\boldsymbol{\beta}}(k)$, is

$$\tilde{\boldsymbol{\beta}}(k) = (\boldsymbol{Z}'\boldsymbol{Z} + k\boldsymbol{I})^{-1}\boldsymbol{Z}'\boldsymbol{Y} \qquad [12.7]$$

where $k \geq 0$. The k is included in the notation for $\tilde{\boldsymbol{\beta}}(k)$ as a reminder that the estimate is dependent on the chosen k. The variance–covariance matrix of the estimates is

$$\mathbf{Var}[\tilde{\boldsymbol{\beta}}(k)] = (\boldsymbol{Z}'\boldsymbol{Z} + k\boldsymbol{I})^{-1}(\boldsymbol{Z}'\boldsymbol{Z})(\boldsymbol{Z}'\boldsymbol{Z} + k\boldsymbol{I})^{-1}\sigma^2 \qquad [12.8]$$

Choice of k

The goal is to choose k so as to minimize the mean squared error (MSE). However, MSE cannot be evaluated and, therefore, the choice of k is somewhat subjective. Usually k is very close to zero, .1 or even smaller. If $k = 0$, the ridge regression solution is the ordinary least squares solution.

As k is increased from zero, several quantities in addition to $\tilde{\boldsymbol{\beta}}(k)$ change. All of the following will monotonically decrease as k increases: maximum variance inflation factor, VIF_{max}, which is defined in ridge regression as the largest diagonal element of $\mathbf{Var}[\tilde{\boldsymbol{\beta}}(k)]/\sigma^2$; the sum of variances of estimated regression coefficients, $\text{tr}\{\mathbf{Var}[\tilde{\boldsymbol{\beta}}(k)]\}$; the length of $\tilde{\boldsymbol{\beta}}(k)$, $[\tilde{\boldsymbol{\beta}}(k)'\tilde{\boldsymbol{\beta}}(k)]^{1/2}$; and R^2. The bias of $\tilde{\boldsymbol{\beta}}(k)$, on the other hand, monotonically increases with k. The MSE for any particular regression coefficient is expected to decrease initially, because of initial rapid decreases in variance, and then increase as bias begins to dominate. The choice of k is a compromise between reducing variance and increasing bias. The effect on variances as k increases can be observed, but not the effect on the biases. Therefore, the strategy is to choose the smallest k that appears to be producing stable estimates of the regression coefficients.

Procedures for Choosing k

One procedure for determining the value of k for the ridge regression solution involves obtaining the ridge regression results for several choices of k near zero, say .005 to .2. The estimates of the individual regression coefficients are plotted against k to give the ridge traces. Plots of R^2, $\text{tr}\{\mathbf{Var}[\tilde{\boldsymbol{\beta}}(k)]\}/\sigma^2$, and $[\tilde{\boldsymbol{\beta}}(k)'\tilde{\boldsymbol{\beta}}(k)]^{1/2}$ against k are also helpful. The value of k chosen is the smallest value where the major changes in $\tilde{\boldsymbol{\beta}}(k)$ and $\mathbf{Var}[\tilde{\boldsymbol{\beta}}(k)]$ have been realized and where the coefficient of determination R^2 has not decreased too much. This approach is subjective and the appearances of stability will depend to some degree on the scaling of the plots.

Alternatively, Hoerl, Kennard, and Baldwin (1975) suggest the use of

$$k = ps^2/[\tilde{\boldsymbol{\beta}}(0)'\tilde{\boldsymbol{\beta}}(0)] \qquad [12.9]$$

where p is the number of parameters excluding β_0, and s^2 is the residual mean square estimated from the ordinary least squares regression ($k = 0$). The denominator of equation 12.9 is the sum of squares of the ordinary least squares regression coefficients $\tilde{\boldsymbol{\beta}}(0)$, excluding the intercept, computed with centered and scaled independent variables, $\boldsymbol{Z}$.

There is a tendency to use too large a value of k when choosing k based on the behavior of the ridge traces (Van Nostrand, 1980). Thus, it is probably best to give greater weight to the value of k determined from equation 12.9. The regression results from the other values of k, however, should strengthen one's confidence in the choice.

Ridge Regression Equation

Given the chosen value of k, the ridge regression equation becomes

$$\hat{Y} = \mathbf{1}\bar{Y} + \mathbf{Z}\tilde{\boldsymbol{\beta}}(k) \qquad [12.10]$$

The key steps in ridge regression are summarized in the box.

Ridge Regression Procedure

1. Center and scale the independent variables to obtain $\mathbf{Z}$ by subtracting the mean of each variable and dividing each by the square root of its sum of squares.
2. Compute $\mathbf{Z}'\mathbf{Z}$ and $\mathbf{Z}'\mathbf{Y}$, the sums of squares and products matrices.
3. Compute the ordinary least squares results in terms of $\mathbf{Z}$ and $\mathbf{Y}$ and compute the Hoerl, Kennard, and Baldwin value of k (equation 12.9).
4. For a sequence of constants, say $k = .005, .01, .02, \ldots$, including the value from equation 12.9, compute
   a. $\tilde{\boldsymbol{\beta}}(k) = (\mathbf{Z}'\mathbf{Z} + k\mathbf{I})^{-1}\mathbf{Z}'\mathbf{Y}$;
   b. $\mathbf{Var}[\tilde{\boldsymbol{\beta}}(k)] = (\mathbf{Z}'\mathbf{Z} + k\mathbf{I})^{-1}(\mathbf{Z}'\mathbf{Z})(\mathbf{Z}'\mathbf{Z} + k\mathbf{I})^{-1}\sigma^2$; and
   c. $\text{SS(Regr)} = \tilde{\boldsymbol{\beta}}(k)'\mathbf{Z}'\mathbf{Y}$.
5. Plot ridge traces, $\tilde{\beta}(k)_j$ versus k, and possibly SS(Regr), R^2, or $\text{tr}\{\mathbf{Var}[\tilde{\boldsymbol{\beta}}(k)]\}$ versus k.
6. Choose the value of k
   a. by equation 12.9; or
   b. where the ridge traces have stabilized, the major decreases in the variances have been realized, and before R^2 has decreased too much. The results for the chosen value of k are the ridge regression solution.

Bayesian Interpretation

The most meaningful interpretation of ridge regression is a Bayesian one. Under the Bayesian interpretation, the *prior* probability distribution of $\boldsymbol{\beta}$ is assumed to have zero mean and variance–covariance matrix $\mathbf{I}(\sigma^2/k)$ (Hoerl and Kennard, 1970a; Marquardt and Snee, 1975; Smith and Campbell, 1980). Thus, the choice of k specifies the prior belief regarding the variances of the distributions of the true regression coefficients. The zero mean for the prior implies that the ridge regression estimates will be "shrunk" toward zero from the ordinary least squares estimates; the larger the value of k, the greater will be the shrinkage. (Recall that one of the effects of increasing k is decreasing $[\tilde{\boldsymbol{\beta}}(k)'\tilde{\boldsymbol{\beta}}(k)]^{1/2}$, the length of the vector of estimates.)

Relationship to Ordinary Least Squares

The relationship between the ridge regression estimates and the ordinary least squares estimates is given by

$$\tilde{\boldsymbol{\beta}} = (\mathbf{Z}'\mathbf{Z} + k\mathbf{I})^{-1}\mathbf{Z}'\mathbf{Z}\hat{\boldsymbol{\beta}} \qquad [12.11]$$

This relationship will be used in the numerical examples to demonstrate that the ridge estimates are weighted averages of the least squares estimates $\hat{\boldsymbol{\beta}}$ with primary weight given to the regression coefficients of the variables involved in the near-singularities. Since $\hat{\boldsymbol{\beta}}$ is unbiased, $\mathscr{E}(\hat{\boldsymbol{\beta}}) = \boldsymbol{\beta}$, it follows from equation 12.11 that $\tilde{\boldsymbol{\beta}}$ is biased unless $k = 0$, $\mathscr{E}(\tilde{\boldsymbol{\beta}}) = (\boldsymbol{Z}'\boldsymbol{Z} + k\boldsymbol{I})^{-1}\boldsymbol{Z}'\boldsymbol{Z}\boldsymbol{\beta}$.

Example 12.7 In order to illustrate biased regression with the artificial data from Example 12.1, a vector of observations for the dependent variable is needed. (Only the matrix of independent variables was used for the correlation matrix, the singular value decomposition, and the biplot.) Artificial values for the dependent variable were generated using the model

$$Y_i = 10 + .4X_{i1} - .2X_{i2} + .4X_{i3} + \varepsilon_i$$

where the ε_i were generated from a normal distribution with zero mean and unit variance. Expressing the model in terms of the centered and standardized variables gives

$$Y_i = 21.28 + 5.138Z_{i1} - 2.440Z_{i2} + 2.683Z_{i3} + \varepsilon_i$$

The vector of observations obtained from the simulation was

$$\boldsymbol{Y} = \begin{bmatrix} 21.607 \\ 21.409 \\ 20.386 \\ 20.946 \\ 17.942 \\ 21.808 \\ 21.358 \\ 21.883 \\ 22.596 \\ 22.483 \\ 20.389 \\ 21.351 \\ 21.374 \\ 19.734 \\ 19.817 \\ 20.337 \\ 19.761 \\ 23.266 \\ 21.600 \\ 23.477 \end{bmatrix}$$

Table 12.3 Ridge regression results for the artificial data set showing changes in $\tilde{\beta}(k)_j$ and their standard errors with choices of k.

	Parameter Estimates			*Standard Errors/s*[a]		
k	$\tilde{\beta}(k)_1$	$\tilde{\beta}(k)_2$	$\tilde{\beta}(k)_3$	$s[\tilde{\beta}(k)_1]$	$s[\tilde{\beta}(k)_2]$	$s[\tilde{\beta}(k)_3]$
0 = OLS	14.48	−13.18	4.49	13.01	13.25	1.30
.005	5.74	−4.28	3.96	4.80	4.89	1.08
.0058[b]	5.28	−3.81	3.93	4.37	4.44	1.08
.01	3.78	−2.28	3.83	2.97	3.02	1.05
.02	2.43	−.90	3.71	1.73	1.75	1.03
.04	1.62	−.07	3.59	1.02	1.03	1.01
.06	1.33	.23	3.51	.79	.79	.98
.08	1.18	.39	3.43	.69	.67	.96
.10	1.09	.49	3.36	.63	.61	.94
.20	.91	.68	3.07	.51	.50	.85

[a] Standard errors have been scaled by dividing by s, the estimate of σ.
[b] The Hoerl, Kennard, Baldwin (1975) estimate of k.

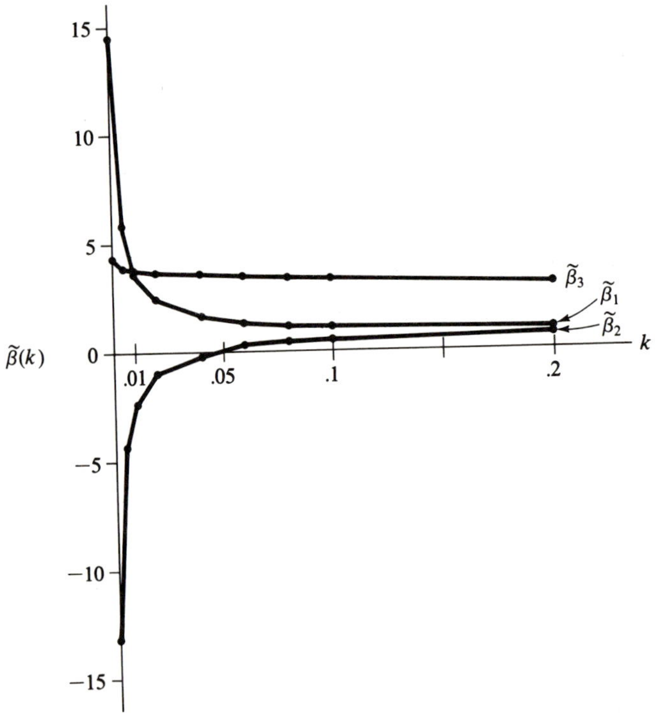

Figure 12.4 Ridge traces for Example 12.7.

The ridge regression results from this example are shown in Table 12.3 and the ridge traces are shown in Figure 12.4. The first row of Table 12.3, corresponding to $k = 0$, gives the ordinary least squares estimates and their standard errors. Notice that $\tilde{\beta}(0)_1$ and $\tilde{\beta}(0)_2$ are of similar magnitude but opposite in sign. Their standard errors are about the same size and large so that neither partial regres-

Table 12.4 Mean squared errors for the ridge regression estimates of the regression coefficients for the artificial data.

	Ridge Regression k				
Variable	0	.0058	.01	.2	.3
Z_1	169.4	25.53	17.58	14.86	15.21
Z_2	176.7	26.43	18.16	14.27	14.06
Z_3	1.7	1.18	1.14	1.08	1.15

sion coefficient is significantly different from zero. Both $\tilde{\beta}(k)_1$ and $\tilde{\beta}(k)_2$ decrease rapidly in absolute value as k is increased. On the other hand, $\tilde{\beta}(k)_3$ changes very little, reflecting the fact that Z_3 is not involved in the near-singularity.

The increased stability of the ridge regression estimates is shown by the rapid decrease in the standard errors of the estimates with increasing k. The objective is to choose the k that realizes most of the decrease in variance without introducing too much bias. If the ridge traces shown in Figure 12.4 are used to pick k, it would appear that k in the vicinity of .01 to .02 is a reasonable choice. At $k = .01$, R^2 has decreased from .626 to .585. On the other hand, the suggestion of Hoerl, Kennard, and Baldwin (1975) leads to a much smaller value of $k = .0058$ in this example.

Since this example was generated from a known model, the bias and the mean squared errors can be evaluated for the different choices of k (Table 12.4). The mean squared errors for $k = 0$ are the variances of the ordinary least squares estimators. For $k > 0$, however, the mean squared errors include the square of the bias in each case. There is a dramatic drop in the variances of the regression coefficients for the first two variables (see the standard errors in Table 12.3). Table 12.4 shows that this drop in variance much more than compensates for the increase in bias; the mean squared error, even with k as small as $k = .0058$, is less than one-sixth that at $k = 0$ for the two variables involved in the near-singularity. Comparison of the mean squared errors with the variances for the corresponding values of k (obtained by squaring the standard errors in Table 12.3) shows, in this example, that most of the mean squared error is due to the bias term for $\tilde{\beta}(k)_1$ and $\tilde{\beta}(k)_2$. Notice also that the mean squared errors do not start to increase in this example until k exceeds .2.

To demonstrate the danger of attempting to assign relative importance to variables from the ridge regression estimates, the coefficient matrix in equation 12.11 is evaluated for $k = .0058$ and $k = .20$. These matrices also demonstrate the "shrinkage" of the least squares estimators toward zero. For $k = .0058$,

$$\begin{pmatrix} \tilde{\beta}(k)_1 \\ \tilde{\beta}(k)_2 \\ \tilde{\beta}(k)_3 \end{pmatrix} = \begin{bmatrix} .672 & .331 & -.018 \\ .331 & .660 & .020 \\ -.018 & .020 & .992 \end{bmatrix} \begin{pmatrix} \hat{\beta}_1 \\ \hat{\beta}_2 \\ \hat{\beta}_3 \end{pmatrix}$$

For $k = .20$, the coefficient matrix is

$$\begin{pmatrix} \tilde{\beta}(k)_1 \\ \tilde{\beta}(k)_2 \\ \tilde{\beta}(k)_3 \end{pmatrix} = \begin{bmatrix} .465 & .443 & .003 \\ .443 & .452 & .049 \\ .003 & .049 & .819 \end{bmatrix} \begin{pmatrix} \hat{\beta}_1 \\ \hat{\beta}_2 \\ \hat{\beta}_3 \end{pmatrix}$$

The ridge estimates of the regression coefficients of Z_1 and Z_2, the two variables causing the collinearity problem in this example, are essentially averages of the least squares estimates. The weights on the two coefficients become more nearly equal as k increases. Even with k as small as $k = .0058$, there is approximately a 6:4 relative weighting of the two coefficients. It is known in this example that the ridge estimates have smaller MSE than the least squares estimates, and in that sense the ridge estimates are better estimates of the individual parameters. However, it is evident that neither the least squares nor ridge regression coefficients provide useful information on the relative importance of Z_1 and Z_2. The least squares estimates are unbiased but are too unreliable to provide meaningful information, and the two ridge estimates are very similar averages of the same quantities. This is reflecting the fact that there is essentially no information in this sample with which to estimate the difference between β_1 and β_2; the dispersion in the $(Z_1 - Z_2)$ dimension of the Z-space is extremely limited.

The ridge regression coefficient for Z_3, the variable not involved in the collinearity, is essentially a constant times the least squares estimate. The constant decreases from 1.0 as k moves away from 0. Likewise, the coefficients in the weighted average of $\hat{\beta}_1$ and $\hat{\beta}_2$ become smaller as k increases. These changes are reflecting the "shrinkage" of the ridge estimates toward zero. ■

12.2.2 Principal Component Regression

Principal component regression approaches the collinearity problem from the point of view of eliminating from consideration those dimensions of the X-space that are causing the collinearity problem. This is similar, in concept, to dropping an independent variable from the model when there is insufficient dispersion in that variable to contribute meaningful information on Y. However, in principal component regression the dimension dropped from consideration is defined by a linear combination of the variables rather than by a single independent variable.

Eigenvectors of Z

Principal component regression builds on the principal component analysis of the matrix of independent variables. As with ridge regression, principal component regression operates on the centered and scaled independent variables, $\boldsymbol{Z}$. The SVD of $\boldsymbol{Z}$ has been used in the analysis of the correlational structure of the X-space and in Gabriel's biplot. This section continues with the notation and results defined earlier. The SVD of $\boldsymbol{Z}$ is used to give

$$\boldsymbol{Z} = \boldsymbol{U}\boldsymbol{L}^{1/2}\boldsymbol{V}' \qquad [12.12]$$

where $\boldsymbol{U}\,(n \times p)$ and $\boldsymbol{V}\,(p \times p)$ are matrices containing the left and right eigenvectors, respectively, and $\boldsymbol{L}^{1/2}$ is the diagonal matrix of singular values. The singular values and their eigenvectors are ordered so that $\lambda_1 > \lambda_2 > \cdots > \lambda_p$. The eigenvectors are pairwise orthogonal and scaled to have unit length so that

$$U'U = V'V = I \qquad [12.13]$$

Principal Components of Z

The principal components of Z are defined as the linear functions of the Z_j specified by the coefficients in the column vectors of V. The first eigenvector in V (first column) defines the first principal component, the second eigenvector in V defines the second principal component, and so forth. Each principal component is a linear function of all independent variables. The principal components W are also given by the columns of U multiplied by the corresponding λ_j. Thus,

$$W = ZV$$

or

$$W = UL^{1/2} \qquad [12.14]$$

is the matrix of principal component variables. Each column in W gives the values for the n observations for one of the principal components.

Sums of Squares and Products of the Principal Components

The sum of squares and products matrix of the principal component variables W is the diagonal matrix of the eigenvalues,

$$W'W = (UL^{1/2})'(UL^{1/2}) = L^{1/2}U'UL^{1/2} = L \qquad [12.15]$$

where $L = \text{Diag}(\lambda_1^2\ \lambda_2^2\ \ldots\ \lambda_p^2)$. Thus, the principal components are orthogonal to each other, since all sums of products are zero, and the sum of squares of each principal component is equal to the corresponding eigenvalue λ_j^2. The first principal component has the largest sum of squares, λ_1^2. The principal components corresponding to the smaller eigenvalues are the dimensions of the Z-space having the least dispersion. These dimensions of the Z-space with limited dispersion are responsible for the collinearity problem if one exists.

Linear Model

The linear model

$$Y = 1\beta_0 + Z\beta + \varepsilon \qquad [12.16]$$

can be written in terms of the principal components W as

$$Y = 1\beta_0 + W\gamma + \varepsilon \qquad [12.17]$$

This uses the fact that $VV' = I$ to transform $Z\beta$ into $W\gamma$:

$$Z\beta = ZVV'\beta = W\gamma \qquad [12.18]$$

Notice that $\gamma = V'\beta$ is the vector of regression coefficients for the principal components; β is the vector of regression coefficients for the Z's. The translation of γ back to $\boldsymbol{\beta}$ is

$$\boldsymbol{\beta} = \boldsymbol{V}\gamma \qquad [12.19]$$

Solution

Ordinary least squares using the principal components as the independent variables gives

$$\hat{\gamma} = (\boldsymbol{W}'\boldsymbol{W})^{-1}\boldsymbol{W}'\boldsymbol{Y} = \boldsymbol{L}^{-1}\boldsymbol{W}'\boldsymbol{Y}$$

$$= \begin{bmatrix} \left(\sum_i W_{i1} Y_i\right) \Big/ \lambda_1^2 \\ \left(\sum_i W_{i2} Y_i\right) \Big/ \lambda_2^2 \\ \vdots \\ \left(\sum_i W_{ip} Y_i\right) \Big/ \lambda_p^2 \end{bmatrix} = \begin{pmatrix} \hat{\gamma}_1 \\ \hat{\gamma}_2 \\ \vdots \\ \hat{\gamma}_p \end{pmatrix} \qquad [12.20]$$

The regression coefficients for the principal components can be computed individually because the principal components are orthogonal; $\boldsymbol{W}'\boldsymbol{W}$ is a diagonal matrix. Likewise, the variance–covariance matrix of $\hat{\gamma}$ is the diagonal matrix

$$\mathbf{Var}(\hat{\gamma}) = \boldsymbol{L}^{-1}\sigma^2 \qquad [12.21]$$

That is, the variance of $\hat{\gamma}_j$ is $s^2(\hat{\gamma}_j) = \sigma^2/\lambda_j^2$, and all covariances are zero. Because of the orthogonality of the principal components, the partial and sequential sums of squares for each principal component are equal and each regression sum of squares can be computed individually as

$$\mathrm{SS}(\gamma_j) = \hat{\gamma}_j^2 \lambda_j^2 \qquad [12.22]$$

Relationship to Ordinary Least Squares

If all principal components are used, the results are equivalent to ordinary least squares regression. The estimate of $\boldsymbol{\beta}$ is obtained from $\hat{\gamma}$ as

$$\hat{\boldsymbol{\beta}} = \boldsymbol{V}\hat{\gamma} \qquad [12.23]$$

and the regression equation can be written as either

$$\hat{\boldsymbol{Y}} = \mathbf{1}\bar{Y} + \boldsymbol{W}\hat{\gamma}$$

or

$$\hat{\boldsymbol{Y}} = \mathbf{1}\bar{Y} + \boldsymbol{Z}\hat{\boldsymbol{\beta}} \qquad [12.24]$$

Eliminating Principal Components

The idea behind principal component regression, however, is to eliminate those dimensions that are causing the collinearity problem, those dimensions for which the λ_j are very small. Assume it has been decided to eliminate s principal components, usually those having the s smallest eigenvalues, and retain g principal components for the analysis ($g + s = p$). The subscript (g) will be used on $\boldsymbol{V}$, $\boldsymbol{L}$, $\boldsymbol{W}$, and $\hat{\boldsymbol{\gamma}}$ to designate the partitions of the corresponding matrices that relate to the g principal component dimensions retained in the analysis. Thus, $\boldsymbol{V}_{(g)}$ is the $p \times g$ matrix of retained eigenvectors, $\boldsymbol{W}_{(g)}$ is the $n \times g$ matrix of the corresponding principal components, and $\hat{\boldsymbol{\gamma}}_{(g)}$ is the vector of their estimated regression coefficients. The subscript (g) will be used on other results to designate the number of principal components retained in the analysis.

Recall that the principal component regression coefficients, their variances, and the sums of squares attributable to each can be computed independently because the principal components are orthogonal. Therefore, $\hat{\boldsymbol{\gamma}}_{(g)}$ is obtained from $\hat{\boldsymbol{\gamma}}$ by simply extracting the g elements corresponding to the retained principal components. The **principal component regression** estimate of $\boldsymbol{\beta}$, the vector of regression coefficients for the Z's, is given by

$$\underset{(p \times 1)}{\boldsymbol{\beta}^{+}_{(g)}} = \underset{(p \times g)}{\boldsymbol{V}_{(g)}} \cdot \underset{(g \times 1)}{\hat{\boldsymbol{\gamma}}_{(g)}} \qquad [12.25]$$

The notation $\boldsymbol{\beta}^{+}$ is used in place of $\hat{\boldsymbol{\beta}}$ to distinguish the principal component estimates of $\boldsymbol{\beta}$ from the least squares estimates. Notice that there are p elements in $\boldsymbol{\beta}^{+}_{(g)}$, even though there are only g elements in $\hat{\boldsymbol{\gamma}}$.

The variance of $\boldsymbol{\beta}^{+}_{(g)}$ is

$$\mathbf{Var}[\boldsymbol{\beta}^{+}_{(g)}] = \boldsymbol{V}_{(g)}\boldsymbol{L}^{-1}_{(g)}\boldsymbol{V}'_{(g)}\sigma^2 \qquad [12.26]$$

These variances involve the reciprocals of only the larger eigenvalues. The smaller ones causing the variance inflation in the ordinary least squares solution have been eliminated.

SS(Regr)

The sum of squares due to regression is the sum of the contributions from the g principal components retained and has g degrees of freedom:

$$\text{SS(Regr)} = \sum_{j \in \{g\}} \text{SS}(\gamma_j) \qquad [12.27]$$

where summation is over the subset of g principal components retained in the model.

Regression Equation

The regression equation can be written either as

$$\hat{\boldsymbol{Y}}_{(g)} = \mathbf{1}\bar{Y} + \boldsymbol{Z}\boldsymbol{\beta}^{+}_{(g)}$$

or

$$\hat{\boldsymbol{Y}}_{(g)} = \mathbf{1}\bar{Y} + \boldsymbol{W}_{(g)}\hat{\boldsymbol{\gamma}}_{(g)} \qquad [12.28]$$

where $\boldsymbol{W}_{(g)}$ is the matrix of retained principal components; $\hat{\beta}_0 = \bar{Y}$ and is orthogonal to each $\beta^{+}_{(g)j}$.

Var($\hat{Y}_{(g)}$)

The variance of $\hat{Y}_{(g)}$ can be written in several forms. Perhaps the simplest is

$$\mathbf{Var}[\hat{\mathbf{Y}}_{(g)}] = \left[\frac{\mathbf{J}}{n} + \mathbf{W}_{(g)}\mathbf{L}_{(g)}^{-1}\mathbf{W}_{(g)}'\right]\sigma^2 \qquad [12.29]$$

Relationship of $\boldsymbol{\beta}_{(g)}^+$ to $\hat{\boldsymbol{\beta}}$

As with ridge regression, the principal component regression coefficients can be expressed as linear functions of the least squares estimates:

$$\begin{aligned}\boldsymbol{\beta}_{(g)}^+ &= \mathbf{V}_{(g)}\mathbf{V}_{(g)}'\hat{\boldsymbol{\beta}} \\ &= [\mathbf{I} - \mathbf{V}_{(s)}\mathbf{V}_{(s)}']\hat{\boldsymbol{\beta}} \qquad [12.30]\end{aligned}$$

where $\mathbf{V}_{(s)}$ is the matrix of s eigenvectors that were dropped from the analysis. Since $\hat{\boldsymbol{\beta}}$ is unbiased, the expectation and bias of the principal component regression coefficients follow from equation 12.30;

$$\mathscr{E}[\boldsymbol{\beta}_{(g)}^+] = \boldsymbol{\beta} - \mathbf{V}_{(s)}\mathbf{V}_{(s)}'\boldsymbol{\beta}$$

or the bias is

$$\text{Bias} = \mathscr{E}[\boldsymbol{\beta}_{(g)}^+] - \boldsymbol{\beta} = -\mathbf{V}_{(s)}\mathbf{V}_{(s)}'\boldsymbol{\beta} \qquad [12.31]$$

Linear Restrictions on $\boldsymbol{\beta}_{(g)}^+$

The fact that $\boldsymbol{\beta}_{(g)}^+$ has p elements, a regression coefficient for each independent variable, even though only g regression coefficients $\hat{\gamma}_{(g)}$ were estimated, implies that there are linear restrictions on $\boldsymbol{\beta}_{(g)}^+$. There is one linear restriction for each eliminated principal component. The linear restrictions on $\boldsymbol{\beta}_{(g)}^+$ are defined by $\mathbf{V}_{(s)}$ as

$$\mathbf{V}_{(s)}'\boldsymbol{\beta}_{(g)}^+ = \mathbf{0} \qquad [12.32]$$

Choosing the Principal Components to Eliminate

It is best to be conservative in eliminating principal components because each one eliminated introduces another constraint on the estimates and another increment of bias. The bias term, equation 12.31, can also be expressed as $-\mathbf{V}_{(s)}\boldsymbol{\gamma}_{(s)}$, where $\boldsymbol{\gamma}_{(s)}$ is the set of principal component regression coefficients dropped. Hence, one does not want to eliminate a principal component for which γ_j is very different from zero. A good working rule seems to be to eliminate only those principal components that

1. have small enough eigenvalues to cause serious variance inflation (see Section 10.3), *and*
2. for which the estimated regression coefficient $\hat{\gamma}_j$ is not significantly different from zero.

One may wish to use a somewhat lower level of significance (say, $\alpha = .10$ or $.20$) for testing the principal component regression coefficients in order to allow for the low power that is likely to be present for the dimensions that have limited dispersion.

The key steps in principal component regression are listed in the box.

Principal Component Regression Procedure

1. Obtain the singular value decomposition on the matrix of centered and scaled independent variables, $\boldsymbol{Z} = \boldsymbol{U}\boldsymbol{L}^{1/2}\boldsymbol{V}'$.
2. The principal components are given by $\boldsymbol{W} = \boldsymbol{Z}\boldsymbol{V}$ or $\boldsymbol{W} = \boldsymbol{U}\boldsymbol{L}^{1/2}$.
3. Regress $\boldsymbol{Y}$ on $\boldsymbol{W}$ to obtain the estimates of the regression coefficients for the p principal components, $\hat{\boldsymbol{\gamma}}$; their estimated variances, $s^2(\hat{\boldsymbol{\gamma}})$; and the sums of squares due to regression, $\text{SS}(\hat{\gamma}_j)$. The residual mean square from the full model is used as the estimate of σ^2 in $s^2(\hat{\boldsymbol{\gamma}})$.
4. Test $H_0\colon \gamma_j = 0$ for each j using Student's t or F. Eliminate from the regression all principal components that
   a. are causing a collinearity problem (condition index > 10, for example), *and*
   b. do not make a significant contribution to the regression.
5. $\hat{\boldsymbol{\gamma}}_{(g)}$ is the vector of estimated regression coefficients retained. SS(Regr) $= \sum \text{SS}(\gamma_j)$, where summation is over the g components retained. SS(Regr) has g degrees of freedom.
6. Convert the regression coefficients for the principal components to the regression coefficients for the original independent variables (centered and scaled) by

$$\boldsymbol{\beta}^{+}_{(g)} = \boldsymbol{V}_{(g)}\hat{\boldsymbol{\gamma}}_{(g)}$$

   which has estimated variance

$$s^2[\boldsymbol{\beta}^{+}_{(g)}] = \boldsymbol{V}_{(g)}\boldsymbol{L}^{-1}_{(g)}\boldsymbol{V}'_{(g)}s^2$$

7. The regression equation is either

$$\hat{\boldsymbol{Y}}_{(g)} = \mathbf{1}\bar{Y} + \boldsymbol{Z}\boldsymbol{\beta}^{+}_{(g)} \quad \text{or} \quad \hat{\boldsymbol{Y}}_{(g)} = \mathbf{1}\bar{Y} + \boldsymbol{W}\hat{\boldsymbol{\gamma}}_{(g)}$$

Example 12.8 The principal component regression analysis for the example begins with the principal component analysis using the data from Example 12.2. The singular values, λ_j, showed that the dimension defined by the third principal component accounted for less than .1% of the total dispersion of the centered and standardized variables, $\boldsymbol{Z}$. The second dimension accounted for 28% of the total dispersion.

The estimates of the regression coefficients for the principal components and the sum of squares attributable to each are shown in Table 12.5 (page 350). The total sum of squares accounted for by the three principal components equals the total sum of squares due to regression of the original variables, SS(Regr) =

Table 12.5 Estimated regression coefficients for the principal components, their standard errors, and the sum of squares attributable to each.

Principal Component j	*Regression Coefficients* $\hat{\gamma}_j$	*Standard Errors*	*Sum of Squares*
1	2.3473	.598	11.940**
2	3.0491	.967	7.718**
3	19.7132	16.361	1.126

** The sum of squares is significant at the .01 level of probability. Each sum of squares has 1 degree of freedom and was tested against the residual mean square from the full model, $s^2 = .776$ with 16 degrees of freedom.

Table 12.6 Principal component regression estimates of the regression coefficients for the original variables using all principal components ($g = 3$) and omitting the third principal component ($g = 2$)[a].

Scaled Variable Z_j	*Regression Coefficients Using g Principal Components*		*Mean Squared Error*
	$g = 3$	$g = 2$	
1	14.477 (11.463)	.678 (.478)	14.83
2	−13.177 (11.674)	.878 (.453)	15.34
3	4.493 (1.144)	3.685 (.927)	1.16

[a] Standard errors are given in parentheses. The mean squared errors are for the $g = 2$ principal component solution.

20.784. The regression coefficients for the first two principal components are highly significant; the regression coefficient for the third component is not significant even at $\alpha = .20$. Consequently, no important information on Y would be lost if the third principal component were to be dropped from the regression. The very large standard error on $\hat{\gamma}_3$ reflects the extremely small amount of variation in the dimension defined by the third principal component.

The principal component analysis and Gabriel's biplot showed that the first principal component is defined primarily by Z_1 and Z_2, with a much smaller contribution from Z_3. This particular linear function of Z_1, Z_2, and Z_3 contains information on Y as shown by its significance. Likewise, the second principal component dominated by Z_3 is important for Y. However, the third principal component, essentially the difference between Z_1 and Z_2, does not make a significant contribution to the regression. This does *not* imply that the difference between Z_1 and Z_2 is unimportant in the process being studied. In fact, the equation used to generate $\mathbf{Y}$ in this artificial example gives greater weight to the difference than it gives to the sum of Z_1 and Z_2. In this particular set of data, however, Z_1 and Z_2 are so nearly collinear that their difference is always very close to being a constant and, therefore, the impact of the difference is estimated only with very low precision.

The principal component regression estimate of $\boldsymbol{\beta}$ (Table 12.6) using all principal components ($g = 3$) reproduces the ordinary least squares result. The estimate of $\boldsymbol{\beta}$ using only the first two principal components, $\boldsymbol{\beta}^{+}_{(2)}$, shows a marked change toward zero in the first two regression coefficients, and a marked decrease in their standard errors. The change is small in the third regression coefficient and its standard error. The large changes associated with Z_1 and Z_2 and the small change associated with Z_3 directly reflect the relative involvement of the independent variables in the near-singularity shown by the third principal component. The coefficient of determination for the principal component regression using the first two principal components is $R^2_{(2)} = .592$, only slightly less than $R^2 = .626$ for ordinary least squares. The regression equation estimated from principal component regression with $g = 2$ is

$$\hat{Y}_{(g)i} = 21.18 + .678Z_{i1} + .878Z_{i2} + 3.685Z_{i3}$$

Since the parameters $\boldsymbol{\beta}$ are known in this artificial example, the mean squared errors for the principal component regression are computed and given in the last column of Table 12.6. As with ridge regression, the mean squared errors for the variables involved in the near-singularity are an order of magnitude smaller than for ordinary least squares. Comparison with the variances of the estimated regression coefficients shows that most of MSE for $\beta^{+}_{(2)1}$ and $\beta^{+}_{(2)2}$ is due to bias.

The relationship between the principal component regression estimates and the least squares estimates for this example are shown by evaluating equation 12.30:

$$\begin{pmatrix} \beta^{+}_{(2)1} \\ \beta^{+}_{(2)2} \\ \beta^{+}_{(2)3} \end{pmatrix} = \begin{bmatrix} .510 & .499 & -.029 \\ .499 & .492 & .029 \\ -.029 & .029 & .998 \end{bmatrix} \begin{pmatrix} \hat{\beta}_1 \\ \hat{\beta}_2 \\ \hat{\beta}_3 \end{pmatrix}$$

The principal component estimates of β_1 and β_2 are very nearly simple averages of the corresponding least squares estimates; this is similar to the ridge regression result. The principal component estimate of β_3 is nearly identical to the least squares estimate. This illustrates a general result of principal component regression: The estimated coefficients for any variables that are nearly orthogonal to the axes causing the collinearity problems are nearly identical to the least squares estimates. However, for variables involved in the collinearity problem, their estimates given by principal component regression are weighted averages of the least squares regression coefficients of *all* variables involved in the collinearity. As with ridge regression, principal component regression provides no information on the relative contribution (to the response variable) of variables involved in the collinearity.

For illustration, it is helpful to follow up on the obvious suggestion from the principal component analysis and the biplot that Z_1 and Z_2, for all practical purposes, present the same information. If the two variables are redundant, a logical course of action is to use only one of the two or their average. The regression analysis was repeated using $\bar{Z} = (Z_1 + Z_2)/2$ as one variable, rescaled to have unit length, and Z_3 as the second variable. Of course, the collinearity problem disappeared. This regression analysis gave $R^2 = .593$, essentially the same as the principal component regression result. The least squares regression coefficient for $\bar{Z}$ was 1.55 (with a standard error of .93). This is almost exactly the sum of the two regression coefficients for Z_1 and Z_2 estimated from the principal component regression using $g = 2$. Thus, the principal component regression analysis replaces the correlated complex of variables causing the near-singularity with a surrogate variable, the principal component, and then "parcels out" the estimated effect of the surrogate variable among the variables that made up the complex. ■

12.3 General Comments

Cases When Collinearity Does Not Cause Serious Problems

The course of action in the presence of collinearity depends on the nature and origin of the collinearity and on the purpose of the regression analysis. If the regression analysis is intended solely for prediction of the dependent variable, the presence of near-singularities in the data does not create serious problems as long as three very important conditions are met:

1. The collinearity shown in the data is a reflection of the correlational structure of the X-space. It must not be an artifact of the sampling process or due to outliers in the data. [Mason and Gunst (1985) discuss the effects and detection of collinearities induced by outliers.]
2. The system continues to operate in the same manner as when the data were generated so that the correlational structure of the X-space remains consistent. This implies that the regression equation is not to be used to predict the response to some modification of the system even if the prediction point is in the sample X-space (condition 3).
3. Prediction is restricted to points within the sample X-space. Extrapolation beyond the data is dangerous in any case but can quickly lead to serious errors of prediction when the regression equation has been estimated from highly collinear data.

These conditions are very limiting and simply reflect the extreme sensitivity of ordinary least squares when collinearity is present. Nevertheless, the impact of collinearity for prediction is much less than it is for estimation (Thisted, 1980). Any variable selection process for model building will tend to select one independent variable from each correlated set to act as a surrogate variable for the complex. The remaining variables in that complex will be dropped. It does not matter for prediction purposes whether the retained variable is a causal variable in the process; it is important only that the system continue to "act" as it did when the data were collected so that the surrogate variable continues to adequately represent the complex of variables.

Cases When Collinearity Is Serious

On the other hand, collinearity creates serious problems if the purpose of the regression is to understand the process, to identify important variables in the process, or to obtain meaningful estimates of the regression coefficients. The ordinary least squares estimates can be far from the true values. In the numerical example we have used throughout, the true values of the regression coefficients were 5.138, -2.440, and 2.683 compared to the estimated values of 14.5, -13.2, and 4.49. While there is always uncertainty with observational data regarding the *true* importance of a variable in the process being studied, the presence of collinearity almost guarantees that the identification of important variables will be wrong. If all potentially important variables are retained in the model, all variables in any correlated complex will appear to be unimportant because any one of them, whether important to the process or not, can usurp the function of the others in the regression equation. Further, any variable selection process to choose the best subset of variables will almost certainly "discard" important variables and the variable retained to represent each correlated complex may very well be unimportant to the process. For these purposes, it is extremely important that the presence of collinearity be recognized and its nature understood.

Some degree of collinearity is expected with observational data. "Seeing" the correlational structure should alert the researcher to the cases where the collinearity is the result of inadequate or erroneous data. The solution to the problem is obvious for these cases; near-singularities that result from inadequate sampling or errors in the data will disappear with more and better data. It may be necessary to change the sampling strategy to obtain data points in regions of the X-space not previously represented. Correlations inherent to the system will persist. Analysis of the correlational structure should provide insight to the researcher on how the system operates and may suggest alternative parameterizations and models. In the final analysis, it will probably be necessary to resort to controlled experimentation to separate the effects of highly collinear variables. Collinearity should seldom be a problem in controlled experiments. The choice of treatment levels for the experiment should be such that the factors are orthogonal, or nearly so.

12.4 Summary

The purposes of this chapter were to emphasize the importance of understanding the nature of any near-singularities in the data that might cause problems with the ordinary least squares regression, to introduce principal component analysis and Gabriel's biplots as tools for aiding this understanding, and to acquaint the reader with two (of the several) biased regression methods. All of the biased regression methods are developed on the premise that estimators with smaller mean squared errors can be found if unbiasedness of the estimators is not required. As with many regression techniques, the reader is cautioned against indiscriminate use of biased regression methods. Every effort should be made to understand the nature and origin of the problem and to correct it with better data if possible.

EXERCISES

12.1. Use the definition of mean squared error in equation 12.4 to show that MSE is the variance of the estimator plus the square of the bias.

12.2. Consider the matrix

$$X = \begin{bmatrix} 3 & 6 \\ 12 & 24 \\ 9 & 18 \\ 4 & 8 \end{bmatrix}$$

a. Show that $X'X$ is singular.

b. Convince yourself (numerically) that $(X'X + kI)$ is not singular. Use any small value of k.

12.3. Use the variance of linear functions to derive $\mathbf{Var}[\tilde{\beta}(k)]$, equation 12.8, assuming that $\mathbf{Var}(\varepsilon) = I\sigma^2$.

12.4. Use the variance of linear functions and $\hat{\gamma} = L^{-1}W'Y$ to show that $\mathbf{Var}(\hat{\gamma}) = L^{-1}\sigma^2$, equation 12.21.

12.5. Use equation 12.25 and the variance of linear functions to derive $\mathbf{Var}(\beta^+_{(g)})$, equation 12.26.

12.6. Show that the sum of the variances of $\beta^+_{(g)j}$ is equal to the sum of the variances of $\hat{\gamma}_j$. That is, show that $\mathrm{tr}\{\mathbf{Var}[\beta^+_{(g)}]\} = \mathrm{tr}\{\mathbf{Var}[\hat{\gamma}_{(g)}]\}$.

* 12.7. Show that the length of the $\beta^+_{(g)}$ vector is the same as the length of $\hat{\gamma}_{(g)}$.

* 12.8. Use the logarithms of the nine independent variables in the peak flow runoff data from exercise 5.1.

a. Center and scale the independent variables to obtain Z and $Z'Z$, the correlation matrix.

b. Do the singular value decomposition on Z and construct the biplot for the first and second principal component dimensions. What proportion of the dispersion in the X-space is accounted for by these first two dimensions?

c. Use the correlation matrix and the biplot to describe the correlational structure of the independent variables.

* 12.9. Do ridge regression on the peak flow runoff data (exercise 5.1) to estimate the regression equation using the logarithms of all independent variables and $\ln(Q)$ as the dependent variable. Use k as determined by Hoerl, Kennard, and Baldwin (equation 12.9). Compare the ridge regression results with ordinary least squares regression and interpret the analyses.

12.10. Do principal component regression on the peak flow runoff data (exercise 5.1) to estimate the regression equation using the logarithms of all independent variables and $\ln(Q)$ as the dependent variable.

a. Which principal components are causing a collinearity problem?

b. Test the significance of the individual principal component regression coefficients. Which principal components will you retain for your regression?

c. Convert the results to $\beta^+_{(g)}$, compute estimates of their variances, and give the final regression equation (in terms of the Z's).

d. Compute R^2 and compare the results with the results of ridge regression (exercise 12.9).

*12.11. Use the data from Andrews and Herzberg (1985) on percentages of sand, silt, and clay in soil at 20 sites given in exercise 10.11.

a. Do the singular value decomposition on Z, the centered and scaled variables, and construct Gabriel's biplot of the data.

b. How many principal components must be used in order to account for 80% of the dispersion?
c. Interpret the results of the biplot (of the first and second principal components) in terms of (i) which variable vectors are not well represented by the biplot, (ii) the correlational structure of the variables, (iii) how the 20 sites tend to cluster, and (iv) which site has very low sand content at depths 1 and 2 but moderately high sand content at depth 3.

12.12. This exercise is a continuation of the Laurie-Alberg experiment on relating fruitfly activity to four enzymes (exercise 10.9). The results of the SVD on $\boldsymbol{Z}$ are given in exercise 10.9. Some of the results from principal component regression are given in the accompanying tables.

Estimates of the regression coefficients (for Z's) retaining the indicated principal components

	Principal Components Retained			
Variable	*All*	1, 2, 3	1, 2	1
Intercept	13.118	13.118	13.118	13.118
SDH	−1.594	2.700	2.472	4.817
FUM	10.153	5.560	5.229	5.444
GH	4.610	6.676	6.400	2.018
GO	4.547	4.580	5.340	4.543

Variances of estimated regression coefficients

	Principal Components Retained			
Variable	*All*	1, 2, 3	1, 2	1
Intercept	.05	.05	.05	.05
SDH	56.8	9.0	6.74	2.72
FUM	63.1	8.4	3.51	3.48
GH	29.0	17.9	14.50	.48
GO	28.8	28.8	2.89	2.42
$\text{tr}\{\mathbf{Var}[\boldsymbol{\beta}^{+}_{(g)}]\}$	177.6	64.1	27.64	9.10

a. From the SVD in exercise 10.9, are any principal components cause for concern in variance inflation? Which Z's are heavily involved in the fourth principal component?
b. From inspection of the behavior of the variances as the principal components are dropped, which variables are heavily involved in the fourth principal component? Which are involved in the third principal component?
c. Which principal component regression solution would you use? The variances continue to decrease as more principal components are dropped from the solution. Why would you not use the solution with *only* the first principal component?
d. Do a t-test of the regression coefficients for your solution. (There were $n = 21$ observations in the data set.) State your conclusions.

CHAPTER THIRTEEN

Case Study: Collinearity Problems

Chapter 12 discussed methods of handling the collinearity problem. This chapter uses the Linthurst data to illustrate the behavior of ordinary least squares when collinearity is a problem. The correlational structure is then analyzed using principal component analysis and Gabriel's biplots. Finally, ridge regression and principal component regression are used and their limitations for the objective of this study are discussed.

This chapter gives the analysis of a set of observational data where collinearity is a problem. The purposes of this case study are (1) to demonstrate the inadequacies of ordinary least squares in the presence of collinearity, (2) to show the value of analyzing the correlational structure of the data, and (3) to demonstrate the use and limitations of ridge regression and principal component regression.

13.1 The Problem

This analysis is a continuation of the first case study (Chapter 5), which used five variables from the September sampling of the Linthurst data on Spartina biomass production in the Cape Fear Estuary of North Carolina.* The objective of the study was to identify physical and chemical properties of the substrate that are influential in determining the widely varying aerial biomass production of Spartina in the Cape Fear Estuary. The sampling plan included three marshes in the estuary and three sites in each marsh representing three ecosystems: an area where the Spartina had previously died but had recently regenerated, an area consisting of short Spartina, and an area consisting of tall Spartina. In each of the nine sites, five random sampling points were chosen from which aerial *BIOMASS* and the following 14 physico-chemical properties of the substrate were measured on a monthly schedule:

*The author appreciates Dr. Rick A. Linthurst's permission to use the data and his contributions to this discussion.

1. Free sulfide (H_2S), moles
2. Salinity (SAL), o/oo
3. Redox potentials at pH 7 ($Eh7$), mv
4. Soil pH in water (pH), 1:1 soil/water
5. Buffer acidity at pH 6.6 (BUF), meg/100 cm^3
6. Phosphorus concentration (P), ppm
7. Potassium concentration (K), ppm
8. Calcium concentration (Ca), ppm
9. Magnesium concentration (Mg), ppm
10. Sodium concentration (Na), ppm
11. Manganese concentration (Mn), ppm
12. Zinc concentration (Zn), ppm
13. Copper concentration (Cu), ppm
14. Ammonium concentration (NH_4), ppm

Table 13.1 contains the data for *BIOMASS* and the 14 substrate variables for the September sampling date. The "Loc" code identifies the three islands in the Cape Fear Estuary. The "Type" code identifies the nature of the Spartina vegetation at each sampling site: DVEG labels the recently regenerated areas, and TALL and SHRT identify the commonly labeled tall and short Spartina areas, respectively.

Analysis of the full data set showed a serious collinearity problem in the data for every sampling date. The five variables used in Chapter 5—*SAL*, *pH*, *K*, *Na*, and *Zn*—were chosen from the larger data set to preserve some of the collinearity problem and yet reduce the dimension of the problem to a more convenient size for presentation. The multiple regression analysis of that subset of data with the five variables in the model showed significance only for *pH*. Backward elimination of one variable at a time led to a final model containing *pH* and *K*. In Chapter 7, all possible regressions showed *pH* and *Na* to be the best two-variable model. Section 10.4 gave the residuals analysis, influence statistics, and the collinearity diagnostics for the model with these five variables.

In this chapter, *BIOMASS* is used as the dependent variable but all 14 physico-chemical variables are investigated as independent variables. The primary objective of this research was to study the observed relationships between *BIOMASS* and the substrate variables with the purpose of identifying substrate variables that with further study might prove to be causal. As in Chapter 5, this analysis concentrates on the *total* variation over the nine sites. The analysis of the "among-site" variation is left to exercises at the end of this chapter. The "within-site" variation can be studied in a similar manner.

Ordinary least squares, perhaps the most commonly used statistical tool for assessing importance of variables, was the first method applied by the researcher. The results obtained, and reported here for the September data, were typical of ordinary least squares results in the presence of collinearity; the inadequacies of

Table 13.1 Aerial biomass and 14 physico-chemical properties of the substrate in the Cape Fear Estuary of North Carolina. The data presented are from the September sampling date and were collected as part of the doctoral research of Dr. R. A. Linthurst, North Carolina State University. (Used with permission.)

Obs.	*Loc.*	*Type*	*BIO*	H_2S	*SAL*	*Eh7*	*pH*	*BUF*	*P*	*K*	*Ca*	*Mg*	*Na*	*Mn*	*Zn*	*Cu*	NH_4
1	OI	DVEG	676	−610	33	−290	5.00	2.34	20.238	1,441.67	2,150.00	5,169.05	35,184.5	14.2857	16.4524	5.02381	59.524
2	OI	DVEG	516	−570	35	−268	4.75	2.66	15.591	1,299.19	1,844.76	4,358.03	28,170.4	7.7285	13.9852	4.19019	51.378
3	OI	DVEG	1,052	−610	32	−282	4.20	4.18	18.716	1,154.27	1,750.36	4,041.27	26,455.0	17.8066	15.3276	4.79221	68.788
4	OI	DVEG	868	−560	30	−232	4.40	3.60	22.821	1,045.15	1,674.36	3,966.08	25,072.9	49.1538	17.3128	4.09487	82.256
5	OI	DVEG	1,008	−610	33	−318	5.55	1.90	37.843	521.62	3,360.02	4,609.39	31,664.2	30.5229	22.3312	4.60131	70.904
6	OI	SHRT	436	−620	33	−308	5.05	3.22	27.381	1,273.02	1,811.11	4,389.84	25,491.7	9.7619	12.2778	4.50794	54.206
7	OI	SHRT	544	−590	36	−264	4.25	4.50	21.284	1,346.35	1,906.63	4,579.33	20,877.3	25.7371	17.8225	4.91093	84.982
8	OI	SHRT	680	−610	30	−340	4.45	3.50	16.511	1,253.88	1,860.29	3,983.09	25,621.3	10.0267	14.3516	5.11364	53.275
9	OI	SHRT	640	−580	38	−252	4.75	2.62	18.199	1,242.65	1,799.02	4,142.40	27,587.3	9.0074	13.6826	4.64461	47.733
10	OI	SHRT	492	−610	30	−288	4.60	3.04	19.321	1,281.95	1,796.66	4,263.93	26,511.7	12.7140	11.7566	4.58761	60.674
11	OI	TALL	984	−540	30	−294	4.10	4.66	16.622	553.69	1,019.56	1,965.95	7,886.5	31.4815	9.8820	1.74582	65.875
12	OI	TALL	1,400	−560	37	−278	3.45	5.24	22.629	494.74	1,373.89	2,366.73	14,596.0	64.4393	16.6752	3.21729	104.550
13	OI	TALL	1,276	−570	33	−248	3.45	6.32	13.015	525.97	1,057.40	2,093.10	9,826.8	48.2886	12.3730	2.97695	75.612
14	OI	TALL	1,736	−580	36	−314	4.10	4.88	13.678	571.14	1,111.29	2,108.47	11,978.4	22.5500	9.4058	2.71841	59.888
15	OI	TALL	1,004	−640	30	−328	3.50	4.70	14.663	408.64	843.50	1,711.42	10,368.6	33.4330	14.9302	1.85407	77.572
16	SI	DVEG	396	−610	30	−328	3.25	6.26	60.862	646.65	1,694.01	3,018.60	17,307.4	52.7993	31.2865	3.72767	102.196
17	SI	DVEG	352	−600	27	−374	3.35	6.36	77.311	514.03	1,667.42	2,444.52	12,822.0	60.4025	30.1652	2.99087	96.418
18	SI	DVEG	328	−630	29	−356	3.20	5.34	73.513	350.73	1,455.84	2,372.91	8,582.6	66.3797	28.5901	2.41503	88.484
19	SI	DVEG	392	−640	34	−354	3.35	4.44	56.762	496.29	2,002.44	2,241.30	12,369.5	56.8681	19.8795	2.45754	91.758
20	SI	DVEG	236	−600	36	−348	3.30	5.90	39.531	580.92	1,427.89	2,778.22	14,731.9	64.5076	18.5056	2.82948	101.712
21	SI	SHRT	392	−640	30	−390	3.25	7.06	39.723	535.82	1,339.26	2,807.64	15,060.6	56.2912	22.1344	3.43709	179.809
22	SI	SHRT	268	−650	28	−358	3.25	7.90	55.566	490.34	1,468.69	2,643.62	11,056.3	58.5863	28.6101	3.47090	168.098
23	SI	SHRT	252	−630	31	−332	3.20	7.72	35.279	552.39	1,377.06	2,674.65	8,118.9	56.7497	23.1908	3.60202	210.316
24	SI	SHRT	236	−640	31	−314	3.20	8.14	97.695	661.32	1,747.56	3,060.10	13,009.5	57.8526	24.6917	3.92552	211.050
25	SI	SHRT	340	−630	35	−332	3.35	7.44	99.169	672.15	1,526.85	2,696.80	15,003.7	45.0128	22.6758	4.23913	185.454
26	SI	TALL	2,436	−620	29	−338	7.10	−.42	3.718	528.65	6,857.39	1,778.77	10,225.0	16.4856	.3729	3.41143	16.497
27	SI	TALL	2,216	−620	35	−268	7.35	−1.04	2.703	563.13	7,178.00	1,837.54	8,024.2	11.4075	.2703	3.43998	13.655
28	SI	TALL	2,096	−570	35	−300	7.45	−1.12	2.633	497.96	6,934.67	1,586.49	10,393.0	7.9561	.3205	3.29673	17.627

29	SI	TALL	1,660	−620	30	−328	7.45	−.86	3.148	458.38	6,911.54	1,483.41	8,711.6	10.4945	.2648	3.11813	15.291
30	SI	TALL	2,272	−570	30	−374	7.40	−.90	2.626	498.25	6,839.54	1,631.32	10,239.6	9.4637	.2105	2.79145	14.750
31	SM	DVEG	824	−620	26	−336	4.85	3.72	16.715	936.26	1,564.84	3,828.75	20,436.0	10.3375	18.9875	5.76402	95.721
32	SM	DVEG	1,196	−630	29	−342	4.60	4.90	16.377	894.79	1,644.37	3,486.84	12,519.9	21.6672	20.9687	5.36276	86.955
33	SM	DVEG	1,960	−630	25	−328	5.20	2.78	21.593	941.36	1,811.00	3,517.16	18,979.0	13.0967	23.9841	5.48042	83.935
34	SM	DVEG	2,080	−630	26	−332	4.75	3.90	18.030	1,038.79	1,706.36	4,096.67	22,986.1	15.6061	19.9727	5.27273	104.439
35	SM	DVEG	1,764	−610	26	−322	5.20	3.60	34.693	898.05	1,642.51	3,593.05	11,704.5	6.9786	21.3864	5.71123	79.773
36	SM	SHRT	412	−640	25	−290	4.55	3.58	28.956	989.87	2,171.35	3,553.17	17,721.0	57.5856	23.7063	3.68392	118.178
37	SM	SHRT	416	−610	26	−352	3.95	5.58	25.741	951.28	1,767.63	3,359.17	16,485.2	72.5160	30.5589	3.91827	123.538
38	SM	SHRT	504	−600	26	−280	3.70	6.58	25.366	939.83	1,654.63	3,545.32	17,101.3	64.4146	26.8415	4.06829	135.268
39	SM	SHRT	492	−620	27	−290	3.75	6.80	17.917	925.42	1,620.83	3,467.92	17,849.0	53.9583	27.7292	3.89583	115.417
40	SM	SHRT	636	−590	27	−328	4.15	5.30	20.259	954.11	1,446.30	3,170.65	16,949.6	22.6657	21.5699	4.70368	108.406
41	SM	TALL	1,756	−560	24	−332	5.60	1.22	134.426	720.72	2,576.08	2,467.52	11,344.6	51.9258	19.6531	4.11065	57.315
42	SM	TALL	1,232	−550	27	−276	5.35	1.82	35.909	782.09	2,659.36	2,772.99	14,752.4	75.1471	20.3295	4.09826	77.193
43	SM	TALL	1,400	−550	26	−282	5.50	1.60	38.719	773.30	2,093.57	2,665.02	13,649.8	71.0254	19.5880	4.31487	68.294
44	SM	TALL	1,620	−540	28	−370	5.50	1.26	33.562	829.26	2,834.25	2,991.99	14,533.0	70.1465	20.1328	6.09432	71.337
45	SM	TALL	1,560	−570	28	−290	5.40	1.56	36.346	856.96	3,459.26	3,059.73	16,892.2	89.2593	19.2420	4.87407	79.383

the method were evident. Principal component analysis and Gabriel's biplot are then used here to develop an understanding of the correlational structure of the independent variables. To complete the case study, ridge regression and principal component regression are applied to the data to illustrate their use, and to show that biased regression methods suffer some of the same inadequacies as least squares when the purpose of the analysis is to identify "important" variables.

While more and better data is the method of first choice for solving the collinearity problem, there will be situations where (1) it is not economically feasible with observational studies to obtain the kind of data needed to disrupt the near-singularities, or (2) the near-singularities are a product of the system and will persist regardless of the amount of data collected. One purpose of this case study is to raise flags of caution on the use of least squares and biased regression methods in such cases. Biased regression methods can have advantages over least squares for estimation of the individual parameters, in terms of mean squared error, but suffer from the same problems as least squares when the purpose is identification of "important" variables.

13.2 Multiple Regression: Ordinary Least Squares

The purpose of presenting this analysis is to illustrate the behavior of ordinary least squares in the presence of collinearity and to demonstrate the misleading nature of the results both for estimation of regression coefficients and for identification of important variables in the system.

Full Model Results

Ordinary least squares regression of *BIOMASS* on all 14 variables gave $R^2 = .807$. The regression coefficients and their standard errors are given in the first three columns of Table 13.2. Only two variables, K and Cu, have regression coefficients differing from zero by more than twice their standard error. Taken at face value, these results would seem to suggest that K and Cu are the only important variables in "determining" *BIOMASS*. However, the magnitudes of regression coefficients and their standard errors in any nonorthogonal data set depend on which other variables are included in the model. (Recall that pH was the only significant variable in the regression on the five variables SAL, pH, K, Na, and Zn.) The conclusion that K and Cu are the only important variables is not warranted.

Stepwise Regression

To demonstrate the dependence of the least squares results on the method used, two stepwise regression procedures, the "maximum R-square" and "stepwise" options in PROC REG (SAS Institute, Inc., 1985d, 1985f), were used to simplify the model and to select "important" variables. These least squares procedures suggested a four-variable model consisting of pH, Mg, Ca, and Cu (Table 13.2, columns 4–11). The "stepwise" procedure terminated at this point, using the default options of the program. In the "maximum R-square" option, the fourth step, at which Cu was added, was the first step for which the C_p statistic was less than p' ($C_p = 4.0$ with $p' = 5$). The major increases in R^2 had been realized at this point ($R^2 = .75$). The two variables that were the only significant variables in the full model, Cu and K, entered in the fourth and sixth steps in the "maximum R-square" stepwise regression.

Table 13.2 Ordinary least squares regression of aerial biomass on 14 soil variables, and stepwise regression results from the "maximum R-square" and "stepwise" options in SAS. All independent variables are centered and standardized.

Multiple Regression			*Maximum R-Square*				*Stepwise*			
Soil Variable X_j	$\hat{\beta}_j$[a]	$s(\hat{\beta})$	*Step*	*Variable Added (+) Deleted (−)*	C_p	R^2	*Step*	*Variable Added (+) Deleted (−)*	$\hat{\beta}_j$[b]	$s(\hat{\beta})$
H_2S	88	610	1	$+pH$	21.4	.599	1	$+pH$	4,793	906
SAL	−591	645	2	$+Mg$	14.1	.659	2	$+Mg$	−2,592	571
$Eh7$	626	493	3	$+Ca$	5.7	.726	3	$+Ca$	−2,350	920
pH	2,005	2,763	4	$+Cu$	4.0	.750	4	$+Cu$	1,121	580
BUF	−117	2,058	5	$+P$	3.8	.764	5	$+P$		
P	−312	483	6	$+K, -P, +Zn$	3.8	.777	6	$-P$		
K	−2,069	952	7	$+NH_4$	4.1	.788	terminated			
Ca	−1,325	1,431	8	$+Eh7, -Zn, +P$	4.7	.797				
Mg	−1,744	1,709	9	$+Zn, -P, +SAL$	5.6	.804				
Na	203	1,128	10	$+P$	7.2	.806				
Mn	−274	872	11	$+Mn$	9.1	.807				
Zn	−1,031	1,195	12	$+Na$	11.0	.807				
Cu	2,374	771	13	$+H_2S$	13.0	.807				
NH_4	−847	1,015	14	$+BUF$	15.0	.807				

[a] Regression coefficients for full model with independent variables standardized.
[b] Regression coefficients for four-variate model with independent variables standardized.

With the other ten variables dropped, the magnitudes of the regression coefficients for the four retained variables and their standard errors changed considerably. The coefficient for pH more than doubled, the coefficients for Mg and Ca increased 50 to 100% (in absolute value), and the coefficient for Cu was halved. The standard errors for pH and Mg were reduced by $\frac{2}{3}$, and the standard errors for Cu and Ca by $\frac{1}{4}$ and $\frac{1}{3}$, respectively. Of these four variables, pH and Mg appear to be the more important, as judged by their early entry into the model and the ratio of their coefficients to their standard errors.

Correlation Matrix

Inspection of the correlation matrix, Table 13.3 (page 362), reveals five variables with reasonably high correlations with $BIOMASS$: pH, BUF, Ca, Zn, and NH_4. Each of these five variables would appear important if used as the only independent variable, but none of these five was identified in the full model, and only pH and Ca were revealed as important in stepwise regression. The other two variables declared important by the stepwise procedure, Mg and Cu, had correlations with $BIOMASS$ of only −.38 and .09, respectively. The second most highly correlated variable with $BIOMASS$, buffer acidity (BUF), was the last of the 14 variables to enter the model in the "maximum R-square" variable selection option.

Inconsistencies

The two stepwise regression methods suggest that future studies concentrate on pH, Mg, Ca, and Cu. On the other hand, ordinary least squares regression using all variables identified only K and Cu as the important variables, and simple regressions on one variable at a time identify pH, BUF, Ca, Zn, and NH_4. None of the results was satisfying to the biologist; the inconsistencies of the results were confusing and variables expected to be biologically important were not showing significant effects.

Table 13.3 Product moment correlations among all variables in the Linthurst September data.

	BIO	H_2S	*SAL*	*Eh7*	*pH*	*BUF*	*P*	*K*	*Ca*	*Mg*	*Na*	*Mn*	*Zn*	*Cu*	NH_4
BIO	1.00														
H_2S	.33	1.00													
SAL	−.10	.10	1.00												
Eh7	.05	.40	.31	1.00											
pH	.77	.27	−.05	.09	1.00										
BUF	−.73	−.37	−.01	−.15	−.95	1.00									
P	−.35	−.12	−.19	−.31	−.40	.38	1.00								
K	−.20	.07	−.02	.42	.02	−.07	−.23	1.00							
Ca	.64	.09	.09	−.04	.88	−.79	−.31	−.26	1.00						
Mg	−.38	−.11	−.01	.30	−.18	.13	−.06	.86	−.42	1.00					
Na	−.27	−.00	.16	.34	−.04	−.06	−.16	.79	−.25	.90	1.00				
Mn	−.35	.14	−.25	−.11	−.48	.42	.50	−.35	−.31	−.22	−.31	1.00			
Zn	−.62	−.27	−.42	−.23	−.72	.71	.56	.07	−.70	.35	.12	.60	1.00		
Cu	.09	.01	−.27	.09	.18	−.14	−.05	.69	−.11	.71	.56	−.23	.21	1.00	
NH_4	−.63	−.43	−.16	−.24	−.75	.85	.49	−.12	−.58	.11	−.11	.53	.72	.93	1.00

Ordinary least squares regression tends either to indicate that none of the variables in a correlated complex is important when all variables are in the model, or to arbitrarily choose one of the variables to represent the complex when an automated variable selection technique is used. A truly important variable may appear unimportant because its contribution is being usurped by variables with which it is correlated. Conversely, unimportant variables may appear important because of their associations with the real causal factors. It is particularly dangerous in the presence of collinearity to use the regression results to impart a "relative importance," whether in a causal sense or not, to the independent variables.

These seemingly inconsistent results are typical of ordinary least squares regression when there are high correlations or, more generally, near-linear dependencies among the independent variables. Inspection of the correlation matrix shows several pairs of independent variables with reasonably high correlations and three with $|r| \geq .90$. The largest absolute correlation, $r = -.95$, is between *pH* and *BUF*, the first and last variables to enter the model in the "maximum *R*-square" stepwise analysis. Any inference that *pH* is an important variable while buffer acidity is not is clearly an unacceptable conclusion. Other less obvious near-linear dependencies among the independent variables may also be influencing the inclusion or exclusion of variables from the model. The correlational structure of the independent variables makes any simple interpretation of the regression analyses unacceptable.

13.3 Analysis of the Correlational Structure

Purpose

The purpose of the analysis of the correlational structure is to gain insight into the relationships among the variables being studied and the causes of the collinearity problem. The analysis may suggest ways of removing some of the collinearity by obtaining more data or redefining variables. The improved understanding will identify the systems of variables that are closely related to the variation in the dependent variable and, hence, which sets of variables merit further study.

Principal Component Analysis

Inspection of the correlations among the independent variables in Table 13.3 reveals several reasonably high correlations. However, correlations reveal only pairwise associations and will provide an adequate picture of the correlational structure only in the simplest cases. A more complete understanding is obtained by using principal component analysis, or singular value decomposition (SVD), of the $n \times p$ matrix of the independent variables. For this purpose, the independent variables are centered and scaled so that the sum of squares of each independent variable is 1.0; the vectors have unit length in n-space. (Refer to Sections 2.7, 2.8, and 12.2 for review of eigenanalysis, singular value decomposition, and construction of the principal component variables.)

The eigenvalues (λ_j^2) and eigenvectors ($\boldsymbol{v}_j$) for these data are given in Table 13.4 (page 364). The first principal component accounts for 35% of the dispersion in X-space, $\lambda_1^2/\sum \lambda_j^2 = .35$, and is defined primarily by the complex of variables pH, BUF, Ca, Zn, and NH_4; these are the variables with the largest coefficients in the first eigenvector, $\boldsymbol{v}_1$. The second principal component, defined primarily by K, Mg, Na, and Cu, accounts for 26% of the dispersion. The four dimensions with eigenvalues greater than 1.0 account for 83% of the dispersion. (If all independent variables had been orthogonal, all eigenvalues would have been 1.0 and each would have accounted for 7% of the dispersion.)

With the singular value decomposition, the measures of collinearity can be used to assess the extent of the collinearity problem. The full impact will not be seen from the singular value decomposition of the centered and scaled matrix because collinearities involving the intercept have been eliminated. Nevertheless, the smaller eigenvalues (Table 13.4) show that there is very little dispersion in several dimensions. The last four principal component dimensions together account for only 1.2% of the dispersion in the X-space; the last six principal components account for 3.4% of the total dispersion in the X-space. Thus, there is very little dispersion in at least six dimensions of a nominal 14-dimensional space. The dimension with the least dispersion, $\lambda_{14}^2 = .0095$, is due primarily to a linear restriction on pH, BUF, and Ca. The correlation between pH and BUF, $-.95$, is the highest correlation among the independent variables (Table 13.3).

Based on a result of Hoerl and Kennard (1970a), the lower bound on the sum of the variances of estimated coefficients is $\sigma^2/\lambda_{14}^2 = 105\sigma^2$. This is compared to $14\sigma^2$ if all independent variables had been pairwise orthogonal. The condition number for the matrix of centered variables is 22.8, above the value of 10 suggested as the point above which collinearity can be expected to cause problems. Thisted's (1980) measure of collinearity is

$$mci = \sum_{j=1}^{14} \lambda_j^{-4}\lambda_{14}^4 = 1.17$$

indicating severe collinearity. (Values of mci near 1.0 indicate high collinearity;

Table 13.4 Eigenvalues and eigenvectors of the $\boldsymbol{X'X}$ matrix for the 14 independent variables in the Linthurst September data. All variables were centered and standardized so that $\boldsymbol{X'X}$ is the correlation matrix.

Eigenvalues	λ_1^2	λ_2^2	λ_3^2	λ_4^2	λ_5^2	λ_6^2	λ_7^2	λ_8^2	λ_9^2	λ_{10}^2	λ_{11}^2	λ_{12}^2	λ_{13}^2	λ_{14}^2
	4.925	3.696	1.607	1.335	.692	.500	.385	.381	.166	.143	.0867	.0451	.0298	.0095
Eigenvectors[a]	v_1	v_2	v_3	v_4	v_5	v_6	v_7	v_8	v_9	v_{10}	v_{11}	v_{12}	v_{13}	v_{14}
H_2S	.164	−.009	.232	−.690	.014	.422	−.293	.087	.169	.296	.221	−.015	−.007	.080
SAL	.108	−.017	.606	.271	.509	−.008	−.389	−.081	−.174	−.227	.090	−.155	.095	−.089
$Eh7$	.124	−.225	.458	−.301	−.166	−.598	.308	.299	−.225	.084	−.023	.055	.033	.023
pH	.408	.028	−.283	−.082	.092	−.190	−.056	.033	.024	.147	.042	−.332	−.025	−.750
BUF	−.412	−.000	.205	.166	−.162	.024	−.110	.159	.097	.103	.340	.455	−.354	−.478
P	−.273	.111	−.160	−.200	.747	.018	.357	.381	.077	−.018	−.035	.064	−.066	−.015
K	.034	−.488	−.023	−.043	−.062	.016	.073	.112	.560	−.554	.219	−.029	.249	−.073
Ca	.358	.181	−.207	.054	.206	−.427	−.117	−.179	.189	.076	.508	.348	−.082	.306
Mg	−.078	−.499	−.050	.037	.103	−.034	.036	−.173	−.012	.111	.119	−.400	−.689	.193
Na	.018	−.470	.051	.055	.240	.059	.160	−.459	.088	.439	−.219	.363	.275	−.144
Mn	−.277	.182	.020	−.483	.039	−.300	−.152	−.524	.086	−.363	−.270	.076	−.172	−.141
Zn	−.404	−.089	−.176	−.150	−.008	−.036	.062	−.211	−.438	.016	.572	−.217	.396	−.042
Cu	.011	−.392	−.377	−.102	.064	−.075	−.549	.305	−.376	−.129	−.194	.304	.000	.043
NH_4	−.399	.026	−.011	.104	−.005	−.378	−.388	.165	.420	.394	.132	.303	.232	.118

[a] The sum of squares of the elements in each eigenvector is 1.0. Thus, if a particular variable's contribution were spread equally over all components, the coefficients would be approximately $\pm 1/\sqrt{14} = \pm .27$.

values greater than 2.0 indicate little or no collinearity.) The variance inflation factors (VIF), the diagonal elements of $(X'X)^{-1}$, also show the effects of collinearity. The largest VIF is 62 for pH, followed by 34.5 for BUF, 23.8 for Mg, 16.6 for Ca, and 11.6 for Zn. The smallest are 1.9 for P and 2.0 for $Eh7$; these two variables are not seriously involved in the near-singularities. (If all independent variables were orthogonal, each VIF would be 1.0.)

In summary, the dispersion of the sample points in at least four principal component dimensions is trivial, accounting for only 1.2% of the total dispersion. This limited dispersion in these principal component dimensions inflates the variances of regression coefficients for *all* independent variables involved in the near-singularities. The observed instability of the least squares regression estimates was to be expected.

Gabriel's Biplot

The major patterns of variation in the X-space can be displayed by plotting the information contained in the major principal components. Gabriel's (1971) biplot using the first two principal components shows the structure of the X matrix as "seen" in these two dimensions (Figure 13.1). This biplot of the first and second principal components accounts for 61% of the dispersion in the original 14-dimensional X-space.

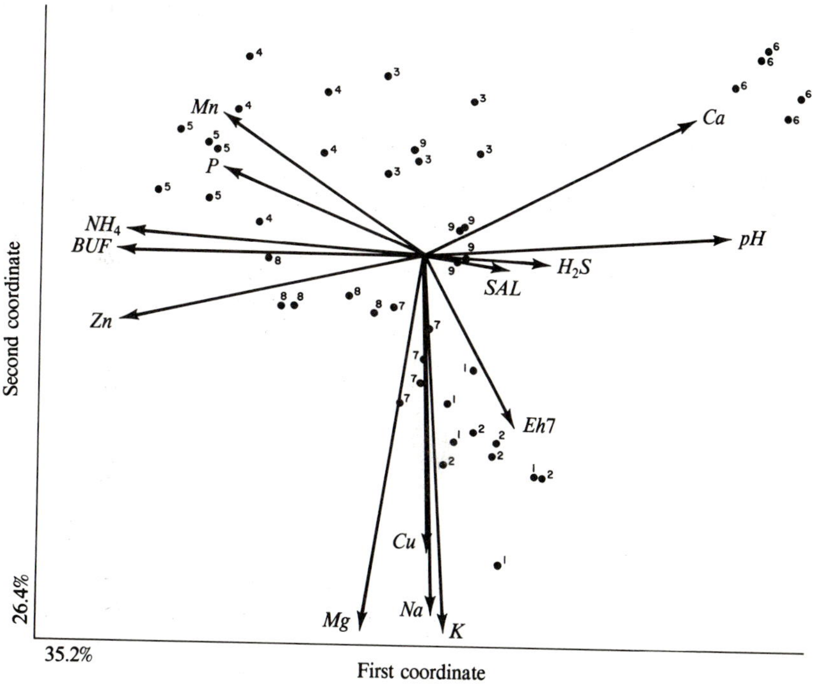

Figure 13.1 Gabriel's biplot of the first and second principal components of the 14 marsh substrate variables. The variables have been centered and scaled so that all vectors have unit length in the original 14-dimensional X-space. The first and second components account for 35.2% and 26.4% of the dispersion in the X-space. Column markers are shown with the vectors, row markers with "•".

Vectors in the biplot are projections of the original variable vectors (in 14-dimensional space) onto the plane defined by the first two principal components. The original vectors were scaled to have unit length. Therefore, the length of each *projected* vector is its correlation with the original vector and reflects the closeness of the original vector to the plane. Thus, the longest vectors, *Ca, pH, Mg, Na, K, Zn*, buffer acidity (*BUF*), and NH_4, indicate the variables are close to the plane being plotted and, consequently, their relationships are well represented by the biplot. The shorter vectors in the biplot, H_2S, salinity (*SAL*), and *Eh*7, identify variables that are more nearly orthogonal to this plane and, therefore, not well represented by this biplot. The other vectors, *Mn, P*, and *Cu*, are intermediate and relationships in this biplot involving these variables should be interpreted with caution.

The near-zero angle between the *Ca* and *pH* vectors, Figure 13.1, shows that the two variables are highly positively correlated ($r = .88$, Table 13.3), as are the three variables NH_4, *BUF*, and *Zn* ($r \geq .71$) and the three variables *Mg, Na*, and *K* ($r \geq .79$). *Ca* is highly negatively correlated with *BUF* and *Zn* ($r \leq -.70$), as is *pH* with *BUF, Zn*, and NH_4 ($r \leq -.72$); the angles are nearly 180°. On the other hand, *pH*, NH_4, *BUF*, and *Zn* are nearly orthogonal to *K* and *Na*. The angles between these vectors are close to 90° and the highest correlation is $r = .12$. The *Cu* and *Na* vectors illustrate the caution needed in interpreting associations between vectors that are not close to unity in length. Even though the angle between the two vectors is close to zero in this biplot, the correlation between *Cu* and *Na* is only .56 (Table 13.3). This apparent inconsistency is because the *Cu* vector is not well represented by this biplot, as indicated by the projected *Cu* vector being appreciably shorter than unity.

More important than the pairwise associations are the two systems of variables revealed by this biplot. The five variables *Ca, pH*, NH_4, *BUF*, and *Zn* strongly associated with the first principal component axis behave as one system; the three variables *Mg, Na*, and *K*, which are strongly associated with the second principal component axis, behave as another. The two sets of variables are nearly orthogonal to each other.

The points in the biplot reflect the relative spatial similarities of the observations (or rows) of the **X** matrix. The number label indicates the sampling site. This biplot indicates that the five samples labeled "6" are very similar to each other and very different from all other samples. The other points also show a distinct tendency to group according to sampling site. The perpendicular projection of the points onto each variable vector shows the relative values of the observations for that variable. Thus, the observations labeled "6" differ from the other points primarily because of their much higher values of *Ca* and *pH* and lower values of NH_4, *BUF*, and *Zn*. On the other hand, observations labeled "1" and "2" tend to be high and observations "3", "4", "5", and "6" tend to be low in *Mg, Na*, and *K*.

Since the first two dimensions account for only 62% of the dispersion in the *X*-space, it is of interest to study the behavior in the third dimension. The first three dimensions account for 73% of the dispersion. Gabriel's biplot of the first and third dimensions (Figure 13.2) shows how the vectors in Figure 13.1 deviate

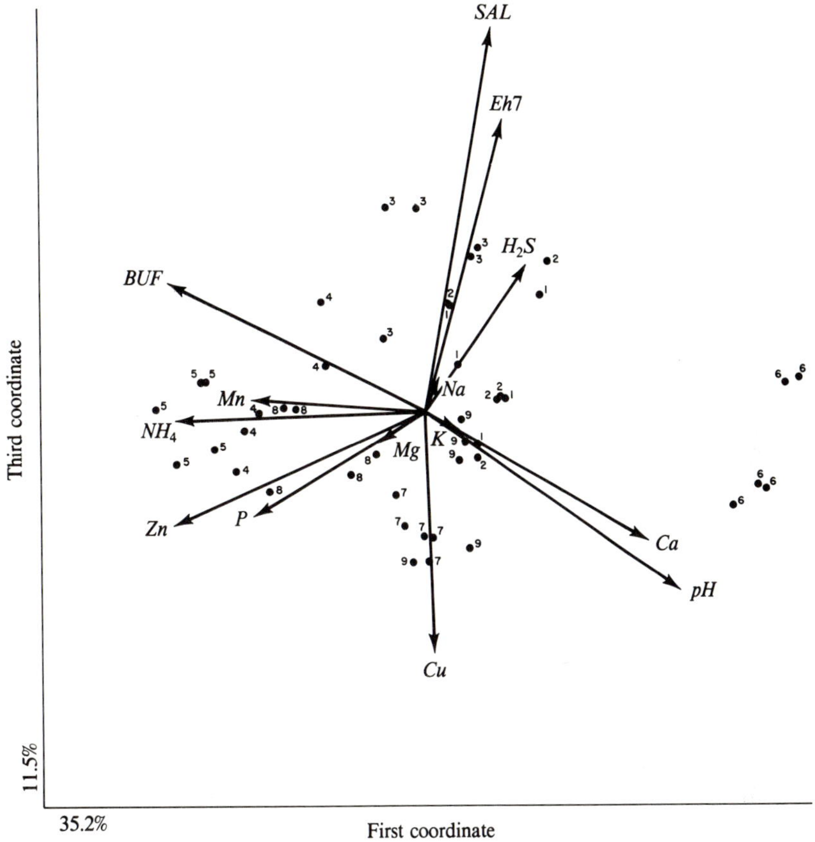

Figure 13.2 Gabriel's biplot of the first and third principal components of the 14 marsh substrate variables. The variables are centered and scaled so that all vectors have unit length in 14-dimensional X-space. The third principal component accounted for 11.5% of the dispersion in the X-space. Column markers are shown with the vectors, row markers with the "•".

above and below the plane representing the first two dimensions. The vectors primarily responsible for defining the second dimension now appear very short because the perspective in Figure 13.2 is down the second axis; only the deviations from the plane of the *Na*, *K*, and *Mg* vectors are observed. The third dimension is defined primarily by salinity (*SAL*), with some impact from *Eh*7 and *Cu*.

The fourth principal component is dominated by H_2S and *Mn* and accounts for 10% of the dispersion. The fifth is dominated by *P* and *SAL* and accounts for 5% of the dispersion, and so on. Gabriel's biplot could also be used to view these dimensions. The principal component analysis and Gabriel's biplots show that the major variation in the X-space is accounted for by relatively few complexes of substrate variables, variables that have a strong tendency to vary together. Interpretation of the associations with *BIOMASS* should focus on these complexes rather than on the individual variables.

Table 13.5 Regression of aerial biomass on individual principal components W_j. (Linthurst September data.)

W_j	SS(Regr)	F (df = 1, 30)
1	10,117,269	82.2*
2	1,018,472	8.3*
3	1,254,969	10.2*
4	496,967	4.0
5	215,196	1.8
6	10,505	.1
7	267,907	2.2
8	675,595	5.5*
9	803,786	6.5*
10	110,826	.9
11	430,865	3.5
12	40,518	.3
13	2,160	.0
14	34,892	.3

* Significant at $\alpha = .05$ using as error the residual mean square from the full model in ordinary least squares (OLS).

Regression of Biomass on Principal Components

The relationship of biomass to the principal component variables can be determined by regressing *BIOMASS* on the principal components. (This is the first step of principal component regression but is presented here to see how *BIOMASS* fits into the principal component structure. Conversion of the regression coefficients for the principal components to the regression coefficients for the original variables will be completed in Section 13.4.) The sums of squares due to regression and the tests of significance of the principal components are given in Table 13.5. The first principal component W_1 dominates the regression, accounting for 65% of the regression sum of squares. (Note that the first principal component is defined so as to account for the greatest dispersion in the X-space, but it does not follow that W_1 will necessarily be the best predictor of *BIOMASS*.) W_2, W_3, W_8, and W_9 also account for significant ($\alpha = .05$) amounts of variation in *BIOMASS*.

Correlations of Independent Variables with Principal Components

For ease of relating the principal components to the original variables, the correlations of each of the 14 original independent variables with these five principal components are given in Table 13.6. The importance of the first principal component in the regression strongly suggests that the *pH–BUF–Ca–Zn–*NH_4 complex of five variables be given primary consideration in future studies of the causes of variation in *BIOMASS* production of Spartina. Perhaps *P* and *Mn* should be included in this set for consideration because of their reasonably high correlations with W_1. Variables of secondary importance are *K*, *Mg*, *Na*, and *Cu*, which are highly correlated with W_2, and salinity and *Eh*7, which are reasonably highly correlated with W_3. W_2 and W_3 account for 7% and 8%, respectively, of the regression sum of squares for *BIOMASS*. H_2S is the only variable not highly correlated with at least one of the three most predictive principal components.

Table 13.6 Correlations between original independent variables, X_k, and the significant principal components, W_j: $\hat{\rho}(X_k, W_j) = v_{jk}\lambda_j$.

Variable	*Principal Component*				
X_k	W_1	W_2	W_3	W_8	W_9
H_2S	.364	−.017	.294	.054	.069
SAL	.240	−.033	.768	−.050	−.071
Eh7	.275	−.433	.581	.185	−.092
pH	.905	.054	−.359	.020	.010
BUF	−.914	.000	.260	.098	.040
P	.606	.213	−.203	.235	.031
K	.075	−.938	−.029	.069	.228
Ca	.794	.348	−.262	−.110	.077
Mg	−.173	−.959	−.063	−.107	−.005
Na	.040	−.904	.065	−.283	.036
Mn	−.615	.350	.025	−.323	.035
Zn	−.897	−.171	−.223	−.130	−.178
Cu	.024	−.754	−.478	.188	−.153
NH_4	−.885	.050	−.014	.102	.171

Dispersion in X-Space

The principal component analysis of the centered and standardized independent variables has demonstrated that most of the dispersion in the X-space can be described by a few complexes of correlated variables. One of these systems, W_1, accounts for a major part of the variation in *BIOMASS*. This complex includes the five variables most highly correlated individually with *BIOMASS*. Three other complexes are accounting for significant but much smaller amounts of variation in *BIOMASS*. The analysis does *not* identify which variable in the complex is responsible for the association. The principal component analysis shows that the data do not contain information that will allow separation of the effects of the individual variables in each complex.

Eliminating Variables to Control Collinearity

A pseudo-solution to the collinearity problem would be to eliminate from the regression model enough independent variables to remove the collinearity. This would be equivalent to retaining one independent variable to represent each major dimension of the original X-space. Variable selection techniques in ordinary least squares regression are, in effect, doing this in a somewhat arbitrary manner. Eliminating variables is not a viable solution when the primary interest is in identifying the important variables. The correlated complexes of variables still exist in nature, but they are no longer "seen" by the regression analysis. It is likely that some of the truly important variables will be lost with such a procedure.

13.4 Biased Regression

Biased regression methods have been suggested as a means of obtaining estimates with smaller mean squared errors in the presence of collinearity. Results from ridge regression and principal component regression are presented for this example to illustrate the impact the methods have on stability of the estimates and the inadequacy of the methods for assigning relative importance to the independent variables. The reader is referred to Section 12.2 for review of these methods.

13.4.1 Ridge Regression

Summary results from ridge regression for several choices of k are shown in Table 13.7. Inspection of the changes in R^2 and the maximum variance inflation factors suggests that k should be between .04 and .10. The Hoerl, Kennard, and Baldwin (1975) suggestion, equation 12.9, gives $k = .08$. However, the ridge traces, plots of the estimated regression coefficients against k (not shown), show four regression coefficients, Cu, Ca, K, and Zn, continuing to change appreciably when $k > .10$; the coefficient for Ca even changed sign at $k = .1$. Thus, ridge regression results for $k = .04, .10$, and $.20$ are included for comparison in Table 13.8. Recall that ridge regression estimates for $k = 0$ are the ordinary least squares estimates.

The dramatic increase in stability of the ridge regression estimates is evident by noting the decrease in VIF_{max} from 62.1 (for $k = 0$) to 4.3 (for $k = .04$), the reduction in $\text{tr}[\mathbf{Var}(\tilde{\boldsymbol{\beta}})]/\sigma^2$, Table 13.7, or the changes in standard errors for several of the individual regression coefficients, Table 13.8. In particular, note the changes for pH and BUF, the two most highly correlated variables, and for Ca and Mg. This gain in stability is realized at a cost of .031 in R^2 and an unknown increase in bias.

The results at $k = .04$, in terms of identifying significant variables, are not very different from ordinary least squares. The coefficients for K and Cu are the only ones differing from zero by more than twice their standard errors. At $k = .10$, pH and Mg are added to the list, and the coefficient for BUF is approaching

Table 13.7 Ridge regression of aerial biomass on 14 substrate variables, Linthurst September data, showing the coefficient of determination, R^2; maximum variance inflation factor, VIF_{max}; sum of all variance inflation factors, $\text{tr}[\mathbf{Var}(\tilde{\boldsymbol{\beta}})]/\sigma^2$; and the length of $\tilde{\boldsymbol{\beta}}$, $(\tilde{\boldsymbol{\beta}}'\tilde{\boldsymbol{\beta}})^{1/2}$, for several choices of k.

k	$100R^2$	VIF_{max} [a]	$\text{tr}[\mathbf{Var}(\tilde{\boldsymbol{\beta}})]/\sigma^2$	$(\tilde{\boldsymbol{\beta}}'\tilde{\boldsymbol{\beta}})^{1/2}$
0	80.7	62.1	196	4,636
.01	79.8	16.0	89	4,078
.02	79.0	8.1	61	3,816
.03	78.3	5.8	47	3,626
.04	77.6	4.3	39	3,473
.06	76.4	2.9	29	3,232
.08	75.4	2.2	23	3,047
.10	74.5	1.8	19	2,899
.20	70.8	.9	11	2,436
.50	63.8	.5	4	1,894

[a] VIF_{max} is the maximum of the diagonal elements of

$$\mathbf{Var}(\tilde{\boldsymbol{\beta}})/\sigma^2 = (X'X + kI)^{-1}(X'X)(X'X + kI)^{-1}$$

Table 13.8 Estimates of regression coefficients and their standard errors for ordinary least squares (OLS) and ridge regression with all independent variables centered and standardized. (Linthurst September data.)

	OLS[a]		Ridge Regression					
			$k = .04$		$k = .10$		$k = .20$	
Variable	$\hat{\beta}$	$s(\hat{\beta})$	$\tilde{\beta}$	$s(\tilde{\beta})$	$\tilde{\beta}$	$s(\tilde{\beta})$	$\tilde{\beta}$	$s(\tilde{\beta})$
H_2S	88	610	343	452	429	376	461	310
SAL	−591	645	−630	438	−588	366	−544	306
$Eh7$	626	493	495	431	362	379	242	322
pH	2,005	2,763	996	621	899*	330	850*	210
BUF	−117	2,058	−759	688	−764	408	−726*	262
P	−312	483	−343	423	−339	377	−324	325
K	−2,069*	952	−1,715*	638	−1,322*	472	−970*	339
Ca	−1,325	1,431	−552	632	−228	433	19	308
Mg	−1,744	1,709	−1,077	729	−846*	416	−694*	257
Na	203	1,128	−341	620	−507	437	−558	317
Mn	−274	872	−529	528	−538	413	−490	324
Zn	−1,031	1,195	−816	697	−662	469	−565	318
Cu	2,374*	771	1,947*	543	1,565*	422	1,211*	327
NH_4	−847	1,015	−598	613	−536	452	−500	335
R^2	.807		.776		.745		.708	
VIF_{max}	62.1		4.3		1.8		.9	

* Estimates exceed twice their standard error.

[a] Ridge regression with $k = 0$ gives OLS results.

twice its standard error. In addition, VIF_{max} is 1.8 and R^2 has decreased another .031. At $k = .2$, the coefficient for BUF also now exceeds twice its standard error. The coefficients for SAL, Na, and Zn are approximately 1.7 times their standard errors. The coefficient of determination has decreased another .037 to $R^2 = .708$ and VIF_{max} is .9, below the cutoff point suggested by Marquardt and Snee (1975). Thus, the ridge regression results would suggest that the important variables, in terms of their association with aerial *BIOMASS*, are K, Cu, pH, Mg, and possibly BUF (buffer acidity).

13.4.2 Principal Component Regression

Deleting Principal Components

The principal component analysis for these data revealed that the six dimensions of the X-space having the *least* dispersion accounted for only 3.4% of the total dispersion in the X-space. Regression of *BIOMASS* on the principal components and tests of significance of the *principal component* regression coefficients revealed that, of these six, only W_9 had significant predictive value for *BIOMASS* (Table 13.5). We then applied the rule that principal components which have small eigenvalues *and* which contain no predictive information for Y should be eliminated. The five principal components corresponding to the five smallest eigenvalues, W_{10} to W_{14}, were deleted for the principal component regression; the first nine principal components ($g = 9$) were retained.

Table 13.9 Cumulative effect of deleting principal components in principal component regression starting with the principal component with the least dispersion, W_{14}. (Linthurst September data.)

Component Deleted	*Information Loss in* $X'X$ (%)	$100R^2$	$\text{tr}[\mathbf{Var}(\boldsymbol{\beta}^+)]/\sigma^2$	$(\boldsymbol{\beta}^{+\prime}\boldsymbol{\beta}^+)^{1/2}$
None = OLS	.0	80.7	196	4,636
14	.1	80.6	91	4,226
13	.3	80.6	57	4,218
12	.6	80.3	35	4,111
11	1.2	78.1	23	3,451
10	2.2	77.5	17	3,333
9	3.4	73.3	10	2,507
8	6.2	69.8	8	2,151
7	8.9	68.4	5	1,952
6	12.5	68.3	3	1,948
5	17.4	67.2	1.8	1,866
4	26.8	64.6	1.1	1,763
3	38.4	58.1	.5	1,526
2	64.8	52.8	.2	1,434
1	100.0	—	—	—

Deleting these five principal components resulted in a loss of 2.2% of the dispersion in the X-space, a loss in predictive value of Y from $R^2 = .807$ to $R^2 = .775$, a decrease in $\text{tr}[\mathbf{Var}(\boldsymbol{\beta}^+)]/\sigma^2$ from 196 to 17, and a decrease in $(\boldsymbol{\beta}^{+\prime}\boldsymbol{\beta}^+)^{1/2}$ from 4,636 to 3,333 (Table 13.9). The stability of the regression estimates increased greatly with an acceptable loss in apparent predictability of *BIOMASS*.

It is of interest to follow the sequential change in these quantities as individual principal components are deleted from the regression (Table 13.9). There is virtually no loss in predictability when W_{12}, W_{13}, and W_{14} are deleted (see R^2, Table 13.9). The variances of the estimates decrease dramatically, particularly with elimination of the 14th principal component (see $\text{tr}[\mathbf{Var}(\boldsymbol{\beta}^+)]/\sigma^2$, Table 13.9). Since W_9 and W_8 are significant, none of the results where W_9 to W_1 have been eliminated would be used. They are presented here only to show the entire pattern.

Regression with Nine Principal Components

The first nine principal components, $g = 9$, were used in principal component regression. The regression coefficients for the nine principal components were converted to estimates of the regression coefficients for the 14 original variables, $\boldsymbol{\beta}^+_{(g)} = V_{(g)}\hat{\boldsymbol{\gamma}}_{(g)}$. The results are given in the last two columns of Table 13.10. Eight of the 14 regression coefficients for the independent variables are significant: pH, BUF, K, Mg, Na, Mn, Cu, and NH_4. (Results from ordinary least squares, $g = 14$, and from the first 11 principal components, $g = 11$, are included for comparison.) The variables pH, BUF, and NH_4 are significant primarily because of their contribution to W_1; K, Mg, and Na are significant primarily through W_2. The significance of Cu and Mn appears to come through their contributions to several principal components. On the other hand, even though

Table 13.10 Principal component regression estimates of regression coefficients and their standard errors using $g = 14$ (OLS), $g = 11$, and $g = 9$ principal components. (Linthurst September data.)

	$g = 14$ (OLS)		$g = 11$		$g = 9$	
Variable	$\hat{\beta}$	$s(\hat{\beta})$	β^+	$s(\beta^+)$	β^+	$s(\beta^+)$
H_2S	88	610	257	538	489	379
SAL	−591	645	−639	458	−238	393
*Eh*7	626	493	609	473	482	465
pH	2,005	2,763	896*	210	858*	152
BUF	−117	2,058	−1,364*	459	−685*	183
P	−312	483	−383	449	−445	446
K	−2,069*	952	−2,247*	761	−1,260*	495
Ca	−1,325	1,431	−1,046	690	30	317
Mg	−1,744	1,709	−817*	228	−652*	145
Na	203	1,128	−488	577	−1,365*	317
Mn	−274	872	−570	604	−848*	385
Zn	−1,031	1,195	−1,005	791	251	410
Cu	2,374*	771	2,168*	563	1,852*	500
NH_4	−847	1,015	−400	621	−1,043*	479
R^2	.807		.803		.775	
VIF_{max}	62.1		5.1		2.0	

* Estimates exceed twice their standard error.

Ca and *Zn* are major components of W_1 and salinity (*SAL*) is a major component of W_3, their contributions to *BIOMASS* through several W_j apparently tend to cancel and make them nonsignificant.

Comparison with Ordinary Least Squares

The increased stability of the principal component regression estimates compared with ordinary least squares is evident in Table 13.10. The cost of the increased stability is a loss in R^2 from .807 to .775, and an introduction of an unknown amount of bias. Hopefully, the decrease in variance is sufficient to more than compensate for the bias so that the principal component estimates will have smaller mean squared error. The large decreases in variance for several of the coefficients make this a reasonable expectation.

The principal component regression has little impact on the regression coefficients for the variables that are not involved in the near-singularities. The regression coefficients and standard errors for *Eh*7 and *P* change relatively little. These two variables have small coefficients for all five principal components eliminated from the principal component regression. All other variables are involved in one or more of the near-singularities.

Judging Importance of Variables

The purpose of this study was to identify "important" variables for further study of the causal mechanisms of *BIOMASS* production. It is dangerous to attempt to assign "relative importance" to the variables based on the relative magnitudes of their partial regression coefficients. This is the case whether the estimates are from ordinary least squares, ridge regression, or principal component regression. The least squares estimates are too unstable in this example to give meaningful results. And both ridge regression and principal component regression estimates are a pooling of the least squares estimates for all variables involved in the strong collinearities (see equations 12.11 and 12.30). The greater stability of the biased regression estimates can be viewed as coming from this "averaging" of information from the correlated variables. However, this does not prove helpful in judging the relative importance of variables in the same correlated complex.

Complexes of Variables

Principal component analysis has shown that the independent variables in this set of data behave as correlated complexes of variables with meaningful variation in only nine dimensions of the 14-dimensional space. The W_1 complex of variables, for example, behaves more or less as a unit in this data set, and it would be inappropriate to designate any one of the five variables as "the variable of importance." It is the complex that, for the moment at least, must be considered of primary importance insofar as *BIOMASS* is concerned. Further research under controlled conditions, where the effect of the individual variables in the complex can be disassociated, is needed before specific causal relationships can be defined.

13.5 Summary

The classical results of ordinary least squares regression in the presence of collinearity are demonstrated with the Linthurst data; either all variables of a correlated complex appear insignificant, if a full multiple regression model is fit, or only one variable in each correlated complex is retained if some stepwise regression procedure is used. In either case, any inference as to which are the important variables can be very misleading. The apparent insignificance of the variables arises from the fact that the near-singularities in X, reflected in the near-zero eigenvalues, cause the ordinary least squares estimates of the regression coefficients to be very unstable. Geometrically, there is only trivial dispersion of the data in one or more dimensions of the X-space and, consequently, the impact of these dimensions on the dependent variable is determined only with very low precision. Conversely, the dimensions of the X-space showing major dispersion are defined by one or more correlated variables. Ordinary least squares somewhat arbitrarily picks one of the variables to represent the complex. If the objective is simply to predict *BIOMASS*, such a procedure is satisfactory as long as care is taken in making predictions. However, when the objective is to identify important variables, such a procedure will be misleading.

Principal component analysis and Gabriel's biplot clarify the complex relationships among the independent variables. Correlated complexes of variables can be identified and their associations with the dependent variable assessed. The primary variables in the complexes that have predictive value can then be studied under controlled conditions to determine their effects on the dependent variable. Biased regression methods, while they may be useful in some cases for estimating regression coefficients, do not prove helpful in assigning relative importance to the independent variables involved in the near-singularities.

EXERCISES

The singular value decomposition of the Linthurst data in this case study was run on the 45 × 14 matrix of individual observations on the 14 independent variables. That analysis operated on the total variation within and among sampling sites. The following exercises study the correlational structure among the independent variables and their relationship to biomass production using only the variation *among* sampling sites. The data to be used are the sampling site means for all variables computed from the data in Table 13.1. The "Loc–Type" codes identify the nine sampling sites.

*13.1. Compute the 9 × 15 matrix of sampling site means for *BIOMASS* and the 14 independent variables. Center and standardize the matrix of means and compute the correlation matrix of all 15 variables. Which independent variables appear to be most highly correlated with *BIOMASS*? Identify insofar as possible the subsets of independent variables that are highly correlated with each other. Are there any independent variables that are nearly independent of the others?

13.2. Extract from the 9 × 15 matrix of centered and standardized variables the 14 independent variables to obtain $\boldsymbol{Z}$. Do the principal component analysis on this matrix. Explain why only eight eigenvalues are nonzero. Describe the composition (in terms of the original variables) of the three principal components that account for the most dispersion. What proportion of the dispersion do they account for? Compare these principal components to those given for the case study using all observations.

*13.3. Drop *BUF* and NH_4 from the data set and repeat exercise 13.2. Describe how the principal components change with these two variables omitted. Notice that the two variables dropped were primary variables in the first principal component computed with all variables, exercise 13.2.

13.4. Use the principal components defined in exercise 13.2 to construct Gabriel's biplot. Use enough dimensions to account for 75% of the dispersion in Z-space. Interpret the biplots with respect to the correlation structure of the variables, the similarity of the sampling sites, and the major differences in the sampling sites.

*13.5. Use the first eight principal components defined in exercise 13.2 as independent variables and the sampling site means for *BIOMASS* as the dependent variable. Regress *BIOMASS* on the principal components (plus an intercept) and compute the sum of squares attributable to each principal component. These sums of squares, multiplied by 5 to put them on a "per-observation" basis, are an orthogonal partitioning of the "among-site" sum of squares. Compute the analysis of variance for the original data to obtain the "among-site" and "within-site" sums of squares. Verify that the "among-site" sums of squares computed by the two methods agree. Test the significance of each principal component using the "within-site" mean square as the estimate of σ^2. Which principal component dominates the regression and which variables does this result suggest might be most important? Which principal component is nearly orthogonal to *BIOMASS* and what does this imply, if anything, about some of the variables?

CHAPTER FOURTEEN

Response Curve Modeling

Chapter 13 completed the series of chapters devoted to problem areas in least squares regression.

This chapter is devoted to response relationships that cannot be represented by a straight line. The extensively used polynomial response models are discussed. Models nonlinear in the parameters are introduced as more realistic models and several examples are given. Then least squares methods for fitting nonlinear models are presented.

Linear Models

Most models previously considered (1) have specified a linear relationship between the dependent variable and each independent variable and (2) have been linear in the parameters. The linear relationship results from each independent variable appearing only to the first degree and in only one term of the model; no terms are included that contain powers or products of independent variables. This restriction forces the rate of change in the mean of the dependent variable with respect to an independent variable to be constant over all values of that and every other independent variable in the model. Linearity in the parameters means that each (additive) term in the model contains only one parameter and only as a multiplicative constant on the independent variable. This restriction excludes many useful mathematical forms, including nearly all models developed from principles of behavior of the system. These are very restrictive models and should be viewed as first-order approximations to true relationships.

Nonlinear Models

In this chapter, the class of models is extended to allow greater flexibility and realism. First, higher-degree polynomial models are considered. These are still to be regarded in most situations as approximations to the true models. Then, the potentially more realistic models that are **nonlinear** in the parameters are introduced. Emphasis will be on the functional form of the model, the part of the model that gives the relationship between the expectation of the dependent variable and the independent variables. Although this chapter does not dwell on the behavior of the residuals, it is important that the assumptions of least squares be continually checked. Growth data, for example, often will not satisfy the homogeneous variance assumption, and will contain correlated errors if the data are collected as repeated measurements over time on the same experimental units. For discussion on experimental designs for fitting response surfaces, the reader is referred to design texts such as Box, Hunter, and Hunter (1978).

14.1 Considerations in Specifying the Functional Form

Regression to Summarize Data

The degree of realism that needs to be incorporated into a model will depend on the purpose of the regression analysis. The least demanding purpose is the simple use of a regression model to *summarize* the observed relationships in a particular set of data. There is no interest in the functional form of the model per se or in predictions to other sets of data or situations. The most demanding is the more esoteric development of mathematical models to describe the physical, chemical, and biological processes in the system. The goal of the latter is to make the model as realistic as the state of knowledge will permit.

The use of regression models simply to summarize observed relationships places no priority on realism because no inference, even to other samples, is intended. The overriding concern is that the model adequately portray the observed relationships. In practice, however, readers will often attach a predictive inference to the presentation of regression results, even if the intent of the author is simply to summarize the data.

Regression for Prediction

When the regression equation is to be used for prediction, it is beneficial to incorporate into the model prior information on the behavior of the system. This serves two goals. First, other things being equal, the more realistic model would be expected to provide better predictions for unobserved points in the X-space, either interpolations or extrapolations. (While extrapolations are always dangerous and are to be avoided, it is not always easy, particularly with observational data, to identify points outside the sample space. Realistic models will tend to provide more protection against large errors in unintentional extrapolations than purely approximating models.) Second, incorporating current beliefs about the behavior of the system into the model provides an opportunity to test and update these theories.

Use of Prior Information

The prior information used in the model may be nothing more than recognizing the general shape the response curve should take. For example, it may be that the response variable should not take negative values, or the response should approach an asymptote for high or low values of an independent variable. Recognizing such constraints on the behavior of the system will often lead to the use of nonlinear models. In some cases, these (presumably) more realistic models will also be simpler models in terms of the number of parameters to be estimated. A response with a plateau, for example, may require several terms of a polynomial model to fit the plateau but might be characterized very well with a two-parameter exponential model. Polynomial models should not a priori be considered the simpler and nonlinear models the more complex.

At the other extreme, prior information on the behavior of a system may include minute details on the physical and chemical interactions in each of several different components of the system and on how these components interact to produce the final product. Such models can become extremely complex and most likely cannot be written as a single functional relationship between $\mathscr{E}(Y)$ and the independent variables. Numerical integration may be required to evaluate and combine the effects of the different components. The detailed crop growth models that predict crop yields based on daily, or even hourly, data on the environmental and cultural conditions during the growing season are examples of such models. (The development of such models will not be pursued in this text. They are mentioned here as an indication of the natural progression of the use of prior information in model building.)

14.2 Polynomial Response Surface Models

The models previously considered have been first-degree polynomial models, models in which each term contains only one independent variable to the first power. The first-degree polynomial model in two variables is

$$Y_i = \beta_0 + \beta_1 X_{i1} + \beta_2 X_{i2} + \varepsilon_i \qquad [14.1]$$

A second-degree polynomial model includes terms, in addition to the first-degree terms, that contain squares or products of the independent variables. The full second-degree polynomial model in two variables is

$$Y_i = \beta_0 + \beta_1 X_{i1} + \beta_2 X_{i2} + \beta_{11} X_{i1}^2 + \beta_{22} X_{i2}^2 + \beta_{12} X_{i1} X_{i2} + \varepsilon_i \qquad [14.2]$$

Degree of a Polynomial

The **degree** (or **order**) of an individual term in a polynomial is defined as the *sum* of the powers of the independent variables in the term. The degree of the entire polynomial is defined as the degree of the highest-degree term. All polynomial models, regardless of their degree, are linear in the parameters. (For the higher-degree polynomial models, the subscript notation on the β's is expanded to reflect the degree of the polynomial term. In general, the number of 1's and the number of 2's in the subscript identify the powers of X_1 and X_2, respectively, in the polynomial term. For example, the two 1's identify β_{11} as the regression coefficient for the second-degree term in X_1.)

The higher-degree polynomial models provide greatly increased flexibility in the response surface. While it is unlikely that any complex process will be truly polynomial in form, the flexibility of the higher-degree polynomials allows any true model to be approximated to any desired degree of precision.

First-Degree Polynomial

The increased flexibility of the higher-degree polynomial models is illustrated with a sequence of polynomial models containing two independent variables. The first-degree polynomial model, equation 14.1, uses a plane to represent $\mathscr{E}(Y_i)$. This surface is a "table top" tilted to give the slopes $\hat{\beta}_1$ in the X_1 direction and $\hat{\beta}_2$ in the X_2 direction (Figure 14.1).

The properties of any response equation can be determined by observing how $\mathscr{E}(Y)$ changes as the values of the independent variables change. For the first-degree polynomial, equation 14.1, the rate of change in $\mathscr{E}(Y)$ as X_1 is changed is the constant β_1, regardless of the values of X_1 and X_2. Similarly, the rate of

Figure 14.1 A first-degree bivariate polynomial response surface.

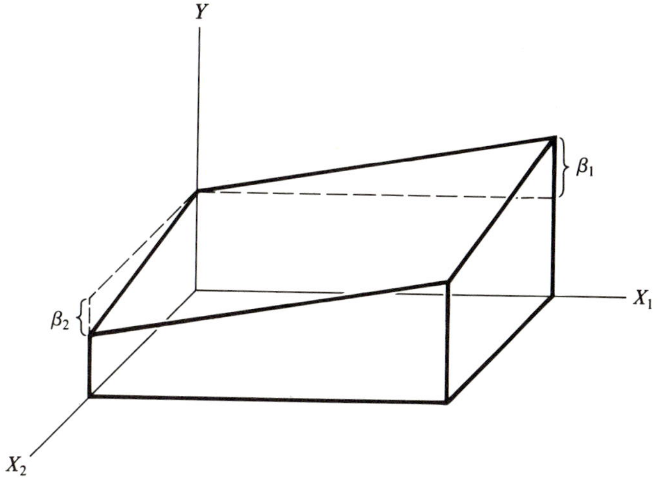

change in $\mathscr{E}(Y)$ as X_2 changes is determined solely by β_2. The changes in $\mathscr{E}(Y)$ as the independent variables change are given by the partial derivatives of $\mathscr{E}(Y)$ with respect to each of the independent variables. For the first-degree polynomial, the partial derivatives are the constants β_1 and β_2:

$$\frac{\partial \mathscr{E}(Y)}{\partial X_1} = \beta_1$$
$$\frac{\partial \mathscr{E}(Y)}{\partial X_2} = \beta_2 \qquad [14.3]$$

The partial derivative with respect to X_j gives the slope of the surface, or the rate of change in $\mathscr{E}(Y)$, in the X_j direction.

Second-Degree Polynomial

The polynomial model is expanded to allow the rate of change in $\mathscr{E}(Y)$ with respect to one independent variable to be dependent on the value of that variable by including a term that contains the square of the variable. For example, adding a second-degree term in X_1 to equation 14.1 gives

$$Y_i = \beta_0 + \beta_1 X_{i1} + \beta_{11} X_{i1}^2 + \beta_2 X_{i2} + \varepsilon_i \qquad [14.4]$$

The partial derivatives for this model are

$$\frac{\partial \mathscr{E}(Y)}{\partial X_1} = \beta_1 + 2\beta_{11} X_{i1}$$
$$\frac{\partial \mathscr{E}(Y)}{\partial X_2} = \beta_2 \qquad [14.5]$$

Now the rate of change in $\mathscr{E}(Y)$ with respect to X_1 is a linear function of X_1, increasing or decreasing according to the sign of β_{11}. The rate of change in $\mathscr{E}(Y)$

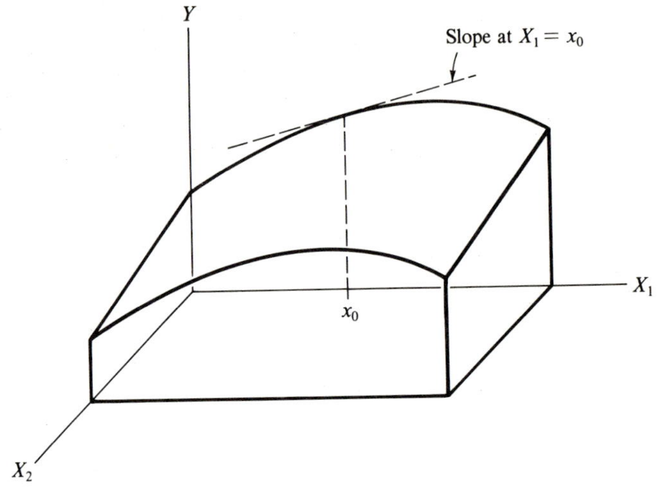

Figure 14.2 A polynomial response surface that is of second degree in X_1 and first degree in X_2.

with respect to X_2 remains a constant β_2. Notice that the meaning of β_1 is not the same in equation 14.4 as it was in the first-degree polynomial, equation 14.1. Here β_1 is the slope of the surface in the X_1 direction *only* where $X_1 = 0$. The nature of this response surface is illustrated in Figure 14.2.

Interaction Term

The rate of change in $\mathscr{E}(Y)$ with respect to one independent variable can be made dependent on another independent variable by including the product of the two variables as a term in the model:

$$Y_i = \beta_0 + \beta_1 X_{i1} + \beta_2 X_{i2} + \beta_{12} X_{i1} X_{i2} + \varepsilon_i \qquad [14.6]$$

The product term, $\beta_{12}X_{i1}X_{i2}$, is referred to as an **interaction term**. It allows one independent variable to influence the impact of another. The derivatives are

$$\frac{\partial \mathscr{E}(Y)}{\partial X_1} = \beta_1 + \beta_{12} X_{i2}$$

$$\frac{\partial \mathscr{E}(Y)}{\partial X_2} = \beta_2 + \beta_{12} X_{i1} \qquad [14.7]$$

The rate of change in $\mathscr{E}(Y)$ with respect to X_1 is now dependent on X_2 but not on X_1, and vice versa. Notice the symmetry of the interaction effect; both partial derivatives are influenced in the same manner by changes in the other variable. This particular type of interaction term is referred to as the **linear-by-linear interaction**, because the linear slope in one variable is changed linearly (at a constant rate) by changes in the other variable and vice versa. This response function gives a "twisted plane" where the response in $\mathscr{E}(Y)$ to changes in either variable is always linear but the slope is dependent on the value of the other variable. This linear-by-linear interaction is illustrated in Figure 14.3 with the three-dimensional figure in part (a) and a two-dimensional representation show-

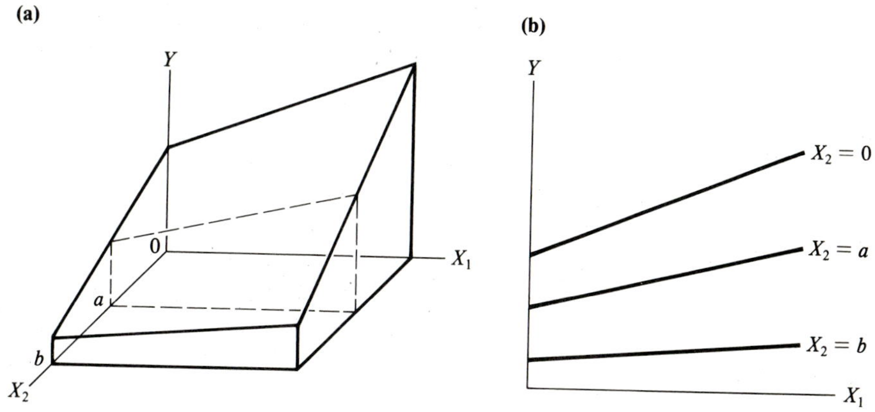

Figure 14.3 Bivariate response surface (a) with interaction and (b) a two-dimensional representation of the surface.

ing the relationship between Y and X_1 for given values of X_2. The interaction is shown by the failure of the three lines in part (b) to be parallel.

Full Second-Degree Bivariate Model

The full second-degree bivariate model includes all possible second-degree terms as shown in equation 14.2. The derivatives with respect to each independent variable are now functions of both independent variables:

$$\frac{\partial \mathscr{E}(Y)}{\partial X_1} = \beta_1 + 2\beta_{11}X_{i1} + \beta_{12}X_{i2}$$

$$\frac{\partial \mathscr{E}(Y)}{\partial X_2} = \beta_2 + 2\beta_{22}X_{i2} + \beta_{12}X_{i1} \qquad [14.8]$$

The squared terms allow for a curved response in each variable. The product term allows for the surface to be "twisted" (Figure 14.4). β_1 and β_2 are the slopes of the response surface in the X_1 and X_2 directions, respectively, only at the point

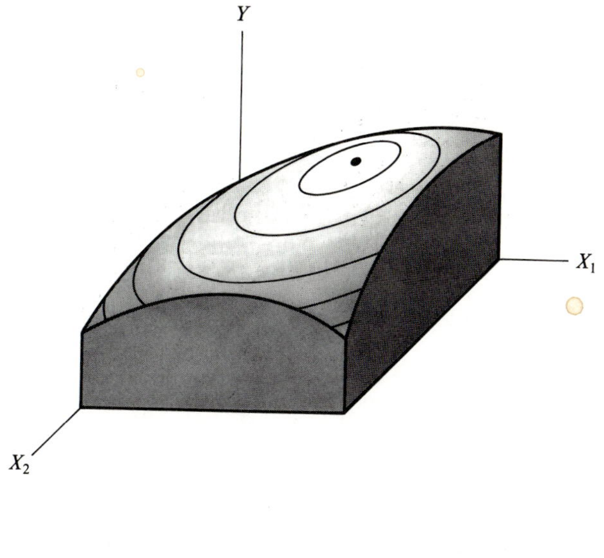

Figure 14.4 A bivariate quadratic response surface with a maximum.

$X_1 = 0$ and $X_2 = 0$. A quadratic response surface will have a maximum, a minimum, or a saddle point, depending on the coefficients in the regression equation. The reader is referred to Box and Draper (1987) for a discussion of the analysis of the properties of quadratic response surfaces. The computer program PROC RSREG (SAS Institute, Inc., 1985f) fits a full quadratic model to a set of data and provides an analysis of the properties of the response surface.

Third-Degree Polynomial

The flexibility of the polynomial models is demonstrated by showing the effects of a third-degree term for one of the variables. For example, consider the model

$$Y_i = \beta_0 + \beta_1 X_{i1} + \beta_2 X_{i2} + \beta_{11} X_{i1}^2 + \beta_{111} X_{i1}^3 + \varepsilon_i \qquad [14.9]$$

The partial derivative with respect to X_1 is now a quadratic function of X_1:

$$\frac{\partial \mathscr{E}(Y)}{\partial X_1} = \beta_1 + 2\beta_{11} X_{i1} + 3\beta_{111} X_{i1}^2 \qquad [14.10]$$

The derivative with respect to X_2 is still β_2. An example of this response surface is shown in Figure 14.5. The full third-degree model in two variables would include all combinations of X_1 and X_2 with sums of the exponents equal to 3 or less.

Flexibility of Polynomials

Increasingly higher-degree terms can be added to the polynomial response model to give an arbitrary degree of flexibility. Any continuous response function can be approximated to any level of precision desired by a polynomial of appropriate degree. Thus, an excellent fit of a polynomial model (or, for that matter, any model) *cannot* be interpreted as an indication that it is in fact the true model. Due to this extreme flexibility, some caution is needed in the use of

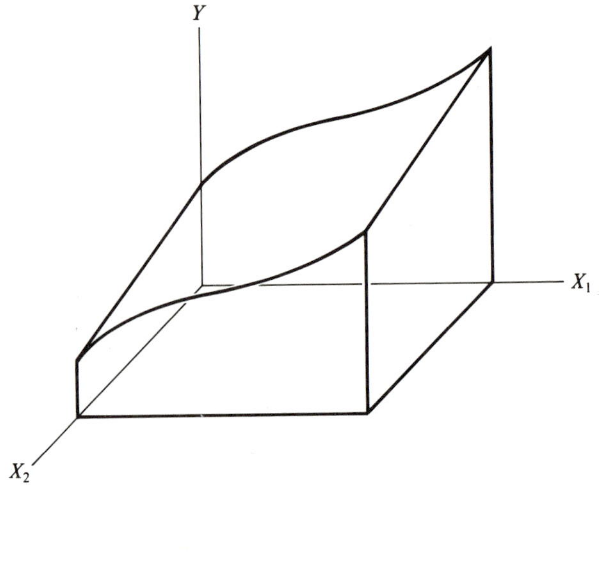

Figure 14.5 A polynomial response surface with a third-degree term in X_1.

polynomial models; it is easy to "overfit" a set of data with polynomial models. Nevertheless, polynomial response models have proven to be extremely useful for summarizing relationships.

Presenting the Response Surface

Polynomial models can be extended to include any number of independent variables. Presenting a multivariate response surface so it can be visualized, however, becomes increasingly difficult. Key features of the response surface (maxima, minima, inflection points) can be determined with the help of calculus. Two- or three-dimensional plots of "slices" of the multivariate surface can be obtained by evaluating the response surface equation at specific values for all independent variables other than the ones of interest.

Caution with Extrapolations

Extrapolation is particularly dangerous when higher-degree polynomial models are being used. The highest-degree term in each independent variable eventually dominates the response in that dimension and the surface will "shoot off" in either the positive or negative direction, depending on the sign of the regression coefficient on that term. Thus, minor extrapolations can have serious errors.

Fitting Polynomials

Fitting polynomial response models with least squares introduces no new conceptual problems. The model is still linear in the parameters and, as long as the usual assumptions on ε are appropriate, ordinary least squares can be used. The higher-degree terms are included in the model by augmenting $\boldsymbol{X}$ with columns of new variables defined as the appropriate powers and products of the independent variables and by augmenting $\boldsymbol{\beta}$ with the respective parameters. The computational problems associated with collinearity are aggravated by the presence of the higher-degree terms because X, X^2, X^3, etc. are often highly collinear. To help alleviate this problem, orthogonal polynomial coefficients can be used (Steel and Torrie, 1980) or each independent variable can be centered *before* the higher-degree terms are included in $\boldsymbol{X}$. For example, the quadratic model

$$Y_i = \beta_0 + \beta_1 X_{i1} + \beta_2 X_{i2} + \beta_{11} X_{i1}^2 + \beta_{22} X_{i2}^2 + \beta_{12} X_{i1} X_{i2} + \varepsilon_i \qquad [14.11]$$

becomes

$$\begin{aligned} Y_i = {} & \gamma_0 + \gamma_1(X_{i1} - \bar{X}_{.1}) + \gamma_2(X_{i2} - \bar{X}_{.2}) + \gamma_{11}(X_{i1} - \bar{X}_{.1})^2 \\ & + \gamma_{22}(X_{i2} - \bar{X}_{.2})^2 + \gamma_{12}(X_{i1} - \bar{X}_{.1})(X_{i2} - \bar{X}_{.2}) + \varepsilon_i \end{aligned} \qquad [14.12]$$

Centering the independent variables changes the definition of the regression coefficients for all but the highest-degree terms. For example, γ_1 and γ_2 are the rates of change in $\mathscr{E}(Y)$ in the X_1 and X_2 directions, respectively, at $X_1 = \bar{X}_{.1}$ and $X_2 = \bar{X}_{.2}$, whereas β_1 and β_2 are the rates of change at $X_1 = X_2 = 0$. The relationship between the two sets of regression coefficients is obtained by expanding the square and product terms in the centered model, equation 14.12, and comparing the coefficients for similar polynomial terms with those in the original model, equation 14.11. Thus,

$$\beta_0 = \gamma_0 - \gamma_1 \bar{X}_{.1} - \gamma_2 \bar{X}_{.2} + \gamma_{11} \bar{X}_{.1}^2 + \gamma_{22} \bar{X}_{.2}^2 + \gamma_{12} \bar{X}_{.1} \bar{X}_{.2}$$

$$\beta_1 = \gamma_1 - 2\gamma_{11} \bar{X}_{.1} - \gamma_{12} \bar{X}_{.2} \qquad [14.13]$$

$$\beta_2 = \gamma_2 - 2\gamma_{22} \bar{X}_{.2} - \gamma_{12} \bar{X}_{.1}$$

$$\beta_{11} = \gamma_{11} \qquad \beta_{22} = \gamma_{22} \qquad \beta_{12} = \gamma_{12}.$$

When the sample X-space does not include the origin, the parameters for the centered model are more meaningful because they relate more directly to the behavior of the surface in the region of interest.

Building the Model

The polynomial model is built sequentially, starting either with a first-degree polynomial and adding progressively higher-order terms as needed, or with a high-degree polynomial and eliminating the unneeded higher-degree terms. The lowest-degree polynomial that accomplishes the degree of approximation needed or warranted by the data is adopted. The error term for the tests of significance at each stage must be an appropriate independent estimate of error, preferably estimated from true replication if available. Otherwise, the residual mean square from a model that contains at least all the terms in the more complex model being considered is used as the estimate of error.

Retaining Lower-Order Terms

It is common practice to retain in the model all lower-degree terms, regardless of their significance, that are contained in, or are subsets of, any significant term. For example, if a second-degree term is significant, the first-degree term in the same variable would be retained even if its partial regression coefficient is not significantly different from zero. If the $X_1^2X_2^2$ term is significant, the X_1, X_2, $X_1^2X_2$, $X_1X_2^2$, and X_1X_2 terms would be retained even if nonsignificant.

The argument for retaining lower-order terms even if not significant is based on two points. First, the meanings and values of the regression coefficients on all except the highest-degree terms change with a simple shift in origin of the independent variables. Recall that reexpressing the independent variables as deviations from their means in a quadratic model changed the meaning of the coefficient for each first-degree term. Thus, the significance or nonsignificance of a lower-order term will depend on the choice of origin for the independent variable during the analysis. A lower-order term that might have been eliminated from a regression equation because it was nonsignificant could "reappear," as a function of the higher-order regression coefficients, when the regression equation is reexpressed with different origins for the independent variables.

Second, eliminating lower-order terms from a polynomial tends to give biased interpretations of the nature of the response surface when the resulting regression equation is studied. For example, eliminating the first-degree term from a second-degree polynomial forces the critical point (maximum, minimum, or saddle point) of the fitted response surface to occur precisely at $X = 0$. (The critical point on a quadratic response surface is found by setting the partial derivatives equal to zero and solving for the values of the independent variable.) For the second-degree polynomial in one variable, the critical point is $X = -\beta_1/(2\beta_{11})$, which is forced to be zero if the first-degree term has been dropped from the model ($\beta_1 = 0$). Even though β_1 may not be significantly different from zero, it would be more informative to investigate the nature of the response surface before such constraints are imposed. The position of the critical point could then be estimated with its standard error and appropriate inferences made.

These arguments for retaining all lower-degree polynomial terms apply when the polynomial model is being used as an approximation of some unknown model. They are not meant to apply to the case where there is a meaningful basis for a model that contains a higher-order term but not the lower-order terms. The development of a prediction equation for the volume of timber from information on diameter and height of the trees provides an illustration. Geometry would suggest that volume should be nearly proportional to the product of $(\text{diameter})^2$ and height. Consequently, a model *without* the lower-order terms, diameter and diameter $\times$ height, would be realistic and appropriate.

Example 14.1

A study of the effects of salinity, temperature, and dissolved oxygen on the resistance of young coho salmon to pentachlorophenate will be used to illustrate the use of polynomial models (Alderdice, 1963, used with permission). The study used a three-factor composite design in two stages to estimate the response surface for median survival time (Y) following exposure to 3 mg/l of sodium pentachlorophenate. The treatment variables were water salinity, temperature, and dissolved oxygen content. The first 15 trials (two replicates) used a 2^3 design of the three factors plus the six axial points and the center point (Table 14.1, page 386). The last 10 trials were a second-stage study to improve the definition of the center of the response surface. The basic levels of the three factors were 9, 5, and 1% salinity; 13, 10, and 7°C temperature; and 7.5, 5.5, and 3.5 mg/l dissolved oxygen. The independent variables were coded as follows:

$$X_1 = (\text{salinity} - 5\%)/4$$

$$X_2 = (\text{temperature} - 10°\text{C})/3$$

$$X_3 = (\text{dissolved oxygen} - 5.5 \text{ mg/l})/2$$

The dependent variable, median lethal time, was computed on samples of ten individuals per experimental unit. The treatment combinations and the observed responses are given in Table 14.1.

It was verified by Alderdice (1963), using the first 15 trials for which there was replication, that a quadratic polynomial response model in the three independent variables was adequate for characterizing the response surface. The replication provided an unbiased estimate of experimental error, which was to test the lack of fit of the quadratic polynomial. Alderdice then fit the full quadratic or second-degree polynomial model to all the data and presented interpretations of the trivariate response surface equation.

Quadratic Model

For this example, the full set of data will be used to develop the simplest polynomial response surface model that adequately represents the data. Since the full quadratic model appears to be more than adequate, that model will be used as the starting point and higher-degree terms will be eliminated if non-

Table 14.1 Treatment combinations of salinity (X_1), temperature (X_2), and dissolved oxygen (X_3), and median lethal time for exposure to 3 mg/l of sodium pentachlorophenate. [Data from Alderdice (1963), and used with permission.]

Trial	*Sal.* X_1	*Temp.* X_2	*Oxy.* X_3	*Median Lethal Time* *Rep* 1	*Rep* 2
1	−1	−1	−1	53	50
2	−1	−1	1	54	42
3	−1	1	−1	40	31
4	−1	1	1	37	28
5	1	−1	−1	84	57
6	1	−1	1	76	78
7	1	1	−1	40	49
8	1	1	1	50	54
9	0	0	0	50	50
10	1.215	0	0	61	76
11	−1.215	0	0	54	45
12	0	1.215	0	39	33
13	0	−1.215	0	67	54
14	0	0	1.215	44	45
15	0	0	−1.215	61	38
16	−1.2500	−1.8867	−.6350	46	
17	.8600	−2.2200	−.4250	66	
18	1.0000	−2.2400	−.3100	68	
19	2.1165	−2.4167	−.1450	75	
20	2.5825	−2.4900	−.0800	75	
21	3.2475	−2.6667	.0800	68	
22	1.1760	−1.3333	0	78	
23	1.4700	−1.6667	0	93	
24	1.7640	−2.0000	0	96	
25	2.0580	−2.3333	0	66	

significant. In addition to the polynomial terms, the model must include a class variable "*REP*" to account for the differences between the two replications in the first trial and between the first and second trials. Thus, the full quadratic model is

$$Y_{ij} = \mu + \rho_i + \beta_1 X_{ij1} + \beta_2 X_{ij2} + \beta_3 X_{ij3} + \beta_{11} X_{ij1}^2 + \beta_{22} X_{ij2}^2 + \beta_{33} X_{ij3}^2 + \beta_{12} X_{ij1} X_{ij2} + \beta_{13} X_{ij1} X_{ij3} + \beta_{23} X_{ij2} X_{ij3} + \varepsilon_{ij} \quad [14.14]$$

where ρ_i is the effect of the ith "rep," $i = 1, 2, 3$, and j designates the observation within the replication. This model allows each "rep" to have its own level of performance but requires the response surface to be the same over replications. The presence of the replication effects creates a singularity that can be handled in any of the ways discussed in Chapter 8. Most convenient in this case is the means reparameterization. Let $\mu_i = \mu + \rho_i$, $i = 1, 2, 3$. Thus, $\boldsymbol{X}$ for the full-rank model consists of three columns of dummy variables for the three replications,

Table 14.2 Partial regression coefficients for the full second-degree polynomial model in three variables for the Alderdice (1963) data.

Term	$\hat{\beta}_j$	$s(\hat{\beta}_j)$	*Student's t*[a]
X_1	9.127	1.772	5.151
X_2	−9.852	1.855	−5.312
X_3	.263	1.862	.141
X_1^2	−1.260	1.464	−.861
X_2^2	−6.498	2.014	−3.225
X_3^2	−2.985	2.952	−1.011
X_1X_2	−.934	1.510	−.618
X_1X_3	2.242	2.150	1.042
X_2X_3	−.139	2.138	−.065

[a] The estimate of σ^2 from this model was $s^2 = 76.533$ with 28 degrees of freedom.

followed by nine columns of X_1, X_2, X_3, and their squares and products. The partial regression coefficients, their standard errors, and the t-statistics for this full model are given in Table 14.2.

Several of the partial regression coefficients do not approach significance, $t_{(.05,28)} = 2.048$; at least some terms can be eliminated from the model. It is not a safe practice, however, to delete all nonsignificant terms in one step unless the columns of the X matrix are orthogonal. The common practice with polynomial models is to eliminate the least important of the highest-degree terms at each step. In this example, the X_2X_3 term would be dropped first. Notice that X_3 is retained in the model at this stage, even though it has the smallest t-value, because there are higher-order terms in the model that contain X_3.

Final Model

The subsequent steps consist of dropping X_1X_2, X_1X_3, X_1^2, X_3^2, and, finally, X_3 in turn. The final polynomial model is

$$Y_{ij} = \mu_i + \beta_1 X_{ij1} + \beta_2 X_{ij2} + \beta_{22} X_{ij2}^2 + \varepsilon_{ij} \qquad [14.15]$$

The residual mean square for this model is $s^2 = 69.09$ with 34 degrees of freedom. (The estimate of experimental error from the replicated data is 62.84 with 14 degrees of freedom.) The regression equation, using the weighted average of the estimates of μ_i, is

$$\hat{Y} = \underset{(2.85)}{59.92} + \underset{(1.47)}{9.21X_1} - \underset{(1.76)}{9.82X_2} - \underset{(1.56)}{6.896X_2^2} \qquad [14.16]$$

The standard errors of the estimates are shown in parentheses. Thus, within the limits of the observed values of the independent variables, survival time of coho salmon with exposure to sodium pentachlorophenate is well represented by a linear response to salinity, and a quadratic response to temperature (Figure 14.6, page 388). There is no significant effect of dissolved oxygen on survival time and there appear to be no interactions among the three environmental factors. The linear effect of salinity is to increase survival time 9.2 minutes per coded unit of salinity, or $9.2/4 = 2.3$ minutes per percent increase in salinity. The quadratic

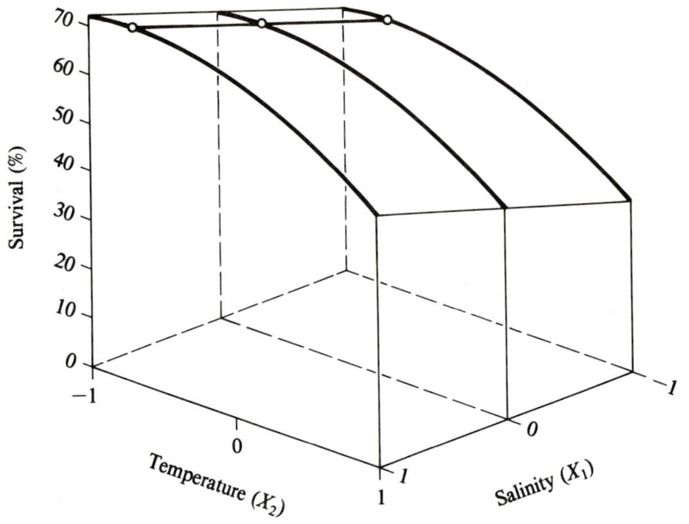

Figure 14.6 Bivariate response surface relating survival time of coho salmon exposed to 3 mg/l of sodium pentachlorophenate to water temperature and water salinity. There was no significant effect of dissolved oxygen (X_3). (Data from Alderdice, 1963; used with permission.)

response to temperature has a maximum at $X_2 = -\hat{\beta}_2/(2\hat{\beta}_{22}) = -.71$, which is 7.9°C on the original temperature scale. (The variance of the estimated maximum point is obtained by using the linear approximation of the ratio of two random variables. This will be discussed in Section 14.4, for the more general case of any nonlinear function with nonlinear models.)

The maximum survival times with respect to temperature for given values of salinity are shown with the line on the surface connecting the open circles at $X_2 = -.71$. The investigated region appears to contain the maximum with respect to temperature, but the results suggest even higher salinities will produce greater survival. The linear response to salinity cannot continue without limit. Using the original *full* quadratic model to investigate the critical points on the response surface, Alderdice (1963) found a maximum at $X_1 = 3.2$ (salinity = 17.8‰), $X_2 = -1.7$ (temperature = 4.9°C), and $X_3 = 1.1$ (dissolved oxygen = 7.7 mg/l). These critical points are near the limits of the sample X-space and should be used with caution. Tests of significance indicate that the data are not adequate to support a statement on curvature with respect to salinity or on even a linear response with respect to dissolved oxygen. ■

14.3 Models Nonlinear in the Parameters

Form of the Model

The more general class of models that are nonlinear in the parameters allows the mean of the dependent variable to be expressed in terms of any function $f(\boldsymbol{x}_i', \boldsymbol{\theta})$

of the independent variables and the parameters. The model becomes

$$Y_i = f(\boldsymbol{x}_i', \boldsymbol{\theta}) + \varepsilon_i \qquad [14.17]$$

where $f(\boldsymbol{x}_i', \boldsymbol{\theta})$ is the nonlinear function relating $\mathscr{E}(Y)$ to the independent variable(s), $\boldsymbol{x}_i'$ is the row vector of observations on k independent variables for the ith observational unit, and $\boldsymbol{\theta}$ is the vector of p parameters. (It is common in nonlinear least squares to use $\boldsymbol{\theta}$ as the vector of parameters rather than $\boldsymbol{\beta}$.) The usual assumptions are made on the random errors.

A sample of nonlinear models is presented to illustrate the types of functions that have proven useful and to show how information on the system can be used to develop more realistic models. Nonlinear models are usually chosen because they are more realistic in some sense or because the functional form of the model allows the response to be better characterized, perhaps with fewer parameters. The procedures for estimating the parameters, using the least squares criterion, are discussed in Section 14.4.

Exponential Decay Model

In many cases the rate of change in the mean level of a response variable at any given point in time (or value of the independent variable) is expected to be proportional to its value or some function of its value. Such information can be used to develop a response model. Models developed in this manner often involve exponentials in some form. For example, assume that the concentration of a drug in the bloodstream is being measured at fixed time points after the drug was injected. The response variable is the concentration of the drug; the independent variable is time (t) after injection. If the rate at which the drug leaves the bloodstream is assumed to be proportional to the mean concentration of the drug in the bloodstream at that point in time, the derivative of $\mathscr{E}(Y)$, drug concentration, with respect to t, time, is

$$\frac{\partial \mathscr{E}(Y)}{\partial t} = -\beta \mathscr{E}(Y) \qquad [14.18]$$

Integrating this differential equation, and imposing the condition that the concentration of the drug at the beginning ($t = 0$) was α, gives

$$\mathscr{E}(Y) = \alpha e^{-\beta t} \qquad [14.19]$$

This is the **exponential decay curve**. If additive errors are assumed, the nonlinear model for a process that operates in this manner would be

$$Y_i = \alpha e^{-\beta t_i} + \varepsilon_i \qquad [14.20]$$

This is a two-parameter model with $\boldsymbol{\theta}' = (\alpha \;\; \beta)$. If multiplicative errors are assumed,

$$Y_i = \alpha(e^{-\beta t_i})\varepsilon_i \qquad [14.21]$$

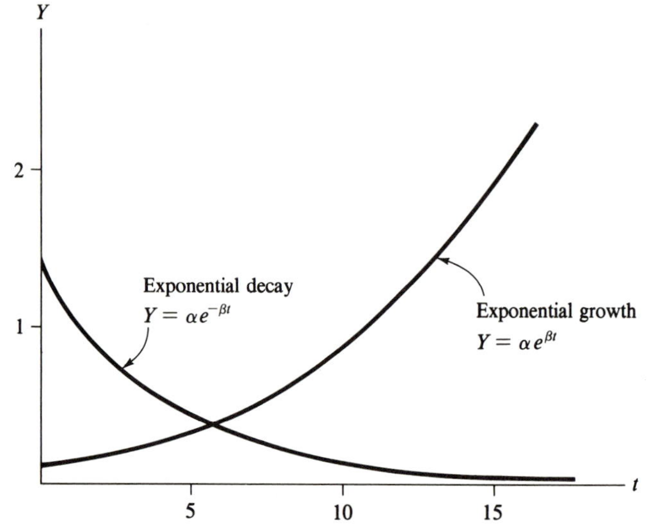

Figure 14.7 Typical forms for the exponential decay model and the exponential growth model. The parameter β is positive in both cases.

The latter is intrinsically linear and is linearized by taking logarithms as discussed in Section 11.2. The model with additive errors, however, cannot be linearized with any transformation and, hence, is intrinsically nonlinear. The remaining discussions in this chapter assume the errors are additive.

Exponential Growth Model

The rate of growth of bacterial colonies might be expected to be proportional to the size of the colony if all cells are actively dividing. The partial derivative in this case would be

$$\frac{\partial \mathscr{E}(Y)}{\partial t} = \beta \mathscr{E}(Y) \qquad [14.22]$$

This is the positive version of equation 14.18, reflecting the expected *growth* of this system. This differential equation yields the **exponential growth model**

$$Y_i = \alpha e^{\beta t_i} + \varepsilon_i \qquad [14.23]$$

where α is the size of the colony at $t = 0$. In both models β is positive; the sign in front of β indicates whether it is an exponential *decay* process or an exponential *growth* process. Their general shapes are shown in Figure 14.7.

Two-Term Exponential Model

A **two-term exponential model** results when, for example, a drug in the bloodstream is being monitored and its concentration in the bloodstream depends on two processes, the movement into the bloodstream from muscle tissue or the digestive system and removal from the bloodstream by, say, the kidneys. Let the mean concentration of the drug in the source tissue be $\mathscr{E}(Y_m)$ and that in the blood be $\mathscr{E}(Y_b)$. Suppose the drug moves into the bloodstream from the muscle at a rate proportional to its concentration in the muscle, $\theta_1 \mathscr{E}(Y_m)$, and is removed from the bloodstream by the kidneys at a rate proportional to its concentration in the bloodstream, $-\theta_2 \mathscr{E}(Y_b)$. Assume $\theta_1 > \theta_2 > 0$. The net rate of change of the drug in the bloodstream is

$$\frac{\partial \mathscr{E}(Y_b)}{\partial t} = \theta_1 \mathscr{E}(Y_m) - \theta_2 \mathscr{E}(Y_b) \qquad [14.24]$$

This process models the concentration in the bloodstream as

$$Y_{bi} = \frac{\theta_1}{\theta_1 - \theta_2}(e^{-\theta_2 t_i} - e^{-\theta_1 t_i}) + \varepsilon_i \qquad [14.25]$$

This response curve shows an increasing concentration of the drug in the blood in the early stages, which reaches a maximum and then declines asymptotically toward zero as the remnants of the drug are removed. This model would also apply to a process where one chemical is being formed by the decay of another, at reaction rate θ_1, and is itself decaying at reaction rate θ_2. If $\theta_1 = \theta_2$, the solution to the differential equations gives the model

$$Y_i = \theta_1 t_i e^{-\theta_1 t_i} + \varepsilon_i \qquad [14.26]$$

The forms of these models are shown in Figure 14.8.

Mitscherlich and Monomolecular Growth Models

When the increase in yield (of a crop) per unit of added nutrient, X, is proportional to the difference between the maximum attainable yield, α, and the

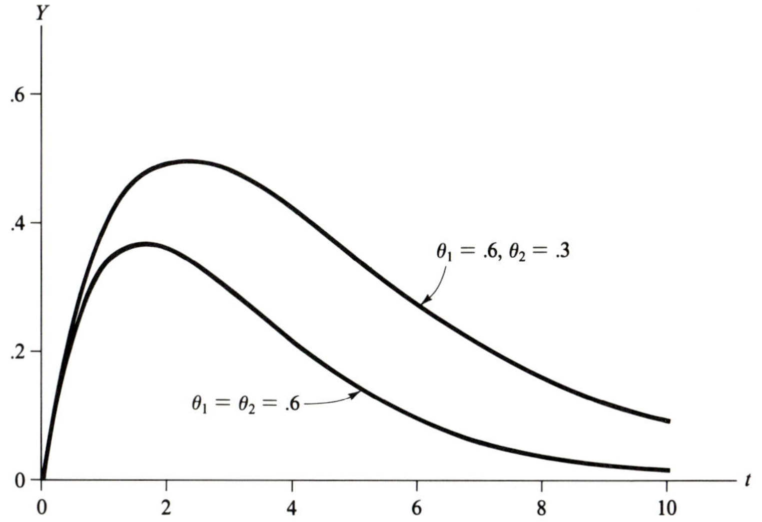

Figure 14.8 The two-term exponential model with $\theta_1 = .6$ and $\theta_2 = .3$, equation 14.25; and its simpler form with $\theta_1 = \theta_2 = .6$, equation 14.26.

actual yield, the partial derivative of Y with respect to X is

$$\frac{\partial \mathscr{E}(Y)}{\partial X} = \beta[\alpha - \mathscr{E}(Y)] \qquad [14.27]$$

This partial derivative generates the model known as the **Mitscherlich equation** (Mombiela and Nelson, 1981):

$$Y_i = \alpha[1 - e^{-\beta(X_i+\delta)}] + \varepsilon_i \qquad [14.28]$$

where δ is the equivalent nutrient value of the soil. This model gives an estimated mean yield of

$$\hat{Y} = \hat{\alpha}(1 - e^{-\hat{\beta}\hat{\delta}}) \qquad [14.29]$$

with no added fertilizer and an asymptotic mean yield of $Y = \hat{\alpha}$ when the amount of added fertilizer is very high. If $\gamma = e^{-\beta\delta}$ is substituted in equation 14.28, this model takes the more familiar form known as **monomolecular growth model**. The form of the Mitscherlich equation is shown in Figure 14.9.

Inverse Polynomial Model

If the rate of increase in yield is postulated to be proportional to the square of $[\alpha - \mathscr{E}(Y)]$, one obtains the **inverse polynomial model** (Nelder, 1966),

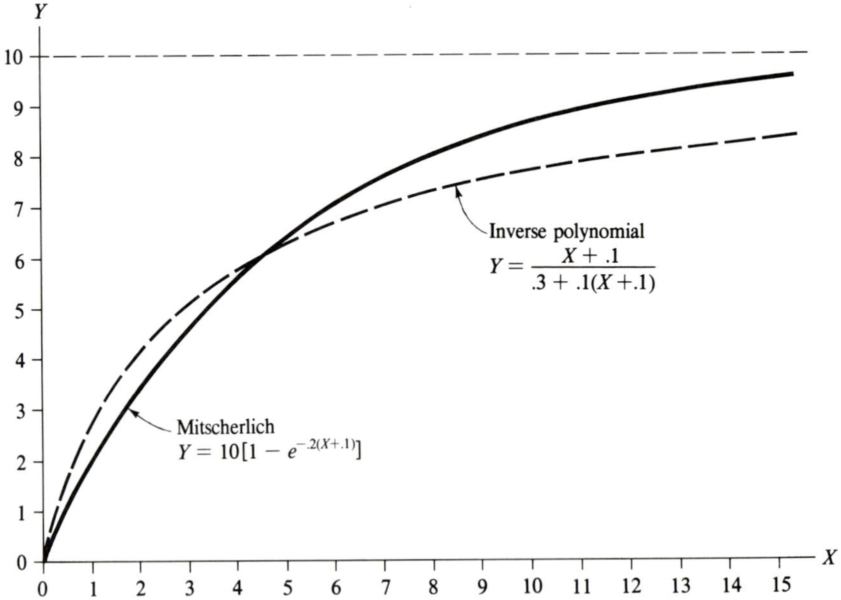

Figure 14.9 The form of the Mitscherlich and inverse polynomial models. The parameter α in the Mitscherlich equation is the upper asymptote and β controls the rate at which the asymptote is approached. The inverse polynomial model approaches its asymptote of $1/\beta_1$ very slowly and at a decreasing rate determined by $\beta_0/[\beta_0 + \beta_1(X + \delta)]^2$.

$$Y_i = \frac{X_i + \delta}{\beta_0 + \beta_1(X_i + \delta)} + \varepsilon_i \qquad [14.30]$$

The inverse polynomial model is also shown in Figure 14.9.

Logistic Growth Model

The **logistic** or **autocatalytic growth function** results when the rate of growth is proportional to the product of the size at the time and the amount of growth remaining:

$$\frac{\partial \mathscr{E}(Y)}{\partial t} = \frac{\beta \mathscr{E}(Y)[\alpha - \mathscr{E}(Y)]}{\alpha} \qquad [14.31]$$

This differential equation gives the model

$$Y_i = \frac{\alpha}{1 + \gamma e^{-\beta t_i}} + \varepsilon_i \qquad [14.32]$$

which has the familiar S-shape associated with growth curves. The curve starts at $\alpha/(1 + \gamma)$ when $t = 0$ and increases to an upper limit of α when t is large.

Gompertz Growth Model

The **Gompertz growth model** results from a rate of growth given by

$$\frac{\partial \mathscr{E}(Y)}{\partial t} = \beta \mathscr{E}(Y)\left\{\ln\left[\frac{\alpha}{\mathscr{E}(Y)}\right]\right\} \qquad [14.33]$$

and has the double exponential form

$$Y_i = \alpha e^{-\gamma e^{-\beta t_i}} + \varepsilon_i \qquad [14.34]$$

Examples of the logistic and Gompertz curves are given in Figure 14.10.

Figure 14.10 The general form of the logistic function and the Gompertz growth model.

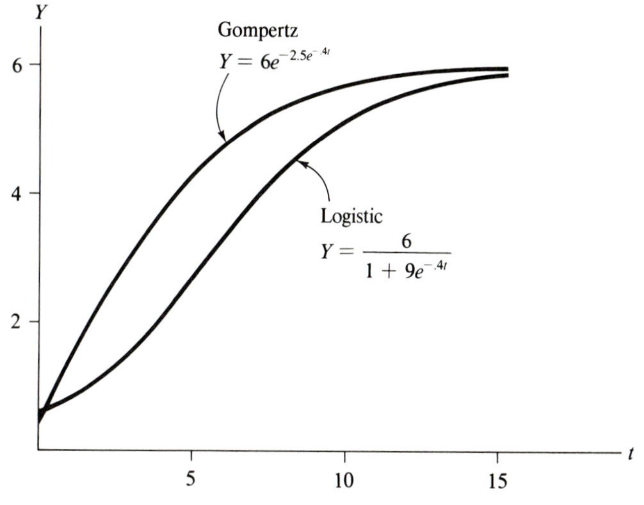

Von Bertalanffy's Model

Von Bertalanffy's model is a more general four-parameter model that yields three of the previous models by appropriate choice of values for the parameter m:

$$Y_i = (\alpha^{1-m} - \theta e^{-\beta t_i})^{1/(1-m)} \qquad [14.35]$$

When $m = 0$, this becomes the monomolecular model with $\theta = \alpha e^{-\beta\delta}$, equation 14.28. When $m = 2$, it simplifies to the logistic model with $\theta = -\gamma/\alpha$, equation 14.32, and if m is allowed to go to unity, the limiting form of Von Bertalanffy's model is the Gompertz model, equation 14.34.

Toxicity Studies

Another class of nonlinear models arises when individuals in a population are being scored for their reaction to some substance, and the individuals differ in their sensitivities to the substance. Such models have been developed most extensively in toxicity studies where it is of interest to determine the dose of a substance that causes a certain proportion of injuries or deaths. It is assumed that there is an underlying probability distribution, called the threshold distribution, of sensitivities of individuals to the toxin. The response curve for the proportion of individuals affected at various doses then follows the cumulative probability distribution of the underlying threshold distribution.

Probit and Logit Models

If the threshold distribution is the normal probability distribution, the proportion of individuals affected at dose X follows the cumulative normal distribution. This model leads to the **probit analysis** common in toxicology. Frequently, the response data are better characterized by the normal distribution after dose has been transformed to ln(dose). Thus, the threshold distribution on the original dose metric is the log-normal distribution. The **logit transformation** results when the underlying threshold distribution is the logistic probability distribution. The probit and logit transformations linearize the corresponding response curves. Alternatively, nonlinear least squares can be used to estimate the parameters of the logistic function. (Weighting should be used to take into account the heterogeneous variances of percentage data.) The cumulative normal distribution has no closed form so that nonlinear least squares cannot be applied directly to estimate the normal parameters.

Weibull Model

The **Weibull probability distribution** is common as the underlying distribution for time-to-failure studies of, for example, electrical systems. The distribution is generated by postulating that a number of individual components must fail, or a number of independent "hits" are needed, before the system fails. Recently, the cumulative form of the Weibull probability distribution has been found to be useful for modeling plant disease progression (Pennypacker et al., 1980) and crop responses to air pollution (Rawlings and Cure, 1985; Heck et al., 1984). The cumulative Weibull probability function is

$$F(X; \mu, \gamma, \sigma) = 1 - e^{-[(X_i-\mu)/\sigma]^{\gamma}} \qquad [14.36]$$

where μ is the lower limit on X. The two parameters σ and γ control the shape of the curve. This function is an increasing function approaching the upper limit of $F = 1$ when X is large.

As a response model, the asymptote can be made arbitrary by introducing another parameter α as a multiplicative constant, and the function can be turned into a monotonically decreasing function by subtracting from α. Thus, the form of the Weibull function used to model crop response to increasing levels of pollution is

$$Y_i = \alpha e^{-(X_i/\sigma)^\gamma} + \varepsilon_i \qquad [14.37]$$

This form assumes that the minimum level of X is zero. The vector of parameters is $\boldsymbol{\theta}' = (\alpha \ \sigma \ \gamma)$. Other experimental design effects such as block effects, cultivar effects, and covariates can be introduced into the Weibull model by expanding the α parameter to include a series of additive terms (Rawlings and Cure, 1985).

Choosing a Nonlinear Model

These examples of nonlinear models illustrate the variety of functional forms available when one is not restricted to linear additive models. There are many other mathematical functions that might serve as useful models. Ideally, the functional form of a model has some theoretical basis as illustrated with the partial derivatives. On the other hand, a nonlinear model might be adopted for no other reason than that it is a simple, convenient representation of the responses being observed. The Weibull model was adopted for characterizing crop losses from ozone pollution because it had a biologically realistic form and its flexibility allowed the use of a common model for all studies at different sites and on different crop species.

Segmented Polynomial Models

In some cases, it is simpler to model a complicated response by using different polynomial equations in different regions of the X-space. Usually constraints are imposed on the polynomials to ensure that they meet in the appropriate way at the "join" points. Such models are called **segmented polynomial models**. When the join points are known, the segmented polynomial models are linear in the parameters and can be fitted using ordinary least squares. However, when the join points must be estimated, the models become nonlinear.

Quadratic–Linear Segmented Polynomial

This class of models is illustrated with the quadratic–linear segmented polynomial model. Assume the first part of the response curve is adequately represented by a quadratic or second-degree polynomial, but at some point the response continues in a linear manner. The value of X at which the two polynomials meet, the "join" point, is labeled θ (Figure 14.11, page 396). Thus, the quadratic–linear model is

$$Y_i = \begin{cases} \beta_0 + \beta_1 X_i + \beta_2 X_i^2 + \varepsilon_i & \text{if } X_i \le \theta \\ \gamma_0 + \gamma_1 X_i + \varepsilon_i & \text{if } X_i > \theta \end{cases} \qquad [14.38]$$

This equation contains six parameters, β_0, β_1, β_2, γ_0, γ_1, and θ. Estimating all six parameters, however, puts no constraints on how, or even if, the two segments meet at the join point. It is common to impose two constraints. The two polynomials should meet when $X = \theta$ and the transition from one polynomial to the other should be smooth. The first requirement implies that

$$\beta_0 + \beta_1\theta + \beta_2\theta^2 = \gamma_0 + \gamma_1\theta \qquad [14.39]$$

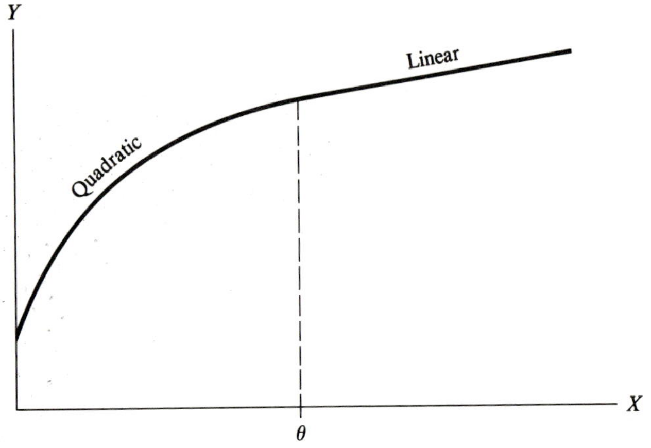

Figure 14.11 An illustration of a quadratic–linear segmented polynomial response curve.

The second condition requires the first derivatives of the two functions to be equal at $X = \theta$; that is, the slopes of both segments must be the same at the join point. Thus,

$$\left.\frac{\partial Y(X \le \theta)}{\partial X}\right|_{X=\theta} = \left.\frac{\partial Y(X > \theta)}{\partial X}\right|_{X=\theta}$$

or

$$\beta_1 + 2\beta_2\theta = \gamma_1 \qquad [14.40]$$

The second constraint requires that γ_1 be a function of β_1 and β_2. Substituting this result into the first constraint and solving for γ_0 gives

$$\gamma_0 = \beta_0 - \beta_2\theta^2 \qquad [14.41]$$

Imposing these two constraints on the original model gives

$$Y_i = \begin{cases} \beta_0 + \beta_1 X_i + \beta_2 X_i^2 + \varepsilon_i & \text{if } X \le \theta \\ (\beta_0 - \beta_2\theta^2) + (\beta_1 + 2\beta_2\theta)X_i + \varepsilon_i & \text{if } X > \theta \end{cases} \qquad [14.42]$$

There are four parameters to be estimated.

This model can be written in one statement if a dummy variable is defined to identify when X is less than θ or greater than θ. Let $T = 0$ if $X \le \theta$ and $T = 1$ if $X > \theta$. Then,

$$\begin{aligned} Y_i &= (1 - T)(\beta_0 + \beta_1 X_i + \beta_2 X_i^2) + T[(\beta_0 - \beta_2\theta^2) + (\beta_1 + 2\beta_2\theta)X_i] \\ &= \beta_0 + \beta_1 X_i + \beta_2[X_i^2 - T(X_i - \theta)^2] \end{aligned} \qquad [14.43]$$

This model is nonlinear in the parameters because the products $\beta_2\theta$ and $\beta_2\theta^2$ are present. If θ is known, the model becomes linear in the parameters. The reader is referred to Anderson and Nelson (1975) and Gallant and Fuller (1973) for more discussion on segmented polynomial models.

14.4 Fitting Models Nonlinear in the Parameters

Least Squares Principle

The least squares principle is used to estimate the parameters in nonlinear models just as in the linear models case. The least squares estimate of $\boldsymbol{\theta}$, labeled $\hat{\boldsymbol{\theta}}$, is the choice of parameters that minimizes the sum of squared residuals

$$\text{SS}[\text{Res}(\hat{\boldsymbol{\theta}})] = \sum_{i=1}^{n} [Y_i - f(\boldsymbol{x}_i', \hat{\boldsymbol{\theta}})]^2$$

or, in matrix notation,

$$\text{SS}[\text{Res}(\hat{\boldsymbol{\theta}})] = [\boldsymbol{Y} - \boldsymbol{f}(\hat{\boldsymbol{\theta}})]'[\boldsymbol{Y} - \boldsymbol{f}(\hat{\boldsymbol{\theta}})] \qquad [14.44]$$

where $\boldsymbol{f}(\hat{\boldsymbol{\theta}})$ is the $n \times 1$ vector of $f(\boldsymbol{x}_i', \hat{\boldsymbol{\theta}})$ evaluated at the n values of $\boldsymbol{x}_i'$. The partial derivatives of $\text{SS}[\text{Res}(\hat{\boldsymbol{\theta}})]$, with respect to each $\hat{\theta}_j$ in turn, are set equal to zero to obtain the p "normal equations." The solution to the normal equations gives the least squares estimate of $\boldsymbol{\theta}$.

Form of the Normal Equations

Each normal equation has the general form

$$\frac{\partial\{\text{SS}[\text{Res}(\hat{\boldsymbol{\theta}})]\}}{\partial\hat{\theta}_j} = -\sum_{i=1}^{n} [Y_i - f(\boldsymbol{x}_i', \hat{\boldsymbol{\theta}})]\left[\frac{\partial f(\boldsymbol{x}_i', \hat{\boldsymbol{\theta}})}{\partial\hat{\theta}_j}\right] = 0 \qquad [14.45]$$

where the second set of brackets contains the partial derivative of the functional form of the model. Unlike linear models, the partial derivatives of a nonlinear model are functions of the parameters. The resulting equations are nonlinear equations and, in general, cannot be solved to obtain explicit solutions for $\hat{\boldsymbol{\theta}}$.

The normal equations for a nonlinear model are illustrated using the exponential growth model, $Y_i = \alpha[\exp(\beta t_i)] + \varepsilon_i$, equation 14.23. The partial derivatives of the model with respect to the two parameters are

$$\frac{\partial f}{\partial \alpha} = \frac{\partial(\alpha e^{\beta t_i})}{\partial \alpha} = e^{\beta t_i}$$

and [14.46]

$$\frac{\partial f}{\partial \beta} = \frac{\partial(\alpha e^{\beta t_i})}{\partial \beta} = \alpha t_i e^{\beta t_i}$$

The two normal equations for this model are

$$\sum_{i=1}^{n} (Y_i - \hat{\alpha} e^{\hat{\beta} t_i})(e^{\hat{\beta} t_i}) = 0$$

and

$$\sum_{i=1}^{n} (Y_i - \hat{\alpha} e^{\hat{\beta} t_i})(\hat{\alpha} t_i e^{\hat{\beta} t_i}) = 0 \qquad [14.47]$$

Solving the Normal Equations

A difficulty with nonlinear least squares arises in trying to solve the normal equations for $\hat{\boldsymbol{\theta}}$. There is no explicit solution even in this simple example. Since explicit solutions cannot be obtained, iterative numerical methods are used. These methods require initial guesses, or starting values, for the parameters; the starting values will be labeled $\boldsymbol{\theta}^0$. The initial guesses are substituted for $\boldsymbol{\theta}$ to compute the residual sum of squares *and* to compute adjustments to $\boldsymbol{\theta}^0$ that will reduce SS(Res) and (hopefully) move $\boldsymbol{\theta}^0$ closer to the least squares solution. The new estimates of the parameters are then used to repeat the process until a sufficiently small adjustment is being made at each step. When this happens, the process is said to have converged to a solution.

Grid Search Method

Several methods for finding a solution to the normal equations are used in various nonlinear least squares computer programs. The simplest conceptual method of finding the solution is a **grid search** over the region of possible values of the parameters for the combination of values that gives the smallest residual sum of squares. This method can be used to provide reasonable starting values for other methods or, if repeated on successively finer grids, to provide the final solution. Such a procedure is not efficient.

Gauss–Newton Method

Four other methods of solving the normal equations are commonly used. The **Gauss–Newton method** uses a Taylor's expansion of $f(\boldsymbol{x}_i', \boldsymbol{\theta})$ about the starting values $\boldsymbol{\theta}^0$ to obtain a linear approximation of the model in the region near the starting values. That is, $f(\boldsymbol{x}_i', \boldsymbol{\theta})$ is replaced with

$$f(\boldsymbol{x}_i', \boldsymbol{\theta}) \doteq f(\boldsymbol{x}_i', \boldsymbol{\theta}^0) + \sum_{j=1}^{p} \left(\frac{\partial f(\boldsymbol{x}_i', \boldsymbol{\theta}^0)}{\partial \theta_j} \right) (\theta_j - \theta_j^0)$$

or

$$\boldsymbol{f}(\boldsymbol{\theta}) \doteq \boldsymbol{f}(\boldsymbol{\theta}^0) + \boldsymbol{F}(\boldsymbol{\theta}^0)(\boldsymbol{\theta} - \boldsymbol{\theta}^0) \qquad [14.48]$$

where $\boldsymbol{F}(\boldsymbol{\theta}^0)$ is the $n \times p$ matrix of partial derivatives evaluated at $\boldsymbol{\theta}^0$ and the n data points $\boldsymbol{x}_i'$. $\boldsymbol{F}(\boldsymbol{\theta}^0)$ has the form

$$\boldsymbol{F}(\boldsymbol{\theta}^0) = \begin{bmatrix} \dfrac{\partial[f(\boldsymbol{x}_1', \boldsymbol{\theta}^0)]}{\partial \theta_1} & \dfrac{\partial[f(\boldsymbol{x}_1', \boldsymbol{\theta}^0)]}{\partial \theta_2} & \cdots & \dfrac{\partial[f(\boldsymbol{x}_1', \boldsymbol{\theta}^0)]}{\partial \theta_p} \\ \dfrac{\partial[f(\boldsymbol{x}_2', \boldsymbol{\theta}^0)]}{\partial \theta_1} & \dfrac{\partial[f(\boldsymbol{x}_2', \boldsymbol{\theta}^0)]}{\partial \theta_2} & \cdots & \dfrac{\partial[f(\boldsymbol{x}_2', \boldsymbol{\theta}^0)]}{\partial \theta_p} \\ \vdots & \vdots & & \vdots \\ \dfrac{\partial[f(\boldsymbol{x}_n', \boldsymbol{\theta}^0)]}{\partial \theta_1} & \dfrac{\partial[f(\boldsymbol{x}_n', \boldsymbol{\theta}^0)]}{\partial \theta_2} & \cdots & \dfrac{\partial[f(\boldsymbol{x}_n', \boldsymbol{\theta}^0)]}{\partial \theta_p} \end{bmatrix} \qquad [14.49]$$

Linear least squares is used on the linearized model to estimate the shift in the parameters, or the amount to adjust the starting values. The model is then linearized about the new values of the parameters and linear least squares is again applied to find the second set of adjustments, and so forth, until the desired degree of convergence is attained. The adjustments obtained from the Gauss–Newton method can be too large and bypass the solution, in which case the residual sum of squares may increase at that step rather than decrease. When this happens, a modified Gauss–Newton method can be used that successively halves the adjustment until the residual sum of squares is smaller than in the previous step (Hartley, 1961).

Method of Steepest Descent

A second method, the **method of steepest descent**, finds the path for amending the initial estimates of the parameters that gives the most rapid decrease in the residual sum of squares (as approximated by the linearization). After each change in the parameter values, the residual sum of squares surface is again approximated in the vicinity of the new solution and a new path is determined. While the method of steepest descent may move rapidly in the initial stages, it can be slow to converge (Draper and Smith, 1981).

Marquardt's Compromise

The third method, called **Marquardt's compromise** (Marquardt, 1963), is designed to capitalize on the best features of the previous two methods. The adjustment computed by Marquardt's method tends toward the Gauss–Newton adjustment if the residual sum of squares is reduced at each step, and toward the steepest descent adjustment if the residual sum of squares increases in any step. This method appears to work well in most cases.

Derivative-Free Method

These three methods require the partial derivatives of the model with respect to each of the parameters. Alternatively, a derivative-free method (Ralston and Jennrich, 1978) can be used in which numerical estimates of the derivatives are computed from observed shifts in $\hat{Y}$ as the values of the θ_j are changed. The derivative-free method appears to work well as long as the data are "rich enough" for the model being fit. There have been cases with relatively limited data where the derivative-free method did not appear to work as well as the derivative methods. Convergence either was not obtained, was not as fast, or the "solution" did not appear to be as good.

Summary of Methods

The details of the numerical methods for finding the least squares solution are not discussed in this text. Gallant (1987) presents a thorough discussion of the theory and methods of nonlinear least squares including the methods of estimation. It is sufficient for now to understand that (1) the least squares principle is being used to find the estimates of the parameters, (2) the nonlinear least squares methods are iterative and use various numerical methods to arrive at the solution, and (3) apparent convergence of the program to a solution does not necessarily imply that the solution is, in fact, the optimum. The methods differ in their rates of convergence to a solution and, in some cases, whether or not a solution is obtained. No one method can be proclaimed as universally best and it may be desirable in some difficult cases to try more than one method.

Starting Values

It is important that the starting values in nonlinear regression be reasonably good. Otherwise, convergence may be slow or not attained. In addition, there may be local minima on the residual sum of squares surface, and poor starting values for the parameters increase the chances that the iterative process will converge to a local minimum rather than the global minimum. To protect against convergence to a local minimum, different sets of starting values can be used to see if convergence is to the same solution in all cases. Plotting the resulting response function with the data superimposed is particularly important in nonlinear regression to ensure that the solution is reasonable.

Nonconvergence

Convergence to a solution may not be obtained in some cases. One reason for nonconvergence is that the functional form of the model is inconsistent with the observed response. For example, an exponential decay model cannot be made to adequately characterize a logistic growth model. "Convergence" in such cases, if attained, would be meaningless. Errors in specification of the derivatives is a common reason for lack of convergence.

Even with an appropriate form for the model and correct derivatives, convergence may not be attained. The reason for lack of convergence can be stated in several ways. (1) The model may be overdefined, meaning the model has more parameters or is more complex than need be for the process. The two-term exponential in the previous section is an overdefined model if the two rate constants are nearly the same. (2) There may not be sufficient data to fully characterize the response curve. This implies that the model is correct; it is the data that are lacking. Of course, a model may appear to be overdefined because there are not sufficient data to show the complete response curve. (3) The model may be poorly parameterized with two (or more) parameters playing very similar roles in the nonlinear function. Thus, very nearly the same fitted response curve can be obtained by very different combinations of values of the parameters. These situations are reflected in the estimates of the parameters being very highly correlated, perhaps .98 or higher. This is the nonlinear models analogue of the collinearity problem in ordinary least squares and has similar effects.

Distribution of $\hat{\theta}$ and SS(Res)

Most properties of linear least squares apply only approximately or asymptotically for nonlinear least squares. The matrix $\boldsymbol{F}(\boldsymbol{\theta}) = \boldsymbol{F}$, equation 14.49, plays the role in nonlinear least squares that $\boldsymbol{X}$ plays in linear least squares. Gallant (1987) shows that, if $\varepsilon \sim \mathrm{N}(\mathbf{0}, \boldsymbol{I}\sigma^2)$, $\hat{\boldsymbol{\theta}}$ is *approximately* normally distributed with mean $\boldsymbol{\theta}$ and $\mathbf{Var}(\hat{\boldsymbol{\theta}}) = (\boldsymbol{F}'\boldsymbol{F})^{-1}\sigma^2$:

$$\hat{\boldsymbol{\theta}} \mathrel{\dot{\sim}} \mathrm{N}[\boldsymbol{\theta}, (\boldsymbol{F}'\boldsymbol{F})^{-1}\sigma^2] \qquad [14.50]$$

where the symbol "$\dot{\sim}$" is read "approximately distributed." The residual sum of squares, $\mathrm{SS}[\mathrm{Res}(\hat{\boldsymbol{\theta}})]$, has *approximately* a chi-squared distribution with $(n - p)$ degrees of freedom. Alternatively, asymptotic arguments can be used to show asymptotic normality of $\hat{\boldsymbol{\theta}}$ as n gets large, without the normality assumption on ε (Gallant, 1987).

Variance of $\hat{\theta}$

In practice, $\boldsymbol{F}(\boldsymbol{\theta})$ is computed as $\boldsymbol{F}(\hat{\boldsymbol{\theta}})$, which is labeled $\hat{\boldsymbol{F}}$ for brevity, and σ^2 is estimated with $s^2 = \mathrm{SS}[\mathrm{Res}(\hat{\boldsymbol{\theta}})]/(n - p)$, so that the estimated asymptotic variance–covariance matrix for $\hat{\boldsymbol{\theta}}$ is

$$\boldsymbol{s}^2(\hat{\boldsymbol{\theta}}) = (\hat{\boldsymbol{F}}'\hat{\boldsymbol{F}})^{-1}s^2 \qquad [14.51]$$

Standard errors given in computer programs are based on this approximation. (Some computer programs for nonlinear least squares give only the standard errors of $\hat{\theta}_j$ and the estimated correlation matrix for $\hat{\boldsymbol{\theta}}$, labeled $\hat{\boldsymbol{\rho}}$. The variance–covariance matrix can be recovered as

$$s^2(\hat{\boldsymbol{\theta}}) = \boldsymbol{S}\hat{\boldsymbol{\rho}}\boldsymbol{S} \qquad [14.52]$$

where $\boldsymbol{S}$ is the $p \times p$ diagonal matrix of standard errors of $\hat{\boldsymbol{\theta}}$.)

Confidence Intervals and Tests of Significance

The approximate normality of $\hat{\boldsymbol{\theta}}$ and chi-squared distribution of $(n - p)s^2$ (and their independence; Gallant, 1987) permit the usual computations of confidence limits and tests of significance of the $\hat{\theta}_j$ and functions of $\hat{\theta}_j$. Let $C = \boldsymbol{K}'\boldsymbol{\theta}$ be any *linear* function of interest. The point estimate of C is $\hat{C} = \boldsymbol{K}'\hat{\boldsymbol{\theta}}$ with (approximate) standard error $s(\hat{C}) = \{\boldsymbol{K}'[s^2(\hat{\boldsymbol{\theta}})]\boldsymbol{K}\}^{1/2}$. The (approximate) 95% confidence interval estimate of C is

$$\hat{C} \pm t_{[\alpha,(n-p)]}s(\hat{C}) \qquad [14.53]$$

The test of the null hypothesis that $C = C_0$ is

$$t = \frac{\hat{C} - C_0}{s(\hat{C})} \qquad [14.54]$$

and is distributed approximately as Student's t with $(n - p)$ degrees of freedom.

Nonlinear Functions of $\hat{\theta}$

Usually the function of interest in nonlinear regression is a *nonlinear* function of $\boldsymbol{\theta}$, which is estimated with the same nonlinear function of $\hat{\boldsymbol{\theta}}$. For example, the fitted values of the response variable, $\hat{\boldsymbol{Y}} = \boldsymbol{f}(\hat{\boldsymbol{\theta}})$, are nonlinear functions of $\hat{\boldsymbol{\theta}}$. Let $h(\boldsymbol{\theta})$ be any nonlinear function of interest. Gallant (1987) shows that $h(\hat{\boldsymbol{\theta}})$ is approximately normally distributed with mean $h(\boldsymbol{\theta})$ and variance $\boldsymbol{H}(\boldsymbol{F}'\boldsymbol{F})^{-1}\boldsymbol{H}'\sigma^2$; that is,

$$h(\hat{\boldsymbol{\theta}}) \mathrel{\dot{\sim}} \mathrm{N}[h(\boldsymbol{\theta}), \boldsymbol{H}(\boldsymbol{F}'\boldsymbol{F})^{-1}\boldsymbol{H}'\sigma^2] \qquad [14.55]$$

where

$$\boldsymbol{H} = \left(\frac{\partial[h(\boldsymbol{\theta})]}{\partial\theta_1} \; \frac{\partial[h(\boldsymbol{\theta})]}{\partial\theta_2} \; \cdots \; \frac{\partial[h(\boldsymbol{\theta})]}{\partial\theta_p}\right) \qquad [14.56]$$

is the row vector of partial derivatives of the function $h(\boldsymbol{\theta})$ with respect to each of the parameters. This result uses the first-order terms of a Taylor's series expansion to approximate $h(\boldsymbol{\theta})$ with a linear function. Thus, $h(\hat{\boldsymbol{\theta}})$ is (approximately) an unbiased estimate of $h(\boldsymbol{\theta})$. Letting $\hat{\boldsymbol{H}} = \boldsymbol{H}(\hat{\boldsymbol{\theta}})$ and $\hat{\boldsymbol{F}} = \boldsymbol{F}(\hat{\boldsymbol{\theta}})$, we can estimate the variance of $h(\hat{\boldsymbol{\theta}})$ by

$$s^2[h(\hat{\boldsymbol{\theta}})] = [\hat{\boldsymbol{H}}(\hat{\boldsymbol{F}}'\hat{\boldsymbol{F}})^{-1}\hat{\boldsymbol{H}}']s^2 \qquad [14.57]$$

The approximate $100(1 - \alpha)\%$ confidence interval estimate of $h(\boldsymbol{\theta})$ is

$$h(\hat{\boldsymbol{\theta}}) \pm t_{[\alpha,(n-p)]}[\hat{\boldsymbol{H}}(\hat{\boldsymbol{F}}'\hat{\boldsymbol{F}})^{-1}\hat{\boldsymbol{H}}'s^2]^{1/2} \qquad [14.58]$$

and the approximate t-test of the null hypothesis that $h(\boldsymbol{\theta}) = h_0$ is

$$t = \frac{h(\hat{\boldsymbol{\theta}}) - h_0}{s[h(\hat{\boldsymbol{\theta}})]} \qquad [14.59]$$

and has $(n - p)$ degrees of freedom.

Several Functions

If there are q functions of interest, $h(\boldsymbol{\theta})$ becomes a vector of order q and **H** becomes a $q \times p$ matrix of partial derivatives with each row being the derivatives for one of the functions. The composite hypothesis

$$H_0\text{:} \quad \boldsymbol{h}(\boldsymbol{\theta}) = \boldsymbol{0}$$

is tested against the two-tailed alternative hypothesis with an approximate test referred to as the **Wald statistic** (Gallant, 1987):

$$W = \frac{\boldsymbol{h}(\hat{\boldsymbol{\theta}})'[\hat{\boldsymbol{H}}(\hat{\boldsymbol{F}}'\hat{\boldsymbol{F}})^{-1}\hat{\boldsymbol{H}}']^{-1}\boldsymbol{h}(\hat{\boldsymbol{\theta}})}{qs^2} \qquad [14.60]$$

Notice the similarity in form of W to the F-statistic in general linear hypotheses. W is approximately distributed as F with q and $(n - p)$ degrees of freedom.

$\hat{Y}_i$

Note that, if the functions of interest are the n values of $\hat{Y}_i$, then $\boldsymbol{h}(\hat{\boldsymbol{\theta}}) = \boldsymbol{f}(\hat{\boldsymbol{\theta}})$ and $\boldsymbol{H}(\hat{\boldsymbol{\theta}}) = \boldsymbol{F}(\hat{\boldsymbol{\theta}})$, so that

$$s^2(\hat{\boldsymbol{Y}}) = [\hat{\boldsymbol{F}}(\hat{\boldsymbol{F}}'\hat{\boldsymbol{F}})^{-1}\hat{\boldsymbol{F}}']s^2 \qquad [14.61]$$

The matrix $[\hat{\boldsymbol{F}}(\hat{\boldsymbol{F}}'\hat{\boldsymbol{F}})^{-1}\hat{\boldsymbol{F}}']$ is analogous to $\boldsymbol{P}$, the projection matrix, in linear least squares.

Wald and Likelihood Ratio

In general, the confidence limits in equation 14.58, the t-test in equation 14.59, and W in equation 14.60 are referred to as the Wald methodology. The Wald approximation appears to work well in most cases in that the stated probability levels are sufficiently close to the true levels (Gallant, 1987). However, Gallant has shown cases where the Wald approach can be seriously wrong. Tests and confidence intervals based on the more difficult likelihood ratio test, however, gave results consistent with the stated probabilities in all cases investigated. For this reason, Gallant recommends that the Wald results be compared to the likelihood ratio results for some cases in each problem to verify that the simpler Wald approach is adequate.

The approximate joint $100(1 - \alpha)\%$ confidence region for $\boldsymbol{\theta}$, based on likelihood ratio theory, is defined as that set of $\boldsymbol{\theta}$ for which

$$\text{SS}[\text{Res}(\boldsymbol{\theta})] - \text{SS}[\text{Res}(\hat{\boldsymbol{\theta}})] \leq p\hat{\sigma}^2 F_{(\alpha;\, p,\, \nu)} \qquad [14.62]$$

where $\hat{\sigma}$ is an estimate of σ^2 based on v degrees of freedom. The reader is referred to Gallant (1987) for discussion of the likelihood ratio procedure.

The validity of the Wald approach depends on how well $f(\boldsymbol{x}_i', \boldsymbol{\theta})$ is represented by the linear approximation in $\boldsymbol{\theta}$. This depends on the parameterization of the model and is referred to as **parameter effects curvature**. Clarke (1987) defined components of overall parameter effects curvature that could be identified with each parameter. These component measures of curvature are then used to define severe curvature, cases in which the Wald methodology may not be adequate for the particular parameters, and to provide higher-order correction terms for the confidence interval estimates. The reader is referred to Clarke (1987) for details.

Example 14.2

The example to illustrate nonlinear regression comes from calcium ion experiments for biochemical analysis of intracellular storage and transport of Ca^{++} across the plasma membrane. The study was run by Howard Grimes, Botany Department, North Carolina State University, and is used with his permission. The data consist of amount of radioactive calcium in cells (nmole/mg) that had been in "hot" calcium suspension for given periods of time (minutes). Data were obtained on 27 independent cell suspensions with times ranging from .45 to 15.00 minutes (Table 14.3).

Table 14.3 Calcium uptake of cells suspended in a solution of radioactive calcium. (Data from H. Grimes, North Carolina State University, and used with permission.)

Suspen. Number	*Time (min)*	*Calcium (nmoles/mg)*	*Suspen. Number*	*Time (min)*	*Calcium (nmoles/mg)*
1	.45	.34170	15	6.10	2.67061
2	.45	−.00438	16	8.05	3.05959
3	.45	.82531	17	8.05	3.94321
4	1.30	1.77967	18	8.05	3.43726
5	1.30	.95384	19	11.15	4.80735
6	1.30	.64080	20	11.15	3.35583
7	2.40	1.75136	21	11.15	2.78309
8	2.40	1.27497	22	13.15	5.13825
9	2.40	1.17332	23	13.15	4.70274
10	4.00	3.12273	24	13.15	4.25702
11	4.00	2.60958	25	15.00	3.60407
12	4.00	2.57429	26	15.00	4.15029
13	6.10	3.17881	27	15.00	3.42484
14	6.10	3.00782			

Proposed Model

The kinetics involved led the researchers to postulate that the response would follow the nonlinear model

$$Y_i = \alpha_1[1 - \exp(-\lambda_1 t_i)] + \alpha_2[1 - \exp(-\lambda_2 t_i)] + \varepsilon_i \qquad [14.63]$$

where Y is nmoles/mg of Ca^{++}. (This model is referred to as the Michaelis–Menten model.) The partial derivatives for this model are

$$\frac{\partial f}{\partial \alpha_1} = 1 - e^{-\lambda_1 t}$$

$$\frac{\partial f}{\partial \lambda_1} = t\alpha_1 e^{-\lambda_1 t}$$

$$\frac{\partial f}{\partial \alpha_2} = 1 - e^{-\lambda_2 t}$$

$$\frac{\partial f}{\partial \lambda_2} = t\alpha_2 e^{-\lambda_2 t}$$

In this case, a derivative-free method was used to fit the data in Table 14.3 to the two-term exponential model. The starting values used for the four parameters were

$$\boldsymbol{\theta}^0 = \begin{pmatrix} \alpha_1^0 \\ \lambda_1^0 \\ \alpha_2^0 \\ \lambda_2^0 \end{pmatrix} = \begin{pmatrix} .05 \\ .09 \\ .20 \\ .20 \end{pmatrix}$$

These were not well-chosen starting values because α_1 and α_2 are the upper asymptotes of the two exponential functions and their sum should be near the upper limits of the data, approximately 4.5. Likewise, the rate constants were chosen quite arbitrarily. This was simply an expedient; if there appeared to be convergence problems or a logical inconsistency in the final model, more effort would be devoted to choice of starting values.

Convergence Not Attained

A solution *appeared* to have been obtained. The residual sum of squares decreased from $\text{SS}[\text{Res}(\boldsymbol{\theta}^0)] = 223$ with the starting values $\boldsymbol{\theta}^0$ to $\text{SS}[\text{Res}(\hat{\boldsymbol{\theta}})] = 7.4645$ with the final solution $\hat{\boldsymbol{\theta}}$. A plot of $\hat{\mathbf{Y}}$ superimposed on the data appeared reasonable. However, the results raised several flags. First, a program message "CONVERGENCE ASSUMED" indicated that the convergence criterion had not been attained. The iterations had terminated because no further progress in reducing the residual sum of squares had been realized during a sequence of halving the size of the parameter changes. Further, the estimates of the parameters and their correlation matrix revealed an overdefined model (Table 14.4). $\hat{\alpha}_1$ converged to the lower bound imposed to keep the estimate positive, .0001, and its standard error was 0.0. $\hat{\lambda}_1$ converged to a very high value with an extremely large standard error and confidence interval. The correlation matrix for the parameter estimates showed other peculiarities. The zeros for the first row and column of the correlation matrix are reflections of the zero *approximated* variance for $\hat{\alpha}_1$. The correlation matrix showed $\hat{\lambda}_1$ and $\hat{\lambda}_2$ to be perfectly negatively correlated, and the correlations of $\hat{\alpha}_2$ with $\hat{\lambda}_1$ and $\hat{\lambda}_2$ were identical in magnitude.

These results are a reflection of the model being overly complex for the response shown in the data. The first exponential component of the model, when

Table 14.4 Nonlinear regression results from Grimes' data using the two-term exponential model.

Analysis of Variance:

Source	d.f.	*Sum of Squares*	*Mean Square*
Model	3[a]	240.78865	80.26288
Residual	24	7.46451	.31102
Uncorr. total	27	248.25315	
Corr. total	26	53.23359	

Parameter	*Estimate*	*Asymptotic Std. Error*	*Asymptotic 95%* *Lower*	*Upper*
$\hat{\alpha}_1$	.000100	.0000000	.0000000	.0000000
$\hat{\lambda}_1$	4,629.250	12,091.767	−20,326.728	29,585.229
$\hat{\alpha}_2$	4.310418	.9179295	2.4159195	6.2049156
$\hat{\lambda}_2$	.208303	.0667369	.0705656	.3460400

Asymptotic Correlation Matrix of the Parameters

	$\hat{\alpha}_1$	$\hat{\lambda}_1$	$\hat{\alpha}_2$	$\hat{\lambda}_2$
$\hat{\alpha}_1$	.0000	.0000	.0000	.0000
$\hat{\lambda}_1$	.0000	1.0000	−.5751	−1.0000
$\hat{\alpha}_2$	.0000	−.5751	1.0000	.5751
$\hat{\lambda}_2$	.0000	−1.0000	.5751	1.0000

[a] Although the model contained four parameters, the convergence of $\hat{\alpha}_1$ to the lower bound of .0001 has effectively removed it as a parameter to be estimated.

evaluated using the parameter estimates, goes to $\hat{\alpha}_1$ for extremely small values of t. For all practical purposes, the first term is contributing only a constant to the overall response curve. This suggests that a single-term exponential model would adequately characterize the behavior of these data.

To verify that these results were not a consequence of the particular starting values, another analysis was run with $\boldsymbol{\theta}^{0\prime} = (1.0 \ .39 \ .50 \ .046)$. Again, the "CONVERGENCE ASSUMED" message was obtained and the residual sum of squares was slightly larger, SS(Res) = 7.4652. The solution, however, was very different (results not given). Now, the estimate of λ_2 and its standard error were exceptionally large, but the correlation matrix appeared quite reasonable. Evaluation of the first exponential term produced very nearly the same numerical results as the second did in the first analysis, and the second exponential term converged to $\hat{\alpha}_2$ for very small values of t. This reversal of roles of the two exponential terms is, again, reflecting an overdefined model.

Simplified Model

The model was simplified to contain only one exponential process,

$$Y_i = \alpha\{1 - \exp[-(t/\sigma)^{\gamma}]\}$$

This is the Weibull growth model, with an upper asymptote of α, and reduces to the exponential growth model if $\gamma = 1.0$. The presence of γ in the model permits greater flexibility than the simple exponential and can be used to test the

Table 14.5 Nonlinear regression results from the Weibull growth model applied to Grimes' data.

Analysis of Variance: *Source*	d.f.	*Sum of Squares*	*Mean Square*
Model	3	240.79017	80.2634
Residual	24	7.46297	.3110
Uncorr. total	27	248.25315	
Corr. total	26	53.23359	

Parameter	*Estimate*	*Asymptotic Std. Error*	*Asymptotic 95%* *Lower*	*Upper*
$\hat{\alpha}$	4.283429	.4743339	3.3044593	5.2623977
$\hat{\sigma}$	4.732545	1.2700253	2.1113631	7.3537277
$\hat{\gamma}$	1.015634	.2272543	.5466084	1.4846603

Asymptotic Correlation Matrix of the Parameters

	$\hat{\alpha}$	$\hat{\sigma}$	$\hat{\gamma}$
$\hat{\alpha}$	1.0000	.9329	−.7774
$\hat{\sigma}$	.9329	1.0000	−.7166
$\hat{\gamma}$	−.7774	−.7166	1.0000

hypothesis that the exponential growth model is adequate, $H_0\colon \gamma = 1.0$. (Notice that σ in this model is equivalent to $1/\lambda$ in the previous exponential models.)

The convergence criterion was met for this model with $\text{SS}[\text{Res}(\hat{\boldsymbol{\theta}})] = 7.4630$, even slightly smaller than obtained with the two-term exponential model. The key results are shown in Table 14.5. There are no indications of any problems with the model. The standard errors and confidence limits on the parameter estimates are reasonable and the correlation matrix shows no extremely high correlations. The Wald t-test of the null hypothesis $H_0\colon \gamma = 1.0$ can be inferred from the confidence limits on γ; $\hat{\gamma}$ is very close to 1.0 and the 95% confidence interval, (.55, 1.48), overlaps 1.0. These results indicate that a simple exponential growth model would suffice.

The logical next step in fitting this model would be to set $\gamma = 1.0$ and fit the simple one-term exponential model

$$Y_i = \alpha[1 - \exp(-t_i/\sigma)] + \varepsilon_i$$

Rather than proceed with that analysis, we will use the present analysis to show the recovery of $s^2(\hat{\boldsymbol{\theta}})$ from the correlation matrix, and the computation of approximate variances and standard errors for nonlinear functions of the parameters.

$s^2(\hat{\boldsymbol{\theta}})$ The estimated variance–covariance matrix of the parameter estimates, equation 14.51, is recovered from the correlation matrix $\hat{\boldsymbol{\rho}}$ by

$$s^2(\hat{\boldsymbol{\theta}}) = \boldsymbol{S}(\hat{\boldsymbol{\rho}})\boldsymbol{S}$$

where $\boldsymbol{S}$ is the diagonal matrix of standard errors of the estimates from Table 14.5,

$$S = \begin{bmatrix} .47433392 & 0 & 0 \\ 0 & 1.27002532 & 0 \\ 0 & 0 & .22725425 \end{bmatrix}$$

The resulting asymptotic variance–covariance matrix, $s^2(\hat{\boldsymbol{\theta}}) = (\hat{\boldsymbol{F}}'\hat{\boldsymbol{F}})^{-1}s^2$, equation 14.51, is

$$s^2(\hat{\boldsymbol{\theta}}) = \begin{bmatrix} .2250 & .5620 & -.0838 \\ .5620 & 1.6130 & -.2068 \\ -.0838 & -.2068 & .0516 \end{bmatrix}$$

Proportional Response Estimates

To illustrate the computation of approximate variances and confidence limits for nonlinear functions of the parameters, assume that the functions of interest are the estimated responses as a proportion of the upper asymptote, α, for $t = 1, 5$, and 15 minutes. That is, the function of interest is

$$h(t, \boldsymbol{\theta}) = 1 - \exp[-(t/\sigma)^{\gamma}]$$

evaluated at $t = 1, 5$, and 15. Writing $h(t, \boldsymbol{\theta})$ for the three values of t as a column vector and substituting $\hat{\boldsymbol{\theta}}' = (4.2834 \;\; 4.7325 \;\; 1.0156)$ from Table 14.5 for $\boldsymbol{\theta}$ gives

$$\boldsymbol{h}(\hat{\boldsymbol{\theta}}) = \begin{pmatrix} h(1, \hat{\boldsymbol{\theta}}) \\ h(5, \hat{\boldsymbol{\theta}}) \\ h(15, \hat{\boldsymbol{\theta}}) \end{pmatrix} = \begin{pmatrix} .1864 \\ .6527 \\ .9603 \end{pmatrix}$$

as the point estimates of the proportional responses.

The partial derivatives of $\boldsymbol{h}(\boldsymbol{\theta})$ are needed to obtain the variance–covariance matrix for $\boldsymbol{h}(\hat{\boldsymbol{\theta}})$, equations 14.56 and 14.57. The partial derivatives are

$$\frac{\partial h}{\partial \alpha} = 0$$

$$\frac{\partial h}{\partial \sigma} = \alpha\left(\frac{\gamma}{\sigma}\right)\left(\frac{t}{\sigma}\right)^{\gamma}\left\{\exp\left[-\left(\frac{t}{\sigma}\right)^{\gamma}\right]\right\}$$

$$\frac{\partial h}{\partial \gamma} = -\alpha\left(\frac{t}{\sigma}\right)^{\gamma}\left[\ln\left(\frac{t}{\sigma}\right)\right]\left\{\exp\left[-\left(\frac{t}{\sigma}\right)^{\gamma}\right]\right\}$$

Writing the partial derivatives as a row vector, equation 14.56, substituting $\hat{\boldsymbol{\theta}}$ for $\boldsymbol{\theta}$, and evaluating the vector for each value of t gives the matrix $\hat{\boldsymbol{H}}$:

$$\hat{\boldsymbol{H}} = \begin{bmatrix} 0 & .36011 & .26084 \\ 0 & .07882 & -.02019 \\ 0 & .02747 & -.14768 \end{bmatrix}$$

The variance–covariance matrix for the predictions, equation 14.57, is

$$s^2(\boldsymbol{h}) = \hat{\boldsymbol{H}}[s^2(\hat{\theta})]\hat{\boldsymbol{H}}'$$

$$= \begin{bmatrix} .001720 & .000204 & -.000776 \\ .000204 & .010701 & .006169 \\ -.000776 & .006169 & .004002 \end{bmatrix}$$

The square roots of the diagonal elements are the standard errors of the estimated levels of Ca^{++} relative to the upper limit at $t = 1$, 5, and 15 minutes. The 95% confidence interval estimates of these increases are given by the Wald approximation as $h(\hat{\theta}) \pm s[h(\hat{\theta})]t_{(.05, 24)}$ since the residual mean square had 24 degrees of freedom. The Wald confidence limits are summarized as follows:

t	$h(\hat{\theta})$	**Lower Limit**	**Upper Limit**
1	.186	.101	.272
5	.653	.439	.866
15	.960	.829	1.091

Note that the upper limit on the interval for $t = 15$ exceeds 1.0, the logical upper bound on a proportion. This reflects inadequacies in the Wald approximation as the limits are approached. ■

EXERCISES

* 14.1. The critical point (maximum or minimum) on a quadratic response curve is that point where the tangent to the curve has slope zero. Plot the equation

$$Y = 10 + 2.5X - .5X^2$$

and find the value of X where the tangent to the curve has slope zero. Is the point on the response curve a maximum or a minimum? The derivative of Y with respect to X is $dY/dX = 2.5 - 1.0X$. Solve for the value of X that makes the derivative equal to zero. How does this point relate to the value of X where the tangent was zero?

14.2. Change the quadratic equation in exercise 14.1 to

$$Y = 10 + 2.5X + .5X^2$$

Again, plot the equation and find the value of X where the tangent to the curve has slope zero. Is this point a maximum or minimum? What characteristic in the quadratic equation determines whether the critical point is a maximum or a minimum?

* 14.3. The critical point on a bivariate quadratic response surface is a maximum, minimum, or saddle point. Plot the bivariate polynomial

$$Y = 10 - X_1 + 4X_2 + .25X_1^2 - .5X_2^2$$

over the region $0 < X_1 < 5$ and $2 < X_2 < 6$. Find the critical point where the slopes of the tangent lines in the X_1 direction *and* the X_2 direction are zero. Is this point a maximum, a minimum, or a saddle point? Now use the partial derivatives to find this critical point.

* 14.4. Assume you have fit the following cubic polynomial to a set of growth data where X ranged from 6 to 20:

$$Y = 50 - 20X + 2.5X^2 - .0667X^3$$

Plot the response equation over the interval of the data. Does it appear to have a reasonable "growth" form? Demonstrate the sensitivity of the polynomial model to extrapolation by plotting the equation over the interval $X = 0$ to $X = 30$.

*14.5. You have obtained the regression equation $Y = 40 - .5X^2$ over the interval $-5 < X < 5$, where X = (temperature in °F − 95). Assume the partial regression coefficient for the linear term was not significant and was dropped from the model. Reexpress the regression equation in degrees centigrade, °C = 5(°F − 32)/9. Find the conversion of X = (°F − 95) to °C and convert the regression equation. What is the linear regression coefficient in the converted equation? What do you conclude about this linear regression coefficient being different from zero *if* the coefficient on X^2, the .5, in the original equation is significantly different from zero?

14.6. You are given the accompanying response data on concentration of a chemical as a function of time. The six sets of observations Y_1 to Y_6 represent different environmental conditions.

Time (h)	Y_1	Y_2	Y_3	Y_4	Y_5	Y_6
6	.38	.20	.34	.43	.10	.26
12	.74	.34	.69	.82	.16	.48
24	.84	.51	.74	.87	.18	.51
48	.70	.41	.62	.69	.19	.44
72	.43	.29	.43	.60	.15	.33

a. Use cubic polynomial models to relate Y = concentration to X = time, where each environment is allowed to have its own intercept and response curve. Is the cubic term significant for any of the environments? [For the purposes of testing homogeneity in part (c) below, retain the minimum-degree polynomial model that describes all responses.]

b. Your knowledge of the process tells you that Y must be zero when $X = 0$. Test the composite null hypothesis that the six intercepts are zero using the model in part (a) as the full model. What model do you adopt based on this test?

c. Use the model determined from the test in part (b) and test the homogeneity of the six response curves. State the conclusion of the test and give the model you have adopted at this stage.

* 14.7. The data in the accompanying table were taken to develop standardized soil moisture curves for each of six soil types. Percent soil moisture is determined at

each of six pressures. The objective is to develop a response curve for prediction of soil moisture from pressure readings. (Data courtesy of Joanne Rebbeck, North Carolina State University.)

Pressure (bars)	Soil Type I	II	III	IV	V	VI
.10	15.31	17.32	14.13	16.75	14.07	14.15
.33	11.59	14.88	10.58	14.20	11.39	10.57
.50	9.74	13.17	8.71	12.07	9.40	9.27
1.00	9.50	12.44	7.62	11.38	8.62	8.73
5.00	6.09	10.08	5.30	9.62	5.17	5.32
15.00	4.49	8.75	4.09	8.59	3.92	4.08

a. Plot percent moisture against pressure for each soil type. Search for a transformation on X or Y or both that linearizes the relationship for all soils. Fit your transformed data and test homogeneity of the responses over the six soil types.

b. Use the nonlinear model $Y = \alpha_j + \beta_j X^\gamma$ to summarize the relationship between Y and X on the original scale, where Y = moisture and X = pressure. [*Caution*: Your nonlinear program may not be able to iterate γ across $\gamma = 0$ and you may have to try both $\gamma > 0$ and $\gamma < 0$.] The full model allows for a value of α_j and β_j for each soil. Fit the reduced model for H_0: $\beta_1 = \beta_2 = \beta_3 = \beta_4 = \beta_5 = \beta_6$ and test this composite null hypothesis. Plot the residuals for your adopted model and summarize the results. What does the estimate of γ suggest about the adequacy in part (a) of only a logarithmic transformation on pressure?

14.8. What model is obtained if $\theta_2 = 0$ in the two-term exponential model, equation 14.25?

* 14.9. The data in the table are from a growth experiment with blue-green algae *Spirulina platensis* conducted by Linda Shurtleff, North Carolina State University (data used with permission). There were four treatments determined by the amount of "aeration" of the cultures:

1. No shaking and no CO_2 aeration;
2. CO_2 bubbled through the culture;
3. Continuous shaking of the culture but no CO_2; and
4. CO_2 bubbled through the culture and continuous shaking of the culture.

There were two replicates for each treatment, each consisting of 14 independent solutions. The 14 solutions in each replicate and treatment were randomly assigned for measurement to one of each of the 14 days of the study. The dependent variable reported is a log-scale measurement of the increased absorbance of light by the solution, which is interpreted as a measure of algae density. The readings for $DAYS = 0$ are a constant zero and are to be omitted from the analyses.

Time (days)	Control		CO_2		Shaking		CO_0 & Shaking	
	Rep 1	Rep 2	Rep 1	Rep 2	Rep 1	Rep 2	Rep 1	Rep 2
0	0	0	0	0	0	0	0	0
1	.220	.482	.530	.184	.536	.531	.740	.638
2	.555	.801	1.183	.664	.974	.926	1.251	1.143
3	1.246	1.483	1.603	1.553	1.707	1.758	2.432	2.058
4	1.456	1.717	1.994	1.910	2.032	2.021	3.054	2.451
5	1.878	2.128	2.708	2.585	2.395	2.374	3.545	2.836
6	2.153	2.194	3.006	3.009	2.706	2.933	4.213	3.296
7	2.245	2.639	3.867	3.403	3.009	3.094	4.570	3.594
8	2.542	2.960	4.059	3.892	3.268	3.402	4.833	3.790
9	2.748	3.203	4.349	4.367	3.485	3.564	5.074	3.898
10	2.937	3.390	4.699	4.551	3.620	3.695	5.268	4.028
11	3.132	3.626	4.983	4.656	3.873	3.852	5.391	4.150
12	3.283	4.003	5.100	4.754	4.042	3.960	5.427	4.253
13	3.397	4.167	5.288	4.842	4.149	4.054	5.549	4.314
14	3.456	4.243	5.374	4.969	4.149	4.168	5.594	4.446

a. Do the analysis of variance on these data (omitting $DAYS = 0$) accounting for treatments, reps within treatments, time, and time × treatment interaction. Plot the residuals from the analysis of variance against time. How do you explain the dispersion in the residuals being smaller for intermediate days? What do these results say about the analysis of variance model?

b. Use quadratic polynomials to represent the response over time. Fit a model that allows each treatment to have its own intercept and quadratic response. Then fit a model that allows each treatment to have its own intercept but forces all to have the same quadratic response. Use the results to test the homogeneity of the responses for the four treatments. [*Note:* Use the residual mean square from the analysis of variance as your estimate of σ^2.] Use the quadratic model you have adopted at this point and define a reduced model that will test the null hypothesis that all intercepts are zero. Complete the test and state your conclusions.

c. The test of zero intercepts in part (b) used quadratic polynomials. Repeat the test of zero intercepts using cubic polynomials for each treatment. Summarize the results.

d. Use the nonlinear Mitscherlich model, equation 14.28 with $\delta = 0$, to describe the change in algae density with time. Allow each treatment to have its own response. Then fit reduced models to test (1) the composite hypothesis that all β_j are equal, and (2) the composite hypothesis that all α_j are equal. Summarize the results and state your conclusions.

14.10. Assigning a visual volume score to vegetation is a nondestructive method of obtaining measures of biomass. The volume score is the volume of space occupied by the plant computed according to an extensive set of rules involving different geometric shapes. The accompanying data on volume scores and biomass dry weights for grasses were obtained for the purpose of developing a prediction equation for dry weight biomass based on the nondestructive volume score. (Data were provided by Steve Byrne, North Carolina State University, and are used with permission.)

Volume	*Dry Wt.*	*Volume*	*Dry Wt.*
5	.8	1,753	3.4
1,201	2.2	70,300	107.6
108,936	87.5	62,000	42.3
105,000	94.4	369	1.0
1,060	4.2	4,100	6.9
1,036	.5	177,500	205.5
33,907	67.7	91,000	120.9
48,500	72.4	2,025	5.5
314	.6	80	1.3
1,400	3.9	54,800	110.3
46,200	87.7	51,000	26.0
76,800	86.8	55	3.4
24,000	57.6	1,605	3.4
1,575	.5	15,262	32.1
9,788	20.7	1,362	1.5
5,650	15.1	57,176	85.1
17,731	26.5	25,000	50.5
38,059	9.3		

a. Use a polynomial response model to develop a prediction equation for $Y =$ (Dry weight)$^{1/2}$ on $X = \ln(\text{volume} + 1)$. What degree polynomial do you need? Plot the residuals, including a normal plot, to see if there are any problems. Would it make sense in this case to force the origin to be zero? Will your fit to the data still be satisfactory if you do?

b. Suppose you want to keep Y on the original scale, $Y =$ Dry weight. Use the nonlinear model $Y = \alpha X^{\gamma} + \varepsilon$ to represent the relationship. (Divide "Volume" by 1,000 to make the numbers more manageable.) Fit the model, plot the residuals, and summarize the results. Define a reduced model that will test $H_0: \gamma = 1$. Is this reduced model *nonlinear*? Complete the test and state your conclusion.

*14.11. Use the data in exercise 14.6 to fit the two-term exponential model (equation 14.25) to the data from each environment separately. Use the derivative-free method and $\theta_1 = .2$ and $\theta_2 = .02$ as starting values. Do you get convergence with all six data sets? Plot the response curves and the data. Do the solutions appear reasonable?

14.12. The data given here are from a study of the colony-forming activity of six bacterial strains (only strain 3 reported here) under exposure to three pH levels (4.5, 6.5, 8.5) and three concentrations of chlorine dioxide in phosphate buffer (20, 50, 80 ppm). (Chlorine dioxide is important in sanitation for controlling bacterial growth.) After suspension of bacteria in the solutions, colony counts were taken on samples from the solutions at recorded time intervals. Use $Y = \ln(\text{count})$ in all analyses. (The data were collected by Vipa Hemstapat, North Carolina State University, and are used with permission.)

ClO_2	pH = 4.5		pH = 6.5		pH = 8.5	
(ppm)	*Time (min)*	*Colony Count*	*Time (min)*	*Colony Count*	*Time (min)*	*Colony Count*
80	0	2,700,000	0	3,100,000	0	2,400,000
80	5	2,300,000	6	1,700,000	5	2,100,000
80	10	610,000	11	180,000	10	730,000
80	15	140,000	15	13,000	15	130,000
80	20	142	20	1	20	186
50	0	7,500,000	0	2,900,000	0	720,000
50	10	2,800,000	10	2,600,000	10	220,000
50	20	670,000	20	1,300,000	20	8,000
50	30	89,000	30	400,000	30	260
50	40	20	40	94	40	1
50	50	2	50	1		
20	0	16,000,000	0	2,400,000	0	2,100,000
20	10	13,000,000	10	2,800,000	10	2,500,000
20	20	11,000,000	20	2,400,000	20	2,300,000
20	30	6,300,000	30	2,500,000	30	2,000,000
20	40	5,900,000	40	1,800,000	50	440,000
20	50	3,400,000	50	970,000	60	260,000
20	60	1,500,000	60	250,000	70	120,000
20	70	340,000	70	240,000	80	46
20	80	1	80	840	90	24
			90	12		

a. Characterize the response of the bacterial strain to ClO_2 for each of the nine pH × ClO_2 combinations by fitting the Weibull model using $Y = \ln(\text{count})$ as the dependent variable and time as the independent variable. You should get convergence in all cases with reasonable starting values; try $\alpha = 20$, $\sigma = 20$, and $\gamma = 2$. Summarize your results with a 3 × 3 table of the estimates of the parameters.

b. Verify algebraically that the time to 50% decline in the colony is given by

$$t_{50} = \hat{\sigma}(.693)^{(1/\hat{\gamma})}$$

Use your fitted Weibull response curves to find t_{50} in each case. Do an analysis of variance of the 3 × 3 table of times to 50% count. (You do not have an estimate of error with which to test the main effects of concentration and pH, but the analysis will show the major patterns.) Summarize the results.

c. The nonlinear function of interest in part (b) is t_{50}. Use the Wald procedure to find the approximate standard error and 95% confidence interval estimate of t_{50} for the middle cell of your 3 × 3 table. You will have to obtain the partial derivatives of t_{50} with respect to the three parameters and recover the variance–covariance matrix for $\hat{\theta}$, and then use these results in equation 14.57.

*14.13. Fit a polynomial model to Grimes' data, Table 14.3, where Y = Calcium (nmoles/mg) and X = time. (The description of the study is given in Example 14.2.) Is there reason to force β_0 to be zero in this case? Plot your polynomial response curve and the Weibull response curve given in the text, and superimpose the observed data. Compare the two curves. Does one appear to provide a "better fit" than the other? If so, in what ways?

CHAPTER
FIFTEEN

Case Study: Response Surface Modeling

The previous chapter discussed the use of polynomial and nonlinear response models.

This chapter uses polynomial models and the nonlinear Weibull model to characterize the seed yield response of soybeans to levels of ozone pollution in one experiment. Then data from four experiments on yield response to ozone are combined, the residuals are inspected, and the response variable is transformed as indicated by the analysis. The response models are fit to the transformed data.

The data used in this case study came from research on the effects of air pollutants on crop yields conducted by Dr. A. S. Heagle, Professor of Plant Pathology, North Carolina State University and USDA; all data are used with his permission. The pollutant of primary interest is ozone. Ozone has been shown to cause crop yield losses and the purpose of this research, as part of a nationwide program, was to quantify the effects of air pollutants on the agricultural industry. Of critical importance in the assessment is the possible interactive effects of ozone with other pollutants and environmental factors. The data from the 1981–1984 studies on soybeans, cultivar Davis, are used in this case study. The studies included effects of sulfur dioxide in 1981, different methods of dispensing ozone in 1982, and different levels of moisture stress in 1983 and 1984.*

Description of Experiments

The pollution studies are conducted in the field using open-top chambers to partially contain the pollutants so that higher than ambient levels of the pollutant gases can be maintained. The air flow through the open-top chamber is sufficient to avoid temperature buildup; plant growth within the chambers is normal. There are measurable chamber effects but they are relatively small. The pollutant levels are controlled by dispensing the gas for 7 hours daily, 10:00 A.M. to 5:00 P.M., into the airstream being forced through the chamber. The level of pollutant in the chamber is continuously monitored and dispensing is adjusted to meet the target value. Since the target value of pollutant is never precisely met, treatments with the same target level have slightly different levels of the gas in different replicates.

* Some of the analyses in this case study were done by V. M. Lesser, North Carolina State University.

The basic details of the four experiments are as follows:

1981. The purpose of the 1981 study was to investigate the bivariate response surface of two pollutant gases—ozone and sulfur dioxide. The experimental design was a randomized complete block design with two blocks and 24 treatments per block. The 24 treatments were all combinations of six levels of ozone and four levels of sulfur dioxide. The six levels of ozone were charcoal-filtered air (CF) which gives about .025 ppm ozone; nonfiltered air (NF), which gives the ambient level of ozone; and constant additions to ambient levels of ozone of .020, .030, .050, and .070 ppm. The constant addition treatments are labeled CA20, CA30, CA50, and CA70, respectively. The four levels of SO_2 were ambient air (NF) and constant additions of .030, .090, and .350 ppm, which are labeled S1, S2, and S3, respectively.

1982. The 1982 study had the purposes of developing more information on the ozone dose–response curve and of investigating possible effects of different methods of dispensing pollutant into the chambers. Prior to 1982, the target with ozone dispensing was to add a constant amount to the ambient levels at any given time. It was believed by some that a proportional increase in the gas at any given time would give more realistic distributions of the pollutant and that differences in distributions of the pollutant might affect plant response. Therefore, the treatments in this study included, in addition to CF and NF, both constant additions of .020, .040, and .060 ppm, and proportional increases of 30%, 60%, and 90% of ambient. The proportional treatments are labeled P13, P16, and P19, respectively. There were a total of eight treatments in a randomized complete block design with two blocks.

1983. The purpose of the 1983 study was to investigate the effects of moisture stress to the plants on their response to ozone. In addition, physiological data were taken on half the plants in each plot so that yield is reported for only one-half plot per chamber. There were two levels of moisture stress and four levels of ozone, CF, NF, CA30, and CA60, giving eight treatments. The experimental design was a randomized complete block design with three blocks.

1984. This was a continuation of the 1983 moisture stress study with, again, only half the plot being used for yield measurement. There were two levels of moisture and six levels of ozone, CF, NF, CA15, CA30, CA45, and CA60, giving 12 treatments in a randomized complete block design with two replications.

Two distinct analyses are presented in this case study. First, the 1981 data alone are analyzed. The bivariate response surface is fit using a polynomial response model and a nonlinear response model. Then, data from all 4 years are combined in an analysis of the residuals. The residuals analysis suggests a transformation of the data, and a nonlinear response model involving ozone, sulfur dioxide, and moisture level is fit to the transformed data.

15.1 The Ozone–Sulfur Dioxide Response Surface (1981)

The objective is to develop a bivariate response surface model to characterize the 1981 yield response of soybeans, cv Davis, to pollutant mixtures of ozone and sulfur dioxide. The yield data and the observed seasonal averages of ozone and sulfur dioxide for each experimental unit are given in Table 15.1. The north and south halves of the experimental plots are recorded separately as Y_1 and Y_2, respectively. This was done to investigate the possibility of an effect of position within the chamber on the response to the pollutant. Preliminary analyses indicated that, while there was a north–south position effect within the chambers, there was no position by treatment interaction effect. Therefore, all analyses reported in this section use the average of Y_1 and Y_2 for each experimental unit.

Analysis of Variance

The analysis of variance for the 1981 soybean data is given in Table 15.2 (page 418). The model for this analysis is

$$Y_{ijk} = \mu + \rho_i + \tau_j + \gamma_k + (\tau\gamma)_{jk} + \varepsilon_{ijk} \qquad [15.1]$$

Table 15.1 Yields of soybean following exposure to ozone (O_3) and sulfur dioxide (SO_2) for 7 hours daily during the growing season. Ozone and sulfur dioxide levels (ppm) are seasonal averages during the exposure period. (Data courtesy A. S. Heagle, plant pathologist, North Carolina State University and USDA; data used with permission.)

Treatment		*Block 1*				*Block 2*			
O_3	SO_2	O_3	SO_2	Y_1[a]	Y_2	O_3	SO_2	Y_1	Y_2
CF	NF	.025	.000	516.5	519.5	.025	.000	603.0	635.0
CF	S1	.023	.022	552.0	596.0	.022	.015	796.0	454.5
CF	S2	.028	.075	569.0	500.5	.018	.100	597.5	697.0
CF	S3	.029	.389	419.0	358.5	.025	.380	458.0	365.5
NF	NF	.059	.000	503.5	449.5	.051	.000	652.0	496.0
NF	S1	.058	.016	411.0	484.0	.052	.028	590.5	292.5
NF	S2	.058	.070	502.5	477.0	.055	.092	440.0	427.5
NF	S3	.058	.350	353.0	338.5	.051	.341	487.0	284.0
CA20	NF	.068	.000	449.5	480.5	.067	.000	533.5	321.5
CA20	S1	.073	.016	472.5	478.0	.066	.023	486.0	317.0
CA20	S2	.072	.085	382.5	411.5	.069	.104	420.5	456.0
CA20	S3	.068	.395	291.0	266.5	.068	.377	271.0	280.5
CA30	NF	.084	.000	399.0	414.5	.089	.000	390.5	324.5
CA30	S1	.086	.034	321.5	336.5	.087	.040	373.0	320.5
CA30	S2	.082	.067	373.0	384.5	.085	.091	321.0	246.0
CA30	S3	.090	.350	269.0	303.0	.083	.379	246.5	274.0
CA50	NF	.105	.000	438.0	345.0	.110	.000	307.0	281.5
CA50	S1	.111	.018	346.5	347.5	.107	.047	387.5	329.5
CA50	S2	.108	.084	297.0	316.5	.100	.098	270.0	246.0
CA50	S3	.106	.369	242.5	244.0	.100	.362	197.5	196.0
CA70	NF	.123	.000	342.5	331.5	.121	.000	275.0	278.5
CA70	S1	.131	.021	269.0	298.5	.125	.028	266.0	243.5
CA70	S2	.126	.056	297.5	308.5	.127	.099	303.0	215.5
CA70	S3	.123	.345	211.0	227.0	.122	.355	283.5	208.0

[a] Y_1 and Y_2 are the yields from the north and south halves of the plot, respectively.

Table 15.2 Analysis of variance of 1981 soybean yield following exposure to ozone and sulfur dioxide pollutants.

Source	d.f.	*Sum of Squares*	*Mean Square*	*F*	Prob > *F*
Total	47	606,481			
Block	1	467	467		
Ozone	5	408,117	81,623	46.01	.0001
Sulfur	3	126,697	42,232	23.81	.0001
Ozone × Sulfur	15	30,400	2,027	1.14	.3766
Error	23	40,799	1,774		

Table 15.3 Soybean mean yields (grams per meter row) for the 1981 ozone by sulfur dioxide study.[a]

SO_2 *Trt.*	*Ozone Treatment* CF	NF	CA20	CA30	CA50	CA70	*Mean*
NF	568.5	525.3	446.3	382.1	342.9	306.9	428.6
S1	599.6	444.5	438.4	337.9	352.8	269.3	407.1
S2	591.0	461.8	417.6	331.1	282.4	281.1	394.2
S3	400.3	365.6	277.3	273.1	220.0	232.4	294.8
Mean	539.8	449.3	394.9	331.1	299.5	272.4	

[a] $s(\bar{Y}_{.ij}) = 29.8$ is the standard error for the cell means. $s(\bar{Y}_{.i.}) = 14.9$ and $s(\bar{Y}_{..j}) = 12.2$ are the standard errors for the ozone and sulfur dioxide marginal means, respectively.

where ρ_i, τ_j, and γ_k are the block, ozone treatment, and sulfur dioxide treatment effects, respectively. All effects are assumed to be fixed; ε_{ijk} are assumed to be normally and independently distributed with zero mean and common variance σ^2.

The analysis of variance shows that there are highly significant ozone and sulfur dioxide effects on soybean seed yield but gives no indication that the two pollutants interact (Table 15.2). The treatment means, Table 15.3, show a 30% change in yield over the sulfur dioxide treatments and a 45% change over the ozone treatments. The standard errors of the treatment means are given in the footnote of Table 15.3. It is this joint response to the two pollutants that is to be characterized with an appropriate response model. For this purpose, the quantitative levels of the pollutants for each plot, rather than the treatment codes, are used.

Pollutant Levels

Using the quantitative levels of the pollutant in each plot introduces a problem that is somewhat unique to these studies. The specified treatments are target levels of the pollutant to be added to ambient air levels. Due to some imprecision in both the monitoring and the dispensing systems, the target levels are not precisely attained. These small discrepancies cause a slight imbalance in the study when the treatments are viewed in terms of the quantitative levels attained. (The effects of imbalance are discussed in Chapter 16. In general, imbalance in an experiment causes the analysis of variance to be inappropriate in that the sum of squares due to one factor will contain effects of other factors.)

In this particular case, the discrepancies in the pollutant levels are relatively minor (Table 15.1) and the analysis of variance can be viewed as a close approximation of the effects of the pollutants. Nevertheless, the ozone treatment sum of squares may contain some bias due to differences in sulfur dioxide levels and vice versa, the ozone by sulfur dioxide interaction sum of squares may contain main effects of the two pollutants, and experimental error may be biased upward by the effects of the pollutants. Thus, the analysis of variance will be used only as a guide to what to expect in the response surface modeling. The lack of fit of the polynomial model cannot be judged solely on how much of the treatment sums of squares are not explained, and experimental error from the analysis of variance will not be used as the unbiased estimate of σ^2.

15.1.1 Polynomial Response Model

Second-Degree Polynomial

The analysis of variance showed significant main effects for both ozone and sulfur dioxide but no indication of an interaction between the two gases (Table 15.2). Therefore, the first polynomial model tried was a second-degree polynomial in both pollutants but with no product, or interaction, term:

$$Y_{ijk} = \beta_0 + \rho D_i + \beta_1 X_{ijk1} + \beta_{11} X_{ijk1}^2 + \beta_2 X_{ijk2} + \beta_{22} X_{ijk2}^2 + \varepsilon_{ijk} \quad [15.2]$$

where D_i is a dummy variable coded $+1$ and -1 to identify the two blocks, ρ is the regression coefficient to account for the block effect, and X_{ijk1} and X_{ijk2} are the observed seasonal averages of ozone and sulfur dioxide, respectively, for the ijkth experimental unit. X for this model is of order 48×6 and consists of the column of ones for the intercept, a column for the dummy variable D_i, and the four columns of X_1, X_1^2, X_2, and X_2^2. The analysis for this model is summarized in Table 15.4. [The analysis was obtained using PROC GLM (SAS Institute, Inc., 1985f).]

Table 15.4 Analysis of variance for the second-degree polynomial model in both gases with no interaction.

Source	d.f.	*Sum of Squares*	*Mean Square*	F	Prob > F
Total	47	606,481			
Regression	5	543,713	108,743	72.76	.0001
Residual	42	62,768	1,494		

SS(Regr) partition:

Source	d.f.	*Type* I *SS*	F	Prob > F	*Type* III *SS*	F	Prob > F
Rep	1	467	.31	.5791	1,792	1.20	.2798
O_3 linear	1	397,665	266.09	.0001	54,922	36.75	.0001
SO_2 linear	1	135,161	90.44	.0001	2,613	1.75	.1933
O_3 quadratic	1	10,281	6.88	.0121	10,295	6.89	.0120
SO_2 quadratic	1	138	.09	.7630	138	.09	.7630

Nonorthogonality

The nonorthogonality of the data, due to the variable treatment levels, is evident. The "SO_2 linear" sum of squares exceeds the total treatment sum of squares for sulfur dioxide, and the residual mean square from the regression analysis is appreciably smaller than experimental error in the analysis of variance, 1,494 versus 1,774. Neither result is possible in the balanced case. Also, differences between Type I (sequential) and Type III (partial) sums of squares for "Rep" and "O_3 quadratic" show that the replication effects are not orthogonal to the realized levels of ozone and sulfur dioxide, and that ozone levels are not orthogonal to sulfur dioxide levels.

Summary

The key results from the analysis of the first polynomial model (Table 15.4) can be summarized as follows:

1. The quadratic term for sulfur dioxide makes no significant contribution and can be dropped from the model.
2. The quadratic term for ozone is significant in both the sequential and partial sums of squares and, consequently, will remain significant even after the "SO_2 quadratic" term is dropped.
3. The Type I sum of squares for "SO_2 linear" is highly significant, and even exceeds the total sulfur dioxide treatment sum of squares. While it is very likely that "SO_2 linear" will remain significant after "SO_2 quadratic" has been dropped, one cannot be certain that it will from this analysis (because the sequential sum of squares for "SO_2 linear" has not been adjusted for "O_3 quadratic"). The nonsignificant partial sum of squares for "SO_2 linear" should be ignored; remember that it has been adjusted for the higher-degree "SO_2 quadratic" term.
4. Block effects are nonsignificant but, since they were part of the basic experimental design, they will be retained in the model. Dropping the block effects, in this case, causes only trivial changes in the final model.

Modifying the Model

Comparison of the sums of squares for the polynomial model with the corresponding treatment sums of squares for ozone and sulfur (remember that in these data they are not precisely comparable) suggests that there is nothing to be gained by expanding the polynomial model to include cubic terms in either variable. On the other hand, there may be some improvement in the model from a second-degree product term, the "O_3 linear × SO_2 linear" interaction term. Even though the interaction sum of squares in the analysis of variance was not significant, it is possible for a single-degree-of-freedom contrast to be significant. Hence, the second polynomial model to be fitted dropped the quadratic term for sulfur dioxide and added the linear-by-linear product term:

$$Y_{ijk} = \beta_0 + \rho D_i + \beta_1 X_{ijk1} + \beta_2 X_{ijk2} + \beta_{11} X_{ijk1}^2 + \beta_{12} X_{ijk1} X_{ijk2} + \varepsilon_{ijk} \qquad [15.3]$$

The analysis of this model is summarized in Table 15.5.

All terms in this model are significant and will be retained. There remains

Table 15.5 Analysis of variance for the polynomial model allowing a quadratic response for ozone, linear response for sulfur dioxide, and a linear-by-linear interaction.

Source	d.f.	*Sum of Squares*	*Mean Square*	*F*	Prob > *F*
Total	47	606,481			
Regression	5	550,004	110,001	81.80	.0001
Residual	42	56,477	1,345		

SS(Regr) partition:

Source	d.f.	*Type* I *SS*	*F*	Prob > *F*	*Type* III *SS*	*F*	Prob > *F*
Rep	1	467	.35	.5587	1,709	1.27	.2659
O_3 linear	1	397,665	295.73	.0001	60,087	44.69	.0001
SO_2 linear	1	135,161	100.52	.0001	46,385	34.50	.0001
O_3 quadratic	1	10,281	7.65	.0084	10,756	8.00	.0071
Linear × linear	1	6,429	4.79	.0344	6,429	4.78	.0344

the possibility that a higher-order product term would contribute significantly to the model. The most logical possibility is the "O_3 quadratic" × "SO_2 linear" interaction term, $X_1^2X_2$, since there is significant quadratic response to ozone and the analysis of variance interaction sum of squares is the largest partition not explained by the present model. It is left as an exercise for the student to show whether this term is needed. While a plot of the data superimposed on the response surface showed considerable dispersion about the surface, there was no apparent pattern suggesting inadequacies in this model. Likewise, the plot of the residuals versus $\hat{Y}$ and the normal plot appeared reasonable.

Final Response Surface

This polynomial model, equation 15.3, is adopted as a reasonable characterization of the ozone–sulfur dioxide response surface in these data. The final response surface equation is

$$\underset{(28)}{\hat{Y} = 724} - \underset{(771)}{5{,}152X_1} + \underset{(4{,}930)}{13{,}944X_1^2} - \underset{(92)}{543X_2} + \underset{(1{,}126)}{2{,}463X_1X_2} \qquad [15.4]$$

The standard errors of the regression coefficients are shown in parentheses. The response surface is shown in Figure 15.1 (page 422) as a series of three response curves for ozone at three levels of SO_2.

Understanding the Response

The response surface has a negative slope with respect to both ozone and sulfur dioxide at near-zero pollution. Thus, there is evidence that increasing levels of either pollutant cause yield of Davis soybean to decline in this environment. The positive sign of the quadratic regression coefficient, $\hat{\beta}_{11}$, indicates that the rate of decline in yield is decreasing with increasing ozone and the polynomial response curve will eventually reach a minimum with yield appearing to increase for levels of ozone beyond that point. The minimum point on the ozone response curve for a given level of sulfur dioxide is obtained by setting the partial derivative of Y with respect to X_1 equal to zero and solving for X_1. The partial derivative is

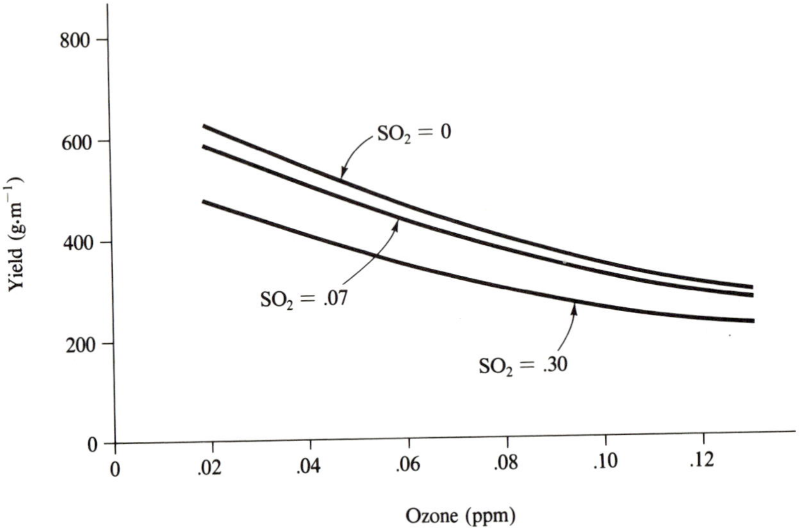

Figure 15.1 The bivariate polynomial response surface for yield of soybeans exposed to chronic doses of ozone and sulfur dioxide. The surface is represented by three traces from the surface for different levels of SO_2.

$$\frac{\partial Y}{\partial X_1} = -5{,}152 + 2(13{,}944)X_1 + 2{,}463X_2 \qquad [15.5]$$

Setting this equation equal to zero and solving for $X_{1\min}$ gives

$$X_{1\min} = \frac{5{,}152 - 2{,}463X_2}{2(13{,}944)}$$

$X_{1\min}$ ranges from .1847 for $X_2 = 0$ to .1582 for $X_2 = .3$. These levels of ozone are beyond the limits of the experiment since the average ozone level for CA70 was .125, and, consequently, any inference that sufficiently high levels of ozone would cause yield to increase would be an inappropriate extrapolation.

The interaction term has the effect of decreasing the rate of decline in yield as the level of the other pollutant increases. The impact of SO_2 at the highest level of O_3 is approximately half, in absolute terms, what it is at the low level of O_3. This diminished effect of one pollutant at higher levels of the other is reasonable because there is less yield to be lost at the higher levels.

Extrapolations

Within the limits of the levels of pollutant in this experiment, the polynomial model provides a reasonable characterization of the response surface. Any extrapolation beyond the limits of the experiment encounters biologically inconsistent predictions: minimum yield in the vicinity of .16 ppm ozone with predictions of increasing yields at higher levels, and predictions of negative yields when SO_2 is sufficiently high, approximately 1.3 ppm.

15.1.2 Nonlinear Weibull Response Model

A nonlinear response model based on the functional form of the Weibull probability distribution has been used as a dose–response model in the ozone pollution research simply because it has a biologically realistic form with sufficient flexibility to cover the range of responses encountered for the various crop species and environmental conditions. A single, flexible form facilitates comparing responses and summarizing the results with a minimum number of response equations.

Form of the Model

The Weibull model in its simplest form was given in equation 14.37. For this experiment, the α term in that model must be extended to account for additional effects—block effects and the effect of sulfur dioxide. Thus, the Weibull model takes the form

$$Y_{ijk} = (\alpha_1 + \alpha_2 D_i + \beta X_{ijk2})e^{-(X_{ijk1}/\sigma)^\gamma} + \varepsilon_{ijk} \qquad [15.6]$$

where the exponential term controls the relative response to ozone, decreasing from 1 at $X_1 = 0$ to a limit of zero when X_1 is large. If $\gamma = 1$, this becomes the exponential decay curve. The three terms in parentheses in front of the exponential term control the yield level under the hypothetical situation of $X_1 = 0$, which is expressed here as an overall constant α_1, a block effect α_2, and a linear adjustment for the level of sulfur dioxide, βX_2. The dummy variable D is defined as 1 if the observation is from block 1 and -1 if the observation is from block 2. Thus, setting $D = 0$ gives an average result for the two blocks so that α_1 is the expected yield for this environment with $X_1 = X_2 = 0$. (On the basis of the polynomial results, SO_2 is handled with a linear response in this model.)

Fitting the Model

The derivative-free method of PROC NLIN in SAS (SAS Institute, Inc., 1985f) was used to fit this model. The program statements that generated the analysis are as follows:

```
PROC NLIN METHOD=DUD;
  PARMS A1=700 A2=0 B=-0.5 SIGMA=0.14 GAMMA=1;
  MODEL PODWT
    =(A1 + A2*D + B*X2)*EXP(-(X1/SIGMA)**GAMMA);
  OUTPUT OUT=OUT.R5 P=PWHAT R=PWRESID;
```

(A1, A2, B, SIGMA, and GAMMA are used in place of α_1, α_2, β, σ, and γ, respectively, because the programming language will not accommodate Greek letters.) The starting values for the parameters are given in the PARMS statement. These values were chosen on the basis of a preliminary plot of the data. The highest yields for the low ozone treatment were in the vicinity of 700; thus, $\alpha_1^0 = 700$. The "rep" effects were small, suggesting $\alpha_2^0 = 0$. The starting value for β, $\beta^0 = -.5$, resulted from a visual assessment of the change in yield per unit change in SO_2 but contained an error in placement of the decimal. The value should have been $\beta^0 = -500$. The parameter σ is interpreted as the dose at which yield has been reduced to the fraction e^{-1} of what it is at zero ozone. The starting value was read off a plot of the data as $\sigma^0 = .14$. Finally, $\gamma^0 = 1$

Table 15.6 Nonlinear regression results from fitting the Weibull model to the 1981 yield data of soybeans following exposure to ozone and sulfur dioxide.

Source	d.f.	*Sum of Squares*	*Mean Square*
Model	5	7,521,067	1,504,213
Residual	43	59,049	1,373
Uncorrected total	48	7,580,116	
(Corrected total)	47	606,481	

Parameter	*Estimate*	*Asymptotic Std. Error*	*Asymptotic 95% Confidence Interval* Lower	Upper
α_1	759.4479	88.2776	581.4198	937.4761
α_2	3.6723	9.4117	−15.3082	22.6529
β	−631.2867	93.9163	−820.6862	−441.8871
σ	.1336	.0145	.1044	.1629
γ	.8788	.2248	.4255	1.3320

Table 15.7 Estimated responses for the nonlinear model and the polynomial model for the 1981 soybean yield response to ozone and sulfur dioxide.

Ozone (ppm)	$SO_2 = 0$ *ppm* Nonlinear	$SO_2 = 0$ *ppm* Polynomial	$SO_2 = .30$ *ppm* Nonlinear	$SO_2 = .30$ *ppm* Polynomial
.02	629.0	626.2	472.2	478.1
.04	537.1	539.8	403.1	406.5
.06	463.1	464.7	347.6	346.2
.08	401.6	400.7	301.5	296.9
.10	349.9	347.8	262.6	258.8
.12	305.8	306.1	229.5	231.9

was chosen because the plot appeared to be similar in shape to an exponential decay curve.

Solution

In spite of a very poor starting value for β, convergence was quickly attained. The summary of this analysis is given in Table 15.6. The residual sum of squares is SS(Res) = 59,049 with 43 degrees of freedom, compared to SS(Res) = 56,477 with 42 degrees of freedom for the final polynomial model. The corresponding mean squares are 1,373 and 1,345. Thus, the nonlinear model with five parameters fits the data nearly as well as the polynomial model with six parameters. [*Note:* The difference in the residual sums of squares for the two models cannot be tested as previously done because neither model is "nested" in the other.] The resulting nonlinear response equation is

$$\hat{Y} = (759.4 + 3.7D - 631X_2)e^{-(X_1/.134)^{.88}} \qquad [15.7]$$

Checking the Equation

The plot of this response equation (not given here) is almost indistinguishable, within the limits of the design space, from the plot for the polynomial response model given in Figure 15.1. Estimated responses for the two equations are compared in Table 15.7. The nonlinear equation has slightly less curvature except

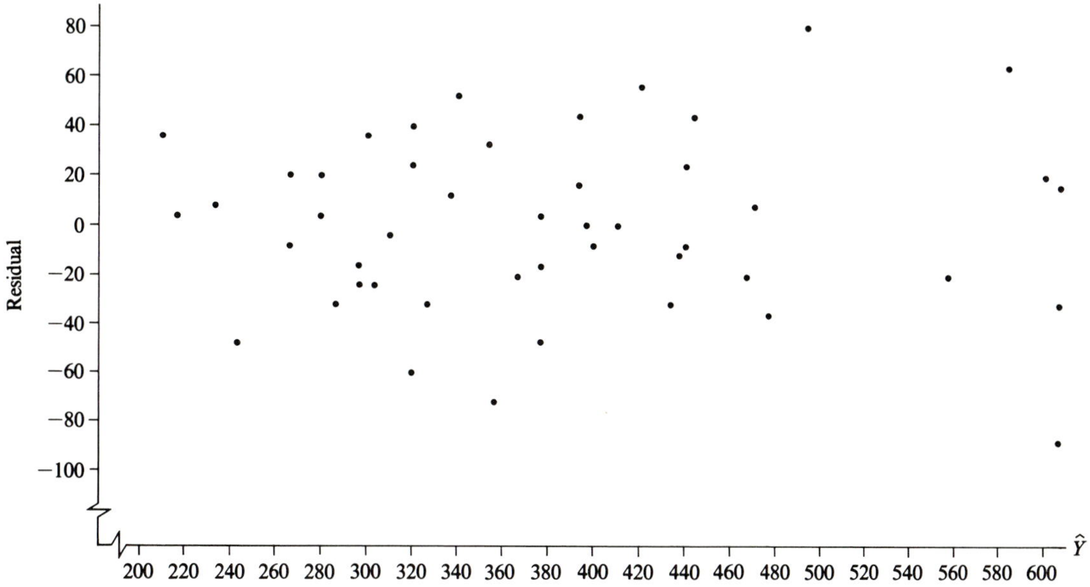

Figure 15.2 The residuals from the nonlinear model for the 1981 soybean response to ozone and sulfur dioxide plotted against the estimated yield.

at the low levels of ozone when sulfur dioxide is near zero. The plot of the residuals against $\hat{Y}$ (Figure 15.2) and the normal plot of the residuals (Figure 15.3, page 426) give no reason for concern about the adequacy of the model. (These plots are very similar to the corresponding plots for the polynomial model. For that reason, the plots are given only for the nonlinear model.)

***Setting* $\gamma = 1$**

The standard error on $\hat{\gamma}$ and the confidence interval estimate of γ (Table 15.6) suggest that the exponential decay model for ozone effects, $\gamma = 1$, would be adequate. The next step in the model building process would be to fit the model with $\gamma = 1$. The nonlinear model would be reduced to four parameters which, it appears, would provide nearly the same fit as the polynomial model with six parameters. This step of the model building is left as an exercise for the student, and the current five-parameter nonlinear response equation will be used for interpretation.

$\hat{\rho}$ *and* $s^2(\hat{\theta})$

The correlation matrix for the estimates of the parameters, $\hat{\theta}' = (\hat{\alpha}_1 \ \hat{\alpha}_2 \ \hat{\beta} \ \hat{\sigma} \ \hat{\gamma})$, is

$$\hat{\rho} = \begin{bmatrix} 1 & .121322 & -.790652 & -.914682 & -.960015 \\ .121322 & 1 & -.072519 & -.128904 & -.108028 \\ -.790652 & -.072519 & 1 & .683393 & .716938 \\ -.914682 & -.128904 & .683393 & 1 & .799745 \\ -.960015 & -.108028 & .716938 & .799745 & 1 \end{bmatrix}$$

The variance–covariance matrix for the estimates of the parameters, reconstructed

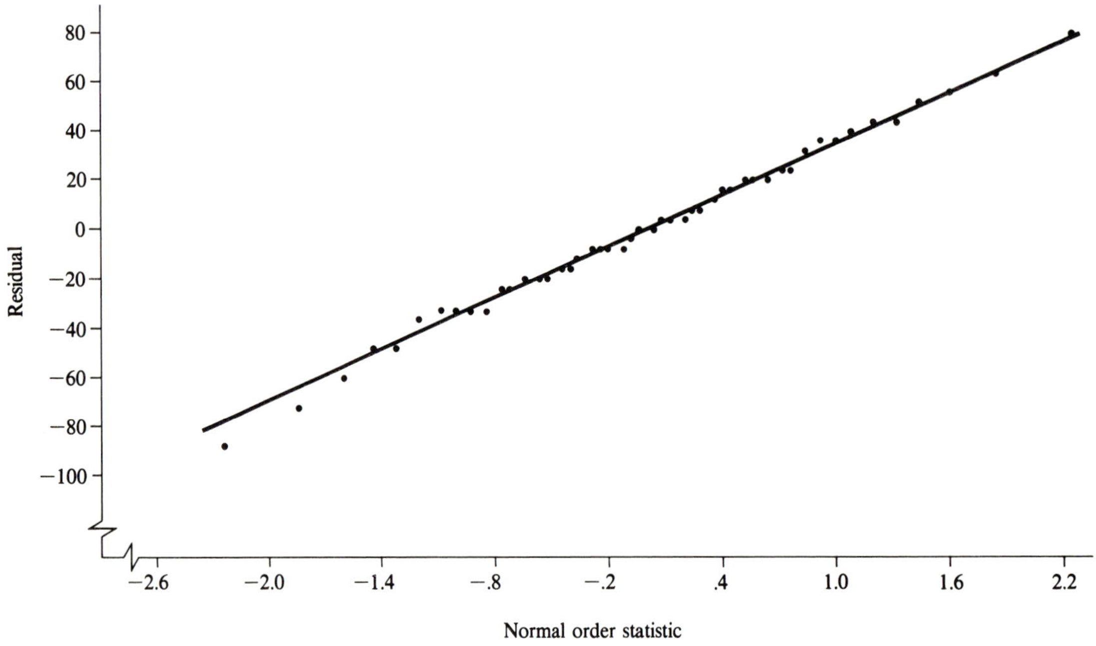

Figure 15.3 The normal plot of the residuals from the nonlinear model for the 1981 soybean response to ozone and sulfur dioxide.

from the correlation matrix, is

$$s^2(\hat{\theta}) = \begin{bmatrix} 7{,}792.938 & 100.8000 & -6{,}555.065 & -1.170042 & -19.04732 \\ 100.8000 & 88.58096 & -64.10073 & -.017580 & -.228514 \\ -6{,}555.065 & -64.10073 & 8{,}820.273 & .930020 & 15.13310 \\ -1.170042 & -.017580 & .930020 & .000210 & .002605 \\ -19.04732 & -.228514 & 15.13310 & .002605 & .050514 \end{bmatrix}$$

The variance–covariance matrix is needed to compute standard errors of any quantities computed from the regression results.

Estimated Yields and Yield Losses

The quantities of particular interest are the estimated yields at specific levels of ozone and sulfur dioxide, and the relative yield losses for given changes in the level of ozone or sulfur dioxide pollution. The use of the regression equation and the determination of variances of the estimated quantities will be illustrated for

1. the estimated yield level for $X_1 = .05$ ppm and $X_2 = .10$ ppm, and
2. the relative yield losses expected from a *change* in the ozone level from $X_{1r} = .025$ ppm to $X_{1o} = .06$ ppm and from $X_{1r} = .025$ ppm to $X_{1o} = .08$ ppm. (X_{1r} and X_{1o} designate the reference level and the postulated new level of ozone, respectively.)

The estimated yield level for $X_1 = .05$ and $X_2 = .10$ is obtained by substitution of these values in the regression equation, along with $D = 0$, to give the average for the two replications. This gives $\hat{Y} = 456.83$ gm^{-1}. The variance is approximated by applying equation 14.57. This requires the partial derivatives of the nonlinear function with respect to each parameter, which for $\hat{Y}$ (with $D = 0$) are

$$\frac{\partial \hat{Y}}{\partial \alpha_1} = E$$

$$\frac{\partial \hat{Y}}{\partial \alpha_2} = 0$$

$$\frac{\partial \hat{Y}}{\partial \beta} = X_2 E \qquad [15.8]$$

$$\frac{\partial \hat{Y}}{\partial \sigma} = (\alpha_1 + \beta X_2) E \left(\frac{\gamma}{\sigma}\right)\left(\frac{X_1}{\sigma}\right)^{\gamma}$$

$$\frac{\partial \hat{Y}}{\partial \gamma} = -(\alpha_1 + \beta X_2) E \left(\frac{X_1}{\sigma}\right)^{\gamma}\left[\ln\left(\frac{X_1}{\sigma}\right)\right]$$

where

$$E = \exp\left[-\left(\frac{X_1}{\sigma}\right)^{\gamma}\right]$$

Evaluating the partial derivatives by substituting the estimates of the parameters, $X_1 = .05$, and $X_2 = .10$, and arranging them in a row vector, gives

$$\boldsymbol{H} = (.65606 \;\; 0 \;\; .065606 \;\; 1{,}266.172 \;\; 189.3025)$$

Thus, the variance of $\hat{Y}$ is approximated by

$$s^2(\hat{Y}) = \boldsymbol{H}[s^2(\hat{\theta})]\boldsymbol{H}'$$
$$= 78.6769$$

or the estimated standard error is $s(\hat{Y}) = 8.87$.

The estimated relative yield loss (RYL) resulting from a change in ozone pollution from X_{1r} to X_{1o} is

$$RYL(X_{1r}, X_{1o}) = \frac{\hat{Y}(X_{1r}) - \hat{Y}(X_{1o})}{\hat{Y}(X_{1r})}$$

$$= 1 - \frac{\exp[-(X_{1o}/\hat{\sigma})^{\hat{\gamma}}]}{\exp[-(X_{1r}/\hat{\sigma})^{\hat{\gamma}}]}$$

$$= 1 - \exp(-DIF) \qquad [15.9]$$

where

$$DIF = \left(\frac{X_{1o}}{\hat{\sigma}}\right)^{\hat{\gamma}} - \left(\frac{X_{1r}}{\hat{\sigma}}\right)^{\hat{\gamma}}$$

For $(X_{1r}, X_{1o}) = (.025, .06)$, $RYL = .233$. Or, there is estimated to be a 23% loss in yield associated with an increase in ozone level from .025 ppm to .06 ppm. For $(X_{1r}, X_{1o}) = (.025, .08)$, $RYL = .335$ or a 34% loss.

Variances of Relative Yield Losses

The partial derivatives of RYL are needed to obtain approximate variances of the estimated relative yield losses. The partial derivatives with respect to α_1, α_2, and β are zero because the function does not involve these parameters. The partial derivatives with respect to σ and γ are

$$\frac{\partial RYL}{\partial \sigma} = -\left(\frac{\gamma}{\sigma}\right)(DIF)\exp(-DIF)$$

$$\frac{\partial RYL}{\partial \gamma} = \exp(-DIF)\left\{\left(\frac{X_{1o}}{\sigma}\right)^{\gamma}\left[\ln\left(\frac{X_{1o}}{\sigma}\right)\right] - \left(\frac{X_{1r}}{\sigma}\right)^{\gamma}\left[\ln\left(\frac{X_{1r}}{\sigma}\right)\right]\right\} \qquad [15.10]$$

where DIF is as defined following equation 15.9. Evaluating the derivatives at $\hat{\theta}$ with $X_{1r} = .025$ and $X_{1o} = .06$ gives

$$\boldsymbol{\hat{H}} = (0 \ 0 \ 0 \ -1.338825 \ -.0091626)$$

and

$$s^2(RYL) = \boldsymbol{\hat{H}}[s^2(\hat{\theta})]\boldsymbol{\hat{H}}'$$
$$= .0004445$$

or an estimated standard error of

$$s(RYL) = .0211$$

For estimated relative yield loss for the $(X_{1r}, X_{1o}) = (.025, .08)$ interval,

$$\boldsymbol{\hat{H}} = (0 \ 0 \ 0 \ -1.783608 \ .0381499)$$

and $s(RYL) = .0197$. These estimated relative yield losses are summarized in the following table:

X_{1r}	X_{1o}	RYL	$s(RYL)$	95% *Confidence Interval*
.025	.06	.233	.0211	(.191, .276)
.025	.08	.335	.0197	(.295, .375)

15.2 Analysis of the Combined Soybean Data

The purpose of this analysis is to use the combined information from the experiments over 4 years, 1981 to 1984, to produce a response equation characterizing the response of Davis soybeans to ozone pollution, sulfur dioxide pollution, and moisture stress. First, the combined data are used to check the validity of the assumptions of normality and constant variance. The 1981 data were given in Table 15.1. The 1982, 1983, and 1984 data are given in Table 15.8 (page 430).

Checking Normality and Constancy of Variance

The individual yearly experiments do not provide sufficient information to critically check normality and constancy of variance. Therefore, data from all experiments were combined to check these assumptions. In 1983 and 1984, half of each chambered plot was used destructively for physiological measurements and, consequently, yield was measured on only the remaining half. In order to keep plot sizes comparable over years, all analyses used the "half-plot" yield as the basic unit. Thus, the north (N) and south (S) halves of each plot in 1981 and 1982 were used as different data sets. (The correlations between the subsets of data in the two experiments were ignored for this analysis of residuals.) The appropriate analysis of variance was run on each data set and the residuals from all analyses were combined to study their behavior. The combined data set has a total of 174 observations and the pooled residuals have 80 degrees of freedom. A missing observation in each of 1983 and 1984 made the data unbalanced from the analysis of variance point of view. (The analysis of unbalanced data is discussed in Chapter 16.) For present purposes, the effects and dummy variables are defined so as to give a full-rank model and regression analyses are used.

The plot of the residuals from the analyses of variance versus $\hat{Y}$, Figure 15.4 (page 431), showed a tendency for increased dispersion at the higher values of $\hat{Y}$. The normal plot of the residuals, Figure 15.5, showed a very slight S-shaped curvature. On the basis of these graphical results, the Box–Cox method was used to find a transformation on Y that would improve normality and constancy of variance.

Logarithmic Transformation Used

The criterion used for choice of power transformation was minimum pooled residual sum of squares from the analyses of variance for the 4 years of data. The pooled residual sums of squares for several choices of λ in the Box–Cox transformation are given in Table 15.9 (page 432). Quadratic interpolation using the three middle points indicated that the minimum was near $\lambda = -.05$ with $SS[Res(\lambda)] = 198{,}471$. The plot of these residual sums of squares and the confidence interval estimate of λ, presented in Chapter 11, Figure 11.4, suggested a logarithmic transformation. The analyses of variance were repeated using $\ln(Y)$ as the dependent variable. The pooled residuals obtained from the analyses on $\ln(Y)$ showed better behavior with respect to both constancy of variance (Figure 15.6) and normality (Figure 15.7, page 433). Consequently, the response model for the combined data is developed using $\ln(Y)$ as the dependent variable.

Full Model

A complete model for the combined 1981–1984 soybean experiments needs to account for differences among years, differences among blocks in years, the joint ozone and sulfur dioxide response in 1981, the joint ozone and method of dispensing effects in 1982, the joint ozone and moisture stress effects in 1983

Table 15.8 Soybean yield data, cultivar Davis, from the 1982, 1983, and 1984 studies on the effects of ozone, dispensing method, and moisture stress. (Data courtesy of Dr. A. S. Heagle, plant pathologist, North Carolina State University and USDA; used with permission.)

1982:

	Block 1			Block 2		
Treatment	*Ozone*	Y_1	Y_2	*Ozone*	Y_1	Y_2
CA20	.0674	487.80	476.40	.0637	511.15	423.00
CA40	.0866	499.95	377.20	.0863	479.50	382.45
CA60	.1135	398.95	283.00	.1051	344.25	266.40
CF	.0149	653.30	583.40	.0222	652.70	600.70
NF	.0406	671.75	525.30	.0483	724.70	627.45
P13	.0635	599.65	412.15	.0672	620.85	513.55
P16	.0798	395.40	378.40	.0817	518.20	438.35
P19	.0933	354.55	288.85	.0902	419.25	325.50

1983:

		Block 1		Block 2		Block 3	
Moisture Stress	*Ozone Trt.*	*Ozone*	*Y*	*Ozone*	*Y*	*Ozone*	*Y*
W	CA30	.0755	477.9	.0773	512.6	.0756	487.2
W	CA60	.0975	395.7	.1010	415.6	.1025	498.0
W	CF	.0299	535.9	.0277	642.0	.0255	639.5
W	NF	.0526	565.4	.0517	493.4	.0488	706.4
D	CA30	.0779	344.0	.0758	225.6	.0753	238.3
D	CA60	.0980	248.4	.1004	237.1	.0947	299.0
D	CF	—	—	.0314	448.8	.0293	282.5
D	NF	.0523	271.9	.0533	211.2	.0520	255.3

1984:

		Block 1		Block 2	
Moisture Stress	*Ozone Trt.*	*Ozone*	*Y*	*Ozone*	*Y*
W	CF	.024	344	.024	416
W	NF	.043	438	.045	428
W	CA15	.065	268	.069	283
W	CA30	.082	293	.082	344
W	CA45	.087	297	.095	231
W	CA60	.104	249	.112	214
D	CF	—	—	.027	297
D	NF	.043	279	.047	330
D	CA15	.066	254	.064	363
D	CA30	.077	202	.081	213
D	CA45	.095	215	.093	229
D	CA60	.107	138	.105	216

and 1984, and possible ozone by year, ozone by dispensing method, ozone by moisture, and ozone by sulfur dioxide interaction effects. However, previous analyses had shown the main and interaction effects due to ozone dispensing methods not to be significant and, consequently, these effects are not included. The year, block, and moisture stress effects are incorporated in the model with

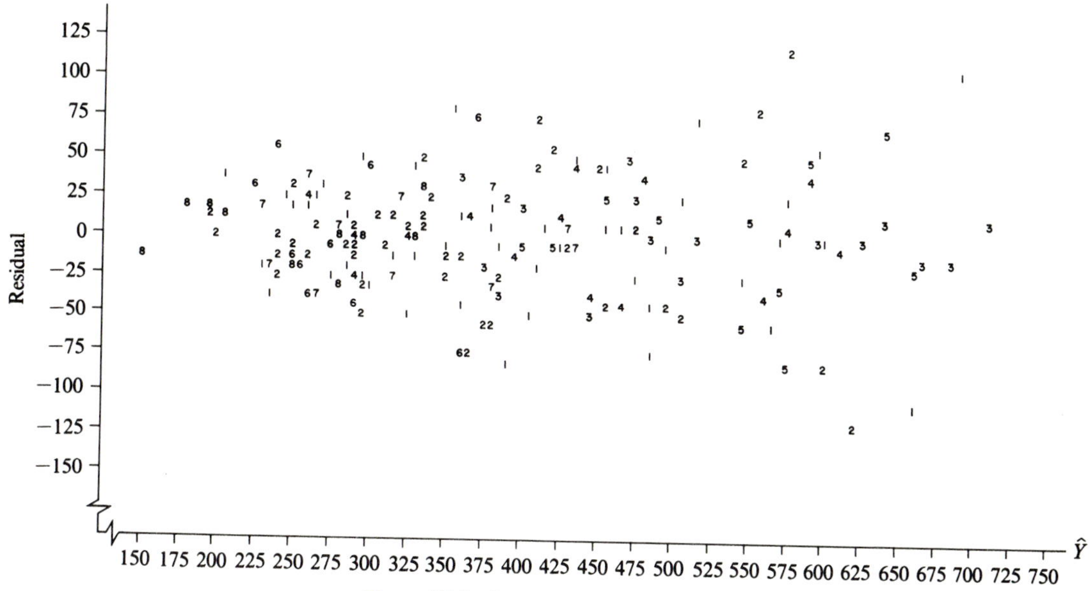

Figure 15.4 Pooled residuals from the separate analyses of variance of yield for the 1981–1984 soybean studies plotted against $\hat{Y}$.

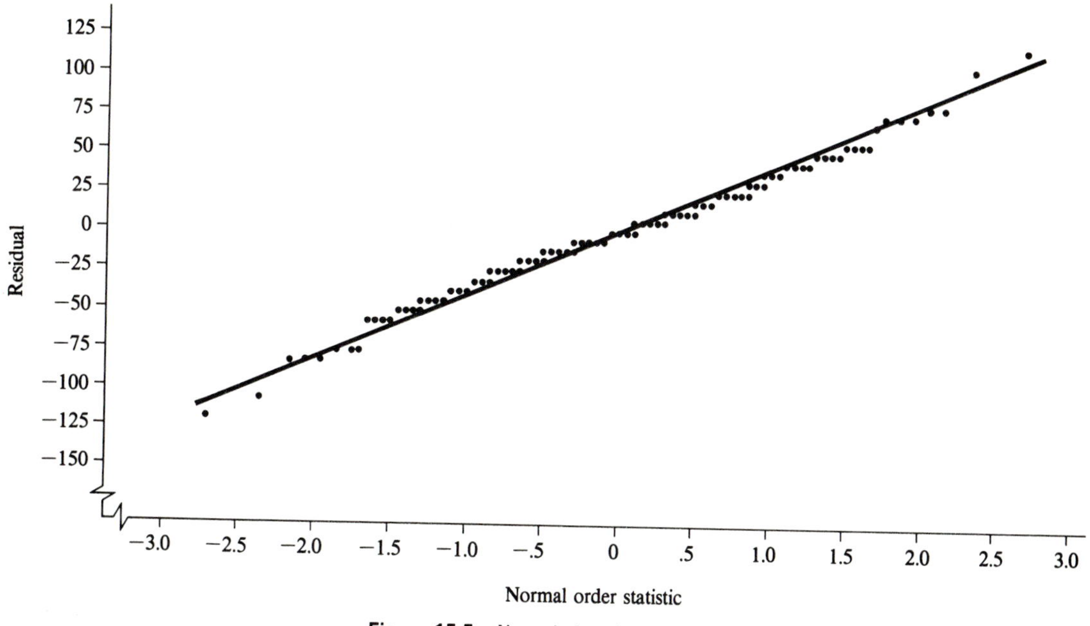

Figure 15.5 Normal plot of the pooled residuals from the analyses of variance of yield for the 1981–1984 soybean studies.

Table 15.9 Pooled residual sums of squares for several choices of λ for the Box–Cox transformation on the 1981–1984 soybean experiments.

λ	*Pooled* SS
−1	243,433
−.5	207,201
0 = ln(*Y*)	198,633
.5	212,027
1	249,122

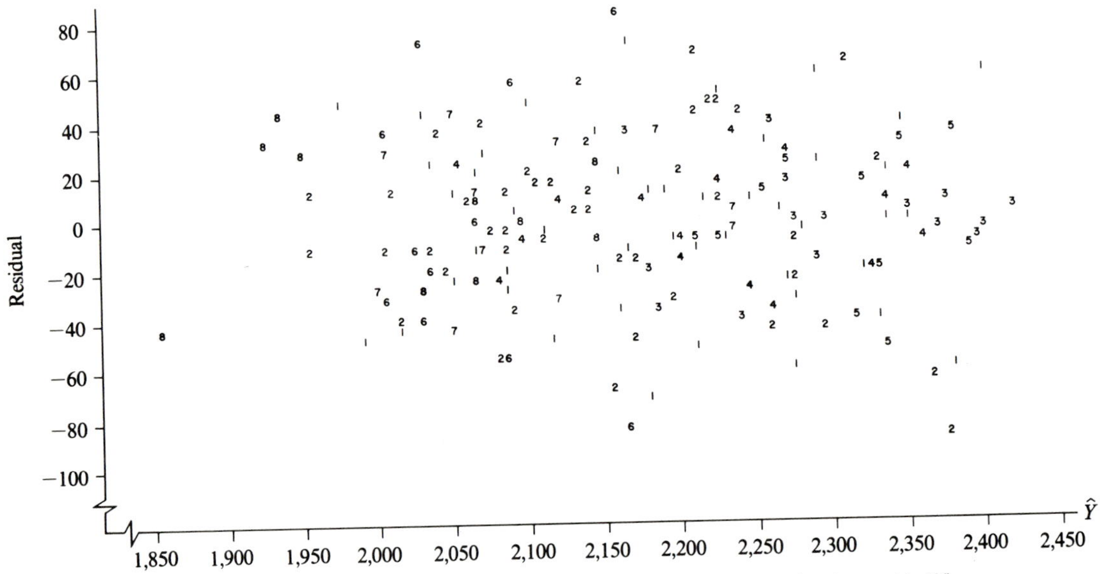

Figure 15.6 Pooled residuals from the analyses of variance of ln(*Y*) for the 1981–1984 soybean studies plotted against $\hat{Y}$.

the use of dummy variables. A plot of the data suggested that a linear regression term would adequately account for the average sulfur dioxide effects. The logarithm of the exponential component in the original Weibull model gives $-(X_1/\sigma)^\gamma$, suggesting that the ozone response on the logarithmic scale can be characterized by a nonlinear term $\beta(X_1)^\gamma$, where $\beta = -(1/\sigma)^\gamma$. Thus, a power parameter γ on the level of ozone is included in the full model. The interaction effects are incorporated as product terms in the usual way.

Let T_1, T_2, T_3, and T_4 be dummy variables identifying the years, by taking the value 1 if the observation is from the year indicated by the subscript and 0 otherwise. Let R_{11}, R_{21}, R_{31}, R_{32}, and R_{41} be dummy variables to account for block differences within each year. Each R_{ij} takes the value 1 if the observation is from the jth block in the ith year and 0 otherwise. Notice that there is one less R_{ij} dummy variable for each year than the number of blocks in that year. The

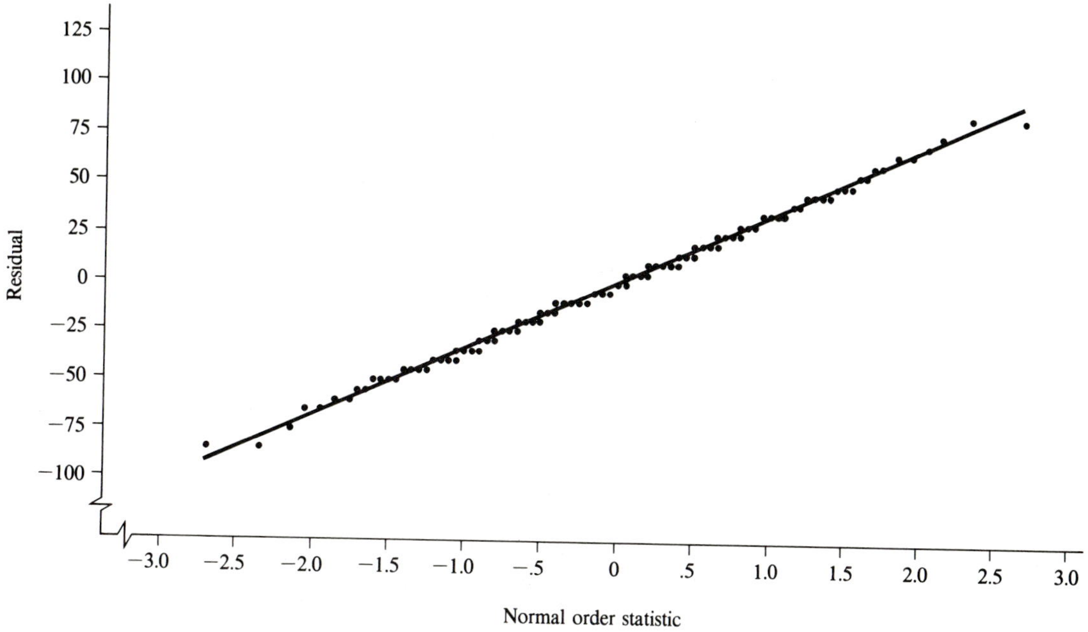

Figure 15.7 Normal plot of the pooled residuals from the analyses of variance of ln(Y) for the 1981–1984 soybean studies.

moisture-stressed plots are identified by $M = 1$, the well-watered plots with $M = 0$. Let MI be a dummy variable to allow for a moisture stress by year interaction between 1983 and 1984, taking the value of 1 if the plot is a moisture-stressed plot in 1983, −1 if it is a moisture-stressed plot in 1984, and 0 otherwise. Thus, the full model, without subscripts to identify the experimental unit, is

$$\begin{aligned} \ln(Y) = {} & \beta_1 T_1 + \beta_2 T_2 + \beta_3 T_3 + \beta_4 T_4 \\ & + \beta_5 R_{11} + \beta_6 R_{21} + \beta_7 R_{31} + \beta_8 R_{32} + \beta_9 R_{41} \\ & + \beta_{10} M + \beta_{11} MI + \beta_{12} X_2 + \beta_{13} X_1^{\gamma} \\ & + \beta_{14} X_2 X_1^{\gamma} + \beta_{15} M X_1^{\gamma} + \beta_{16} T_1 X_1^{\gamma} \\ & + \beta_{17} T_2 X_1^{\gamma} + \beta_{18} T_3 X_1^{\gamma} + \varepsilon \end{aligned} \qquad [15.11]$$

where X_2 is the level of sulfur dioxide and X_1 is the level of ozone. The product term MX_1^{γ} allows the moisture-stressed plots to have a different response to ozone, and the last three terms allow for year by ozone interactions. This is a nonlinear model only because of the power parameter on X_1.

Fitting the Model

This model was fitted using the derivative-free option in PROC NLIN (SAS Institute, Inc., 1985f). The starting values for the parameters were $\beta_1 = \beta_2 = \beta_3 = \beta_4 = 6.5, \beta_{12} = -1, \beta_{13} = -5, \gamma = 1$, and all others zero. Although convergence was obtained, the derivative-free method appeared to be inefficient.

Table 15.10 Summary of the nonlinear least squares analysis of ln(seed yield) for the 1981–1984 soybean data using the full model.

Source	d.f.	*Sum of Squares*	*Mean Square*
Model	19	6,089.9388	320.5231
Residual	155	3.6824	.0238
Uncorrected total	174	6,093.6212	
(Corrected total)	173	20.0321	

Parameter		*Estimate*	*Asymptotic Std. Error*	*Asymptotic 95% Confidence Interval*
β_1	Years	6.4828[a]	.0948	(6.2956, 6.6700)
β_2	Years	6.6811[a]	.1092	(6.4654, 6.8968)
β_3	Years	6.5359[a]	.1129	(6.3129, 6.7589)
β_4	Years	6.2472[a]	.1278	(5.9948, 6.4997)
β_5	Blocks/years	.0604	.0315	(−.0018, .1227)
β_6	Blocks/years	−.0614	.0546	(−.1693, .0465)
β_7	Blocks/years	−.0346	.0805	(−.1936, .1245)
β_8	Blocks/years	−.0552	.0771	(−.2075, .0971)
β_9	Blocks/years	−.1005	.0647	(−.2284, .0273)
β_{10}:	M	−.4712[a]	.1206	(−.7094, −.2330)
β_{11}:	M × Yr	−.2059[a]	.0460	(−.2967, −.1151)
β_{12}:	SO_2	−1.0194[a]	.2563	(−1.5257, −.5130)
β_{13}:	O_3	−9.3216[a]	4.4734	(−18.1584, −.4848)
β_{14}:	$SO_2 \times O_3$	.3332	4.0286	(−7.6249, 8.2913)
β_{15}:	M × O_3	.7626	2.1813	(−3.5464, 5.0716)
β_{16}	Yr × O_3	.4954	1.9172	(−3.2918, 4.2827)
β_{17}	Yr × O_3	−.9537	2.2033	(−5.3061, 3.3987)
β_{18}	Yr × O_3	3.9134	2.7087	(−1.4375, 9.2642)
γ:	O_3 power	1.1287[a]	.2330	(.6684, 1.5889)

[a] 95% confidence interval does not overlap zero.

With 19 parameters in the model, 20 iterations are required with the derivative-free method before the numerical estimates of all derivatives can be computed. In this particular case, 7 additional iterations were made and then iterations were restarted with a smaller grid around the current estimates. This required an additional 20 iterations to recompute the numerical derivatives and a final 5 iterations to reach convergence. Thus, 52 iterations were needed to find the solution. Except for the terms involving X_1^γ, this model is linear in the parameters. In models that are "nearly" linear in the parameters, convergence is usually fairly rapid when the derivatives are specified. It is left as an exercise for the reader to fit this model using derivatives.

Summary of the Analysis

The summary of this analysis is shown in Table 15.10. The asymptotic confidence intervals can be used as guides to the significance of the various parameters. This is equivalent to testing the corresponding hypotheses using the Wald statistics. The year parameters, β_1 to β_4, are different from zero, as expected, and will be retained in the model. The block differences within years are not significantly different from zero, as shown by the confidence intervals for β_5 to β_9 overlapping zero. However, the block effects are part of the original experimental designs and will be kept in the model. The average moisture stress effect, β_{10}, the moisture stress by year interaction effect, β_{11}, and the regression coefficients for sulfur dioxide, β_{12}, and ozone, β_{13}, are significantly different from zero. The analysis gives no indication of an ozone by sulfur dioxide interaction, β_{14}, a moisture stress by ozone interaction, β_{15}, or any year by ozone interactions, β_{16}, β_{17}, and β_{18}.

Rather than dropping all nonsignificant interaction terms at one time, the analysis will proceed more cautiously by dropping first the year by ozone interaction effects and then dropping other interaction effects if they remain unimportant. This protects against dropping effects that may become significant after other effects in the model have been dropped, and it provides the opportunity to test the significance of the effects with the likelihood ratio test using the difference in residual sums of squares from the two models.

Dropping Year by Ozone Interactions

The residual sum of squares from the model in which all year by ozone interaction effects, β_{16}, β_{17}, and β_{18}, are set equal to zero is SS(Res) = 3.8088 with 158 degrees of freedom. Comparing this to the residual sum of squares from the full model, Table 15.10, and computing the F-statistic gives

$$F = \frac{[\text{SS}(\text{Res}_{\text{reduced}}) - \text{SS}(\text{Res}_{\text{full}})]/q}{\text{SS}(\text{Res}_{\text{full}})/(n-p)}$$

$$= \frac{(3.8088 - 3.6824)/3}{3.6824/155} = 1.77$$

where $q = 3$ is the number of constraints placed on the parameters. This is an approximate F-test with q and $(n - p)$ degrees of freedom and is nonsignificant. Gallant (1987) shows that this is equivalent to the likelihood ratio test. This confirms the decision based on the Wald statistic that β_{16}, β_{17}, and β_{18} are not different from zero. The reduced model continues to show that β_{14} and β_{15}, the sulfur dioxide by ozone interaction and the moisture stress by ozone interaction, are not different from zero.

Moisture Stress by Ozone Interaction

The model *without* β_{16}, β_{17}, and β_{18} is adopted as the full model for testing the significance of β_{15}. The reduced model, with β_{15} set equal to zero, gives $\text{SS}(\text{Res}_{\text{reduced}}) = 3.8447$ with 159 degrees of freedom. The likelihood ratio test of H_0: $\beta_{15} = 0$ gives

$$F = \frac{3.8447 - 3.8088}{3.8088/158} = 1.49$$

which, with 1 and 158 degrees of freedom, is not significant and β_{15} will be dropped from the model.

Sulfur Dioxide by Ozone Interaction

The Wald confidence interval for the model with β_{15} dropped continues to indicate that β_{14} is not significantly different from zero. The model was further reduced by setting $\beta_{14} = 0$. This gives $\text{SS}(\text{Res}_{\text{reduced}}) = 3.8449$ with 160 degrees

Table 15.11 Summary of the nonlinear least squares analysis of ln(seed yield) for the 1981–1984 soybean data using the reduced model.

Source	d.f.	*Sum of Squares*	*Mean Square*
Model	14	6,089.7763	434.9840
Residual	160	3.8449	.0240
Uncorrected total	174	6,093.6212	
(Corrected total)	173	20.0321	

Parameter		*Estimate*	*Asymptotic Std. Error*	*Asymptotic 95% Confidence Interval*
β_1	Years	6.4910[a]	.0870	(6.3191, 6.6629)
β_2	Years	6.6180[a]	.0986	(6.4233, 6.8127)
β_3	Years	6.6931[a]	.1065	(6.4828, 6.9034)
β_4	Years	6.2316[a]	.1022	(6.0299, 6.4333)
β_5	Blocks/years	.0603	.0317	(−.0023, .1228)
β_6	Blocks/years	−.0616	.0549	(−.1701, .0469)
β_7	Blocks/years	−.0143	.0805	(−.1732, .1446)
β_8	Blocks/years	−.0496	.0775	(−.2027, .1035)
β_9	Blocks/years	−.0987	.0648	(−.2267, .0293)
β_{10}: M		−.4283[a]	.0459	(−.5190, −.3376)
β_{11}: M × Yr		−.2020[a]	.0458	(−.2925, −.1114)
β_{12}: SO_2		−.9996[a]	.1083	(−1.2135, −.7856)
β_{13}: O_3		−7.9230[a]	3.2129	(−14.2682, −1.5778)
γ: O_3 power		1.0778[a]	.2292	(.6253, 1.5304)

[a] 95% confidence interval does not overlap zero.

of freedom (Table 15.11). Comparing this to the residual sum of squares for the previous model gives

$$F = \frac{3.8449 - 3.8447}{3.8447/159} = .01$$

which is nonsignificant. Thus, the sulfur dioxide by ozone interaction effect is also not important and can be dropped from the model. The only interaction effect remaining is the moisture stress by year interaction, β_{11}, which is significant in this reduced model. Likewise, the moisture stress effect, the sulfur dioxide effect, and the ozone effect remain significant as judged by their 95% approximate confidence interval estimates.

Setting $\gamma = 1$

The final stage in simplifying this model relates to γ, the power parameter on X_1. The logical null hypothesis for γ is H_0: $\gamma = 1.0$ which, if true, removes the nonlinearity of the model. The point estimate of γ (in the last reduced model) is $\hat{\gamma} = 1.078$ and the 95% confidence interval estimate is (.625, 1.530). There appears to be no reason to reject the null hypothesis that $\gamma = 1.0$. Since the model with $\gamma = 1$ is linear in the parameters, PROC GLM with the no-intercept option is used to fit this final reduced model. The results are summarized in Table 15.12. The likelihood ratio test of the null hypothesis that $\gamma = 1.0$ gives

Table 15.12 Summary of the analysis of ln(seed yield) for the 1981–1984 soybean data using the final linear model.

Source	d.f.	*Sum of Squares*	*Mean Square*
Model	13	6,089.7733	468.4441
Residual	161	3.8479	.0239
Uncorrected total	174	6,093.6212	
(Corrected total)	173	20.0321	

Parameter		*Estimate*	*Asymptotic Std. Error*	*Asymptotic 95% Confidence Interval*
β_1	Years	6.5193[a]	.0394	(6.4421, 6.5966)
β_2	Years	6.6486[a]	.0473	(6.5559, 6.7414)
β_3	Years	6.7227[a]	.0678	(6.5899, 6.8555)
β_4	Years	6.2611[a]	.0611	(6.1413, 6.3809)
β_5	Blocks/years	.0605	.0316	(−.0014, .1224)
β_6	Blocks/years	−.0627	.0547	(−.1699, .0444)
β_7	Blocks/years	−.0133	.0802	(−.1704, .1439)
β_8	Blocks/years	−.0494	.0773	(−.2009, .1021)
β_9	Blocks/years	−.0981	.0646	(−.2247, .0285)
β_{10}: M		−.4275[a]	.0457	(−.5171, −.3378)
β_{11}: M × Yr		−.2019[a]	.0457	(−.2915, −.1123)
β_{12}: SO_2		−.9977[a]	.1079	(−1.2091, −.7862)
β_{13}: O_3		−6.9170[a]	.3869	(−7.6753, −6.1587)

[a] 95% confidence interval does not overlap zero.

$$F = \frac{3.8479 - 3.8449}{3.8449/160} = .12$$

which is clearly nonsignificant. All the remaining terms in this model, except the block effects, are significant. The plot of the residuals versus $\hat{Y}$ (Figure 15.8, page 438) and the normal plot of the residuals (Figure 15.9) give no reason for concern about inadequacies in the model.

Relative Yield Losses

Thus, the final model to represent the 1981–1984 soybean response to sulfur dioxide and ozone shows a decline in $\ln(Y)$ of 6.9 units per ppm increase in ozone and a decline of 1.0 unit per ppm increase in sulfur dioxide. Translating this regression equation back to the original scale, by taking the antilogarithm, and computing the relative yield loss (RYL) for changes in ozone gives

$$\begin{aligned} RYL &= 1 - \exp[\hat{\beta}_{13}(X_{1o} - X_{1r})] \\ &= 1 - \exp[-6.917(X_{1o} - X_{1r})] \end{aligned}$$

The partial derivatives of RYL with respect to the parameters in the model are all zero except for the partial derivative with respect to β_{13},

$$\frac{\partial RYL}{\partial \beta_{13}} = \{-\exp[\beta_{13}(X_{1o} - X_{1r})]\}(X_{1o} - X_{1r})$$

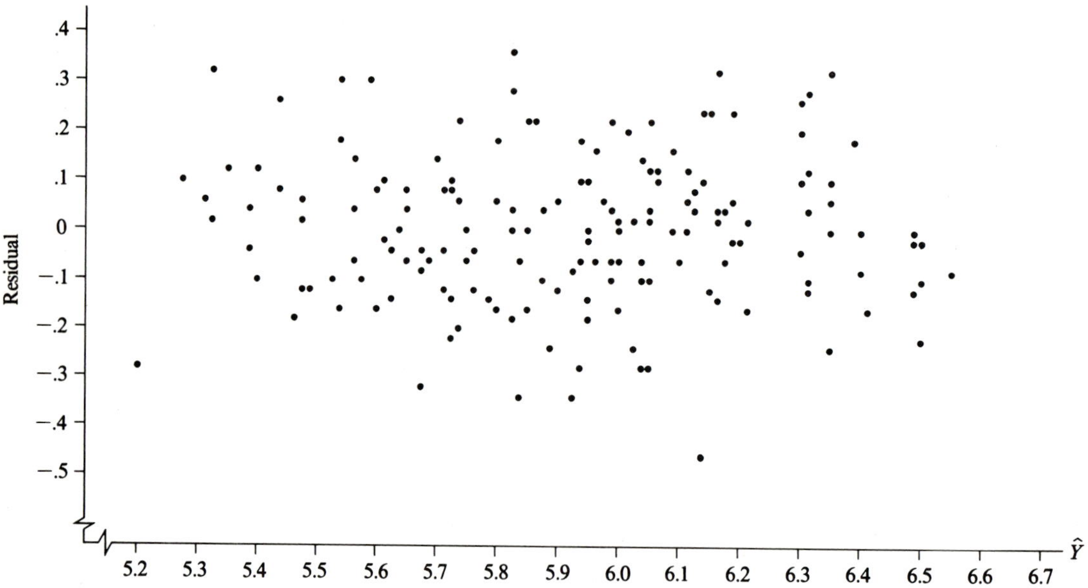

Figure 15.8 Pooled residuals from the final response model for the 1981–1984 soybean data plotted against $\hat{Y}$.

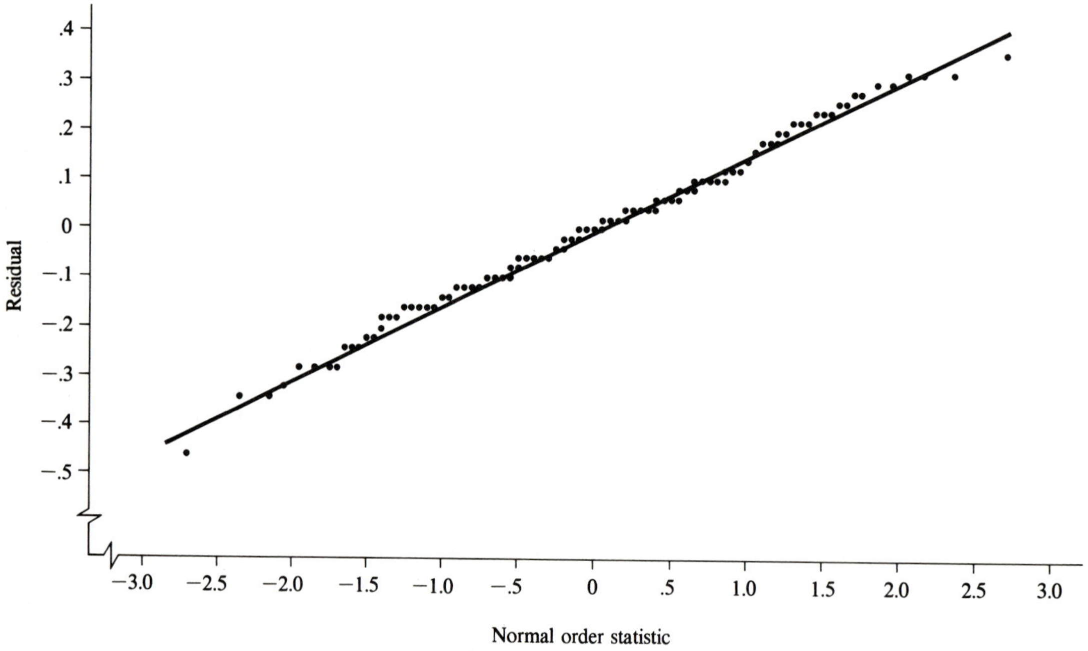

Figure 15.9 Normal plot of the residuals from the final response model for the 1981–1984 soybean data.

Table 15.13 Estimates of relative yield losses (*RYL*), their approximate standard errors, and approximate 95% confidence interval estimates (using $t = 1.975$).

Est. Interval X_{1r}	X_{1o}	*RYL*	*Approx.* $s(RYL)$	*Approximate 95% Confidence Interval*
.025	.03	.034	.0019	(.030, .038)
	.04	.099	.0052	(.088, .109)
	.05	.158	.0081	(.143, .175)
	.06	.215	.0106	(.194, .236)
	.07	.267	.0128	(.242, .293)
	.08	.316	.0145	(.288, .345)

Thus, $s^2(RYL)$ involves only the one variance, $s^2(\hat{\beta}_{13})$, multiplied by the square of the partial derivative evaluated at $\hat{\beta}_{13}$. The estimated relative yield losses (RYL), their approximate standard errors, and the 95% approximate confidence interval estimates for several choices of X_{1o} are given in Table 15.13.

An alternative approach to obtain confidence interval estimates of RYL in this example is to first compute the confidence interval estimates of

$$\ln(1 - RYL) = \beta_{13}(X_{1o} - X_{1r})$$

as

$$[\hat{\beta}_{13} \pm t_{(\alpha,\nu)} s(\hat{\beta}_{13})](X_{1o} - X_{1r})$$

and then transform the limits. The antilogarithms of these limits subtracted from unity give the limits on RYL. In this example, the limits obtained in this way agreed to the third decimal with those in Table 15.13 in all cases except for a difference of 1 in the third decimal when $X_{1o} = .08$.

The estimates of relative yield losses are very similar to those obtained from the 1981 data alone, .215 versus .233 for $X_{1o} = .06$, and .316 versus .335 for $X_{1o} = .08$. The standard errors are appreciably smaller as expected from the use of additional information, .011 versus .018 and .015 versus .033.

Consequences of Setting $\gamma = 1$

Most of the point estimates of the parameters changed only slightly when γ was set equal to 1.0 in the last step of developing this model. The estimate of β_{13} changed most noticeably, from -7.92 to -6.92, but this was to be expected because β_{13} is now the coefficient on X_1, not X_1^{γ}. The standard error on β_{13}, however, decreased to only one-tenth its previous value when γ was set equal to 1.0. This greatly increased precision in the estimate of β_{13} is the result of eliminating a collinearity problem; the correlation between $\hat{\beta}_{13}$ and $\hat{\gamma}$ was $-.990$. This high negative correlation means that changes in one parameter could be offset by compensating changes in the other parameter; the joint confidence region for the two parameters would be a very elongate ellipse.

EXERCISES

*15.1. The polynomial response model adopted for the 1981 soybean data did not use the O_3 quadratic $\times$ SO_2 linear interaction term but the text suggested that it would be the next most logical term to test. Add the term $X_1^2X_2$ to the model shown in equation 15.4 and fit the 1981 soybean data (Table 15.1). Compare these results to those obtained from the model shown in equation 15.4 and test the significance of the new term. State your conclusions.

15.2. Determine whether cubic terms in either ozone or sulfur dioxide would have significantly improved the polynomial response model, equation 15.4, for the 1981 soybean data.

15.3. The sums of squares due to the polynomial terms in the analysis of the 1981 data were not partitions of the analysis of variance due to the fact that a given pollutant treatment was not constant over the levels of the other pollutant and the two replications. Rerun the polynomial analysis using the *mean* ozone level for each ozone treatment and the *mean* sulfur dioxide level for each sulfur dioxide treatment; that is, use $\bar{X}_{i..1}$ and $\bar{X}_{.j.2}$. How does this change your results? What polynomial model do you adopt? Are the sums of squares due to the polynomial terms in ozone level partitions of the ozone treatment sum of squares? Are the Type I (sequential) sums of squares due to the polynomial terms in sulfur dioxide level partitions of the sulfur dioxide treatment sum of squares?

15.4. Refit the Weibull model, equation 15.6, to the 1981 soybean data using one of the methods that require derivatives. Compare your results to those reported in the text for the derivative-free method (Table 15.6).

*15.5. Use the likelihood ratio test with the 1981 data to test the null hypothesis that the parameter γ in the Weibull model is equal to 1. (Refit the nonlinear model you obtain from the Weibull model by setting $\gamma = 1$. Test the increase in residual sum of squares of this "reduced" model over that of the "full" model against the residual mean square from the "full" model using an F-test.) Is the result of this test consistent with the conclusion you reach if you use the Wald test?

15.6. The nonlinear model used in relating $\ln(Y)$ to the treatment variables in the combined 1981–1984 data, equation 15.11, was fit using the derivative-free method. Convergence was slow because of the large number of parameters in the model. Refit the model using one of the methods requiring derivatives. Use the same starting values used in the text. Was convergence obtained or assumed? How many iterations were required? Does the solution agree with that from the derivative-free method, Table 15.10? Does it appear reasonable from these results to set $\gamma = 1$? On what do you base your answer?

*15.7. The nonlinear model used in relating $\ln(Y)$ to the treatment variables in the combined 1981–1984 data, equation 15.11, can also be fit using linear least squares. If γ is fixed at some value, the model is linear in the parameters. Fitting this linear model gives a residual sum of squares that is conditional on the chosen value of γ. Repeating the analysis for a series of values of γ from which the one with the minimum residual sum of squares is chosen will eventually lead to the least squares solution if small enough steps in γ are used. Obtain the least squares solution by this grid search method and compare your results with those obtained from nonlinear least squares. Use $\gamma = 1.0, 1.1, 1.12, 1.13, 1.14, 1.20$ as trial values.

CHAPTER
SIXTEEN

Analysis of Unbalanced Data

Chapter 8 introduced the use of class variables, with which the classical analyses of variance for balanced data became special cases of least squares regression.

This chapter discusses the analysis of unbalanced data using least squares regression with class variables. Emphasis is on understanding the estimable functions of the parameters being tested by the various sums of squares. Treatment means adjusted for the effects of imbalance are defined. Finally, models that contain more than one random effect are introduced.

Definition of "Balance"

The classical analyses of variance for the standard experimental designs are appropriate only for data from balanced experiments. The common definition of balance is that an experiment is **balanced** if all cells of the data table have equal numbers of observations. Critical to this definition is the understanding, which is often not stated, that the "cells" of the data table must include a cell for every possible combination of the levels of all treatment factors, and, if blocking is used, for each combination of treatments and blocks. These conditions imply that every possible multiway table involving different treatment factors (and blocks) will have the same number of observations in all cells of the table.

The balance in the data allows contrasts, and sums of squares associated with the contrasts, to be computed directly from corresponding marginal data tables. (Marginal data tables are constructed by summing across factors not involved in the contrast of immediate interest.) Without balance, contrasts on the marginal sums (or means) will include unwanted effects of other treatment factors. This leads to a "working" definition of balance:

> Data are *balanced* if the contrasts of interest, and sums of squares for the contrasts, can be computed directly from the marginal sums (or means) for the factors involved in the contrast.

[There are other definitions of balance; see, for example, Basson (1965). The definition given here is more restrictive than necessary. Unequal but proportional numbers, for example, may be sufficient for some cases.]

Methods for Unbalanced Data

In this chapter, methods of analyzing unbalanced data are discussed. The first two methods attempt to avoid the effects of imbalance by applying least squares analysis to cell means. (The analysis of cell means is not to be confused with the use of the means model.) The third method applies least squares principles to obtain estimates of estimable functions of the parameters and sums of squares for relevant testable hypotheses. The emphasis in this text is on the application of least squares to the classical effects models. The reader is referred to Hocking (1985) for a thorough discussion of the alternative of using means models.

Many procedures for the analysis of unbalanced data concentrate more on partitioning sums of squares than on the hypotheses being tested. Consequently, the hypotheses often are not the most meaningful and may not even be clearly specified. [See Hocking and Speed (1975), Speed and Hocking (1976), and Speed, Hocking, and Hackney (1978) for extensive discussions on analysis of unbalanced data.] The emphasis in this text is on estimable functions and testable hypotheses in order to enhance the reader's understanding of the analyses. The general linear models procedure, PROC GLM (SAS Institute, Inc., 1985d, 1985f) is used extensively. This procedure computes four types of sums of squares, which include most of the options usually considered, and provides the estimable functions of the parameters being tested by these sums of squares. [The reader is referred to Freund, Littell, and Spector (1986) and Searle and Henderson (1979) for more discussion on PROC GLM.]

16.1 Sources of Imbalance

Imbalance in data can arise for different reasons and at different "levels" in the experiment. The imbalance may be deliberate in the design of the experiment or it may be the result of failure to give adequate consideration to the design. Certain treatment combinations, such as simultaneous high temperature and high pressure, may not be possible for the particular system being studied, or limited resources may restrict the number of treatment combinations that can be handled.

Most often, however, unequal numbers arise due to accidents during the experiment; contamination of material or mortality of animals or plants causes the loss of experimental units, sample material is lost or handled incorrectly before it can be analyzed and data recorded, or data are recorded incorrectly and subsequently have to be discarded. The loss of data may occur at the sampling unit level (if sampling units are used), at the experimental unit level, or at the treatment level. The loss of an entire treatment will cause confounding of effects if the treatment is one of a factorial set of treatments.

While imbalance is occasionally deemed necessary because of the nature of the system being studied and often occurs accidentally, the availability of computing power and general analysis programs such as PROC GLM should never be the justification for conducting an unbalanced experiment. As will be seen, the analysis and interpretation of results are much more difficult for unbalanced data and, frequently, the imbalance will result in the loss of important information.

16.2 Effects of Imbalance

Two-Way Model

The confounding effects of imbalance are illustrated with a 2 × 3 factorial set of treatments in a completely random experimental design. The effects model for this case is

$$Y_{ijk} = \mu + \alpha_i + \beta_j + \gamma_{ij} + \varepsilon_{ijk} \quad [16.1]$$

where α_i and β_j are the effects of the ith and jth levels of treatment factors A and B, respectively; γ_{ij} is the interaction effect between the ith level of A and the jth level of B; and ε_{ijk} is the random error associated with the observation from the kth experimental unit receiving the ijth treatment combination.

Balanced Data: Expectations of Cell Means

When the data are balanced, the sums of squares for the standard analysis of variance are computed directly from contrasts on the treatment means. Functions of the squared differences among the A treatment means generate the sum of squares for the A treatment factor unconfounded by the effects of factor B, and vice versa. The simplicity of the analysis of variance is a direct result of the balance in the data. The reason is evident from the expectations of the cell and marginal means (Table 16.1). Expectations of the cell means are obtained by averaging the fixed effects in the model, equation 16.1, over subscript k, the observations within each cell. In this case, the fixed effects do not involve the subscript k, so the expectation for the ijth cell mean is

$$\mathscr{E}(\bar{Y}_{ij.}) = \mu + \alpha_i + \beta_j + \gamma_{ij}$$

The expectations of the marginal means are obtained by averaging the cell expectations over each row or column, as the case may be, giving equal weight to each cell. The equal weight for each cell simulates the averaging one would do if all cells had the same number of observations.

Table 16.1 The expectations of the cell means and the marginal means for a 2 × 3 factorial in a completely random experimental design. The marginal means are computed assuming equal numbers of observations in each cell.[a]

		B			
		1	2	3	$\mathscr{E}(\bar{Y}_{i..})$
A	1	$\mu + \alpha_1 + \beta_1 + \gamma_{11}$	$\mu + \alpha_1 + \beta_2 + \gamma_{12}$	$\mu + \alpha_1 + \beta_3 + \gamma_{13}$	$\mu + \alpha_1 + \bar{\beta}_. + \bar{\gamma}_{1.}$
	2	$\mu + \alpha_2 + \beta_1 + \gamma_{21}$	$\mu + \alpha_2 + \beta_2 + \gamma_{22}$	$\mu + \alpha_2 + \beta_3 + \gamma_{23}$	$\mu + \alpha_2 + \bar{\beta}_. + \bar{\gamma}_{2.}$
$\mathscr{E}(\bar{Y}_{.j.})$		$\mu + \bar{\alpha}_. + \beta_1 + \bar{\gamma}_{.1}$	$\mu + \bar{\alpha}_. + \beta_2 + \bar{\gamma}_{.2}$	$\mu + \bar{\alpha}_. + \beta_3 + \bar{\gamma}_{.3}$	$\mu + \bar{\alpha}_. + \bar{\beta}_. + \bar{\gamma}_{..}$

[a] The bar over the symbol indicates the average over the subscript that has been replaced with a dot.

Balanced Data: Expectations of Contrasts

The expectations of all marginal means for the B factor contain exactly the same function of the α_i effects (Table 16.1). Thus, all α_i effects will cancel in the expectation of any *contrast* on the marginal means for the B factor. For example, the contrast between levels 1 and 2 for the B factor has expectation

$$\mathscr{E}(\bar{Y}_{.1.} - \bar{Y}_{.2.}) = \beta_1 - \beta_2 + (\bar{\gamma}_{.1} - \bar{\gamma}_{.2}) \qquad [16.2]$$

which involves no α_i. The result is that any contrast of interest on the β_j effects is estimated with the same contrast on the marginal means for the B factor and is not *confounded* with the effects of the A factor. Similarly, any contrast of interest on the α_i effects is estimated with the same contrast on the marginal means for the A factor without being confounded with β_j effects. It follows that the sums of squares for contrasts among the A factor means will not involve the β_j effects and sums of squares for contrasts among the B factor means will not involve the α_i effects when the data are balanced.

Balanced Data: Interaction Effects

The interaction effects, γ_{ij}, do not cancel in contrasts on the marginal means in balanced data, but they are present in very specific ways. The expectation of any contrast on marginal means in balanced data involves the *same* contrast on the simple marginal averages of the γ_{ij} effects. Notice that, if one imposed the commonly used constraint on the parameters that the interaction effects sum to zero over any row or column, the $\bar{\gamma}_{i.}$ and $\bar{\gamma}_{.j}$ would be zero and would disappear from the marginal means and from contrasts on the marginal means. (There is no function of the data that will estimate a contrast on main effects without involving interaction effects, if the model contains interaction effects, unless constraints are imposed on the parameters. In this discussion, all results are presented in terms of the full model without constraints. Thus, contrasts involving *only* main effects, $\alpha_1 - \alpha_2$, for example, are nonestimable.)

Unbalanced Data: Expectations

The effect of imbalance is illustrated by considering the same set of factorial treatments but with unequal cell numbers. Let

$$n_{11} = 1, \quad n_{12} = 2, \quad n_{13} = 1$$
$$n_{21} = 3, \quad n_{22} = 1, \quad n_{23} = 1$$

The expectations of the cell means remain as shown in Table 16.1. However, the expectations of the marginal means now are weighted averages of the expectations of the cell means, where the weighting is by n_{ij}. Thus,

$$\mathscr{E}(\bar{Y}_{1..}) = \frac{\mathscr{E}(\bar{Y}_{11.}) + 2\mathscr{E}(\bar{Y}_{12.}) + \mathscr{E}(\bar{Y}_{13.})}{4}$$

$$= \mu + \alpha_1 + \frac{\beta_1 + 2\beta_2 + \beta_3}{4} + \frac{\gamma_{11} + 2\gamma_{12} + \gamma_{13}}{4} \qquad [16.3]$$

and

$$\mathscr{E}(\bar{Y}_{2..}) = \frac{3\mathscr{E}(\bar{Y}_{21.}) + \mathscr{E}(\bar{Y}_{22.}) + \mathscr{E}(\bar{Y}_{23.})}{5}$$

$$= \mu + \alpha_2 + \frac{3\beta_1 + \beta_2 + \beta_3}{5} + \frac{3\gamma_{21} + \gamma_{22} + \gamma_{23}}{5} \quad [16.4]$$

The marginal means for the A factor now involve different functions of the β_j so that they will not cancel in a contrast on the A treatment means:

$$\mathscr{E}(\bar{Y}_{1..} - \bar{Y}_{2..}) = \alpha_1 - \alpha_2 + \frac{(-7\beta_1 + 6\beta_2 + \beta_3)}{20}$$

$$+ \left[\frac{(\gamma_{11} + 2\gamma_{12} + \gamma_{13})}{4} - \frac{(3\gamma_{21} + \gamma_{22} + \gamma_{23})}{5}\right] \quad [16.5]$$

Similarly, contrasts on the B treatment means will be confounded with α_i effects. Further, the expectations contain different functions of the interaction effects from the balanced case. Simple contrasts on the treatment means, and sums of squares for these contrasts, no longer provide direct estimates of the appropriate functions of the parameters. Other approaches become necessary.

Empty Cells

This illustration assumed that the unequal numbers did not create any empty cells, cells with $n_{ij} = 0$. As long as there are no empty cells, all functions of the parameters that were estimable with balanced data remain estimable in the unbalanced data. However, when there are empty cells, some additional functions become nonestimable and it may be impossible to obtain estimates of some functions of interest.

16.3 Unweighted Analysis of Cell Means

Averaging

The method of **unweighted analysis of cell means** is an attempt to avoid the effects of imbalance by replacing the unequal numbers of observations with their cell means. The method is dependent on there being no empty cells. If the imbalance arises from unequal numbers of sampling units within experimental units, the available sampling observations from each experimental unit are averaged to obtain a mean response for each experimental unit. The analysis is then conducted on these experimental unit means, as if there had been no sampling. If the imbalance arises from experimental units being lost, data from the available experimental units for each treatment are averaged and then used for the analysis of treatment effects.

Caution

The analysis of cell means will avoid the confounding of effects associated with imbalance only in those cases where the averaging is over observations that have the same expectation. Or, equivalently, the averaging must be over observations that differ only in random elements. Averaging over unequal numbers of sampling units always provides unbiased estimates of treatment comparisons. Averaging over experimental units to obtain cell means, however, requires care to avoid confounding fixed effects in the final analysis. If the experimental design is a completely random design or if the experimental design is a randomized complete block design with *random* block effects, the analysis of cell means will yield unbiased comparisons of treatment effects. However, some of the efficiency of blocking will be lost because variances of treatment comparisons will involve the component of variance due to random block effects. If the block effects are *fixed* effects, treatment comparisons based on unweighted means will be confounded with block effects.

Inefficiency

Although the unweighted analysis of cell means is simple, it is not an efficient analysis because unequal variances (of the cell means) are being ignored. Further, the sums of squares that are generated are not distributed as chi-squared random variables and, hence, the conventional tests of significance are only approximate. With the computing facilities generally available, the simplicity of the unweighted analysis of cell means does not justify its use (Speed, Hocking, and Hackney, 1978). A brief outline of the analysis is given to set the stage for the weighted analysis of cell means, Section 16.4.

Model for Observations

The analysis of cell means is described in terms of a completely random experimental design with an $A \times B$ factorial set of treatments. Let n_{ij} be the number of experimental units receiving the ijth treatment combination. The effects model for the individual observations is

$$Y_{ijk} = \mu + \alpha_i + \beta_j + \gamma_{ij} + \varepsilon_{ijk} \qquad [16.6]$$

where α_i ($i = 1, \ldots, a$) is the effect of the ith level of factor A, β_j ($j = 1, \ldots, b$) is the effect of the jth level of factor B, and γ_{ij} is the interaction effect between the ith level of factor A and the jth level of factor B. The subscript k designates the observation receiving the ijth treatment combination ($k = 1, \ldots, n_{ij}$). The usual least squares assumptions apply to ε_{ijk}. The data are unbalanced if the n_{ij} are not equal.

Model for Cell Means

The cell means are obtained by averaging over the n_{ij} observations receiving the ijth treatment,

$$\bar{Y}_{ij.} = \frac{1}{n_{ij}}\left(\sum_{k=1}^{n_{ij}} Y_{ijk}\right) \qquad [16.7]$$

The model in terms of these cell means is

$$\bar{Y}_{ij.} = \mu + \alpha_i + \beta_j + \gamma_{ij} + \bar{\varepsilon}_{ij.} \qquad [16.8]$$

If the variance–covariance matrix of the ε_{ijk} in the original data is $\mathbf{Var}(\varepsilon) = \boldsymbol{I}\sigma^2$, the variance–covariance matrix for the $\bar{\varepsilon}_{ij.}$ in the cell means model will be

$$\mathbf{Var}(\bar{\varepsilon}) = \begin{bmatrix} 1/n_{11} & 0 & \cdots & 0 \\ 0 & 1/n_{12} & \cdots & 0 \\ \vdots & \vdots & \ddots & \vdots \\ 0 & 0 & \cdots & 1/n_{ab} \end{bmatrix} \sigma^2 \qquad [16.9]$$

Table 16.2 Degrees of freedom and mean square expectations for the unweighted analysis of cell means for an $A \times B$ factorial with n_{ij} observations per treatment in a completely random design; all $n_{ij} > 0$.

Source	d.f.	$\mathcal{E}$(mean square)[a]
Total	$ab - 1$	
A	$a - 1$	$\sigma^2 + \bar{n}_h\theta_\gamma^2 + b\bar{n}_h\theta_\alpha^2$
B	$b - 1$	$\sigma^2 + \bar{n}_h\theta_\gamma^2 + a\bar{n}_h\theta_\beta^2$
$A \times B$	$(a - 1)(b - 1)$	$\sigma^2 + \bar{n}_h\theta_\gamma^2$
Exp. error	$n_{..} - ab$	σ^2

[a] The θ^2 terms are quadratic forms of the fixed effects indicated by the subscript.

The *unweighted* analysis of cell means ignores these unequal variances and proceeds as if $\mathbf{Var}(\bar{\varepsilon}) = I\sigma^2$.

Expectations of Means

The expectations of the cell means, given by the first four terms in the model, equation 16.8, and the expectations of the marginal means, obtained by unweighted averaging of the cell means, have the same composition of all fixed effects as with balanced data (Table 16.1).

Analysis of Variance

The analysis of variance of the $a \times b$ table of cell means, with each sum of squares multiplied by $\bar{n}_h$, the harmonic mean of the numbers of observations per cell, gives SS(A), SS(B), and SS(AB). The harmonic mean is

$$\bar{n}_h = \frac{ab}{\sum_{i=1}^{a}\sum_{j=1}^{b}\frac{1}{n_{ij}}} \qquad [16.10]$$

which simplifies to n when all $n_{ij} = n$. The mean squares estimate the same functions of the fixed effects as the corresponding analysis with balanced data except the coefficient n is replaced with $\bar{n}_h$ (Table 16.2).

Variances

The estimate of σ^2 is obtained from a separate computation of the variances among experimental units within treatments and pooled over the ab treatments. Thus,

$$\text{MS(Error)} = \frac{\sum_{i=1}^{a}\sum_{j=1}^{b}\left[\sum_{k=1}^{n_{ij}}(Y_{ijk} - \bar{Y}_{ij.})^2\right]}{\nu} \qquad [16.11]$$

where

$$\nu = \sum_{i=1}^{a}\sum_{j=1}^{b}(n_{ij} - 1) \qquad [16.12]$$

is the degrees of freedom.

The variance of the ijth treatment mean is σ^2/n_{ij} as shown in equation 16.9. The variance of a marginal treatment mean, computed as the unweighted average of cell means, is σ^2/k, where the divisor k is the product of the number of cell means in the average and the harmonic mean of the n_{ij} *for those cells.* The variance of the difference between two unweighted marginal treatment means is the sum of the variances of the two means.

16.4 Weighted Analysis of Cell Means

The **weighted analysis of cell means** uses weighted least squares to take into account the unequal variances of the cell means. The relative sizes of the variances of the cell means are determined by $1/n_{ij}$, equation 16.9, so that the appropriate weighting matrix is a diagonal matrix of the n_{ij}. The estimates of estimable functions of the parameters are best linear unbiased estimates. The sums of squares obtained correspond to those obtained from the general linear models analysis of the original observations, to be discussed in the next section.

16.5 General Linear Models Analysis of Unbalanced Data

Least squares regression with linear models containing class variables reproduces the analyses of variance for the standard experimental designs when the data are balanced (Chapter 8). The general linear models approach, however, does not require balanced data. As long as the parametric functions of interest remain estimable, the general linear models approach will provide estimates of the functions and sums of squares for tests of significance of any testable hypotheses. This section discusses the use of least squares regression with class variables for the analysis of unbalanced data.

General Procedure

The general procedure is as discussed in Chapter 8. To review briefly, a linear model is constructed using dummy variables in $\boldsymbol{X}$ to bring in the effects of class variables, such as treatments. Each set of dummy variables introduces at least one linear dependency among the columns of $\boldsymbol{X}$ so that the model is not of full rank and the unique inverse does not exist. The general linear models approach uses a generalized inverse of $\boldsymbol{X}'\boldsymbol{X}$ to obtain one of the nonunique solutions to the normal equations,

$$\boldsymbol{\beta}^0 = (\boldsymbol{X}'\boldsymbol{X})^-\boldsymbol{X}'\boldsymbol{Y} \qquad [16.13]$$

where $(\boldsymbol{X}'\boldsymbol{X})^-$ is a generalized inverse of $\boldsymbol{X}'\boldsymbol{X}$. Even though $\boldsymbol{\beta}^0$ is not unique, it can be used to obtain a unique estimate of any estimable function of the parameters and a unique sum of squares for any testable hypothesis. That is, if $\boldsymbol{K}'\boldsymbol{\beta}$ is an estimable function of $\boldsymbol{\beta}$, it is uniquely estimated with $\boldsymbol{K}'\boldsymbol{\beta}^0$, where $\boldsymbol{\beta}^0$ is one of the nonunique solutions. Further, if $\boldsymbol{K}'\boldsymbol{\beta}$ is estimable, $\boldsymbol{K}'\boldsymbol{\beta} = \boldsymbol{0}$ is a testable hypothesis for which the unique sum of squares is

$$Q = (\boldsymbol{K}'\boldsymbol{\beta}^0)'[\boldsymbol{K}'(\boldsymbol{X}'\boldsymbol{X})^-\boldsymbol{K}]^{-1}(\boldsymbol{K}'\boldsymbol{\beta}^0) \qquad [16.14]$$

with $r(\boldsymbol{K}')$ degrees of freedom.

Estimable Functions

The specific linear functions of parameters that are estimable play a dominant role in the analysis of models of less than full rank. This was indicated in the discussion of the analysis of balanced data (Chapter 8), but the specific form of the estimable functions was not critical to that discussion and was not pursued at that time. In the analysis of unbalanced data, however, the form of the estimable functions *defines* different types of sums of squares that might be computed and serves as a convenient vehicle for describing these differences. First, and for background, the general form of the estimable functions and the specific forms that generate the sums of squares in the analysis of variance of *balanced data* are presented. Then, the estimable functions that generate the sums of squares for four classes of hypotheses with unbalanced data are discussed. These four classes of hypotheses are labeled Types I, II, III, and IV in the general linear models program PROC GLM (SAS Institute, Inc., 1985d, 1985f) and include all but H4 and H8 of the nine hypotheses discussed by Speed, Hocking, and Hackney (1978, Table 7) for the two-way classified model.

16.5.1 Estimable Functions with Balanced Data

General Form

A general form, $\boldsymbol{L}'\boldsymbol{\beta}$, that encompasses all linear estimable functions can be obtained from the $\boldsymbol{X}$ matrix. The coefficients in each row of $\boldsymbol{X}$ define an estimable function of $\boldsymbol{\beta}$. This follows from the fact that each observation in $\boldsymbol{Y}$ is an unbiased estimate of the particular function of $\boldsymbol{\beta}$ defined by the corresponding row of $\boldsymbol{X}$. That is, $\mathscr{E}(Y_i) = \boldsymbol{x}_i'\boldsymbol{\beta}$, where $\boldsymbol{x}_i'$ is the ith row of $\boldsymbol{X}$. It also follows that any linear function of the rows of $\boldsymbol{X}$ also defines an estimable function of $\boldsymbol{\beta}$.

This principle is used to generate, by row operations on $\boldsymbol{X}$, a general form that encompasses all estimable functions for a given model and set of data. Only the unique rows of $\boldsymbol{X}$ need be considered. That is, no new estimable function is generated by an additional observation that has the same expectation (identical values of $\boldsymbol{X}$) as a previously considered observation. (A corollary of this statement is that imbalance in data does not change the set of estimable functions as long as none of the unique rows of $\boldsymbol{X}$ has been lost. This requires that there be at least one observation in every cell.)

Illustration with CRD

Derivation of the general form of the estimable function is illustrated for the completely random experimental design with $t = 4$ treatments. The general linear model is

$$Y_{ij} = \mu + \tau_i + \varepsilon_{ij} \quad (i = 1, \ldots, 4)$$

from which the *unique* rows of $\boldsymbol{X}$ are

$$\boldsymbol{X} = \begin{bmatrix} 1 & 1 & 0 & 0 & 0 \\ 1 & 0 & 1 & 0 & 0 \\ 1 & 0 & 0 & 1 & 0 \\ 1 & 0 & 0 & 0 & 1 \end{bmatrix}$$

The linear functions of the parameters defined by $\boldsymbol{X\beta}$ are estimable. To obtain the general form of estimable functions as given by PROC GLM, row operations on X are used to reduce it to a form as nearly like the identity matrix as possible. In this case, we obtain

$$\boldsymbol{X}^* = \begin{bmatrix} 1 & 0 & 0 & 0 & 1 \\ 0 & 1 & 0 & 0 & -1 \\ 0 & 0 & 1 & 0 & -1 \\ 0 & 0 & 0 & 1 & -1 \end{bmatrix}$$

The row operations on X are linear operators so that all linear functions defined by $\boldsymbol{X}^*\boldsymbol{\beta}$ are also estimable. The first row of $\boldsymbol{X}^*$ says that $(\mu + \tau_4)$ is estimable, the second row says that $(\tau_1 - \tau_4)$ is estimable, and so forth.

Further, any arbitrary linear function of these estimable functions will be estimable. Let the arbitrary linear function be defined by the coefficients

$$\boldsymbol{C} = (L_1 \; L_2 \; L_3 \; L_4)$$

Thus, the general form that encompasses all estimable functions for this example is

$$\boldsymbol{CX}^*\boldsymbol{\beta} = L_1\mu + L_2\tau_1 + L_3\tau_2 + L_4\tau_3 + (L_1 - L_2 - L_3 - L_4)\tau_4$$

or, letting $\boldsymbol{L}' = \boldsymbol{CX}^*$,

$$\boldsymbol{L}' = [L_1, \; L_2, \; L_3, \; L_4, \; (L_1 - L_2 - L_3 - L_4)] \qquad [16.15]$$

Notice that the fifth element of $\boldsymbol{L}$, the coefficient of τ_4, is a linear function of other L_j. This reflects the overparameterization of the model.

Any choice of values for the L_j yields an estimable function of the parameters. For example, setting $L_1 = 1, L_2 = 1$, and all others equal to zero gives $(\mu + \tau_1)$, which is the expectation of the mean of the first treatment. Setting $L_1 = 1$ and $L_2 = L_3 = L_4 = \frac{1}{4}$ shows that $(\mu + \bar{\tau}_.)$ is estimable.

Estimable Contrasts for CRD

To obtain an estimable *contrast* on the treatment effects, L_1 must be set to zero to avoid having μ involved. There are three remaining "free" coefficients in $\boldsymbol{L}$ involving only the τ_i so that there are a maximum of three linearly independent estimable functions of the τ_i. (This is why 3 degrees of freedom are assigned to the treatment sum of squares.) Setting $L_2 = 1$ and all others equal to zero gives $(\tau_1 - \tau_4)$ as one estimable contrast. Similarly, setting $L_3 = 1$ and all others zero gives $(\tau_2 - \tau_4)$, and setting $L_4 = 1$ and all others zero gives $(\tau_3 - \tau_4)$. If these three choices of the L_j are combined in one matrix,

$$\boldsymbol{K}' = \begin{bmatrix} 0 & 1 & 0 & 0 & -1 \\ 0 & 0 & 1 & 0 & -1 \\ 0 & 0 & 0 & 1 & -1 \end{bmatrix} \qquad [16.16]$$

then $\boldsymbol{K}'\boldsymbol{\beta}$ is a set of linearly independent *estimable* functions (contrasts) involving

Table 16.3 The general form for estimable functions in a 2 × 3 factorial (with no empty cells) and choices of L_k that give the conventional analysis of variance results with balanced data.

Parameters	*Coefficients for General Form*[a]	*Specific Estimable Functions* for α's	for β's (1)	for β's (2)	for γ's (1)	for γ's (2)
μ	L_1	0	0	0	0	0
α_1	L_2	1	0	0	0	0
α_2	$L_1 - L_2$	-1	0	0	0	0
β_1	L_4	0	1	0	0	0
β_2	L_5	0	0	1	0	0
β_3	$L_1 - L_4 - L_5$	0	-1	-1	0	0
γ_{11}	L_7	$\frac{1}{3}$	$\frac{1}{2}$	0	1	0
γ_{12}	L_8	$\frac{1}{3}$	0	$\frac{1}{2}$	0	1
γ_{13}	$L_2 - L_7 - L_8$	$\frac{1}{3}$	$-\frac{1}{2}$	$-\frac{1}{2}$	-1	-1
γ_{21}	$L_4 - L_7$	$-\frac{1}{3}$	$\frac{1}{2}$	0	-1	0
γ_{22}	$L_5 - L_8$	$-\frac{1}{3}$	0	$\frac{1}{2}$	0	-1
γ_{23}	$L_1 - L_2 - L_4 - L_5 + L_7 + L_8$	$-\frac{1}{3}$	$-\frac{1}{2}$	$-\frac{1}{2}$	1	1

[a] The subscripts on the L coefficients correspond to the sequence of the parameters in $\boldsymbol{\beta}$. The missing coefficients, L_3 and L_6, for example, are constrained by the design and the model to be equal to the linear function of the coefficients shown in their place.

the τ_i. The composite hypothesis that all τ_i are equal, or that there are no differences among the treatments, can be written as H_0: $\boldsymbol{K'\beta} = \mathbf{0}$. This is a *testable* hypothesis because each row vector in $\boldsymbol{K'}$ defines an estimable function of $\boldsymbol{\beta}$.

General Form for 2 × 3 Factorial

Return now to the 2 × 3 factorial in a completely random experimental design, which was used to illustrate the effects of imbalance (Section 16.2). The general form for all estimable functions for the 2 × 3 factorial with interaction is given in the second column of Table 16.3. The last five columns give the specific estimable functions that generate the sums of squares for the conventional analysis of variance with *balanced data.*

SS(*A*)

The estimable function (contrast) of the α's that generates SS(A) (column 3 of Table 16.3) is obtained by setting L_1 equal to zero to remove μ from the contrast, the remaining free coefficient on the α_i, L_2, equal to unity, and L_4 and L_5 equal to zero to remove the β_j effects from the contrast. This leaves L_7 and L_8 to be determined. When the data are balanced, comparisons on the marginal means for the A factor involve the same comparisons on the row averages of the γ_{ij} effects. That result is obtained by setting $L_7 = L_8 = \frac{1}{3}L_2$ (see Table 16.1). The divisor of 3 comes from the number of levels of the B factor being averaged across.

SS(*B*)

Two linearly independent contrasts, 2 degrees of freedom, are required to generate SS(B), the variation due to the β_j. This is evident in the general form by the two "free" coefficients, L_4 and L_5, associated with the β_j. There are several ways to define contrasts whenever more than 1 degree of freedom is involved. It is necessary only that the contrasts be linearly independent. The contrasts on β_j require that $L_1 = L_2 = 0$ to avoid confounding the contrast with μ and

α_i. The first contrast in Table 16.3 sets $L_4 = 1$ and $L_5 = 0$; the second contrast is the converse. The L_7 and L_8 coefficients are chosen in each case to give the same contrast on the column averages of the interaction effects. The sum of squares due to the composite hypothesis that both contrasts are zero is SS(B) in the analysis of variance of balanced data.

SS(AB)

Finally, contrasts for the γ_{ij} effects require that all L_k except L_7 and L_8 be zero to avoid confounding the interaction contrasts with μ, α_i, or β_j. This leaves two free coefficients, L_7 and L_8, and hence two linearly independent contrasts to be defined. The first contrast in Table 16.3 uses $L_7 = 1$ and $L_8 = 0$; the second uses $L_7 = 0$ and $L_8 = 1$. The sum of squares due to the composite hypothesis that both contrasts are zero is SS(AB) in the analysis of variance of balanced data.

Properties for Balanced Data

This illustrates the general nature of the estimable functions or the testable hypotheses that generate the sums of squares in the conventional analyses of variance for balanced data (Table 16.3). These linear functions define the hypotheses being tested with balanced data and they provide a guide for the kinds of hypotheses that might be considered in the analysis of unbalanced data. They possess the following three properties that can be used to define various types of hypotheses, and their sums of squares, for unbalanced data:

Property 1 No estimable function for generating a main effect sum of squares, such as the contrast on α_i or the contrasts on the β_j, involves main effects of the other factor. However, each does contain a contrast on higher-order interaction effects involving the same factor. This illustrates the more general result:

> Estimable functions for the sum of squares for any one class of effects, main effects or interaction effects, will not involve any other class of effects *except* those that are higher-order interaction effects or higher-level nested effects of the same factor.

For example, the estimable functions for the $A \times B$ interaction sum of squares in a *three*-factor factorial will have zero coefficients on all main effects and the $A \times C$ and $B \times C$ interaction effects. They will have nonzero coefficients on the $A \times B \times C$ interaction effects because this is a higher-level interaction effect involving $A \times B$. The $A \times B \times C$ interaction effect is said to "contain" (in notation) the $A \times B$ interaction effect. Thus, estimable functions for any class of effects will have zero coefficients on all other classes of effects that do not *contain* the effects being contrasted.

Property 2 An estimable function for the sum of squares for one class of main effects includes the same contrast on averages of the corresponding interaction effects. In effect, the coefficient on each main effect is divided and *equitably distributed* over the interaction effects associated with the same cells as the main effect. For example, the "-1" coefficient on α_2 in the first contrast (Table 16.3) is distributed equally over the three interaction effects γ_{21}, γ_{22}, and γ_{23}, with a coefficient of $-\frac{1}{3}$ on each. In multifactor experiments, this property of "equitable distribution" of coefficients extends to all higher-order interaction effects that *contain* the class of effects on which the estimable function is being constructed. This is referred to as the **equitable distribution property** of the coefficients and is always obtained in balanced data.

Property 3 The estimable function for the sum of squares for the α_i effects is *orthogonal* to both estimable functions constructed for the sum of squares for the γ_{ij} effects. Similarly, the two estimable functions constructed for the sum of squares for the β_j effects are pairwise orthogonal to the two estimable functions constructed for the γ_{ij} effects. [The sum of products of the coefficients in any one of columns 3, 4, or 5 with the coefficients in either one of columns 6 or 7 is zero (Table 16.3).] This is referred to as the **orthogonality property** and is always obtained in balanced data. More generally, the orthogonality property states that:

> The estimable functions, or the testable hypotheses, constructed for the sum of squares for any class of effects are pairwise orthogonal to the estimable functions constructed for the sum of squares for any class of effects that contain them.

16.5.2 Estimable Functions with Unbalanced Data

Effect of Imbalance on Estimable Functions

Imbalance in the data does not change the general form of estimable functions as long as all cells of the table have at least one observation. When there are empty cells, the general form of estimable functions will change, and some additional linear functions will become nonestimable if the missing data has caused the loss of one or more of the unique rows of X. Even if the general form of estimable functions has not changed, imbalance does change the functions being estimated by the "analysis of variance" sums of squares, and there are different methods of adjusting for the confounding of effects that results. These different methods of adjusting are equivalent to imposing different conditions on the choice of coefficients in the general form of estimable functions.

Four Types of SS

Four types of sums of squares can be computed for unbalanced data. Each might be considered a logical extension in one way or another of the analysis of variance for balanced data to the unbalanced case. In all cases, the sums of squares are conveniently described in terms of the testable hypotheses they represent. Type I and Type II sums of squares can be described solely in terms of the other effects in the model for which the sum of squares has been adjusted. If a sum of squares has been adjusted for a particular class of effects, the testable hypotheses for that sum of squares have zero coefficients on that class of effects. For both the Type I and Type II sums of squares, no control is exercised over the coefficients on classes of effects for which the sum of squares has *not* been adjusted. The Type III and Type IV sums of squares differ from Type I and Type II in that regard; constraints are imposed on the coefficients of the classes of effects for which the sum of squares has not been adjusted. Constraints are imposed so that the underlying hypotheses possess the *orthogonality* property (Type III), the *equitable distribution* property (Type IV), or both.

While these are the four types of sums of squares computed by PROC GLM (SAS Institute, Inc., 1985d, 1985f), various ones of these are also computed by other programs. The reader is referred to Speed, Hocking, and Hackney (1978) for a summary of the hypotheses being tested by the sums of squares from various programs. Speed, Hocking, and Hackney specify their hypotheses in terms of the full-rank means model, but there is an equivalence to the classical effects model (Speed and Hocking, 1976).

16.5.2.1 Sequential Sums of Squares: Type I

***Type* I: *Sequential* SS**

The Type I sums of squares are the classical sequential sums of squares obtained from adding the terms to the model in some logical sequence. The sum of squares for each class of effects is adjusted for only those effects that precede it in the model. Thus, the sums of squares and their expectations are dependent on the order in which the model is specified. Using the 2×3 factorial for illustration, adding the terms to the model in the order A, B, AB would generate Type I sums of squares described with the R-notation as

$$\mathrm{SS}(A) = R(\boldsymbol{\alpha}|\mu)$$

$$\mathrm{SS}(B) = R(\boldsymbol{\beta}|\boldsymbol{\alpha}, \mu)$$

$$\mathrm{SS}(AB) = R(\gamma|\boldsymbol{\alpha}, \boldsymbol{\beta}, \mu)$$

SS(*A*)

The sum of squares for the α effects, SS(A), has been adjusted only for μ. It is computed as the (corrected) sum of squares among the A treatment totals giving no consideration to the β_j and γ_{ij} effects. The estimable function that generates this Type I sum of squares is obtained from the general form, Table 16.3, by setting $L_2 = 1$, to give a contrast on α_1 and α_2, and $L_1 = 0$, to remove the effect of μ. All other coefficients in the general estimable form are allowed to take whatever values result from computing the contrast on the α_i adjusted for μ. These coefficients will be functions of the n_{ij}, the numbers of observations in the cells. SS(A) will almost certainly be confounded with β_j effects in unbalanced data. It is often referred to as the sum of squares for A "ignoring" B.

SS(*B*)

The Type I sum of squares for the β effects, SS(B), is adjusted for both μ and the α_i effects, because these effects precede B in the model statement. It is computed as the sum of squares for differences among the levels of the B factor, but further adjusted to remove any α_i effects. The two estimable functions that generate this sum of squares have $L_1 = L_2 = 0$, to remove μ and the α_i, and L_4 and L_5 chosen to specify two contrasts on the β_j as in Table 16.3. The coefficients on the γ_{ij}, L_7 and L_8, however, take whatever values they happen to have after adjustment for μ and the α_i, and are functions of the numbers of observations. Thus, Type I SS(B) is not confounded with the α_i effects but the function of the γ_{ij} effects contained in the contrasts is not as shown in Table 16.3.

SS(*AB*)

The Type I sum of squares for interaction, SS(AB), is adjusted for all other effects in the model because it occurs last in the model statement. The estimable functions that generate this sum of squares are the same as those shown in Table 16.3 for balanced data.

Uses of Type I SS

Because of the sequential manner in which the Type I sums of squares are adjusted, they are not appropriate for many hypotheses used in analysis of variance problems. They are appropriate sums of squares for testing hypotheses when there is some logic in the particular sequence of adjustments such as, for example, the contributions of successively higher-degree terms in a polynomial model or the sequential terms in a purely nested model. Sums of Type I sums of squares are useful for testing composite hypotheses of several class effects if appropriately ordered in the model. In general, however, the Type I sums of squares should be used with caution.

16.5.2.2 Partial Sums of Squares: Types II, III, IV

Adjustments for Types II, III, and IV

The remaining three types of sums of squares, Types II, III, and IV, are "partial" sums of squares in the sense that each is adjusted for all other classes of effects in the model, but each is adjusted according to different rules. One general rule applies to all three types: The estimable functions that generate the sum of squares for one class of effects will not involve any other classes of effects except those that "contain" the class of effects in question. This is the first general property noted in Section 16.5.1 on the nature of estimable functions in *balanced* data. Thus, Types II, III, and IV sums of squares are defined so as to test hypotheses that contain the same classes of effects as the corresponding hypotheses in balanced data. For example, the estimable functions that generate SS(AB) in a *three*-factor factorial will have zero coefficients on all main effects and the $A \times C$ and $B \times C$ interaction effects. They will contain nonzero coefficients on the $A \times B \times C$ interaction effects, because the $A \times B \times C$ interaction "contains" the $A \times B$ interaction.

Type II Adjustments

Types II, III, and IV sums of squares differ from each other in how the coefficients are determined for the classes of effects that do not have zero coefficients—those that contain the class of effects in question. The estimable functions for the Type II sums of squares impose no restrictions on the values of the nonzero coefficients on the remaining effects; they are allowed to take whatever values result from the computations adjusting for effects that are required to have zero coefficients. Thus, the coefficients on the higher-order interaction effects and higher-level nesting effects are functions of the numbers of observations in the data. The Type II sums of squares for the $A \times B$ factorial can be described in the R-notation as

$$\text{SS}(A) = R(\boldsymbol{\alpha}|\boldsymbol{\beta}, \mu)$$

$$\text{SS}(B) = R(\boldsymbol{\beta}|\boldsymbol{\alpha}, \mu)$$

$$\text{SS}(AB) = R(\gamma|\boldsymbol{\alpha}, \boldsymbol{\beta}, \mu)$$

In general, the Type II sums of squares do not possess the properties of equitable distribution and orthogonality characteristic of balanced data.

Type III, IV Adjustments

The Type III and Type IV sums of squares differ from the Type II sums of squares in that the coefficients on the higher-order interaction or nested effects that contain the effects in question are also adjusted so as to satisfy either the orthogonality condition (Type III) or the equitable distribution property (Type IV). The coefficients on these effects are no longer functions of the n_{ij} and, consequently, are the same for all designs with the same general form of estimable functions. If there are no empty cells (no $n_{ij} = 0$), both conditions can be satisfied at the same time and Type III and Type IV sums of squares are equal. The hypotheses being tested are the same as when the data are balanced.

***Type* III ≠ Type IV**

When there are empty cells, the hypotheses being tested by the Type III and Type IV sums of squares may differ. The Type III criterion of orthogonality reproduces the same hypotheses one obtains if effects are assumed to add to zero. (This is one of the constraints frequently imposed to reparameterize a model of less than full rank.) When there are empty cells this is modified to "the effects that are present are assumed to add to zero." The Type IV hypotheses utilize balanced subsets of nonempty cells and may not be unique. (PROC GLM warns the user when other Type IV estimable functions exist and the Type IV sums of squares are not unique.)

Which Type Is Appropriate?

When data are balanced, the four types of sums of squares are the same and identical to the conventional analysis of variance for the particular design. When the data are unbalanced, the four types of sums of squares and the hypotheses being tested may differ. Decisions as to which are the "appropriate" sums of squares to use should be based on which sums of squares test the most meaningful hypotheses. The Type I sums of squares, being sequential sums of squares adjusted *only* for effects that precede it in the model, are usually not appropriate for the classical analysis of variance hypotheses. They are appropriate in special cases as noted in Section 16.5.2.1.

The Types II, III, and IV sums of squares are adjusted so that the classes of effects involved (those that have nonzero coefficients) in each sum of squares are the same as in the sums of squares for balanced data. Type II sums of squares allow the nonzero coefficients on the higher-order interaction effects that contain the effects in question to depend on the numbers of observations in the cells. Hence, the hypotheses being tested by the Type II sums of squares are dependent on the results of that particular experiment. Since hypotheses should be independent of the experimental results, the hypotheses tested by Type II sums of squares usually are not meaningful hypotheses. They might be considered meaningful if, for example, the unequal numbers represented relative frequencies of occurrence in natural populations being studied, or if the models contained only main effects.

Types III and IV sums of squares adjust the nonzero coefficients on the higher-order effects to satisfy either the orthogonality principle or the equitable distribution principle, both of which are satisfied when data are balanced. The hypotheses being tested are no longer dependent on the particular n_{ij}. The Types III and IV sums of squares differ only when there are empty cells. The Type IV sums of squares test hypotheses constructed on balanced subsets of the data and, in general, are not unique. The Type III sums of squares would appear to be the most appropriate for testing the usual hypotheses associated with analysis of variance problems. [The reader is referred to Freund, Littell, and Spector (1986), *SAS/STAT Guide for Personal Computers* (1985d), and *SAS*

User's Guide: Statistics (1985f) for more discussion of the four types of sums of squares.]

16.5.3 Unbalanced Data: An Example

Example 16.1 The differences in the four types of sums of squares and their estimable functions are illustrated using a specific unbalanced case of the 2×3 factorial. The example is taken from Searle and Henderson (1979), and is used with their permission. The data and numbers of observations per cell are as follows:

Data:

		Factor B: 1	2	3
Factor A:	1	2, 4, 6	4, 6	5
	2	12, 8	11, 7	—

$\boldsymbol{n_{ij}}$**:**

		Factor B: 1	2	3	
Factor A:	1	3	2	1	6
	2	2	2	0	4
		5	4	1	10

The data contain one missing cell: $n_{23} = 0$. The numbers of observations for the other cells vary from $n_{13} = 1$ to $n_{11} = 3$. The model is the same as used earlier, equation 16.6:

$$Y_{ijk} = \mu + \alpha_i + \beta_j + \gamma_{ij} + \varepsilon_{ijk}$$

where α_i and β_j are the main effects and the γ_{ij} are the interaction effects. The difference is that the $(i, j) = (2, 3)$ combination does not occur because that cell is empty. The four types of sums of squares (computed using PROC GLM with the model specified in the order A, B, AB) are given in Table 16.4 (page 458). In this example, Type III = Type IV for SS(A), Type I = Type II for SS(B), and all four types are the same for SS(AB).

General Form The general form of the estimable functions for this set of data differs from that for the balanced 2×3 factorial, Table 16.3, only because of the empty cell. The general form for the estimable functions is obtained by row operations on the unique rows of $\boldsymbol{X}$. The absence of an observation in cell (2, 3) caused the loss of the row of $\boldsymbol{X}$ containing γ_{23} and, consequently, must affect the estimable functions. The general coefficients for the α_i and β_j effects remain as shown in

Table 16.4 Analysis of data for an unbalanced 2 × 3 factorial with one empty cell. (From Searle and Henderson, 1979; used with permission.)

Source	d.f.	*Sum of Squares*	*Mean Square*
Model	4	62.5	15.625
Error	5	26.0	5.200
Total	9	88.5	

		Sum of Squares			
Sources	d.f.	*Type* I	*Type* II	*Type* III	*Type* IV
A	1	60.00	57.02	54.55	54.55
B	2	.32	.32	.21	1.50
$A \times B$	1	2.18	2.18	2.18	2.18

Table 16.5 Estimable functions for the sum of squares for factor A from a 2 × 3 factorial with cell (2, 3) missing.

Type SS	*Parameter* μ	α_1	α_2	β_1	β_2	β_3	γ_{11}	γ_{12}	γ_{13}	γ_{21}	γ_{22}
I	0	1	-1	0	$-\frac{1}{6}$	$\frac{1}{6}$	$\frac{3}{6}$	$\frac{2}{6}$	$\frac{1}{6}$	$-\frac{1}{2}$	$-\frac{1}{2}$
II	0	1	-1	0	0	0	.546	.454	0	$-.546$	$-.454$
III	0	1	-1	0	0	0	$\frac{1}{2}$	$\frac{1}{2}$	0	$-\frac{1}{2}$	$-\frac{1}{2}$
IV	0	1	-1	0	0	0	$\frac{1}{2}$	$\frac{1}{2}$	0	$-\frac{1}{2}$	$-\frac{1}{2}$

Table 16.3; the general coefficients on the interaction effects change to the following:

$$
\begin{aligned}
&\gamma_{11}\colon\; L_7 \\
&\gamma_{12}\colon\; -L_1 + L_2 + L_4 + L_5 - L_7 \\
&\gamma_{13}\colon\; L_1 - L_4 - L_5 \\
&\gamma_{21}\colon\; L_4 - L_7 \\
&\gamma_{22}\colon\; L_1 - L_2 - L_4 + L_7
\end{aligned}
\qquad [16.17]
$$

The absence of γ_{23} in this list should be interpreted as the coefficient on γ_{23} always being zero.

***Estimable Functions for* SS**

The differences among the four types of sums of squares are illustrated by the estimable function(s) being considered in each case. The estimable functions for the A sum of squares (Table 16.5) show the following:

1. The Type I sum of squares involves β_j effects, whereas the other three types involve only contrasts on the α_i and the γ_{ij}.
2. The coefficients on the γ_{ij} in the Types I and II sums of squares are functions of the n_{ij}, whereas those in Types III and IV are not.

3. The Types III and IV estimable functions for SS(A) are equal, as is seen numerically in the sums of squares in Table 16.4. The equitable distribution and orthogonality properties of Type III and Type IV sums of squares happen to be satisfied with the same choice of coefficients. The equitable distribution property of coefficients characteristic of Type IV sums of squares is evident. The orthogonality property for the Type III sums of squares will be seen later when the contrasts for the interaction sum of squares are presented.

SS(A)

The Type I SS(A) is inappropriate for testing hypotheses about α_i; it is confounded with the β_j effects. The Type II sum of squares does not involve the β_j but has the undesirable property that the coefficients on the γ_{ij} have been determined by the n_{ij} obtained in this particular experiment. The Types III and IV sums of squares for A are based on an estimable function similar in form to that in the balanced case. They differ from the balanced case in that there is no information on γ_{23}, and the coefficient on γ_{13} has been forced to zero to meet the orthogonality and equitable distribution properties.

SS(B)

The estimable functions for SS(B) sums of squares are shown in Table 16.6. There are 2 degrees of freedom for SS(B), there are two "free" coefficients in the general form, so that two linear contrasts are required. The Type I sum of squares for B does not involve α_i effects, whereas the Type I sum of squares for A did involve β_j effects. This results from B occurring after A in the model and reflects the sequential nature of the Type I sums of squares. The Type I and Type II sums of squares for B are identical because both have been adjusted for all effects that do not contain the β_j, the α_i in this example. The coefficients on the γ_{ij} remain functions of the n_{ij}. The orthogonality and equitable distribution properties for the coefficients on the higher-order interaction effects lead to different coefficients on the γ_{ij} for the Types III and IV sums of squares for B. It is left as an exercise for the student to show numerically that the Type IV sums of squares are not unique. Relabeling the treatment levels so that the empty cell occurs in a different position may lead to different Type IV sums of squares.

Table 16.6 The estimable functions for the sum of squares for the B factor in a 2×3 factorial with cell (2, 3) missing.

Type	*Parameter*										
SS	μ	α_1	α_2	β_1	β_2	β_3	γ_{11}	γ_{12}	γ_{13}	γ_{21}	γ_{22}
I = II	0	0	0	1	0	-1	$\frac{9}{11}$	$\frac{2}{11}$	-1	$\frac{2}{11}$	$-\frac{2}{11}$
	0	0	0	0	1	-1	$\frac{3}{11}$	$\frac{8}{11}$	-1	$-\frac{3}{11}$	$\frac{3}{11}$
III	0	0	0	1	0	-1	$\frac{3}{4}$	$\frac{1}{4}$	-1	$\frac{1}{4}$	$-\frac{1}{4}$
	0	0	0	0	1	-1	$\frac{1}{4}$	$\frac{3}{4}$	-1	$-\frac{1}{4}$	$\frac{1}{4}$
IV	0	0	0	1	0	-1	1	0	-1	0	0
	0	0	0	0	1	-1	0	1	-1	0	0

SS(AB)

The estimable functions for SS(AB) are the same for all four types. The contrast is

$$\gamma_{11} - \gamma_{12} - \gamma_{21} + \gamma_{22}$$

This contrast involves only the effects in the 2 × 2 part of the table that does not involve the missing cell. The orthogonality criterion of the Type III sums of squares can be verified by computing the sum of products of the coefficients for the A and B Type III (Tables 16.5 and 16.6) contrasts with the $A \times B$ Type III contrast. The Type IV contrasts in SS(B) are not orthogonal to the $A \times B$ contrast. [The Type IV contrast in SS(A) is orthogonal to the $A \times B$ contrast in this particular case because both the equitable distribution and orthogonality properties are satisfied with the same choice of coefficients in the SS(A) contrast.] ■

This discussion and Example 16.1 have centered on the factorial model. Models with nested effects or both nested and cross-classified effects follow much the same rules. The general form of the estimable functions for any specific case can be determined from the unique rows of the X matrix before reparameterization [see *SAS User's Guide: Statistics, Version* 5 (1985f), or *SAS/STAT Guide for Personal Computers* (1985d)] and can be requested as the E option in the model statement in PROC GLM. The specific estimable functions for each type of sum of squares can also be obtained as the E1, E2, E3, and E4 options.

16.5.4 Least Squares Means

Definition

The marginal means in an unbalanced set of data do not in general provide meaningful comparisons. The least squares solution to the normal equations, however, can be used to obtain estimates of the *same* linear functions of effects as provided by the corresponding means in balanced data *if* these functions are estimable. The estimates can be thought of as adjusted means, adjusted to remove the unwanted, confounding effects. They are called the **least squares means** and will be designated with "*LS*" in front of the usual mean notation.

The particular linear functions of $\boldsymbol{\beta}$ that must be estimated to obtain the least squares means are defined by the expectations of the corresponding means for balanced data. These expectations are called **population marginal means** (Searle, Speed, and Milliken, 1980). The population marginal means are obtained by averaging the fixed effects in the model in the manner specified by the particular mean being considered. Thus, the expectation of the mean is defined by the subscript–dot notation used to define the mean. The rules for writing the expectation for a particular mean when the data are balanced are given in the box.

Rules for Obtaining Population Marginal Means

1. Specify the desired mean using the dot notation.
2. Include in its expectation a term for each class of fixed effects in the model. Drop all random effects.

3. On each fixed effects term, replace each subscript in the model with the specific number or dot consistent with the notation for the particular mean of interest.
4. Any fixed effect that contains a dot in its subscript is an average of effects as indicated by the dot notation. Place a "bar" over the effect to denote a mean.

Illustration with 2 × 3 Factorial

To illustrate, consider the expectation of the marginal means for an $A \times B$ balanced factorial with interaction effects in the model. The model is

$$Y_{ijk} = \mu + \alpha_i + \beta_j + \gamma_{ij} + \varepsilon_{ijk}$$

Assume there are two levels of factor A and three levels of factor B. To obtain $\mathscr{E}(\bar{Y}_{1..})$, drop the random effects term, ε_{ijk}, replace the subscript i on α_i and γ_{ij} with "1" and the subscript j on β_j and γ_{ij} with "$.$", and place a "bar" over all terms with a dot. Thus,

$$\mathscr{E}(\bar{Y}_{1..}) = \mu + \alpha_1 + \bar{\beta}_. + \bar{\gamma}_{1.}$$

Similarly,

$$\begin{aligned}
\mathscr{E}(\bar{Y}_{2..}) &= \mu + \alpha_2 + \bar{\beta}_. + \bar{\gamma}_{2.} \\
\mathscr{E}(\bar{Y}_{.1.}) &= \mu + \bar{\alpha}_. + \beta_1 + \bar{\gamma}_{.1} \\
\mathscr{E}(\bar{Y}_{.2.}) &= \mu + \bar{\alpha}_. + \beta_2 + \bar{\gamma}_{.2} \\
\mathscr{E}(\bar{Y}_{.3.}) &= \mu + \bar{\alpha}_. + \beta_3 + \bar{\gamma}_{.3}
\end{aligned} \qquad [16.18]$$

Marginal Means

The least squares marginal treatment means for this model, $LS\bar{Y}_{i..}$ and $LS\bar{Y}_{.j.}$, are defined as the best linear unbiased estimates of the corresponding linear functions of the parameters in $\mathscr{E}(\bar{Y}_{i..})$ and $\mathscr{E}(\bar{Y}_{.j.})$, equation 16.18. All are estimable if there are no empty cells. When cell (2, 3) is empty, as in Example 16.1, there is no information on γ_{23} and, therefore, any expectation involving γ_{23} must be a nonestimable function. Thus, it is not possible in Example 16.1 to compute $LS\bar{Y}_{2..}$ and $LS\bar{Y}_{.3.}$ because their expectations involve γ_{23}. While the concept of estimability applies to linear functions of the parameters, for convenience the terms "estimable" and "nonestimable" will be attached to the least squares means according to whether the corresponding population marginal means are estimable or nonestimable. [SAS (1985f) defines the expectation to be estimated by the least squares means as the average of the expectations over *only* the cells that contain data.]

Determining Estimability of Population Marginal Means

If a population marginal mean is estimable, its expectation can be obtained from the general form of estimable functions for that specific case with proper choice of coefficients.

Example 16.2

The estimability of population marginal means is illustrated with the 2 × 3 factorial with cell (2, 3) empty (Example 16.1). The particular linear function of the parameters contained in the expectation of $\bar{Y}_{1..}$, equation 16.18, is obtained from the general form by setting $L_1 = L_2 = 1$ and $L_4 = L_5 = L_7 = \frac{1}{3}$. [Combine equation 16.17 with Table 16.3 to obtain the general linear form for the case with cell (2, 3) empty.] Therefore, $LS\bar{Y}_{1..}$ is an estimable least squares mean in this example. $\mathscr{E}(\bar{Y}_{.1.})$ is obtained by setting $L_1 = L_4 = 1$, $L_2 = L_7 = \frac{1}{2}$, and $L_5 = 0$ and, therefore, $LS\bar{Y}_{.1.}$ is estimable. On the other hand, $\mathscr{E}(\bar{Y}_{2..})$ cannot be obtained by any choice of coefficients, and therefore $LS\bar{Y}_{2..}$ is nonestimable. (PROC GLM informs the user when least squares means are nonestimable.) The population means for the individual cells of the table have expectations

$$\mathscr{E}(LS\bar{Y}_{ij.}) = \mu + \alpha_i + \beta_j + \gamma_{ij}$$

which can be obtained from the general form for all (i, j) except $(i = 2, j = 3)$. Therefore, all $LS\bar{Y}_{ij.}$ except $LS\bar{Y}_{23.}$ are estimable. ■

Estimability Dependent on Model

The estimability of the population marginal means for a particular set of data is dependent on the model being used. This can be seen in the 2 × 3 example by noting that all marginal means become estimable if the model does not contain interaction effects, γ_{ij}, even though cell (2, 3) is empty. The general form of estimable functions is as before but with the γ_{ij} coefficients dropped. $\mathscr{E}(Y_{.3.}) = \mu + \bar{\alpha}_. + \beta_3$ is estimable and is obtained from the general linear form by setting $L_1 = 1$ and $L_2 = L_4 = L_5 = 0$.

Computation

The least squares means are computed as linear functions of one of the nonunique solutions $\boldsymbol{\beta}^0$ to the normal equations. The least squares estimate $\boldsymbol{\beta}^0$ is biased, $\mathscr{E}(\boldsymbol{\beta}^0) \neq \boldsymbol{\beta}$, since $\boldsymbol{X}$ is not of full rank. However, the best linear unbiased estimate of any *estimable function* of $\boldsymbol{\beta}$ is given by the same linear function of the least squares solution.

Example 16.3

The vector of parameters in Example 16.1 is

$$\boldsymbol{\beta}' = (\mu \ \alpha_1 \ \alpha_2 \ \beta_1 \ \beta_2 \ \beta_3 \ \gamma_{11} \ \gamma_{12} \ \gamma_{13} \ \gamma_{21} \ \gamma_{22} \ \text{—})$$

Notice that γ_{23} is missing because cell (2, 3) is empty. A dash has been inserted in its place so that it is not forgotten. The estimates $\boldsymbol{\beta}^0$ computed by PROC GLM and their expectations are given in Table 16.7. The $\mathscr{E}(LS\bar{Y}_{1..})$ in equation 16.18 is written in vector notation, $\mathscr{E}(LS\bar{Y}_{1..}) = \boldsymbol{K}_1'\boldsymbol{\beta}$, where $\boldsymbol{K}_1'$ is

$$\boldsymbol{K}_1' = (1 \ 1 \ 0 \ \tfrac{1}{3} \ \tfrac{1}{3} \ \tfrac{1}{3} \ \tfrac{1}{3} \ \tfrac{1}{3} \ \tfrac{1}{3} \ 0 \ 0 \ 0)$$

Thus, the least squares mean for the first level of factor A is computed from Table 16.7 as

$$\begin{aligned} LS\bar{Y}_{1..} &= \boldsymbol{K}_1'\boldsymbol{\beta}^0 \\ &= 9 + (-4) + (\tfrac{1}{3}) + (-\tfrac{2}{3}) = 4.667 \end{aligned}$$

Table 16.7 The GLM solution to the 2 × 3 factorial with cell (2, 3) empty and the expectations of the estimates.

GLM Results		
Parameter	*Estimate*	*Expectation of the Estimate*
Intercept	9.0 B[a]	$\mu + \alpha_2 + \beta_3 - \gamma_{12} + \gamma_{13} + \gamma_{22}$
A 1	−4.0 B	$\alpha_1 - \alpha_2 + \gamma_{12} - \gamma_{22}$
2	0 B	0
B 1	1.0 B	$\beta_1 - \beta_3 + \gamma_{12} - \gamma_{13} + \gamma_{21} - \gamma_{22}$
2	0 B	$\beta_2 - \beta_3 + \gamma_{12} - \gamma_{13}$
3	0 B	0
AB 11	−2.0 B	$\gamma_{11} - \gamma_{12} - \gamma_{21} + \gamma_{22}$
12	0 B	0
13	0 B	0
21	0 B	0
22	0 B	0

[a] The "B" is part of the SAS output to remind the user that the estimates are biased.

$LS\bar{Y}_{2..}$ would be computed as $\boldsymbol{K}_2'\boldsymbol{\beta}^0$ where

$$\boldsymbol{K}_2' = (1\ 0\ 1\ \tfrac{1}{3}\ \tfrac{1}{3}\ \tfrac{1}{3}\ 0\ 0\ 0\ \tfrac{1}{3}\ \tfrac{1}{3}\ \tfrac{1}{3})$$

except for the fact that the last element in $\boldsymbol{K}_2'$ is the coefficient on the missing γ_{23}. Therefore, $LS\bar{Y}_{2..}$ cannot be computed, or $LS\bar{Y}_{2..}$ is nonestimable. Any least squares mean that has a nonzero coefficient on γ_{23} is nonestimable in this example. ■

Variances of LS Means

The variances for the least squares means that are estimable are obtained by applying the rule for variances of linear functions, and using $\mathbf{Var}(\boldsymbol{\beta}^0) = (\boldsymbol{X}'\boldsymbol{X})^-\sigma^2$. For example, $\text{Var}(LS\bar{Y}_{1..}) = \boldsymbol{K}_1'(\boldsymbol{X}'\boldsymbol{X})^-\boldsymbol{K}_1\sigma^2$. The estimate of the variance is obtained by substituting s^2 for σ^2. The standard deviations of the least squares means are available on request as one of the options in PROC GLM.

Example 16.4

(*Continuation of Example* 16.1) Table 16.8 (page 464) gives the estimates of the estimable least squares means, their standard errors, and the linear functions of the parameters being estimated. The least squares means for the individual cells of the table, the $A \times B$ means, are the same as the unadjusted means and their variances are σ^2/n_{ij}. This will always be the case for the smallest subdivision of a factorial table. ■

Summary

Section 16.5 has been a general introduction to the analysis of unbalanced data and has concentrated on the results obtained from PROC GLM. Freund, Littell, and Spector (1986) and Searle and Henderson (1979) are recommended reading for other applications and more detailed discussions of the use of PROC GLM. There are other computer programs that treat the analysis of unbalanced data, for example, BMDP (Dixon, 1981) and SPSS (Norusis, 1985). It is not always clear which sums of squares are being computed by the various programs

Table 16.8 Least squares means, their standard errors, and expectations for the 2 × 3 factorial example with cell (2, 3) empty.

LSMEAN			Std. Error	Expectation
LSMEANS for A factor:				
A	1	4.667	1.029	$\mu + \alpha_1 + \bar{\beta}_{.} + \bar{\gamma}_{1.}$
	2	Nonest.	—	
LSMEANS for B factor:				
B	1	7.00	1.041	$\mu + \bar{\alpha}_{.} + \beta_1 + \bar{\gamma}_{.1}$
	2	7.00	1.140	$\mu + \bar{\alpha}_{.} + \beta_2 + \bar{\gamma}_{.2}$
	3	Nonest.	—	
LSMEANS for $A \times B$:				
A	B			
1	1	4.00	1.317	$\mu + \alpha_1 + \beta_1 + \gamma_{11}$
1	2	5.00	1.612	$\mu + \alpha_1 + \beta_2 + \gamma_{12}$
1	3	5.00	2.280	$\mu + \alpha_1 + \beta_3 + \gamma_{13}$
2	1	10.00	1.612	$\mu + \alpha_2 + \beta_1 + \gamma_{21}$
2	2	9.00	1.612	$\mu + \alpha_2 + \beta_2 + \gamma_{22}$

and, therefore, which hypotheses are being tested. It is important that the user understand the program and its output in order to avoid misinterpretation of the results. Other recent references are Myers (1986) and Searle (1986).

Warning

Programs such as PROC GLM are powerful programs that can accommodate a wide variety of data analysis problems. Their generality, however, will also tend to make them inefficient for specific problems that can be handled with less sophistication. For example, analysis of balanced data using PROC GLM will be more costly than using an analysis of variance program, such as PROC ANOVA (SAS Institute, Inc., 1985d, 1985f). Likewise, regression analyses with no class variables will be more efficiently run on PROC REG (SAS Institute, Inc., 1985d, 1985f), which is designed to handle only full-rank models. The generality of PROC GLM also causes computer space problems to be encountered for problems that might be handled with modest space in an analysis of variance program. GLM uses the $\boldsymbol{X}$ matrix *without* reparameterization. This can be a very large matrix and require a large amount of computer memory even for relatively modest problems. Analyses using the means model (Hocking, 1985) require less computer memory, because it is a full-rank model, and may be the preferred approach in such cases.

16.6 Models with More Than One Random Effect

Fixed Models

The classical least squares model contains only one random element, the random error; all other effects are assumed to be fixed constants. For this class of models, the assumption of independence of the ε_i implies independence of the Y_i. That is, if $\mathbf{Var}(\varepsilon) = \boldsymbol{I}\sigma^2$, then $\mathbf{Var}(Y) = \boldsymbol{I}\sigma^2$ also. Such models are called **fixed effects models**, or more simply, **fixed models**.

Random and Mixed Models

Many situations call for models in which there is more than one random term. The classical **variance components problems**, in which the purpose is to estimate components of variance rather than specific treatment effects, is one example. In these cases, the "treatment effects" are assumed to be a random sample from a population of such effects and the goal of the study is to estimate the variance among these effects in the population. The individual effects that happen to be observed in the study are not of any particular interest except for the information they provide on the variance component. Models in which all effects are assumed to be random effects are called **random models**.

Observational studies often involve a hierarchy of nested effects that represent "levels" of random sampling of some population, such as random homes in random counties in random states of the United States. The sampling of environments in which controlled experiments are conducted, locations and years, often is regarded as a random sampling of environmental conditions. The purpose is to infer behavior of the fixed treatments over some population of environments, rather than just to the fixed set of environments encountered in the experiments. Models that contain both fixed and random effects are called **mixed models**. Even the commonly used split-plot experimental design specifies two random terms in the model—the whole-plot error and the subplot error.

$\mathrm{Var}(Y) \neq I\sigma^2$

The net effect of more than one random term in the model is that $\mathbf{Var}(\boldsymbol{Y}) \neq \boldsymbol{I}\sigma^2$ even if $\mathbf{Var}(\boldsymbol{\varepsilon}) = \boldsymbol{I}\sigma^2$. The random elements shared by observations introduce nonzero covariances among all observations having common "levels" of the random effects. This is illustrated with the model for the split-plot experiment in a randomized complete block experimental design:

$$Y_{ijk} = \mu + \rho_i + \alpha_j + \delta_{ij} + \beta_k + \gamma_{jk} + \varepsilon_{ijk}$$

where

ρ_i is the effect of the ith block;

α_j is the effect of the jth whole-plot treatment;

β_k is the effect of the kth subplot treatment;

γ_{jk} is the interaction effect due to the jth and kth levels of the treatments;

δ_{ij} is the whole-plot error, $N(0, \sigma_\delta^2)$; and

ε_{ijk} is the subplot error, $N(0, \sigma^2)$.

All effects are assumed to be fixed effects except δ_{ij} and ε_{ijk}. With this model, the variance of each Y_{ijk} is

$$\sigma^2(Y_{ijk}) = \sigma^2 + \sigma_\delta^2$$

Any two observations from the same whole-plot, same ij subscripts, share the same random δ_{ij} and, consequently, have covariance

$$\mathrm{Cov}(Y_{ijk}, Y_{ijk'}) = \sigma_\delta^2$$

Thus, the variance–covariance matrix for Y will consist of nonzero off-diagonal elements for any pair of observations that come from the same whole-plot. By least squares theory, the best linear unbiased estimates of the fixed effects are obtained only if generalized least squares is used (see Section 11.5.2). Some computer programs—for example, program P3V in the BMDP system (Dixon, 1981)—can provide maximum likelihood and restricted maximum likelihood estimates for mixed models. The user is warned about the heavy computational load in this program for large problems.

Analysis of Variance Approach

The conventional least squares approach, sometimes called the **analysis of variance approach**, to models with more than one random term is to assume initially that all effects, other than the term that assigns a unique random element to each observation, are fixed effects. Least squares is applied to this "fixed" model to obtain relevant partitions of the sums of squares. The δ_{ij} in the split-plot model, for example, are regarded as parameters to be estimated—in this case, as fixed interaction effects between blocks and the whole-plot treatment. The whole-plot error for the split-plot analysis is computed as the sum of squares for the hypothesis that these fixed interaction effects are zero. Then, the model containing the random effects is reinstated and *expectations* of the mean squares are derived. The mean square expectations determine how tests of significance are to be made and how variance components are to be estimated.

Limitations of Analysis of Variance Approach

With balanced data, this approach generates the conventional analysis of variance for the design and, with the appropriate adjustment of the mean square expectations for the random effects, gives the same results as would be obtained with a full generalized least squares analysis. The generalized least squares analysis is not obtained by this method, however, when the data are not balanced. Nevertheless, this is still the common method of analysis of models with more than one random element, undoubtedly due to the much greater computational load associated with generalized least squares.

All computations in the classical approach to the analysis of mixed models assume the model is fixed; there is only one random element. As already noted, the estimates of $\boldsymbol{\beta}$, and all linear functions of $\boldsymbol{\beta}$, obtained by this approach will not be the best linear unbiased estimates because the true variance–covariance structure is not being taken into account. The estimates are unbiased but there will be some loss in precision. In addition and perhaps more critically, the tests of significance and the computed measures of precision, $s(\boldsymbol{\beta}^0)$, $s(\hat{Y})$, and standard errors of the least squares means, will be incorrect. The measures of precision are computed as if $\boldsymbol{I}\sigma^2$ were the true variance–covariance matrix of Y rather than the more general $\mathbf{Var}(\boldsymbol{Y})$. If ordinary least squares is to be used for the analysis of models with more than one random component, adjustments to the tests of significance and the estimates of the standard errors must be considered.

Expectations of Mean Squares

Adjustments to tests of significance are made by "constructing" an error mean square that has the proper expectation with respect to the random elements. This requires the expectations of the mean squares under the random model. For balanced data the mean square expectations are easily obtained and are reported in many places (e.g., Searle, 1971, 1986; Steel and Torrie, 1980). For unbalanced data, computer programs provide the expectations. The "RANDOM" statement in PROC GLM prompts the program to provide the mean square expectations under a mixed model in which the random effects are specified in the "RANDOM" statement. (The "MODEL" statement in GLM specifies all classes of effects, fixed and random, except for the unique random element associated with each observation.) The expectations are given for any of the four types of sums of squares and all contrasts used in the analysis. The expectations are expressed in terms of a linear function of the variance components for the random effects plus a general statement of the classes of fixed effects involved in the quadratic function. The specific quadratic functions of the fixed effects can also be obtained, if needed.

Example 16.5 To illustrate the use of the results provided by the "RANDOM" statement, the mean square expectations are given here for the Type III sums of squares for the whole-plot treatment factor and the whole-plot error for the unbalanced data to be analyzed in Chapter 17. The experiment is a split-plot experiment with the whole-plot treatments (a factorial set of treatments involving factors A and B) arranged in a randomized complete block experimental design. The estimate of the whole-plot error was computed as the "Block $\times$ A $\times$ B" interaction, where "Block $\times$ A" and "Block $\times$ B" are not included in the model. (You are referred to Chapter 17 for the details of the experiment.) The expectations for the Type III mean squares for treatment factor A and the Block $\times$ A $\times$ B interaction are as follows:

Source	*Type* III *Expected Mean Square*
A	$\sigma^2 + 1.0909\sigma_\delta^2 + Q(A, A \times B)$
Block $\times$ A $\times$ B	$\sigma^2 + 1.9048\sigma_\delta^2$

The expectation of the residual mean square is σ^2. The $Q(\cdot)$ function indicates that the mean square expectation is a quadratic function of the "A" and "$A \times B$" treatment effects. For simplicity, let E_a = MS(Block $\times$ A $\times$ B) and E_b = MS(Res).

If these data had been balanced, the coefficient on σ_δ^2 would have been 2 in each case, the number of levels of the subplot treatment factor, and E_a would have been the appropriate error for testing the null hypothesis that $Q(A, A \times B) = 0$. With the imbalance, the coefficients on σ_δ^2 differ and, consequently, E_a is not the appropriate error for the test. An approximate test is obtained by constructing a mean square that has the correct expectation. The test for H_0: $Q(A, A \times B) = 0$ requires a denominator mean square whose expectation is $\sigma^2 + 1.0909\sigma_\delta^2$. Such a mean square is constructed as a linear function of E_a and E_b as follows:

$$E' = \left(\frac{1.0909}{1.9048}\right)E_a + \left(1 - \frac{1.0909}{1.9048}\right)E_b$$

The approximate test of H_0: $Q(A, A \times B)$ is then

$$F' = \frac{\text{MS}(A \times B)}{E'}$$ ■

Distribution of F′

The constructed variance ratio, F', in Example 16.5 is only approximately distributed as an F-statistic for two reasons. First, a linear function of mean squares does not behave quite like a chi-squared random variable as is required for the F-test. The degrees of freedom for E' are determined so as to minimize this problem (Satterthwaite, 1946). Second, the Types II, III, and IV sums of squares, in general, are not orthogonal partitions of the model sum of squares and, hence, the numerator and denominator mean squares in F' are not independent. This will be the case for all tests in which the constructed denominator mean square involves one of the partitions. This lack of independence is ignored in the test of significance. [MS(Residual) is orthogonal to all partitions of the model sum of squares.]

The Satterthwaite (1946) approximation for the degrees of freedom for a linear function of mean squares, $\sum a_i \text{MS}_i$, is

$$f' = \frac{\left(\sum_i a_i \text{MS}_i\right)^2}{\sum_i \left(\dfrac{a_i^2 \text{MS}_i^2}{f_i}\right)} \qquad [16.19]$$

where f_i is the degrees of freedom of MS_i.

Differences in Defining Interaction Effects

A word of warning is needed on the use of the mean square expectations. There are differences of opinion on how interaction effects between a fixed and a random factor are to be handled in deriving mean square expectations. Some argue that if one of the factors involved in the interaction is a random factor, the interaction effects should be treated as completely random variables with no "constraints" imposed on their behavior. In such cases, the interaction component of variance is present in the expectations of the interaction mean square and *both* main effects mean squares. SAS uses this procedure in deriving expectations in the "RANDOM" option.

The classical approach to handling interaction effects is to impose the constraint that the interaction effects sum to zero over the levels of the *fixed* factor; that is, the effects sum to zero in the fixed direction of the two-way table of effects. This causes the interaction component of variance to "drop out" of the mean square expectation for the *random* main effect. These expectations are consistent with those derived under a two-dimensional finite sampling model in which the samples of effects for factor A and factor B are assumed to have resulted from taking random samples from the two finite populations of effects. Let N_a and N_b be the two population sizes, and n_a and n_b be the respective sample sizes, $n_a < N_a$ and $n_b < N_b$. The mean square expectations for the *mixed* model are then obtained from this finite model by letting the population size go to infinity for the random factor and to the sample size for the fixed factor. The covariances among the effects due to the finiteness of the population cause the interaction effects to drop out of the mean square expectation for the random factor.

These differences in philosophy do not enter into the present split-plot example because all treatment factors are assumed to be fixed. The differences will affect the choice of error in many cases and the reader needs to be aware of the problem. Speed and Hocking (1976) provide more discussion on this point.

Adjusting Measures of Precision

Two methods of adjusting the measures of precision obtained from the standard least squares analysis might be used. If the generalized inverse of $\boldsymbol{X'X}$ is available from the computer program, the correct variance–covariance matrix for any linear function of $\boldsymbol{\beta}^0$ can be computed using a matrix program such as IML (SAS Institute, Inc., 1985a). Let $\boldsymbol{L'\beta}^0$ be k linear functions of $\boldsymbol{\beta}^0$ of interest and let $s^2(\boldsymbol{Y})$ be an estimate of $\mathbf{Var}(\boldsymbol{Y})$, the variance–covariance matrix of $\boldsymbol{Y}$. The true variance of $\boldsymbol{L'\beta}^0$, when $\boldsymbol{\beta}^0$ has been computed *assuming* $\mathbf{Var}(\boldsymbol{Y}) = \boldsymbol{I}\sigma^2$, is

$$\mathbf{Var}(\boldsymbol{L'\beta}^0) = \boldsymbol{L'}(\boldsymbol{X'X})^{-}\boldsymbol{X'}[\mathbf{Var}(\boldsymbol{Y})]\boldsymbol{X}(\boldsymbol{X'X})^{-}\boldsymbol{L} \qquad [16.20]$$

and is estimated by substituting $s^2(\boldsymbol{Y})$ for $\mathbf{Var}(\boldsymbol{Y})$. This gives $s^2(\boldsymbol{\beta}^0)$ if $\boldsymbol{L'}$ is the identity matrix, $s^2(\hat{\boldsymbol{Y}})$ if $\boldsymbol{L'}$ is $\boldsymbol{X}$, and s^2(LSMEANS) if $\boldsymbol{L'}$ consists of row vectors of the estimable functions for the least squares means.

As an alternative to computing the exact variances, expectations of the mean squares can be used to make approximate adjustments to standard errors of the least squares means. The expectation for the random elements of a particular mean square provides an *average* variance for the class of means involved in that mean square. As with the tests of significance, a mean square can be constructed which has this expectation. Multiplication of the standard errors reported for any particular class of means by

$$\text{Ratio} = \left[\frac{\text{Constructed MS}}{\text{MS(Residual)}}\right]^{1/2}$$

provides reasonable approximations of the standard errors. (A comparison of the two methods will be given for the case study in Chapter 17.)

A General Method for Mixed Models

Recently, methods have been developed that reduce the computational load of the more appropriate generalized least squares analysis of models with more than one random element. MIXMOD, for example, is a program designed to analyze data in the context of a general mixed model (Giesbrecht, 1983, 1984, 1986; Giesbrecht and Burns, 1985). It is a two-stage procedure that estimates the variance components due to the random effects and then uses these estimates to construct the variance–covariance matrix for $\boldsymbol{Y}$ and compute the generalized least squares estimates of the fixed effects. Since the estimates of the variance components and the fixed effects are interdependent in unbalanced data, the procedure is iterative. Experience has shown that it converges rapidly. The details of this method are beyond the scope of this text, but the reader should be aware of the existence as such options to the more conventional methods. For comparison with GLM, results from MIXMOD will be presented for the case study in Chapter 17.

EXERCISES

16.1. Unequal numbers of observations may be designed into an experiment. Discuss two situations in which it might be desirable to have unequal numbers of observations. For each situation, discuss whether Type II and Type III hypotheses would be more meaningful.

16.2. Table 16.1 gives the expectations of the cell means for a 2 × 3 factorial in a completely random experimental design. Construct a similar $A \times B$ table but for a randomized complete block design with balanced data. Assume that the block effects are fixed effects. Include $A \times B$ interactions but do not include interactions with blocks. Demonstrate that the expectation of any contrast on treatment means, cell means or marginal means, does not involve block effects.

* 16.3. Reconstruct the table developed in exercise 16.2 assuming there are three blocks but that treatment (2, 3) is missing in block 3. Identify the contrasts on cell treatment means and on marginal treatment means that are "free" of block effects. Would the analysis of cell means be appropriate for these data? Show why or why not.

16.4. In exercise 16.2 fixed block effects were assumed. Show how the expectations change if block effects are assumed to be random variables with zero mean and variance σ_b^2. Show how this changes your conclusions when the numbers are unequal as in exercise 16.3.

* 16.5. Exercise 8.12 used data on survival time of patients with different types of cancer (Cameron and Pauling, 1978). The data are cross-classified with unequal numbers if both sex of patient and cancer type are considered. Use the logarithm of the ratio of days survival of the treated patient to the mean days survival of his control group as the dependent variable. (In the following analyses, include interaction effects between sex of patient and type of cancer in your models, but ignore differences in age.)

a. Do an unweighted analysis of cell means to investigate the effects of sex, cancer type, and their interaction. Compute the within-cell variance and the harmonic mean of the numbers of observations, and summarize the results in an analysis of variance table. Note that the four types of sums of squares are equal and that the ordinary means are equal to the least squares means. Can you explain why?

b. Do a weighted analysis of cell means, with weights n_{ij}, using PROC GLM or a similar program. Request that all four types of sums of squares be computed. Do any of the four types equal the sums of squares obtained from the unweighted analysis of cell means? Do the ordinary means or the least squares means agree with those from the unweighted analysis?

c. Use the general linear models approach (PROC GLM or similar program) to analyze the data. Request all four types of sums of squares. Compare the analysis with the weighted analysis of cell means. Compare the least squares means with those from the unweighted and weighted analyses of cell means.

16.6. Repeat exercise 16.5 with the "Type × Sex" interactions omitted from all models. Compare the sums of squares, the ordinary means, and the least squares means with those obtained based on models with interaction effects.

* 16.7. In the weighted analysis of cell means, weighting was determined by the n_{ij}. This resulted from the assumption of constant variance for the ε_{ijk}; that is, $\mathbf{Var}(\boldsymbol{\varepsilon}) = \boldsymbol{I}\sigma^2$. (See equation 16.9.) Suppose the variances for the observations as well as the numbers of observations differed from cell to cell. Let the variance of

cell ij be σ_{ij}^2. What would be an appropriate weighting for the weighted analysis of cell means? How would you determine numerical values for the weights?

16.8. Construct the general form of estimable functions $\boldsymbol{L}'\boldsymbol{\beta}$ for the nested model

$$Y_{ijk} = \mu + \alpha_i + \beta_{ij} + \varepsilon_{ijk}$$

where $i = 1, 2$ and $j = 1, 2$. Assume all effects are fixed effects and

$$\boldsymbol{\beta}' = (\mu \quad \alpha_1 \quad \alpha_2 \quad \beta_{11} \quad \beta_{12} \quad \beta_{21} \quad \beta_{22})$$

(You need to define $\boldsymbol{X}$ for this model, eliminate any nonunique rows, and then use row operations to reduce $\boldsymbol{X}$ to the "near-identity" form.) You should obtain

$$\boldsymbol{L}' = (L_1, \quad L_2, \quad (L_1 - L_2), \quad L_4, \quad (L_2 - L_4), \quad L_6, \quad (L_1 - L_2 - L_6))$$

* 16.9. Use the general form of estimable functions in exercise 16.8 to determine if each of the following is an estimable function. Give the choice of coefficients that generates the linear function if it is an estimable function.

a. $\mu + \alpha_1 + \beta_{11}$
b. $\beta_{11} - \beta_{12}$
c. $\beta_{11} + \beta_{21}$
d. $\alpha_1 - \alpha_2$
e. $\beta_{11} + \beta_{21} - \beta_{12} - \beta_{22}$
f. $\mu + \alpha_1 + \frac{1}{2}(\beta_{11} + \beta_{12})$
g. $\alpha_1 - \alpha_2 + \frac{1}{2}(\beta_{11} + \beta_{12} - \beta_{21} - \beta_{22})$

16.10. Use the model and the general form of estimable functions in exercise 16.8 to answer each of the following. In each case, explain how you arrived at your answer. [*Note:* In the nested model, the nested effects β_{ij} "contain" the α_i effects.]

a. How many degrees of freedom are there for SS(A)?
b. How many degrees of freedom are there for SS[$B(A)$]?
c. Which coefficients will be zero for the Type I sum of squares SS(A)? For the Type II SS(A)?
d. Which coefficients will be zero for the Type I sum of squares SS[$B(A)$]? For the Type II SS[$B(A)$]?
e. If there are unequal numbers of observations in the ij cells, explain how the coefficients for the Type II and Type III sums of squares for SS(A) will differ. (Do not try to give numbers.)

*16.11. Construct the population marginal means for A and $B(A)$ for the nested model in exercise 16.8. Are they all estimable if there are no empty cells? How do the population marginal means for A change if β_{ij} effects are random effects?

16.12. Construct an artificial set of data for the nested model in exercise 16.8 with $n_{11} = 2$, $n_{12} = 3$, $n_{21} = 1$, and $n_{22} = 2$, and use PROC GLM or a similar program to obtain the general linear form and the specific estimable functions for the sums of squares. Request the LSMEANS for A and $B(A)$. Compare the results with your answers to exercises 16.9–16.11. (It does not matter what you use for the values of the dependent variable because the estimable functions depend only on $\boldsymbol{X}$.) Use PROC GLM to determine how SAS defines the least squares mean for the first level of factor A when cell (1, 2) is empty.

*16.13. The numerical example of unbalanced data used in Section 16.5.3 had $n_{23} = 0$. Elsewhere it was stated that the Type IV sums of squares may not be unique. Relabel the first level of factor A as level 3 and the first level of factor B as level 4. [This moves the empty cell to the $(i, j) = (1, 2)$ position in the table.] Use PROC GLM or a similar program to obtain the analysis of the relabeled data. Be sure to request Type IV sums of squares. Are the Type IV sums of squares different from the results reported in Table 16.4? Do the estimates of the least squares means change?

16.14. The 1983 soybean data from Heagle (Table 15.8, page 430) contain one missing observation. Do the analysis of variance, using the general linear models approach (PROC GLM). Include block, ozone, moisture, and ozone × moisture interaction effects in the model. (Use the ozone treatment codes and ignore the slight differences in realized ozone levels.) Use the Type III sums of squares to interpret the results. Use the estimable functions to interpret the differences in the Types II and III sums of squares for ozone and moisture. Are all relevant least squares means estimable?

*16.15. Do exercise 16.14 using the 1984 soybean data from Heagle (Table 15.8).

16.16. In the split-plot example in Section 16.6 it was stated that $\sigma^2(Y_{ijk}) = \sigma^2 + \sigma_\delta^2$. Derive this result using the definition of variance

$$\sigma^2(Y_{ijk}) = \mathscr{E}\{[Y_{ijk} - \mathscr{E}(Y_{ijk})]^2\}$$

and the split-plot model given in the text. Derive the covariance of Y_{ijk} and $Y_{ijk'}$ using the definition

$$\text{Cov}(Y_{ijk}, Y_{ijk'}) = \mathscr{E}\{[Y_{ijk} - \mathscr{E}(Y_{ijk})][Y_{ijk'} - \mathscr{E}(Y_{ijk'})]\}$$

*16.17. You have a completely random experimental design with t treatments and r experimental units per treatment. The response of each experimental unit was determined by measuring the response variable on each of s random samples. This gives the model

$$Y_{ijk} = \mu + \tau_i + \varepsilon_{ij} + \delta_{ijk}$$

where τ_i are fixed treatment effects and ε_{ij} and δ_{ijk} are random experimental unit and sampling unit effects with zero means and variances σ^2 and σ_δ^2, respectively.

a. What is $\sigma^2(Y_{ijk})$? What is $\text{Cov}(Y_{ijk}, Y_{ijk'})$? Show the form of the variance–covariance matrix $\mathbf{Var}(\boldsymbol{Y})$.
b. What is the form of $\mathbf{Var}(\boldsymbol{Y})$ *if* the mean of all samples within each experimental unit is used as the response variable?
c. If the Y_{ijk} are used in the analysis using GLM, how are the standard errors of the treatment means as given by GLM computed? Are they correct? If not, how can they be corrected?
d. If the experimental unit means are used in the analysis, how are the standard errors of the treatment means computed in PROC GLM? Are they correct? What if the numbers of samples per experimental unit are not constant?
e. Explain the differences in assumptions between doing the analysis with a general linear models program such as GLM and with a program such as MIXMOD.

16.18. Use the corn borer data in exercise 8.4. Make the data unbalanced by assuming the first two observations in Days $= 3$ (the 17 and 22) and in Days $= 6$ (the 37 and 26) are missing. Analyze the data using (a) unweighted analysis of cell means, (b) weighted analysis of cell means, and (c) a general linear models procedure such as PROC GLM. Obtain the simple treatment means and the least squares treatment means. Do they differ? Why or why not?

*16.19. The Weber data, exercise 8.11, comprise a $2 \times 2 \times 5$ factorial in a randomized complete block design with $r = 2$ blocks. Make the data unbalanced by assuming that the two highest concentrations (80 and 100) of herbicide B could not be used at the high temperature (55°C). (Call all treatment factors class variables.) Include block effects, treatment main effects, and treatment interaction effects in the model. Use PROC GLM to analyze the data and obtain the simple and least squares treatment means.

a. Which sums of squares will you use for testing hypotheses about the treatment effects? Explain why you choose the particular set you do.
b. Which least squares means are nonestimable? Explain why these particular means are nonestimable. Do the results of the analysis let you simplify the model so that all relevant means are estimable?
c. Summarize the results with tables of relevant least squares means and their standard errors.

CHAPTER
SEVENTEEN

Case Study: Analysis of Unbalanced Data

Chapter 16 discussed the analysis of unbalanced data and introduced models with more than one random effect.

This case study illustrates the analysis of unbalanced data where the model contains more than one random effect. The classical less-than-full-rank effects model is used, and the results are compared to results obtained from MIXMOD, a program designed to handle models with more than one random effect.

The data for this case study are from a study of several management systems for corn production (courtesy of Dr. Gar House, North Carolina State University). The set of treatments was intended to be the $2 \times 2 \times 2$ factorial from the three factors method of tillage ($TILL$), herbicide application ($HERB$), and additional removal of weeds by hand ($CULT$). The levels of the treatment factors were conventional tillage (CT) and no tillage (NT) for the factor $TILL$, a recommended level of herbicide (H) and no herbicide (NOH) for the factor $HERB$, and hand weeding (C) and no hand weeding (NOC) for the factor $CULT$. The experimental design was a split-plot design with whole-plots in a randomized complete block design with four blocks. The whole-plot treatments were the four $TILL$–$HERB$ treatment combinations; the subplot treatments were the two levels of $CULT$. There are a total of $2^3 \times 4 = 32$ experimental units.

Cause of Imbalance

The data are unbalanced because the hand weeding (C) was not done on the no-tillage plots (NT) and, hence, the C level became an NOC treatment for those plots. In addition, the NOC–H–CT observation in block 1 is missing. (This observation was dropped for this case study to introduce more imbalance.) The number of observations per treatment are as follows:

	TILL:	*CT*		*NT*	
	HERB:	*H*	*NOH*	*H*	*NOH*
CULT	*C*	4	4	0	0
	NOC	3	4	8	8

Table 17.1 Yield in bushels per acre for the unbalanced 2 × 2 × 2 factorial study of cultural practices on corn yield. (Data courtesy of Dr. Gar House, North Carolina State University; used with permission.)

Treatment			*BLOCK*			
TILL	*HERB*	*CULT*	1	2	3	4
CT	*H*	*C*	75.38	92.11	79.59	94.22
CT	*H*	*NOC*	—	39.80	51.54	51.05
CT	*NOH*	*C*	16.59	61.88	68.06	94.50
CT	*NOH*	*NOC*	5.34	25.88	8.57	39.24
NT	*H*	*NOC*	51.47	71.16	45.84	77.06
NT	*H*	*NOC*	55.13	55.13	63.84	74.40
NT	*NOH*	*NOC*	0.00[a]	7.31	0.00	58.22
NT	*NOH*	*NOC*	0.00	0.00	0.00	31.78

[a] The zeros represent zero yield and not missing values.

This missing observation in the lower left-hand cell is from block 1. Otherwise all treatments were equally represented in each block. The data, yield of corn in bushels per acre, are given in Table 17.1.

Model

The linear effects model for the full 2 × 2 × 2 factorial in a split-plot arrangement is

$$Y_{ijkl} = \mu + B_i + T_j + H_k + TH_{jk} + \delta_{ijk} + C_l + TC_{jl} + HC_{kl} + THC_{jkl} + \varepsilon_{ijkl} \qquad [17.1]$$

where B_i, T_j, H_k, and C_l are block, tillage, herbicide, and cultivation effects, respectively, and products designate the respective interaction effects; $i = 1, 2, 3, 4$; $j = k = l = 1, 2$. In this study, however, the absence of the C level of the cultivation treatment factor when the tillage treatment is NT makes it impossible to estimate any *TILL* × *CULT* or *TILL* × *HERB* × *CULT* interactions. Therefore, the TC_{jl} and THC_{jkl} terms are dropped from the model, which is equivalent to imposing the constraints that these effects are zero. These constraints are reflected in the analysis. In this case, the *full* 2 × 2 × 2 factorial model gives somewhat larger Types III and IV SS(*HERB*) than the simpler model, and most of the least squares means are nonestimable, because the required two- and three-factor interaction effects are nonestimable.

Random Errors

The random error associated with subplots is designated by ε_{ijkl} and the whole-plot error is designated by δ_{ijk}. Both are assumed to be normally distributed with variances σ^2 and σ_δ^2, respectively. The presence of several zero yields in the *NT–NOH* treatment (five out of the eight are zero) raises the possibility that assumptions of normality and common variance over all treatments may not be satisfied. The large readings for the fourth block, however, show that the variation for these two treatments is comparable to that for the others. It is likely that, with the wide range in yields observed in this study, the variance will be associated with the mean yield level. For the purpose of demonstrating the analysis of unbalanced data, common variance and normality will be assumed.

It is left as an exercise for the student to investigate the need for a transformation to stabilize the variance.

Logical Comparisons

Due to the empty cells, the treatments are more appropriately described as the 2 × 2 factorial for *HERB* and *CULT* conducted at *TILL* = *CT*, and the 2 × 2 factorial for *TILL* and *HERB* conducted at *CULT* = *NOC*, with two treatments being common to the two sets. From this perspective, it is clear that the *HERB* effect, *CULT* effect, and *HERB* × *CULT* interaction effect can be estimated from the two-way table for *TILL* = *CT*, and the *TILL* effect, *HERB* effect, and *TILL* × *HERB* interaction effect can be estimated at the *NOC* level of the factor *CULT*. Notice that the *HERB* effect is estimated in both tables. These are logical contrasts one might generate if the analysis is approached from the cell means model point of view (Hocking, 1985). This case study will emphasize the analysis using the effects model.

Outline of the Analysis

The general linear model analysis for fixed models, assuming for the moment that the δ_{ijk} are fixed effects, will be used to partition the sums of squares and obtain the least squares means. (This ignores the covariance structure that exists among the Y_{ijkl} due to observations having common δ_{ijk}.) Then, the expectations of the mean squares will be determined with δ_{ijk} and ε_{ijkl} assumed to be random variables. The mean square expectations will be used to determine appropriate (approximate) tests of significance and to obtain better approximations of the standard errors of the least squares means. PROC GLM (SAS Institute, Inc., 1985d, 1985f) is used for the analysis with the RANDOM option providing the expectations of the Type III sums of squares. An interactive matrix language program (IML, SAS Institute, Inc., 1985a) is used to determine the correct variances of the least squares means.

17.1 The Analysis of Variance

The CLASS and MODEL statements for PROC GLM are

```
PROC GLM; CLASS BLOCK TILL HERB CULT;
MODEL Y = BLOCK TILL HERB TILL*HERB
          BLOCK*TILL*HERB CULT HERB * CULT /
          E E1 E2 E3 E4 SS1 SS2 SS3 SS4;
```

The sum of squares for the whole-plot error is computed as the three-factor interaction *BLOCK* × *TILL* × *HERB*. The sum of squares for the subplot error will appear as the residual sum of squares (labeled ERROR in PROC GLM). The options E, E1, E2, E3, and E4 request the general form of the estimable functions and the specific form of the estimable functions for each of the four types of sums of squares, respectively. (These options generate several pages of results and should not be requested unless needed for understanding the analysis.) The SS1, SS2, SS3, and SS4 options request that all four types of sums of squares be computed. Again, this is for illustration of this problem. Normally, the default option, which gives the Type I and Type III sums of squares, is adequate.

Table 17.2 Analysis of variance from the cultural practices study on yield, from PROC GLM, SAS.

Source	d.f.	*Sum of Squares*	*Mean Square*
Model	17	27,760.81	1,632.99
Error	13	1,554.23	119.56

		Sum of Squares			
Source	d.f.	*Type* I	*Type* II	*Type* III	*Type* IV
BLOCK	3	5,213.91	5,209.93	4,830.67	4,830.67
TILL	1	1,890.70	93.81	109.35	109.35
HERB	1	11,621.47	10,672.85	7,431.66	7,431.66
TILL × *HERB*	1	961.45	868.06	692.73	692.73
BLOCK × *TILL* × *HERB*	9	2,249.50	1,677.09	1,677.09	1,677.09
CULT	1	5,823.38	5,823.38	5,718.07	5,718.07
HERB × *CULT*	1	.39	.39	.39	.39

Analysis of Variance

The results of this analysis are summarized in Table 17.2. The sum of squares denoted "MODEL" by SAS is SS(Regr) in the notation of this text. The sum of squares labeled "ERROR" is the residual sum of squares, which in the split-plot analysis is an estimate of the subplot error. The bottom portion of Table 17.2 gives the four types of sums of squares for each class of effects in the model. The discussion in Chapter 16 noted that Type III sums of squares tested the most reasonable hypotheses in most cases of unbalanced data. However, for pedagogical purposes the results for all four types of sums of squares are discussed.

Estimating Whole-Plot Error

The degrees of freedom for *BLOCK* × *TILL* × *HERB* and Error sources of variation need explanation. Usually, an interaction sum of squares has degrees of freedom equal to the corresponding product of the degrees of freedom of the component main effects, which, in this case, would be 3 for the *BLOCK* × *TILL* × *HERB* interaction. However, the two-factor interactions *BLOCK* × *TILL* and *BLOCK* × *HERB* are not specified in the model and both are *contained* in the three-factor interaction. Consequently, the degrees of freedom and sums of squares for these two-factor interactions are absorbed by the three-factor interaction. The interactions of the whole-plot treatments with blocks in the split-plot model are estimates of whole-plot error and this specification of the model is a convenient technique of pooling these sums of squares.

Degrees of Freedom for Subplot Error

The residual sum of squares in the conventional split-plot design would have degrees of freedom determined by the pooling of the sums of squares for the interactions between block effects and subplot treatment and interaction effects. This would give 12 degrees of freedom if the data were balanced. In this case, the residual sum of squares is the pooling of *CULT* × *BLOCK*, with 3 degrees of freedom, *HERB* × *CULT* × *BLOCK*, with 3 degrees of freedom, and differences between duplicate plots of the *NT–NC* treatment in each level of *HERB* in each of the four blocks, 8 degrees of freedom, minus 1 degree of freedom for the missing plot.

Comparison of Sums of Squares

It is evident from the sums of squares that the data are not balanced because the four types of sums of squares differ. The largest adjustments in the sums of squares are for SS(*TILL*) (compare Type I with the others), and for SS(*HERB*) (compare Types I and II with Types III and IV). The difference between the simple averages of all plots receiving the *CT* treatment and all plots receiving the *NT* treatment is reflecting primarily the confounded cultivation effect, *C* versus *NOC*. Recall that none of the *NT* treated plots received the *C* cultivation treatment.

In this particular case Type III and Type IV sums of squares are the same for all sources of variation. This is because the imbalance did not give empty cells for any of the multiway tables defined by interaction terms in the model. For example, every cell in the $BLOCK \times TILL \times HERB$ table has two observations except the *CT–H* cell in block 1, which has one observation. If that one observation had also been missing, the Type III and Type IV sums of squares would have differed.

General Form of Estimable Functions

The estimable functions explicitly define the differences in the types of sums of squares. The general form for estimable functions for this model and this set of data is given in Table 17.3. The specific forms for the estimable functions for the four types of sums of squares are given for each source of variation in Tables 17.4–17.9 (pages 480–485).

Degrees of Freedom

The number of *free* coefficients in the general form of estimable functions, Table 17.3, for any particular class of effects shows the number of linearly independent contrasts for that class and the number of degrees of freedom for its sum of squares. The free coefficients for any class of effects are those coefficients in that class that are not involved in any other classes of effects *except* those that "contain" the effects in question. Thus, there are three "free" coefficients for the *BLOCK* effects, L_2, L_3, and L_4; the other coefficient in that set, L_1, is involved in the Intercept and, therefore, is not a free coefficient. L_2, L_3, and L_4 are involved in the $BLOCK \times TILL \times HERB$ interaction, but this is a class of effects that contains the *BLOCK* effects. There are nine free coefficients for the $BLOCK \times TILL \times HERB$ effects, L_{14} to L_{24} excluding L_{17} and L_{21}, and, hence, nine linearly independent contrasts and 9 degrees of freedom for its sums of squares. The remaining coefficients in the $BLOCK \times TILL \times HERB$ effects *must* be set equal to zero to remove all other effects. There are no other classes of effects that contain this class of effects.

Specific Estimable Functions for SS

The specific estimable functions in Tables 17.4–17.9 are determined from this general form. For example, the Type I estimable function for *BLOCK* sum of squares, Table 17.4, is obtained by

1. setting $L_1 = 0$ (to remove the intercept);
2. leaving L_2, L_3, and L_4 general as the free coefficients; and
3. setting all other coefficients to multiples of L_2, L_3, and L_4; $L_6 = L_8 = -.0714L_2$, $L_{10} = -.1071L_2$, $L_{14} = .1429L_2$, and so forth. These nonzero coefficients are functions of the numbers of observations and result from the computations adjusting the *BLOCK* sum of squares for μ. It is

Table 17.3 The general form of estimable functions for the unbalanced split-plot study.

Effect		*Coefficients*
Intercept		L_1
BLOCK	1	L_2
	2	L_3
	3	L_4
	4	$L_1 - L_2 - L_3 - L_4$
TILL	*CT*	L_6
	NT	$L_1 - L_6$
HERB	*H*	L_8
	NOH	$L_1 - L_8$
TILL × *HERB*	*CT H*	L_{10}
	CT NOH	$L_6 - L_{10}$
	NT H	$L_8 - L_{10}$
	NT NOH	$L_1 - L_6 - L_8 + L_{10}$
BLOCK × *TILL* × *HERB*	1 *CT H*	L_{14}
	1 *CT NOH*	L_{15}
	1 *NT H*	L_{16}
	1 *NT NOH*	$L_2 - L_{14} - L_{15} - L_{16}$
	2 *CT H*	L_{18}
	2 *CT NOH*	L_{19}
	2 *NT H*	L_{20}
	2 *NT NOH*	$L_3 - L_{18} - L_{19} - L_{20}$
	3 *CT H*	L_{22}
	3 *CT NOH*	L_{23}
	3 *NT H*	L_{24}
	3 *NT NOH*	$L_4 - L_{22} - L_{23} - L_{24}$
	4 *CT H*	$L_{10} - L_{14} - L_{18} - L_{22}$
	4 *CT NOH*	$L_6 - L_{10} - L_{15} - L_{19} - L_{23}$
	4 *NT H*	$L_8 - L_{10} - L_{16} - L_{20} - L_{24}$
	4 *NT NOH*	$L_1 - L_2 - L_3 - L_4 - L_6 - L_8 + L_{10}$ $+ L_{14} + L_{15} + L_{16} + L_{18} + L_{19}$ $+ L_{20} + L_{22} + L_{23} + L_{24}$
CULT	*C*	L_{30}
	NOC	$L_1 - L_{30}$
HERB × *CULT*	*H C*	L_{32}
	H NOC	$L_8 - L_{32}$
	NOH C	$L_{30} - L_{32}$
	NOH NOC	$L_1 - L_8 - L_{30} + L_{32}$

important to note which coefficients are nonzero and that they are functions of the numbers of observations.

The Type II and Type III (= Type IV) estimable functions for *BLOCK* sum of squares are obtained by

1. setting $L_1 = L_6 = L_8 = L_{10} = L_{30} = L_{32} = 0$ (to remove all other effects that do not contain *BLOCK* effects);
2. leaving L_2, L_3, and L_4 general; and

Table 17.4 The estimable functions for *BLOCK* sums of squares.

Effect		Coefficients: Type I	Type II	Type III = Type IV
Intercept		0	0	0
BLOCK	1	L_2	L_2	L_2
	2	L_3	L_3	L_3
	3	L_4	L_4	L_4
	4	$-L_2 - L_3 - L_4$	$-L_2 - L_3 - L_4$	$-L_2 - L_3 - L_4$
TILL	*CT*	$-.0714L_2$	0	0
	NT	$.0714L_2$	0	0
HERB	*H*	$-.0714L_2$	0	0
	NOH	$.0714L_2$	0	0
TILL × *HERB*	*CT H*	$-.1071L_2$	0	0
	CT NOH	$.0357L_2$	0	0
	NT H	$.0357L_2$	0	0
	NT NOH	$.0357L_2$	0	0
BLOCK × *TILL* × *HERB*	1 *CT H*	$.1429L_2$	$.1429L_2$	$.25L_2$
	1 *CT NOH*	$.2857L_2$	$.2857L_2$	$.25L_2$
	1 *NT H*	$.2857L_2$	$.2857L_2$	$.25L_2$
	1 *NT NOH*	$.2857L_2$	$.2857L_2$	$.25L_2$
	2 *CT H*	$.25L_3$	$.0357L_2 + .25L_3$	$.25L_3$
	2 *CT NOH*	$.25L_3$	$-.0119L_2 + .25L_3$	$.25L_3$
	2 *NT H*	$.25L_3$	$-.0119L_2 + .25L_3$	$.25L_3$
	2 *NT NOH*	$.25L_3$	$-.0119L_2 + .25L_3$	$.25L_3$
	3 *CT H*	$.25L_4$	$.0357L_2 + .25L_4$	$.25L_4$
	3 *CT NOH*	$.25L_4$	$-.0119L_2 + .25L_4$	$.25L_4$
	3 *NT H*	$.25L_4$	$-.0119L_2 + .25L_4$	$.25L_4$
	3 *NT NOH*	$.25L_4$	$-.0119L_2 + .25L_4$	$.25L_4$
	4 *CT H*	$-.25L_2 - .25L_3 - .25L_4$	$-.2143L_2 - .25L_3 - .25L_4$	$-.25L_2 - .25L_3 - .25L_4$
	4 *CT NOH*	$-.25L_2 - .25L_3 - .25L_4$	$-.2619L_2 - .25L_3 - .25L_4$	$-.25L_2 - .25L_3 - .25L_4$
	4 *NT H*	$-.25L_2 - .25L_3 - .25L_4$	$-.2619L_2 - .25L_3 - .25L_4$	$-.25L_2 - .25L_3 - .25L_4$
	4 *NT NOH*	$-.25L_2 - .25L_3 - .25L_4$	$-.2619L_2 - .25L_3 - .25L_4$	$-.25L_2 - .25L_3 - .25L_4$
CULT	*C*	$.0375L_2$	0	0
	NOC	$-.0375L_2$	0	0
HERB × *CULT*	*H C*	$.0179L_2$	0	0
	H NOC	$-.0893L_2$	0	0
	NOH C	$.0179L_2$	0	0
	NOH NOC	$.0536L_2$	0	0

Table 17.5 The estimable functions for *TILL* sums of squares.

Effect		*Coefficients* Type I	Type II	Type III = Type IV
Intercept		0	0	0
BLOCK	1	0	0	0
	2	0	0	0
	3	0	0	0
	4	0	0	0
TILL	*CT*	L_6	L_6	L_6
	NT	$-L_6$	$-L_6$	$-L_6$
HERB	*H*	$-.037L_6$	0	0
	NOH	$.037L_6$	0	0
TILL × *HERB*	*CT H*	$.463L_6$	$.4437L_6$	$.5L_6$
	CT NOH	$.537L_6$	$.5563L_6$	$.5L_6$
	NT H	$-.5L_6$	$-.4437L_6$	$-.5L_6$
	NT NOH	$-.5L_6$	$-.5563L_6$	$-.5L_6$
BLOCK × *TILL* × *HERB*	1 *CT H*	$.0741L_6$	$.0158L_6$	$.125L_6$
	1 *CT NOH*	$.1481L_6$	$.1708L_6$	$.125L_6$
	1 *NT H*	$-.1111L_6$	$-.0792L_6$	$-.125L_6$
	1 *NT NOH*	$-.1111L_6$	$-.1074L_6$	$-.125L_6$
	2 *CT H*	$.1296L_6$	$.1426L_6$	$.125L_6$
	2 *CT NOH*	$.1296L_6$	$.1285L_6$	$.125L_6$
	2 *NT H*	$-.1296L_6$	$-.1215L_6$	$-.125L_6$
	2 *NT NOH*	$-.1296L_6$	$-.1496L_6$	$-.125L_6$
	3 *CT H*	$.1296L_6$	$.1426L_6$	$.125L_6$
	3 *CT NOH*	$.1296L_6$	$.1285L_6$	$.125L_6$
	3 *NT H*	$-.1296L_6$	$-.1215L_6$	$-.125L_6$
	3 *NT NOH*	$-.1296L_6$	$-.1496L_6$	$-.125L_6$
	4 *CT H*	$.1296L_6$	$.1426L_6$	$.125L_6$
	4 *CT NOH*	$.1296L_6$	$.1285L_6$	$.125L_6$
	4 *NT H*	$-.1296L_6$	$-.1215L_6$	$-.125L_6$
	4 *NT NOH*	$-.1296L_6$	$-.1496L_6$	$-.125L_6$
CULT	*C*	$.537L_6$	0	0
	NOC	$-.537L_6$	0	0
HERB × *CULT*	*H C*	$.2685L_6$	0	0
	H NOC	$-.3056L_6$	0	0
	NOH C	$.2685L_6$	0	0
	NOH NOC	$-.2315L_6$	0	0

3. setting all other coefficients, L_{14} to L_{24}, to multiples of L_2, L_3, and L_4. The multiples for the Type II estimable functions are determined by the computations involved in forcing L_1, L_6, L_8, L_{10}, L_{30}, and L_{32} to be zero. The multiples for the Type III estimable functions are chosen to satisfy the orthogonality and equitable distribution properties.

Nonzero coefficients for *TILL*, *HERB*, *TILL* × *HERB*, *CULT*, and *HERB* × *CULT* effects in the estimable function for the Type I sum of squares for *BLOCK* (Table 17.4) result from the *sequential* nature of the Type I sums of

Table 17.6 The estimable functions for *HERB* sums of squares.

Effect		Coefficients		
		Type I	*Type* II	*Type* III = *Type* IV
Intercept		0	0	0
BLOCK	1	0	0	0
	2	0	0	0
	3	0	0	0
	4	0	0	0
TILL	*CT*	0	0	0
	NT	0	0	0
HERB	*H*	L_8	L_8	L_8
	NOH	$-L_8$	$-L_8$	$-L_8$
TILL × *HERB*	*CT H*	$.4808L_8$	$.4792L_8$	$.5L_8$
	CT NOH	$-.4808L_8$	$-.4792L_8$	$-.5L_8$
	NT H	$.5192L_8$	$.5208L_8$	$.5L_8$
	NT NOH	$-.5192L_8$	$-.5208L_8$	$-.5L_8$
BLOCK × *TILL* × *HERB*	1 *CT H*	$.0769L_8$	$.0729L_8$	$.125L_8$
	1 *CT NOH*	$-.1058L_8$	$-.1042L_8$	$-.125L_8$
	1 *NT H*	$.1442L_8$	$.1458L_8$	$.125L_8$
	1 *NT NOH*	$-.1154L_8$	$-.1146L_8$	$-.125L_8$
	2 *CT H*	$.1346L_8$	$.1354L_8$	$.125L_8$
	2 *CT NOH*	$-.125L_8$	$-.125L_8$	$-.125L_8$
	2 *NT H*	$.125L_8$	$.125L_8$	$.125L_8$
	2 *NT NOH*	$-.1346L_8$	$-.1354L_8$	$-.125L_8$
	3 *CT H*	$.1346L_8$	$.1354L_8$	$.125L_8$
	3 *CT NOH*	$-.125L_8$	$-.125L_8$	$-.125L_8$
	3 *NT H*	$.125L_8$	$.125L_8$	$.125L_8$
	3 *NT NOH*	$-.1346L_8$	$-.1354L_8$	$-.125L_8$
	4 *CT H*	$.1346L_8$	$.1354L_8$	$.125L_8$
	4 *CT NOH*	$-.125L_8$	$-.125L_8$	$-.125L_8$
	4 *NT H*	$.125L_8$	$.125L_8$	$.125L_8$
	4 *NT NOH*	$-.1346L_8$	$-.1354L_8$	$-.125L_8$
CULT	*C*	$.0385L_8$	0	0
	NOC	$-.0385L_8$	0	0
HERB × *CULT*	*H C*	$.2788L_8$	$.2604L_8$	$.5L_8$
	H NOC	$.7212L_8$	$.7396L_8$	$.5L_8$
	NOH C	$-.2404L_8$	$-.2604L_8$	$-.5L_8$
	NOH NOC	$-.7596L_8$	$-.7396L_8$	$-.5L_8$

squares. Type I sums of squares for a particular effect are adjusted *only* for effects that precede it in the model. Consequently, the *BLOCK* sum of squares, being first in the model statement, is adjusted only for μ. Clearly, the Type I *BLOCK* sum of squares is confounded with all other effects in the model. On the other hand, the Type II and Type III *BLOCK* sums of squares have been *adjusted* for all effects that do not contain the *BLOCK* effects by setting L_6, L_8, L_{10}, L_{30}, and L_{32} equal to zero. The Type II and Type III sums of squares differ from each other only in how the multiples of L_2, L_3, and L_4 are chosen for L_{14} to L_{24}. For the Type II sums of squares, the coefficients are allowed to take whatever values

Table 17.7 The estimable functions for *TILL* × *HERB* sums of squares.

Effect		*Coefficients* *Type* I	*Type* II	*Type* III = *Type* IV
Intercept		0	0	0
BLOCK	1	0	0	0
	2	0	0	0
	3	0	0	0
	4	0	0	0
TILL	*CT*	0	0	0
	NT	0	0	0
HERB	*H*	0	0	0
	NOH	0	0	0
TILL × *HERB*	*CT H*	L_{10}	L_{10}	L_{10}
	CT NOH	$-L_{10}$	$-L_{10}$	$-L_{10}$
	NT H	$-L_{10}$	$-L_{10}$	$-L_{10}$
	NT NOH	L_{10}	L_{10}	L_{10}
BLOCK × *TILL* × *HERB*	1 *CT H*	$.16L_{10}$	$.0375L_{10}$	$.25L_{10}$
	1 *CT NOH*	$-.22L_{10}$	$-.1786L_{10}$	$-.25L_{10}$
	1 *NT H*	$-.22L_{10}$	$-.1786L_{10}$	$-.25L_{10}$
	1 *NT NOH*	$.28L_{10}$	$.3214L_{10}$	$.25L_{10}$
	2 *CT H*	$.28L_{10}$	$.3214L_{10}$	$.25L_{10}$
	2 *CT NOH*	$-.26L_{10}$	$-.2738L_{10}$	$-.25L_{10}$
	2 *NT H*	$-.26L_{10}$	$-.2738L_{10}$	$-.25L_{10}$
	2 *NT NOH*	$.24L_{10}$	$.2262L_{10}$	$.25L_{10}$
	3 *CT H*	$.28L_{10}$	$.3214L_{10}$	$.25L_{10}$
	3 *CT NOH*	$-.26L_{10}$	$-.2738L_{10}$	$-.25L_{10}$
	3 *NT H*	$-.26L_{10}$	$-.2738L_{10}$	$-.25L_{10}$
	3 *NT NOH*	$.24L_{10}$	$.2262L_{10}$	$.25L_{10}$
	4 *CT H*	$.28L_{10}$	$.3214L_{10}$	$.25L_{10}$
	4 *CT NOH*	$-.26L_{10}$	$-.2738L_{10}$	$-.25L_{10}$
	4 *NT H*	$-.26L_{10}$	$-.2738L_{10}$	$-.25L_{10}$
	4 *NT NOH*	$.24L_{10}$	$.2262L_{10}$	$.25L_{10}$
CULT	*C*	$.08L_{10}$	0	0
	NOC	$-.08L_{10}$	0	0
HERB × *CULT*	*H C*	$.58L_{10}$	0	0
	H NOC	$-.58L_{10}$	0	0
	NOH C	$-.5L_{10}$	0	0
	NOH NOC	$.5L_{10}$	0	0

they happen to have after the adjustment for other effects is completed. In general, these are functions of the numbers of observations in the cells. For Type III and Type IV, the multiples of L_2, L_3, and L_4 are chosen to satisfy the orthogonality property (Type III) or the equitable distribution property (Type IV) for the higher-order interaction effects that contain *BLOCK* effects. In this case, both properties are satisfied with the same choices and, consequently, Type III = Type IV.

Using Estimable Functions to Compare SS

Each of Tables 17.4–17.9 contains the estimable functions for all four types of sums of squares for one source of variation. The sequence of the tables

Table 17.8 The estimable functions for *BLOCK* × *TILL* × *HERB* sums of squares.

Effect		*Coefficients*	
		Type I	*Type* II = *Type* III = *Type* IV
Intercept		0	0
BLOCK	1	0	0
	2	0	0
	3	0	0
	4	0	0
TILL	*CT*	0	0
	NT	0	0
HERB	*H*	0	0
	NOH	0	0
TILL × *HERB*	*CT H*	0	0
	CT NOH	0	0
	NT H	0	0
	NT NOH	0	0
BLOCK × *TILL* × *HERB*	1 *CT H*	L_{14}	L_{14}
	1 *CT NOH*	L_{15}	L_{15}
	1 *NT H*	L_{16}	L_{16}
	1 *NT NOH*	$-L_{14} - L_{15} - L_{16}$	$-L_{14} - L_{15} - L_{16}$
	2 *CT H*	L_{18}	L_{18}
	2 *CT NOH*	L_{19}	L_{19}
	2 *NT H*	L_{20}	L_{20}
	2 *NT NOH*	$-L_{18} - L_{19} - L_{20}$	$-L_{18} - L_{19} - L_{20}$
	3 *CT H*	L_{22}	L_{22}
	3 *CT NOH*	L_{23}	L_{23}
	3 *NT H*	L_{24}	L_{24}
	3 *NT NOH*	$-L_{22} - L_{23} - L_{24}$	$-L_{22} - L_{23} - L_{24}$
	4 *CT H*	$-L_{14} - L_{18} - L_{22}$	$-L_{14} - L_{18} - L_{22}$
	4 *CT NOH*	$-L_{15} - L_{19} - L_{23}$	$-L_{15} - L_{19} - L_{23}$
	4 *NT H*	$-L_{16} - L_{20} - L_{24}$	$-L_{16} - L_{20} - L_{24}$
	4 *NT NOH*	$L_{14} + L_{15} + L_{16} + L_{18} + L_{19}$ $+ L_{20} + L_{22} + L_{23} + L_{24}$	$L_{14} + L_{15} + L_{16} + L_{18} + L_{19}$ $+ L_{20} + L_{22} + L_{23} + L_{24}$
CULT	*C*	$.5L_{14}$	0
	NOC	$-.5L_{14}$	0
HERB × *CULT*	*H C*	$.5L_{14}$	0
	H NOC	$-.5L_{14}$	0
	NOH C	0	0
	NOH NOC	0	0

corresponds to the order in which the class variables were entered into the model statement. Thus, comparison of the Type I estimable functions from table to table shows the sequential nature of the Type I sums of squares. The Type I estimable function for *BLOCK* sum of squares, Table 17.4, contains nonzero coefficients for all effects other than the Intercept; it is confounded with all other effects. The Type I estimable function for *TILL*, Table 17.5, has zero coefficients for *BLOCK* effects but nonzero coefficients for all succeeding classes of effects; this sum of squares is adjusted for *BLOCK* effects but is confounded with all classes of effects that follow *TILL* in the model statement. Inspection of the remaining tables

Table 17.9 The estimable functions for *CULT* sums of squares.

Effect		*Coefficients*	
		Type I = *Type* II	*Type* III = *Type* IV
Intercept		0	0
BLOCK	1	0	0
	2	0	0
	3	0	0
	4	0	0
TILL	*CT*	0	0
	NT	0	0
HERB	*H*	0	0
	NOH	0	0
TILL × *HERB*	*CT H*	0	0
	CT NOH	0	0
	NT H	0	0
	NT NOH	0	0
BLOCK × *TILL* × *HERB*	1 *CT H*	0	0
	1 *CT NOH*	0	0
	1 *NT H*	0	0
	1 *NT NOH*	0	0
	2 *CT H*	0	0
	2 *CT NOH*	0	0
	2 *NT H*	0	0
	2 *NT NOH*	0	0
	3 *CT H*	0	0
	3 *CT NOH*	0	0
	3 *NT H*	0	0
	3 *NT NOH*	0	0
	4 *CT H*	0	0
	4 *CT NOH*	0	0
	4 *NT H*	0	0
	4 *NT NOH*	0	0
CULT	*C*	L_{30}	L_{30}
	NOC	$-L_{30}$	$-L_{30}$
HERB × *CULT*	*H C*	$.4286L_{30}$	$.5L_{30}$
	H NOC	$-.4286L_{30}$	$-.5L_{30}$
	NOH C	$.5714L_{30}$	$.5L_{30}$
	NOH NOC	$-.5714L_{30}$	$-.5L_{30}$

shows that this pattern continues for successive terms in the model. The estimable function for the last term in the model, *HERB* × *CULT*, is the same for all types of sums of squares and is not given in a separate table. Being the last term in the model, the Type I *HERB* × *CULT* estimable function has zero coefficients for *all* other effects.

To reiterate, the Type I estimable function for each class of effects:

1. is adjusted only for other classes of effects that precede it in the model statement and, consequently, is confounded with all classes of effects that follow it in the model;

2. has coefficients on the effects for which it is not adjusted that are dependent on the cell numbers; and
3. has coefficients that have neither the orthogonality property nor the equitable distribution property. This confounding of different classes of effects and the dependence of the coefficients on the numbers of observations make the Type I sums of squares inappropriate for testing hypotheses in this example.

Comparisons of the Type II and Type III estimable functions across Tables 17.4–17.9 show the following:

1. The same classes of effects are involved (have nonzero coefficients) in both Type II and Type III estimable functions in all cases. The Types II, III, and IV estimable functions always have zero coefficients for all effects that do not contain the effects in question.
2. The coefficients on the Type II effects are functions of the cell numbers, whereas the Type III coefficients have the orthogonality and equitable distribution properties. In effect, there have been additional constraints imposed on the Type III (and Type IV) estimable functions to reproduce conditions like those that would be obtained with balanced data. Since there are no empty cells in this example, the equitable distribution property and the orthogonality property characteristic of balanced data are satisfied by the same choice of coefficients and, hence, Type III and Type IV estimable functions are the same.

Summary

In summary, the Type I sums of squares are not appropriate for testing hypotheses in analysis of variance problems. The hypotheses being tested by the Type II sums of squares suffer from being dependent on the particular set of n_{ij} that occurred in that particular experiment. The Type III and Type IV hypotheses are the same, when there are no empty cells (in the multiway tables defined by the effects in the model), and are the same as the hypotheses being tested by the analysis of variance sums of squares in balanced data. Thus, the Type III (and IV) sums of squares would seem to be the most appropriate in this example for testing hypotheses that various classes of effects are zero. When there are empty cells (not illustrated with this example), the Type IV hypotheses may not be unique and for this reason the Type III sums of squares would be preferred.

17.2 Mean Square Expectations and Choice of Errors

Before we turn to interpretation of the analysis of variance, the analysis based on a fixed effects model must be reconciled with the fact that the correct model contains two random effects—the whole-plot effect, δ_{ijk}, and the subplot effect, ε_{ijkl}. With balanced data, the whole-plot error is estimated with the interaction mean square between blocks and the whole-plot treatments, in this case, the $BLOCK \times TILL \times HERB$ mean square. With unbalanced data, the expectations of the mean squares must be used to determine proper error terms. The

Table 17.10 Expectations of Type III mean squares for the split-plot experiment using the RANDOM option in PROC GLM.

Mean Square	*Expectation of Mean Square*[a]
BLOCK	$\sigma^2 + 1.8667\sigma_\delta^2 + Q(BLOCK)$
TILL	$\sigma^2 + 1.0909\sigma_\delta^2 + Q(TILL, TILL \times HERB)$
HERB	$\sigma^2 + 1.3333\sigma_\delta^2 + Q(HERB, TILL \times HERB, HERB \times CULT)$
TILL × *HERB*	$\sigma^2 + 1.0909\sigma_\delta^2 + Q(TILL \times HERB)$
ERROR A[b]	$\sigma^2 + 1.9048\sigma_\delta^2$
CULT	$\sigma^2 + Q(CULT, HERB \times CULT)$
HERB × *CULT*	$\sigma^2 + Q(HERB \times CULT)$
ERROR B	σ^2

[a] $Q(\cdot)$ is a quadratic function of the effects in parentheses.
[b] ERROR A = MS(*BLOCK* × *TILL* × *HERB*).

RANDOM statement in PROC GLM was used to obtain these expectations. The residual mean square always has expectation σ^2, where σ^2 is the true variance of the unique random element in the model, ε_{ijkl} in this case. Thus, $s^2 = 119.56$ with 13 degrees of freedom is the estimate of the subplot error variance.

Expectations Involving Only σ^2

The expectations of the Type III mean squares in the analysis are given in Table 17.10. The random component in the expectations of *CULT* and *HERB* × *CULT* mean squares is σ^2. This confirms that the subplot error, ERROR B, is the appropriate error term for testing hypotheses about *CULT* and *HERB* × *CULT* effects, the subplot treatment comparisons, as is the case with balanced data.

Tests Using Error B

The variance ratio for *HERB* × *CULT* interaction is less than unity, suggesting that the herbicide effects and the cultivation effects are additive. The variance ratio for *CULT* effects is $F = 5{,}718.07/119.56 = 47.8$, which is highly significant; that is, the average difference in yield between the *CULT* treatments is too large to be explained by random variation. The absence of a *HERB* × *CULT* interaction indicates that this effect of hand weeding is consistent over both herbicide levels. Recall that the information on the *HERB* × *CULT* interaction effects and the *CULT* effects comes only from data on conventional tillage, *TILL* = *CT*. These conclusions can be extended to the *TILL* = *NT* treatment only if there is no interaction of these effects with *TILL*. This was implicitly assumed when the *TILL* × *CULT* and *TILL* × *CULT* × *HERB* interaction effects were dropped from the model, but these assumptions cannot be tested with these data.

Expectations Involving Both σ^2 and σ_δ^2

The random components in the expectations of the remaining mean squares are not the same as in balanced data. If the data were balanced, the expectation of the mean square for the whole-plot error (ERROR A) would contain $\sigma^2 + k\sigma_\delta^2$, where k is the number of subplots per whole-plot. The expectations of all whole-plot treatment mean squares also would contain $\sigma^2 + k\sigma_\delta^2$, plus a quadratic function of fixed effects, so that ERROR A would be the appropriate error mean square for all tests of whole-plot treatment effects. With this unbalanced example, the coefficients on σ_δ^2 for *TILL*, *HERB*, and *TILL* × *HERB* differ from that for ERROR A (Table 17.10). Thus, ERROR A is not the appropriate error for tests of significance. (If the coefficients were very similar, one might be content to use ERROR A in approximate tests of hypotheses about whole-plot treatment effects. In this case, the coefficients are quite different, 1.0909 versus 1.9048, so that tests using ERROR A could be seriously biased unless σ_δ^2 is close to zero.)

Constructed Error Mean Squares

When the coefficients are more than trivially different, it is better to construct for each F-test an error mean square that has the same expectation for the random elements as the numerator mean square. The constructed error mean square for testing *TILL* and *TILL* × *HERB* effects is that linear function of ERROR A (E_a) and ERROR B (E_b) that has expectation $\sigma^2 + 1.0909\sigma_\delta^2$. Thus,

$$E' = \frac{1.0909}{1.9048}E_a + \left(1 - \frac{1.0909}{1.9048}\right)E_b = 157.81$$

The degrees of freedom for this estimate of error are approximated with Satterthwaite's approximation as

$$\begin{aligned} f' &= \frac{(\sum a_i \text{MS}_i)^2}{\sum (a_i^2 \text{MS}_i^2/f_i)} \\ &= \frac{(157.8080)^2}{\left(\frac{1.0909}{1.9048}\right)^2 \frac{(186.34)^2}{9} + \left(1 - \frac{1.0909}{1.9048}\right)^2 \frac{(119.56)^2}{13}} \\ &= 16.98 \quad \text{or 17 degrees of freedom} \end{aligned}$$

With this constructed error term, the variance ratio for *TILL* × *HERB* is $F' = 4.39$, which just misses being significant at $\alpha = .05$, $F_{(.05;1,17)} = 4.45$. If one adheres strictly to the chosen α, the interaction effect between *TILL* and *HERB* would be declared unimportant. However, one would probably report the herbicide effects at each tillage level and then point out that the differences were not quite significant (at $\alpha = .05$). The variance ratio for the test of *TILL* effects *averaged* over the levels of *HERB* is $F' = .69$, which is not significant. This does not imply that the tillage effects are negligible *within* each herbicide treatment.

The constructed error term for testing *HERB* effects is $E' = 166.31$ with approximate degrees of freedom $f' = 14$. The variance ratio for this test is $F' = 44.68$, far exceeding the critical level for $\alpha = .01$. Unlike the *TILL* and *CULT* main effects, information on the *HERB* effect comes from both two-way tables. This *average* herbicide effect, averaged over *TILL* and *CULT* treatments, is significantly different from zero but the (nearly) significant *TILL* × *HERB* interaction suggests that the herbicide effect is not the same for the two tillage levels.

Summary of Analysis of Variance

To summarize the results of the analysis of variance, the near significance of the interaction between *TILL* and *HERB* suggests that the yield response to herbicide depends on whether conventional tillage or no tillage is used. The *average* herbicide effect is significant but is somewhat difficult to interpret because it is an average from the two two-way factorials, one of which shows an interaction. The average cultivation effect is different from zero and its effects are relatively constant over levels of *HERB* as observed under the *TILL* = *CT* treatment. These results suggest that the effects of the treatments can be summarized in the two-way table of *TILL* × *HERB* means and the marginal means for *CULT*.

17.3 Least Squares Means and Standard Errors

The least squares means are computed as the linear functions of $\boldsymbol{\beta}^0$ that have the same expectations as the corresponding means in balanced data, the population marginal means. In the tillage–herbicide–cultivation study, there are no empty cells for any of the effects defined in the model, so all population marginal means are estimable. (It was recognized that there was no information in the data on two of the interactions, and their effects were dropped from the model. If these effects had been retained, many of the population marginal means would not have been estimable.)

Expectations of Least Squares Means

The expectations of the least squares marginal means for the herbicide treatments, *HERB*, and the cultivation treatments, *CULT*, are given in Table 17.11 (page 490). (For comparison, the expectation of the *unadjusted* mean for the *C* level of *CULT* is also given. The differences in coefficients between the last column and the third column show the nature of the confounding in this unadjusted mean. The coefficient of 1.0 on the *CT* effect of the *TILL* factor shows that the unadjusted *C* mean is completely confounded with the *CT* effect.) The estimable functions for the two-way table of *TILL* × *HERB* means are given in Table 17.12 (page 491). The coefficients in each column of Tables 17.11 and 17.12 define the linear functions of $\boldsymbol{\beta}^0$ that must be computed to obtain the least squares mean.

Estimates of Means and Interpretations

The least squares marginal means for all three treatment factors and the two-way *TILL* × *HERB* treatment means are given in column 2 in Table 17.13 (page 492). The *unadjusted* treatment means are given in the last column of the table for comparison only. All interpretations should be based on the least squares means. The tests of significance have indicated that the *CT* and the *NT* means for tillage are not different. (The unadjusted tillage means, on the other hand, were very different—53.58 versus 36.96. The adjustment is primarily on the *NT* treatment mean and is reflecting its total confounding with the *NOC* treatment. The *NT* treatment did not involve any plots on which there was additional hand weeding.)

The difference between the herbicide treatment means is significant; the presence of herbicide more than doubled yield in this experiment. Similarly, additional hand weeding, *C*, doubled yield. It must not be overlooked, however, that there was no measure of the interaction between *CULT* and *TILL* because hand weeding, *C*, was done only on the no-tillage, *NT*, plots. Thus, it would be an extrapolation to imply that hand weeding would have this same effect on the conventional-tillage plots.

The two-way *TILL* × *HERB* means are given because the interaction was close to significance at $\alpha = .05$. The pattern of the means in this two-way table

Table 17.11 The estimable functions for the least squares means for levels of herbicide (*HERB*) and cultivation (*CULT*). The unadjusted *C* mean is given for comparison.

		CULT		*HERB*		
Effect		C	*NOC*	*H*	*NOH*	*Unadj.* C *Mean*
Intercept		1	1	1	1	1
BLOCK	1	$\frac{1}{4}$	$\frac{1}{4}$	$\frac{1}{4}$	$\frac{1}{4}$	$\frac{1}{4}$
	2	$\frac{1}{4}$	$\frac{1}{4}$	$\frac{1}{4}$	$\frac{1}{4}$	$\frac{1}{4}$
	3	$\frac{1}{4}$	$\frac{1}{4}$	$\frac{1}{4}$	$\frac{1}{4}$	$\frac{1}{4}$
	4	$\frac{1}{4}$	$\frac{1}{4}$	$\frac{1}{4}$	$\frac{1}{4}$	$\frac{1}{4}$
TILL	*CT*	$\frac{1}{2}$	$\frac{1}{2}$	$\frac{1}{2}$	$\frac{1}{2}$	1
	NT	$\frac{1}{2}$	$\frac{1}{2}$	$\frac{1}{2}$	$\frac{1}{2}$	0
HERB	*H*	$\frac{1}{2}$	$\frac{1}{2}$	1	0	$\frac{1}{2}$
	NOH	$\frac{1}{2}$	$\frac{1}{2}$	0	1	$\frac{1}{2}$
TILL × *HERB*	*CT H*	$\frac{1}{4}$	$\frac{1}{4}$	$\frac{1}{2}$	0	$\frac{1}{2}$
	CT NOH	$\frac{1}{4}$	$\frac{1}{4}$	0	$\frac{1}{2}$	$\frac{1}{2}$
	NT H	$\frac{1}{4}$	$\frac{1}{4}$	$\frac{1}{2}$	0	0
	NT NOH	$\frac{1}{4}$	$\frac{1}{4}$	0	$\frac{1}{2}$	0
BLOCK × *TILL* × *HERB*	1 *CT H*	$\frac{1}{16}$	$\frac{1}{16}$	$\frac{1}{8}$	0	$\frac{1}{8}$
	1 *CT NOH*	$\frac{1}{16}$	$\frac{1}{16}$	0	$\frac{1}{8}$	$\frac{1}{8}$
	1 *NT H*	$\frac{1}{16}$	$\frac{1}{16}$	$\frac{1}{8}$	0	0
	1 *NT NOH*	$\frac{1}{16}$	$\frac{1}{16}$	0	$\frac{1}{8}$	0
	2 *CT H*	$\frac{1}{16}$	$\frac{1}{16}$	$\frac{1}{8}$	0	$\frac{1}{8}$
	2 *CT NOH*	$\frac{1}{16}$	$\frac{1}{16}$	0	$\frac{1}{8}$	$\frac{1}{8}$
	2 *NT H*	$\frac{1}{16}$	$\frac{1}{16}$	$\frac{1}{8}$	0	0
	2 *NT NOH*	$\frac{1}{16}$	$\frac{1}{16}$	0	$\frac{1}{8}$	0
	3 *CT H*	$\frac{1}{16}$	$\frac{1}{16}$	$\frac{1}{8}$	0	$\frac{1}{8}$
	3 *CT NOH*	$\frac{1}{16}$	$\frac{1}{16}$	0	$\frac{1}{8}$	$\frac{1}{8}$
	3 *NT H*	$\frac{1}{16}$	$\frac{1}{16}$	$\frac{1}{8}$	0	0
	3 *NT NOH*	$\frac{1}{16}$	$\frac{1}{16}$	0	$\frac{1}{8}$	0
	4 *CT H*	$\frac{1}{16}$	$\frac{1}{16}$	$\frac{1}{8}$	0	$\frac{1}{8}$
	4 *CT NOH*	$\frac{1}{16}$	$\frac{1}{16}$	0	$\frac{1}{8}$	$\frac{1}{8}$
	4 *NT H*	$\frac{1}{16}$	$\frac{1}{16}$	$\frac{1}{8}$	0	0
	4 *NT NOH*	$\frac{1}{16}$	$\frac{1}{16}$	0	$\frac{1}{8}$	0
CULT	*C*	1	0	$\frac{1}{2}$	$\frac{1}{2}$	1
	NOC	0	1	$\frac{1}{2}$	$\frac{1}{2}$	0
HERB × *CULT*	*H C*	$\frac{1}{2}$	0	$\frac{1}{2}$	0	$\frac{1}{2}$
	H NOC	0	$\frac{1}{2}$	$\frac{1}{2}$	0	0
	NOH C	$\frac{1}{2}$	0	0	$\frac{1}{2}$	$\frac{1}{2}$
	NOH NOC	0	$\frac{1}{2}$	0	$\frac{1}{2}$	0

suggests that no tillage, *NT*, is better than conventional tillage, *CT*, when herbicide is being used, but is slightly worse if no herbicide is used. The herbicide effect is positive under both types of tillage but the difference is much larger in the *NT* treatment. It appears from this study that it is better to use herbicide and, if herbicide is to be used, to also use the no-tillage method.

Standard Errors

Columns 3–5 in Table 17.13 give standard errors of the least squares means computed according to different rules. The first column of standard errors, labeled "GLM", are as given by PROC GLM. The GLM standard errors are

Table 17.12 The estimable functions for the two-way table of least squares means for levels of tillage (*TILL*) and herbicide (*HERB*).

Effect		*CT H*	*CT NOH*	*NT H*	*NT NOH*
Intercept		1	1	1	1
BLOCK	1	$\frac{1}{4}$	$\frac{1}{4}$	$\frac{1}{4}$	$\frac{1}{4}$
	2	$\frac{1}{4}$	$\frac{1}{4}$	$\frac{1}{4}$	$\frac{1}{4}$
	3	$\frac{1}{4}$	$\frac{1}{4}$	$\frac{1}{4}$	$\frac{1}{4}$
	4	$\frac{1}{4}$	$\frac{1}{4}$	$\frac{1}{4}$	$\frac{1}{4}$
TILL	*CT*	1	1	0	0
	NT	0	0	1	1
HERB	*H*	1	0	1	0
	NOH	0	1	0	1
TILL × *HERB*	*CT H*	1	0	0	0
	CT NOH	0	1	0	0
	NT H	0	0	1	0
	NT NOH	0	0	0	1
BLOCK × *TILL* × *HERB*	1 *CT H*	$\frac{1}{4}$	0	0	0
	1 *CT NOH*	0	$\frac{1}{4}$	0	0
	1 *NT H*	0	0	$\frac{1}{4}$	0
	1 *NT NOH*	0	0	0	$\frac{1}{4}$
	2 *CT H*	$\frac{1}{4}$	0	0	0
	2 *CT NOH*	0	$\frac{1}{4}$	0	0
	2 *NT H*	0	0	$\frac{1}{4}$	0
	2 *NT NOH*	0	0	0	$\frac{1}{4}$
	3 *CT H*	$\frac{1}{4}$	0	0	0
	3 *CT NOH*	0	$\frac{1}{4}$	0	0
	3 *NT H*	0	0	$\frac{1}{4}$	0
	3 *NT NOH*	0	0	0	$\frac{1}{4}$
	4 *CT H*	$\frac{1}{4}$	0	0	0
	4 *CT NOH*	0	$\frac{1}{4}$	0	0
	4 *NT H*	0	0	$\frac{1}{4}$	0
	4 *NT NOH*	0	0	0	$\frac{1}{4}$
CULT	*C*	$\frac{1}{2}$	$\frac{1}{2}$	$\frac{1}{2}$	$\frac{1}{2}$
	NOC	$\frac{1}{2}$	$\frac{1}{2}$	$\frac{1}{2}$	$\frac{1}{2}$
HERB × *CULT*	*H C*	$\frac{1}{2}$	0	$\frac{1}{2}$	0
	H NOC	$\frac{1}{2}$	0	$\frac{1}{2}$	0
	NOH C	0	$\frac{1}{2}$	0	$\frac{1}{2}$
	NOH NOC	0	$\frac{1}{2}$	0	$\frac{1}{2}$

computed as if $\mathbf{Var}(\mathbf{Y}) = \mathbf{I}\sigma^2$ and the residual mean square, ERROR B = 119.56, is used as the estimate of σ^2.

Adjusted Standard Errors

The second column of standard errors, labeled "GLM ADJ", has been computed from the "GLM" standard errors by multiplying each by the square root of the ratio of the constructed error mean square to ERROR B. This approach still assumes $\mathbf{Var}(\mathbf{Y}) = \mathbf{I}\sigma^2$, but replaces σ^2 with an average variance of the means in that class of means; the average is taken from the expectations of the Type III mean squares given by the RANDOM option. The estimates of

Table 17.13 Least squares means, standard errors of least squares means as given by GLM, GLM standard errors adjusted for the mean square expectations, exact standard errors of least squares means, standard errors of mean differences, and unadjusted treatment means.

Treatment	*Least Squares Means*	*Standard Errors* GLM	GLM ADJ	EXACT	*S.E. Mean Diff.*[a]	*Unadj. Means*
TILL:						
CT	52.37	2.95	3.39	3.62		53.58
NT	57.38	4.02	4.62	4.54	6.01	36.96
HERB:						
H	73.54	3.35	3.95	3.95		65.18
NOH	36.21	3.35	3.95	3.95	5.58	26.09
CULT:						
C	75.29	4.67	4.67	4.89		72.79
NOC	34.46	2.62	2.62	3.01	5.91	35.34
TILL × *HERB*:						
CT H	64.74	4.46	5.12	5.35		69.10
CT NOH	40.01	3.87	4.47	4.87	7.23	40.01
NT H	82.34	5.91	6.79	6.62		61.75
NT NOH	32.41	5.47	6.29	6.22	9.07	12.16

[a] Standard errors of differences between adjacent pairs of treatment means using the EXACT computations.

the error components of variance are computed from the PROC GLM Type III sums of squares. For example, the GLM standard errors for the *TILL* means have been multiplied by $\sqrt{157.8/119.6} = 1.149$ to obtain GLM ADJ. The 157.8 is the error mean square constructed as the appropriate denominator for the F-test of tillage effects.

"Exact" Standard Errors

The third column of standard errors in Table 17.13, labeled "EXACT", uses the estimated variance–covariance matrix for **Y**, which takes into account the covariances of observations due to the presence of more than one random element, and the PROC GLM algebra to compute correct estimated standard errors of the least squares means (see equation 16.20). The estimates of the variance components used to obtain $s^2(\mathbf{Y})$ were computed from GLM Type III sums of squares.

The standard errors reported by PROC GLM will not in general be correct when the model involves more than one random element (Table 17.13). (This is true whether or not the data are balanced.) In this case study, the GLM standard errors for the whole-plot treatment means (*TILL* and *HERB*) varied from 81% to 89% of the "EXACT" standard errors. The standard errors for the subplot treatment means (*CULT*), which contain only the one variance component, varied from 87% to 96% of the "EXACT" standard errors. The GLM ADJ standard errors provide better agreement with the "EXACT" for the whole-plot treatment means. This adjustment has no effect on the standard errors for the subplot treatment means.

The need for correcting the GLM standard errors will depend on the relative magnitudes of the components of variance in the model. Multiplying by the square root of the ratio of the appropriate error mean squares, GLM ADJ, is a simple adjustment and is recommended in all cases where computation of the "EXACT" standard errors does not seem practical. Adjustments to the standard errors are necessary even when the data are balanced. In the balanced case, the GLM ADJ procedure gives the "EXACT" result.

Standard Errors of Mean Differences

The standard errors of the mean differences, column 6 of Table 17.13, are given to emphasize that, with unbalanced data, variances of differences cannot in general be computed simply as the sum of the variances; the least squares means are *not* independent. The standard errors of the mean differences given in Table 17.13 are computed using the exact method, which takes the covariances into account. The mean difference between the *CULT* treatments, 40.83, has a standard error of 5.91 if computed with the exact method but 5.74 if computed by summing the GLM variances as if the means were independent. Of the marginal treatment means, only the *H* and *NOH* treatment means for the *HERB* treatment factor are independent. The variance of the difference between the *H* and *NOH* means is equal to the sum of the two variances. Within the two-way table of *TILL* × *HERB* means, all means are independent except the *CT–H* mean and the *NT–H* mean.

All least squares means were estimable in this case because it was recognized in advance that the data contained no information on interactions between *CULT* and *TILL*, and these interaction effects were left out of the model. Had this not been done, any least squares means involving the nonestimable higher-order interactions in their expectations would not have been estimable. Nonestimability of least squares means is a common problem in the analysis of unbalanced data when the model includes higher-order interactions. In such cases, it is sometimes necessary to simplify the model by dropping interaction effects to make the means estimable. If the interactions are significant, this creates problems with interpretation.

17.4 MIXMOD Analysis

Description

MIXMOD is a computer program designed to analyze data within the general linear models approach where the model may contain more than one random effect (Giesbrecht, 1984). The program constructs the variance–covariance matrix for $\boldsymbol{Y}$, based on prior "guesses" of the values of the variance components, and then estimates the fixed effects in the model using *generalized* least squares. If the relative sizes of the variance components are correct, this provides the best linear unbiased estimates of the fixed effects. Estimates of the variance components are also provided. Although the estimates depend on the prior values, they are unbiased as long as the choice of prior values does not depend on the data. Experience has shown that setting all prior values to 1.0 is "not unreasonable" (Giesbrecht, 1984).

Iteration to Solution

Usually, the solution is obtained iteratively. The estimates of the variance components from one step are used as the prior values for the next cycle of estimation. Giesbrecht notes that convergence to a stable answer is obtained quickly, in two or three cycles. The user must make a decision at each step on how to treat negative estimates of variance components, if they are obtained. The negative estimate of a component can be used in the next cycle, it can be set to zero, or the random effect can be dropped from the model. If the random elements in the model are all normally distributed and no negative estimates of variance components are obtained, this iterative process gives the modified maximum likelihood or the restricted maximum likelihood estimates of the components. The reader is referred to Giesbrecht (1984) for details on both the theory of the method and the use of the program.

Results from MIXMOD

The results from MIXMOD are presented for this case study for comparison with the more common analysis of variance approach. The $BLOCK \times TILL \times HERB$ effects, which correspond to the δ_{ijkl} in the model, and the ε_{ijk} were treated as random effects. All other effects were assumed to be fixed. Convergence was very rapid. Only trivial changes occurred in the estimates of the variance components from the second to the third iterations. The estimates of the variance components (and their standard errors) given by the third iteration of MIXMOD were

$$\hat{\sigma}_\delta^2\text{:}\quad 37.2373 \pm 52.45$$

$$\hat{\sigma}^2\text{:}\quad 117.597 \pm 45.96$$

The estimates from the Type III sums of squares in SAS were

$$\hat{\sigma}_\delta^2 = 35.0605$$

$$\hat{\sigma}^2 = 119.5562$$

MIXMOD Estimates of Means

MIXMOD provides a solution to the normal equations, $\boldsymbol{\beta}^0$, and the variance–covariance matrix for the solution, $s^2(\boldsymbol{\beta}^0)$. In a separate matrix operation, the MIXMOD solution and its variance–covariance matrix were used to compute the least squares means and their standard errors. This was accomplished by defining a matrix of coefficients, $\boldsymbol{K}'$, such that $\boldsymbol{K}'\boldsymbol{\beta}$ gives the expectations of the least squares means of interest. Each row of $\boldsymbol{K}'$ contains the coefficients of the expectation of one of the least squares means, such as given in Tables 17.11 and 17.12. Then, the vector of least squares means is computed as

$$LSMEANS = \boldsymbol{K}'\boldsymbol{\beta}^0$$

and the estimated variance–covariance matrix for the least squares means is

$$s^2(LSMEANS) = \boldsymbol{K}'[s^2(\boldsymbol{\beta}^0)]\boldsymbol{K}$$

The least squares means and their standard errors are given in Table 17.14.

Comparison to GLM

The differences between the MIXMOD and GLM estimates of the least squares means are small in this example. The estimates of the *HERB* means are identical, as they always will be when the data are balanced. The estimates of the

Table 17.14 Least squares means and standard errors estimated from MIXMOD.

Treatment	*Least Squares Mean*	*Standard Error*
TILL:		
CT	52.03	3.584
NT	57.72	4.494
HERB:		
H	73.54	3.960
NOH	36.21	3.960
CULT:		
C	75.64	4.835
NOC	34.11	2.946
TILL:		
CT H	64.04	5.231
CT NOH	40.01	4.900
NT H	83.04	6.486
NT NOH	32.41	6.222

standard errors are also very similar (to the exact standard errors) in all cases. The small difference in the standard errors of the *HERB* means, 3.95 versus 3.96, is due to the difference in the GLM and MIXMOD estimates of the variance components. Large differences in the estimates are not expected unless the variance components being ignored by GLM are relatively important. In this case, $\hat{\sigma}_\delta^2$ is less than its standard error and only about one-third as large as $\hat{\sigma}^2$. The advantages of the MIXMOD procedure are that it utilizes the variance–covariance information, which will produce more precise estimates if the estimates of the variance components are reliable, and it can accommodate larger models than PROC GLM.

EXERCISES

*17.1. Investigate whether a transformation of the data in this case study might be desirable. Use the Box–Cox transformation on yield for several values of λ. You will need to add a small constant to avoid problems with the zero yields. Run PROC GLM (or a similar program) for each transformed yield variable and plot the residual sums of squares against λ. Construct the confidence interval on λ. What transformation is suggested?

*17.2. Use the Type III sums of squares from the analysis of variance, Table 17.2, and the mean square expectations, Table 17.10, to estimate the two components of variance, σ^2 and σ_δ^2. Compute standard errors for each. [Assume that each mean square is distributed as a chi-squared random variable scaled by $\mathscr{E}(\text{MS})/\text{d.f.}$ so that its variance is $2[\mathscr{E}(\text{MS})]^2/\text{d.f.}$, and that the two mean squares are independent. The estimate of the variance of a chi-squared random variable is obtained by substituting the observed mean square of its expectation.] Compare these standard errors with those given for the MIXMOD solution, Section 17.4.

17.3. Verify that the constructed error mean square for testing *HERB* effects is $E' = 166.31$ and that its approximate degrees of freedom are $f' = 14$.

*17.4. Determine the estimable functions for the population marginal means for a 2×3 factorial set of treatments in a randomized complete block design with $r = 4$ blocks. Include $A \times B$ interactions in your model. Give the estimable functions for the six treatment means and for the marginal treatment means for each treatment factor. How do the estimable functions change if there are no interactions in the model? Suppose cell (1, 2) is empty. Which means become nonestimable if there are interactions in the model? If there are no interactions in the model?

Appendix Tables

Appendix Table A Distribution of *t*.

	Probability								
d.f.	.5	.4	.3	.2	.1	.05	.02	.01	.001
1	1.000	1.376	1.963	3.078	6.314	12.706	31.821	63.657	636.619
2	.816	1.061	1.386	1.886	2.920	4.303	6.965	9.925	31.598
3	.765	.978	1.250	1.638	2.353	3.182	4.541	5.841	12.924
4	.741	.941	1.190	1.533	2.132	2.776	3.747	4.604	8.610
5	.727	.920	1.156	1.476	2.015	2.571	3.365	4.032	6.869
6	.718	.906	1.134	1.440	1.943	2.447	3.143	3.707	5.959
7	.711	.896	1.119	1.415	1.895	2.365	2.998	3.499	5.408
8	.706	.889	1.108	1.397	1.860	2.306	2.896	3.355	5.041
9	.703	.883	1.100	1.383	1.833	2.262	2.821	3.250	4.781
10	.700	.879	1.093	1.372	1.812	2.228	2.764	3.169	4.587
11	.697	.876	1.088	1.363	1.796	2.201	2.718	3.106	4.437
12	.695	.873	1.083	1.356	1.782	2.179	2.681	3.055	4.318
13	.694	.870	1.079	1.350	1.771	2.160	2.650	3.012	4.221
14	.692	.868	1.076	1.345	1.761	2.145	2.624	2.977	4.140
15	.691	.866	1.074	1.341	1.753	2.131	2.602	2.947	4.073
16	.690	.865	1.071	1.337	1.746	2.120	2.583	2.921	4.015
17	.689	.963	1.069	1.333	1.740	2.110	2.567	2.898	3.965
18	.688	.862	1.067	1.330	1.734	2.101	2.552	2.878	3.922
19	.688	.861	1.066	1.328	1.729	2.093	2.539	2.861	3.883
20	.687	.860	1.064	1.325	1.725	2.086	2.528	2.845	3.850
21	.686	.859	1.063	1.323	1.721	2.080	2.518	2.831	3.819
22	.686	.858	1.061	1.321	1.717	2.074	2.508	2.819	3.792
23	.685	.858	1.060	1.319	1.714	2.069	2.500	2.807	3.767
24	.685	.857	1.059	1.318	1.711	2.064	2.492	2.797	3.745
25	.684	.856	1.058	1.316	1.708	2.060	2.485	2.787	3.725
26	.684	.856	1.058	1.315	1.706	2.056	2.479	2.779	3.707
27	.684	.855	1.057	1.314	1.703	2.052	2.473	2.771	3.690
28	.683	.855	1.056	1.313	1.701	2.048	2.467	2.763	3.674
29	.683	.854	1.055	1.311	1.699	2.045	2.462	2.756	3.659
30	.683	.854	1.055	1.310	1.697	2.042	2.457	2.750	3.646
40	.681	.851	1.050	1.303	1.684	2.021	2.423	2.704	3.551
60	.679	.848	1.046	1.296	1.671	2.000	2.390	2.660	3.460
120	.677	.845	1.041	1.289	1.658	1.980	2.358	2.617	3.373
∞	.674	.842	1.036	1.282	1.645	1.960	2.326	2.576	3.291

From Table III of *Statistical Tables for Biological, Agricultural and Medical Research* by R. A. Fisher and F. Yates, published by Longman Group UK Ltd, London (previously published by Oliver and Boyd Ltd, Edinburgh) and reproduced by permission of the authors and publishers.

Appendix Table B Percentage points for the *F*-distribution (variance ratio).
Upper 10% points

ν_2 \ ν_1	1	2	3	4	5	6	7	8	9	10	12	15	20	24	30	40	60	120	∞
1	39.86	49.50	53.59	55.83	57.24	58.20	58.91	59.44	59.86	60.19	60.71	61.22	61.74	62.00	62.26	62.53	62.79	63.06	63.33
2	8.53	9.00	9.16	9.24	9.29	9.33	9.35	9.37	9.38	9.39	9.41	9.42	9.44	9.45	9.46	9.47	9.47	9.48	9.49
3	5.54	5.46	5.39	5.34	5.31	5.28	5.27	5.25	5.24	5.23	5.22	5.20	5.18	5.18	5.17	5.16	5.15	5.14	5.13
4	4.54	4.32	4.19	4.11	4.05	4.01	3.98	3.95	3.94	3.92	3.90	3.87	3.84	3.83	3.82	3.80	3.79	3.78	3.76
5	4.06	3.78	3.62	3.52	3.45	3.40	3.37	3.34	3.32	3.30	3.27	3.24	3.21	3.19	3.17	3.16	3.14	3.12	3.10
6	3.78	3.46	3.29	3.18	3.11	3.05	3.01	2.98	2.96	2.94	2.90	2.87	2.84	2.82	2.80	2.78	2.76	2.74	2.72
7	3.59	3.26	3.07	2.96	2.88	2.83	2.78	2.75	2.72	2.70	2.67	2.63	2.59	2.58	2.56	2.54	2.51	2.49	2.47
8	3.46	3.11	2.92	2.81	2.73	2.67	2.62	2.59	2.56	2.54	2.50	2.46	2.42	2.40	2.38	2.36	2.34	2.32	2.29
9	3.36	3.01	2.81	2.69	2.61	2.55	2.51	2.47	2.44	2.42	2.38	2.34	2.30	2.28	2.25	2.23	2.21	2.18	2.16
10	3.29	2.92	2.73	2.61	2.52	2.46	2.41	2.38	2.35	2.32	2.28	2.24	2.20	2.18	2.16	2.13	2.11	2.08	2.06
11	3.23	2.86	2.66	2.54	2.45	2.39	2.34	2.30	2.27	2.25	2.21	2.17	2.12	2.10	2.08	2.05	2.03	2.00	1.97
12	3.18	2.81	2.61	2.48	2.39	2.33	2.28	2.24	2.21	2.19	2.15	2.10	2.06	2.04	2.01	1.99	1.96	1.93	1.90
13	3.14	2.76	2.56	2.43	2.35	2.28	2.23	2.20	2.16	2.14	2.10	2.05	2.01	1.98	1.96	1.93	1.90	1.88	1.85
14	3.10	2.73	2.52	2.39	2.31	2.24	2.19	2.15	2.12	2.10	2.05	2.01	1.96	1.94	1.91	1.89	1.86	1.83	1.80
15	3.07	2.70	2.49	2.36	2.27	2.21	2.16	2.12	2.09	2.06	2.02	1.97	1.92	1.90	1.87	1.85	1.82	1.79	1.76
16	3.05	2.67	2.46	2.33	2.24	2.18	2.13	2.09	2.06	2.03	1.99	1.94	1.89	1.87	1.84	1.81	1.78	1.75	1.72
17	3.03	2.64	2.44	2.31	2.22	2.15	2.10	2.06	2.03	2.00	1.96	1.91	1.86	1.84	1.81	1.78	1.75	1.72	1.69
18	3.01	2.62	2.42	2.29	2.20	2.13	2.08	2.04	2.00	1.98	1.93	1.89	1.84	1.81	1.78	1.75	1.72	1.69	1.66
19	2.99	2.61	2.40	2.27	2.18	2.11	2.06	2.02	1.98	1.96	1.91	1.86	1.81	1.79	1.76	1.73	1.70	1.67	1.63
20	2.97	2.59	2.38	2.25	2.16	2.09	2.04	2.00	1.96	1.94	1.89	1.84	1.79	1.77	1.74	1.71	1.68	1.64	1.61
21	2.96	2.57	2.36	2.23	2.14	2.08	2.02	1.98	1.95	1.92	1.87	1.83	1.78	1.75	1.72	1.69	1.66	1.62	1.59
22	2.95	2.56	2.35	2.22	2.13	2.06	2.01	1.97	1.93	1.90	1.86	1.81	1.76	1.73	1.70	1.67	1.64	1.60	1.57
23	2.94	2.55	2.34	2.21	2.11	2.05	1.99	1.95	1.92	1.89	1.84	1.80	1.74	1.72	1.69	1.66	1.62	1.59	1.55
24	2.93	2.54	2.33	2.19	2.10	2.04	1.98	1.94	1.91	1.88	1.83	1.78	1.73	1.70	1.67	1.64	1.61	1.57	1.53
25	2.92	2.53	2.32	2.18	2.09	2.02	1.97	1.93	1.89	1.87	1.82	1.77	1.72	1.69	1.66	1.63	1.59	1.56	1.52
26	2.91	2.52	2.31	2.17	2.08	2.01	1.96	1.92	1.88	1.86	1.81	1.76	1.71	1.68	1.65	1.61	1.58	1.54	1.50
27	2.90	2.51	2.30	2.17	2.07	2.00	1.95	1.91	1.87	1.85	1.80	1.75	1.70	1.67	1.64	1.60	1.57	1.53	1.49
28	2.89	2.50	2.29	2.16	2.06	2.00	1.94	1.90	1.87	1.84	1.79	1.74	1.69	1.66	1.63	1.59	1.56	1.52	1.48
29	2.89	2.50	2.28	2.15	2.06	1.99	1.93	1.89	1.86	1.83	1.78	1.73	1.68	1.65	1.62	1.58	1.55	1.51	1.47
30	2.88	2.49	2.28	2.14	2.05	1.98	1.93	1.88	1.85	1.82	1.77	1.72	1.67	1.64	1.61	1.57	1.54	1.50	1.46
40	2.84	2.44	2.23	2.09	2.00	1.93	1.87	1.83	1.79	1.76	1.71	1.66	1.61	1.57	1.54	1.51	1.47	1.42	1.38
60	2.79	2.39	2.18	2.04	1.95	1.87	1.82	1.77	1.74	1.71	1.66	1.60	1.54	1.51	1.48	1.44	1.40	1.35	1.29
120	2.75	2.35	2.13	1.99	1.90	1.82	1.77	1.72	1.68	1.65	1.60	1.55	1.48	1.45	1.41	1.37	1.32	1.26	1.19
∞	2.71	2.30	2.08	1.94	1.85	1.77	1.72	1.67	1.63	1.60	1.55	1.49	1.42	1.38	1.34	1.30	1.24	1.17	1.00

$F = \frac{s_1^2}{s_2^2} = \frac{S_1}{\nu_1} \Big/ \frac{S_2}{\nu_2}$, where $s_1^2 = S_1/\nu_1$ and $s_2^2 = S_2/\nu_2$ are independent mean squares estimating a common variance σ^2, based on ν_1 and ν_2 degrees of freedom, respectively.

Appendix Table B (*continued*)
Upper 5% points

ν_2 \ ν_1	1	2	3	4	5	6	7	8	9	10	12	15	20	24	30	40	60	120	∞
1	161.4	199.5	215.7	224.6	230.2	234.0	236.8	238.9	240.5	241.9	243.9	245.9	248.0	249.1	250.1	251.1	252.2	253.3	254.3
2	18.51	19.00	19.16	19.25	19.30	19.33	19.35	19.37	19.38	19.40	19.41	19.43	19.45	19.45	19.46	19.47	19.48	19.49	19.50
3	10.13	9.55	9.28	9.12	9.01	8.94	8.89	8.85	8.81	8.79	8.74	8.70	8.66	8.64	8.62	8.59	8.57	8.55	8.53
4	7.71	6.94	6.59	6.39	6.26	6.16	6.09	6.04	6.00	5.96	5.91	5.86	5.80	5.77	5.75	5.72	5.69	5.66	5.63
5	6.61	5.79	5.41	5.19	5.05	4.95	4.88	4.82	4.77	4.74	4.68	4.62	4.56	4.53	4.50	4.46	4.43	4.40	4.36
6	5.99	5.14	4.76	4.53	4.39	4.28	4.21	4.15	4.10	4.06	4.00	3.94	3.87	3.84	3.81	3.77	3.74	3.70	3.67
7	5.59	4.74	4.35	4.12	3.97	3.87	3.79	3.73	3.68	3.64	3.57	3.51	3.44	3.41	3.38	3.34	3.30	3.27	3.23
8	5.32	4.46	4.07	3.84	3.69	3.58	3.50	3.44	3.39	3.35	3.28	3.22	3.15	3.12	3.08	3.04	3.01	2.97	2.93
9	5.12	4.26	3.86	3.63	3.48	3.37	3.29	3.23	3.18	3.14	3.07	3.01	2.94	2.90	2.86	2.83	2.79	2.75	2.71
10	4.96	4.10	3.71	3.48	3.33	3.22	3.14	3.07	3.02	2.98	2.91	2.85	2.77	2.74	2.70	2.66	2.62	2.58	2.54
11	4.84	3.98	3.59	3.36	3.20	3.09	3.01	2.95	2.90	2.85	2.79	2.72	2.65	2.61	2.57	2.53	2.49	2.45	2.40
12	4.75	3.89	3.49	3.26	3.11	3.00	2.91	2.85	2.80	2.75	2.69	2.62	2.54	2.51	2.47	2.43	2.38	2.34	2.30
13	4.67	3.81	3.41	3.18	3.03	2.92	2.83	2.77	2.71	2.67	2.60	2.53	2.46	2.42	2.38	2.34	2.30	2.25	2.21
14	4.60	3.74	3.34	3.11	2.96	2.85	2.76	2.70	2.65	2.60	2.53	2.46	2.39	2.35	2.31	2.27	2.22	2.18	2.13
15	4.54	3.68	3.29	3.06	2.90	2.79	2.71	2.64	2.59	2.54	2.48	2.40	2.33	2.29	2.25	2.20	2.16	2.11	2.07
16	4.49	3.63	3.24	3.01	2.85	2.74	2.66	2.59	2.54	2.49	2.42	2.35	2.28	2.24	2.19	2.15	2.11	2.06	2.01
17	4.45	3.59	3.20	2.96	2.81	2.70	2.61	2.55	2.49	2.45	2.38	2.31	2.23	2.19	2.15	2.10	2.06	2.01	1.96
18	4.41	3.55	3.16	2.93	2.77	2.66	2.58	2.51	2.46	2.41	2.34	2.27	2.19	2.15	2.11	2.06	2.02	1.97	1.92
19	4.28	3.52	3.13	2.90	2.74	2.63	2.54	2.48	2.42	2.38	2.31	2.23	2.16	2.11	2.07	2.03	1.98	1.93	1.88
20	4.35	3.49	3.10	2.87	2.71	2.60	2.51	2.45	2.39	2.35	2.28	2.20	2.12	2.08	2.04	1.99	1.95	1.90	1.84
21	4.32	3.47	3.07	2.84	2.68	2.57	2.49	2.42	2.37	2.32	2.25	2.18	2.10	2.05	2.01	1.96	1.92	1.87	1.81
22	4.30	3.44	3.05	2.82	2.66	2.55	2.46	2.40	2.34	2.30	2.23	2.15	2.07	2.03	1.98	1.94	1.89	1.84	1.78
23	4.28	3.42	3.03	2.80	2.64	2.53	2.44	2.37	2.32	2.27	2.20	2.13	2.05	2.01	1.96	1.91	1.86	1.81	1.76
24	4.26	3.40	3.01	2.78	2.62	2.51	2.42	2.36	2.30	2.25	2.18	2.11	2.03	1.98	1.94	1.89	1.84	1.79	1.73
25	4.24	3.39	2.99	2.76	2.60	2.49	2.40	2.34	2.28	2.24	2.16	2.09	2.01	1.96	1.92	1.87	1.82	1.77	1.71
26	4.23	3.37	2.98	2.74	2.59	2.47	2.39	2.32	2.27	2.22	2.15	2.07	1.99	1.95	1.90	1.85	1.80	1.75	1.69
27	4.21	3.35	2.96	2.73	2.57	2.46	2.37	2.31	2.25	2.20	2.13	2.06	1.97	1.93	1.88	1.84	1.79	1.73	1.67
28	4.20	3.34	2.95	2.71	2.56	2.45	2.36	2.29	2.24	2.19	2.12	2.04	1.96	1.91	1.87	1.82	1.77	1.71	1.65
29	4.18	3.33	2.93	2.70	2.55	2.43	2.35	2.28	2.22	2.18	2.10	2.03	1.94	1.90	1.85	1.81	1.75	1.70	1.64
30	4.17	3.32	2.92	2.69	2.53	2.42	2.33	2.27	2.21	2.16	2.09	2.01	1.93	1.89	1.84	1.79	1.74	1.68	1.62
40	4.08	3.23	2.84	2.61	2.45	2.34	2.25	2.18	2.12	2.08	2.00	1.92	1.84	1.79	1.74	1.69	1.64	1.58	1.51
60	4.00	3.15	2.76	2.53	2.37	2.25	2.17	2.10	2.04	1.99	1.92	1.84	1.75	1.70	1.65	1.59	1.53	1.47	1.39
120	3.92	3.07	2.68	2.45	2.29	2.17	2.09	2.02	1.96	1.91	1.83	1.75	1.66	1.61	1.55	1.50	1.43	1.35	1.25
∞	3.84	3.00	2.60	2.37	2.21	2.10	2.01	1.94	1.88	1.83	1.75	1.67	1.57	1.52	1.46	1.39	1.32	1.22	1.00

$F = \dfrac{s_1^2}{s_2^2} = \dfrac{S_1}{\nu_1}\Big/\dfrac{S_2}{\nu_2}$, where $s_1^2 = S_1/\nu_1$ and $s_2^2 = S_2/\nu_2$ are independent mean squares estimating a common variance σ^2, based on ν_1 and ν_2 degrees of freedom, respectively.

Appendix Table B *(continued)*
Upper 2.5% points

ν_2 \ ν_1	1	2	3	4	5	6	7	8	9	10	12	15	20	24	30	40	60	120	∞
1	647.8	799.5	864.2	899.6	921.8	937.1	948.2	956.7	963.3	968.6	976.7	984.9	993.1	997.2	1,001	1,006	1,010	1,014	1,018
2	38.51	39.00	39.17	39.25	39.30	39.33	39.36	39.37	39.39	39.40	39.41	39.43	39.45	39.46	39.46	39.47	39.48	39.49	39.50
3	17.44	16.04	15.44	15.10	14.88	14.73	14.62	14.54	14.47	14.42	14.34	14.25	14.17	14.12	14.08	14.04	13.99	13.95	13.90
4	12.22	10.65	9.98	9.60	9.36	9.20	9.07	8.98	8.90	8.84	8.75	8.66	8.56	8.51	8.46	8.41	8.36	8.31	8.26
5	10.01	8.43	7.76	7.39	7.15	6.98	6.85	6.76	6.68	6.62	6.52	6.43	6.33	6.28	6.23	6.18	6.12	6.07	6.02
6	8.81	7.26	6.60	6.23	5.99	5.82	5.70	5.60	5.52	5.46	5.37	5.27	5.17	5.12	5.07	5.01	4.96	4.90	4.85
7	8.07	6.54	5.89	5.52	5.29	5.12	4.99	4.90	4.82	4.76	4.67	4.57	4.47	4.42	4.36	4.31	4.25	4.20	4.14
8	7.57	6.06	5.42	5.05	4.82	4.65	4.53	4.43	4.36	4.30	4.20	4.10	4.00	3.95	3.89	3.84	3.78	3.73	3.67
9	7.21	5.71	5.08	4.72	4.48	4.32	4.20	4.10	4.03	3.96	3.87	3.77	3.67	3.61	3.56	3.51	3.45	3.39	3.33
10	6.94	5.46	4.83	4.47	4.24	4.07	3.95	3.85	3.78	3.72	3.62	3.52	3.42	3.37	3.31	3.26	3.20	3.14	3.08
11	6.72	5.26	4.63	4.28	4.04	3.88	3.76	3.66	3.59	3.53	3.43	3.33	3.23	3.17	3.12	3.06	3.00	2.94	2.88
12	6.55	5.10	4.47	4.12	3.89	3.73	3.61	3.51	3.44	3.37	3.28	3.18	3.07	3.02	2.96	2.91	2.85	2.79	2.72
13	6.41	4.97	4.35	4.00	3.77	3.60	3.48	3.39	3.31	3.25	3.15	3.05	2.95	2.89	2.84	2.78	2.72	2.66	2.60
14	6.30	4.86	4.24	3.89	3.66	3.50	3.38	3.29	3.21	3.15	3.05	2.95	2.84	2.79	2.73	2.67	2.61	2.55	2.49
15	6.20	4.77	4.15	3.80	3.58	3.41	3.29	3.20	3.12	3.06	2.96	2.86	2.76	2.70	2.64	2.59	2.52	2.46	2.40
16	6.12	4.69	4.08	3.73	3.50	3.34	3.22	3.12	3.05	2.99	2.89	2.79	2.68	2.63	2.57	2.51	2.45	2.38	2.32
17	6.04	4.62	4.01	3.66	3.44	3.28	3.16	3.06	2.98	2.92	2.82	2.72	2.62	2.56	2.50	2.44	2.38	2.32	2.25
18	5.98	4.56	3.95	3.61	3.38	3.22	3.10	3.01	2.93	2.87	2.77	2.67	2.56	2.50	2.44	2.38	2.32	2.26	2.19
19	5.92	4.51	3.90	3.56	3.33	3.17	3.05	2.96	2.88	2.82	2.72	2.62	2.51	2.45	2.39	2.33	2.27	2.20	2.13
20	5.87	4.46	3.86	3.51	3.29	3.13	3.01	2.91	2.84	2.77	2.68	2.57	2.46	2.41	2.35	2.29	2.22	2.16	2.09
21	5.83	4.42	3.82	3.48	3.25	3.09	2.97	2.87	2.80	2.73	2.64	2.53	2.42	2.37	2.31	2.25	2.18	2.11	2.04
22	5.79	4.38	3.78	3.44	3.22	3.05	2.93	2.84	2.76	2.70	2.60	2.50	2.39	2.33	2.27	2.21	2.14	2.08	2.00
23	5.75	4.35	3.75	3.41	3.18	3.02	2.90	2.81	2.73	2.67	2.57	2.47	2.36	2.30	2.24	2.18	2.11	2.04	1.97
24	5.72	4.32	3.72	3.38	3.15	2.99	2.87	2.78	2.70	2.64	2.54	2.44	2.33	2.27	2.21	2.15	2.08	2.01	1.94
25	5.69	4.29	3.69	3.35	3.13	2.97	2.85	2.75	2.68	2.61	2.51	2.41	2.30	2.24	2.18	2.12	2.05	1.98	1.91
26	5.66	4.27	3.67	3.33	3.10	2.94	2.82	2.73	2.65	2.59	2.49	2.39	2.28	2.22	2.16	2.09	2.03	1.95	1.88
27	5.63	4.24	3.65	3.31	3.08	2.92	2.80	2.71	2.63	2.57	2.47	2.36	2.25	2.19	2.13	2.07	2.00	1.93	1.85
28	5.61	4.22	3.63	3.29	3.06	2.90	2.78	2.69	2.61	2.55	2.45	2.34	2.23	2.17	2.11	2.05	1.98	1.91	1.83
29	5.59	4.20	3.61	3.27	3.04	2.88	2.76	2.67	2.59	2.53	2.43	2.32	2.21	2.15	2.09	2.03	1.96	1.89	1.81
30	5.57	4.18	3.59	3.25	3.03	2.87	2.75	2.65	2.57	2.51	2.41	2.31	2.20	2.14	2.07	2.01	1.94	1.87	1.79
40	5.42	4.05	3.46	3.13	2.90	2.74	2.62	2.53	2.45	2.39	2.29	2.18	2.07	2.01	1.94	1.88	1.80	1.72	1.64
60	5.29	3.93	3.34	3.01	2.79	2.63	2.51	2.41	2.33	2.27	2.17	2.06	1.94	1.88	1.82	1.74	1.67	1.58	1.48
120	5.15	3.80	3.23	2.89	2.67	2.52	2.39	2.30	2.22	2.16	2.05	1.94	1.82	1.76	1.69	1.61	1.53	1.43	1.31
∞	5.02	3.69	3.12	2.79	2.57	2.41	2.29	2.19	2.11	2.05	1.94	1.83	1.71	1.64	1.57	1.48	1.39	1.27	1.00

$F = \frac{s_1^2}{s_2^2} = \frac{S_1}{\nu_1} \Big/ \frac{S_2}{\nu_2}$, where $s_1^2 = S_1/\nu_1$ and $s_2^2 = S_2/\nu_2$ are independent mean squares estimating a common variance σ^2, based on ν_1 and ν_2 degrees of freedom, respectively.

Appendix Table B *(continued)*
Upper 1% points

v_2 \ v_1	1	2	3	4	5	6	7	8	9	10	12	15	20	24	30	40	60	120	∞
1	4,052	4,999.5	5,403	5,625	5,764	5,859	5,928	5,982	6,022	6,056	6,106	6,157	6,209	6,235	6,261	6,287	6,313	6,339	6,366
2	98.50	99.00	99.17	99.25	99.30	99.33	99.36	99.37	99.39	99.40	99.42	99.43	99.45	99.46	99.47	99.47	99.48	99.49	99.50
3	34.12	30.82	29.46	28.71	28.24	27.91	27.67	27.49	27.35	27.23	27.05	26.87	26.69	26.60	26.50	26.41	26.32	26.22	26.13
4	21.20	18.00	16.69	15.98	15.52	15.21	14.98	14.80	14.66	14.55	14.37	14.20	14.02	13.93	13.84	13.75	13.65	13.56	13.46
5	16.26	13.27	12.06	11.39	10.97	10.67	10.46	10.29	10.16	10.05	9.89	9.72	9.55	9.47	9.38	9.29	9.20	9.11	9.02
6	13.75	10.92	9.78	9.15	8.75	8.47	8.26	8.10	7.98	7.87	7.72	7.56	7.40	7.31	7.23	7.14	7.06	6.97	6.88
7	12.25	9.55	8.45	7.85	7.46	7.19	6.99	6.84	6.72	6.62	6.47	6.31	6.16	6.07	5.99	5.91	5.82	5.74	5.65
8	11.26	8.65	7.59	7.01	6.63	6.37	6.18	6.03	5.91	5.81	5.67	5.52	5.36	5.28	5.20	5.12	5.03	4.95	4.86
9	10.56	8.02	6.99	6.42	6.06	5.80	5.61	5.47	5.35	5.26	5.11	4.96	4.81	4.73	4.65	4.57	4.48	4.40	4.31
10	10.04	7.56	6.55	5.99	5.64	5.39	5.20	5.06	4.94	4.85	4.71	4.56	4.41	4.33	4.25	4.17	4.08	4.00	3.91
11	9.65	7.21	6.22	5.67	5.32	5.07	4.89	4.74	4.63	4.54	4.40	4.25	4.10	4.02	3.94	3.86	3.78	3.69	3.60
12	9.33	6.93	5.95	5.41	5.06	4.82	4.64	4.50	4.39	4.30	4.16	4.01	3.86	3.78	3.70	3.62	3.54	3.45	3.36
13	9.07	6.70	5.74	5.21	4.86	4.62	4.44	4.30	4.19	4.10	3.96	3.82	3.66	3.59	3.51	3.43	3.34	3.25	3.17
14	8.86	6.51	5.56	5.04	4.69	4.46	4.28	4.14	4.03	3.94	3.80	3.66	3.51	3.43	3.35	3.27	3.18	3.09	3.00
15	8.68	6.36	5.42	4.89	4.56	4.32	4.14	4.00	3.89	3.80	3.67	3.52	3.37	3.29	3.21	3.13	3.05	2.96	2.87
16	8.53	6.23	5.29	4.77	4.44	4.20	4.03	3.89	3.78	3.69	3.55	3.41	3.26	3.18	3.10	3.02	2.93	2.84	2.75
17	8.40	6.11	5.18	4.67	4.34	4.10	3.93	3.79	3.68	3.59	3.46	3.31	3.16	3.08	3.00	2.92	2.83	2.75	2.65
18	8.29	6.01	5.09	4.58	4.25	4.01	3.84	3.71	3.60	3.51	3.37	3.23	3.08	3.00	2.92	2.84	2.75	2.66	2.57
19	8.18	5.93	5.01	4.50	4.17	3.94	3.77	3.63	3.52	3.43	3.30	3.15	3.00	2.92	2.84	2.76	2.67	2.58	2.49
20	8.10	5.85	4.94	4.43	4.10	3.87	3.70	3.56	3.46	3.37	3.23	3.09	2.94	2.86	2.78	2.69	2.61	2.52	2.42
21	8.02	5.78	4.87	4.37	4.04	3.81	3.64	3.51	3.40	3.31	3.17	3.03	2.88	2.80	2.72	2.64	2.55	2.46	2.36
22	7.95	5.72	4.82	4.31	3.99	3.76	3.59	3.45	3.35	3.26	3.12	2.98	2.83	2.75	2.67	2.58	2.50	2.40	2.31
23	7.88	5.66	4.76	4.26	3.94	3.71	3.54	3.41	3.30	3.21	3.07	2.93	2.78	2.70	2.62	2.54	2.45	2.35	2.26
24	7.82	5.61	4.72	4.22	3.90	3.67	3.50	3.36	3.26	3.17	3.03	2.89	2.74	2.66	2.58	2.49	2.40	2.31	2.21
25	7.77	5.57	4.68	4.18	3.85	3.63	3.46	3.32	3.22	3.13	2.99	2.85	2.70	2.62	2.54	2.45	2.36	2.27	2.17
26	7.72	5.53	4.64	4.14	3.82	3.59	3.42	3.29	3.18	3.09	2.96	2.81	2.66	2.58	2.50	2.42	2.33	2.23	2.13
27	7.68	5.49	4.60	4.11	3.78	3.56	3.30	3.26	3.15	3.06	2.93	2.78	2.63	2.55	2.47	2.38	2.29	2.20	2.10
28	7.64	5.45	4.57	4.07	3.75	3.53	3.36	3.23	3.12	3.03	2.90	2.75	2.60	2.52	2.44	2.35	2.26	2.17	2.06
29	7.60	5.42	4.54	4.04	3.73	3.50	3.33	3.20	3.09	3.00	2.87	2.73	2.57	2.49	2.41	2.33	2.23	2.14	2.03
30	7.56	5.39	4.51	4.02	3.70	3.47	3.30	3.17	3.07	2.98	2.84	2.70	2.55	2.47	2.39	2.30	2.21	2.11	2.01
40	7.31	5.18	4.31	3.83	3.51	3.29	3.12	2.99	2.89	2.80	2.66	2.52	2.37	2.29	2.20	2.11	2.02	1.92	1.80
60	7.08	4.98	4.13	3.65	3.34	3.12	2.95	2.82	2.72	2.63	2.50	2.35	2.20	2.12	2.03	1.94	1.84	1.73	1.60
120	6.85	4.79	3.95	3.48	3.17	2.96	2.79	2.66	2.56	2.47	2.34	2.19	2.03	1.95	1.86	1.76	1.66	1.53	1.38
∞	6.63	4.61	3.78	3.32	3.02	2.80	2.64	2.51	2.41	2.32	2.18	2.04	1.88	1.79	1.70	1.59	1.47	1.32	1.00

$F = \frac{s_1^2}{s_2^2} = \frac{S_1}{v_1} \Big/ \frac{S_2}{v_2}$, where $s_1^2 = S_1/v_1$ and $s_2^2 = S_2/v_2$ are independent mean squares estimating a common variance σ^2, based on v_1 and v_2 degrees of freedom, respectively.

Appendix Table B *(continued)*
Upper .5% points

ν_2 \ ν_1	1	2	3	4	5	6	7	8	9	10	12	15	20	24	30	40	60	120	∞
1	16,211	20,000	21,615	22,500	23,056	23,437	23,715	23,925	24,091	24,224	24,426	24,630	24,836	24,940	25,044	25,148	25,253	25,359	25,465
2	198.5	199.0	199.2	199.2	199.3	199.3	199.4	199.4	199.4	199.4	199.4	199.4	199.4	199.5	199.5	199.5	199.5	199.5	199.5
3	55.55	49.80	47.47	46.19	45.39	44.84	44.43	44.13	43.88	43.69	43.39	43.08	42.78	42.62	42.47	42.31	42.15	41.99	41.83
4	31.33	26.28	24.26	23.15	22.46	21.97	21.62	21.35	21.14	20.97	20.70	20.44	20.17	20.03	19.89	19.75	19.61	19.47	19.32
5	22.78	18.31	16.53	15.56	14.94	14.51	14.20	13.96	13.77	13.62	13.38	13.15	12.90	12.78	12.66	12.53	12.40	12.27	12.14
6	18.63	14.54	12.92	12.03	11.46	11.07	10.79	10.57	10.39	10.25	10.03	9.81	9.59	9.47	9.36	9.24	9.12	9.00	8.88
7	16.24	12.40	10.88	10.05	9.52	9.16	8.89	8.68	8.51	8.38	8.18	7.97	7.75	7.65	7.53	7.42	7.31	7.19	7.08
8	14.69	11.04	9.60	8.81	8.30	7.95	7.69	7.50	7.34	7.21	7.01	6.81	6.61	6.50	6.40	6.29	6.18	6.06	5.95
9	13.61	10.11	8.72	7.96	7.47	7.13	6.88	6.69	6.54	6.42	6.23	6.03	5.83	5.73	5.62	5.52	5.41	5.30	5.19
10	12.83	9.43	8.08	7.34	6.87	6.54	6.30	6.12	5.97	5.85	5.66	5.47	5.27	5.17	5.07	4.97	4.86	4.75	4.64
11	12.23	8.91	7.60	6.88	6.42	6.10	5.86	5.68	5.54	5.42	5.24	5.05	4.86	4.76	4.65	4.55	4.44	4.34	4.23
12	11.75	8.51	7.23	6.52	6.07	5.76	5.52	5.35	5.20	5.09	4.91	4.72	4.53	4.43	4.33	4.23	4.12	4.01	3.90
13	11.37	8.19	6.93	6.23	5.79	5.48	5.25	5.08	4.94	4.82	4.64	4.46	4.27	4.17	4.07	3.97	3.87	3.76	3.65
14	11.06	7.92	6.68	6.00	5.56	5.26	5.03	4.86	4.72	4.60	4.43	4.25	4.06	3.96	3.86	3.76	3.66	3.55	3.44
15	10.80	7.70	6.48	5.80	5.37	5.07	4.85	4.67	4.54	4.42	4.25	4.07	3.88	3.79	3.69	3.58	3.48	3.37	3.26
16	10.58	7.51	6.30	5.64	5.21	4.91	4.69	4.52	4.38	4.27	4.10	3.92	3.73	3.64	3.54	3.44	3.33	3.22	3.11
17	10.38	7.35	6.16	5.50	5.07	4.78	4.56	4.39	4.25	4.14	3.97	3.79	3.61	3.51	3.41	3.31	3.21	3.10	2.98
18	10.22	7.21	6.03	5.37	4.96	4.66	4.44	4.28	4.14	4.03	3.86	3.68	3.50	3.40	3.30	3.20	3.10	2.99	2.87
19	10.07	7.09	5.92	5.27	4.85	4.56	4.34	4.18	4.04	3.93	3.76	3.59	3.40	3.31	3.21	3.11	3.00	2.89	2.78
20	9.94	6.99	5.82	5.17	4.76	4.47	4.26	4.09	3.96	3.85	3.68	3.50	3.32	3.22	3.12	3.02	2.92	2.81	2.69
21	9.83	6.89	5.73	5.09	4.68	4.39	4.18	4.01	3.88	3.77	3.60	3.43	3.24	3.15	3.05	2.95	2.84	2.73	2.61
22	9.73	6.81	5.65	5.02	4.61	4.32	4.11	3.94	3.81	3.70	3.54	3.36	3.18	3.08	2.98	2.88	2.77	2.66	2.55
23	9.63	6.73	5.58	4.95	4.54	4.26	4.05	3.88	3.75	3.64	3.47	3.30	3.12	3.02	2.92	2.82	2.71	2.60	2.48
24	9.55	6.66	5.52	4.89	4.49	4.20	3.99	3.83	3.69	3.59	3.42	3.25	3.06	2.97	2.87	2.77	2.66	2.55	2.43
25	9.48	6.60	5.46	4.84	4.43	4.15	3.94	3.78	3.64	3.54	3.37	3.20	3.01	2.92	2.82	2.72	2.61	2.50	2.38
26	9.41	6.54	5.41	4.79	4.38	4.10	3.89	3.73	3.60	3.49	3.33	3.15	2.97	2.87	2.77	2.67	2.56	2.45	2.33
27	9.34	6.49	5.36	4.74	4.34	4.06	3.85	3.69	3.56	3.45	3.28	3.11	2.93	2.83	2.73	2.63	2.52	2.41	2.25
28	9.28	6.44	5.32	4.70	4.30	4.02	3.81	3.65	3.52	3.41	3.25	3.07	2.89	2.79	2.69	2.59	2.48	2.37	2.29
29	9.23	6.40	5.28	4.66	4.26	3.98	3.77	3.61	3.48	3.38	3.21	3.04	2.86	2.76	2.66	2.56	2.45	2.33	2.24
30	9.18	6.35	5.24	4.62	4.23	3.95	3.74	3.58	3.45	3.34	3.18	3.01	2.82	2.73	2.63	2.52	2.42	2.30	2.18
40	8.83	6.07	4.98	4.37	3.99	3.71	3.51	3.35	3.22	3.12	2.95	2.78	2.60	2.50	2.40	2.30	2.18	2.06	1.93
60	8.49	5.79	4.73	4.14	3.76	3.49	3.29	3.13	3.01	2.90	2.74	2.57	2.39	2.29	2.19	2.08	1.96	1.83	1.69
120	8.18	5.54	4.50	3.92	3.55	3.28	3.09	2.93	2.81	2.71	2.54	2.37	2.19	2.09	1.98	1.87	1.75	1.61	1.43
∞	7.88	5.30	4.28	3.72	3.35	3.09	2.90	2.74	2.62	2.52	2.36	2.19	2.00	1.90	1.79	1.67	1.53	1.36	1.00

$F = \frac{s_1^2}{s_2^2} = \frac{S_1}{\nu_1} \Big/ \frac{S_2}{\nu_2}$, where $s_1^2 = S_1/\nu_1$ and $s_2^2 = S_2/\nu_2$ are independent mean squares estimating a common variance σ^2, based on ν_1 and ν_2 degrees of freedom, respectively.

Appendix Table B is adapted with permission of the Biometrika Trustees from *Biometrika Tables for Statisticians*, Volume II, Table 5, edited by E. S. Pearson and H. O. Hartley, published for Biometrika Trustees, Cambridge University Press, Cambridge, England, 1972.

Appendix Table C Significance points of d_L and d_U for the Durbin–Watson test for correlation.

5%

	$k' = 1$		$k' = 2$		$k' = 3$		$k' = 4$		$k' = 5$	
n	d_L	d_U	d_L	d_U	d_L	d_U	d_L	d_U	d_L	d_U
15	1.08	1.36	.95	1.54	.82	1.75	.69	1.97	.56	2.21
16	1.10	1.37	.98	1.54	.86	1.73	.74	1.93	.62	2.15
17	1.13	1.38	1.02	1.54	.90	1.71	.78	1.90	.67	2.10
18	1.16	1.39	1.05	1.53	.93	1.69	.82	1.87	.71	2.06
19	1.18	1.40	1.08	1.53	.97	1.68	.86	1.85	.75	2.02
20	1.20	1.41	1.10	1.54	1.00	1.68	.90	1.83	.79	1.99
21	1.22	1.42	1.13	1.54	1.03	1.67	.93	1.81	.83	1.96
22	1.24	1.43	1.15	1.54	1.05	1.66	.96	1.80	.86	1.94
23	1.26	1.44	1.17	1.54	1.08	1.66	.99	1.79	.90	1.92
24	1.27	1.45	1.19	1.55	1.10	1.66	1.01	1.78	.93	1.90
25	1.29	1.45	1.21	1.55	1.12	1.66	1.04	1.77	.95	1.89
26	1.30	1.46	1.22	1.55	1.14	1.65	1.06	1.76	.98	1.88
27	1.32	1.47	1.24	1.56	1.16	1.65	1.08	1.76	1.01	1.86
28	1.33	1.48	1.26	1.56	1.18	1.65	1.10	1.75	1.03	1.85
29	1.34	1.48	1.27	1.56	1.20	1.65	1.12	1.74	1.05	1.84
30	1.35	1.49	1.28	1.57	1.21	1.65	1.14	1.74	1.07	1.83
31	1.36	1.50	1.30	1.57	1.23	1.65	1.16	1.74	1.09	1.83
32	1.37	1.50	1.31	1.57	1.24	1.65	1.18	1.73	1.11	1.82
33	1.38	1.51	1.32	1.58	1.26	1.65	1.19	1.73	1.13	1.81
34	1.39	1.51	1.33	1.58	1.27	1.65	1.21	1.73	1.15	1.81
35	1.40	1.52	1.34	1.58	1.28	1.65	1.22	1.73	1.16	1.80
36	1.41	1.52	1.35	1.59	1.29	1.65	1.24	1.73	1.18	1.80
37	1.42	1.53	1.36	1.59	1.31	1.66	1.25	1.72	1.19	1.80
38	1.43	1.54	1.37	1.59	1.32	1.66	1.26	1.72	1.21	1.79
39	1.43	1.54	1.38	1.60	1.33	1.66	1.27	1.72	1.22	1.79
40	1.44	1.54	1.39	1.60	1.34	1.66	1.29	1.72	1.23	1.79
45	1.48	1.57	1.43	1.62	1.38	1.67	1.34	1.72	1.29	1.78
50	1.50	1.59	1.46	1.63	1.42	1.67	1.38	1.72	1.34	1.77
55	1.53	1.60	1.49	1.64	1.45	1.68	1.41	1.72	1.38	1.77
60	1.55	1.62	1.51	1.65	1.48	1.69	1.44	1.73	1.41	1.77
65	1.57	1.63	1.54	1.66	1.50	1.70	1.47	1.73	1.44	1.77
70	1.58	1.64	1.55	1.67	1.52	1.70	1.49	1.74	1.46	1.77
75	1.60	1.65	1.57	1.68	1.54	1.71	1.51	1.74	1.49	1.77
80	1.61	1.66	1.59	1.69	1.56	1.72	1.53	1.74	1.51	1.77
85	1.62	1.67	1.60	1.70	1.57	1.72	1.55	1.75	1.52	1.77
90	1.63	1.68	1.61	1.70	1.59	1.73	1.57	1.75	1.54	1.78
95	1.64	1.69	1.62	1.71	1.60	1.73	1.58	1.75	1.56	1.78
100	1.65	1.69	1.63	1.72	1.61	1.74	1.59	1.76	1.57	1.78

Appendix Table C (*continued*)
2.5%

n	$k' = 1$		$k' = 2$		$k' = 3$		$k' = 4$		$k' = 5$	
	d_L	d_U	d_L	d_U	d_L	d_U	d_L	d_U	d_L	d_U
15	.95	1.23	.83	1.40	.71	1.61	.59	1.84	.48	2.09
16	.98	1.24	.86	1.40	.75	1.59	.64	1.80	.53	2.03
17	1.01	1.25	.90	1.40	.79	1.58	.68	1.77	.57	1.98
18	1.03	1.26	.93	1.40	.82	1.56	.72	1.74	.62	1.93
19	1.06	1.28	.96	1.41	.86	1.55	.76	1.72	.66	1.90
20	1.08	1.28	.99	1.41	.89	1.55	.79	1.70	.70	1.87
21	1.10	1.30	1.01	1.41	.92	1.54	.83	1.69	.73	1.84
22	1.12	1.31	1.04	1.42	.95	1.54	.86	1.68	.77	1.82
23	1.14	1.32	1.06	1.42	.97	1.54	.89	1.67	.80	1.80
24	1.16	1.33	1.08	1.43	1.00	1.54	.91	1.66	.83	1.79
25	1.18	1.34	1.10	1.43	1.02	1.54	.94	1.65	.86	1.77
26	1.19	1.35	1.12	1.44	1.04	1.54	.96	1.65	.88	1.76
27	1.21	1.36	1.13	1.44	1.06	1.54	.99	1.64	.91	1.75
28	1.22	1.37	1.15	1.45	1.08	1.54	1.01	1.64	.93	1.74
29	1.24	1.38	1.17	1.45	1.10	1.54	1.03	1.63	.96	1.73
30	1.25	1.38	1.18	1.46	1.12	1.54	1.05	1.63	.98	1.73
31	1.26	1.39	1.20	1.47	1.13	1.55	1.07	1.63	1.00	1.72
32	1.27	1.40	1.21	1.47	1.15	1.55	1.08	1.63	1.02	1.71
33	1.28	1.41	1.22	1.48	1.16	1.55	1.10	1.63	1.04	1.71
34	1.29	1.41	1.24	1.48	1.17	1.55	1.12	1.63	1.06	1.70
35	1.30	1.42	1.25	1.48	1.19	1.55	1.13	1.63	1.07	1.70
36	1.31	1.43	1.26	1.49	1.20	1.56	1.15	1.63	1.09	1.70
37	1.32	1.43	1.27	1.49	1.21	1.56	1.16	1.62	1.10	1.70
38	1.33	1.44	1.28	1.50	1.23	1.56	1.17	1.62	1.12	1.70
39	1.34	1.44	1.29	1.50	1.24	1.56	1.19	1.63	1.13	1.69
40	1.35	1.45	1.30	1.51	1.25	1.57	1.20	1.63	1.15	1.69
45	1.39	1.48	1.34	1.53	1.30	1.58	1.25	1.63	1.21	1.69
50	1.42	1.50	1.38	1.54	1.34	1.59	1.30	1.64	1.26	1.69
55	1.45	1.52	1.41	1.56	1.37	1.60	1.33	1.64	1.30	1.69
60	1.47	1.54	1.44	1.57	1.40	1.61	1.37	1.65	1.33	1.69
65	1.49	1.55	1.46	1.59	1.43	1.62	1.40	1.66	1.36	1.69
70	1.51	1.57	1.48	1.60	1.45	1.63	1.42	1.66	1.39	1.70
75	1.53	1.58	1.50	1.61	1.47	1.64	1.45	1.67	1.42	1.70
80	1.54	1.59	1.52	1.62	1.49	1.65	1.47	1.67	1.44	1.70
85	1.56	1.60	1.53	1.63	1.51	1.65	1.49	1.68	1.46	1.71
90	1.57	1.61	1.55	1.64	1.53	1.66	1.50	1.69	1.48	1.71
95	1.58	1.62	1.56	1.65	1.54	1.67	1.52	1.69	1.50	1.71
100	1.59	1.63	1.57	1.65	1.55	1.67	1.53	1.70	1.51	1.72

Appendix Table C (*continued*)
1%

n	$k' = 1$		$k' = 2$		$k' = 3$		$k' = 4$		$k' = 5$	
	d_L	d_U	d_L	d_U	d_L	d_U	d_L	d_U	d_L	d_U
15	.81	1.07	.70	1.25	.59	1.46	.49	1.70	.30	1.96
16	.84	1.09	.74	1.25	.63	1.44	.53	1.66	.44	1.90
17	.87	1.10	.77	1.25	.67	1.43	.57	1.63	.48	1.85
18	.90	1.12	.80	1.26	.71	1.42	.61	1.60	.52	1.80
19	.93	1.13	.83	1.26	.74	1.41	.65	1.58	.56	1.77
20	.95	1.15	.86	1.27	.77	1.41	.68	1.57	.60	1.74
21	.97	1.16	.89	1.27	.80	1.41	.72	1.55	.63	1.71
22	1.00	1.17	.91	1.28	.83	1.40	.75	1.54	.66	1.69
23	1.02	1.19	.94	1.29	.86	1.40	.77	1.53	.70	1.67
24	1.04	1.20	.96	1.30	.88	1.41	.80	1.53	.72	1.66
25	1.05	1.21	.98	1.30	.90	1.41	.83	1.52	.75	1.65
26	1.07	1.22	1.00	1.31	.93	1.41	.85	1.52	.78	1.64
27	1.09	1.23	1.02	1.32	.95	1.41	.88	1.51	.81	1.63
28	1.10	1.24	1.04	1.32	.97	1.41	.90	1.51	.83	1.62
29	1.12	1.25	1.05	1.33	.99	1.42	.92	1.51	.85	1.61
30	1.13	1.26	1.07	1.34	1.01	1.42	.94	1.51	.88	1.61
31	1.15	1.27	1.08	1.34	1.02	1.42	.96	1.51	.90	1.60
32	1.16	1.28	1.10	1.35	1.04	1.43	.98	1.51	.92	1.60
33	1.17	1.29	1.11	1.36	1.05	1.43	1.00	1.51	.94	1.59
34	1.18	1.30	1.13	1.36	1.07	1.43	1.01	1.51	.95	1.59
35	1.19	1.31	1.14	1.37	1.08	1.44	1.03	1.51	.97	1.59
36	1.21	1.32	1.15	1.38	1.10	1.44	1.04	1.51	.99	1.59
37	1.22	1.32	1.16	1.38	1.11	1.45	1.06	1.51	1.00	1.59
38	1.23	1.33	1.18	1.39	1.12	1.45	1.07	1.52	1.02	1.58
39	1.24	1.34	1.19	1.39	1.14	1.45	1.09	1.52	1.03	1.58
40	1.25	1.34	1.20	1.40	1.15	1.46	1.10	1.52	1.05	1.58
45	1.29	1.38	1.24	1.42	1.20	1.48	1.16	1.53	1.11	1.58
50	1.32	1.40	1.28	1.45	1.24	1.49	1.20	1.54	1.16	1.59
55	1.36	1.43	1.32	1.47	1.28	1.51	1.25	1.55	1.21	1.59
60	1.38	1.45	1.35	1.48	1.32	1.52	1.28	1.56	1.25	1.60
65	1.41	1.47	1.38	1.50	1.35	1.53	1.31	1.57	1.28	1.61
70	1.43	1.49	1.40	1.52	1.37	1.55	1.34	1.58	1.31	1.61
75	1.45	1.50	1.42	1.53	1.39	1.56	1.37	1.59	1.34	1.62
80	1.47	1.52	1.44	1.54	1.42	1.57	1.39	1.60	1.36	1.62
85	1.48	1.53	1.46	1.55	1.43	1.58	1.41	1.60	1.39	1.63
90	1.50	1.54	1.47	1.56	1.45	1.59	1.43	1.61	1.41	1.64
95	1.51	1.55	1.49	1.57	1.47	1.60	1.45	1.62	1.42	1.64
100	1.52	1.56	1.50	1.58	1.48	1.60	1.46	1.63	1.44	1.65

Reproduced in part from Tables 4, 5, and 6 of J. Durbin and G. S. Watson, Testing for correlation in least squares regression. II. *Biometrika 38*, 159–178 (1951) with permission of the Biometrika Trustees.

Appendix Table D Empirical percentage points of the approximate W' test.

	P										
n	.01	.05	.10	.15	.20	.50	.80	.85	.90	.95	.99
35	.919	.943	.952	.956	.964	.976	.982	.985	.987	.989	.992
50	.935	.953	.963	.968	.971	.981	.987	.988	.990	.991	.994
51	.935	.954	.964	.968	.971	.981	.988	.989	.990	.992	.994
53	.938	.957	.964	.969	.972	.982	.988	.989	.990	.992	.994
55	.940	.958	.965	.971	.973	.983	.988	.990	.991	.992	.994
57	.944	.961	.966	.971	.974	.983	.989	.990	.991	.992	.994
59	.945	.962	.967	.972	.975	.983	.989	.990	.991	.992	.994
61	.947	.963	.968	.973	.975	.984	.990	.990	.991	.992	.994
63	.947	.964	.970	.973	.976	.984	.990	.991	.992	.993	.994
65	.948	.965	.971	.974	.976	.985	.990	.991	.992	.993	.995
67	.950	.966	.971	.974	.977	.985	.990	.991	.992	.993	.995
69	.951	.966	.972	.976	.978	.986	.990	.991	.992	.993	.995
71	.953	.967	.972	.976	.978	.986	.990	.991	.992	.994	.995
73	.956	.968	.973	.976	.979	.986	.991	.992	.993	.994	.995
75	.956	.969	.973	.976	.979	.986	.991	.992	.993	.994	.995
77	.957	.969	.974	.977	.980	.987	.991	.992	.993	.994	.996
79	.957	.970	.975	.978	.980	.987	.991	.992	.993	.994	.996
81	.958	.970	.975	.979	.981	.987	.992	.992	.993	.994	.996
83	.960	.971	.976	.979	.981	.988	.992	.992	.993	.994	.996
85	.961	.972	.977	.980	.981	.988	.992	.992	.993	.994	.996
87	.961	.972	.977	.980	.982	.988	.992	.993	.994	.994	.996
89	.961	.972	.977	.981	.982	.988	.992	.993	.994	.995	.996
91	.962	.973	.978	.981	.983	.989	.992	.993	.994	.995	.996
93	.963	.973	.979	.981	.983	.989	.992	.993	.994	.995	.996
95	.965	.974	.979	.981	.983	.989	.993	.993	.994	.995	.996
97	.965	.975	.979	.982	.984	.989	.993	.993	.994	.995	.996
99	.967	.976	.980	.982	.984	.989	.993	.994	.994	.995	.996

Reproduced with permission from S. S. Shapiro and R. S. Francia, An approximate analysis of variance test for normality. *Journal of the American Statistical Association 67*, 215–216 (1972).

Answers to Selected Exercises

Chapter 1

1.4. a. Graph not given. There appears to be an increasing trend in Y with increasing X.

b. $\hat{Y}_i = 4.7990 + .5947X_i$. The slope coefficient $\hat{\beta}_1 = .5947$ indicates an average increase of .59 beat per minute in heart rate at rest for each kilogram increase in body weight in these data.

c. The point (67, 40) appears to have a major effect on the slope of the regression line. If this point were deleted, there would be little or no apparent trend remaining in the data; $\hat{\beta}_1$ would decrease toward zero (becomes .0788) and $\hat{\beta}_0$ would increase toward $\bar{Y}$ (becomes 49.1641).

d. $\hat{Y}_{(X=88)} = 57.13$. The 95% confidence interval estimate of the mean heart rate for $X = 88$ is (44.53, 69.73).

e. $\hat{Y}_{\text{pred}(X=88)} = 57.13$. The 95% confidence interval estimate of the heart rate for a particular subject weighing $X = 88$ is (30.09, 84.17).

f. Observation 5 with $X = 81$.

1.5. $r(X, Y) = r(Y, \hat{Y}) = .5687$ and $r(X, \hat{Y}) = 1.00$, where $r(\cdot, \cdot)$ denotes the correlation between the two variables in parentheses. Proofs not given.

1.10. a. Analysis of variance:

Source	d.f.	SS	MS
Total	3	32.73	
Regression	1	29.23	29.23
Residual	2	3.50	1.75

b. $R^2 = .89$; 89% of the dispersion in Y is attributable to the dispersion in X.

1.11. a. $s^2(\bar{Y}_1) = 5.8750$, $s^2(\bar{Y}_2) = 3.9167$, $s^2(\bar{Y}_3) = 7.8333$, $s^2(\bar{Y}_4) = 2.6111$.

b. $s^2(\bar{Y}_3 + \bar{Y}_4 - 2\bar{Y}_1) = 33.9444$

c. $s^2[(\bar{Y}_1 + \bar{Y}_2 + \bar{Y}_3)/3] = 1.9583$

d. $s^2[(4\bar{Y}_1 + 6\bar{Y}_2 + 3\bar{Y}_3)/13] = 1.8077$

1.12. Normal equations: $n\mu = \sum Y_i$, $\beta_1 \sum x_i^2 = \sum x_i Y_i$

1.17. $\sigma^2(\hat{Y}_{\text{pred(mean of } q)}) = \left[\frac{1}{q} + \frac{1}{n} + \frac{(X_i - \bar{X})^2}{\sum (X_i - \bar{X})^2}\right]$

1.19. **a.** Low ozone: $\hat{Y}_i = -5.360 + .0405X_i$; high ozone: $\hat{Y}_i = -2.2419 + .0338X_i$; 95% confidence limits:

	On β_0	On β_1
Low ozone	$(-9.6858, -1.0349)$	(.0284, .0526)
High ozone	$(-4.5463, .0626)$	(.0266, .0409)

Strongly overlapping intervals suggest no significant differences.

b. $\hat{Y}_i = -3.7475 + .0371X_i$

1.21. **a.** Plot not given.

b. Plot not given. Subtraction of control group has reduced dispersion.

c. H_0: $\beta_0 = 0$; $t = -.442$, Prob($|t| > .442$) = .67. The intercept can be dropped from the model. Without β_0: $\hat{Y}_i = .000624X_i$, $s(\hat{\beta}_1) = .000032$. SS(Model) = 17.87346, SS(Res) = .55012 with 12 degrees of freedom.

Chapter 2

2.1. **a.** $c'A = (5 \quad 8)$

c.

$$B' + A = \begin{bmatrix} 2 & 0 \\ 4 & 7 \\ -2 & -2 \end{bmatrix}$$

e. Not possible

2.3.

$$D = \begin{bmatrix} 1 & 0 & 0 & 0 \\ 0 & 1 & 0 & 0 \\ 0 & 0 & 1 & 0 \\ 0 & 0 & 0 & 1 \end{bmatrix}, \quad D \text{ is idempotent, } r(D) = 4$$

2.5. $AB = I_2$; therefore, $A^{-1} = B$.

2.7. Plot not given. v_2 and v_3 are pairwise orthogonal since $v_2'v_3 = 0$. $v_1'v_2 = 5$ and $v_1'v_3 = -3$ and, therefore, neither v_1 and v_2 nor v_1 and v_3 are pairwise orthogonal. Three vectors plotted in two-dimensional space must contain a linear dependency.

2.8. **a.** $r(V) = r(V \ v_3)$; equations are consistent.

b. $r(V)$ = number of unknowns; solution is unique.

c. $x' = (-\frac{17}{3} \ \frac{5}{3})$

2.11. **a.** $A = ZLZ'$ where

$$Z = \begin{bmatrix} .81649658 & -.57735027 \\ .57735027 & .81649658 \end{bmatrix} \qquad L = \begin{bmatrix} 4 & 0 \\ 0 & 1 \end{bmatrix}$$

b. $r(A) = 2$ since there are two nonzero eigenvalues.

2.13.

$$A^{-1} = \begin{bmatrix} .2 & 0 & 0 \\ 0 & .11538462 & -.0769231 \\ 0 & -.0769231 & .38461538 \end{bmatrix}$$

2.15. **a.** $X = UL^{1/2}Z'$ where

$$U = \begin{bmatrix} .30781896 & -.5025993 & -.3572695 \\ .33684738 & -.1877123 & -.3553842 \\ .36587579 & .12717465 & -.353499 \\ .3949042 & .44206161 & -.3516137 \\ .30921568 & -.5069614 & .34982249 \\ .33824409 & -.1920745 & .35170774 \\ .3672725 & .12281249 & .35359299 \\ .39630092 & .43769945 & .35547823 \end{bmatrix}$$

$$Z = \begin{bmatrix} .88611956 & -.4606721 & -.0507278 \\ .46132559 & .8872299 & .00133138 \\ .04439387 & -.0245818 & .99871163 \end{bmatrix}$$

$$L^{1/2} = \begin{bmatrix} 3.1784419 & 0 & 0 \\ 0 & .56352281 & 0 \\ 0 & 0 & .14124212 \end{bmatrix}$$

$r(X) = 3$ since all roots are nonzero.

b.

$$X_1 = \lambda_1 u_1 z_1' = \begin{bmatrix} .8669658 & .45135389 & .04343428 \\ .9487237 & .49391816 & .04753028 \\ 1.0304817 & .53648242 & .05162629 \\ 1.1122396 & .57904668 & .05572229 \\ .8708996 & .4534019 & .04363136 \\ .9526576 & .49596616 & .04772736 \\ 1.0344155 & .53853042 & .05182337 \\ 1.1161734 & .58109468 & .05591937 \end{bmatrix};$$

Goodness of fit $= .9677$

c.

$$X_1'X_1 = \lambda_1^2 z_1 z_1' = \begin{bmatrix} 7.9325569 & 4.1297943 & .39741461 \\ 4.1297943 & 2.1500257 & .20689932 \\ .3974146 & .2068993 & .01991015 \end{bmatrix};$$

$\text{tr}(X_1'X_1) = \lambda_1^2 = 10.10249$,
$\text{tr}(X'X) = 10.44$

2.18. $r(A) = r(A \;\; y) = 2$

2.19. Show that $AA^-A = A$ or $A^-AA^- = A^-$

2.21. $x' = (1 \;\; 1 \;\; 1)$

2.23. Distance (y_1, y_2) = Distance $(x_1^*, x_2^*) = 26$ but Distance $(x_1, x_2) = 84$

Chapter 3

3.2. X_1 is of full column rank and therefore there is a unique solution to the normal equations. X_2 and X_3 are not of full rank and will not have unique solutions.

3.4. $n = 6, \sum X_i^2 = 31, \sum x_i^2 = 29.5$

3.5. **a.** Definitions of matrices not given.

b.
$$(X'X) = \begin{bmatrix} 7 & 51 & 32 \\ 51 & 471 & 235 \\ 32 & 235 & 163.84 \end{bmatrix}, \quad (X'Y) = \begin{bmatrix} 652.5 \\ 4{,}915.3 \\ 3{,}103.66 \end{bmatrix}$$

c. $\hat{\boldsymbol{\beta}}' = (51.569698 \;\; 1.4974105 \;\; 6.7232557)$;
$\hat{Y}_i = 51.5697 + 1.4974X_{i1} + 6.7233X_{i2}$.
Interpretation of $\hat{\beta}_j$ not given. Units of measure for $\hat{\beta}_0$, $\hat{\beta}_1$, and $\hat{\beta}_2$ are grams dry weight, grams dry weight per percent organic matter, and grams dry weight per kilogram supplemental soil nitrogen, respectively.

d.
$$\hat{\boldsymbol{Y}} = \begin{bmatrix} 79.532036 \\ 72.56455 \\ 105.69144 \\ 88.883305 \\ 97.012501 \\ 105.01912 \\ 103.79704 \end{bmatrix}, \quad \boldsymbol{e} = \begin{bmatrix} -1.032036 \\ 1.735450 \\ -1.391445 \\ -1.283305 \\ -1.112501 \\ 4.180881 \\ -1.097045 \end{bmatrix}$$

e. $v_{11} = .36431779$, $v_{12} = .33531579$;
$\text{Var}(\hat{\beta}_1) = (.0100774)\sigma^2$, $\quad \text{Var}(\hat{Y}_{\text{pred}_1}) = (1.36431779)\sigma^2$,
$\text{Var}(\hat{Y}_1) = (.36431779)\sigma^2$, $\quad \text{Var}(e_1) = (.63568221)\sigma^2$

3.7. $\boldsymbol{P} = \boldsymbol{P}'$ verifies symmetry, $\boldsymbol{PP} = \boldsymbol{P}$ verifies idempotency. $\hat{Y}_3 = .08625337Y_1 + .20754717Y_2 + .30458221Y_3 + .40161725Y_4$; the coefficients for $\hat{Y}_3$ are the elements of row 3 of $\boldsymbol{P}$.

3.9. a.
$$\boldsymbol{X} = \begin{bmatrix} 1 & -195.708 & -.0225 \\ 1 & -98.908 & -.0225 \\ 1 & -30.208 & -.0225 \\ 1 & 73.792 & -.0225 \\ 1 & 137.392 & -.0225 \\ 1 & 201.492 & -.0225 \\ 1 & -205.008 & .0225 \\ 1 & -114.508 & .0225 \\ 1 & -49.908 & .0225 \\ 1 & 44.092 & .0225 \\ 1 & 99.092 & .0225 \\ 1 & 138.392 & .0225 \end{bmatrix}, \quad \boldsymbol{\beta} = \begin{pmatrix} \beta_0 \\ \beta_1 \\ \beta_2 \end{pmatrix}, \quad \hat{\boldsymbol{\beta}} = \begin{pmatrix} 7.908321 \\ .037549 \\ 22.582618 \end{pmatrix}$$

$\hat{Y}_i = 7.90832 + .037549(X_{i1} - \bar{X}_1) + 22.58262(X_{i2} - \bar{X}_2)$ where X_1 = radiation and X_2 = ozone level.

b. $\hat{Y} = 7.85903 + .037121(X_1 - \bar{X}_1) + 22.30463(X_2 - \bar{X}_2) - .149614(X_1 - \bar{X}_1)(X_2 - \bar{X}_2)$. The presence of the product term X_1X_2 decreased the residual sum of squares from 13.15948 (with 9 degrees of freedom) to 10.93078 (with 8 degrees of freedom).

3.12. a. $\hat{Y}_i = .0873 - .0354X_1 + .0047X_2 - .0019X_3 - .1201X_4$. The standard errors of the regression coefficients are $s(\hat{\beta}_0) = .3119$, $s(\hat{\beta}_1) = .0093$, $s(\hat{\beta}_2) = .0048$, $s(\hat{\beta}_3) = .0042$, and $s(\hat{\beta}_4) = .0238$.

b. $\hat{Y}_i = -.0330X_1 + .0058X_2 - .0008X_3 - .1215X_4$. The standard errors are $s(\hat{\beta}_1) = .0038$, $s(\hat{\beta}_2) = .0028$, $s(\hat{\beta}_3) = .0005$, and $s(\hat{\beta}_4) = .0214$.

c. $\hat{Y}_i = -.0332X_1 - .1243X_4$. The standard errors are $s(\hat{\beta}_1) = .0039$ and $s(\hat{\beta}_4) =$

.0122. The regression coefficients on X_1 and X_4 changed only slightly as the models were changed; X_1 and X_4 must be very nearly orthogonal to X_2 and X_3.

d. Standard errors needed for comparison are given in parts (a)–(c).

3.13. a. $\hat{\boldsymbol{\beta}} = (\boldsymbol{X}'\boldsymbol{X})^{-1}\boldsymbol{X}'\boldsymbol{Y}$ but $\mathscr{E}(\boldsymbol{Y}) = \boldsymbol{X\beta} + \boldsymbol{Z\gamma}$. Therefore, $\mathscr{E}(\hat{\boldsymbol{\beta}}) = \boldsymbol{\beta} + (\boldsymbol{X}'\boldsymbol{X})^{-1}\boldsymbol{X}'\boldsymbol{Z\gamma}$

b. $\hat{\boldsymbol{\beta}}$ is unbiased only if the columns of $\boldsymbol{X}$ and $\boldsymbol{Z}$ are pairwise orthogonal matrices so that $\boldsymbol{X}'\boldsymbol{Z} = \boldsymbol{0}$, or $\boldsymbol{\gamma} = \boldsymbol{0}$.

Chapter 4

4.1. a. $\hat{Y} = 95 + 3.3708X_1 + .31858X_2 - 3.10X_3$

b. Analysis of variance:

Source	d.f.	SS	MS
$\text{Total}_{\text{uncorr}}$	20	185,883	
$R(\mu)$	1	180,500	
$\text{Total}_{\text{corr}}$	19	5,383	
$R(\beta_1\ \beta_2\ \beta_3\|\mu)$	3	5,034.7037	1,678.2345
Residual	16	348.2963	21.7685

c. $s^2 = 21.7685$; standard errors of the regression coefficients are the square roots of the diagonal elements of $s^2(\hat{\boldsymbol{\beta}})$:

$$s^2(\hat{\boldsymbol{\beta}}) = \begin{bmatrix} 1.088426 & 0 & 0 & 0 \\ 0 & .238526 & -.094422 & 0 \\ 0 & -.094422 & .058866 & 0 \\ 0 & 0 & 0 & .170067 \end{bmatrix}$$

$\text{Cov}(\hat{\beta}_1, \hat{\beta}_2) = -.094422 \qquad \text{Cov}(\hat{\beta}_1, \hat{\beta}_3) = 0$

d. $R(\beta_3|\mu\ \beta_1\ \beta_2) = 1{,}230.0800$, $R(\beta_2|\mu\ \beta_1\ \beta_3) = 37.5309$. X_3 makes the greater contribution. $\hat{\beta}_1$ did not change in value when X_3 was dropped because X_1 and X_3 are pairwise orthogonal as shown by the zero sum of products, element (2, 4) in $\boldsymbol{X}'\boldsymbol{X}$. For the full model, $\hat{\beta}_1$ is the change in Y per unit change in X_1 conditional on X_2 and X_3 being constant. When X_3 is dropped, $\hat{\beta}_1$ is the change in Y per unit change in X_1 conditional on X_2 being constant but ignoring the value of X_3. A similar interpretation applies when X_2 is dropped but X_3 is retained.

e. The zero elements in the first row of $\boldsymbol{X}'\boldsymbol{X}$ show that the sums of X_1, X_2, and X_3 are zero and, therefore, their means are zero. If the variables had not been centered, it would have been necessary to invert the full 4×4 matrix. With the centered variables, the largest matrix to be inverted is the 2×2 matrix in the center of $\boldsymbol{X}'\boldsymbol{X}$.

4.3. a. Answer not given.

b. SS(Res) has 4 degrees of freedom. $s^2 = 6.8952$, $s^2(\hat{\beta}_1) = .06949$, $\text{Cov}(\hat{\beta}_1, \hat{\beta}_2) = -.00735$.

c. 95% univariate confidence limits:
For β_1: (.7656, 2.2292); for β_2: (4.9817, 8.4648).
95% Bonferroni confidence intervals:
For β_1: (.5761, 2.4187); for β_2: (4.5307, 8.9159).
95% Scheffé confidence intervals:
For β_1: (.5153, 2.4795); for β_2: (4.3860, 9.0605).

d. H_0: $2\beta_1 \geq \beta_2$, H_a: $2\beta_1 < \beta_2$. $t = -4.4534$, $\text{Prob}(t_{(4)} \leq -4.4534) = .006$.

e. $\boldsymbol{K}' = (0 \ 2 \ -1)$, $\boldsymbol{m} = 0$, $Q = 136.7518$, $F = 19.83$, $\text{Prob}(F_{(1,4)} > 19.83) = .012$. H_a: $2\beta_1 \neq \beta_2$.

f. $Y_i = \beta_0 + \beta X_i^* + \varepsilon_i$, where $X_i^* = X_{i1} + 2X_{i2}$. Test is same as in part (e).

4.5. a. Answer not given.

b. Analysis of variance:

Source	d.f.	SS	MS	*F-Value*	Prob > *F*
$\text{Total}_{\text{corr}}$	8	4.0778			
Regression	5	3.5977	.7195	4.50	.1229
Residual	3	.4801	.1600		

c. $.238 = R(\beta_2 | \beta_1 \ \beta_3 \ \beta_4 \ \beta_5 \ \beta_0)$

d. Statements (i) and (ii) are not valid; statement (iii) is valid. Statement (i) is not valid because a regression analysis does not establish causality. Statement (ii) is not valid because one or more of X_2, X_3, and X_4 may appear unimportant due to correlations among the variables, and become important to the regression when the correlated variable is dropped.

4.7. a. Analysis of variance summary:

Source	d.f.	SS	MS	*F-Value*	Prob > *F*
$\text{Total}_{\text{corr}}$	15	316,148			
Regression	2	134,156	67,078	4.791	.0276
Residual	13	181,992	13,999		

Variable	d.f.	*Sequential* SS	*Partial* SS
Intercept	1	4,449,568	901
X_2 = Vessels	1	19,638	1,781
X_1 = Pressure	1	114,518	114,518

Parameter estimates:

Variable	d.f.	*Parameter Estimate*	*Standard Error*	*t for* $\beta_j = 0$	Prob > $\lvert t \rvert$
Intercept	1	−119.0926	469.3404	−.254	.8037
X_2 = Vessels	1	2.2783	6.3870	.357	.7270
X_1 = Pressure	1	1.0538	.3685	2.860	.0134

95% Bonferroni confidence intervals on β_j:

Parameter	*Lower Limit*	*Upper Limit*
β_0	−1,407.4	1,169.2
β_2	−15.254	19.811
β_1	.04241	2.06527

$\hat{Y}(X_1 = 400, X_2 = 70) = 461.9$; 95% confidence limits on the prediction are (186.0, 737.9).

b. See t-tests of regression coefficients under part (a). Reject $H_0: \beta_1 = 0$ but do not reject $H_0: \beta_2 = 0$.

c. Keep only X_1 = Pressure. X_1 by itself is significant at $\alpha = .01$.

d. Recommend using total fishing pressure, but caution that this may not be very effective.

4.9. a. Analysis of variance:

Source	d.f.	SS	MS	*F-Value*	Prob > F
$\text{Total}_{\text{corr}}$	11	295.00917			
Regression	3	284.07839	94.69280	69.304	.0001
Residual	8	10.93078	$1.36635 = s^2$		

Sequential and partial SS:

Variable	d.f.	*Sequential* SS	*Partial* SS
Intercept	1	750.500833	731.603115
$x_1 = (X_1 - \bar{X}_1)$	1	278.790952	271.014210
$x_2 = (X_2 - \bar{X}_2)$	1	3.058731	2.983270
$x_1 x_2$	1	2.228705	2.228705

X_1 = Total cumulative solar radiation
X_2 = Ozone exposure level

Parameter estimates and confidence intervals:

Variable	d.f.	*Parameter Estimate*	*Standard Error*	*90% Bonferroni Confidence Limits*
Intercept	1	7.8590	.33963	(6.9244, 8.7937)
x_1	1	.0371	.00264	(.02987, .04437)
x_2	1	22.3046	15.09487	(−19.236, 63.846)
$x_1 x_2$	1	−.1496	.11715	(−.4720, .1728)

$t_{(.10/4, 8)} = 2.752$. Only β_0 and β_1 are clearly different from zero based on the 90% Bonferroni confidence limits. For $H_0: \beta_3 = 0$ (β_3 is the regression coefficient for the product term), $t = -1.277$, $\text{Prob}(|t_{(8)}| > 1.277) = .24$. Conclusion agrees with conclusion based on confidence interval. (They need not agree because different Type I error rates are being used.)

b. Analysis of variance:

Source	d.f.	SS	MS	*F-Value*	Prob > F
$\text{Total}_{\text{corr}}$	11	295.00917			
Regression	2	281.84968	140.92484	96.38	.0001
Residual	9	13.15948	1.46216		

Partial and sequential SS:

Variable	d.f.	*Sequential* SS	*Partial* SS
Intercept	1	750.500833	750.498458
x_1	1	278.790952	281.828850
x_2	1	3.058731	3.058731

The difference in SS(Res) is (13.15948 − 10.93078) = 2.22870 with 1 d.f. F = 1.63, Prob($F_{(1,8)} > 1.63$) = .24, the same result as with the t-test.
Parameter estimates:

Variable	d.f.	*Parameter Estimate*	*Standard Error*	*t for* $\beta_j = 0$	Prob > $\lvert t \rvert$
Intercept	1	7.90832	.34907	22.656	.0001
x_1	1	.03755	.00270	13.883	.0001
x_2	1	22.58262	15.61356	1.446	.1820

The estimated seed weight for $X_1 = \bar{X}_1$ and $X_2 = .025$ is $\hat{Y} = 7.4002$; for $X_1 = \bar{X}_1$ and $X_2 = .07$ is $\hat{Y} = 8.4164$; the estimated mean change in seed weight for ozone (X_2) changing from .025 to .07 is −1.0162 grams per plant (a *gain*). To obtain these three linear functions as $\boldsymbol{L} = \boldsymbol{K}'\hat{\boldsymbol{\beta}}$ define $\boldsymbol{K}'$ as

$$\boldsymbol{K}' = \begin{bmatrix} 1 & 0 & -.0225 \\ 1 & 0 & .0225 \\ 0 & 0 & -.045 \end{bmatrix} \quad \text{(Recall that } X\text{'s are centered.)}$$

Then,

$$s^2(\boldsymbol{L}) = \boldsymbol{K}'[s^2(\hat{\boldsymbol{\beta}})]\boldsymbol{K} = \begin{bmatrix} .24526 & -.00157 & .24683 \\ -.00157 & .24526 & -.24683 \\ .24683 & -.24683 & .49366 \end{bmatrix}$$

The coordinates for plotting the joint confidence ellipse for β_1 and β_2 are:

Value of β_2	*Lower Value of* β_1	*Upper Value of* β_1
−15	.0355	.0381
−10	.0334	.0404
−5	.0324	.0416
0	.0318	.0424
5	.0313	.0431
10	.0311	.0435
15	.0309	.0439
20	.0309	.0441
25	.0310	.0442
30	.0312	.0442
35	.0316	.0440
40	.0320	.0438
45	.0326	.0433
50	.0335	.0427
55	.0347	.0417
60	.0369	.0397

4.12.

Numerator SS	*Null Hypothesis*
$R(\beta_3 \mid \beta_0)$	$\beta_3 = 0$
$R(\beta_1 \mid \beta_0\ \beta_3)$	$\beta_{1.3} = 0$
$R(\beta_2 \mid \beta_0\ \beta_1\ \beta_3)$	$\beta_{2.13} = 0$
$R(\beta_3 \mid \beta_0\ \beta_1\ \beta_2)$	$\beta_{3.12} = 0$
$R(\beta_1 \mid \beta_0\ \beta_2\ \beta_3)$	$\beta_{1.23} = 0$
$R(\beta_2 \mid \beta_0\ \beta_1\ \beta_3)$	$\beta_{2.13} = 0$

4.14. a. X_3 would account for the greatest variability. The coefficient of determination for the regression of Y on X_3 would be $R^2 = (.974)^2 = .949$. The R^2 for the multiple regression must be larger.

b. There is a very high correlation between X_1 and X_2 of .983. Any information on Y provided by X_1 will also be provided by X_2 so that one becomes redundant.

Chapter 5

5.1. Correlation coefficients for peak flow data on original scale ($n = 30$):

	X_1	X_2	X_3	X_4	X_5	X_6	X_7	X_8	X_9	Q
X_1	1.00	.80	−.08	.92	−.74	.12	.83	.17	.23	.78
X_2	.80	1.00	−.07	.76	−.49	.07	.83	.11	.13	.67
X_3	−.08	−.07	1.00	.24	−.40	.15	−.31	−.07	−.13	.21
X_4	.92	.76	.24	1.00	−.78	.06	.75	.17	.22	.87
X_5	−.74	−.49	−.40	−.78	1.00	−.27	−.48	−.04	−.11	−.62
X_6	.12	.07	.15	.06	−.27	1.00	.12	.07	−.01	.05
X_7	.83	.83	−.31	.75	−.48	.12	1.00	.17	.24	.67
X_8	.17	.11	−.07	.17	−.04	.07	.17	1.00	.89	.33
X_9	.23	.13	−.13	.22	−.11	−.01	.24	.89	1.00	.28
Q	.78	.67	.21	.87	−.62	.05	.67	.33	.28	1.00

Based on the simple correlations with Q, the variables most likely to contribute significantly to variation in Q are X_1, X_2, X_4, X_5, and X_7. The best single variable would be X_4 (= stream length).

5.3. The test of the composite null hypothesis that all partial regression coefficients are zero: $F = 169.77$ with 9 and 20 degrees of freedom; reject H_0. The tests of the individual partial regression coefficients showed X_1, X_6, X_8, and X_9 to be significantly different from zero ($\alpha = .05$). The variable X_7 had the least significant partial regression coefficient and would be the first eliminated from the full model.

5.5. 95% univariate confidence interval estimates:

Parameter	*95% Univariate*	*95% Bonferroni*
β_0	(5.42, 6.33)	(Not included)
β_1	(.6329, .7295)	(.6157, .7466)
β_3	(.1794, .5712)	(.1099, .6407)
β_6	(−.5040, −.1676)	(−.5638, −.1078)
β_8	(1.3789, 2.0711)	(1.2559, 2.1941)
β_9	(−1.7655, −1.1475)	(−1.8752, −1.0378)

(The Bonferroni intervals are based on only the five independent variables.)

Chapter 6

6.1. $\hat{\beta}_1 \doteq \frac{17}{44}$, $\hat{\beta}_2 \doteq \frac{24}{55}$. $\hat{Y}$ would have to fall left of X_2 for $\hat{\beta}_1$ to be negative, and below X_1 for $\hat{\beta}_2$ to be negative. Both conditions would have to be met for both coefficients to be negative.

6.4. Diagram not given.

a. $\hat{\beta}_1 \doteq -.20$, $\hat{\beta}_2 \doteq -.85$.

b. SS(Regr) $= \hat{Y}'\hat{Y} \doteq 10$, d.f. $= 2$

c. No. Y is needed to compute SS(Res); d.f. $= 8$

d. Yes. Length(Y) cannot be less than length($\hat{Y}$) $= \sqrt{10}$.

6.6. **a.** length(X) $= 1{,}019.776$, length(Y) $= 1{,}234.354$, length($\hat{Y}$) $= 1{,}230.925$, length(e) $= 91.947$.

b. angle(X, Y) $= 4.27°$, angle(X, $\hat{Y}$) $= 0°$, angle(X, e) $= 90°$, angle(Y, $\hat{Y}$) $= 4.27°$, angle(Y, e) $= 85.7°$, angle($\hat{Y}$, e) $= 90°$.

c. Plot not given. The X-space is one-dimensional. As a result, $\hat{Y}$ must fall on the space defined by X and the angle between X and $\hat{Y}$ must be 0°. $\hat{\beta} \doteq 1.21$.

6.8.

$$\cos(\theta) = \frac{\mathbf{1}'X}{[(\mathbf{1}'\mathbf{1})(X'X)]^{1/2}} = \frac{1}{\left[1 + \frac{(n-1)}{n}(CV)^2\right]^{1/2}}$$

where $CV = s_x/\bar{X}$ is the coefficient of variation for the independent variable. Multiplication of X by a constant does not affect CV and, consequently, does not change the angle between **1** and X. Adding or subtracting a constant changes the mean but not the variance, and CV and the angle change accordingly. The larger the mean for a fixed variance, the smaller will be the angle. If the variable is centered by subtracting the mean, the angle between **1** and X will be 90°.

Chapter 7

7.1. $R^2 = 1 - \dfrac{(n-p')\text{MS(Res)}}{\text{SS(Total)}}$

7.5. Only key points are given:

a. Observational data cannot establish causality.

b. Partial regression coefficients in nonorthogonal data will absorb part of the effects of other correlated variables and, hence, may not reflect their true causal effects.

c. Prediction of the effect of stricter government standards may be inappropriate, even if the causality has been established, if the change in standards disrupts the correlational structure of the system.

d. Yes. See comment (b).

7.7. Total number of models is $2^9 - 1 = 511$. The C_p plot is not given. The *two* "best" models for each stage are as follows (× = variables in best model, o = variables in second best model):

No. of Var.	R^2	MSE	C_p	X_1	X_2	X_3	X_4	X_5	X_6	X_7	X_8	X_9
1	.9100	.230	113.4				×					
	.8875	.287	148.1	o								
2	.9161	.222	105.9				×				×	
	.9155	.224	106.9				o	o				

3	.9718	.078	21.7				×				×	×
	.9629	.102	35.5	o							o	o
4	.9808	.055	9.7				×		×		×	×
	.9744	.073	19.7	o					o		o	o
5	.9845	.046	6.0	×		×			×		×	×
	.9834	.049	7.7	o			o		o		o	o
6	.9860	.043	5.7	×		×		×	×		×	×
	.9856	.045	6.3	o		o	o		o		o	o
7	.9865	.044	6.9	×	×	×		×	×		×	×
	.9863	.044	7.2	o		o		o	o	o	o	o
8	.9870	.044	8.1	×	×	×	×	×	×		×	×
	.9866	.045	8.7	o		o	o	o	o	o	o	o
9	.9871	.046	10	×	×	×	×	×	×	×	×	×

7.9. Backward elimination with SLS = .10 stopped with five variables. The model contained the same variables identified as "best" at that stage by R^2 (see answer to exercise 7.7); $C_p = 6.0$. C_p plot not given. Backward elimination did not give same model as found in exercise 7.8.

7.11. Summary of chosen five-variable model, $R^2 = .9845$, $C_p = 6.02$:

Source	d.f.	SS	MS	*F*	Prob > *F*
Regression	5	70.2848	14.0570	304.56	.0001
Error	24	1.1077	.0462		
Total	29	71.3926			

Variable	*Parameter Estimate*	*Std. Error*	*Partial* SS	*F*	Prob > *F*
INTERCEP	5.8759	.2202	32.8638	712.02	.0001
X_1 (LAREA)	.6812	.0234	38.9963	844.89	.0001
X_3 (LAVSLOPE)	.3753	.0949	.7219	15.64	.0006
X_6 (LSTORAGE)	−.3358	.0815	.7834	16.97	.0004
X_8 (LRAIN)	1.7250	.1677	4.8849	105.84	.0001
X_9 (LTIME)	−1.4565	.1497	4.3667	94.61	.0001

Chapter 8

8.3.

$$X^* = \begin{bmatrix} 1 & 0 & 0 \\ \vdots & \vdots & \vdots \\ 1 & 1 & 0 \\ \vdots & \vdots & \vdots \\ 1 & 0 & 1 \\ \vdots & \vdots & \vdots \end{bmatrix} \begin{matrix} \} n_1 \text{ identical rows} \\ \\ \} n_2 \text{ identical rows} \\ \\ \} n_3 \text{ identical rows} \\ \\ \end{matrix} \qquad \boldsymbol{\beta}^* = \begin{pmatrix} \mu^* \\ \delta_1^* \\ \delta_2^* \end{pmatrix}$$

X is full rank. The form of X is equivalent to the reparameterization using $\tau_1 = 0$.

8.5. For H_0: $\tau_1 = \tau_2$, $\boldsymbol{K}' = (0\ 1\ -1\ 0\ 0)$.

For the composite null hypothesis,

$$K' = \begin{bmatrix} 0 & 0 & 0 & 1 & -1 \\ 0 & 1 & -1 & 0 & 0 \\ 0 & 1 & 1 & -1 & -1 \end{bmatrix}$$

These hypotheses are testable. The sum of squares for the composite hypothesis equals the treatment sum of squares.

8.7. With $b = 2$ and $t = 4$, there are 3 degrees of freedom for the block-by-treatment interaction. With $\boldsymbol{\mu}$ defined as

$$\boldsymbol{\mu}' = (\mu_{11}\ \mu_{12}\ \mu_{13}\ \mu_{14}\ \mu_{21}\ \mu_{22}\ \mu_{23}\ \mu_{24})$$

one possible $\boldsymbol{K}'$ is contructed by using the interaction contrast for each adjacent 2×2 table of μ_{ij}:

$$K' = \begin{bmatrix} 1 & -1 & 0 & 0 & -1 & 1 & 0 & 0 \\ 0 & 1 & -1 & 0 & 0 & -1 & 1 & 0 \\ 0 & 0 & 1 & -1 & 0 & 0 & -1 & 1 \end{bmatrix}$$

8.11. a. $\boldsymbol{X}$ is of order 40×23 and contains two singularities; $\boldsymbol{\beta}$ is 23×1.

$$X = \begin{bmatrix} \mathbf{1}_{20} & \mathbf{1}_{20} & \mathbf{0}_{20} & I_{20} \\ \mathbf{1}_{20} & \mathbf{0}_{20} & \mathbf{1}_{20} & I_{20} \end{bmatrix} \qquad \boldsymbol{\beta}' = (\mu\ \gamma_1\ \gamma_2\ \tau_1\ \tau_2\ \cdots\ \tau_{20})$$

where $\mathbf{1}_{20}$ is a 20×1 column vector of ones, $\mathbf{0}_{20}$ is a 20×1 column vector of zeros, and $\boldsymbol{I}_{20}$ is the 20×20 identity matrix. Reparameterization using $\gamma_2 = 0$ and $\tau_{20} = 0$ gives $\boldsymbol{X}^*$ equal to $\boldsymbol{X}$ with columns 3 and 23 dropped. SS(Blocks) = .0000729; SS(Treatments) = .421552.

b. The composite null hypothesis is H_0: $\tau_1 = \tau_{11}$ and $\tau_2 = \tau_{12}$ and, ..., $\tau_{10} = \tau_{20}$. $\boldsymbol{K}'$ for this hypothesis will have 10 rows and 10 degrees of freedom, and have the form

$$K' = (\mathbf{0}_{10}\ \mathbf{0}_{10}\ \mathbf{0}_{10}\ I_{10}\ -I_{10})$$

where $\mathbf{0}_{10}$ is the 10×1 column vector of zeros and $\boldsymbol{I}_{10}$ is the 10×10 identity matrix. The degrees of freedom for this hypothesis equals the sum of the degrees of freedom for Temp, Temp × Herb, Temp × Conc, and Temp × Herb × Conc.

$\boldsymbol{K}'$ for the null hypothesis that the *average* effect of temperature is zero is of order 1×23 and the sum of squares (= SS(Temp)) has 1 degree of freedom;

$$K' = (0\ 0\ 0\ \mathbf{1}'_{10}\ -\mathbf{1}'_{10})$$

c. The factorial model with only the main effects (plus block effects) will be of order 40×12 and will contain 4 singularities.

$$X = \begin{bmatrix} \mathbf{1}_5 & \mathbf{1}_5 & \mathbf{0}_5 & \mathbf{1}_5 & \mathbf{0}_5 & \mathbf{1}_5 & \mathbf{0}_5 & I_5 \\ \mathbf{1}_5 & \mathbf{1}_5 & \mathbf{0}_5 & \mathbf{1}_5 & \mathbf{0}_5 & \mathbf{0}_5 & \mathbf{1}_5 & I_5 \\ \mathbf{1}_5 & \mathbf{1}_5 & \mathbf{0}_5 & \mathbf{0}_5 & \mathbf{1}_5 & \mathbf{1}_5 & \mathbf{0}_5 & I_5 \\ \mathbf{1}_5 & \mathbf{1}_5 & \mathbf{0}_5 & \mathbf{0}_5 & \mathbf{1}_5 & \mathbf{0}_5 & \mathbf{1}_5 & I_5 \\ \mathbf{1}_5 & \mathbf{0}_5 & \mathbf{1}_5 & \mathbf{1}_5 & \mathbf{0}_5 & \mathbf{1}_5 & \mathbf{0}_5 & I_5 \\ \mathbf{1}_5 & \mathbf{0}_5 & \mathbf{1}_5 & \mathbf{1}_5 & \mathbf{0}_5 & \mathbf{0}_5 & \mathbf{1}_5 & I_5 \\ \mathbf{1}_5 & \mathbf{0}_5 & \mathbf{1}_5 & \mathbf{0}_5 & \mathbf{1}_5 & \mathbf{1}_5 & \mathbf{0}_5 & I_5 \\ \mathbf{1}_5 & \mathbf{0}_5 & \mathbf{1}_5 & \mathbf{0}_5 & \mathbf{1}_5 & \mathbf{0}_5 & \mathbf{1}_5 & I_5 \end{bmatrix} \qquad \boldsymbol{\beta} = \begin{bmatrix} \mu \\ \gamma_1 \\ \gamma_2 \\ T_1 \\ T_2 \\ H_1 \\ H_2 \\ C_1 \\ C_2 \\ C_3 \\ C_4 \end{bmatrix}$$

where $\mathbf{1}_5$ and $\mathbf{0}_5$ are the unity and null vectors of order 5, and $\boldsymbol{I}_5$ is the 5×5

identity matrix. If the sum constraints are used, columns 3, 5, 7, and 12 are deleted in X. The sums of squares are given in part (e) below.

d. Including $(TH)_{jk}$ effects in the model adds 4 columns to X and introduces 3 singularities. Including $(TC)_{jl}$ and $(HC)_{kl}$ effects adds 10 columns each to X and introduces 6 singularities. Including $(THC)_{jkl}$ effects adds 20 columns and introduces 16 singularities.

e. Factorial analysis of variance:

Source	d.f.	SS	MS	*F-Value*	Prob > *F*
Total	39	.42256			
BLOCK	1	.00007	.00007	1.48	.2385
TEMP	1	.00313	.00313	63.66	.0001
HERB	1	.13877	.13877	2,819.59	.0001
CONC	4	.26859	.06715	1,364.37	.0001
TEMP * HERB	1	.00620	.00620	125.98	.0001
TEMP * CONC	4	.00085	.00021	4.30	.0121
HERB * CONC	4	.00309	.00077	15.68	.0001
TEMP * HERB * CONC	4	.00092	.00023	4.69	.0084
Error	19	.00094	.00005		

8.13.

$$X^* = \begin{bmatrix} 1 & 1 & 0 & 1 & 1 & 0 \\ 1 & 1 & 0 & 0 & 0 & 0 \\ 1 & 0 & 1 & 1 & 0 & 1 \\ 1 & 0 & 1 & 0 & 0 & 0 \\ 1 & 0 & 0 & 1 & 0 & 0 \\ 1 & 0 & 0 & 0 & 0 & 0 \end{bmatrix} \qquad \beta^* = \begin{pmatrix} \mu^* \\ \gamma_1^* \\ \gamma_2^* \\ \tau_1^* \\ (\gamma\tau)_{11}^* \\ (\gamma\tau)_{21}^* \end{pmatrix}$$

$$K' = \begin{bmatrix} 0 & 0 & 0 & 0 & 1 & 0 \\ 0 & 0 & 0 & 0 & 0 & 1 \end{bmatrix}$$

8.15. One possible specification of the reduced model: Let $2\mu = \mu_{1.} = \mu_{2.} = \mu_{3.}$. Note that μ is then the overall mean. Then $\mu_{12} = 2\mu - \mu_{11}$, $\mu_{22} = 2\mu - \mu_{21}$, and $\mu_{32} = 2\mu - \mu_{31}$. This reduces the number of parameters to four (μ, μ_{11}, μ_{21}, and μ_{31}) with X having the form

$$X = \begin{bmatrix} 1 & 1 & 0 & 0 \\ 2 & -1 & 0 & 0 \\ 1 & 0 & 1 & 0 \\ 2 & 0 & -1 & 0 \\ 1 & 0 & 0 & 1 \\ 2 & 0 & 0 & -1 \end{bmatrix} \quad \text{where each element is a } 10 \times 1 \text{ vector.}$$

$SS(Res_{reduced}) = 3{,}400.4$ with 56 d.f.
$SS(Res_{full}) = 2{,}491.1$ with 54 d.f.
$Q = 3{,}400.4 - 2{,}491.1 = 909.3$ with 2 d.f.
$F = 9.86$, $Prob(F_{(2,54)} > 9.86) < .001$

8.16. $SS(Res_{reduced}) = 9{,}619{,}402.993$ with 25 d.f.
$SS(Res_{full}) = 4{,}248{,}097.043$ with 12 d.f.
$Q = 5{,}371{,}305.9$ with 13 d.f.
$F = 1.17$, $Prob(F_{(13,12)} > 1.17) > .25$

Chapter 9

9.1. a. The dependent variable would not be expected to satisfy assumptions of normality and homogeneous variances.

b. Plot not given. Data will plot only at $Y = 0$ or $Y = 1$ with $Y = 0$ dominating at the low doses and $Y = 1$ dominating at the high doses. The regression line will violate the natural boundaries of 0 and 1.

c. For $X = 2$, the point estimate of the mean is .104. The 95% confidence interval estimate of the mean at $X = 2$ is $(-.121, .329)$. The true proportion affected cannot be negative and, therefore, the lower bound is unreasonable.

d. Normality would be very nearly satisfied (because of the central limit theorem).

9.3. Variance of simple average = 3.222. Variance of weighted average = 2.3607. No other weighted average will give a lower variance.

9.5. No. Variance often increases with mean level. Measurements from the same plots over time will tend to be positively correlated. Similarly, measurements made at the same time on different plots also will tend to be positively correlated.

9.6. For $X = 0, 1, 2$, and 9, the perturbation in $\hat{\beta}_1$ would be $-\frac{3}{50}, -\frac{2}{50}, -\frac{1}{50}$, and $\frac{6}{50}$. Observation $X = 9$ will have the largest element in $\boldsymbol{P}$.

Chapter 10

10.1. Plot of Studentized residual versus $\hat{Y}$:

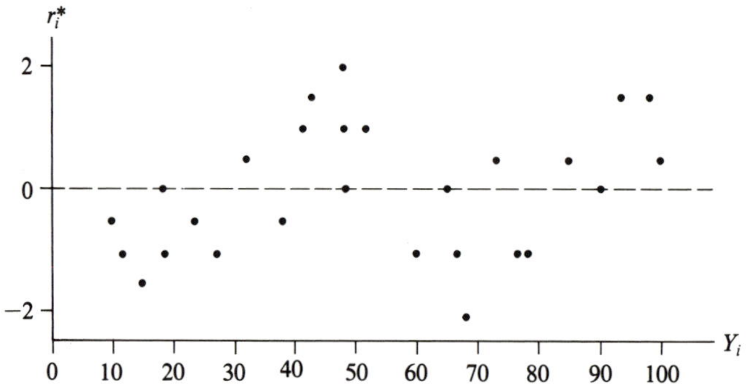

The plot shows two distinct groups of residuals with positive slopes. The model is not properly taking into account the group differences.

10.3. a. Plot the residuals against $\hat{Y}$. Look for any pattern of changing *magnitudes* of the residuals.

b. Look at the influence statistics for observation 11. Cook's D would provide an overall assessment of its influence. A large number would be expected if observation 11 is influential.

c. Plot the residuals against X_3. Any curved pattern of the residuals about the zero line would suggest inadequacies in the linear model. A partial regression leverage plot for X_3 might be helpful.

d. Plot residuals against $\hat{Y}$ and look for residuals that deviate far from zero or that do not follow the pattern of the other residuals.

e. Plot the residuals against $\hat{Y}$ and the independent variables. A pattern in any of the plots may suggest an omitted variable.

10.5. a. $K(X) = \delta_5 = 4.58, \delta_4 = 2.65, \delta_3 = 1.62, \delta_2 = 1.11$

b. $mci = 1.132$. Condition number indicates no problem but *mci* suggests there may be.

10.7. a. $r(\mathbf{X}) = 4$

b. $K(\mathbf{X}) = 24.379 > 10$; there may be a collinearity problem.

c. Large proportions of the variances of $\hat{\beta}_0$ (97%) and $\hat{\beta}_1$ (90%) come from the last principal component. This, in conjunction with the large condition number, indicates that collinearity is probably inflating the variances of these two regression coefficients.

d. The adjusted variance proportions for the third principal component are (.8796, .9327, .8425, .0947). This does not indicate serious variance inflation from collinearity, even though several proportions are greater than 50%, since the condition index for the third principal component does not indicate a collinearity problem.

10.9. a. Proportions of dispersion in the X-space for the four principal components are 54.9%, 27.0%, 13.7%, and 4.4%, respectively.

b. $K(\mathbf{Z}) = 3.53$. The condition indices are $\delta_4 = 3.53$, $\delta_3 = 2.00$, $\delta_2 = 1.43$, and $\delta_1 = 1.0$. There would appear to be no problem from collinearity.

c. The first principal component is nearly an average of the four variables with somewhat less weight given to *GH*. The second principal component is primarily the difference between *GH* and *SDH*.

d. $\text{tr}[\mathbf{Var}(\hat{\boldsymbol{\beta}})] = 8.882\sigma^2 \geq 5.675\sigma^2 = \sigma^2/\lambda_4^2$. If the four variables had been orthogonal, $\lambda_4^2 = 1$ so that $\sigma^2/\lambda_4^2 = \sigma^2$.

10.11. a. The sand, silt, and clay variables for each depth are percentages that very nearly add to 100. Therefore, one can expect three near-singularities from these built-in constraints alone.

b. The SVD indicates the presence of a collinearity problem. The condition number of the centered and scaled matrix is $K(\mathbf{X}) = 779$, and, in addition, $\delta_8 = 312$ and $\delta_7 = 189$. The collinearity problem would have been at least this severe if the variables had not been centered and the intercept included because centering has removed any collinearities with the intercept.

Chapter 11

11.1. a. The plots suggest that variance increases with increasing Y (or X). It appears that a quadratic polynomial might be needed to fit the data.

b. Move up the ladder of transformations for a transformation on Y or down the ladder for a transformation on X. The heterogeneous variances suggest that a transformation on Y would be helpful but moving up the ladder would be in the wrong direction. Therefore, prefer to make the transformation on X or first find the appropriate transformation on Y to stabilize the variance and then find the transformation on X that will straighten the relationship.

c. The residual sums of squares for the Box–Cox analysis using the quadratic model are:

λ	**Residual SS**
0	80.51
.2	45.30
.3	41.08
.4	41.00
.5	43.73
.7	55.97
1.0	93.62

The minimum occurs between $\lambda = .3$ and $\lambda = .4$. Using minimum SS = 41.00, the approximate 95% confidence interval on λ includes all values of λ that give a residual SS less than 43.7. This does not quite include $\lambda = .5$. [Answers in part (e) are based on $\lambda = .4$.]

d. The results of the Box–Cox analysis using a linear model are:

λ	**Residual SS**
0	128.86
.2	77.91
.3	68.04
.4	63.24
.5	61.96
.7	67.71
1.0	98.33

The minimum residual SS is near $\lambda = .5$, suggesting a square-root transformation. The change in the value of λ compared to part (c) results from the "efforts" of the Box–Cox analysis to straighten the relationship when the linear model is being used. The quadratic model allows Box–Cox to concentrate more on homogeneity of variance and normality.

e. Using $\lambda = .4$ and a linear model relating $Y^{(\lambda)}$ to X, the Box–Tidwell analysis estimated the power transformation on X as $\hat{\alpha} = .34$ at the end of the third iteration. The results from the iterations were:

Iteration	$\hat{\beta}$	$\hat{\gamma}$	$\hat{\alpha}$
1	.033535	−.74165	.26
2	1.799768	.28898	1.28898
3	1.118700	.01798	1.01798

The final $\hat{\alpha}$ is the product of the individual $\hat{\alpha}$'s.

11.3. **a.** Plot suggests that σ is approximately proportional to μ. Therefore, $k = 1$, which suggests the logarithmic transformation.

b. The weighting matrix is a 375×375 diagonal matrix where the diagonal elements are the reciprocals of the error variances from each location. The weighting can be accomplished without forming this large matrix by dividing Y and the elements of X for each observation by the square root of the error variance from the location from which the observation was obtained.

c. Discussion not given.

11.5. No

11.7. The Box–Cox analysis using Q and the logarithm of the nine independent variables gives a minimum residual SS near $\lambda = -.2$.

λ	**Res SS**
−.6	2,308,811
−.4	686,648
−.2	270,592
0	312,630
.2	601,770
.4	1,138,183
1.0	6,304,446

The approximate 95% confidence interval on λ is defined by the sum of squares $270{,}592[1 + (2.086)^2/20] = 329{,}465$ and contains $\lambda = 0$.

11.9. The Box–Cox transformation using a linear model in X gives a minimum residual SS near $\lambda = 1.5$. The 95% confidence interval estimate of λ includes $\lambda = 1$ to $\lambda = 2$. The power transformation $\lambda = 1.5$ is suggested.

11.11. The Box–Cox analysis using the linear model with $X^* = \ln(X)$ gives a minimum residual SS near $\lambda = 1.0$. The 95% confidence interval includes $\lambda = .75$ to $\lambda = 1.25$. No transformation is suggested.

11.13. The Box–Cox analysis using the analysis of variance model gives a minimum SS between $\lambda = 0$ and $\lambda = .25$. The approximate 95% confidence interval for λ overlapped $\lambda = 0$ and $\lambda = .5$; $\lambda = 0$ gave a smaller residual SS than did $\lambda = .5$, so the logarithmic transformation would be suggested.

Chapter 12

12.7. From equation 12.25, $\boldsymbol{\beta}^+_{(g)} = \boldsymbol{V}_{(g)}\hat{\boldsymbol{\gamma}}_{(g)}$. Therefore,

$$\begin{aligned}\text{length}[\boldsymbol{\beta}^+_{(g)}] &= [\boldsymbol{\beta}^{+\prime}_{(g)}\boldsymbol{\beta}^+_{(g)}]^{1/2}\\ &= [(\boldsymbol{V}_{(g)}\hat{\boldsymbol{\gamma}}_{(g)})'(\boldsymbol{V}_{(g)}\hat{\boldsymbol{\gamma}}_{(g)})]^{1/2}\\ &= (\hat{\boldsymbol{\gamma}}'_{(g)}\hat{\boldsymbol{\gamma}}_{(g)})^{1/2} = \text{length}(\hat{\boldsymbol{\gamma}}_{(g)})\end{aligned}$$

since $\boldsymbol{V}'_{(g)}\boldsymbol{V}_{(g)} = \boldsymbol{I}$.

12.8. **a.** $\boldsymbol{Z}$ and the correlation matrix are not given.

b. Eigenvectors for singular value decomposition of $\boldsymbol{Z}$ and the biplot are not given. The eigenvalues are as follows:

$$\lambda' = (2.0392\ \ 1.4095\ \ 1.0751\ \ 1.0270\ \ .6075\ \ .4037\ \ .3029\ \ .1301\ \ .0608)$$

The first two dimensions account for 68% of the dispersion.

c. Description of correlational structure not given.

12.9. $k = .00465$ from equation 12.9; $k = .005$ used in answers. The ordinary least squares estimates ($k = 0$) and the ridge regression estimates ($k = .005$) of the regression coefficients, with standard errors in parentheses, are as follows:

Variable	$k = 0$	$k = .005$
1	6.677 (2.754)	4.759 (1.256)
2	−.599 (.730)	−.173 (.597)
3	.802 (.752)	1.075 (.597)
4	2.198 (2.452)	2.988 (1.204)
5	.806 (.722)	.361 (.547)
6	−.743 (.331)	−.866 (.287)
7	.158 (.655)	.525 (.535)
8	4.349 (.489)	4.243 (.457)
9	−4.318 (.497)	−4.160 (.468)
SS(Regr)	70.47	70.09

SS(Total) = 71.39

12.11. **a.** Gabriel's biplot not given.

b. Three principal components must be used to account for 80%.

c. Variables not well represented by the biplot of the first two dimensions are silt content at depths 1 and 3 and, perhaps, clay content at depth 3. The biplot of the first two dimensions suggests that each soil component at depth 1 is reasonably highly positively correlated with the corresponding component at depth 2, but only moderately positively correlated with the component at depth 3. The sand and clay components at each depth are moderately highly negatively correlated, whereas the silt and clay components tend to be slightly positively correlated. The 20 sites tend to cluster as (7, 16, 18, 20), (1, 12, 17, 14), (2, 13, 15), (8, 10), (3, 9), (5, 6), with 4, 11, and 19 appearing alone. Site 4 has very low sand content at depths 1 and 2 but moderately high sand content at depth 3. *Caution:* This biplot accounts for only 76% of the dispersion in the X-space so that these interpretations should be regarded as tentative until the higher dimensions have been studied.

Chapter 13

13.1. Correlation matrix not given. Variables most highly correlated with *BIOMASS* are *pH*, *BUF*, *Ca*, *Zn*, and NH_4. The correlations for these variables exceed the critical value for $\alpha = .05$ for the test of H_0: $\rho = 0$.

The variables *pH*, *BUF*, *Ca*, *Zn*, and NH_4 are all reasonably highly correlated with each other, with *pH* and *BUF* being particularly highly (negatively) correlated. In addition, *K*, *Mg*, and *Na* are highly intercorrelated. Only H_2S, *SAL*, and *Eh*7 are not significantly correlated with at least one other variable.

13.3. Dropping buffer acidity (*BUF*) and NH_4 caused the principal components to be redefined. With some exceptions, the variables that originally dominated the first principal component now dominate the second principal component and vice versa.

13.5. Analysis of site means gives SS(Total) = SS(Regr) = 3,363,635.2; SS(Res) = 0. Five times SS(Total) corresponds to the SS(Among sites) from the analysis of variance of the original 45 observations. The within-site sampling variance is $s^2 =$ 65,355 with 36 degrees of freedom. The sum of squares attributable to each principal component (multiplied by 5), the *F*-ratio for each, and the probability of a larger *F* under the null hypothesis are given in the table.

Principal Component	SS(Regr)	*F*	Prob > *F*
1	9,378,000	143.5	.0001
2	2,003,370	30.7	.0001
3	2,192,052	33.5	.0001
4	62	.0	.9760
5	863,325	13.2	.0009
6	288,755	4.4	.0426
7	1,296,320	19.8	.0001
8	796,290	12.2	.0013

All principal components except the fourth are providing information on *BIOMASS*; the first contributes more than all others combined. The variables primarily involved in the first principal component are those with the large (absolute) coefficients in the first eigenvector. The fourth principal component is nearly orthogonal to *BIOMASS*. The primary variables involved in the fourth principal component are H_2S, *Eh*7, and *Mn* and, as long as these variables are not important in other principal components, they tentatively may be considered unimportant in *BIOMASS* production.

Chapter 14

14.1. Critical point at $X = 2.5$ is a maximum.

14.3. Critical point at $(X_1, X_2) = (2, 4)$ is a saddle point, a minimum with respect to X_1 and a maximum with respect to X_2.

14.4. In the interval $X = 6$ to 20 the response curve is the classical S-shaped growth curve. However, extrapolation from $X = 6$ to $X = 0$ shows an upward turn in the curve with $Y = 50$ at $X = 0$, and extrapolation from $X = 20$ to 30 shows a rapid decrease in the curve with $Y = -101$ at $X = 30$.

14.5. The transformed equation is $Y = -1{,}944.5 + 113.4T_c - 1.62T_c^2$, where T_c is temperature in °C. The linear coefficient is 113.4, which is a constant times the coefficient on X^2 in the original equation, and would be significantly different from zero if the original coefficient is significant.

14.7. **a.** The transformation of X = pressure to X = ln(pressure) appears to give a nearly linear relationship; logarithmic transformation on both percent moisture and pressure does even slightly better. With Y = ln(percent moisture) and X = ln(pressure), the full model allowing each soil to have its own intercept and slope gives $SS(Res_{full}) = .05018$ with 24 degrees of freedom. The reduced model forcing all soils to have a common slope gives $SS(Res_{reduced}) = .35763$ with 29 degrees of freedom. The test of homogeneity of slopes gives $F = 29.41$, which is significant ($\alpha = .01$). Reject the null hypothesis of common slopes.

b. The full nonlinear model, requiring a common γ but allowing α_j and β_j to differ with soils, gives $SS(Res_{full}) = 4.4209$ with 23 degrees of freedom. The reduced model requiring all β_j to be equal gives $SS(Res_{reduced}) = 7.7201$ with 28 degrees of freedom; $F = 3.4331$, which exceeds $F_{(.05;5,23)} = 2.64$ and the null hypothesis is rejected. The estimate of γ is $-.2337$ with the Wald 95% confidence limits extending from $-.30$ to $-.17$. This interval does not overlap zero and indicates that the logarithmic transformation on pressure *only* would not be adequate to straighten the relationship.

14.9. Analysis of variance and polynomial regression results:

Source	d.f.	SS	MS	*F*
Total	111	223.4322		
Treatments	3	23.6661	7.8887	4.64
Rep(Trt.)	4	6.7980	1.6995	
Days	13	185.7240	14.2865	435.56
Linear	1	173.6874	173.6874	5,295.35
Quadratic	1	11.6966	11.6966	356.60
Deviations	11	.3400	.0309	.94
Days × Trt.	39	5.5394	.1420	4.33
Lin. × Trt.	3	3.7374	1.2458	37.98
Quad. × Trt.	3	1.3065	.4355	13.27
Dev. × Trt.	33	2.4309	.0737	2.25
Error	52	1.7048	.0328	

a. The unusual pattern in the analysis of variance residuals results from the implicit assumption that the replication effects are the same for all "Days"—that is, that there are no true "Day by Rep" interaction effects—whereas the replication differences are small at day 1 and tend to increase with time.

b. The full quadratic model that allows each treatment to have its own intercept and quadratic response gives $SS(Res_{full}) = 9.3383$ with 100 degrees of freedom. The reduced model that allows each to have its own intercept but forces all to have the same quadratic response gives $SS(Res_{reduced}) = 14.3822$ with 106 degrees of freedom. The test of homogeneity of the quadratic responses gives $F = 25.6$ with 6 and 52 degrees of freedom; the null hypothesis is rejected. The reduced model for testing the null hypothesis that all intercepts are zero, conditional on a quadratic response for each treatment, gives $SS(Res_{reduced}) =$

9.8423 with 104 degrees of freedom. The test of all intercepts being zero is $F = 3.84$ with 4 and 52 degrees of freedom, which is significant at $\alpha = .01$. The full quadratic model allowing each treatment to have its own intercept and response would be adopted at this stage.

c. The test of zero intercepts in the presence of cubic models gives $F = .60$ with 4 and 52 degrees of freedom. The null hypothesis of zero intercepts is adopted.

d. The full model using the Mitscherlich response model for each treatment gave $SS(Res_{full}) = 10.3929$ with 104 degrees of freedom. The reduced Mitscherlich model with all treatments having common β_j gave $SS(Res_{reduced}) = 12.5168$ with 107 degrees of freedom. The reduced Mitscherlich model with all treatments having common α_j gave $SS(Res_{reduced}) = 12.1294$ with 107 degrees of freedom. Each of these changes in residual sums of squares is highly significant, indicating that different equations are needed if the Mitscherlich model is to be used to summarize the responses.

14.11. Convergence is attained for all environments. The estimates of the parameters for the six environments and their approximate standard errors are given in the table.

Environment	θ_1	$s(\theta_1)$	θ_2	$s(\theta_2)$
1	.1107	.0206	.0110	.0021
2	.0450	.0030	.0245	.0015
3	.0956	.0123	.0130	.0017
4	.1269	.0205	.0080	.0015
5	.0181	.0016	.0489	.0038
6	.0579	.0054	.0222	.0018

The plots of the response curves and the data appear reasonable.

14.13. It is reasonable to expect β_0 to be zero. The quadratic polynomial with zero intercept was adopted:

$$\hat{Y} = .6838X - .028306X^2$$

The polynomial shows a distinct maximum near Time = 12. The Weibull shows a monotonic increase to an upper asymptote. The quadratic appears to fit the data slightly better for the last few time periods, but which is "better" would be decided on whether a maximum and then a drop in calcium uptake with time is realistic.

Chapter 15

15.1. The sum of squares attributable to $X_1^2X_2$ is 406.2339. The F-test of the hypothesis H_0: $\beta_{112} = 0$ is $F = .23$ (using $s^2 = 1{,}774$ from the analysis of variance), which is nonsignificant. There is no need to add this term to the model.

15.3. The use of the mean ozone and mean sulfur dioxide levels over replications, rather than the individual plot values, caused relatively minor changes in the regression sums of squares. None of the conclusions is altered and the same-degree polynomial model is adopted with small changes in the coefficients:

$$\hat{Y} = 721 - 4{,}955X_1 + 12{,}127X_1^2 - 595X_2 + 3{,}151X_1X_2$$

Since the levels of X_1 and X_2 are constant over replicates of the same treatment in this analysis and the experiment is balanced, the sequential sums of squares for the polynomial terms in X_1 are partitions of the ozone treatment sums of squares in the analysis of variance, and similarly for the polynomial terms in X_2.

15.5. $SS(Res_{reduced}) = 59{,}431$ with 44 degrees of freedom. $SS(Res_{full}) = 59{,}049$ with 43 degrees of freedom. The F-test of H_0: $\gamma = 1$ is $F = 382/1{,}373 = .28$, which is nonsignificant. The yield loss in these data does not differ significantly from an exponential decay function. In this case, the conclusion is consistent with that obtained by the Wald test; the confidence interval for γ overlaps $\gamma = 1$.

15.7. The residual sums of squares obtained from linear least squares for specific choices of γ are as follows:

γ	**SS(Res)**
1.0	3.689997
1.05	3.685206
1.10	3.682756
1.12	3.682415
1.13	3.682380
1.14	3.682434
1.20	3.684597

The minimum among these choices of γ is at $\gamma = 1.13$. Quadratic interpolation will give a minimum slightly below 1.13. The nonlinear least squares solution was $\hat{\gamma} = 1.1289$.

Chapter 16

16.3. If treatment (2, 3) is missing from block 3, $\bar{\rho}_. = (\rho_1 + \rho_2 + \rho_3)/3$ for all cells in the table except cell (2, 3) and for all marginal means except those involving cell (2, 3). For cell (2, 3), $\bar{\rho}_. = (\rho_1 + \rho_2)/2$. Therefore, any contrasts *not* involving cell (2, 3) or means computed from cell (2, 3) will be free of block effects. Analysis of cell means will not produce unbiased estimates because contrasts on the treatment means will involve block effects.

16.5. **a.** s^2(within cells) $= 1.6210$ with 41 degrees of freedom, $\bar{n}_h = 5.3202$. The analysis of unweighted means with all sums of squares multiplied by $\bar{n}_h$ is:

Source	d.f.	SS	MS	*F*
Sex	1	.1756	.1756	.11
Type cancer	2	3.4622	1.7311	1.07
Sex × Type	2	.3126	.1563	.10
Within	41	66.4609	1.6210	

The four types of sums of squares are equal and the ordinary means are equal to the least squares means because the cell means data are balanced.

b. Sums of squares not given. None of the four types of sums of squares equals the sums of squares obtained from the unweighted analysis of cell means. The ordinary means from the weighted analysis are the weighted averages of the appropriate cell means and differ from the means obtained from the unweighted analysis. The least squares means reproduce the ordinary means from the unweighted analysis.

c. Sums of squares not given. All four types of sums of squares and all means are identical to those obtained from the weighted analysis of cell means.

16.7. The weighting matrix would be a diagonal matrix of n_{ij}/σ_{ij}^2. Estimates of the weights would be obtained from estimates of the within-cell variances.

16.9. **a.** Estimable; let $(L_1 \ L_2 \ L_4 \ L_6) = (1 \ 1 \ 1 \ 0)$.

b. Estimable: let $(L_1 \ L_2 \ L_4 \ L_6) = (0 \ 0 \ 1 \ 0)$.

c. Not estimable.

d. Not estimable.

e. Estimable; let $(L_1\ L_2\ L_4\ L_6) = (0\ 0\ 1\ 1)$.
f. Estimable; let $(L_1\ L_2\ L_4\ L_6) = (1\ 1\ \frac{1}{2}\ 0)$.
g. Estimable; let $(L_1\ L_2\ L_4\ L_6) = (0\ 1\ \frac{1}{2}\ -\frac{1}{2})$.

16.11. The population marginal means are as follows (the numbers are the coefficients on the model effects in the order $\mu, \alpha_1, \alpha_2, \beta_{11}, \beta_{12}, \beta_{21}, \beta_{22}$):

A_1: 1 1 0 $\frac{1}{2}$ $\frac{1}{2}$ 0 0
A_2: 1 0 1 0 0 $\frac{1}{2}$ $\frac{1}{2}$
B_{11}: 1 1 0 1 0 0 0
B_{12}: 1 1 0 0 1 0 0
B_{21}: 1 0 1 0 0 1 0
B_{22}: 1 0 1 0 0 0 1

All are estimable. If the B_{ij} are random effects, the population marginal means for A contain only the μ and α_i terms.

16.13. The Type IV sums of squares for B change to .7857; all other Type IV sums of squares are the same. The least squares means are not changed.

16.15. Analysis of variance:

Source	d.f.	*Type* III SS	MS	*F*	Prob > *F*
Block	1	3,823	3,823	2.84	.1229
Ozone	5	68,626	13,725	10.19	.0011
Moisture	1	24,345	24,345	18.08	.0017
Ozone × Moist.	5	17,020	3,404	2.53	.0994
Error	10	13,464	1,346		

Estimable functions for Type II and Type III sums of squares for ozone treatments:

		Type II	*Type* III
Intercept		0	0
Block		0	0
	2	0	0
Ozone	CF	L_4	L_4
	NF	L_5	L_5
	15	L_6	L_6
	30	L_7	L_7
	45	$-L_4 - L_5 - L_6 - L_7 - L_8$	$-L_4 - L_5 - L_6 - L_7 - L_8$
Moisture	D	0	0
	W	0	0
Ozone × Moist.	CF D	$.3438L_4$	$.1667L_{10}$
	CF W	$.6563L_4$	$.1667L_{10}$
	NF D	$.0313L_4 + \frac{1}{2}L_5$	$\frac{1}{2}L_5$
	NF W	$-.0313L_4 + \frac{1}{2}L_5$	$\frac{1}{2}L_5$
	15 D	$.0313L_4 + \frac{1}{2}L_6$	$\frac{1}{2}L_6$
	15 W	$-.0313L_4 + \frac{1}{2}L_6$	$\frac{1}{2}L_6$
	30 D	$.0313L_4 + \frac{1}{2}L_7$	$\frac{1}{2}L_7$
	30 W	$-.0313L_4 + \frac{1}{2}L_7$	$\frac{1}{2}L_7$
	45 D	$.0313L_4 + \frac{1}{2}L_8$	$\frac{1}{2}L_8$
	45 W	$-.0313L_4 + \frac{1}{2}L_8$	$\frac{1}{2}L_8$
	60 D	$-.4687L_4 - \frac{1}{2}(L_5 + L_6 + L_7 + L_8)$	$-\frac{1}{2}(L_4 + L_5 + L_6 + L_7 + L_8)$
	60 W	$-.5313L_4 - \frac{1}{2}(L_5 + L_6 + L_7 + L_8)$	$-\frac{1}{2}(L_4 + L_5 + L_6 + L_7 + L_8)$

Estimable functions for Type II and Type III sums of squares for moisture treatments:

		Type II	*Type* III
Intercept		0	0
Block	1	0	0
	2	0	0
Ozone	CF	0	0
	NF	0	0
	15	0	0
	30	0	0
	45	0	0
Moisture	D	L_{10}	L_{10}
	W	$-L_{10}$	$-L_{10}$
Ozone × Moist.	CF D	$.1146L_{10}$	$.1667L_{10}$
	CF W	$-.1146L_{10}$	$-.1667L_{10}$
	NF D	$.1771L_{10}$	$.1667L_{10}$
	NF W	$-.1771L_{10}$	$-.1667L_{10}$
	15 D	$.1771L_{10}$	$.1667L_{10}$
	15 W	$-.1771L_{10}$	$-.1667L_{10}$
	30 D	$.1771L_{10}$	$.1667L_{10}$
	30 W	$-.1771L_{10}$	$-.1667L_{10}$
	45 D	$.1771L_{10}$	$.1667L_{10}$
	45 W	$-.1771L_{10}$	$-.1667L_{10}$
	60 D	$.1771L_{10}$	$.1667L_{10}$
	60 W	$-.1771L_{10}$	$-.1667L_{10}$

All least squares means are estimable.

16.17. **a.** $\sigma^2(Y_{ijk}) = \sigma^2 + \sigma_\delta^2$; $\text{Cov}(Y_{ijk}, Y_{ijk'}) = \sigma^2$. Thus, $\mathbf{Var}(Y) = \text{Diag}(\mathbf{A}\ \mathbf{A}\ \ldots\ \mathbf{A})$, of order rts, where A is an $s \times s$ matrix consisting of $\sigma^2 + \sigma_\delta^2$ on the diagonal and σ^2 in all off-diagonal positions.

b. If means over samples are used, $\mathbf{Var}(Y) = I(\sigma^2 + \sigma_\delta^2/s)$, where I is of order tr.

c. PROC GLM uses the residual mean square from the analysis to compute standard errors, which in this case estimates σ_δ^2 and, consequently, is incorrect. The standard errors can be adjusted by multiplying by [MS(Among experimental units)/MS(Residual)]$^{1/2}$.

d. The residual mean square is now an estimate of $\sigma^2 + \sigma_\delta^2/s$ and the standard errors of the means are computed correctly. If the numbers of samples are not constant, the residual mean square is an estimate of $\sigma^2 + \sigma_\delta^2/\bar{s}_h$, where $\bar{s}_h$ is the harmonic mean of the numbers of samples per experimental unit. Dividing the residual mean square by r will not give precisely the correct divisor on σ_δ^2 for the variance of any treatment mean, although the discrepancies may be small.

e. The general linear model assumes that only the δ_{ijk} effects are random, whereas a more general approach such as MIXMOD can accommodate several random effects and, in this case, would treat ε_{ij} and δ_{ijk} as random.

16.19. **a.** Analysis and interpretation not given.

b. The least squares main effect means for $TEMP = 55$, $HERB = \mathbf{B}$, and $CONC = 80$ or 100 are not estimable. In addition, all two-factor or three-factor means that involve any of these levels of the three factors are not estimable. If the model is simplified by deleting the three-factor interaction, all main effect means and two-factor means become estimable. The three-factor interaction is significant and therefore could not be eliminated from the model without some loss of information. However, its contribution to the sums of squares is minor relative to that of the main effects and the temperature × herbicide interaction effects so that one might be willing to accept this simplification in order to obtain estimates of all means.

c. The analysis of variance without the three-factor interaction in the model shows the two-factor interactions temperature × herbicide and herbicide × concentration to be significant. Therefore, the corresponding two-way tables of least squares means summarize the results (standard errors are given in parentheses):

		$HERB = A$	$HERB = B$
$TEMP =$	10	.4174 (.0027)	.5103 (.0027)
	55	.3748 (.0027)	.5147 (.0052)
		$HERB = A$	$HERB = B$
$CONC =$	20	.2705 (.0043)	.3558 (.0043)
	40	.3595 (.0043)	.4833 (.0043)
	60	.4185 (.0043)	.5440 (.0043)
	80	.4555 (.0043)	.5690 (.0082)
	100	.4765 (.0043)	.6105 (.0082)

Chapter 17

17.1. A constant 1 was added to all yield values to avoid the zero readings. The Box–Cox transformation was made for $\lambda = 0, .25, .50, .75$, and 1.0 with the following results:

λ	**SS(Res)**
0	1,771
.25	986
.50	867
.75	1,074
1.00	1,554

It appears that the minimum will fall slightly below $\lambda = .5$. Assuming SS(Res) $= 867$ is the minimum, the 95% confidence interval on λ is established as

$$\text{SS(Res)}[1 + (2.160^2)/13] = 1{,}178$$

The limits on λ determined by all λ that provide SS(Res) $< 1{,}178$ include all values from approximately $\lambda = .2$ to $\lambda = .8$. The interval does not include $\lambda = 1.0$, no transformation, and indicates that the square-root transformation would be desirable.

17.2. ERROR B provides a direct estimate of σ^2, $\hat{\sigma}^2 = 119.56$. The standard error of this estimate is $s(\hat{\sigma}^2) = \{[2(119.56)^2]/13\}^{1/2} = 46.90$. The estimate of σ_δ^2 is obtained by equating ERROR A (the $BLOCK \times TILL \times HERB$ mean square) and ERROR B to their expectations and solving for σ_δ^2. This gives

$$\hat{\sigma}_\delta^2 = (186.34 - 119.56)/1.9048 = 35.06$$

The standard error of this estimate is

$$s(\hat{\sigma}_\delta^2) = [2(186.34)^2/9 + 2(119.56)^2/13]^{1/2}/1.9048 = 52.28$$

These results are very similar to those given by MIXMOD.

17.4. Let ρ_i be the block effects and γ_{ij} be the interaction effects between α_i and β_j. The estimable functions for the population marginal means are:

Mean:	A_1	A_2	B_1	B_2	B_3	A_1B_1	A_1B_2	A_1B_3	A_2B_1	A_2B_2	A_2B_3
μ	1	1	1	1	1	1	1	1	1	1	1
ρ_1	$\frac{1}{4}$	$\frac{1}{4}$	$\frac{1}{4}$	$\frac{1}{4}$	$\frac{1}{4}$	$\frac{1}{4}$	$\frac{1}{4}$	$\frac{1}{4}$	$\frac{1}{4}$	$\frac{1}{4}$	$\frac{1}{4}$
ρ_2	$\frac{1}{4}$	$\frac{1}{4}$	$\frac{1}{4}$	$\frac{1}{4}$	$\frac{1}{4}$	$\frac{1}{4}$	$\frac{1}{4}$	$\frac{1}{4}$	$\frac{1}{4}$	$\frac{1}{4}$	$\frac{1}{4}$
ρ_3	$\frac{1}{4}$	$\frac{1}{4}$	$\frac{1}{4}$	$\frac{1}{4}$	$\frac{1}{4}$	$\frac{1}{4}$	$\frac{1}{4}$	$\frac{1}{4}$	$\frac{1}{4}$	$\frac{1}{4}$	$\frac{1}{4}$
ρ_4	$\frac{1}{4}$	$\frac{1}{4}$	$\frac{1}{4}$	$\frac{1}{4}$	$\frac{1}{4}$	$\frac{1}{4}$	$\frac{1}{4}$	$\frac{1}{4}$	$\frac{1}{4}$	$\frac{1}{4}$	$\frac{1}{4}$
α_1	1	0	$\frac{1}{2}$	$\frac{1}{2}$	$\frac{1}{2}$	1	1	1	0	0	0
α_2	0	1	$\frac{1}{2}$	$\frac{1}{2}$	$\frac{1}{2}$	0	0	0	1	1	1
β_1	$\frac{1}{3}$	$\frac{1}{3}$	1	0	0	1	0	0	1	0	0
β_2	$\frac{1}{3}$	$\frac{1}{3}$	0	1	0	0	1	0	0	1	0
β_3	$\frac{1}{3}$	$\frac{1}{3}$	0	0	1	0	0	1	0	0	1
γ_{11}	$\frac{1}{3}$	0	$\frac{1}{2}$	0	0	1	0	0	0	0	0
γ_{12}	$\frac{1}{3}$	0	0	$\frac{1}{2}$	0	0	1	0	0	0	0
γ_{13}	$\frac{1}{3}$	0	0	0	$\frac{1}{2}$	0	0	1	0	0	0
γ_{21}	0	$\frac{1}{3}$	$\frac{1}{2}$	0	0	0	0	0	1	0	0
γ_{22}	0	$\frac{1}{3}$	0	$\frac{1}{2}$	0	0	0	0	0	1	0
γ_{23}	0	$\frac{1}{3}$	0	0	$\frac{1}{2}$	0	0	0	0	0	1

If cell (1, 2) is empty, there is no information on γ_{12} and all population marginal means that have nonzero coefficient on γ_{12} are nonestimable—namely, A_1, B_2, and A_1B_2. If the model contains no interaction, all least squares means become estimable.

Bibliography

(The following list includes some references not cited in the text.)

Alderdice, D. F. (1963). Some effects of simultaneous variation in salinity, temperature and dissolved oxygen on the resistance of young coho salmon to a toxic substance. *Journal of the Fisheries Research Board of Canada 20*, 525–550.

Allen, D. M. (1971a). Mean square error of prediction as a criterion for selecting variables. *Technometrics 13*, 469–475.

Allen, D. M. (1971b). The prediction sum of squares as a criterion for selection of predictor variables. Technical Report No. 23, Department of Statistics, University of Kentucky.

Anderson, R. L. and Nelson, L. A. (1975). A family of models involving intersecting straight lines and concomitant experimental designs useful in evaluating response to fertilizer nutrients. *Biometrics 31*, 303–318.

Andrews, D. F. (1971). A note on the selection of data transformations. *Biometrika 58*, 249–254.

Andrews, D. F. (1974). A robust method for multiple linear regression. *Technometrics 16*, 523–531.

Andrews, D. F. and Herzberg, A. M. (1985). *Data: A Collection of Problems from Many Fields for the Student and Research Worker*. New York: Springer-Verlag.

Andrews, D. F. and Pregibon, D. (1978). Finding the outliers that matter. *Journal of the Royal Statistical Society, Series B 40*, 85–93.

Anscombe, F. J. (1973). Graphs in statistical analysis. *The American Statistician 27*, 17–21.

Atkinson, A. C. (1981). Two graphical displays for outlying and influential observations in regression. *Biometrika 68*, 13–20.

Atkinson, A. C. (1982). Regression diagnostics, transformations and constructed variables. *Journal of the Royal Statistical Society, Series B 44*, 1–36.

Atkinson, A. C. (1983). Diagnostic regression analysis and shifted power transformations. *Technometrics 25*, 23–33.

Atkinson, A. C. (1986). Diagnostic tests for transformations. *Technometrics 28*, 29–37.

Bancroft, T. A. (1964). Analysis and inference for incompletely specified models involving the use of preliminary test(s) of significance. *Biometrics 20*, 427–442.

Bartlett, M. S. (1947). The use of transformations. *Biometrics 3*, 39–53.

Bartlett, M. S. (1949). Fitting a straight line when both variables are subject to error. *Biometrics 5*, 207–212.

Baskerville, J. C. and Toogood, J. H. (1982). Guided regression modeling for prediction and exploration of structure with many explanatory variables. *Technometrics 24*, 9–17.

Basson, R. P. (1965). On unbiased estimation in variance component models. Ph.D. thesis, Iowa State University of Science and Technology.

Belsley, D. A. (1984). Demeaning conditioning diagnostics through centering (with Discussion). *The American Statistician 38*, 73–77.

Belsley, D. A., Kuh, E., and Welsch, R. E. (1980). *Regression Diagnostics: Identifying Influential Data and Sources of Collinearity*. New York: Wiley.

Bendel, R. B. and Afifi, A. A. (1977). Comparison of stopping rules in forward "stepwise" regression. *Journal of the American Statistical Association 72*, 46–53.

Berk, K. N. (1977). Tolerance and condition in regression computations. *Journal of the American Statistical Association 72*, 863–866.

Berk, K. N. (1978). Comparing subset regression procedures. *Technometrics 20*, 1–6.

Berk, K. N. (1984). Validating regression procedures with new data. *Technometrics 26*, 331–338.

Bickel, P. J. (1973). On some analogues to linear combinations of order statistics in the linear model. *Annals of Statistics 1*, 597–616.

Blom, G. (1958). *Statistical Estimates and Transformed Beta Variates*. New York: Wiley.

Bloomfield, P. (1976). *Fourier Analysis of Time Series: An Introduction*. New York: Wiley.

Box, G. E. P. (1980). Sampling and Bayes' inference in scientific modelling and robustness. *Journal of the Royal Statistical Society, Series A 143*, 383–430.

Box, G. E. P. and Cox, D. R. (1964). An analysis of transformations. *Journal of the Royal Statistical Society, Series B 26*, 211–243.

Box, G. E. P. and Draper, N. R. (1987). *Empirical Model-Building and Response Surfaces*. New York: Wiley.

Box, G. E. P., Hunter, W. G., and Hunter, J. S. (1978). *Statistics for Experimenters: An Introduction to Design, Data Analysis, and Model Building*. New York: Wiley.

Box, G. E. P. and Tidwell, P. W. (1962). Transformation of the independent variables. *Technometrics 4*, 531–550.

Box, G. E. P. and Watson, G. S. (1962). Robustness to nonnormality of regression tests. *Biometrika 49*, 93–106.

Bradley, R. A. and Srivastava, S. S. (1979). Correlation in polynomial regression. *The American Statistician 33*, 11–14.

Bradu, D. and Gabriel, K. R. (1974). Simultaneous statistical inference on interactions in two-way analysis of variance. *Journal of the American Statistical Association 69*, 428–436.

Bradu, D. and Gabriel, K. R. (1978). The biplot as a diagnostic tool for models of two-way tables. *Technometrics 20*, 47–68.

Brown, R. L., Durbin, J., and Evans, J. M. (1975). Techniques for testing the constancy of regression relationships over time. *Journal of the Royal Statistical Society, Series B 37*, 149–192.

Bryant, P. (1984). Geometry, statistics, probability: Variations on a common theme. *The American Statistician 38*, 38–48.

Bunke, O. and Droge, B. (1984). Estimators of the mean squared error of prediction in linear regression. *Technometrics 26*, 145–155.

Buonagurio, D. A., Nakada, S., Parvin, J. D., Krystal, M., Palese, P., and Fitch, W. M. (1986). Evolution of human influenza A viruses over 50 years: Rapid, uniform rate of change in NS gene. *Science 232*, 980–982.

Cameron, E. and Pauling, L. (1978). Supplemental ascorbate in the supportive treatment of cancer: Reevaluation of prolongation of survival times in terminal human cancer. *Proceedings of the National Academy of Sciences U.S.A. 75*, 4538–4542.

Carroll, R. J. and Ruppert, D. (1985). Transformations in regression: A robust analysis. *Technometrics 27*, 1–12.

Carter, R. L. and Fuller, W. A. (1980). Instrumental variable estimation of the simple errors-in-variables model. *Journal of the American Statistical Association 75*, 687–692.

Clarke, G. P. Y. (1987). Marginal curvatures in the analysis of nonlinear regression models. *Journal of the American Statistical Association 82*, 844–850.

Cochran, W. G. (1983). *Planning and Analysis of Observational Studies.* New York: Wiley.

Cochran, W. G. and Cox, G. M. (1957). *Experimental Designs*, 2nd edition. New York: Wiley.

Cook, R. D. (1977). Detection of influential observations in linear regression. *Technometrics 19*, 15–18.

Cook, R. D. (1979). Influential observations in linear regression. *Journal of the American Statistical Association 74*, 169–174.

Cook, R. D. (1984). Comment [to Belsley, D. A. (1984)]. *The American Statistician 38*, 78–79.

Cook, R. D. and Prescott, P. (1981). On the accuracy of Bonferroni significance levels for detecting outliers in linear models. *Technometrics 23*, 59–63.

Cook, R. D. and Wang, P. C. (1983). Transformations and influential cases in regression. *Technometrics 25*, 337–343.

Cook, R. D. and Weisberg, S. (1982). *Residuals and Influence in Regression.* London: Chapman and Hall.

Corsten, L. C. A. and Gabriel, K. R. (1976). Graphical exploration in comparing variance matrices. *Biometrics 32*, 851–863.

Cramér, H. (1946). *Mathematical Methods of Statistics.* Princeton, New Jersey: Princeton University Press.

Daniel, C. and Wood, F. S. (1980). *Fitting Equations to Data: Computer Analysis of Multifactor Data*, 2nd edition. New York: Wiley.

Dixon, W. J. (1981). *BMDP Statistical Software* 1981, W. J. Dixon, chief editor, Berkeley, California: University of California Press.

Draper, N. and Smith, H. (1981). *Applied Regression Analysis*, 2nd edition. New York: Wiley.

Durbin, J. and Watson, G. S. (1951). Testing for serial correlation in least squares regression. II. *Biometrika 38*, 159–178.

Durbin, J. and Watson, G. S. (1971). Testing for serial correlation in least squares regression. III. *Biometrika 58*, 1–19.

Feldstein, M. (1974). Errors in variables: A consistent estimator with smaller MSE in finite samples. *Journal of the American Statistical Association 69*, 990–996.

Filliben, J. J. (1975). The probability plot correlation coefficient test for normality. *Technometrics 17*, 111–118.

Freund, R. J., Littell, R. C., and Spector, P. C. (1986). *SAS System for Linear Models*, 2nd edition. Cary, North Carolina: SAS Institute, Inc.

Fuller, W. A. (1976). *Introduction to Statistical Time Series.* New York: Wiley.

Furnival, G. M. (1971). All possible regressions with less computation. *Technometrics 13*, 403–408.

Furnival, G. M. and Wilson, R. B. (1974). Regressions by leaps and bounds. *Technometrics 16*, 499–511.

Gabriel, K. R. (1971). The biplot graphic display of matrices with application to principal component analysis. *Biometrika 58*, 453–467.

Gabriel, K. R. (1972). Analysis of meteorological data by means of canonical decomposition and biplots. *Journal of Applied Meteorology 11*, 1071–1077.

Gabriel, K. R. (1978). Least squares approximation of matrices by additive and multiplicative models. *Journal of the Royal Statistical Society, Series B 40*, 186–196.

Gallant, A. R. (1987). *Nonlinear Statistical Models*. New York: Wiley.

Gallant, A. R. and Fuller, W. A. (1973). Fitting segmented polynomial models whose join points have to be estimated. *Journal of the American Statistical Association 68*, 144–147.

Galpin, J. S. and Hawkins, D. M. (1984). The use of recursive residuals in checking model fit in linear regression. *The American Statistician 38*, 94–105.

Giesbrecht, F. G. (1983). An efficient procedure for computing MINQUE of variance components and generalized least squares estimates of fixed effects. *Communications in Statistics—Theory and Methods 12*(18), 2169–2177.

Giesbrecht, F. G. (1984). MIXMOD, a SAS procedure for analysing mixed models. North Carolina State University, Institute of Statistics Mimeograph Series No. 1659.

Giesbrecht, F. G. (1986). Analysis of data from incomplete block designs. *Biometrics 42*, 437–448.

Giesbrecht, F. G. and Burns, J. C. (1985). Two-stage analysis based on a mixed model: Large-sample asymptotic theory and small-sample simulation results. *Biometrics 41*, 477–486.

Golub, B. H. and Reinsch, C. (1970). Singular value decomposition and least squares solution. *Numerische Mathematik 14*, 403–420.

Good, I. J. (1969). Some applications of the singular value decomposition of a matrix. *Technometrics 11*, 823–831.

Graybill, F. A. (1961). *An Introduction to Linear Statistical Models*. New York: McGraw-Hill.

Graybill, F. A. (1983). *Matrices with Applications in Statistics*, 2nd edition. Belmont, California: Wadsworth.

Gunst, R. F. (1983). Regression analysis with multicollinear predictor variables: Definition, detection, and effects. *Communications in Statistics—Theory and Methods 12*, 2217–2260.

Gunst, R. F. (1984). Comment: Toward a balanced assessment of collinearity diagnostics. *The American Statistician 38*, 79–82.

Gunst, R. F. and Mason, R. L. (1980). *Regression Analysis and Its Applications: A Data-Oriented Approach*. New York: Marcel Dekker.

Hampel, F. R., Ronchetti, E. M., Rousseeuw, P. J., and Stahel, W. A. (1986). *Robust Statistics, The Approach Based on Influence Functions*. New York: Wiley.

Hartley, H. O. (1961). The modified Gauss–Newton method for the fitting of nonlinear regression functions by least-squares. *Technometrics 3*, 269–280.

Harvey, A. C. and Collier, P. (1977). Testing for functional misspecification in regression analysis. *Journal of Econometrics 6*, 103–120.

Harvey, A. C. and Phillips, G. D. A. (1974). A comparison of the power of some tests for heteroscedasticity in the general linear model. *Journal of Econometrics 2*, 307–316.

Hawkins, C. M. (1973). On the investigation of alternative regressions by principal component analysis. *Applied Statistics 22*, 275–286.

Heck, W. W., Cure, W. W., Rawlings, J. O., Zaragosa, L. J., Heagle, A. S., Heggestad, H. E., Kohut, R. J., Kress, L. W., and Temple, P. J. (1984). Assessing impacts of ozone on agricultural crops: II. *Journal of the Air Pollution Control Association 34*, 810–817.

Hedayat, A. and Robson, D. S. (1970). Independent stepwise residuals for testing homoscedasticity. *Journal of the American Statistical Association 65*, 1573–1581.

Hernandez, F. and Johnson, R. A. (1980). The large-sample behavior of transformations to normality. *Journal of the American Statistical Association 75*, 855–861.

Hoaglin, D. C. and Welsch, R. E. (1978). The hat matrix in regression and ANOVA. *The American Statistician 32*, 17–22.

Hocking, R. R. (1973). A discussion of the two-way mixed model. *The American Statistician 27*, 148–152.

Hocking, R. R. (1976). The analysis and selection of variables in linear regression. *Biometrics 32*, 1–49.

Hocking, R. R. (1983). Developments in linear regression methodology: 1959–1982 (with Discussion). *Technometrics 25*, 219–249.

Hocking, R. R. (1985). *The Analysis of Linear Models.* Monterey, California: Brooks/Cole.

Hocking, R. R. and Pendleton, O. J. (1983). The regression dilemma. *Communications in Statistics A12*, 497–527.

Hocking, R. R. and Speed, F. M. (1975). A full-rank analysis of some linear model problems. *Journal of the American Statistical Association 70*, 706–712.

Hocking, R. R., Speed, F. M., and Lynn, M. J. (1976). A class of biased estimators in linear regression. *Technometrics 18*, 425–437.

Hoerl, A. E. (1962). Application of ridge analysis to regression problems. *Chemical Engineering Progress 58*, 54–59.

Hoerl, A. E. and Kennard, R. W. (1970a). Ridge regression: Biased estimation for nonorthogonal problems. *Technometrics 12*, 55–67.

Hoerl, A. E. and Kennard, R. W. (1970b). Ridge regression: Applications to nonorthogonal problems. *Technometrics 12*, 69–82.

Hoerl, A. E., Kennard, R. W., and Baldwin, K. F. (1975). Ridge regression: Some simulations. *Communications in Statistics 4*, 105–124.

Hogg, R. V. and Randles, R. H. (1975). Adaptive distribution-free regression methods and their applications. *Technometrics 17*, 399–408.

Hotelling, H. (1957). Relation of the newer multivariate statistical methods to factor analysis. *British Journal of Statistical Psychology 10*, 69–79.

Householder, A. S. and Young, G. (1938). Matrix approximation and latent roots. *American Mathematical Monthly 45*, 165–171.

Huang, C. J. and Bolch, B. W. (1974). On testing of regression disturbances for normality. *Journal of the American Statistical Association 69*, 330–335.

Huber, P. J. (1973). Robust regression: Asymptotics, conjectures and Monte Carlo. *Annals of Statistics 1*, 799–821.

Huber, P. J. (1981). *Robust Statistics.* New York: Wiley.

Jensen, B. C. and McDonald, J. B. (1976). A pedagogical example of heteroscedasticity and autocorrelation. *The American Statistician 30*, 192–193.

Joshi, P. C. and Lalitha, S. (1986). Tests for two outliers in a linear model. *Biometrika 73*, 236–239.

Kennedy, W. J. and Bancroft, T. A. (1971). Model-building for prediction in regression based on repeated significance tests. *Annals of Mathematical Statistics 42*, 1273–1284.

Ketellapper, R. H. (1983). On estimating parameters in a simple linear errors-in-variables model. *Technometrics 25*, 43–47.

Kleinbaum, D. G., Kupper, L. L., and Muller, K. E. (1988). *Applied Regression Analysis and Other Multivariable Methods*, 2nd edition. Boston: PWS-Kent.

Koenker, R. and Bassett, G. (1978). Regression quantiles. *Econometrica 46*, 33–50.

Land, S. B. (1973). Sea water flood tolerance of some southern pines. Ph.D. thesis,

Department of Forestry and Department of Genetics, North Carolina State University.

Lawson, C. L. and Hanson, R. J. (1974). *Solving Least-Squares Problems*. Englewood Cliffs, New Jersey: Prentice-Hall.

Linthurst, R. A. (1979). Aeration, nitrogen, pH and salinity as factors affecting *Spartina Alterniflora* growth and dieback. Ph.D. thesis, North Carolina State University.

Lott, W. F. (1973). The optimal set of principal component restrictions on a least squares regression. *Communications in Statistics 2*, 449–464.

Madansky, A. (1959). The fitting of straight lines when both variables are subject to error. *Journal of the American Statistical Association 54*, 173–205.

Mallows, C. L. (1973a). Some comments on C_p. *Technometrics 15*, 661–675.

Mallows, C. L. (1973b). Data analysis in a regression context. In *Proceedings of University of Kentucky Conference on Regression with a Large Number of Predictor Variables*, W. O. Thompson and F. B. Cady (eds.). Department of Statistics, University of Kentucky.

Marquardt, D. W. (1963). An algorithm for least-squares estimation of nonlinear parameters. *Journal of the Society for Industrial and Applied Mathematics 11*, 431–441.

Marquardt, D. W. (1970). Generalized inverses, ridge regression, biased linear estimation, and nonlinear estimation. *Technometrics 12*, 591–612.

Marquardt, D. W. (1980). Comment: You should standardize the predictor variables in your regression models. *Journal of the American Statistical Association 75*, 87–91.

Marquardt, D. W. and Snee, R. D. (1975). Ridge regression in practice. *The American Statistician 29*, 3–19.

Mason, R. L. and Gunst, R. F. (1985). Outlier-induced collinearities. *Technometrics 27*, 401–407.

Massy, W. F. (1965). Principal components regression in exploratory statistical research. *Journal of the American Statistical Association 60*, 234–256.

Miller, R. G., Jr. (1981). *Simultaneous Statistical Inference*, 2nd edition. New York: Springer-Verlag.

Mombiela, F. A. and Nelson, L. A. (1981). Relationships among some biological and empirical fertilizer response models and use of the power family of transformations to identify an appropriate model. *Agronomy Journal 73*, 353–356.

Mosteller, F. and Tukey, J. W. (1977). *Data Analysis and Regression: A Second Course in Statistics*. Reading, Massachusetts: Addison-Wesley.

Mullett, G. M. (1976). Why regression coefficients have the wrong sign. *Journal of Quality Technology 8*, 121–126.

Myers, R. H. (1986). *Classical and Modern Regression with Applications*. Boston: Duxbury Press.

Nelder, J. A. (1966). Inverse polynomials, a useful group of multifactor response functions. *Biometrics 22*, 128–140.

Nelson, W. R. and Ahrenholz, D. W. (1986). Population and fishery characteristics of Gulf Menhaden, *Brevoortia patronus*. *Fishery Bulletin 84*, 311–325.

Neter, J., Wasserman, W., and Kutner, M. H. (1983). *Applied Linear Regression Models*. Homewood, Illinois: Richard D. Irwin.

Neyman, J. (1937). "Smooth" test for goodness of fit. *Skandinavisk Aktuarietidskrift 20*, 149–199.

Nielsen, D. R., Biggar, J. W., and Erh, E. T. (1973). Spatial variability of field-measured soil-water properties. *Hilgardia 42*, 215–259.

Norusis, M. J. (1985). *SPSS-X Advanced Statistics Guide*. Chicago: McGraw-Hill.

Park, S. H. (1981). Collinearity and optimal restrictions on regression parameters for estimating responses. *Technometrics 23*, 289–295.

Pearson, E. S. and Hartley, H. O. (1966). *Biometrika Tables for Statisticians*, Volume 1, 3rd edition. London: Cambridge University Press.

Pearson, E. S. and Please, N. W. (1975). Relation between the shape of population distribution and the robustness of four simple statistical tests. *Biometrika 62*, 223–242.

Pearson, E. S. and Stephens, M. A. (1964). The ratio of range to standard deviation in the same normal sample. *Biometrika 51*, 484–487.

Pennypacker, S. P., Knoble, H. D., Antle, C. E., and Madden, L. V. (1980). A flexible model for studying plant disease progression. *Phytopathology 70*, 232–235.

Pierce, D. A. and Gray, R. J. (1982). Testing normality of errors in regression models. *Biometrika 69*, 233–236.

Pierce, D. A. and Kopecky, K. J. (1979). Testing goodness of fit for the distribution of errors in regression models. *Biometrika 66*, 1–5.

Plackett, R. L. (1950). Some theorems in least squares. *Biometrika 37*, 149–157.

Quesenberry, C. P. (1986). Some transformation methods in goodness-of-fit. Chapter 6 in *Goodness of Fit Techniques*, R. B. D'Agostino and M. A. Stephens (eds.). New York: Marcel Dekker.

Quesenberry, C. P. and Quesenberry, C., Jr. (1982). On the distribution of residuals from fitted parametric models. *Journal of Statistical Computation and Simulation 15*, 129–140.

Ralston, M. L. and Jennrich, R. I. (1978). Dud, a derivative-free algorithm for nonlinear least squares. *Technometrics 20*, 7–14.

Rao, C. R. (1973). *Linear Statistical Inference and Its Applications*, 2nd edition. New York: Wiley.

Rawlings, J. O. and Cure, W. W. (1985). The Weibull function as a dose–response model for air pollution effects on crop yields. *Crop Science 25*, 807–814.

Riggs, D. S., Guarnieri, J. A., and Addelman, S. (1978). Fitting straight lines when both variables are subject to error. *Life Sciences 22*, 1305–1360.

Rohlf, F. J. and Sokal, R. R. (1981). *Statistical Tables*, 2nd edition. San Francisco: W. H. Freeman.

Saeed, M. and Francis, C. A. (1984). Association of weather variables with genotype × environment interactions in grain sorghum. *Crop Science 24*, 13–16.

SAS Institute, Inc. (1985a). *SAS/IML User's Guide for Personal Computers*, Version 6 edition. Cary, North Carolina: SAS Institute, Inc.

SAS Institute, Inc. (1985b). *SAS Language Guide for Personal Computers*, Version 6 edition. Cary, North Carolina: SAS Institute, Inc.

SAS Institute, Inc. (1985c). *SAS Procedures Guide for Personal Computers*, Version 6 edition. Cary, North Carolina: SAS Institute, Inc.

SAS Institute, Inc. (1985d). *SAS/STAT Guide for Personal Computers*, Version 6 edition. Cary, North Carolina: SAS Institute, Inc.

SAS Institute, Inc. (1985e). *SAS User's Guide: Basics*, Version 5 edition. Cary, North Carolina: SAS Institute, Inc.

SAS Institute, Inc. (1985f). *SAS User's Guide: Statistics*, Version 5 edition. Cary, North Carolina: SAS Institute, Inc.

Satterthwaite, F. E. (1946). An approximate distribution of estimates of variance components. *Biometrics Bulletin*, *2*, 110–114.

Scheffé, H. (1953). A method for judging all contrasts in the analysis of variance. *Biometrika 40*, 87–104.

Scheffé, H. (1959). *The Analysis of Variance*. New York: Wiley.

Schneeweiss, H. (1976). Consistent estimation of a regression with errors in the variables. *Metrika 23*, 101–116.

Searle, S. R. (1971). *Linear Models*. New York: Wiley.

Searle, S. R. (1982). *Matrix Algebra Useful for Statistics*. New York: Wiley.

Searle, S. R. (1986). *Linear Models for Unbalanced Data*. New York: Wiley.

Searle, S. R. and Hausman, W. H. (1970). *Matrix Algebra for Business and Economics*. New York: Wiley.

Searle, S. R. and Henderson, H. V. (1979). Annotated computer output for analyses of unbalanced data: SAS GLM. Technical Report BU-641-M, Biometrics Unit, Cornell University.

Searle, S. R., Speed, F. M., and Milliken, G. A. (1980). Population marginal means in the linear model: An alternative to least squares means. *The American Statistician 34*, 216–221.

Shapiro, S. S. and Francia, R. S. (1972). An approximate analysis of variance test for normality. *Journal of the American Statistical Association 67*, 215–216.

Shapiro, S. S. and Wilk, M. B. (1965). An analysis of variance test for normality (complete samples). *Biometrika 52*, 591–611.

Shapiro, S. S., Wilk, M. B., and Chen, H. J. (1968). A comparative study of various tests of normality. *Journal of the American Statistical Association 63*, 1343–1372.

Shy-Modjeska, J. S., Riviere, J. E., and Rawlings, J. O. (1984). Application of biplot methods to the multivariate analysis of toxicological and pharmacokinetic data. *Toxicology and Applied Pharmacology 72*, 91–101.

Smith, G. and Campbell, F. (1980). A critique of some ridge regression methods. *Journal of the American Statistical Association 75*, 74–81.

Snedecor, G. W. and Cochran, W. G. (1980). *Statistical Methods*, 7th edition. Ames, Iowa: Iowa State University Press.

Snee, R. D. (1973). Some aspects of nonorthogonal data analysis, Part I. Developing prediction equations. *Journal of Quality Technology 5*, 67–79.

Snee, R. D. (1977). Validation of regression models: Methods and examples. *Technometrics 19*, 415–428.

Snee, R. D. (1983). Discussion [of Hocking, R. R. (1983)]. *Technometrics 25*, 230–237.

Snee, R. D. and Marquardt, D. W. (1984). Comment: Collinearity diagnostics depend on the domain of prediction, the model, and the data. *The American Statistician 38*, 83–87.

Speed, F. M. and Hocking, R. R. (1976). The use of the $R(\cdot)$-notation with unbalanced data. *The American Statistician 30*, 30–33.

Speed, F. M., Hocking, R. R., and Hackney, O. P. (1978). Methods of analysis of linear models with unbalanced data. *Journal of the American Statistical Association 73*, 105–112.

Steel, R. G. D. and Torrie, J. H. (1980). *Principles and Procedures of Statistics: A Biometrical Approach*, 2nd edition. New York: McGraw-Hill.

Stein, C. M. (1960). Multiple regression. *In Contributions to Probability and Statistics, Essays in Honor of Harold Hotelling*, I. Olkin (ed.), 424–443. Stanford, California: Stanford University Press.

Stephens, M. A. (1970). Use of the Kolmogorov–Smirnov, Cramér–von Mises and related statistics without extensive tables. *Journal of the Royal Statistical Society, Series B 32*, 115–122.

Stewart, G. W. (1973). *Introduction to Matrix Computations*. New York: Academic Press.

Swed, F. S. and Eisenhart, C. (1943). Tables for testing randomness of grouping in a sequence of alternatives. *Annals of Mathematical Statistics 14*, 66–87.

Theil, H. (1971). *Principles of Econometrics.* New York: Wiley.

Thisted, R. A. (1976). Ridge regression, minimax estimation, and empirical Bayes methods. Stanford University Division of Biostatistics Technical Report No. 28.

Thisted, R. A. (1980). Comment: A critique of some ridge regression methods. *Journal of the American Statistical Association 75*, 81–86.

Thisted, R. A. and Morris, C. N. (1979). Theoretical results for adaptive ordinary ridge regression estimators. Technical Report No. 94, University of Chicago, Department of Statistics.

Tukey, J. W. (1977). *Exploratory Data Analysis.* Reading, Massachusetts: Addison-Wesley.

Van Nostrand, R. C. (1980). Comment: A critique of some ridge regression methods. *Journal of the American Statistical Association 75*, 92–94.

Wald, A. (1940). The fitting of straight lines if both variables are subject to error. *Annals of Mathematical Statistics 11*, 284–300.

Watson, G. S. (1961). Goodness-of-fit tests on a circle. *Biometrika 48*, 109–114.

Webster, J. T., Gunst, R. F., and Mason, R. L. (1974). Latent root regression analysis. *Technometrics 16*, 513–522.

Weisberg, S. (1974). An empirical comparison of the percentage points of W and W'. *Biometrika 61*, 644–646.

Weisberg, S. (1980). Comment [on White and MacDonald (1980)]. *Journal of the American Statistical Association 75*, 28–31.

Weisberg, S. (1981). A statistic for allocating C_p to individual cases. *Technometrics 23*, 27–31.

Weisberg, S. (1985). *Applied Linear Regression*, 2nd edition. New York: Wiley.

White, H. and MacDonald, G. M. (1980). Some large-sample tests for nonnormality in the linear regression model (with Comment by S. Weisberg). *Journal of the American Statistical Association 75*, 16–31.

Wilkinson, J. H. (1965). *The Algebraic Eigenvalue Problem.* Oxford: Clarendon Press.

Wilkinson, L. and Dallal, G. E. (1981). Tests of significance in forward selection regression with an F-to-enter stopping rule. *Technometrics 23*, 377–380.

Willan, A. R. and Watts, D. G. (1978). Meaningful multicollinearity measures. *Technometrics 20*, 407–412.

Wood, F. S. (1973). The use of individual effects and residuals in fitting equations to data. *Technometrics 15*, 677–695.

Wood, F. S. (1984). Comment: Effect of centering on collinearity and interpretation of the constant. *The American Statistician 38*, 88–90.

Working, H. and Hotelling, H. (1929). Application of the theory of error to the interpretation of trends. *Journal of the American Statistical Association, Supplement (Proceedings) 24*, 73–85.

Yale, C. and Forsythe, A. B. (1976). Winsorized regression. *Technometrics 18*, 291–300.

Author Index

Subject Index